T0200665

An Introduction to
MATHEMATICAL
LOGIC

An Introduction to
MATHEMATICAL
LOGIC

Richard E. Hodel

Duke University

DOVER PUBLICATIONS, INC.
Garden City, New York

Bibliographical Note

This Dover edition, first published in 2013, is a slightly corrected republication
of the edition originally published by PWS Publishing, Boston, in 1995. The
charts originally located on the endpapers have been moved to pages xv–xviii.

International Standard Book Number

ISBN-13: 978-0-486-49785-3
ISBN-10: 0-486-49785-2

Manufactured in the United States by LSC Communications
49785208   2020
www.doverpublications.com

To Margaret, Richie, and Katie
and
To Mother, and the memory of my Father

Contents

Preface

T his book is designed for a variety of courses in mathematical logic and recursion theory. Although primarily written with the mathematics major in mind, the book, with its emphasis on algorithms and on the validity of argument forms, should also be of interest to students in computer science and philosophy. A distinguishing characteristic of the text is the rather large number of exercises. Before discussing possible uses of the text, let me briefly outline its contents.

Chapter 1 serves as background for the remainder of the text; I recommend that the material in this chapter be covered as needed. The topics are induction, formal systems, set theory, functions and relations, countable sets, uncountable sets and Cantor diagonal arguments, axiom systems (for example, the Peano postulates and the axioms for a group), computable functions and decidable relations, recursive functions and recursive relations, and Church's thesis.

Chapters 2 and 3 cover propositional logic. In Chapter 2, the emphasis is on the language and semantics of propositional logic, the fundamental notion of tautological consequence, and adequate sets of connectives. This material is motivated by questions on the validity of arguments that can be expressed in the language of propositional logic.

In Chapter 3, the emphasis shifts from logic to metalogic. Here we introduce a formal system for propositional logic and then study the interplay between semantic and syntactic concepts; the soundness and adequacy theorems are the high points of this chapter. At the end of Chapter 3, there are two sections (3.5 and 3.6) that survey other formal systems for propositional logic. This material can be used as an alternative to the formal system we use (thereby giving flexibility to the text), or as projects for independent study or seminars in logic. The discussion in Section 3.5 on intuitionistic propositional logic should be of interest to philosophy students; Section 3.6, on Gentzen-style proof systems, is written with the computer science major in mind. In any case, these two sections need not be covered before proceeding to Chapter 4.

The material in Chapter 1 might be used in conjunction with Chapters 2 and 3 as follows: Cover Section 1.2, on induction, before starting Chapter 2, on informal propositional logic; cover Section 1.3, on formal systems, before starting Chapter 3, on the formal approach to propositional logic. A mathematically mature class could probably skip Sections 1.2 and 1.3, or could be assigned this material as outside reading. It might also be a good idea to discuss some of the topics in Section 1.4, on set theory, as background for the proof of the adequacy theorem (note especially Exercises 6 and 7 in Section 1.4). Section 1.5 includes a proof that a language with a countable number of symbols has a countable number of formulas.

Chapters 4 and 5 do for first-order logic what Chapters 2 and 3 do for propositional logic. Chapter 4 covers first-order languages, interpretations and models, and Tarski's definition of truth. The approach is fairly leisurely; in particular, we start off with a *specific* first-order language L_{NN} for arithmetic. This has two advantages: Students, at an early stage, become comfortable with the language used in Gödel's incompleteness theorem; students also get practice in expressing mathematical ideas in a first-order language (this idea is used very effectively by Hofstadter in *Gödel, Escher, Bach* 1979). The last section of Chapter 4 covers substitution for free variables. This is the most technical section of the chapter, but it does pave the way for the formal approach to first-order logic in Chapter 5. In particular, in Section 4.4 we show that substitution axioms and equality axioms are logically valid; these are key steps in the proof of the soundness theorem for first-order logic in Chapter 5.

In Chapter 5, we introduce a formal system for first-order logic. The most difficult theorem in this chapter comes in Section 5.4, where we use Henkin's ideas to show that a consistent set of sentences has a model. The proof is eased by the fact that this approach is also used earlier—in Chapter 3, on propositional logic. The completeness theorems of Gödel are the high points of Chapter 5, but in addition we discuss the decidability and listability of the theorems of a set Γ of sentences of a first-order language. These ideas are fundamental in later chapters. The material in Chapter 1 might be used in conjunction with that of Chapters 4 and 5 as follows: Cover Section 1.4, on relations, and Section 1.6, on axiom systems, before starting Chapter 4, on first-order languages; this will help students understand the role of the nonlogical symbols in a first-order language. One might also want to cover Section 1.7, on decidable and semidecidable relations, before discussing the material in Section 5.5 on decidability and listability.

In Chapters 6 and beyond, the emphasis shifts from logic approached from a mathematical point of view to the interplay between mathematics and logic. Chapter 6 begins with a discussion of mathematical theories (including Peano arithmetic); we then ask questions about these theories that obviously involve logic. A number of these questions set the stage for the theorems of Gödel, Church, and Tarski in Chapter 7. This section also has a discussion of Hilbert's program. Chapter 6 continues with applications of the Löwenheim-Skolem and compactness theorems (especially on the shortcomings of first-order logic), decidable theories, and the axioms of Zermelo-Frankel set theory.

Chapter 7 covers the theorems of Gödel, Church, and Tarski on incompleteness, undecidability, and indefinability. The approach is informal in the sense that we establish the recursiveness (and hence the expressibility) of certain relations by showing that the relation in question is decidable (and then use Church's thesis). The hardest section of this chapter is the rigorous proof in Section 7.3 of Gödel's result that every recursive relation is expressible. Background for this chapter is provided in Sections 1.7 and 1.8, on decidable relations and recursive relations; Section 6.1 might also be useful. The proof of Gödel's theorem is modeled on his original idea: Construct a sentence whose interpretation is *I am not a theorem.* Gödel's second incompleteness theorem is also discussed in this chapter.

Chapter 8 is a rigorous treatment of recursive functions and recursive relations. In particular, we develop more than enough recursion theory to prove the theorems of Gödel, Church, and Tarski in the following form: *The 1-ary relation* THM_Γ *is RE but not recursive* (Γ a consistent recursively axiomatized extension of arithmetic); *the 1-ary relation TR is not even definable.*

Chapter 9 covers computability theory. A major goal is to introduce a new model of computation and then to prove it equivalent to the recursive function approach. The model we use is the unlimited register machine of Shepherdson and Sturgis. Key theorems of this chapter are the recursiveness of the Kleene computation relation T_n and the parameter theorem. Chapter 9 ends with a discussion of word problems and the proof of the Markov-Post theorem that there is a finitely presented semigroup with an unsolvable word problem. The book is organized so that the following sections constitute a self-contained proof of this result: Sections 1.7 and 1.8; Sections 8.1–8.3 and 8.5; Sections 9.1–9.3 and 9.6.

Chapter 10 covers Hilbert's tenth problem. Here we see another application of logic to mathematics: the nonexistence of an algorithm for the solution of certain Diophantine equations. The book is organized so that the following sections constitute a self-contained presentation of Hilbert's tenth problem: Sections 1.7 and 1.8; Sections 8.1–8.3 and 8.5; Chapter 10.

If there is a common thread in Chapters 8–10, it is this: In each chapter, we prove the extremely important result that *there is an RE relation that is not recursive.*

Now I will suggest some ways to use the text. The following topics are appropriate for a one-semester course for upper-level undergraduates in mathematics, computer science, and philosophy:

Sections

Propositional logic: 1.2; 2.1–2.3;
1.3 (and perhaps 1.4 and 1.5); 3.1–3.4.
First-order logic: 1.4 and 1.6; 4.1–4.4; 5.1–5.5.
Incompleteness: 1.7 and 1.8; Chapter 7 (skim Section 7.3).

Some instructors may wish to include topics from Chapter 6 (for example, Hilbert's program, the compactness theorem, and Zermelo-Frankel set theory). A one-semester course in logic at the graduate level might cover Chapters 1 through 8, with some material assigned for outside reading (for example, selected parts of Chapter 7). Sections 1.7 and 1.8 and Chapters 8, 9, and 10 are appropriate for a one-semester course in recursion theory. Finally, there is enough material for a one-year course in logic and recursion theory.

Anyone familiar with the logic text *Mathematical Logic* 1967 by Joe Shoenfield, will know that I am deeply indebted to him for many of the ideas that appear in this text. Perhaps most conspicuous is the use of his set of axioms for propositional logic; these are, it seems to me, ideal for a first introduction to mathematic logic. The axioms have intuitive appeal and are very easy to work with from the point of view of writing out formal proofs. The one drawback is the proof of the deduction theorem; I give a traditional proof of this result, but it is not quite so easily obtained as with a Hilbert-style proof system whose only rule of inference is modus ponens. In the long run, however, the advantages greatly outweigh this disadvantage. An instructor who prefers a standard Hilbert-style proof system with language $\neg$, $\rightarrow$ can use the system L or LF presented in Section 3.5 with only minor inconvenience.

A solid-box symbol ($\blacksquare$) denotes the end of a proof; a hollow-box symbol ($\square$) denotes the end of an example. Throughout the text, I use $\Leftrightarrow$ for "if and only if" and $\Rightarrow$ for "implies." (There are two exceptions to this convention: Section 3.6, on Gentzen systems, and Section 9.6, on semi-Thue systems.)

—————————————— *Acknowledgments* ——————————————

Let me first express my appreciation to Joe Shoenfield for the opportunity to attend his stimulating and crystal-clear lectures at Duke University on model theory, logic, set theory, and recursion theory. Although J. H. Roberts is my mathematical father, Joe Shoenfield is surely my mathematical uncle. I am also indebted to two former graduate students in logic at Duke for their helpful comments: Bill Mueller and Kyriakos Kontostathis. There are too many undergraduates to mention individually, but I especially want to thank Marc Borkan and Jennifer Fleischer for their help. In addition, I have profited from a number of discussions with two of my colleagues in the computer science department at Duke, Don Loveland and Gopalan Nadathur.

In the references, there is a list of logic texts that I have found useful (in addition to the Shoenfield text). Of these, I have probably referred to Mendelson's book most often.

I would also like to express my appreciation to the following reviewers: Peter G. Hinman, University of Michigan; Ned I. Rosen, Boston College; Judy Roitman, The University of Kansas; Samuel D. Shore, University of New Hampshire; Rick L. Smith, University of Florida; Thomas W. Wieting, Reed College. All made suggestions that have proved to be very valuable.

I also wish to thank Steve Quigley, the senior mathematics editor at PWS Publishing Company, and John Ward, editorial assistant, for their patience, help, and encouragement. Thanks also to Monique Calello, the production editor, and to Judith Abrahms, the copyeditor, for their hard work and careful attention to detail.

Richard Hodel

Axioms and Rules for Propositional Logic

Prop Axiom **(AXIOM)**	$\neg A \vee A$
Associative **(ASSOC)**	$\dfrac{A \vee (B \vee C)}{(A \vee B) \vee C} \qquad \dfrac{(A \vee B) \vee C}{A \vee (B \vee C)}$
Cut **(CUT)**	$\dfrac{A \vee B, \neg A \vee C}{B \vee C}$
Expansion **(EXP)**	$\dfrac{A}{B \vee A} \qquad \dfrac{A}{A \vee B}$
Contraction **(CTN)**	$\dfrac{A \vee A}{A}$
Disjunctive Syllogism **(DS)**	$\dfrac{A \vee B, \neg A}{B}$
Commutative **(CM)**	$\dfrac{A \vee B}{B \vee A}$
$\neg\neg$ *RULES*	$\dfrac{A}{\neg\neg A} \qquad \dfrac{A \vee B}{\neg\neg A \vee B}$
	$\dfrac{\neg\neg A}{A} \qquad \dfrac{\neg\neg A \vee B}{A \vee B}$
Modus Ponens **(MP)**	$\dfrac{A \rightarrow B, A}{B}$
Modus Tollens **(MT)**	$\dfrac{A \rightarrow B, \neg B}{\neg A}$
Hypothetical Syllogism **(HS)**	$\dfrac{A \rightarrow B, B \rightarrow C}{A \rightarrow C}$
Contrapositive **(CP)**	$\dfrac{A \rightarrow B}{\neg B \rightarrow \neg A}$
Join **(JOIN)**	$\dfrac{A, B}{A \wedge B}$
Separation rules **(SEP)**	$\dfrac{A \wedge B}{A} \qquad \dfrac{A \wedge B}{B}$

(In all cases, $A_x[t]$ requires that t is substitutable for x in A.)

Substitution Axioms (SUBST AX) $\forall xA \rightarrow A_x[t]$

                                       $\forall xA \rightarrow A$ (special case)

Equality Axioms (EQ AX) $(x = t) \rightarrow (A \leftrightarrow A_x[t])$

Reflexive Axiom (REF AX) $\forall x_1(x_1 = x_1)$

Substitution Theorems (SUBST THM) $A_x[t] \rightarrow \exists xA$

                                       $A \rightarrow \exists xA$ (special case)

∀-Elimination (∀-ELIM) $\dfrac{\forall xA}{A_x[t]} \qquad \dfrac{\forall xA}{A}$ (special case)

Existential Generalization (∃-GEN) $\dfrac{A_x[t]}{\exists xA} \qquad \dfrac{A}{\exists xA}$ (special case)

Add ∀-Rule (ADD ∀) $\dfrac{A \vee B}{A \vee \forall xB}$ (x not free in A)

Generalization (GEN) $\dfrac{A}{\forall xA}$

Substitution Rule (SUBST RULE) $\dfrac{A}{A_x[t]}$

∀-Introduction Rule (∀-INTRO) $\dfrac{A \rightarrow B}{A \rightarrow \forall xB}$ (x not free in A)

∃-Introduction Rule (∃-INTRO) $\dfrac{A \rightarrow B}{\exists xA \rightarrow B}$ (x not free in B)

∀-Distribution Rule (∀-DIST) $\dfrac{A \rightarrow B}{\forall xA \rightarrow \forall xB}$

∃-Distribution Rule (∃-DIST) $\dfrac{A \rightarrow B}{\exists xA \rightarrow \exists xB}$

List of Symbols

First-order languages

Mathematical theories

Starting functions: $+$, $\times$, $U_{n,k}$, $(n \geq 1, 1 \leq K \leq n)$, $K_<$

Composition: $F(a_1, \ldots, a_n) = G(H_1(a_1, \ldots, a_n), \ldots, H_k(a_1, \ldots, a_n))$

Minimalization: $F(a_1, \ldots, a_n) = \mu x[G(a_1, \ldots, a_n, x) = 0]$ (G regular)

Background

T his chapter contains background material for the remainder of the text and should be covered as required. Topics include induction, formal systems, set theory, functions, relations, countable and uncountable sets, Cantor diagonal arguments, axiom systems, algorithms, computable functions and decidable relations, recursive functions and recursive relations, and Church's thesis.

1.1
Overview of Mathematical Logic

Logic is often described as the analysis of reasoning. The first systematic study of logic is due to the Greek philosopher Aristotle; his methods, however, were not mathematical. In mathematical logic we study logic using mathematical tools; in addition, there is an emphasis on the language of mathematics and the type of reasoning done by mathematicians. Thus concepts such as *axiom, proof, theorem, consistency, truth, logical validity,* and *tautological consequence* are given precise definitions, and theorems are proved about these concepts.

Let us give a brief overview of classical two-valued logic. The simplest branch of logic is *propositional logic,* the study of the propositional connectives *not, or, and, if...then,* and *if and only if* (denoted $\neg$, $\vee$, $\wedge$, $\rightarrow$, and $\leftrightarrow$ respectively). Chapters 2 and 3 are devoted to propositional logic. But propositional logic is not adequate for analyzing all arguments in ordinary discourse or in mathematics, so there is a more advanced branch of logic, called *first-order logic.* First-order logic (studied in Chapters 4 and 5) includes propositional logic, but in addition uses the quantifiers *for all* (denoted $\forall$) and *there exists* (denoted $\exists$).

By 1850, logic consisted mainly of propositional logic and the classification of syllogisms, a special case of first-order logic. The German philosopher Immanuel Kant stated in 1787 that since Aristotle, logic *"has not been able to advance a single step, and to all appearances is a closed and completed body of study."* The modern, mathematical approach to logic begins with Boole's publication of *The Mathematical Analysis of Logic* in 1847 and Frege's publication of *Begriffsschrift* (literally, "concept writing") in 1879. In his famous treatise, Boole describes an algebra that can be interpreted either as propositional logic or as the logic of syllogisms. To Frege we owe the development of first-order logic. First-order logic differs from propositional logic not only in the use of the quantifiers *for all* and *there exists,* but also by the inclusion of symbols for constants, n-ary relations, and n-ary functions.

An interesting feature of mathematical logic is the interplay between *syntactic* and *semantic* concepts. The distinction between the two will become clear as we proceed, but for now we can say that semantic concepts have to do with meaning and interpretation, whereas syntactic concepts have to do with form. To illustrate, in first-order logic there is the syntactic concept of *theorem* and the semantic concept of *logical validity.* A major theorem of first-order logic states that a formula is a theorem if and only if it is logically valid. Again in first-order logic, we have the syntactic concept of *consistency* and the semantic concept of a *model.* In Chapter 5 we will prove that an extension of first-order logic is consistent if and only if it has a model. These two results are instances of Gödel's completeness theorem (1930), arguably the most important theorem in mathematical logic.

Mathematical logic is a branch of mathematics; in this respect, it is similar to Euclidean geometry and to number theory. Euclidean geometry is the study of points and lines; number theory is the study of properties of the integers (for example, primality). To give some idea of the scope of mathematical logic, we will discuss a number of questions that have greatly influenced its development in the twentieth century; most of these questions were raised by the German mathematician David Hilbert between 1900 and 1930.

***Question 1** Is arithmetic consistent?* By arithmetic, we mean that branch of mathematics which studies addition and multiplication of the natural numbers $\mathbb{N} = \{0, 1, 2, \ldots\}$; that is, elementary number theory. Questions about the consistency of arithmetic and of other mathematical systems were motivated in part by various paradoxes in set theory that surfaced around 1900 (see Russell's paradox, in Section 4 of this chapter). Some related, but even more fundamental, questions are the following: What do we mean by consistency? What methods are acceptable in proving consistency? Later we will prove that both propositional logic and first-order logic are consistent (Chapters 3 and 5 respectively).

***Question 2** Is arithmetic complete?* To elaborate, suppose we divide all mathematical statements about arithmetic into two categories: *true* and *false.* Next, select an easily recognizable list of axioms for arithmetic; these axioms should be mathematical statements that are obviously true (e.g., $x + y = y + x$). We now ask: Can we select this list of axioms so that all true statements of arithmetic are derivable from the axioms? In 1931, Kurt Gödel published his celebrated incompleteness theorem; this remarkable result asserts that the answer is *no:* It is impossible to select a consistent, easily recognizable list of axioms for arithmetic and then derive all true statements of arithmetic from these axioms. Gödel's theorem dramatizes the fact that there is a distinction between truth (semantic concept) and provability (syntactic concept). Gödel's incompleteness theorem is one of the deepest and most important theorems in twentieth-century mathematics, and the techniques developed by Gödel are fundamental to modern mathematical logic. The proof of this famous result is outlined in Chapter 7 and complete details are given in Chapter 8.

In mathematical logic, there is an emphasis on formal languages. One reason for this is suggested by Question 2, in which we refer to mathematical statements about arithmetic. What do we mean by this? Consider, for example, the two statements (1) 5 is prime; (2) 5 is green. By choosing an appropriate formal language, we can eliminate (2) as a mathematical statement about arithmetic.

Question 3 Is there a concise list of logical axioms and rules of inference that summarizes all of the logic used in mathematical reasoning? The answer is *yes*; this is a consequence of Gödel's completeness theorem (see Chapter 5).

Question 4 Is there an algorithm that, given an arbitrary polynomial $p(x_1, \ldots, x_n)$ with integer coefficients, decides (YES or NO, in finitely many steps) whether the polynomial equation $p(x_1, \ldots, x_n) = 0$ has a solution in integers? This is Hilbert's tenth problem; it is one of 23 problems raised by Hilbert in 1900 at the International Congress of Mathematicians. Although this appears to be a problem in number theory, mathematical logic plays a key role in its solution. Hilbert's tenth problem was not completely solved until 1970. The proof that no such algorithm exists is one of the outstanding achievements of mathematical logic (see Chapter 10).

Question 4 asks for an algorithm to solve an infinite list of problems. Such problems are called *decision problems,* and come within the scope of mathematical logic. This suggests the following, more fundamental question.

Question 5 What is an algorithm? To focus the question somewhat, suppose we have a function from the set $\mathbb{N}$ of natural numbers to the set of natural numbers (that is, $F: \mathbb{N} \to \mathbb{N}$). We say that F is *computable* if there is an algorithm such that given $a \in \mathbb{N}$, the algorithm computes the value $F(a)$ in a finite number of steps. The successor function $S(a) = a + 1$ is an example of a computable function. This definition of a computable function is not precise, because it depends upon the notation of an algorithm, which is itself not a precisely defined concept. A major contribution of mathematical logic both to mathematics and to computer science is the clarification of the notion of an algorithm and a precise definition of a computable function. In particular, mathematical logic gives us a general framework for proving the *unsolvability* of certain decision problems.

Question 6 Are we justified in using infinite sets in mathematics? Consider: Research in physics suggests that the infinite does not exist in nature; careless use of infinite sets can lead to a contradiction; Brouwer claims that the logic used in mathematics was designed to reason about finite sets, not infinite sets. Is it inevitable that the use of infinite sets will lead to contradictions? Or can we introduce an appropriate axiom system for infinite sets and prove, without recourse to infinity, the consistency of this axiom system? These and related ideas will be discussed in Section 6.4.

— 1.2 —

Induction

The principle of finite induction is a fundamental tool for proving results in mathematical logic. We begin with the simplest version.

Induction: Let $S(n)$ be a statement that depends on the positive integer n. To prove that $S(n)$ is true for all $n \geq 1$, it suffices to prove the following:

Beginning step $S(1)$ is true.
Induction step For $n \geq 1$, if $S(n)$ is true, then $S(n + 1)$ is true.

Induction is a fundamental property of the positive integers. As such, it is regarded as an axiom or a basic principle rather than something which requires proof. Its plausibility, however, can be shown as follows: Abbreviate the induction step by $S(n) \rightarrow S(n + 1)$. We now argue that $S(1)$, $S(2)$, $S(3)$, ... are all true. By the beginning step, $S(1)$ is true. Now take $n = 1$ in the induction step $S(n) \rightarrow S(n + 1)$ to obtain $S(1) \rightarrow S(2)$. By modus ponens, we conclude that $S(2)$ is true. Next, take $n = 2$ in the induction step $S(n) \rightarrow S(n + 1)$ to obtain $S(2) \rightarrow S(3)$. Since $S(2)$ is true, we again use modus ponens to conclude that $S(3)$ is true. Continuing this process, we see that each of $S(1)$, $S(2)$, $S(3)$, ... is true. Let us illustrate induction with an example.

Example 1:

For $n \geq 1$, $1 + 2 + \cdots + n = n(n + 1)/2$. Let $S(n)$ say: The sum of the first n positive integers is $n(n + 1)/2$.

Beginning step $S(1)$ states that $1 = (1)(2)/2$, and this is certainly true. Thus the beginning step has been checked.

Induction step Let $n \geq 1$ and assume that $S(n)$ is true. We must show that $S(n + 1)$ is true; that is, that $1 + 2 + \cdots + (n + 1) = (n + 1)(n + 2)/2$. We have:

$$\sum_{k=1}^{n+1} k = \sum_{k=1}^{n} k + (n + 1)$$

$$= \frac{n(n + 1)}{2} + (n + 1) \quad \text{induction hypothesis}$$

$$= \frac{(n + 1)}{2}[n + 2] \quad \text{factor out } \frac{(n + 1)}{2}$$

The second line is the key step. We replace $1 + 2 + \cdots + n$ with $n(n + 1)/2$. This is allowed because we are *assuming* that $S(n)$ is true. The assumption that $S(n)$ is true is called the *induction hypothesis*. □

In proving the induction step, that is, in proving that $S(n + 1)$ is true—we may need to know that several, perhaps all, of the statements $S(1), \ldots, S(n)$ are true. We extend induction to cover this situation as follows.

Complete induction: Let $S(n)$ be a statement that depends on the positive integer n. To prove that $S(n)$ is true for all $n \geq 1$, it suffices to prove the following:

Beginning step $S(1)$ is true.

Induction step For $n \geq 1$, if $S(1), \ldots, S(n)$ are all true, then $S(n + 1)$ is true.

We point out that induction or complete induction can start at any integer a. In the case of complete induction, the beginning step is to show that $S(a)$ is true, and the induction step consists of showing that if $S(a), \ldots, S(n)$ are true, then $S(n + 1)$ is true. Many induction arguments start at 0.

We illustrate these ideas with a complete induction argument that begins at 2. Recall that a positive integer $n > 1$ is *composite* if it can be written as $n = a \times b$, where a and b are integers with $1 < a < n$ and $1 < b < n$. A positive integer $n > 1$ that is not composite is *prime*. For example, $2, 3, 5$, and 7 are prime numbers; $4, 6, 8$, and 9 are composite numbers. The number 1 is neither prime nor composite. Thus the set of positive integers partitions into three sets: 1, primes, and composites.

Theorem 1 (Factorization into primes): Every integer $n > 1$ is a product of primes.

Proof: Let $S(n)$ say: n is a product of primes. We use complete induction to show that $S(n)$ is true for all $n \geq 2$.

Beginning step $S(2)$ states that 2 is a product of primes. Since 2 is prime, this is certainly true. (By convention, a prime is considered a product of primes.)

Induction step Now let $n \geq 2$ and assume that $S(2), \ldots, S(n)$ are all true. We are required to prove that $n + 1$ is a product of primes. If $n + 1$ is prime, there is nothing to prove, by our convention. So assume that $n + 1$ is composite and write $n + 1 = a \times b$ with $1 < a < n + 1$ and $1 < b < n + 1$. By the induction hypothesis $S(a)$, a is a product of primes, say $a = p_1 \times \cdots \times p_r$. Likewise, by the induction hypothesis $S(b)$, b is a product of primes, say $b = q_1 \times \cdots \times q_s$. But $n + 1 = a \times b$, so we have

$$n + 1 = (p_1 \times \cdots \times p_r) \times (q_1 \times \cdots \times q_s)$$

and $n + 1$ is a product of primes as required. ■

There is another version of induction, called the *well-ordering principle,* or the *least number principle.*

Well-ordering principle: Every nonempty set X of positive integers has a smallest element; in other words, there is a positive integer a in X such that $a \leq b$ for all b in X.

We have not proved induction but instead have treated it as a basic property of the positive integers. On the other hand, induction, complete induction, and the well-ordering principle are closely related. Namely, using basic set theory and familiar properties of the positive integers, one can assume any one of these and prove the other two; see Exercises 9 –11 and Exercise 3 of Section 1.4.

───────────── *Exercises for Section 1.2* ─────────────

1. Use induction to prove the following:

   (1) $\sum_{k=1}^{n} k^2 = \dfrac{n(n + 1)(2n + 1)}{6}$.   (3) $\sum_{k=1}^{n} k(k + 1) = \dfrac{n(n + 1)(n + 2)}{3}$.

   (2) $\sum_{k=1}^{n} k^3 = \left[\dfrac{n(n + 1)}{2}\right]^2$.   (4) For $n \geq 0$, $(2n)! \geq (n!)(n!)2^n$.

2. Let $H_n = \sum_{k=1}^{n} 1/k$. Show that $\sum_{k=1}^{n} H_k = (n + 1)H_n - n$.

3. Prove deMoivre's theorem: $(\cos \theta + i \sin \theta)^n = \cos n\theta + i \sin n\theta$ (all $n \geq 1$). (*Note:* $i^2 = -1$.)

4. (*VPI&SU Mathematics Contest*) Let $x_1 = 1$, $x_2 = 3$, and for $n > 1$ let $x_{n+1} = \dfrac{1}{n + 1} \sum_{k=1}^{n} x_k$. Find $\lim_{n \to \infty} x_n$ and justify your answer.

5. (*VPI&SU Mathematics Contest*) Let f be a function such that $f(1) = 1$ and, for $n > 1$, $f(n) = f(f(n - 1)) + f(n - f(n - 1))$. Find $f(10,000)$.

6. Let $a_0 = a_1 = 1$ and $a_{n+1} = a_n + 2a_{n-1}$ for $n \geq 1$. Show that $a_n = [2^{n+1} + (-1)^n]/3$ for all $n \geq 0$.

7. The Fibonacci numbers are defined by $F_0 = F_1 = 1$ and $F_{n+1} = F_n + F_{n-1}$ for $n \geq 1$. Thus the first few Fibonacci numbers are $1, 1, 2, 3, 5, 8, \ldots$.

   (1) Let $\phi = (1 + \sqrt{5})/2$, $\psi = (1 - \sqrt{5})/2$. Show that $\phi^2 = 1 + \phi$ and $\psi^2 = 1 + \psi$.

   (2) Use induction to show that $F_n = (\phi^{n+1} - \psi^{n+1})/\sqrt{5}$ for all $n \geq 0$.

8. Let $\mathbb{N} = \{0, 1, 2, \ldots\}$ and let ψ be a function from $\mathbb{N} \times \mathbb{N}$ into $\mathbb{N}$ such that

   R1 $\psi(a, 0) = a + 1$;

   R2 $\psi(0, b + 1) = \psi(1, b)$;

   R3 $\psi(a + 1, b + 1) = \psi(\psi(a, b + 1), b)$.

   (1) Find $\psi(2, 1)$.

   (2) Show that for all $a, b \in \mathbb{N}$, $\psi(a, b)$ can be calculated in a finite number of steps using (R1), (R2), and (R3). (*Hint:* Use double induction. Let $S(b)$ say: For all $a \in \mathbb{N}$, $\psi(a, b)$ can be calculated in a finite number of steps using (R1), (R2), and (R3). Prove $S(b)$ by induction. In proving $S(b + 1)$, let $T(a)$ say: $\psi(a, b + 1)$ can be calculated in a finite number of steps using (R1), (R2), and (R3). Prove $T(a)$ by induction.)

   (3) Guess at, and then prove by induction, a formula for $\psi(a, 2)$.

   (4) Find $\psi(5, 4)$.

9. Derive complete induction from induction. (*Hint:* Assume that $S(1)$ is true and that for all $n \geq 1$, if $S(1), \ldots, S(n)$ are true, then $S(n + 1)$ is true. Let $T(n)$ say: $S(1), \ldots, S(n)$

are all true. Use induction to show that $T(n)$ holds for all $n \geq 1$; conclude that $S(n)$ holds for all $n \geq 1$.)

10. Use complete induction to prove the well-ordering principle.
11. Use the well-ordering principle to prove induction.

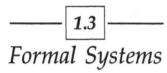

Formal Systems

A formal system is the appropriate mathematical framework for defining and clarifying such fundamental concepts as *mathematical statement, axiom, proof, theorem*, and *consistency*. We begin by introducing some terminology that is used in the study of formal languages.

Definition 1: Let S be a set of symbols. An *expression in S* (or *word in S*) is a finite sequence of symbols of S. For example, if the set of symbols is $\{a, b, c\}$, then *aabc* and *cba* are both expressions in S.

To describe a formal language, one first specifies a set S of symbols and then selects certain expressions as the *formulas*. In the most important cases, the formulas are those expressions that have a meaningful interpretation. For example, if S is the set of the 26 letters of the alphabet—that is, if $S = \{a, b, c, \ldots, z\}$—then *bird* and *ffmg* are both expressions in S. If we declare the formulas to be those expressions that appear in some standard dictionary, then *bird* is a formula but *ffmg* is not.

Definition 2: A *formal system F* has three components:

• **Formal language** (symbols + certain expressions called *formulas*);
• **Axioms** (certain formulas);
• **Rules of inference**.

The language should have the property that there is an algorithm—that is, a mechanical procedure—for deciding (YES or NO in a finite number of steps) whether an arbitrary expression is a formula. All of the languages we study will have this property. Throughout this text, the letters A, B, C, D, and E are used to denote formulas of a formal system, and Γ and Δ are used to denote sets of formulas.

The *axioms* of a formal system F are certain carefully selected formulas. In many cases the language of the formal system is selected with some interpretation in mind, and the axioms are then certain formulas that are "obviously true." A formal system is required to have an algorithm that decides whether an arbitrary formula is an axiom.

A *rule of inference* is a rule that states that one formula (called the *conclusion*) follows from a finite set of formulas (called the *hypotheses*). Suppose $A_1, \ldots, A_k, B$ are formulas (in most cases $k = 1$ or $k = 2$). A rule of inference with hypotheses $A_1, \ldots, A_k$ and conclusion B is often denoted schematically by $A_1, \ldots, A_k/B$ or by

$$\frac{A_1, \ldots, A_k}{B}.$$

In any formal system, the number of rules is finite. Before proceeding further, let us give an example of a formal system.

Example 1:

The formal system ADD.

Language There are three symbols: +, =, and |. A formula is any expression of the form $x + y = z$, where x, y, and z are expressions that use just the symbol |. For example, ||| + || = ||||| is a formula, but || + + = || and || = | + | are not formulas.

Axioms The only axiom is the formula | + | = ||.

Rules of inference There are two rules of inference:

$$\text{R1:} \quad \frac{x + y = z}{x| + y = z|} \qquad \text{R2:} \quad \frac{x + y = z}{y + x = z}$$

□

A formal system is usually introduced with an interpretation in mind; in other words, the language and axioms are selected to study specific subject matter such as the logic of propositions, arithmetic, set theory, and so on. An interpretation is not, however, a required part of a formal system.

Those concepts of a formal system that are defined with no reference to an interpretation are called *syntactic* concepts. On the other hand, concepts that depend on the meaning or interpretation of the symbols are *semantic* concepts. The interplay between syntactic and semantic concepts is fundamental in mathematical logic. For example, in propositional logic we have the syntactic concept of a *theorem* and the semantic concept of a *tautology*. We will spend considerable effort in showing that a formula of propositional logic is a theorem if and only if it is a tautology.

Now that the concept of a formal system is available, we can give precise definitions of *proof* and *theorem*. Both are syntactic concepts.

Definition 3: Let F be a formal system. A *proof in* F is a finite sequence $A_1, \ldots, A_n$ of formulas of F such that for $1 \le k \le n$, one of the following conditions is satisfied: (1) A_k is an axiom of F; (2) $k > 1$ and A_k is the conclusion of a rule of inference whose hypotheses are among the previous formulas $A_1, \ldots, A_{k-1}$. A proof in F can be displayed schematically as follows:

(1) A_1

  ⋮
  ⋮
  ⋮

(n) A_n.

If $A_1, \ldots, A_n$ is a proof in F with $A_n = A$, we say that $A_1, \ldots, A_n$ is a *proof of* A and that A is a *theorem of* F; we denote this by $\vdash_F A$. When the formal system F under discussion is understood, we often omit the subscript F and simply write $\vdash A$.

Example 2:

The formula $\| + \||| = \|||||$ is a theorem of the formal system ADD. To show this, it suffices to write out a formal proof:

(1)	$\| + \| = \|\|$	AXIOM
(2)	$\|\| + \| = \|\|\|$	R1
(3)	$\|\|\| + \| = \|\|\|\|$	R1
(4)	$\| + \|\|\| = \|\|\|\|$	R2
(5)	$\|\| + \|\|\| = \|\|\|\|\|$	R1

In summary, $\vdash_{\text{ADD}} \| + \||| = \|||||$ as required. □

The notion of a formal proof is very important. Indeed, formal systems were introduced by Hilbert so that the notion of a mathematical proof could be studied using mathematical tools. Just as geometry is the study of points and lines and number theory is the study of the natural numbers, *proof theory* is the study of mathematical proofs. A formal system is the framework in which proof theory takes place.

Given a formal system F and a property Q of formulas, we often want to prove that every theorem of F has property Q. Let $\vdash_F A$; then there is a proof of A in n steps for some $n \geq 1$, and we can picture this proof of A as follows:

(1) A_1

(2) A_2

$\vdots$

$\vdots$

(n) $A_n = A$.

Now assume that property Q satisfies these two conditions: (1) Each axiom of F has property Q; (2) for each rule of inference of F, if each hypothesis of the rule has property Q, then the conclusion of the rule also has property Q. With these assumptions, we claim that each of the formulas $A_1, \ldots, A_n$ has property Q; since the last formula A_n is A, the formula A has property Q as required.

The formula A_1 is an axiom, since it is the first line of a proof. By (1) we conclude that A_1 has property Q. Now consider the second formula A_2. If A_2 is also an axiom, then A_2 has property Q as before. Otherwise, A_2 is the conclusion of a rule of inference whose hypothesis is A_1. But we have already seen that A_1 has property Q, so by (2), A_2 has property Q. Continuing in this way we see that each of the formulas $A_1, \ldots, A_n$ has property Q.

We have just given an informal proof of an important technique known as *induction on theorems.* We will now give a rigorous proof of this result, mainly to emphasize the role that complete induction plays in its justification.

Theorem 1 (Induction on theorems): Let F be a formal system and let Q be a property of formulas of F. To prove that every theorem of F has property Q, it suffices to prove the following: (1) Every axiom of F has property Q; (2) for each rule of inference of F, if each hypothesis of the rule has property Q, then the conclusion also has property Q.

Proof: Let $S(n)$ say: Every theorem with a proof in n steps has property Q. We use complete induction to show that $S(n)$ is true for all $n \geq 1$; since every theorem has a proof in n steps for some $n \geq 1$, it follows that every theorem has property Q as required.

Beginning step $S(1)$ states that a theorem with a proof in one step has property Q. Now a one-step proof is actually an axiom, and every axiom has property Q by (1).

Induction step Let $n \geq 1$ and assume that $S(1), \ldots, S(n)$ are all true; we are required to prove that every theorem with a proof in $n + 1$ steps has property Q. Let A have a proof in $n + 1$ steps, say $A_1, \ldots, A_n, A_{n+1}$, where $A_{n+1} = A$. We consider the case in which A_{n+1} is the conclusion of a rule of inference whose hypotheses are among $A_1, \ldots, A_n$. By (2), A_{n+1} has property Q provided that each hypothesis (among $A_1, \ldots, A_n$) has property Q. Now for $1 \leq k \leq n$, A_k is a theorem with a proof in k steps; it follows by the induction hypothesis $S(k)$ that A_k has property Q. Thus each hypothesis has property Q; hence A_{n+1} has property Q as required. ∎

It is not surprising that an application of induction on theorems is often referred to as *proof by induction on the number of steps in the proof.* We now illustrate its use with an example.

Example 3:

Call a formula $x + y = z$ of the formal system ADD *true* if the total number of occurrences of the symbol | on the left side of = is the same as the number of occurrences on the right side. For example, || + ||| = ||||| is true, because the number of occurrences of | in each case is 5. On the other hand, || + ||| = |||| is not true, because the count for the left side is 5 but the count for the right side is 4. We now use induction on theorems to show that every theorem of ADD is true. In this example, Q is the property of being true. The only axiom is | + | = ||, and this formula is true (the count on each side of = is 2). There are two rules; we discuss R1 and leave R2 to the reader. Assume the hypothesis $x + y = z$ of R1 is true and let i, j, k be the number of occurrences of | in $x, y,$ and z respectively. Since $x + y = z$ is true, $i + j = k$. Now the number of occurrences of | in $x|$ is $i + 1$ and the number of occurrences of | in $z|$ is $k + 1$. Since $(i + 1) + j = (k + 1)$, the conclusion $x| + y = z|$ of R1 is true, as required. □

Definition 4: A formal system F is *decidable* if there is an algorithm that, given an arbitrary formula A of F, decides (YES or NO in a finite number of steps) whether $\vdash_F A$. If there is no such algorithm, we say that F is *undecidable*.

One of the most important questions we can ask about a formal system is whether it is decidable. Later we will see that propositional logic is decidable and that first-order logic is undecidable. As a rather trivial example for now, the formal system ADD is decidable (see Exercise 8).

To avoid false impressions about the role of formal systems, we emphasize that formal systems provide us with a framework in which we can define such fundamental concepts as *axiom, theorem, proof,* and *consistency* and then prove mathematical results about these concepts. Formal systems are *not* introduced for the sake of writing out formal proofs of specific formulas such as the proof of $|| + ||| = |||||$ (Example 2). Rather, the emphasis is on proving results *about* formal systems. A major portion of this text is devoted to the study of these important formal systems: P (propositional logic); FO_L (first-order logic); NN (formal arithmetic), the setting of Gödel's incompleteness theorem; and ZF (a formal system in which it is possible to develop all of classical mathematics).

───────── *Exercises for Section 1.3* ─────────

1. (*A* Tag *game that tags itself*) List all theorems of the following formal system. (This exercise and the next appear in Yasuhara's *Recursive Function Theory and Logic*.)

   Language There are three symbols: a, b, and c. Any expression is a formula.
   Axiom cabcba
   Rules of inference There are three rules of inference:

   > a Rule: If a formula begins with a, add *cac* to the right and then erase the first three symbols.
   > b Rule: If a formula begins with b, add *bab* to the right and then erase the first three symbols.
   > c Rule: If a formula begins with c, add *ca* to the right and then erase the first three symbols.

2. (*A* Tag *game that cycles*) List all theorems of the following formal system:

   Language There are three symbols: a, b, and c. Any expression is a formula.
   Axiom abccba
   Rules of inference There are three rules of inference:

   > a Rule: If a formula begins with a, add *bab* to the right and then erase the first three symbols.
   > b Rule: If a formula begins with b, add *abba* to the right and then erase the first three symbols.
   > c Rule: If a formula begins with c, add *ca* to the right and then erase the first three symbols.

3. Invent a formal system, MULT, that is similar to ADD but whose theorems are true statements about multiplication of positive integers. Then show that the formula $|| \times ||| = ||||||$ is a theorem of the system.

4. We describe a formal system PR, whose theorems are well-formed strings of parentheses. The language has two symbols, namely (and); any expression is a formula; ( ) is the only axiom. Let A and B denote formulas (=expressions). There are three rules of inference:

$$\text{Add Rule: } \frac{A}{(A)} \qquad \text{Double Rule: } \frac{A}{AA} \qquad \text{Omit Rule: } \frac{A()B}{AB}.$$

(1) Show that the following formulas are theorems of PR:

    (a) $((( ))( ))$;

    (b) $(( ))( )( )$;

    (c) $( )(( )( ))( )$.

(2) Prove that every theorem of PR has the property that the number of left parentheses is the same as the number of right parentheses.

5. We describe another formal system, PR, whose theorems are well-formed strings of parentheses. The language and axiom are the same as in Exercise 4. There are two rules of inference, as follows:

$$\text{Add Rule: } \frac{A}{(A)} \qquad \text{Join Rule: } \frac{A, B}{AB}.$$

Show that the formulas in Exercise 4(1) are theorems of PR.

6. Let F be a formal system, let FOR be the set of all formulas of F, and let f be a function from FOR into $\{0, 1\}$ having these two properties: (1) $f(A) = 0$ for every axiom A of F; (2) if $A_1, \ldots, A_k/B$ is any rule of inference of F and $f(A_1) = f(A_n) = 0$, then $f(B) = 0$. Prove that $f(A) = 0$ for every theorem A of F.

7. We describe a formal system EVEN, whose theorems are even positive integers. The only symbol is | and any expression is a formula. The only axiom is the formula ||. There is just one rule of inference, namely $A/A||$.

(1) Show that every theorem of EVEN has an even number of occurrences of the symbol |.

(2) Show that every formula with an even number of occurrences of the symbol | is a theorem of EVEN.

(3) Show that EVEN is decidable.

8. Refer to the formal system ADD.

(1) Show that every true formula of ADD is a theorem of ADD.

(2) Show that ADD is decidable.

9. Let F be the formal system described as follows: The language has two symbols, a and b; any expression is a formula; the axiom is ab; there is one rule of inference, namely A/aAb.

(1) Prove that for all $n \geq 1$, $a^n b^n$ is a theorem of F. (*Note:* a^n denotes a string of n a's.)

(2) Let A be a theorem of F. Prove that there exists $n \geq 1$ such that A is $a^n b^n$.

(3) Show that F is decidable.

10. We describe a formal system PAL, whose theorems are all palindromes on three letters. The language has three symbols, a, b, c; any expression is a formula; there are six axioms: $a, b, c, aa, bb,$ and cc; there are three rules of inference:

$$\text{Add } a \text{ Rule: } \frac{A}{aAa} \qquad \text{Add } b \text{ Rule: } \frac{A}{bAb} \qquad \text{Add } c \text{ Rule: } \frac{A}{cAc}$$

(1) Show that $abccccba$ and $abcccba$ are theorems of PAL.

(2) Show that every theorem of PAL is a palindrome on the three letters a, b, c.

(3) Show that every palindrome on the three letters a, b, c is a theorem of PAL.

(4) Show that F is decidable. (Note: See Example 1 in Section 1.7.)

11. Let F be a formal system and let Γ be a set of formulas of F. A *proof in* F *using* Γ is a finite sequence $A_1, \ldots, A_n$ of formulas of F such that for $1 \le k \le n$, one of the following conditions is satisfied: (1) A_k is an axiom of F; (2) $A_k \in \Gamma$; (3) $k > 1$ and A_k is the conclusion of a rule of inference whose hypotheses are among the previous formulas $A_1, \ldots, A_{k-1}$. Informally, the formulas in Γ are "additional" axioms that may be used in the proof $A_1, \ldots, A_n$. If $A_1, \ldots, A_n$ is a proof in F using Γ with $A_n = A$, we say that $A_1, \ldots, A_n$ is a *proof of A using* Γ and that A is a *theorem of* Γ; we denote this by $\Gamma \vdash_F A$ (more simply, $\Gamma \vdash A$). Now let Γ and Δ be sets of formulas in F and let A be a formula in F. Prove the following:
    (1) If $\Gamma \subseteq \Delta$ and $\Gamma \vdash A$, then $\Delta \vdash A$.
    (2) If $\Gamma \vdash A$, then there is a finite $\Gamma_0 \subseteq \Gamma$ such that $\Gamma_0 \vdash A$.
    (3) If $\Gamma \vdash A$ and $\Delta \vdash B$ for every formula $B \in \Gamma$, then $\Delta \vdash A$.
12. (*Consistency*) A formal system F is said to be *absolutely consistent* if there is at least one formula that is not a theorem of F. Now let F be a formal system whose language has a symbol $\neg$ and assume that for every formula A, $\neg A$ is also a formula. (The intended interpretation of $\neg$ is *not*.) We say that F is *consistent* if there is no formula A such that both A and $\neg A$ are theorems of F. Prove the following:
    (1) If F is consistent, then F is absolutely consistent.
    (2) Assume that for any formulas A and B, $\{A, \neg A\} \vdash_F B$ (see preceding exercise). Show that if F is absolutely consistent, then F is consistent.
13. (*Top-down = bottom-up*) Our definition of a theorem of a formal system is sometimes described as a *bottom-up* approach, since we start with axioms and apply rules of inference to build up longer and longer proofs, and thus more and more theorems. In this problem we outline another approach in which we obtain the theorems *top-down*. Let F be a formal system, let FOR be the set of formulas of F, and let THM be the set of theorems of F. A subset I of FOR is said to be *inductive* if it satisfies these two conditions: (a) Every axiom of the formal system F is in I; (b) If $A_1, \ldots, A_n/B$ is a rule of inference of F, and each of the hypotheses $A_1, \ldots, A_n$ are in I, then B is also in I.
    (1) Show that both FOR and THM are inductive.
    (2) Let I be any inductive set. Show that THM $\subseteq I$.
    (3) Let $\{I_t: t \in T\}$ be a collection of subsets of FOR, each of which is inductive. Show that $\bigcap_{t \in T} I_t$ is inductive.
    (4) Let $\{I_t: t \in T\}$ be the collection of all inductive subsets of FOR. Show that THM $= \bigcap_{t \in T} I_t$. In other words, the set of theorems is the smallest inductive subset of FOR. (*Note*: See Section 1.4 for set-theoretic terminology and for the definition of $\bigcap_{t \in T} I_t$.)
14. (*Hofstadter's MIU-system*) We describe a formal system MIU that appears in Hofstadter's *Gödel, Escher, Bach* 1979, 33–41. (This book is highly recommended to the reader.) The language has three symbols, M, I, and U; any expression is a formula; the only axiom is MI; there are four rules (x denotes any expression):

    R1 If xI is a theorem, so is xIU.
    R2 If Mx is a theorem, so is Mxx.
    R3 In any theorem, III can be replaced by U.
    R4 UU can be omitted from any theorem.

    (1) Show that $MUII$ is a theorem of MIU.
    (2) Show that the number of I's in a theorem of MIU is never a multiple of 3 (that is, $0, 3, 6, \ldots$ or $3k$, $k = 0, 1, 2, \ldots$).
    (3) Is MU a theorem of MIU?
15. (*MIU-system is decidable*) This exercise outlines a proof, due to Swanson and McEliece (*Mathematical Intelligencer* 10(#2) 1988, 48–49), that the MIU-system of Hofstadter is decidable. (See the preceding exercise for the description of MIU.) Let I^n be defined inductively by $I^1 = I$, $I^{n+1} = I^n I$; in other words, I^n is a string of n I's. Prove the

following:

(1) MI^{2^n} is a theorem of MIU.

(2) If $xIII$ is a theorem of MIU, then x is also a theorem of MIU.

(3) If n is a positive integer of the form $2^m - 3k(k \geq 0)$, then MI^n is a theorem of MIU.

(4) MI^n is a theorem of MIU for any positive integer n that is not a multiple of 3. (*Hint:* By (3), it suffices to show that any positive integer n not a multiple of 3 can be written in the form $2^m - 3k(k \geq 0)$. Choose m such that $2^m \geq n$; then show that $2^m - n$ or $2^{m+1} - n$ is a multiple of 3. *Note:* n is of the form $3t + 1$ or $3t + 2$; likewise for 2^m.)

(5) An expression x of the MIU-system is a theorem of MIU if and only if x is of the form My, where y is a string of I's and U's and the number of I's is not a multiple of 3.

Set Theory, Functions, and Relations

Basic Concepts of Set Theory

Set theory begins with two primitive, or undefined, concepts: *set* and *is an element of.* These two concepts are not defined because they do not seem to be explainable in more elementary terms. Roughly speaking, *sets* are the objects we are interested in and *is an element of* is a binary relation that either does or does not hold between any two given sets. Thus, if a and X are sets, exactly one of the following holds:

- a is an element of X, written $a \in X$;
- a is *not* an element of X, written $a \notin X$.

To avoid monotony, we sometimes use *collection* instead of *set* and often express $a \in X$ as *a is in X* or *a is a member of X*. We illustrate these ideas with familiar examples from mathematics.

$\mathbb{N}$ = set of natural numbers $0, 1, 2, 3, \ldots$

$\mathbb{N}^+$ = set of positive integers $1, 2, 3, \ldots$

$\mathbb{Z}$ = set of integers $0, \pm 1, \pm 2, \pm 3, \ldots$

$\mathbb{Q}$ = set of rational numbers r/s (r and s integers with $s > 0$)

$\mathbb{R}$ = set of real numbers

The number 1 is an element of all of these sets; $1/2 \in \mathbb{Q}$ but $1/2 \notin \mathbb{Z}$; $\sqrt{2} \in \mathbb{R}$ but $\sqrt{2} \notin \mathbb{Q}$. Throughout this text $\mathbb{N}, \mathbb{N}^+, \mathbb{Z}, \mathbb{Q}$, and $\mathbb{R}$ are consistently used to denote these five important sets. We now use the membership relation $\in$ to define a new binary relation between sets.

Definition 1: Given two sets A and B, we say that A is a *subset* of B, written $A \subseteq B$, if every element of A is also an element of B.

For example, $\mathbb{N}^+ \subseteq \mathbb{N}, \mathbb{N} \subseteq \mathbb{Z}, \mathbb{Z} \subseteq \mathbb{Q}$, and $\mathbb{Q} \subseteq \mathbb{R}$. The set that has no elements is called the *empty set* and is denoted $\varnothing$; for any set A, we have

$\emptyset \subseteq A$. In mathematics, *equals* means *is identical with*. In particular, two sets A and B are *equal*, written $A = B$, if they are one and the same set. If A and B are not equal, we write $A \neq B$. In axiomatic set theory, it is necessary to have an axiom that establishes a connection between the membership relation $\in$ and the equality relation $=$; this is called the *axiom of extensionality* and can be stated in terms of subsets as follows: *If $A \subseteq B$ and $B \subseteq A$, then $A = B$.*

Specification of sets: There are several ways of describing a set. One way is to list the elements of the set in braces $\{\ldots\}$; for example, $\{1, 2\}$ is the set having the first two positive integers as elements. For any two sets a and X, $a \in X$ if and only if $\{a\} \subseteq X$. Here is another way of specifying sets: Given a set X and a property Q of sets, we construct a subset A of X by

$$A = \{x : x \in X \text{ and } x \text{ has property } Q\}.$$

For example, $(a, b) = \{x : x \in \mathbb{R} \text{ and } a < x < b\}$.

Russell's paradox: Careless use of specification can lead to a contradiction, as illustrated by this famous example, due to Bertrand Russell in 1902. Let $A = \{x : x \text{ is a set and } x \notin x\}$, and assume that A is a "legitimate" set. Then for any set x, we have $x \in A \Leftrightarrow x$ is a set and $x \notin x$. In particular, if we let x be the set A, we obtain

$$A \in A \Leftrightarrow A \text{ is a set and } A \notin A.$$

But we are assuming that A is a set, so this simplifies to $A \in A \Leftrightarrow A \notin A$, a contradiction. One can avoid this paradox by a careful statement of the operations that are allowed in the construction of sets. The reader is referred to the discussion of Zermelo-Frankel set theory in Section 6.4.

We now discuss the following operations for constructing new sets from given sets: Boolean operations; arbitrary unions and intersections; power set of a set; and Cartesian products.

Boolean operations: Given a set X and subsets A and B of X, there are subsets $A \cup B$, $A \cap B$, A^c, and $A - B$ of X as follows:

- the *union of A and B* is the set $A \cup B = \{x : x \in X; x \in A \text{ or } x \in B\}$;
- the *intersection of A and B* is the set $A \cap B = \{x : x \in X; x \in A \text{ and } x \in B\}$;
- the *complement of A in X* is the set $A^c = \{x : x \in X \text{ and } x \notin A\}$;
- the *difference of A and B* is the set $A - B = \{x : x \in X; x \in A \text{ and } x \notin B\}$.

Digression: Application of set theory to the analysis of syllogisms: A syllogism is a special type of argument form, first studied by Aristotle, the founder of logic. Set theory and Venn diagrams easily decide the validity of a syllogism. We illustrate with two examples.

Example 1:

All *A* are *B*.

Some *C* are not *B*.

∴ Some *C* are not *A*.

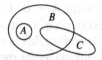

This translates into set theory as $A \subseteq B$, $C \cap B^c \neq \emptyset$, ∴ $C \cap A^c \neq \emptyset$. The proof proceeds as follows: Let $x \in (C \cap B^c)$. Then $x \in C$ and $x \in B^c$. Now $x \in B^c$ implies $x \notin B$; but $A \subseteq B$, so $x \notin A$. We now have $x \in C$ and $x \in A^c$; hence $x \in (C \cap A^c)$ and $C \cap A^c \neq \emptyset$ as required. This shows that the syllogism is valid.   ☐

Example 2:

All *A* are *B*.

Some *B* are *C*.

∴ Some *A* are *C*.

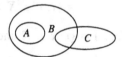

This translates into set theory as $A \subseteq B$, $B \cap C \neq \emptyset$, ∴ $A \cap C \neq \emptyset$. This is not always the case; for example, take $A = \{1\}$, $B = \{1, 2\}$, and $C = \{2, 3\}$. For this choice of sets, both $A \subseteq B$ and $B \cap C \neq \emptyset$ hold, but $A \cap C \neq \emptyset$ does not hold. This shows that the syllogism is invalid.   ☐

Arbitrary unions and intersections: The operations of union and intersection extend to arbitrary collections. Let X and T be sets, and suppose that for each $t \in T$ we have $A_t \subseteq X$. We denote this collection of subsets of X by $\{A_t : t \in T\}$ and call T an *index set* for the collection. The *union* and *intersection* of $\{A_t : t \in T\}$ are defined by

$$\bigcup_{t \in T} A_t = \{x : x \in X \text{ and } x \in A_t \text{ for at least one } t \in T\};$$

$$\bigcap_{t \in T} A_t = \{x : x \in X \text{ and } x \in A_t \text{ for all } t \in T\}.$$

The special case in which the index set is $\mathbb{N}$ is of interest: Given a collection $\{A_n : n \in \mathbb{N}\}$ of subsets of X,

$$x \in \bigcup_{n \in N} A_n \text{ if and only if } x \in A_n \text{ for some } n \in \mathbb{N};$$

$$x \in \bigcap_{n \in N} A_n \text{ if and only if } x \in A_n \text{ for all } n \in \mathbb{N}.$$

Power set of a set: The *power set* of A is defined by $P(A) = \{B : B \subseteq A\}$. For example, if $A = \{a, b\}$, then $P(A) = \{\emptyset, \{a\}, \{b\}, \{a, b\}\}$. If the set A has n elements, then $P(A)$ has 2^n elements (proof by induction).

Cartesian products: Yet another technique for constructing new sets from given sets is that of taking their Cartesian product. First we need to discuss ordered n-tuples. Let a and b be sets. The set consisting of a and b can be specified as $\{a, b\}$ or $\{b, a\}$; that is, the order is unimportant. If the order is important, we write $\langle a, b \rangle$ and call this the *ordered pair of a and b* with *first coordinate a* and *second coordinate b*. For ordered pairs, we have $\langle a, b \rangle = \langle c, d \rangle$ if and only if $a = c$ and $b = d$. Thus, $\{1, 2\} = \{2, 1\}$ but $\langle 1, 2 \rangle \neq \langle 2, 1 \rangle$. More generally, for ordered n-tuples we have $\langle a_1, \ldots, a_n \rangle = \langle b_1, \ldots, b_n \rangle$ if and only if $a_k = b_k$ for $1 \leq k \leq n$. Now let $A_1, \ldots, A_n$ be n sets. The *Cartesian product* of these sets is the set

$$A_1 \times \cdots \times A_n = \{\langle a_1, \ldots, a_n \rangle : a_k \in A_k \text{ for } 1 \leq k \leq n\}.$$

In many cases $A_1, \ldots, A_n$ are all the same set, say A, in which case we write A^n. We are especially interested in the sets $\mathbb{N}^n$ and $\{T, F\}^n$. In this second example T and F are two symbols that intuitively stand for *true* and *false*; note that $\{T, F\}^2 = \{\langle T, T \rangle, \langle T, F \rangle, \langle F, T \rangle, \langle F, F \rangle\}$ has four elements and in general $\{T, F\}^n$ has 2^n elements.

Cartesian products are defined in terms of ordered n-tuples. For the record, ordered n-tuples can be defined rigorously in terms of previous concepts as follows: First, ordered pairs are defined by $\langle a, b \rangle = \{\{a\}, \{a, b\}\}$ (with this definition one can prove the important property that $\langle a, b \rangle = \langle c, d \rangle$ if and only if $a = c$ and $b = d$); ordered n-tuples are defined inductively by $\langle a_1, \ldots, a_n \rangle = \langle \langle a_1, \ldots, a_{n-1} \rangle, a_n \rangle$.

Functions

Definition 2: A *function f from X into Y*, written $f : X \rightarrow Y$, is a rule that assigns to each $a \in X$ a *unique* $b \in Y$; this assignment is written $b = f(a)$, and b is called the *value of f at a*. The sets X and Y are called the *domain* and *codomain* of f respectively, and the set of all values of f is called the *range* of f and is denoted $f(X)$. Thus,

$$f(X) = \{f(a) : a \in X\}$$
$$= \{b : b \in Y \text{ and there exists } a \in X \text{ such that } f(a) = b\}.$$

Note that $f(X) \subseteq Y$ always holds. The function f is *one-to-one* if the following holds for all $a, b \in X$: If $a \neq b$, then $f(a) \neq f(b)$. It is often easier to work with the contrapositive: If $f(a) = f(b)$, then $a = b$. The function f is *onto* if for each $b \in Y$, there is at least one $a \in X$ such that $f(a) = b$. Thus, f is onto if and only if $f(X) = Y$; that is, the range and codomain of f are equal. Given $f : X \rightarrow Y$ and $g : Y \rightarrow Z$, the *composition* of f and g is the function $g \circ f : X \rightarrow Z$ defined by $(g \circ f)(a) = g(f(a))$. The following are basic results about these properties; the proofs are left as exercises for the reader.

Theorem 1 (Existence of an inverse): Let $f : X \rightarrow Y$ be both one-to-one and onto. Then there is a one-to-one function $f^{-1} : Y \rightarrow X$ from Y onto X (called the *inverse* of f) such that for all $a \in X$ and $b \in Y$, $f(a) = b$ if and only if $f^{-1}(b) = a$.

Theorem 2 (Properties of composition) Let $f : X \rightarrow Y$ and $g : Y \rightarrow Z$ be functions. (1) If f and g are both one-to-one, then $g \circ f : X \rightarrow Z$ is one-to-one; (2) if f and g are both onto, then $g \circ f : X \rightarrow Z$ is onto.

Our informal approach to functions can be made rigorous as follows: A *function from X into Y* is a subset f of $X \times Y$ satisfying these two conditions: (1) For each $a \in X$, there exists $b \in Y$ such that $\langle a, b \rangle \in f$; (2) if $\langle a, b \rangle \in f$ and $\langle a, c \rangle \in f$, then $b = c$. Note that the term *rule* is clarified by stating that f

is a subset of $X \times Y$; part (1) of the definition states that f assigns an element of Y to each element of X; part (2) states that this assignment is unique.

Relations

Functions and the more general notion of a relation are absolutely fundamental in mathematics; indeed, all of classical mathematics can be reduced to a discussion of these two concepts. This point of view will be important in our discussion of first-order languages. We begin with the important special case of a binary relation.

Definition 3: Let X be a set. A *binary relation on* X is a subset R of $X \times X$.

For a binary relation R it is customary to write $x R y$ instead of the proper $\langle x, y \rangle \in R$. For example, in working with real numbers, one writes $2 \leq 3$ rather than $\langle 2, 3 \rangle \in \leq$. For any set X, equality is a binary relation on X and $\subseteq$ is a binary relation on $P(X)$. The usual *less than* relation among natural numbers is a binary relation on $\mathbb{N}$ that can be defined in terms of addition by $a < b \Leftrightarrow$ there exists $c \in \mathbb{N}^+$ such that $a + c = b$. We now define three important properties of a binary relation.

Definition 4: Let R be a binary relation on X; R is *reflexive* if $a R a$ for all $a \in X$; R is *transitive* if for all $a, b, c \in R$, $a R b$ and $b R c$ imply $a R c$; R is *symmetric* if for all $a, b \in R$, $a R b$ implies $b R a$.

A relation that satisfies all three of these properties is called an *equivalence relation*; $\sim$ is often used to denote an equivalence relation. For any set X, equality is an equivalence relation on X and equipotence is an equivalence relation on $P(X)$ (see Section 1.5). Given an equivalence relation $\sim$ on X and $a \in X$, $[a] = \{b : b \in X \text{ and } a \sim b\}$; $[a]$ is a subset of X called an *equivalence class* of $\sim$. The set of all equivalence classes is denoted $X/\sim$; thus $X/\sim = \{[a] : a \in X\}$. Each $x \in [a]$ is called a *representative element* of $[a]$. By the reflexive property, $a \in [a]$, so $[a]$ is always nonempty and a is a representative element of $[a]$. The following theorem gives further basic information about equivalence classes. (The proof is left as an exercise for the reader.)

Theorem 3: Let $\sim$ be an equivalence relation on a set X and let $a, b \in X$. Then (1) $a \sim b$ if and only if $[a] = [b]$; (2) $[a] = [b]$ or $[a] \cap [b] = \varnothing$.

Suppose $\sim$ is an equivalence relation on a set X. The reflexive property implies that $a \in [a]$ for all $a \in X$. So every element of X is in some equivalence class and every equivalence class is nonempty. By Theorem 3(2), any two equivalence classes are either disjoint or equal. Thus an equivalence relation on X partitions X into a pairwise disjoint collection of nonempty subsets of X.

We now generalize the idea of a binary relation.

Definition 5: Let X be a set and let $n \geq 1$. An *n-ary relation* on X is a subset R of X^n. Note that a 2-ary relation is the same as a binary relation and a 1-ary relation on X is simply a subset of X.

Example 3:

We give several examples of relations on $\mathbb{N}$.

(1) $R_1(a) \Leftrightarrow a \in \mathbb{N}$ and a is a prime number.

(2) $R_2(a) \Leftrightarrow a \in \mathbb{N}$, a is an even number ≥ 4 and the sum of two primes.

(3) $R_3(a, b, c) \Leftrightarrow a, b, c \in \mathbb{N}^+$ and $a^2 + b^2 = c^2$.

(4) $R_4(a, b, c, k) \Leftrightarrow a, b, c, k \in \mathbb{N}^+$ with $k > 2$ and $a^k + b^k = c^k$.

R_1 and R_2 are 1-ary relations on $\mathbb{N}$. The number of primes is infinite, so R_1 is actually an infinite subset of $\mathbb{N}$. The relation R_2 is related to Goldbach's conjecture. This conjecture, which goes back to 1742, states that every even integer ≥ 4 is the sum of two primes. This conjecture is still unproved; if it is correct, then $R_2 = \{n: n \text{ even and } n \geq 4\}$. The relation R_3 is a 3-ary relation on $\mathbb{N}$ and $\langle 3, 4, 5 \rangle$, $\langle 7, 12, 13 \rangle \in R_3$; in fact, R_3 is infinite. Elements of R_3 are called *Pythagorean triples*. The relation R_4 is a 4-ary relation on $\mathbb{N}$ and is related to Fermat's last theorem, which asserts that for all $k \geq 3$, the equation $x^k + y^k = z^k$ has no solution in positive integers. This result was not actually proved by Fermat, and for more than 300 years it has been a famous unsolved problem of number theory. If it is correct, then $R_4 = \emptyset$. (*Note*: A. Wyles (1993) announced a proof of Fermat's result).   □

Exercises for Section 1.4

1. (*$P(X)$ is a Boolean algebra*) Let X be a set and let A, B, C be subsets of X. Prove the following (verification of (1)–(4) shows that $P(X)$ is a Boolean algebra; see Section 1.6).
   (1) $\left.\begin{array}{l} A \cup B = B \cup A \\ A \cap B = B \cap A \end{array}\right\}$ (commutative laws);
   (2) $\left.\begin{array}{l} A \cup (B \cap C) = (A \cup B) \cap (A \cup C) \\ A \cap (B \cup C) = (A \cap B) \cup (A \cap C) \end{array}\right\}$ (distributive laws);
   (3) $\left.\begin{array}{l} A \cup \emptyset = A \\ A \cap \emptyset = \emptyset \end{array}\right\}$ (identity laws);
   (4) $\left.\begin{array}{l} A \cup A^c = X \\ A \cap A^c = \emptyset \end{array}\right\}$ (complement laws);
   (5) $\left.\begin{array}{l} (A \cup B)^c = A^c \cap B^c \\ (A \cap B)^c = A^c \cup B^c \end{array}\right\}$ (deMorgan's laws).
2. Let A be a set with n elements. Show that the power set $P(A)$ of A has 2^n elements.
3. Use basic set theory and familiar properties of the natural numbers to prove that induction (**I**), complete induction (**CI**), and the well-ordering principle (**WOP**) are equivalent.
   I Let $S \subseteq \mathbb{N}$. Then $S = \mathbb{N}$ if these two conditions are satisfied: (1) $0 \in S$; (2) if $n \in S$, then $n + 1 \in S$.
   CI Let $S \subseteq \mathbb{N}$. Then $S = \mathbb{N}$ if these two conditions are satisfied: (1) $0 \in S$; (2) if $\{0, 1, \ldots, n\} \subseteq S$, then $n + 1 \in S$.

WOP Every nonempty subset of $\mathbb{N}$ has a smallest element.

4. Use set theory to classify the following syllogisms as valid or invalid.
   (1) All A are B, all B are C, $\therefore$ all A are C.
   (2) Some A are B, all B are C, $\therefore$ some A are C.
   (3) Some A are B, some B are C, $\therefore$ some A are C.
   (4) No A are B, all C are B, $\therefore$ no A are C.
   (5) Some A are not B, all C are B, $\therefore$ no A are C.

5. Consider the syllogism *No A are B, all B are C, $\therefore$ some C are not A.*
   (1) Show the syllogism valid under the assumption that $B \neq \varnothing$.
   (2) Show the syllogism invalid without this additional assumption.

6. Let X be a set, let $\{A_n : n \in \mathbb{N}\}$ be a collection of subsets of X such that $A_0 \subseteq A_1 \subseteq A_2 \subseteq \ldots$ (i.e., the sets are *increasing*), and let F be a finite subset of X such that $F \subseteq \bigcup_{n \in \mathbb{N}} A_n$. Show that there is some $k \in \mathbb{N}$ such that $F \subseteq A_k$.

7. Let F be a formal system, let $\{\Gamma_n : n \in \mathbb{N}\}$ be sets of formulas of F such that $\Gamma_0 \subseteq \Gamma_1 \subseteq \Gamma_2 \ldots$ (i.e., increasing), and let $\Delta = \bigcup_{n \in \mathbb{N}} \Gamma_n$. Suppose A is a formula such that $\Delta \vdash A$. Show that there is some $k \in \mathbb{N}$ such that $\Gamma_k \vdash A$. (See Exercise 6 and Exercise 11(2) in Section 1.3.)

8. Let X be a set and let $\{A_n : n \in \mathbb{N}\}$ be a collection of subsets of X. Prove

   (1) $X - \bigcup_{n \in \mathbb{N}} A_n = \bigcap_{n \in \mathbb{N}} (X - A_n);$

   (2) $X - \bigcap_{n \in \mathbb{N}} A_n = \bigcup_{n \in \mathbb{N}} (X - A_n).$

9. The following exercise is a variation of a puzzle that appears in C. R. Wylie, Jr., *101 Puzzles in Thought and Logic* (New York: Dover Publications, Inc., 1957). Determine all functions $f : \{a, b, c\} \rightarrow \{x, y, z\}$ satisfying these four conditions: (1) f is one-to-one and onto; (2) if $f(a) = x$, then $f(c) \neq z$; (3) if $f(b) = x$ or $f(b) = z$, then $f(a) = y$; (4) if $f(a) = y$ or $f(a) = z$, then $f(b) = x$.

10. Use the informal definition of a function to prove Theorem 1.

11. Use the formal definition of a function to prove Theorem 1.

12. Prove Theorem 2.

13. Let $X = \{a, b, c\}$. For each binary relation below, decide whether the relation is reflexive or not, transitive or not, symmetric or not.
    (1) $R_1 = \{\langle a, a \rangle, \langle b, b \rangle, \langle c, c \rangle, \langle b, c \rangle\};$
    (2) $R_2 = \{\langle a, a \rangle, \langle b, b \rangle, \langle a, b \rangle, \langle b, a \rangle\};$
    (3) $R_3 = \{\langle a, a \rangle, \langle b, b \rangle, \langle c, c \rangle, \langle a, b \rangle, \langle b, a \rangle, \langle b, c \rangle, \langle c, b \rangle\}.$

14. The divisibility relation $\mid$ on $\mathbb{N}^+$ is defined by $a \mid b \Leftrightarrow b = a \times c$ for some $c \in \mathbb{N}^+$. Show that $\mid$ is reflexive and transitive but not symmetric.

15. Define a relation $\sim$ on $\mathbb{Z}$ by $a \sim b \Leftrightarrow a - b$ is even. Show that $\sim$ is an equivalence relation on $\mathbb{Z}$ and that $\mathbb{Z}/\sim = \{[0], [1]\}$.

16. (*Construction of integers $\mathbb{Z}$ from positive integers $\mathbb{N}^+$*) Show that the following relation is an equivalence relation on $\mathbb{N}^+ \times \mathbb{N}^+$:

$$\langle a, b \rangle \sim \langle c, d \rangle \quad \text{if and only if} \quad a + d = b + c.$$

17. (*Construction of rationals $\mathbb{Q}$ from integers $\mathbb{Z}$*) Define a relation $\sim$ on the set $X = \{\langle a, b \rangle : a, b \in \mathbb{Z}, b \neq 0\}$ by $\langle a, b \rangle \sim \langle c, d \rangle \Leftrightarrow ad = bc$. Show that $\sim$ is an equivalence relation on X.

18. Define a relation $\sim$ on $\mathbb{R} \times \mathbb{R}$ by $\langle a, b \rangle \sim \langle c, d \rangle \Leftrightarrow a^2 + b^2 = c^2 + d^2$.
    (1) Show that $\sim$ is an equivalence relation on X.
    (2) Give a geometric description of the equivalence class $[\langle 1, 0 \rangle]$.

19. Given $f : X \rightarrow Y$, define a relation $\sim$ on X by $a \sim b \Leftrightarrow f(a) = f(b)$. Show that $\sim$ is an equivalence relation on X and that the function $g : X/\sim \rightarrow Y$ defined by $g([a]) = f(a)$ is one-to-one.

20. Let $\sim$ be an equivalence relation on a set X and let $a, b \in X$. Show that the following are equivalent: (1) $a \sim b$; (2) $a \in [b]$; (3) $b \in [a]$; (4) $[a] \cap [b] \neq \varnothing$; (5) $[a] = [b]$.

——————————— $\boxed{1.5}$ ———————————

Countable and Uncountable Sets

Informally speaking, a set is *countable* if it either is finite or can be put into a one-to-one correspondence with the set $\mathbb{N}$ of natural numbers. In this section, we will make this definition precise and establish basic properties of countable sets. We have two main goals: Prove that if a set S of symbols is countable, then the set of all expressions in S is also countable; discuss Cantor diagonal arguments, the basic technique for showing that a set is *not* countable.

Definition 1: Two sets A and B are *equipotent*, written $A \approx B$, if there is a one-to-one function from A onto B. Intuitively speaking, two sets A and B are equipotent if they have the same number of elements. For example, $\mathbb{N} \approx \mathbb{N}^+$ and $\mathbb{Z} \approx \mathbb{N}$. (The proof is left as an exercise for the reader.)

Theorem 1 (Equipotence is an equivalence relation): Let A, B, and C be sets. Then (1) $A \approx A$; (2) if $A \approx B$, then $B \approx A$; (3) if $A \approx B$ and $B \approx C$, then $A \approx C$.

Proof: To prove (1), define $f: A \rightarrow A$ by $f(a) = a$ (the identity function on A); clearly f is both one-to-one and onto; (2) follows from Theorem 1 in Section 1.4 and (3) follows from Theorem 2 in Section 1.4. ∎

Definition 2: A set A is *finite* if $A = \varnothing$ or if there exists $n \in \mathbb{N}$ such that $A \approx \{0, \ldots, n\}$; a set that is not finite is *infinite*. A set A is *countable* if it is finite or $A \approx \mathbb{N}$. In other words, A is countable if one of the following holds: $A = \varnothing$; $A \approx \{0, \ldots, n\}$; $A \approx \mathbb{N}$. Finally, a set A is *uncountable* if it is not countable.

Example 1:

Both $\mathbb{N}^+$ and $\mathbb{Z}$ are countable, since $\mathbb{N} \approx \mathbb{N}^+$ and $\mathbb{Z} \approx \mathbb{N}$. The set $\mathbb{Q}$ of rational numbers is also countable; a proof of this fact is outlined in the exercises. On the other hand, $P(\mathbb{N})$ and $\mathbb{R}$ are uncountable; proofs of these two results will be given later when we discuss Cantor diagonal arguments. □

Lemma 1: If A is a countable set and $A \approx B$, then B is countable.

Proof: We consider the case in which $A \approx \mathbb{N}$ and leave the other cases to the reader. From $A \approx \mathbb{N}$ and $A \approx B$, we obtain $B \approx \mathbb{N}$ (use (2) and (3) of Theorem 1); hence B is countable as required. ∎

As a consequence of Lemma 1, any of the sets $\mathbb{N}^+$, $\mathbb{Z}$, and $\mathbb{Q}$ can be used to prove that a set is countable. A countable set has the important property that it can be indexed by $\mathbb{N}$ or by any set equipotent to $\mathbb{N}$; let us state and prove the precise result for $\mathbb{N}$.

Theorem 2 (Enumeration): Let A be a nonempty but countable set. Then the elements of A can be enumerated as $a_0, a_1, a_2, \ldots$; in other words, $A = \{a_n : n \in \mathbb{N}\}$. Moreover, for A infinite, we can assume that $a_m \neq a_n$ whenever $m \neq n$.

Proof: First assume $A \approx \mathbb{N}$ and let $f : \mathbb{N} \to A$ be one-to-one and onto. By the onto property, $A = \{f(n) : n \in \mathbb{N}\}$; hence if we let $a_n = f(n)$ for all $n \in \mathbb{N}$, then $A = \{a_n : n \in \mathbb{N}\}$. By the one-to-one property of f, $a_m \neq a_n$ for $m \neq n$. Now assume that A is nonempty and finite. In this case, $A \approx \{0, \ldots, k\}$ for some $k \in \mathbb{N}$; hence $A = \{a_n : 0 \le n \le k\}$ (use an argument similar to the one above); if we let $a_n = a_k$ for all $n \ge k$, we also have $A = \{a_n : n \in \mathbb{N}\}$, as required. ■

The next two theorems are fundamental tools for proving countability.

Theorem 3 (Subset): Every subset of a countable set is countable.

Proof: Let $A \subseteq B$ with B countable. We first prove the result for the special case in which $B = \mathbb{N}$. The proof is somewhat informal; for a more detailed proof, see Exercise 8. If A is finite, then A is countable by definition and we are finished. So we may suppose that A is an infinite subset of $\mathbb{N}$. Construct an increasing sequence $a_0 < a_1 < \ldots < a_n < \ldots$ of natural numbers inductively as follows:

$$a_0 = \text{smallest element of } A;$$

$$a_{n+1} = \text{smallest element of } A - \{a_0, a_1, \ldots, a_n\}.$$

Since A is infinite, at each stage the set $A - \{a_0, a_1, \ldots, a_n\}$ is nonempty; by the well-ordering principle, this set has a smallest element a_{n+1}. One can show that $A = \{a_n : n \in \mathbb{N}\}$ and that $a_n \neq a_m$ for $n \neq m$. From this it follows that the function $f : \mathbb{N} \to A$ defined by $f(n) = a_n$ is both one-to-one and onto. It follows that $A \approx \mathbb{N}$, so A is countable as required.

Now consider the general case $A \subseteq B$ with B countable. We may assume that B is infinite. Then $B \approx \mathbb{N}$ and there is a one-to-one and onto function $f : B \to \mathbb{N}$. Let $C = f(A) = \{f(a) : a \in A\}$; now $C \subseteq \mathbb{N}$, so the special case discussed above applies and C is countable. Moreover, by restricting the domain of f to A, we obtain a one-to-one function from A onto C; hence $A \approx C$. In summary, $A \approx C$ with C countable; so by Lemma 1 the set A is countable as required. ■

Theorem 4 (One-to-one): Let A be a set. If there is a countable set B and a one-to-one function $f : A \to B$, then A is countable.

Proof: Let $C = f(A)$; we can regard f as a one-to-one function from A onto C, hence $A \approx C$. But C is a subset of B and B is countable; hence C is countable (subset theorem). In summary, we have $A \approx C$ with C countable; Lemma 1 applies and A is countable as required. ■

Example 2:

In many applications of Theorem 4, the codomain B is $\mathbb{N}$. We illustrate by showing that $\mathbb{N} \times \mathbb{N}$ is countable. Define $f : \mathbb{N} \times \mathbb{N} \to \mathbb{N}$ by $f(\langle i, j \rangle) = 2^i \times 3^j$; we are required to show that f is one-to-one. Let $f(\langle i, j \rangle) = f(\langle m, n \rangle)$; that is, $2^i \times 3^j = 2^m \times 3^n$. Suppose $i \neq m$, say $i < m$. Divide by 2^i to obtain $3^j = 2^{m-i} \times 3^n$. This is impossible, since the left side is odd and the right side is even. Hence $i = m$ and we have $3^j = 3^n$. From this it follows that $j = n$, so $\langle i, j \rangle = \langle m, n \rangle$, as required. $\qquad\square$

Theorem 5 (Number of expressions): Let S be a countable set of symbols and let EXP be the set of all expressions in S. Then EXP is countable.

Proof: To simplify the discussion, assume that the set S of symbols is infinite. By the enumeration theorem, we can write $S = \{s_n : n \in \mathbb{N}^+\}$ with $s_n \neq s_m$ for $n \neq m$. A typical element of EXP is of the form $s_{i_1} \ldots s_{i_k}$. Now let $p_1 < p_2 < \ldots < p_n < \ldots$ denote the prime numbers in increasing order ($p_1 = 2$, $p_2 = 3$, $p_3 = 5$, and so on). Define $f : \text{EXP} \to \mathbb{N}$ by

$$f(s_{i_1} \ldots s_{i_k}) = p_1^{i_1} \times p_2^{i_2} \times \cdots \times p_k^{i_k}.$$

(For example, $s_3 s_7 s_2$ is an expression in S and $f(s_3 s_7 s_2) = 2^3 3^7 5^2$.) We now use the unique factorization theorem (see the appendix on number theory) to show that f is a one-to-one function. Assuming this is so, it follows from Theorem 4 (one-to-one theorem) that EXP is countable.

Let u and v be expressions in S such that $f(u) = f(v)$; our goal is to show that $u = v$. Let $u = s_{i_1} \ldots s_{i_k}$, $v = s_{j_1} \ldots s_{j_n}$. From $f(u) = f(v)$, we have $p_1^{i_1} \times \cdots \times p_k^{i_k} = p_1^{j_1} \times \cdots \times p_n^{j_n}$. By the unique factorization theorem, we have $k = n$ and $i_r = j_r$ for $1 \leq r \leq k$. It easily follows that $u = v$, as required. ∎

Corollary 1 (Number of formulas): Let F be a formal system whose language has a countable number of symbols. Then the set FOR of all formulas of F is also countable.

Proof: The set EXP is countable, and FOR $\subseteq$ EXP; hence FOR is countable by Theorem 3 (subset theorem). ∎

——————— Cantor Diagonal Argument ———————

We use a technique called a *Cantor diagonal argument* to show that certain sets are uncountable. Georg Cantor (1845–1918) is the founder of set theory.

Theorem 6 (Cantor, uncountability of $P(\mathbb{N})$): The set $P(\mathbb{N})$ is uncountable (i.e., the set of all subsets of $\mathbb{N}$ cannot be put into a one-to-one correspondence with $\mathbb{N}$).

Proof: We first give a somewhat intuitive argument to motivate the method. Suppose $P(\mathbb{N})$ is countable. By the enumeration theorem, $P(\mathbb{N}) = \{A_n : n \in \mathbb{N}\}$. We now construct a set B such that

- $B \subseteq \mathbb{N}$;
- $B \neq A_n$ for all $n \in \mathbb{N}$.

We then have $B \in P(\mathbb{N})$ and $B \notin \{A_n : n \in \mathbb{N}\}$, a contradiction of $P(\mathbb{N}) = \{A_n : n \in \mathbb{N}\}$.

To ensure that $B \neq A_0$, we decide what to do with 0 as follows: if $0 \in A_0$, then $0 \notin B$; but if $0 \notin A_0$, then $0 \in B$. This guarantees that $B \neq A_0$. Next, to ensure that $B \neq A_1$, we decide what to do with 1 as follows: if $1 \in A_1$, then $1 \notin B$; but if $1 \notin A_1$, then $1 \in B$. This guarantees that $B \neq A_1$. Continuing in this way, we construct $B \subseteq \mathbb{N}$ so that $B \neq A_n$ for all $n \in \mathbb{N}$.

We now give a more direct proof. Let $B = \{n : n \in \mathbb{N}$ and $n \notin A_n\}$. We prove that $B \neq A_n$ for all $n \in \mathbb{N}$. Fix $n \in \mathbb{N}$, and consider two cases. First suppose $n \in A_n$. Then $n \notin B$ (see the definition of B); hence $B \neq A_n$. Now suppose $n \notin A_n$. Then $n \in B$, and again $B \neq A_n$ as required. ∎

We now use a Cantor diagonal argument to prove the uncountability of the set $\mathbb{R}$ of real numbers. In the proof we will use the following fact about decimal expansions: If two numbers are written in their unique nonterminating decimal expansion, and if these expansions differ in the nth place for some $n \geq 1$, then the two numbers are not equal. (Recall that some numbers can be written in decimal form in two different ways; for example, $1/2 = .5$ and $1/2 = .49999\ldots$. We will always choose the nonterminating expansion.)

Theorem 7 (Cantor, uncountability of $\mathbb{R}$): The set $\mathbb{R}$ of real numbers is uncountable.

Proof: By Theorem 3 (subset theorem), it suffices to prove that the set $(0, 1)$ of all real numbers between 0 and 1 is uncountable. Suppose this is false; that is, suppose that $(0, 1)$ is countable. By the enumeration theorem, $(0, 1) = \{x_n : n \in \mathbb{N}^+\}$. We obtain a contradiction by constructing a real number y such that $y \in (0, 1)$ but $y \neq x_n$ for all $n \in \mathbb{N}^+$. Write each x_n in its unique nonterminating decimal expansion:

$$x_1 = .\boxed{x_{11}}x_{12}x_{13}\ldots x_{1n}\ldots$$
$$x_2 = .x_{21}\boxed{x_{22}}x_{23}\ldots x_{2n}\ldots$$
$$\vdots$$
$$x_n = .x_{n1}x_{n2}x_{n3}\ldots \boxed{x_{nn}}\ldots$$

The idea is to construct a number $y = .y_1 y_2 \ldots y_n \ldots$ in $(0, 1)$ such that for all $n \geq 1$, y_n (the nth position in the nonterminating decimal expansion of y) differs from x_{nn} (the nth position in the nonterminating decimal expansion of x_n). To achieve this, let $y = .y_1 y_2 \ldots y_n \ldots$, where

$$y_n = \begin{cases} 1 & \text{if } x_{nn} \neq 1; \\ 2 & \text{if } x_{nn} = 1. \end{cases}$$

Clearly $y = .y_1 y_2 \ldots y_n \ldots$ is a nonterminating decimal expansion with $y \in$

$(0, 1)$. By construction, $y_n \neq x_{nn}$ for all $n \in \mathbb{N}^+$; hence $y \neq x_n$ for all $n \in \mathbb{N}^+$. Thus y is a number in $(0, 1)$ with $y \neq x_n$ for all $n \in \mathbb{N}^+$ as required. ∎

No discussion of countable and uncountable sets is complete without mention of the continuum hypothesis, or CH. This famous conjecture is due to Cantor and can be stated as follows:

CH   For every infinite subset A of $\mathbb{R}$, either $A \approx \mathbb{N}$ or $A \approx \mathbb{R}$.

Cantor showed that CH holds for infinite *closed* sets of $\mathbb{R}$, but he was unable to settle the original question. The following two theorems of twentieth-century mathematics show why.

Theorem 8 (Gödel, 1938): If set theory is consistent, then set theory plus CH is consistent.

Theorem 9 (Cohen, 1963): If set theory is consistent, then set theory plus the negation of CH is consistent.

Gödel's result tells us that we cannot prove the negation of CH from the usual axioms of set theory (unless set theory is inconsistent, in which case everything is provable!). On the other hand, by Cohen's result we cannot prove CH from the usual axioms of set theory. Mathematicians are still looking for an "obviously true" statement about sets that implies either CH or not CH.

──────────*Exercises for Section 1.5*──────────

1. Prove the following: (1) $\mathbb{N}^+ \approx \mathbb{N}$; (2) $\mathbb{Z} \approx \mathbb{N}$; (3) $[0, 1] \approx [0, 1)$.
2. Let $a, b \in \mathbb{R}$ with $a < b$. Prove the following:
   (1) $[0, 1] \approx [a, b]$;   (4) $(-1, 1) \approx \mathbb{R}$ (*Hint:* $f(x) = x/(1 + |x|)$);
   (2) $(0, 1) \approx (a, b)$;   (5) $(a, b) \approx \mathbb{R}$ (use (2), (4));
   (3) $(0, 1) \approx [0, 1]$;   (6) $[a, b] \approx \mathbb{R}$ (use (1), (3), (5)).
3. Let A, B, C, D be sets such that $A \approx C$ and $B \approx D$. Prove that $A \times B \approx C \times D$.
4. A computer program can ultimately be thought of as a finite string of 0's and 1's—for example, 1110010001110010. Prove that the number of computer programs is countable.
5. Show that $\mathbb{Q}$, the set of rational numbers, is countable. (*Hint:* Construct a one-to-one function $f: \mathbb{Q} \to \mathbb{Z}$; also see the proof that $\mathbb{N} \times \mathbb{N}$ is countable.)
6. Show that $\mathbb{N} \times \mathbb{N} \times \mathbb{N}$ is countable.
7. Let SEQ be the set of all finite sequences of natural numbers; thus,

$$\text{SEQ} = \mathbb{N} \cup (\mathbb{N} \times \mathbb{N}) \cup (\mathbb{N} \times \mathbb{N} \times \mathbb{N}) \cup \dots$$
$$= \{\langle n_1, \dots, n_k \rangle : k \geq 1, n_1, \dots, n_k \in \mathbb{N}\}.$$

   Show that SEQ is countable. (*Hint:* Method of proof in Theorem 5.)
8. (*Subset theorem*) This problem outlines another proof of the special case of Theorem 3. Let A be an infinite subset of $\mathbb{N}$ and define $f: A \to \mathbb{N}^+$ by $f(k) = |\{0, 1, \dots, k\} \cap A|$ (i.e., $f(k)$ is the number of elements in A that are $\leq k$).
   (1) Show that f is one-to-one. More precisely, show that if $i < j$, then $f(i) < f(j)$.
   (2) Show that f is onto. (*Hint:* Use induction. To begin, use the well-ordering principle to choose the smallest number $k \in A$ and show that $f(k) = 1$.)
9. (*Onto theorem*) Let A be a countable set and let $f: A \to B$ be onto. Show that B is countable. (*Hint:* For each $b \in B$, choose $a_b \in A$ such that $f(a_b) = b$. Define $g: B \to A$ by $g(b) = a_b$. Show that g is one-to-one and use Theorem 4 (one-to-one theorem).)
10. (*Union theorem*) Let $\{A_n : n \in \mathbb{N}\}$ be a countable collection of sets such that each set A_n

is countable. Show that $\bigcup_{n \in \mathbb{N}} A_n$ is countable. (*Hint*: Since each A_n is countable, $A_n = \{x_{nk} : k \in \mathbb{N}\}$; define $f : \mathbb{N} \times \mathbb{N} \to \bigcup_{n \in \mathbb{N}} A_n$ by $f(n, k) = x_{nk}$ and use the onto theorem.)

11. (*Finite product theorem*) Prove that the product of a finite number of countable sets is countable. (*Hint*: First show that the product of two countable sets A and B is countable. To do this, find a function from $\mathbb{N} \times \mathbb{N}$ onto $A \times B$ and use the onto theorem. Then proceed by induction to show that any finite product of countable sets is countable.)

12. Let A be the set of all real numbers x in the interval $(0, 1)$ such that every digit in the nonterminating decimal expansion of x is either 2 or 3. Show that A is uncountable.

13. (X is not equipotent to $P(X)$.) Let X be a set.
    (1) Construct a one-to-one function from X into $P(X)$.
    (2) Let $f : X \to P(X)$. Use a Cantor diagonal argument to show that f cannot be onto. Conclude that X is not equipotent to $P(X)$.

14. Let X be a set and let $F(X)$ be the collection of all functions from X into $\{0, 1\}$. Show that $F(X)$ is uncountable.

15. Prove that the collection of all finite subsets of $\mathbb{N}$ is countable.

16. Prove that the collection of all infinite subsets of $\mathbb{N}$ is uncountable.

17. Let $\mathcal{A} = \{A : A \subseteq \mathbb{N}$ and both A and A^c are infinite$\}$. For example, the set of even positive integers is in $\mathcal{A}$. Is $\mathcal{A}$ countable or uncountable? Prove your result.

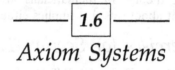

Axiom Systems

Set theory, functions, and relations provide us with an economical, unified method for introducing mathematical structures and axiom systems. We will give a number of examples to support this claim. These examples will be used later, in Chapters 4 through 6, to illustrate important concepts of first-order logic. The following terminology is used: an *n-ary operation* on a set X is a function from X^n into X. For example, subtraction is a binary ($= 2$-ary) operation on $\mathbb{Z}$. We are primarily interested in 1-ary and 2-ary operations. For a 2-ary operation $\circ$ on X, it is customary to write $a \circ b$ rather than the proper $\circ(a, b)$.

Definition 1: The *natural number system* is an ordered 5-tuple $\langle \mathbb{N}, 0, S, +, \times \rangle$, where $\mathbb{N}$ is a set whose elements are called *natural numbers*, 0 is an element of $\mathbb{N}$ called *zero*, S is a 1-ary operation on $\mathbb{N}$ called the *successor function*, $+$ and $\times$ are 2-ary operations on $\mathbb{N}$, called *addition* and *multiplication* respectively, and the following properties hold for all $a, b \in \mathbb{N}$:

P1 $S(a) \neq 0$.

P2 If $S(a) = S(b)$, then $a = b$.

P3 Let $A \subseteq \mathbb{N}$ satisfy (1) $0 \in A$; and (2) if $a \in A$, then $S(a) \in A$. Then $A = \mathbb{N}$.

P4 $a + 0 = a$.

P5 $a + S(b) = S(a + b)$.

P6 $a \times 0 = 0$.

P7 $a \times S(b) = (a \times b) + a$.

Dedekind and Peano were the first mathematicians to study the natural number system from an axiomatic point of view. P1–P3 are known as *Peano's postulates*, although they are originally due to Dedekind. By definition, $1 = S(0)$, $2 = S(S(0))$, and so on. P1 states that 0 is not the successor of any natural number; P2 states that the successor function is one-to-one; P3 is the principle of finite induction. From P1–P7 one can prove familiar properties of $+$ and $\times$ such as the associative, commutative, and distributive laws. It is a remarkable fact that $+$ and $\times$ are actually superfluous in this definition; starting with P1–P3 and using basic set theory, one can construct 2-ary operations $+$ and $\times$ on $\mathbb{N}$ and then prove that P4–P7 hold. In other words, one can define the natural number system as an ordered 3-tuple $\langle \mathbb{N}, 0, S \rangle$ that satisfies P1–P3 and then prove the existence of unique functions $+$ and $\times$ that satisfy P4–P7.

Theorem 1 (Dedekind): The natural number system $\langle \mathbb{N}, 0, S \rangle$ is unique.

A proof of Dedekind's result is outlined in the exercises. To state his theorem precisely, let us introduce the following fundamental definition.

Definition 2: Consider an ordered triple $\langle X, z, f \rangle$, where $z \in X$ and $f: X \rightarrow X$. An *isomorphism from* $\langle \mathbb{N}, 0, S \rangle$ to $\langle X, z, f \rangle$ is a one-to-one and onto function $h: \mathbb{N} \rightarrow X$ such that the following two properties hold: (1) $h(0) = z$; (2) $h(S(a)) = f(h(a))$ for all $a \in \mathbb{N}$. We also say that $\langle \mathbb{N}, 0, S \rangle$ and $\langle X, z, f \rangle$ are *isomorphic*. The concept of an isomorphism is a precise way of stating that $\langle \mathbb{N}, 0, S \rangle$ and $\langle X, z, f \rangle$ are mathematically the same. Dedekind's theorem states: any ordered triple $\langle X, z, f \rangle$ that satisfies P1–P3 (replace $\mathbb{N}$ by X, 0 by z, and S by f) is isomorphic to $\langle \mathbb{N}, 0, S \rangle$.

Definition 3: A *group* is an ordered triple $\langle G, e, \circ \rangle$, where G is a set, $e \in G$, $\circ$ is a binary operation on G, and these three properties hold:

G1 $a \circ (b \circ c) = (a \circ b) \circ c$ for all $a, b, c \in G$ ($\circ$ is *associative*).

G2 $a \circ e = a$ for all $a \in G$ (e is a *right identity* for G).

G3 For all a in G, there exists b in G such that $a \circ b = e$ (each element a of G has a *right inverse* b).

Example 1:

(1) $\langle \mathbb{Z}, 0, + \rangle$ is a group (the underlying set is $\mathbb{Z}$, the right identity is 0, the operation is addition); (2) $\langle \mathbb{R}^+, 1, \times \rangle$ is a group ($\mathbb{R}^+$ is the set of positive real numbers, the right identity is 1, the operation is multiplication); (3) $\langle \mathbb{Z}, 0, - \rangle$ is *not* a group (the operation is subtraction). We leave it to the reader to check that G1–G3 hold for the first two examples and that G2 and G3, but not G1, hold for the third example. $\square$

Definition 4: Consider the following properties of a binary relation $\leq$ on a nonempty set X ($a, b, c \in X$):

PO1 $a \leq a$ (*reflexive*).
PO2 If $a \leq b$ and $b \leq c$, then $a \leq c$ (*transitive*).
PO3 If $a \leq b$ and $b \leq a$, then $a = b$ (*anti-symmetric*).
D $a \leq b$ or $b \leq a$ (*dichotomy*).
WO Every nonempty subset A of X has a smallest element (i.e., there exists $a \in A$ such that $a \leq b$ for all $b \in A$).

The relation $\leq$ is a *partial order* if it satisfies PO1–PO3; a *total order* (or *linear order*) if it satisfies PO1–PO3 and D; a *well-order* if it satisfies PO1–PO3 and WO. Also, a *partially ordered set* is a pair $\langle X, \leq \rangle$, where X is a nonempty set and $\leq$ is a partial order on X; similarly for *totally ordered set* and *well-ordered set*. Every well-ordered set is a totally ordered set.

Example 2:

The relation $\leq$ on $\mathbb{N}$ is a well-order; the relation $\leq$ on $\mathbb{Z}$ is a total order but not a well-order. The divisibility relation $\mid$ on $\mathbb{N}^+$ is a partial order that is not a total order ($a \mid b \Leftrightarrow$ there exists $c \in \mathbb{N}^+$ such that $b = a \times c$). $\quad\square$

Definition 5: A *Boolean algebra* is an ordered 6-tuple $\langle B, +, \times, ', 0, 1 \rangle$, where B is a nonempty set, $+$ and $\times$ are binary operations on B, $'$ is a 1-ary operation on B (i.e., a function from B into B), 0 and 1 are elements of B, and the following properties hold for all $a, b, c \in B$:

BA1 $a + b = b + a$.
BA2 $a \times b = b \times a$.
BA3 $a + 0 = a$.
BA4 $a \times 1 = a$.
BA5 $a \times (b + c) = (a \times b) + (a \times c)$.
BA6 $a + (b \times c) = (a + b) \times (a + c)$.
BA7 $a + a' = 1$.
BA8 $a \times a' = 0$.

Example 3:

$P(X)$, the power set of X, is a Boolean algebra with $+$, $\times$, $'$, 0, and 1 defined as follows (see Exercise 1 of Section 1.4): $A + B = A \cup B$, $A \times B = A \cap B$, $A' = A^c$, $0 = \varnothing$, $1 = X$. $\quad\square$

—————— Exercises for Section 1.6 ——————

1. Refer to the Peano postulates P1–P7. By definition, $1 = S(0)$, $2 = S(1)$ or $2 = S(S(0))$. Prove the following:
   (1) $1 + 1 = 2$ (in other words, $S(0) + S(0) = S(S(0))$);
   (2) $1 + 2 = 2 + 1$;
   (3) $1 \times 1 = 1$ (in other words, $S(0) \times S(0) = S(0)$);
   (4) $1 \times 2 = 2$.
2. Prove that for every $a \in \mathbb{N}$, $a = 0$ or $a = S(b)$ for some $b \in \mathbb{N}$. (*Hint:* Let $A = \{a: a = 0$ or $a = S(b)$ for some $b \in \mathbb{N}\}$ and use P3 to show $A = \mathbb{N}$.)

3. Refer to the Peano postulates P1–P7; by definition, $1 = S(0)$. Prove the following:
   (1) $S(a) = a + 1$ for all $a \in \mathbb{N}$ (in other words, $S(a) = a + S(0)$);
   (2) $S(a) \neq a$ for all $a \in \mathbb{N}$;
   (3) $0 + a = a$ for all $a \in \mathbb{N}$;
   (4) $0 \times a = 0$ for all $a \in \mathbb{N}$;
   (5) $1 \times a = a$ for all $a \in \mathbb{N}$;
   (6) $a \times 1 = a$ for all $a \in \mathbb{N}$.

4. (*Consequences of* P1 *and* P2) Let $\langle X, z, f \rangle$ be an ordered triple, where X is a set, $z \in X$, and f is a 1-ary operation on X. For $n \geq 0$, define z^n inductively by $z^0 = z$ and $z^{n+1} = f(z^n)$. Assume that $\langle X, z, f \rangle$ satisfies P1: There is no $x \in X$ such that $f(x) = z$; P2: If $f(x) = f(y)$, then $x = y$.
   (1) Let $m, n \in \mathbb{N}$. Prove that $z^m = z^n$ implies $m = n$. (*Hint*: Let $T(n)$ say: For any $m \in \mathbb{N}$, if $z^m = z^n$, then $m = n$. Prove that $T(n)$ holds for all $n \in \mathbb{N}$.)
   (2) Define $h: \mathbb{N} \to X$ by $h(n) = z^n$. Show that (a) h is one-to-one; (b) $h(0) = z$; (c) $h(S(n)) = f(h(n))$ for all $n \in \mathbb{N}$.
   (3) Show by example that h need not be onto.

5. (*Dedekind's theorem*) Refer to the Exercise 4 for notation. Let $\langle X, z, f \rangle$ satisfy P1, P2, and P3: Let $A \subseteq X$ satisfy $z \in A$; if $x \in A$, then $f(x) \in A$; then $A = X$. Show that the function h in the previous exercise is an isomorphism; it suffices to show that h is onto.

6. Let $\langle \mathbb{N}, 0, S \rangle$ be isomorphic to $\langle X, z, f \rangle$. Prove that $\langle X, z, f \rangle$ satisfies P1–P3. (Refer to the preceding two exercises for notation.)

7. Show that each of the following ordered 3-tuples is a group.
   (1) $G = \langle \{1, -1\}, 1, \times \rangle$;
   (2) $G = \langle \{1, -1, i, -i\}, 1, \times \rangle$, where $i^2 = -1$;
   (3) $G = \langle \mathbb{R}^+, 1, \times \rangle$.

8. Let X be a nonempty set and let G be the set of all one-to-one functions from X onto X. Show that $\langle G, i, \circ \rangle$ is a group. (The function i is the identity function $i(x) = x$ on X; the operation $\circ$ is composition $f \circ g(x) = f(g(x))$.)

9. Prove the right cancellation law for groups: If $a \circ c = b \circ c$, then $a = b$.

10. Define a 2-ary operation $\circ$ on $\mathbb{Z}$ by $a \circ b = a + b - 5$. Is $\mathbb{Z}$ with this operation a group?

11. Define a 2-ary operation $\circ$ on $\mathbb{Q}^+$ (positive rationals) by $a \circ b = ab/5$. Is $\mathbb{Q}^+$ with this operation a group?

12. Let $G = \{a: a \in \mathbb{R}, a \neq -1\}$ and $a \circ b = a + b + ab$. (1) Show that $\circ$ is a 2-ary operation on G by showing that if $a \neq -1$ and $b \neq -1$, then $a \circ b \neq -1$. (2) Is G with operation $\circ$ a group?

13. (*Isomorphisms of groups*) Two groups $\langle G, e, * \rangle$ and $\langle H, e', \circ \rangle$ are *isomorphic* if there is a one-to-one function h from G onto H such that for all $a, b \in G$, $h(a * b) = h(a) \circ h(b)$. (1) Let $2\mathbb{Z} = \{2k: k \in \mathbb{Z}\}$. Show that $\langle \mathbb{Z}, 0, + \rangle$ and $\langle 2\mathbb{Z}, 0, + \rangle$ are isomorphic. (2) Let $\langle G, e, * \rangle$ be a group and let h be a one-to-one function from G onto H. Define a 2-ary operation $\circ$ on H by $h(a) \circ h(b) = h(a * b)$. Show that $\langle H, h(e), \circ \rangle$ is a group and is isomorphic to G.

14. Show that the axioms for a group are not categorical by giving two examples of groups that are not isomorphic.

15. (*Abelian groups*) A group $\langle G, e, \circ \rangle$ is said to be *abelian* if it satisfies G4: For all $a, b \in G$, $a \circ b = b \circ a$.
   (1) Give an example of an abelian group.
   (2) Let $G = \{\langle a, b \rangle: a, b \in \mathbb{R}$ and $a \neq 0\}$, and define an operation $\circ$ on G by $\langle a, b \rangle \circ \langle c, d \rangle = \langle ac, bc + d \rangle$. Verify that G is a nonabelian group. Conclude that G4 cannot be derived from G1–G3.

16. Let G be a group such that $a \circ a = e$ for every $a \in G$. Prove that G is abelian (see Exercise 15).

17. (*Alternative approach to totally ordered sets*) The axioms for a totally ordered set $\langle L, \leq \rangle$ are PO1–PO3 and D (see Definition 4). Consider the following properties of a binary relation $<$ on L:

L1 $<$ is irreflexive ($a < a$ never holds);

L2 $<$ is transitive ($a < b$ and $b < c$ imply $a < c$);

L3 $<$ is linear (for all $a, b \in L$, either $a < b$, $a = b$, or $b < a$).

Prove the following:

(1) If $a < b$, then $b < a$ does not hold.

(2) At most one of the following holds: $a < b$, $a = b$, $b < a$.

(3) If $\langle L, \leq \rangle$ is a totally ordered set, then $<$ satisfies L1–L3, where $<$ is defined by $x < y \Leftrightarrow x \leq y$ and $x \neq y$.

(4) If $<$ satisfies L1–L3, then $\langle L, \leq \rangle$ is a totally ordered set, where $\leq$ is defined by $x \leq y \Leftrightarrow x < y$ or $x = y$.

18. Let $\langle L, \leq \rangle$ be a well-ordered set. Prove the following:

(1) $\langle L, \leq \rangle$ is a totally ordered set.

(2) Every nonempty subset of L is a well-ordered set.

19. (*Partially ordered set from a reflexive, transitive relation*) Let R be a binary relation on a set X that is reflexive and transitive. Define a binary relation $\sim$ on X by $a \sim b$ if and only if aRb and bRa. Prove the following:

(1) $\sim$ is an equivalence relation on X.

(2) If $a \sim a'$ and $b \sim b'$, then $aRb \Leftrightarrow a'Rb'$.

(3) Define a relation $\leq$ on $X/\sim$ by $[a] \leq [b]$ if and only if aRb. (From (2) it follows that the relation $\leq$ is well defined; its definition does not depend on the choice of representative elements.) Then $\langle X/\sim, \leq \rangle$ is a partially ordered set.

20. Let B be a Boolean algebra and let $a, b, c \in B$. Prove the following:

(1) $a + a = a$
(2) $a \times a = a$ (idempotent);

(5) $a + 1 = 1$
(6) $a \times 0 = 0$ (boundedness).

(3) $a + (a \times b) = a$
(4) $a \times (a + b) = a$ (absorption)

1.7

Decidability and Computability

Algorithms

Algorithms are fundamental in mathematics and computer science. They also form a basic part of mathematical logic; indeed, one of the great achievements of mathematical logic in this century is the clarification of the notion of an algorithm. Examples of algorithms from number theory include the division algorithm and the Euclidean algorithm. Here is an example of an algorithm from formal language theory.

Example 1:

(*Algorithm for palindromes*) Let S be a nonempty but finite set of symbols. Recall that a palindrome on S is an expression on S that reads the same backward and forward; for example, *abcba* and *cccc* are palindromes on $\{a, b, c\}$. We give an algorithm that, given an arbitrary expression on S, decides (YES or NO in finitely many steps) whether the expression is a palindrome on S.

(1) Input an expression E of S.

(2) Is the number of symbols in $E \leq 1$?

(3) If *Yes,* print YES and halt.

(4) Are the first and last symbols of E the same?

(5) If *No,* print NO and halt.

(6) Erase the first and last symbols of E and go to step (2).   □

We will now discuss the characteristics of an algorithm. First of all, an algorithm consists of a *finite* list of instructions, say $I_1, \ldots, I_t$. We require that each instruction be stated in a clear, precise manner so that it can be carried out by a machine. In particular, an instruction should require no ingenuity or judgment on the part of the agent that executes the instruction. Typically, an algorithm is given one or more inputs, and after a finite number of steps produces an output. A carefully written computer program is a paragon of an algorithm.

Decidable Formal Systems and Decidable Relations

We now introduce a number of ideas, each of which is defined in terms of an algorithm. We emphasize that algorithms are not precisely defined, and hence concepts defined in terms of algorithms are also not precisely defined. We will refer to a definition that is stated in terms of an algorithm as an *informal* definition or an *informal* concept. The theory of recursive functions (see Section 1.8 and Chapter 8) allows us to replace informal concepts with precise, or *formal,* concepts.

Definition 1 (Informal): A formal system F is *decidable* if there is an algorithm that, given an arbitrary formula A of F, decides whether $\vdash_F A$; if there is no such algorithm, we say that F is *undecidable.* An example of a decidable formal system is ADD (see Section 1.3).

Although this is an informal concept, it is still useful. Namely, to show that a formal system *is* decidable, it suffices to describe an algorithm and then to show that it works. In Chapter 3 we will show that propositional logic is decidable. The difficulty arises in trying to show that a formal system is undecidable. In this case, one must prove the nonexistence of an algorithm, and for this a more penetrating analysis of algorithms is required. This will be done in Chapter 8, on recursive functions. For now, we simply point out that both arithmetic and first-order logic are undecidable.

We now introduce a closely related concept, that of a *decidable relation on* $\mathbb{N}$. Questions about the decidability of a formal system can be reduced to questions about decidable relations on $\mathbb{N}$ by means of an effective coding of the language of the formal system (see Exercise 12). The advantage of such a reduction is that relations on $\mathbb{N}$ are easy to treat mathematically.

Convention and Notation Unless otherwise stated, *relation* means *relation on* $\mathbb{N}$; in particular, a 1-ary relation is a subset of $\mathbb{N}$. Given an *n*-ary relation R, we write

$$R(a_1, \ldots, a_n) \quad \text{for} \quad \langle a_1, \ldots, a_n \rangle \in \mathbb{N};$$

$$\neg R(a_1, \ldots, a_n) \quad \text{for} \quad \langle a_1, \ldots, a_n \rangle \notin \mathbb{N}.$$

Definition 2 (Informal): An *n*-ary relation R is *decidable* if there is an algorithm such that, given $a_1, \ldots, a_n \in \mathbb{N}$ (i.e., $\langle a_1, \ldots, a_n \rangle \in \mathbb{N}^n$), the algorithm decides (YES or NO in a finite number of steps) whether $R(a_1, \ldots, a_n)$. A relation that is not decidable is said to be *undecidable*.

Example 2:

Each of the relations R_1–R_4 in Example 3 of Section 1.4 is decidable. We write an algorithm to justify this for R_2 on Goldbach's conjecture (we assume from elementary number theory that there is an algorithm for deciding whether a positive integer is prime).

(1) Input $a \in \mathbb{N}$.
(2) Is a even and ≥ 4? If *No*, $\neg R_2(a)$; print NO and halt.
(3) Is there a natural number k, $2 \leq k \leq n - 2$, such that both k and $(n - k)$ are prime?
(4) If *No*, $\neg R_2(a)$; print NO and halt.
(5) Otherwise, $R_2(a)$; print YES and halt. $\square$

It is not easy to describe a *specific* relation on $\mathbb{N}$ that is not decidable. We can, however, use a counting argument to show the existence of such a relation.

Theorem 1 (Existence of an undecidable relation): There is a 1-ary relation on $\mathbb{N}$ that is undecidable.

Proof: We first show that the set $D(\mathbb{N})$ of all 1-ary decidable relations is countable. Fix a language L with a finite number of symbols such that all algorithms can be written in L (English, for example). The collection ALG of all algorithms written in L is countable (Theorem 5 in Section 1.5, on the number of expressions). Define $\Phi: D(\mathbb{N}) \to$ ALG by

$$\Phi(R) = \text{some algorithm that decides the relation } R.$$

The function Φ is one-to-one and ALG is a countable set; hence the one-to-one theorem applies and $D(\mathbb{N})$ is countable (see Theorem 4 in Section 1.5). On the other hand, we have proved that $P(\mathbb{N})$ is uncountable. Thus $D(\mathbb{N}) \neq P(\mathbb{N})$, so there is a subset of $\mathbb{N}$ that is not decidable, as required.

It is worthwhile to give the last part of the argument in a slightly different way so as to highlight the fact that a Cantor diagonal argument is used in the proof. The set $D(\mathbb{N})$ of 1-ary decidable relations is countable; hence, by the enumeration theorem, $D(\mathbb{N}) = \{R_n : n \in \mathbb{N}\}$. Now let $Q = \{n : n \in \mathbb{N}$ and

$n \notin R_n$}; then $Q \neq R_n$ for all $n \in \mathbb{N}$; hence $Q \notin D(\mathbb{N})$ and Q is not decidable, as required. ■

Computable Functions

Decidable relations are better understood when compared with the closely related notion of a *computable function*. But before giving the definition, let us agree on some terminology. By an *n-ary function* we mean a function $F\colon \mathbb{N}^n \to \mathbb{N}$. We emphasize that an *n*-ary function always has $\mathbb{N}^n$ as its domain and $\mathbb{N}$ as its codomain. If F is an *n*-ary function and $a_1, \ldots, a_n \in \mathbb{N}$, we write $F(a_1, \ldots, a_n)$ instead of the proper $F(\langle a_1, \ldots, a_n \rangle)$.

Definition 3 (Informal): An *n*-ary function F is *computable* if there is an algorithm such that, given $a_1, \ldots, a_n \in \mathbb{N}$, the algorithm computes the value $F(a_1, \ldots, a_n)$ in a finite number of steps.

Examples of 2-ary computable functions are the familiar operations of addition and multiplication. Indeed, it is not unreasonable to claim that every *n*-ary function ordinarily encountered in mathematics is a computable function.

We now introduce the notion of the characteristic function of an *n*-ary relation; this allows us to relate the decidability of a relation to the computability of a function.

Definition 4: The *characteristic function* of an *n*-ary relation R is the *n*-ary function K_R defined by

$$K_R(a_1, \ldots, a_n) = \begin{cases} 0 & \text{if } R(a_1, \ldots, a_n); \\ 1 & \text{if } \neg R(a_1, \ldots, a_n). \end{cases}$$

The function $K_R\colon \mathbb{N}^n \to \mathbb{N}$ is a decision function for R; by computing K_R, we can decide whether an element of $\mathbb{N}^n$ belongs to R ($0 = $ YES, $1 = $ NO).

Theorem 2: Let R be an *n*-ary relation on $\mathbb{N}$. Then R is decidable if and only if its characteristic function K_R is computable.

Proof: Assume R is decidable; here is an algorithm for computing K_R.

(1) Input $a_1, \ldots, a_n \in \mathbb{N}$.
(2) Use the algorithm for R to decide whether $R(a_1, \ldots, a_n)$.
(3) If *Yes*, set $K_R(a_1, \ldots, a_n) = 0$ and halt.
(4) Otherwise, set $K_R(a_1, \ldots, a_n) = 1$ and halt.

We leave the proof of the converse to the reader. ■

Although it is difficult to give an explicit example of a noncomputable function, we can prove the existence of such a function using a Cantor diagonal argument.

Theorem 3 (Existence of a noncomputable function): There is a 1-ary function that is not computable.

Proof: As in the proof of Theorem 1, the set of all computable functions is countable. Let $\{F_n : n \in \mathbb{N}\}$ be all 1-ary computable functions (enumeration theorem). We use a Cantor diagonal argument to construct a function $G : \mathbb{N} \to \mathbb{N}$ that is not in this list; such a function cannot be computable. Let $G(n) = F_n(n) + 1$; then for all $n \in \mathbb{N}$, $G(n) \neq F_n(n)$, and so $G \neq F_n$, as required. ∎

Note $G(n) = F_n(n) + 1$ is the way we diagonalize when working with functions.

——— *Semidecidable Relations and Listable Sets* ———

There are some rather natural algorithms that, for some inputs, go into an infinite loop. We will illustrate this with an example.

Example 3

Is there an algorithm that, given a polynomial $p(x) = a_n x^n + \cdots + a_0$ with integer coefficients $a_0, \ldots, a_n$, decides (YES or NO in a finite number of steps) whether the equation $p(x) = 0$ has a solution in integers? Consider the following candidate.

(1) Input a polynomial $p(x) = a_n x^n + \ldots + a_0$ and set $x = 0$.
(2) Compute $p(x)$ and $p(-x)$. Does $p(x) = 0$ or $p(-x) = 0$?
(3) If *Yes*, print YES and halt.
(4) If *NO*, add 1 to x and go to step (2).

This algorithm systematically searches for a solution of $p(x) = 0$ by calculating $p(x)$ at $x = 0, \pm 1, \pm 2, \ldots$. If there is a solution in integers, the algorithm eventually finds the smallest in absolute value, prints YES, and halts. But if there is no solution, the algorithm continues forever. (There is a better algorithm for this problem: Example 2 in Section 10.1 describes an algorithm that, after a finite number of steps, prints either YES (there is a solution in integers) or NO (there is no solution in integers)). □

The idea of an algorithm continuing forever for some inputs is the basis of the following generalization of a decidable relation.

Definition 5 (Informal): An n-ary relation R is *semidecidable* if there is an algorithm such that for all $a_1, \ldots, a_n \in \mathbb{N}$, (1) if $R(a_1, \ldots, a_n)$, the algorithm halts after a finite number of steps and prints YES; (2) if $\neg R(a_1, \ldots, a_n)$, the algorithm continues forever.

Theorem 4: An n-ary relation R is decidable if and only if both R and its complement R^c are semidecidable.

Proof: We leave the easy direction as an exercise for the reader (R decidable implies both R and R^c semidecidable). Conversely, assume that both R and R^c are semidecidable with algorithms Φ and Φ^c respectively. The required algorithm for R uses a technique called *dovetailing*, described as follows: Given inputs $a_1, \ldots, a_n \in \mathbb{N}$, do the first step of the algorithm Φ for R; then do the first step of the algorithm Φ^c for R^c; then do the first two steps of the algorithm Φ for R; then do the first two steps of the algorithm Φ^c for R^c; and so on. Now either $R(a_1, \ldots, a_n)$ or $\neg R(a_1, \ldots, a_n)$; hence this procedure must eventually halt. In the algorithm below, x is the number of steps executed.

(1) Input $a_1, \ldots, a_n \in \mathbb{N}$ and set $x = 1$.
(2) Does the algorithm for R applied to $a_1, \ldots, a_n$ halt in x steps and print YES? If *Yes*, print YES and halt.
(3) Does the algorithm for R^c applied to $a_1, \ldots, a_n$ halt in x steps and print YES? If *Yes*, print NO and halt.
(4) If *No*, add 1 to x and go to step (2). ∎

The next theorem characterizes semidecidability in terms of decidability.

Theorem 5: An n-ary relation R is semidecidable if and only if there is an $n + 1$-ary decidable relation Q such that for all $a_1, \ldots, a_n \in \mathbb{N}$,

$$R(a_1, \ldots, a_n) \Leftrightarrow \text{there exists } x \in \mathbb{N} \text{ such that } Q(a_1, \ldots, a_n, x). \qquad (*)$$

Proof: Assume that there is a decidable relation Q such that $(*)$ holds. The following algorithm shows that R is semidecidable.

(1) Input $a_1, \ldots, a_n \in \mathbb{N}$ and set $x = 0$.
(2) Use the algorithm for Q to decide whether $Q(a_1, \ldots, a_n, x)$.
(3) If *Yes*, then $R(a_1, \ldots, a_n)$ (see $(*)$); print YES and halt.
(4) Otherwise, add 1 to x and go to step (2).

This algorithm systematically searches for the smallest $x \in \mathbb{N}$ such that $Q(a_1, \ldots, a_n, x)$; if there is no such x, the algorithm continues forever.
Now assume that R is semidecidable. Define an $n + 1$-ary relation Q by

$$Q(a_1, \ldots, a_n, x) \Leftrightarrow \text{the algorithm for } R \text{ with inputs } a_1, \ldots, a_n$$
$$\text{halts after} \leq x \text{ steps and prints YES.}$$

We leave it to the reader to check that Q is decidable and that $(*)$ holds as required. ∎

If we start with a relation R, then its characteristic function K_R is a function such that R is decidable if and only if K_R is computable. We now want to go the other way: Starting with a function F, find a relation R such that F is computable if and only if R is decidable.

Definition 6: The *graph* of an n-ary function F is the $n + 1$-ary relation $\mathscr{G}_F$ defined by $\mathscr{G}_F(a_1, \ldots, a_n, b) \Leftrightarrow F(a_1, \ldots, a_n) = b$.

Theorem 6: Let F be an n-ary function. The following are equivalent: (1) F is computable; (2) $\mathscr{G}_F$ is decidable; (3) $\mathscr{G}_F$ is semidecidable.

Proof: We prove (3) $\Rightarrow$ (1) and leave (1) $\Rightarrow$ (2) and (2) $\Rightarrow$ (3) to the reader. Assume $\mathscr{G}_F$ is semidecidable; recall that for all $a_1, \ldots, a_n \in \mathbb{N}$,

$$\mathscr{G}_F(a_1, \ldots, a_n, b) \Leftrightarrow F(a_1, \ldots, a_n) = b. \tag{*}$$

The required algorithm for F is motivated as follows: Let $a_1, \ldots, a_n \in \mathbb{N}$. Since F is a function, there exists $b \in \mathbb{N}$ such that $F(a_1, \ldots, a_n) = b$; in other words, there exists $b \in \mathbb{N}$ such that $\mathscr{G}_F(a_1, \ldots, a_n, b)$ (see (*)). Moreover, the algorithm for $\mathscr{G}_F$ with inputs $a_1, \ldots, a_n, b$ halts in a finite number of steps, say k steps. So the strategy for an algorithm for computing F is to systematically search for the smallest $x \geq 1$ such that $b \leq x$ and $k \leq x$.

(1) Input $a_1, \ldots, a_n \in \mathbb{N}$ and set $x = 1$.
(2) Is there some $b \leq x$ and some $k \leq x$ such that the algorithm for $\mathscr{G}_F$ with inputs $a_1, \ldots, a_n, b$ halts in k steps and prints YES?
(3) If *Yes*, then $\mathscr{G}_F(a_1, \ldots, a_n, b)$; set $F(a_1, \ldots, a_n) = b$ and halt (see (*)).
(4) If *No*, add 1 to x and go to step (2).

It should be clear that the algorithm always halts, and with the correct output. For example, suppose that $F(a_1, \ldots, a_n) = 100$ and that the algorithm for $\mathscr{G}_F$ with inputs $a_1, \ldots, a_n, 100$ halts in 1000 steps; then the above algorithm halts when $x = 1000$ and the output will be 100. Similarly, if $F(a_1, \ldots, a_n) = 500$ and the algorithm for $\mathscr{G}_F$ with inputs $a_1, \ldots, a_n, 500$ halts in 200 steps, then the above algorithm halts when $x = 500$ and the output will be 500. ∎

Now let us restrict our attention to 1-ary relations; that is, to subsets of $\mathbb{N}$. In this special case, semidecidability has an interesting characterization.

Definition 7 (Informal): A subset R of $\mathbb{N}$ is *listable* if $R = \varnothing$ or if there is a 1-ary computable function F such that $R = \{F(n): n \in \mathbb{N}\}$. For example, the set R of even integers that are ≥ 4 is a listable set: $R = \{F(n): n \in \mathbb{N}\}$, where F is the computable function $F(n) = 2n + 4$.

Theorem 7: A subset R of $\mathbb{N}$ is listable if and only if it is semidecidable.

Proof: Assume R is listable. If $R = \varnothing$, then R is obviously decidable and hence semidecidable. So assume $R = \{F(n): n \in \mathbb{N}\}$, where F is a 1-ary computable function. The following algorithm shows that R is semidecidable.

(1) Input $a \in \mathbb{N}$ and set $n = 0$.
(2) Compute $F(n)$. Is $F(n) = a$?
(3) If *Yes*, $a \in R$; print YES and halt.
(4) If *No*, add 1 to n and go to step (2).

Note that if $a \notin R$, then $F(n) \neq a$ for all $n \in \mathbb{N}$; hence the algorithm continues forever, computing $F(0), F(1), \ldots$.

Now assume that R is semidecidable and that $R \neq \varnothing$, say $a_0 \in R$. We are required to find a computable function $F: \mathbb{N} \to \mathbb{N}$ such that $R = \{F(n): n \in \mathbb{N}\}$. For each $b \in R$, the algorithm for R with input b halts in a finite number of steps, say x steps. In this case we let $F(2^b 3^x) = b$. (Informally, the number

$2^b 3^x$ "codes" the situation in which $b \in R$ and the algorithm for R with input b halts in x steps.) In all other cases, $F(n) = a_0$. We leave it to the reader to verify that $R = \{F(n) : n \in \mathbb{N}\}$; it remains to describe an algorithm for computing F.

(1) Input $n \in \mathbb{N}$.

(2) Is $n = 2^b 3^x$, where $b \geq 0$ and $x \geq 1$?

(3) If *No*, set $F(n) = a_0$ and halt.

(4) Otherwise, $n = 2^b 3^x$ for some $b \geq 0$ and some $x \geq 1$. Does the algorithm for R with input b halt in x steps and print YES?

(5) If *No*, set $F(n) = a_0$; if *Yes*, set $F(n) = b$. Then halt. ∎

Decision Problems

The notion of a decision problem can be used to give a unified approach to some of the most important results in mathematical logic.

Definition 8 (Informal): Let X be a countable set and let $Y \subseteq X$. The *decision problem for Y in X* asks whether there is an algorithm that, given $x \in X$, decides (YES or NO) whether $x \in Y$. If such an algorithm exists, we say that the decision problem for Y in X is *solvable*; otherwise, the decision problem is *unsolvable*. Here are four examples of decision problems; others will occur throughout the text.

Decision problem for a formal system: This can be described as the decision problem for Y in X, where $X =$ all formulas of F and $Y =$ all theorems of F.

Validity problem for L: Let L be a first-order language. Is there an algorithm that, given an arbitrary formula A of L, decides (YES or NO) whether A is logically valid? This can be described as the decision problem for Y in X, where $X =$ all formulas of L and $Y =$ all logically valid formulas of L. There is a first-order language for which the validity problem is unsolvable. This result is due independently to Church and Turing, and is proved in Sections 8.6 and 9.3.

Halting problem: This decision problem can be stated in a variety of ways, but to be specific, let us restrict our attention to all computer programs that are written in BASIC and that accept as input a natural number $n \in \mathbb{N}$. Let $\mathcal{P}$ denote the collection of all such programs. Is there an algorithm such that, given any program $P \in \mathcal{P}$ and any $n \in \mathbb{N}$, the algorithm decides (YES or NO) whether program P with input n halts after a finite number of steps? This can be described as the decision problem for Y in X, where $X = \mathcal{P} \times \mathbb{N}$ and $Y = \{\langle P, n \rangle : \langle P, n \rangle \in \mathcal{P} \times \mathbb{N}$ and program P with input n halts$\}$. The halting problem is unsolvable; this result is due to Turing and is proved in Section 9.2.

Hilbert's tenth problem: Is there an algorithm that, given a polynomial $p(x_1, \ldots, x_n)$ with integer coefficients, decides (YES or NO) whether there are integers $k_1, \ldots, k_n$ such that $p(k_1, \ldots, k_n) = 0$? This is the decision problem for Y in X, where X = all polynomials with integer coefficients and Y = all polynomials with integer coefficients that have integer solutions. See Chapter 10 for a proof that Hilbert's tenth problem is unsolvable.

──────────── *Exercises for Section 1.7* ────────────

1. Construct an algorithm that, given an integer $a \in \mathbb{Z}$, decides (YES or NO in a finite number of steps) whether the equation $x^2 + y^3 - a = 0$ has a solution in integers.

2. Show that the following relations on $\mathbb{N}$ are decidable:
   (1) $R(a, b, c, n) \Leftrightarrow a, b, c \in \mathbb{N}^+$, $n > 2$, and $a^n + b^n = c^n$;
   (2) $R(c, n) \Leftrightarrow c \in \mathbb{N}^+$, $n > 2$, and $x^n + y^n = c^n$ has a solution in positive integers.

3. (*Algebra of decidable sets*) Let Q and R be 1-ary decidable relations. Prove that the following are decidable: (1) $Q \cap R$; (2) $Q \cup R$; (3) Q^c; (4) $Q - R$.

4. Let R be an n-ary relation such that the characteristic function K_R of R is computable. Show that R is decidable.

5. Prove: (1) Every finite subset of $\mathbb{N}$ is decidable; (2) if R is a subset of $\mathbb{N}$ whose complement is finite, then R is decidable.

6. Let Q and R be 1-ary decidable relations. Show that the 2-ary relation $Q \times R$ is decidable.

7. Let R be a 2-ary decidable relation on $\mathbb{N}$ and let P be defined by $P(b) \Leftrightarrow$ there exists $a \le b$ such that $R(a, b)$. Show that P is a decidable relation.

8. Let F be an n-ary computable function. Show that $\mathscr{G}_F$ is decidable.

9. (*Algorithm for checking proofs*) Every formal system F is assumed to have these two properties: (1) There is an algorithm that decides whether an arbitrary formula is an axiom of F. (2) There is an algorithm that decides whether B is the conclusion of a rule of inference whose hypotheses are among $A_1, \ldots, A_k$. Show that for every formal system F, there is an algorithm CHKPRF that, given a finite sequence $A_1, \ldots, A_n$ of formulas of F, decides (YES or NO) whether $A_1, \ldots, A_n$ is a proof in F.

10. (*No 2-ary decidable relation enumerates all 1-ary decidable relations.*) Let R be a 2-ary relation with the following property: For each 1-ary decidable relation Q, there exists $n \in \mathbb{N}$ such that for all $a \in \mathbb{N}$, $R(n, a) \Leftrightarrow Q(a)$. (In this case, we say that R *enumerates* the 1-ary decidable relations.) Show that R is *not* decidable. (*Hint*: Use a Cantor diagonal argument: Assume R is decidable and show that $Q(a) \Leftrightarrow \neg R(a, a)$ must also be decidable.)

11. Recall the proof of the existence of an undecidable relation; Q is defined by $Q(n) \Leftrightarrow \neg R_n(n)$, where $\{R_n : n \in \mathbb{N}\}$ is an enumeration of all 1-ary decidable relations. Criticize the following "algorithm" for showing that Q is decidable:
    (1) Input $n \in \mathbb{N}$.
    (2) Use the algorithm for R_n to decide whether $R_n(n)$.
    (3) If $\neg R_n(n)$, print YES and halt; if $R_n(n)$, print NO and halt.

12. (*Effective coding and decidable formal systems*) Let F be a formal system, let FOR be the set of all formulas of F, and let g be a one-to-one function $g:$ FOR $\rightarrow \mathbb{N}$ satisfying the following two conditions: **Gö1** There is an algorithm that, given a formula A of F, finds $g(A)$ in a finite number of steps. **Gö2** There is an algorithm that, given $a \in \mathbb{N}$, decides whether there is a formula A such that $g(A) = a$; if YES, the algorithm finds A such that $g(A) = a$. A one-to-one function g satisfying Gö1 and Gö2 is called an *effective coding* of FOR. Define a 1-ary relation on $\mathbb{N}$ by

$$\text{THM}_F(a) \Leftrightarrow \text{there is a formula } A \text{ of F such that } g(A) = a \text{ and } A \text{ is a theorem of F.}$$

Show that the formal system F is decidable if and only if the 1-ary relation THM_F is decidable. (*Hint*: See Section 7.5.)

13. (*Construction of an effective coding*) Refer to Exercise 12. Let F be a formal system. Construct an effective coding for the set FOR of formulas of F. (*Hint:* See Section 7.2.)

14. Let $F: \mathbb{N} \to \mathbb{N}$ be a computable function that is both one-to-one and onto. Prove that F^{-1} is also computable. (*Hint:* $F^{-1}(b) = a \Leftrightarrow F(a) = b$.)

15. Let $F: \mathbb{N} \times \mathbb{N} \to \mathbb{N}$ be a 2-ary computable function that is onto. Write an algorithm that, given input $c \in \mathbb{N}$, finds $a, b \in \mathbb{N}$ such that $F(a, b) = c$.

16. Let F and G be 1-ary functions defined as follows:

$$F(n) = \begin{cases} 1 & \text{The } n\text{th digit in the decimal expansion of } \pi \text{ is 3;} \\ 0 & \text{otherwise.} \end{cases}$$

$$G(n) = \begin{cases} 1 & \text{There is a block of at least } n \text{ consecutive} \\ & \text{3s in the decimal expansion of } \pi; \\ 0 & \text{otherwise.} \end{cases}$$

(1) Show that F is a computable function.
(2) Show that G has the following property: If $G(n) = 0$, then $G(n + 1) = 0$.
(3) Show that one of the following must hold: $G(n) = 1$ for all $n \in \mathbb{N}$; there is some k such that $G(n) = 1$ for all $n \leq k$ and $G(n) = 0$ for all $n > k$.
(4) Is G computable?

17. Let $F: \mathbb{N} \to \mathbb{N}$ be defined as follows: If Goldbach's conjecture is true, then $F(a) = 0$ for all $a \in \mathbb{N}$; if Goldbach's conjecture is false, then $F(a) = 1$ for all $a \in \mathbb{N}$. Explain why F is a computable function.

18. (*Composition of computable functions is computable.*) Let F be the composition of $G, H_1, \ldots, H_k$ (defined in Section 1.8). Prove that if $G, H_1, \ldots, H_k$ are computable, then F is computable.

19. (*Minimalization of a computable function is computable.*) Let F be the minimalization of an $n + 1$-ary regular function (defined in Section 1.8). Prove that if G is computable, then F is computable.

20. Let F be a computable function, let Q be a 1-ary decidable relation, and define R by $R(n) \Leftrightarrow \neg Q(F(n))$. Show that R is decidable.

21. Let F be a 1-ary computable function and Q a 2-ary decidable relation. Show that R is decidable, where $R(a) \Leftrightarrow$ there exists $x \leq F(a)$ such that $Q(a, x)$.

22. (*No 2-ary computable function enumerates all 1-ary computable functions.*) Let F be a 2-ary function with the following property: For each 1-ary computable function G, there exists $n \in \mathbb{N}$ such that for all $a \in \mathbb{N}$, $G(a) = F(n, a)$. (In this case, we say that F *enumerates* all 1-ary computable functions.) Show that F is *not* computable. (*Hint:* Use a Cantor diagonal argument: Assume F is computable; define $G: \mathbb{N} \to \mathbb{N}$ by $G(a) = F(a, a) + 1$ and show that G must also be computable.)

23. Recall our proof of the existence of a 1-ary function G that is not computable. Criticize the following "algorithm" for computing G:
(1) Input $n \in \mathbb{N}$.
(2) Use the algorithm for F_n to compute $F_n(n)$.
(3) Compute $G(n) = F_n(n) + 1$.

24. Find a one-to-one function from $\mathbb{N} \times \mathbb{N}$ onto $\mathbb{N}$ that is computable. (*Hint:* See the discussion of Cantor's pairing function in Section 8.3.)

25. Show that for all $n \geq 1$, there is a one-to-one function from $\mathbb{N}^n$ onto $\mathbb{N}$ that is computable. (*Hint:* Use induction and the result from exercise 24.)

26. Assume R is a decidable relation. Show that both R and R^c are semidecidable.

27. Let Q and R be listable sets. Prove that the following sets are listable: (1) $Q \cap R$; (2) $Q \cup R$.

28. Let $R = \{n: n \in \mathbb{N}, \text{ there exist } a, b, c \in \mathbb{N}^+ \text{ such that } a^n + b^n = c^n\}$. Show that R is semidecidable.

29. Use a counting argument to prove the following: (1) There is a 1-ary relation that is not

listable; (2) there is a 1-ary relation R such that neither R nor its complement R^c is listable.

30. This exercise outlines an intuitive construction of a set K that is semidecidable but not decidable. We consider algorithms Φ with the following two properties: (1) The input for Φ is any natural number $a \in \mathbb{N}$. (2) If Φ is given an input $a \in \mathbb{N}$, then either Φ halts after a finite number of steps, in which case it prints YES, or else Φ continues to run forever. Let $\Omega = \{\Phi_n : n \in \mathbb{N}\}$ be the collection of all algorithms satisfying (1) and (2). Further assume that we have an algorithm Σ such that for each $n \in \mathbb{N}$, we can completely recover algorithm Φ_n from n; in other words, given $n \in \mathbb{N}$, we can write out the list of instructions of Φ_n and then execute Φ_n for any input $a \in \mathbb{N}$. Let $K = \{n : n \in \mathbb{N}$ and algorithm Φ_n with input n halts$\}$.

(1) Show that K is semidecidable.

(2) Show that K^c is not semidecidable. (*Hint:* Assume K^c is semidecidable and let Φ_n be an algorithm for K^c.) Conclude that K is not decidable.

1.8

Recursive Functions and Recursive Relations

The goal of this section is to give precise definitions of computability (of an n-ary function) and decidability (of an n-ary relation). *Reminder*: All functions are from $\mathbb{N}^n$ into $\mathbb{N}$ and all relations are on $\mathbb{N}$; in other words, $F: \mathbb{N}^n \to \mathbb{N}$ and $R \subseteq \mathbb{N}^n$; generally speaking, F, G, H denote functions and P, Q, R denote relations. We proceed as follows: First introduce the class of *recursive* functions as the precise or formal counterpart of the class of computable functions; then define an n-ary relation to be *recursive* if its characteristic function is a recursive function. In summary:

Formal Concept	*Informal Counterpart*
Recursive function	Computable function
Recursive relation	Decidable relation

The problem of giving a precise definition of computability has been approached in a number of ways. Among other definitions are (1) computable by a Turing machine; (2) computable by a register machine; (3) recursive; (4) Herbrand-Gödel computable; (5) Markov computable; (6) Post computable; (7) λ-definable. One can prove that these different approaches lead to the same class of functions. (In Chapter 9 we show that *recursive = computable by a register machine*.) Each approach has its own advantages and disadvantages. Recursive functions have the advantage of being easy to describe and easy to work with mathematically. A disadvantage, however, is that they are not immediately convincing; that is, it is not clear at the outset that the class of recursive functions is broad enough to capture the concept of computability.

The basic idea in defining the class of recursive functions is to begin with certain *starting* functions ("axioms") and then apply the operations of *composition* and/or *minimalization* ("rules of inference") to construct additional functions in the class. The starting functions and the two operations are as follows.

Starting functions: $+, \times, U_{n,k}$ ($n \geq 1, 1 \leq k \leq n$), and $K_<$. Here, $+$ and $\times$ are the 2-ary functions of addition and multiplication. The function $U_{n,k}$ is called the *n-ary projection function into the kth coordinate* and is defined by $U_{n,k}(a_1, \ldots, a_n) = a_k$, $1 \leq k \leq n$. Note that $U_{1,1}(a) = a$ and so the identity function on $\mathbb{N}$ is a starting function. Finally, $K_<$ is the characteristic function for the relation $<$ on $\mathbb{N}$; in other words, $K_<$ is the 2-ary decision function defined by $K_<(a, b) = 0$ if $a < b$ and $K_<(a, b) = 1$ if $a \geq b$. Each starting function is a computable function (in the intuitive sense).

Composition: Let G be a k-ary function and let $H_1, \ldots, H_k$ be n-ary functions. The *composition of $G, H_1, \ldots, H_k$* is the n-ary function F defined by

$$F(a_1, \ldots, a_n) = G(H_1(a_1, \ldots, a_n), \ldots, H_k(a_1, \ldots, a_n)).$$

Minimalization (of a regular function): Let G be an $n + 1$-ary function such that for all $a_1, \ldots, a_n \in \mathbb{N}$, there is at least one number $x \in \mathbb{N}$ such that $G(a_1, \ldots, a_n, x) = 0$. A function with this property is said to be *regular*. Given that G is regular, we define an n-ary function F by

$$F(a_1, \ldots, a_n) = \mu x[G(a_1, \ldots, a_n, x) = 0],$$

where $\mu x[..x..]$ means the smallest natural number x such that $..x..$ holds. (Recall the well-ordering principle, which states that every nonempty set of natural numbers has a smallest element.) Thus, the value of $F(a_1, \ldots, a_n)$ is the smallest $x \in \mathbb{N}$ such that $G(a_1, \ldots, a_n, x) = 0$. The function F is said to be obtained from G by *minimalization (of a regular function)*.

To better understand composition and minimalization, let us prove the following two lemmas.

Lemma 1: Let G be a k-ary function, let $H_1, \ldots, H_k$ be n-ary functions, and let F be the composition of $G, H_1, \ldots, H_k$. If the functions $G, H_1, \ldots, H_k$ are all computable, then F is computable.

Proof: Here is an algorithm for computing F.

(1) Input $a_1, \ldots, a_n \in \mathbb{N}$.
(2) For $1 \leq i \leq k$, use the algorithm for H_i to compute $b_i = H_i(a_1, \ldots, a_n)$.
(3) Use the algorithm for G to compute $G(b_1, \ldots, b_k)$.
(4) $F(a_1, \ldots, a_n) = G(b_1, \ldots, b_k)$. ∎

Lemma 2: Let G be an $n + 1$-ary regular function and let F be obtained from G by minimalization. If G is computable, then F is computable.

Proof: Here is an algorithm for computing F.

(1) Input $a_1, \ldots, a_n \in \mathbb{N}$ and let $x = 0$.
(2) Compute $G(a_1, \ldots, a_n, x)$ (G is computable).
(3) Is $G(a_1, \ldots, a_n, x) = 0$? If *Yes*, $F(a_1, \ldots, a_n) = x$; halt.
(4) If *No*, replace x with $x + 1$ and go to (2).

This algorithm does not go into an infinite loop, since we have assumed that for all $a_1, \ldots, a_n \in \mathbb{N}$, there is at least one $x \in \mathbb{N}$ such that $G(a_1, \ldots, a_n, x) = 0$. Since $F(a_1, \ldots, a_n)$ is defined as the smallest such x, the algorithm systematically searches for x by calculating $G(a_1, \ldots, a_n, 0)$, $G(a_1, \ldots, a_n, 1)$, and so on, until it finds the first x for which $G(a_1, \ldots, a_n, x) = 0$. ∎

Definition 1: An n-ary function F is *recursive* if there is a finite sequence $F_1, \ldots, F_m$ of functions with $F_m = F$ and such that for $1 \le k \le m$, one of the following conditions holds: (1) F_k is a starting function; (2) $k > 1$ and F_k is obtained from previous functions (i.e., among $F_1, \ldots, F_{k-1}$) by composition or minimalization (of a regular function).

Thus, showing that a function F is recursive proceeds much like a proof in a formal system; starting functions are like axioms and the two operations of composition and minimalization are like rules of inference. However, this procedure is much too cumbersome to carry out in practice, and instead we have the following more efficient method (proof given in Section 8.1).

Lemma 3: Let F be an n-ary function. To show that F is recursive, it suffices to show that F satisfies one of the following three conditions: (1) F is a starting function; (2) F is the composition of recursive functions; (3) F is the minimalization of a regular recursive function.

Perhaps the reader is not impressed with the power of the two operations of composition and minimalization. Indeed, it is not obvious that a simple function such as the zero function $Z(a) = 0$ is recursive. Much less obvious is the fact that the factorial function $F(n) = n!$, which is certainly computable, is recursive.

Nevertheless, a systematic study of recursive functions, as carried out in Chapter 8, shows that the short list of starting functions, together with the operations of composition and minimalization, are powerful enough to prove that a wide range of computable functions are indeed recursive. In fact, there is no known example of a function that is computable but fails to be recursive.

As an example, how do we see that $F(n) = n!$ is recursive? In Section 8.4 we will justify the following operation for constructing recursive functions.

Primitive recursion (Special case): Let $k \in \mathbb{N}$ and let H be a 2-ary recursive function. Then F is recursive, where

$$F(0) = k;$$

$$F(n + 1) = H(n, F(n)).$$

To see how this applies to $F(n) = n!$, take $k = 1$ and $H(n, x) = (n + 1) \times x$ (for details, see Section 8.4). Incidentally, the important operation of primitive recursion explains the term *recursive function*.

The class of recursive functions turns out to be an important and natural class of functions. The following discussion anticipates results we will consider in detail in Chapter 9. Take an ideal computer; the computer we have in mind can perform the same operations as a typical real-life computer but differs in that it has an infinite amount of storage. Specifically, there are an infinite number of registers $R_1, R_2, R_3, \ldots$, and each of these registers can hold any natural number, regardless of size. An n-ary function F is computed by such a computer if there is a program with the following property: Given $a_1, \ldots, a_n \in \mathbb{N}$, put $a_1, \ldots, a_n$ in registers $R_1, \ldots, R_n$ respectively, and then start the program; after a finite number of steps, $F(a_1, \ldots, a_n)$ appears in register R_1. *Question*: What functions can be calculated by such a computer? *Answer*: Let F be an n-ary function; then F can be calculated by this computer if and only if F is recursive.

When working with formal systems, we have available the technique of induction on theorems. There is a similar procedure for showing that all recursive functions have some property Q of functions.

Theorem 1 (Induction on recursive functions): Let Q be a property of functions. To prove that every recursive function has property Q, it suffices to prove the following: (1) Every starting function has property Q; (2) if F is the composition of $G, H_1, \ldots, H_k$, and each of $G, H_1, \ldots, H_k$ has property Q, then F has property Q; (3) if F is the minimalization of a regular function G, and G has property Q, then F has property Q.

Proof: We give an informal argument; a precise proof would use complete induction. Let F be a recursive function. Then there is a finite sequence $F_1, \ldots, F_n$ of functions with $F_n = F$ and such that for $1 \leq k \leq n$, one of the following two conditions holds: F_k is a starting function; $k > 1$ and F_k is obtained from previous functions (i.e., among $F_1, \ldots, F_{k-1}$) by composition or minimalization of a regular function. We now argue that each of the functions $F_1, \ldots, F_n$ has property Q. First of all, F_1 is a starting function and hence F_1 has property Q by (1). Now consider F_2; if F_2 is also a starting function, then F_2 has property Q for the same reason. Otherwise, F_2 is obtained from F_1 by composition or minimalization. By (2) or (3), it follows that F_2 has property Q. Continuing in this way, we see that each of the functions $F_1, \ldots, F_n$ has property Q; in particular, the function F_n has property Q as required. ∎

As an application of Theorem 1, let us show that all recursive functions are computable.

Theorem 2: Every recursive function is computable.

Proof: We use induction on recursive functions with Q the property of being computable (in the intuitive sense). It is obvious that the starting functions $+, \times, U_{n,k}(1 \leq k \leq n)$, and $K_<$ are all computable. Moreover, the two lemmas proved earlier in this section show that the composition or minimalization of computable functions is computable. ∎

The recursive functions are the formal counterpart of the computable functions. Let us now turn to the problem of defining the formal counterpart of the decidable relations. The definition we give is obviously based on Theorem 2 in Section 1.7.

Definition 2: An *n*-ary relation R is *recursive* if its characteristic function K_R is recursive. For example, the 2-ary relation $<$ is recursive since its characteristic function $K_<$ is a starting function and hence is recursive. We have the following consequence of Theorem 2.

Corollary 1: Every recursive relation is decidable.

Proof: Let R be a recursive relation. The following steps show that R is decidable: K_R is a recursive function (definition of R recursive); K_R is a computable function (Theorem 2); R is a decidable relation (Theorem 2 in Section 1.7). ∎

─────────────── *Church's Thesis* ───────────────

Have we succeeded in capturing the informal notion of a computable function? We have already proved that every recursive function is computable. But can we prove the converse? This amounts to showing that *not recursive* ⟹ *not computable*. This statement is not subject to proof. For suppose we have a function F that is not recursive. In order to show that F is not computable, we must show that there does not exist an algorithm that calculates F. Since the concept of an algorithm is not precisely defined, we cannot hope to prove its nonexistence. Nevertheless, there is strong evidence that this converse does indeed hold. This is known as Church's thesis, and is stated as follows.

Church's Thesis Every computable function is recursive.

Although we cannot prove Church's thesis, we can give evidence that it is correct:

- *Verification argument* Every function known to be computable can be shown to be recursive using the powerful techniques developed in Chapter 8.
- *Agreement argument* The class of recursive functions is a very natural and fundamental class of functions. There have been many attempts to capture the notion of a computable function: recursive, register-machine computable, Turing-machine computable, and others. All of these approaches give rise to the same class of functions.

How could Church's thesis be false? Consider the following scenario: A bright mathematician describes, to a panel of experts on computability, a function F and an algorithm for computing F. The experts all agree that the algorithm does indeed compute the function F. The clever mathematician then proceeds to give a precise proof that F is *not* recursive. In other words, a counterexample to Church's thesis is mathematically possible. But most

recursion theorists accept Church's thesis and do not feel that a counterexample is forthcoming. Church's thesis can also be stated in terms of
relations. This is an easy consequence of the following result.

Theorem 3: The following two conditions are equivalent: (1) Every
computable function is recursive; (2) every decidable relation is recursive.

Proof that (1) $\Rightarrow$ (2): Let R be an n-ary relation that is decidable. The
following steps show that R is recursive: K_R is computable (Theorem 2 in
Section 1.7); K_R is recursive (see (1)); R is recursive (definition of R recursive).

Proof that (2) $\Rightarrow$ (1): Let F be an n-ary computable function; we are
required to show that F is a recursive function. Since F is computable, the
graph $\mathcal{G}_F$ of F is decidable (Theorem 6 in Section 1.7). By (2), $\mathcal{G}_F$ is recursive.
Let K be the characteristic function of $\mathcal{G}_F$; in other words,

$$K(a_1, \ldots, a_n, b) = \begin{cases} 0 & \text{if } F(a_1, \ldots, a_n) = b; \\ 1 & \text{if } F(a_1, \ldots, a_n) \neq b. \end{cases}$$

Since $\mathcal{G}_F$ is a recursive relation, K is a recursive function. In addition, the
function K is regular. To see this, let $a_1, \ldots, a_n \in \mathbb{N}$; we are required to show
that there is some $b \in \mathbb{N}$ such that $K(a_1, \ldots, a_n, b) = 0$. Take $b = F(a_1, \ldots, a_n)$. Finally, $F(a_1, \ldots, a_n) = \mu b[K(a_1, \ldots, a_n, b) = 0]$; hence F is
recursive by minimalization of a regular recursive function (see Lemma 3). ∎

Theorem 3 does not assert Church's thesis; rather, it states that if we
accept Church's thesis, then we must also accept the assertion that every
decidable relation is recursive. Conversely, if we accept (2), then we must also
accept Church's thesis. Let us formulate (2) as an alternate version of Church's
thesis.

Church's Thesis (relation version) Every decidable relation is
recursive.

Henceforth, we will refer to either (1) or (2) of Theorem 3 as *Church's thesis.*
Recursion theory clarifies the concept of an algorithm. Let us now discuss
what is undoubtedly the main application of recursion theory to mathematics.
Let X be a countable set, let $Y \subseteq X$, and suppose that we want to show that
the decision problem for Y in X is unsolvable. We proceed as follows: (1) Show
that there is a relation R on $\mathbb{N}$ such that the decision problem for Y in X is
solvable if and only if R is decidable; (2) show that R is not recursive. We then
say that the decision problem for Y in X is *recursively unsolvable.* Moreover, if
we accept Church's thesis, then the relation R is actually undecidable, in which
case we say that the decision problem for Y in X is *algorithmically unsolvable.*
This procedure is used to prove, among other results, the undecidability of
arithmetic (see Sections 7.5 and 8.6) and the unsolvability of the halting
problem (see Section 9.2).

RE Relations

We end this section with a brief survey of some ideas that are discussed in great detail in Section 8.5. Recursive relations are the formal counterpart of the decidable relations. Is there a formal counterpart of the semidecidable relations?

Definition 3: An n-ary relation R is RE (*recursively enumerable*) if there is an $n + 1$-ary recursive relation Q such that for all $a_1, \ldots, a_n \in \mathbb{N}$,

$$R(a_1, \ldots, a_n) \Leftrightarrow \text{there exists } x \in \mathbb{N} \text{ such that } Q(a_1, \ldots, a_n, x).$$

This definition is obviously inspired by the characterization of semidecidable relations in Section 1.7 (see Theorem 5). In Section 8.5, the following fundamental results are proved: (1) Every recursive relation is an RE relation; (2) every RE relation is semidecidable; (3) assuming Church's thesis, every semidecidable relation is RE. In summary:

Formal Concept	*Informal Counterpart*
Recursive function	Computable function
Recursive relation	Decidable relation
RE relation	Semidecidable relation

Finally, the existence of an RE relation that is not recursive is one of the most important and unifying results in mathematical logic. In this text we will give three rather different proofs of this fundamental result: THM_Γ is RE but not recursive (Chapter 8); HLT is RE but not recursive (Chapter 9); SOLN is RE but not recursive (Chapter 10). But for now, let us simply record the precise result.

Theorem 4 (Fundamental theorem of recursion theory): There is an RE relation that is not recursive.

Exercises for Section 1.8

1. Show that the number of recursive functions is countable. Conclude that there is a 1-ary function that is not recursive.
2. Let $\{F_n : n \in \mathbb{N}\}$ be the collection of all 1-ary recursive functions. Define a 1-ary function $G : \mathbb{N} \to \mathbb{N}$ by $G(n) = F_n(n) + 1$ for all $n \geq 0$. Is G recursive?
3. Let $\{R_n : n \in \mathbb{N}\}$ be the collection of all 1-ary recursive relations. Define a 1-ary relation Q by $Q = \{n : n \in \mathbb{N} \text{ and } n \notin R_n\}$. Is Q recursive?
4. Let F and G be 1-ary recursive functions. Show that the following functions are recursive:
   (1) $H_1(a, b) = F(a) + G(b)$;
   (2) $H_2(a, b) = F(a) \times G(b)$.
   (*Hint:* Lemma 3, composition, projection functions, $+$ and $\times$ starting functions.)
5. Let $F(a, b, c) = a \times (b + c)$. Show that F is recursive.
6. Let G be a 3-ary recursive function and let H be a 2-ary recursive function. Show that the function $F(a, b) = G(b, H(a, b), a)$ is recursive. (*Hint:* Lemma 3, composition, projection functions.)
7. Let R be an $n + 1$-ary recursive relation such that for all $a_1, \ldots, a_n \in \mathbb{N}$, there exists $b \in \mathbb{N}$ such that $R(a_1, \ldots, a_n, b)$. Show that the function $F(a_1, \ldots, a_n) = \mu b[R(a_1, \ldots, a_n, b)]$ is recursive.

8. Show that the following functions are recursive:
   (1) The zero function $Z(a) = 0$ (*Hint*: minimalization, $U_{2,2}$).
   (2) The successor function $S(a) = a + 1$ (*Hint*: minimalization, $K_<$).
   (3) The constant function $C(a) = 1$.
   (4) The 1-ary decision functions sg and csg defined by

$$sg(a) = \begin{cases} 0 & a = 0; \\ 1 & a \neq 0; \end{cases} \quad csg(a) = \begin{cases} 1 & a = 0; \\ 0 & a \neq 0. \end{cases}$$

9. Let R be a 1-ary relation, and suppose there is a 1-ary recursive function F such that for all $a \in \mathbb{N}$, $a \in R \Leftrightarrow F(a) = 0$. Show that R is recursive. (*Hint*: Use the function sg in Exercise 8).
10. Prove Lemma 3.

2

Language and Semantics of Propositional Logic

P ropositional logic is the study of the five logical connectives *not, or, and, if... then,* and *if and only if.* In this chapter we describe a language for propositional logic, assign truth functions to the five logical connectives, and define the fundamental concept *tautological consequence.* In Chapter 3 we will introduce a formal system whose intended interpretation is propositional logic. Thus, one may regard Chapter 2 as a discussion of the language of the formal system that we propose to study in Chapter 3; alternatively, Chapter 2 is a semantic approach to propositional logic.

2.1

Language of Propositional Logic

Propositional logic is concerned with the analysis of arguments such as the following:

(1) Snow is white or 1 + 1 = 2.
∴ 1 + 1 = 2.

(2) Snow is white or 1 + 1 = 3.
∴ 1 + 1 = 3.

(3) Snow is red or 1 + 1 = 2.
Snow is not red.
∴ 1 + 1 = 2.

(4) Snow is red or 1 + 1 = 3.
Snow is not red.
∴ 1 + 1 = 3.

First we make some comments about the statements that appear in these four arguments. By a *simple proposition,* we mean a declarative sentence that is either true or false and has no connectives. For example, "Snow is white" and "1 + 1 = 2" are simple propositions. By a *proposition,* we mean a declarative sentence that either is a simple proposition or is built up from simple propositions using one or more of the connectives *not, or, and, if... then,* and *if and only if.*

Examples of propositions are "Snow is red or $1 + 1 = 2$" and "Snow is not red." For later use, we record the truth value of the propositions that appear in the arguments given above:

Snow is white or $1 + 1 = 2$: TRUE

Snow is white or $1 + 1 = 3$: TRUE

Snow is red or $1 + 1 = 2$: TRUE

Snow is red or $1 + 1 = 3$: FALSE

Snow is not red: TRUE

$1 + 1 = 2$: TRUE

$1 + 1 = 3$: FALSE

We now comment on the arguments themselves. Although arguments (1) and (2) involve different propositions, they nevertheless have the same *form*, namely $p \vee q \therefore q$. Likewise, arguments (3) and (4) have the same *form*, namely $p \vee q$, $\neg p$, $\therefore q$. In these forms, one should interpret the symbol $\vee$ as *or* and the symbol $\neg$ as *not*; the letters p and q stand for simple propositions. Arguments are classified as either *valid* or *invalid*; the key to this classification is the form of an argument. Later we will give precise definitions of these concepts. For the time being and for the sake of continuing the present discussion, let us assume that the argument form $p \vee q$, $\therefore q$ is invalid, whereas the argument form $p \vee q$, $\neg p$, $\therefore q$ is valid. This means that arguments (1) and (2) are invalid and arguments (3) and (4) are valid.

Although argument (4) is valid, its conclusion is false. Note, however, that one of its hypotheses is also false. Thus we cannot automatically say that the conclusion of a valid argument is true. In both arguments (1) and (2) the hypothesis is true. In argument (1), however, the conclusion is true whereas in argument (2) the conclusion is false. These two examples show that the conclusion of an invalid argument can be either true or false, even when all hypotheses are true.

Thus it seems that the only time we can be confident that the conclusion of an argument is true is when the argument is valid *and* all hypotheses of the argument are true. This is actually the basis for the definition of a valid argument (we will say more about this later).

We now describe a formal language for propositional logic. The description is given in two steps: symbols of the language and formulas of the language. The formulas are those expressions that can be interpreted as propositions.

Symbols: The symbols of the language are

p_1 p_2 p_3 $\cdots$	(infinite list)
( )	(left and right parentheses)
$\neg$ $\vee$ $\wedge$ $\rightarrow$ $\leftrightarrow$	(logical connectives)

The letters $p_1, p_2, p_3, \ldots$ are called *propositional variables* and are interpreted as simple propositions. The list is infinite to ensure that we never run out of propositional variables when constructing formulas. The two symbols (and) are used for punctuation. Finally, the remaining symbols are called *logical connectives.* Their names and interpretations are as follows:

Symbol	Name	Interpretation
¬	negation	not
∨	disjunction	or
∧	conjunction	and
→	conditional	if ... then
↔	biconditional	if and only if

The symbols (,), ¬, ∨, ∧, →, and ↔ are often called *logical symbols.* In translating English sentences into the language of propositional logic, these symbols are always interpreted in the same way.

Formulas: The formulas are defined inductively by these three rules:

F1 Each propositional variable is a formula.

F2 If A and B are formulas, so are ¬A, $(A \vee B)$, $(A \wedge B)$, $(A \to B)$, and $(A \leftrightarrow B)$.

F3 Every formula is obtained by a finite number of applications of F1 and F2.

In other words, the formulas of propositional logic either are propositional variables or are built up from simpler formulas using the operators ¬, ∨, ∧, →, and ↔. The formula ¬A is called the *negation* of A; the formula $(A \vee B)$ is called a *disjunction* with *disjuncts* A and B; the formula $(A \wedge B)$ is called a *conjunction* with *conjuncts* A and B; the formula $(A \to B)$ is called a *conditional* (or *implication*) with *antecedent* A and *consequent* B; the formula $(A \leftrightarrow B)$ is called the *biconditional* of A and B.

Example 1:

Each of the following expressions is a formula:

(1) p_4;   **(3)** $(\neg p_1 \vee p_2)$;
(2) $\neg(p_1 \vee p_2)$;   **(4)** $((p_1 \to p_2) \wedge \neg p_6)$.

For example, to justify (4): p_1, p_2, p_6 are formulas by F1; $(p_1 \to p_2)$ and $\neg p_6$ are formulas by F2; $((p_1 \to p_2) \wedge \neg p_6)$ is a formula by F2. □

Example 2:

We will show that the expression $(p_1 \wedge \neg(\neg p_3))$ is not a formula. Suppose $(p_1 \wedge \neg(\neg p_3))$ is a formula. By F3, $(p_1 \wedge \neg(\neg p_3))$ is obtained by a finite number of applications of F1 and F2. In particular, F2 applies and p_1 and $\neg(\neg p_3)$ must both be formulas. Now if $\neg(\neg p_3)$ is a formula, then, again by F2, $(\neg p_3)$ must be a formula. Moreover, $(\neg p_3)$ is the result of an application of F2. But according to F2, parentheses are *not* placed about the negation symbol. So $(p_1 \wedge \neg(\neg p_3))$ is not a formula, as required. □

Conventions It is convenient to make some conventions on the writing of formulas. First of all, the letters p, q, r, s, and so on are often used for propositional variables instead of the more precise $p_1, p_2, p_3, \ldots$. Also, outer parentheses in formulas are usually omitted since no ambiguity can occur. With these conventions, formulas (2), (3), and (4) of Example 1 become

(2)′ $\neg(p \vee q)$;   **(3)′** $\neg p \vee q$;   **(4)′** $(p \rightarrow q) \wedge \neg r$.

When $\vee$ is used more than twice in succession without parentheses, it is understood that the parentheses are associated to the right. For example, $p \vee q \vee r$ is used for $(p \vee (q \vee r))$, and $p \vee q \vee r \vee s$ is used for $(p \vee (q \vee (r \vee s)))$. A similar remark applies to $\wedge$. Finally, brackets are occasionally used in place of parentheses to improve readability. For example,

$$[p \rightarrow (q \rightarrow r)] \leftrightarrow [(p \wedge q) \rightarrow r]$$

is acceptable as a formula.

Example 3:

We express the following argument in the language of propositional logic:

The quartet will not play both Haydn and Mozart.

The quartet will not play Haydn.

Therefore, the quartet will play Mozart.

Assign propositional variables to simple propositions as follows: p = The quartet will play Haydn; q = The quartet will play Mozart. We then have $\neg(p \wedge q)$, $\neg p$, $\therefore q$. A more relaxed and suggestive translation is $\neg(H \wedge M)$, $\neg H$, $\therefore M$. The reader is encouraged to use this more informal symbolism when translating English sentences into the language of propositional logic. □

The letters A, B, C, D, and E are used to denote formulas; Γ and Δ denote sets of formulas. It should be stressed that while $(p_1 \rightarrow p_2)$ is an official formula of our language, $(A \rightarrow B)$ is not. Rather, $(A \rightarrow B)$ represents an infinite number of formulas and is an example of what is known as a *formula scheme*. To illustrate, $(p_1 \rightarrow p_2)$ and $(p_3 \rightarrow p_1)$, as well as more complicated formulas such as $(\neg p_2 \rightarrow (p_3 \wedge p_4))$, are all instances of $(A \rightarrow B)$.

In the study of formal languages, it is customary to call the language under study the *object language* and the language used to talk about the object language the *metalanguage*. For example, in this section the object language is the language of propositional logic and the metalanguage is English (supplemented with some additional symbols). In this situation, the letters A, B, C, D, and E, which stand for formulas, are called *metavariables*. They are not a part of the object language but instead stand for certain concepts of the object language, namely formulas.

We will end this section with an inductive procedure for showing that every formula of propositional logic has some property Q. This procedure is often called *induction on the number of connectives in a formula.*

Theorem 1 (Induction on formulas): Let Q be a property of formulas. To show that every formula A of propositional logic has property Q, it suffices to prove the following:

(1) Every propositional variable has property Q;
(2) if A is $\neg B$ and B has property Q, then A has property Q;
(3) if A is one of $(B \vee C)$, $(B \wedge C)$, $(B \rightarrow C)$, and $(B \leftrightarrow C)$, and both B and C have property Q, then A has property Q.

Proof: Let $S(n)$ say: Every formula with n occurrences of the logical connectives $\neg$, $\vee$, $\wedge$, $\rightarrow$, and $\leftrightarrow$ has property Q. We use complete induction to show that $S(n)$ holds for all $n \geq 0$. Since every formula has n connectives for some $n \geq 0$, it follows that every formula has property Q, as required.

Beginning step $S(0)$ states that every formula with no connectives has property Q. A formula with no connectives is a propositional variable, and each propositional variable has property Q by (1).

Induction step Let $n \geq 0$ and assume that $S(0), \ldots, S(n)$ all hold. $S(n + 1)$ states that every formula A with $n + 1$ connectives has property Q. There are five cases: A is $\neg B$, $(B \vee C)$, $(B \wedge C)$, $(B \rightarrow C)$, and $(B \leftrightarrow C)$. Consider the case in which A is $(B \vee C)$. Both B and C have fewer than $n + 1$ connectives, so by the induction hypotheses $S(0), \ldots, S(n)$ it follows that both B and C have property Q. Now apply (3) to conclude that $(B \vee C)$ has property Q. The remaining four cases, namely that A is $\neg B$, $(B \wedge C)$, $(B \rightarrow C)$, or $(B \leftrightarrow C)$, are similar and are left to the reader. ∎

In a proof by induction on formulas, (1) corresponds to the beginning step of an induction proof, and (2) and (3) correspond to the induction step. In (2) and (3), the assumption that B and C have property Q is the *induction hypothesis*.

Example 4:

We show that each formula of propositional logic has the same number of left parentheses as right parentheses. Given a formula A, let

$L(A) = $ # of left parentheses in A and
$R(A) = $ # of right parentheses in A.

We use induction on formulas to show that $L(A) = R(A)$ for each formula A. First of all, $L(p) = R(p) = 0$ for any propositional variable p. Now suppose A is $(B \vee C)$. By induction hypothesis, $L(B) = R(B)$ and $L(C) = R(C)$. Moreover,

$L((B \vee C)) = L(B) + L(C) + 1$ and
$R((B \vee C)) = R(B) + R(C) + 1$;

hence $L(A) = R(A)$, as required. The cases in which A is $\neg B$, $(B \wedge C)$, $(B \rightarrow C)$, or $(B \leftrightarrow C)$ are similar and are left to the reader. □

──────────*Exercises for Section 2.1*──────────

1. Decide whether the following expressions are formulas. If the expression is a formula, derive it from F1 and F2.

   (1) $((p_1 \vee p_2) \wedge p_3)$;   (3) $(p_1 \vee p_2 \wedge p_3)$;
   (2) $(\neg p_1 \vee \neg\neg p_2))$;   (4) $((p_1 \to p_2) \to p_3)$.

2. Decide whether the following expressions are formulas. If the expression is a formula, derive it from F1 and F2. This time, assume conventions in writing formulas (use p, q, r for propositional variables, omit outer parentheses, use brackets).

   (1) $(p \wedge q) \vee r$;   (3) $[(p \to q) \to p] \to p$;
   (2) $(p \wedge q) \leftrightarrow \wedge r$;   (4) $p \to (q \to p)$.

3. Write each of the following arguments in the language of propositional logic and try to classify each as valid or invalid.

   (1) If Fred can understand mathematics, then he can understand logic. Fred cannot understand logic. $\therefore$ Fred cannot understand mathematics.
   (2) If it rains or if it snows, then the electricity will go out. It will rain. $\therefore$ The electricity will go out.
   (3) If it rains or if it snows, then the electricity will go out. The electricity does not go out. $\therefore$ It did not snow.
   (4) If it rains and the fog settles in, then flying will be dangerous. It will rain. $\therefore$ Flying will be dangerous.
   (5) If $\sin x$ is differentiable, then $\sin x$ is continuous. If $\sin x$ is continuous, then $\sin x$ is integrable. The function $\sin x$ is differentiable. $\therefore$ The function $\sin x$ is integrable.
   (6) If Gödel had been president, then Congress would pass logical laws. Gödel was not president. $\therefore$ Congress does not pass logical laws.
   (7) If it rains, then we will not have a picnic. If it snows, then we will not have a picnic. It will rain or it will snow. $\therefore$ We will not have a picnic.

4. Complete the proof of Theorem 1 (induction on formulas) by considering the cases in which A is $\neg B$, $(B \wedge C)$, $(B \to C)$, or $(B \leftrightarrow C)$.

5. Complete the proof of Example 4.

6. Use the well-ordering principle to prove induction on formulas. (*Hint:* Assume (1)–(3) hold, but suppose there is some formula that does not have property Q. Choose one with the least number of occurrences of the connectives $\neg$, $\vee$, $\wedge$, $\to$, and $\leftrightarrow$.)

7. Prove that every formula has at least one occurrence of a propositional variable.

8. Prove that every formula with an occurrence of $\vee$ has at least two propositional variables.

9. Prove that every formula with n occurrences of propositional variables has $n-1$ occurrences of the connectives $\vee$, $\wedge$, $\to$, and $\leftrightarrow$.

10. Let FOR $(\neg, \vee)$ be all formulas of propositional logic that use just $\neg$ and $\vee$. *Note:* FOR $(\neg, \vee)$ is defined inductively by the following: Every propositional variable is in FOR $(\neg, \vee)$; if A and B are in FOR $(\neg, \vee)$, so are $\neg A$ and $(A \vee B)$. Prove that every $A \in$ FOR $(\neg, \vee)$ is of the form p, $\neg p$, $\neg\neg B$, $B \vee C$, or $\neg(B \vee C)$.

11. Let FOR $(\neg, \vee)$ be all formulas that use just $\neg$ and $\vee$ (see Exercise 10) and let $\Gamma \subseteq$ FOR $(\neg, \vee)$ satisfy these four conditions: (1) There is no propositional variable p such that $\{p, \neg p\} \subseteq \Gamma$; (2) if $\neg\neg A \in \Gamma$, then $A \in \Gamma$; (3) if $\neg(A \vee B) \in \Gamma$, then $\neg A \in \Gamma$ and $\neg B \in \Gamma$; (4) if $(A \vee B) \in \Gamma$, then $A \in \Gamma$ or $B \in \Gamma$. Show that $\{A, \neg A\} \not\subseteq \Gamma$ for all $A \in$ FOR $(\neg, \vee)$.

12. (*Top-down = bottom-up*) Our definition of a formula of propositional logic is often described as a *bottom-up* approach, since we start with propositional variables and then build up more complicated formulas using the five operators $\neg$, $\vee$, $\wedge$, $\to$, and $\leftrightarrow$. We will give a precise definition of the bottom-up approach and then show that the formulas of propositional logic can also be described as *top-down*. The symbol set for propositional logic is $S = \{\neg, \vee, \wedge, \to, \leftrightarrow, (,), p_1, p_2, \ldots\}$. Let EXP be the set of all expressions on S. The bottom-up definition is as follows: An expression A of S is a *formula* if there is a finite sequence $A_1, A_2, \ldots, A_n$ of expressions with $A_n = A$ and such that for

$1 \leq k \leq n$, one of the following holds: (a) A_k is a propositional variable; (b) $k > 1$ and A_k has one of the forms $\neg B$, $(B \vee C)$, $(B \wedge C)$, $(B \to C)$, or $(B \leftrightarrow C)$, where B and C are among the previous expressions $A_1, \ldots, A_{k-1}$. Let FOR$^-$ denote the set of formulas according to this definition. Now we give the top-down approach. Define a 1-ary function $F_\neg$ and 2-ary functions $F_\vee$, $F_\wedge$, $F_\to$, and $F_\leftrightarrow$ on EXP by $F_\neg(A) = \neg A$, $F_\to(A, B) = (A \to B)$, $F_\wedge(A, B) = (A \wedge B)$, $F_\vee(A, B) = (A \vee B)$, and $F_\leftrightarrow(A, B) = (A \leftrightarrow B)$. Call a subset I of EXP *F-closed* if whenever A and B are in I, then $F_\neg(A)$, $F_\vee(A, B)$, $F_\wedge(A, B)$, $F_\to(A, B)$, and $F_\leftrightarrow(A, B)$ also belong to I. Call a subset I of EXP *inductive* if it is F-closed and every propositional variable is in I. For example, EXP itself is inductive. Finally, for the top-down approach to formulas, let FOR$^+ = \bigcap\{I : I \subseteq$ EXP and I is inductive\}. Show that FOR$^- =$ FOR$^+$.

2.2

Tautological Consequence

—————— *Truth Assignments* ——————

We begin with two basic properties of propositions: (1) Every simple proposition (e.g., "Snow is white") is either true or false but not both; in other words, every simple proposition has a unique truth value. (2) A proposition built up from simple propositions using connectives (e.g., "If it snows, then we will not have a picnic") is either true or false but not both; this unique truth value depends on the connectives used and the truth values of the simple propositions that make up the proposition.

Actually, we are more interested in the formulas of propositional logic than in propositions, but our approach is motivated by (1) and (2). Select two distinct symbols T and F, called *truth values*. Each propositional variable has two possible truth values, namely T or F. Informally, T stands for true, F for false. This step is based on (1). To each connective we assign a function called its *truth function*; these truth functions are used to assign truth values to formulas. This step is based on (2). We will begin with a general discussion of truth functions.

Definition 1: An *n-ary truth function* is a function $H: \{T, F\}^n \to \{T, F\}$.

We are primarily interested in 1-ary and 2-ary truth functions. A *truth table* is simply a tabular description of a truth function. More generally, truth tables are used to display calculations with truth functions.

Example 1:

Define $H: \{T, F\}^2 \to \{T, F\}$ by $H(T, T) = H(T, F) = T$ and $H(F, T) = H(F, F) = F$; H is a 2-ary truth function with the following truth table:

T	T	T
T	F	T
F	T	F
F	F	F

The first two columns systematically list all elements of the domain $\{T, F\}^2$ of H; the third column gives the value of H at each given 2-tuple.   □

We now assign a truth function to each of the five connectives. In each case, the truth table description is given first.

Negation (Not): truth function denoted by $H_\neg$.

A	¬A
T	F
F	T

$$H_\neg(T) = F; \qquad H_\neg(F) = T.$$

Informally, $¬A$ is false when A is true and vice versa. The column headings A and $¬A$ are not an official part of the truth table; rather, their purpose is to emphasize or highlight the effect of the truth function on a formula.

Conjunction (And): truth function denoted by $H_\wedge$.

A	B	A ∧ B
T	T	T
T	F	F
F	T	F
F	F	F

$$H_\wedge(T, T) = T;$$

$$H_\wedge(T, F) = H_\wedge(F, T) = H_\wedge(F, F) = F.$$

Informally, the only time $A \wedge B$ is true is in the case in which *both* A and B are true. This is the commonly accepted use of *and*.

Disjunction (Or): truth function denoted by $H_\vee$.

A	B	A ∨ B
T	T	T
T	F	T
F	T	T
F	F	F

$$H_\vee(T, T) = H_\vee(T, F) = H_\vee(F, T) = T;$$

$$H_\vee(F, F) = F.$$

Informally, the only time $A \vee B$ is false is when *both* A and B are false. This is the commonly accepted use of *or* in mathematics and corresponds to the inclusive *or* of English.

Conditional (If ... then or implies): truth function denoted by $H_\rightarrow$.

A	B	$A \rightarrow B$
T	T	T
T	F	F
F	T	T
F	F	T

$$H_\rightarrow(T, T) = H_\rightarrow(F, T) = H_\rightarrow(F, F) = T;$$

$$H_\rightarrow(T, F) = F.$$

Informally, the only time $A \rightarrow B$ is false is when A is true but B is false. The choice of the "proper" truth function for the conditional is not as straightforward as in the previous cases. The connective *if...then* is used in a wide variety of ways in everyday discourse. There is, however, a common core to all of these meanings; namely, "*if A, then B*" is false whenever the antecedent A is true but the consequent B is false. This corresponds to row 2 of the table for $\rightarrow$ (disregard the column headings A, B, $A \rightarrow B$). One must still choose a truth value for the other three cases (rows 1, 3, 4), and T is chosen for each case. This choice reflects the use of *if... then* in mathematics. For a mathematician, a statement of the form *if* ***, *then* +++ is true *except* in the one case in which the hypothesis *** is true but the conclusion +++ is false. A mathematician, confronted with a statement of the form *if* ***, *then* +++, proceeds as follows: Assume *** is true; prove +++ must also be true. This mathematical proof shows that in the case in which the antecedent is true, it is row 1 and not row 2 that applies; the mathematician may safely and conveniently ignore the case in which the antecedent is false (rows 3 and 4).

The conditional $\rightarrow$ is sometimes referred to as *material implication* or *truth-functional implication*. There are uses of *if... then* in ordinary discourse that are not truth-functional; consequently, we cannot expect to find a truth function that completely captures this connective. But we can say that the truth table chosen for $\rightarrow$ has these properties: (1) It is the best truth-functional approximation of *if... then*; (2) it captures the use of *if... then* in mathematics. For a further discussion of these ideas, see the book *If P, Then Q: Conditionals and the Foundations of Reasoning* by David Sanford.

Given rows 1 and 2 of the truth table for the conditional, there are three other ways in which one could assign a truth value to $A \rightarrow B$ in rows 3 and 4. But these choices give conjunction or biconditional, or else repeat the truth value of B (see Exercise 5).

Biconditional (If and only if): truth function denoted by $H_\leftrightarrow$.

A	B	$A \leftrightarrow B$
T	T	T
T	F	F
F	T	F
F	F	T

$$H_\leftrightarrow(T, T) = H_\leftrightarrow(F, F) = T;$$

$$H_\leftrightarrow(T, F) = H_\leftrightarrow(F, T) = F.$$

Informally, $A \leftrightarrow B$ is true whenever A and B have the same truth value and false whenever A and B have opposite truth values.

Now let us show how truth functions are used. Let A be a formula, and assume that the propositional variables that occur in A are among $p_1, p_2, \ldots, p_n$. Assign truth values (either T or F) to each of the propositional variables $p_1, p_2, \ldots, p_n$. Then A has a truth value that can be calculated using the truth functions $H_\neg, H_\vee, H_\wedge, H_\rightarrow, H_\leftrightarrow$. This truth value for A is unique and can be computed in a mechanical way. We illustrate with an example.

Example 2:

The formula $(p \rightarrow q) \leftrightarrow \neg(p \wedge \neg q)$ has truth value T for every possible assignment of truth values to the propositional variables p and q. The table below is used to calculate the truth value of $(p \rightarrow q) \leftrightarrow \neg(p \wedge \neg q)$ for various assignments of truth values to p and q. Note that columns 1 and 2 systematically list all possible truth assignments to the propositional variables p and q and column 7 gives the corresponding truth value of $(p \rightarrow q) \leftrightarrow \neg(p \wedge \neg q)$. Columns 3–6 are intermediate calculations; for example, column 3, with heading $p \rightarrow q$, is obtained using the truth function $H_\rightarrow$.

p	q	$p \rightarrow q$	$\neg q$	$p \wedge \neg q$	$\neg(p \wedge \neg q)$	$(p \rightarrow q) \leftrightarrow \neg(p \wedge \neg q)$
T	T	T	F	F	T	T
T	F	F	T	T	F	T
F	T	T	F	F	T	T
F	F	T	T	F	T	T

□

We now introduce an alternate but essentially equivalent way of computing truth values. First, some notation. Let

PROP = set of all propositional variables $p_1, p_2, \ldots$;

FOR = set of all formulas of propositional logic.

Definition 2: A *truth assignment* is a function $\phi: \text{PROP} \rightarrow \{T, F\}$. For example, the function ϕ that assigns T to $p_2, p_4, p_6, \ldots$ and F to $p_1, p_3, p_5, \ldots$ is a truth assignment. The Greek letters ϕ and ψ are used to denote truth assignments.

A truth assignment assigns a truth value T or F to propositional variables. Our goal, however, is to assign truth values to formulas. So we use the truth functions $H_\neg, H_\vee, H_\wedge, H_\rightarrow, H_\leftrightarrow$ to extend each truth assignment ϕ so that it assigns a unique truth value T or F to *every* formula. This extension, *which we continue to denote by ϕ*, is defined inductively by

(1) $\phi(\neg A) = H_\neg(\phi(A))$;

(2) $\phi(A \vee B) = H_\vee(\phi(A), \phi(B))$;

(3) $\phi(A \wedge B) = H_\wedge(\phi(A), \phi(B))$;

(4) $\phi(A \rightarrow B) = H_\rightarrow(\phi(A), \phi(B))$;

(5) $\phi(A \leftrightarrow B) = H_\leftrightarrow(\phi(A), \phi(B))$.

Although (1)–(5) make up the official definition of the extension of a truth assignment ϕ to all formulas, it is convenient to have available the following equivalent conditions:

$\mathbf{TA\neg}$ $\phi(A) \neq \phi(\neg A)$;

$\mathbf{TA\vee}$ $\phi(A \vee B) = T \Leftrightarrow \phi(A) = T$ or $\phi(B) = T$;

$\quad\quad$ $\phi(A \vee B) = F \Leftrightarrow \phi(A) = F$ and $\phi(B) = F$;

$\mathbf{TA\wedge}$ $\phi(A \wedge B) = T \Leftrightarrow \phi(A) = T$ and $\phi(B) = T$;

$\quad\quad$ $\phi(A \wedge B) = F \Leftrightarrow \phi(A) = F$ or $\phi(B) = F$;

$\mathbf{TA\rightarrow}$ $\phi(A \rightarrow B) = F \Leftrightarrow \phi(A) = T$ and $\phi(B) = F$;

$\quad\quad$ $\phi(A \rightarrow B) = T \Leftrightarrow \phi(A) = F$ or $\phi(B) = T$;

$\mathbf{TA\leftrightarrow}$ $\phi(A \leftrightarrow B) = T \Leftrightarrow \phi(A) = \phi(B)$;

$\quad\quad$ $\phi(A \leftrightarrow B) = F \Leftrightarrow \phi(A) \neq \phi(B)$.

Informally, $\mathbf{TA\neg}$ tells us that the symbol $\neg$ is interpreted as *not*, $\mathbf{TA\vee}$ tells us that the symbol $\vee$ is interpreted as *or*, and so on. It is not necessary to remember the inductive definition used to extend a truth assignment ϕ; instead, one should remember the five conditions $\mathbf{TA\neg}$, $\mathbf{TA\vee}$, $\mathbf{TA\wedge}$, $\mathbf{TA\rightarrow}$, $\mathbf{TA\leftrightarrow}$.

Summary: A truth assignment is a function $\phi: \mathrm{PROP} \rightarrow \{T, F\}$; this function has a unique extension $\phi: \mathrm{FOR} \rightarrow \{T, F\}$ that satisfies the five conditions $\mathbf{TA\neg}$, $\mathbf{TA\vee}$, $\mathbf{TA\wedge}$, $\mathbf{TA\rightarrow}$, and $\mathbf{TA\leftrightarrow}$.

Example 3:

We will show that $\phi((p \rightarrow q) \leftrightarrow \neg(p \wedge \neg q)) = T$ for every truth assignment ϕ (this is the same as Example 2). First we have:

$$\phi(p \rightarrow q) = F \Leftrightarrow \phi(p) = T \text{ and } \phi(q) = F \quad\quad \mathbf{TA\rightarrow}$$

$$\Leftrightarrow \phi(p) = T \text{ and } \phi(\neg q) = T \quad\quad \mathbf{TA\neg}$$

$$\Leftrightarrow \phi(p \wedge \neg q) = T \quad\quad\quad\quad\quad \mathbf{TA\wedge}$$

$$\Leftrightarrow \phi(\neg(p \wedge \neg q)) = F \quad\quad\quad\quad \mathbf{TA\neg}$$

From these calculations, it follows that $\phi(p \rightarrow q) = \phi(\neg(p \wedge \neg q))$; $\mathbf{TA\leftrightarrow}$ now applies and $\phi((p \rightarrow q) \leftrightarrow \neg(p \wedge \neg q)) = T$, as required. $\quad\square$

From Examples 2 and 3 we see that truth tables and truth assignments are essentially equivalent ways of calculating a truth value for a formula. There is, however, one difference: Suppose A is a formula with propositional variables $p_1, \ldots, p_n$. When working with truth tables, we assign truth values to $p_1, \ldots, p_n$ and ignore all other propositional variables. On the other hand, with truth assignments *all* propositional variables are assigned truth values.

But as we will now show, for any truth assignment ϕ, the value of $\phi(A)$ does not depend on $\phi(p)$ for those propositional variables p that do not occur in A.

Theorem 1: If two truth assignments ϕ and ψ agree on all propositional variables that occur in A, then $\phi(A) = \psi(A)$.

Proof: We use induction on formulas. The result is obvious whenever A is a propositional variable. Suppose A is $(B \vee C)$. By hypothesis, ϕ and ψ agree on all propositional variables that occur in A. Thus, ϕ and ψ certainly agree on all propositional variables in B and C. By the induction hypothesis, we have $\phi(B) = \psi(B)$ and $\phi(C) = \psi(C)$. It easily follows from TA$\vee$ that $\phi(B \vee C) = \psi(B \vee C)$; in other words, $\phi(A) = \psi(A)$, as required. We leave the cases in which A is $\neg B$, $B \wedge C$, $B \rightarrow C$, or $B \leftrightarrow C$ to the reader. ∎

———— *Tautological Consequence* ————

The proposition "Snow is white" is true by virtue of a fact of nature; it could be false in some other possible world. On the other hand, the proposition "Snow is white or snow is not white" is true by virtue of the meaning of *not* and *or*; it is true in all possible worlds. This second example motivates the following definition.

Definition 3: A formula A of propositional logic is a *tautology* if $\phi(A) = T$ for every truth assignment ϕ.

In other words, A is a tautology if for every assignment of truth values to the propositional variables that occur in A, the truth value of the formula A is T. For example, the formulas $p \vee \neg p$ and $(p \rightarrow q) \leftrightarrow \neg(p \wedge \neg q)$ are tautologies, whereas $p \vee q$ is not a tautology. A tautology is true by virtue of the meaning of its connectives and is independent of the interpretation of its propositional variables.

Example 4:

We will give two proofs that the formula $[\neg A \rightarrow (B \wedge \neg B)] \rightarrow A$ is a tautology.

First proof: Use truth assignments. Let ϕ be a truth assignment; we are required to show that $\phi([\neg A \rightarrow (B \wedge \neg B)] \rightarrow A) = T$. Suppose not; then $\phi([\neg A \rightarrow (B \wedge \neg B)] \rightarrow A) = F$. By TA$\rightarrow$ we have

$$(1)\ \ \phi(\neg A \rightarrow (B \wedge \neg B)) = T; \qquad (2)\ \ \phi(A) = F.$$

Now, by (1) and TA$\rightarrow$, $\phi(\neg A) = F$ or $\phi(B \wedge \neg B) = T$. But $\phi(\neg A) = F$ is impossible by (2), so we must have $\phi(B \wedge \neg B) = T$. It follows from TA$\wedge$ that $\phi(B) = T$ and $\phi(\neg B) = T$, and this contradicts TA$\neg$.

Second proof: Use truth tables. Assign truth values to the propositional variables that occur in A and B; for this assignment, A has a truth value of either T or F, and the same holds for B. So altogether there are four cases, and we can systematically calculate the corresponding truth value of $[\neg A \rightarrow (B \wedge \neg B)] \rightarrow A$ by a truth table; we leave the construction of the table as an exercise for the reader. But in each of the four cases, the truth value of $[\neg A \rightarrow (B \wedge \neg B)] \rightarrow A$ is T, so the formula is a tautology, as required.   □

Let A be a formula of propositional logic, and assume that the propositional variables that occur in A are among $p_1, \ldots, p_n$. To check that A is a tautology, we must show that $\phi(A) = T$ for every truth assignment ϕ. Since there are uncountably many truth assignments, this appears to be a formidable task. But Theorem 1 states that $\phi(A)$ does not depend on $\phi(p_k)$ for $k > n$. Thus we need only check the 2^n possible truth assignments to $p_1, \ldots, p_n$, and this can be done mechanically in a finite number of steps with truth tables. We summarize this discussion as a theorem.

Theorem 2: There is an algorithm that, given an arbitrary formula of propositional logic, decides (**YES** *or* **NO**) whether the formula is a tautology.

Proof: The required algorithm is to write out the truth table of the formula. ∎

The following important ideas are closely related to the notion of a tautology.

Definition 4: A formula A is *satisfiable* if there is at least one truth assignment ϕ such that $\phi(A) = T$; in this case, we also say that the truth assignment ϕ *satisfies* the formula A. If there is no truth assignment that satisfies A, we say that A is *unsatisfiable*.

Note that a formula A is a tautology if every truth assignment satisfies A. In addition, one has the following easy result (proof left to the reader): *A is a tautology if and only if $\neg A$ is unsatisfiable.*

Example 5:

To show that $p_1 \vee \neg p_2$ is not a tautology, it suffices to find a truth assignment that does not satisfy the formula. Let $\phi(p_1) = F$ and $\phi(p_n) = T$ for $n > 1$; then $\phi(p_1 \vee \neg p_2) = F$, as required.   □

We will now generalize the above ideas to obtain the fundamental concept of propositional logic: tautological consequence.

Definition 5: A set Γ of formulas is *satisfiable* if there is at least one truth assignment ϕ that satisfies every formula in Γ (i.e., $\phi(A) = T$ for every $A \in \Gamma$); in this case, we also say that ϕ satisfies Γ, or that ϕ is a *model* of Γ. If there is no truth assignment that satisfies Γ, we say that Γ is *unsatisfiable*.

Definition 6: Let B be a formula and Γ a set of formulas. We say that B is a *tautological consequence* of Γ, written $\Gamma \models B$, if every truth assignment that satisfies Γ also satisfies B. To prove $\Gamma \models B$, proceed as follows: Let ϕ be a truth assignment such that $\phi(A) = T$ for every $A \in \Gamma$; show that $\phi(B) = T$.

In the case in which Γ is the empty set, $\Gamma \models B$ reduces to the statement that B is a tautology and we write $\models B$. More generally, for Γ finite, $\Gamma \models B$ reduces to a tautology as follows.

Theorem 3: Let $A_1, \ldots, A_n, B$ be formulas. The following are equivalent: (1) $\{A_1, \ldots, A_n\} \models B$; (2) $(A_1 \wedge \ldots \wedge A_n) \to B$ is a tautology.

Proof: The key is the observation that for any truth assignment ϕ,

$$\phi(A_1 \wedge \ldots \wedge A_n) = T \Leftrightarrow \phi \text{ satisfies } \{A_1, \ldots, A_n\}.$$

We leave the details to the reader. ∎

Corollary 1: There is an algorithm that, given formulas $A_1, \ldots, A_n, B$, decides (YES or NO) whether $\{A_1, \ldots, A_n\} \models B$.

Proof: Use truth tables to decide whether $(A_1 \wedge \ldots \wedge A_n) \to B$ is a tautology. ∎

Now let us apply these ideas to the problem of defining a valid argument form. Recall the argument in Example 3 of Section 2.1 that translates into the language of propositional logic as $\neg(p \wedge q)$, $\neg p$, $\therefore q$. This translation is called the *form* of the argument. We are going to define the two concepts *valid argument form* and *invalid argument form*. Assuming this is done, we then agree that an argument is *valid* if its argument form is valid and *invalid* if its argument form is invalid.

The notation for an argument form is $A_1, \ldots, A_k \therefore B$. Informally, the formulas $A_1, \ldots, A_k, B$ are the result of translating an argument involving propositions into the language of propositional logic. Formally, $A_1, \ldots, A_k \therefore B$ is a finite sequence of formulas in which $A_1, \ldots, A_k$ are called the *hypotheses* and B is called the *conclusion*.

Definition 7: An argument form $A_1, \ldots, A_k \therefore B$ is *valid* if B is a tautological consequence of $\{A_1, \ldots, A_k\}$—in other words, if $\{A_1, \ldots, A_k\} \models B$. Otherwise, $A_1, \ldots, A_k \therefore B$ is *invalid*.

Let us elaborate on this definition. The argument form $A_1, \ldots, A_k \therefore B$ is *valid* if for every assignment of truth values to the propositional variables that occur in $A_1, \ldots, A_k, B$, if the formulas $A_1, \ldots, A_k$ all have truth value T, then the formula B must also have truth value T. On the other hand, the argument form $A_1, \ldots, A_k \therefore B$ is *invalid* if it is possible to assign truth values to the propositional variables that occur in the formulas $A_1, \ldots, A_k, B$ in such a way that $A_1, \ldots, A_k$ all have truth value T but B has truth value F. By Corollary 1, there is an algorithm that decides (YES or NO in a finite number of steps) whether an arbitrary argument form is valid.

Example 6:

We show that the argument form $A \to B$, $C \to D$, $A \lor C \therefore B \lor D$ is valid by showing that $\{A \to B,\ C \to D,\ A \lor C\} \models B \lor D$. Let ϕ be any truth assignment such that

(1) $\phi(A \to B) = T$,   (2) $\phi(C \to D) = T$,   (3) $\phi(A \lor C) = T$;

we are required to show $\phi(B \lor D) = T$. By (3) and TA$\lor$, $\phi(A) = T$ or $\phi(C) = T$. First, suppose $\phi(A) = T$. By (1) and TA$\to$, $\phi(A) = F$ or $\phi(B) = T$. But $\phi(A) = T$, so $\phi(A) = F$ cannot hold and thus we have $\phi(B) = T$. It follows easily from TA$\lor$ that $\phi(B \lor D) = T$ as required. The case in which $\phi(C) = T$ is similar and is left to the reader; this time, (2) is used.   □

Example 7:

To show that the argument form $\neg(p \land q)$, $\neg p$, $\therefore q$ is invalid, it suffices to find a truth assignment ϕ that satisfies $\{\neg(p \land q), \neg p\}$ and for which $\phi(q) = F$. The choice is obvious: Take $\phi(p) = F$ and $\phi(q) = F$; then $\phi(\neg p) = \phi\neg(p \land q) = T$ and $\phi(q) = F$, as required.   □

—— Compactness Theorem for Propositional Logic ——

We have already noted that for Γ finite, $\Gamma \models B$ reduces to a tautology (see Theorem 3). What about the case in which Γ is infinite? To answer this, we need the compactness theorem for propositional logic. But before we state this important result, let us prove the following.

Theorem 4: $\Gamma \models B$ if and only if $\Gamma \cup \{\neg B\}$ is unsatisfiable.

Proof: First assume $\Gamma \models B$. Suppose, however, that there is some truth assignment ϕ that satisfies $\Gamma \cup \{\neg B\}$. Then ϕ satisfies Γ and $\phi(\neg B) = T$. But $\Gamma \models B$; hence $\phi(B) = T$. This contradicts $\phi(\neg B) = T$.

Now assume $\Gamma \cup \{\neg B\}$ is unsatisfiable. To show $\Gamma \models B$, let ϕ be a truth assignment that satisfies Γ; our goal is to show that $\phi(B) = T$. Now $\Gamma \cup \{\neg B\}$ is unsatisfiable, so ϕ cannot satisfy $\Gamma \cup \{\neg B\}$. But ϕ does satisfy Γ; hence $\phi(\neg B) = F$. It follows from TA$\neg$ that $\phi(B) = T$, as required. ∎

There are actually two versions of the compactness theorem. The first version is conveniently stated in terms of the following property.

Definition 8: A set Γ of formulas is *finitely satisfiable* if every finite subset of Γ is satisfiable.

Theorem 5 (Compactness theorem for propositional logic, Version I): If Γ is finitely satisfiable, then Γ is satisfiable.

Compactness theorem I is an easy consequence of the model existence and consistency theorems in Chapter 3, so we will postpone its proof for now. This later proof takes advantage of certain syntactic concepts, including that of consistency. There is, however, a direct proof of the compactness theorem based entirely on semantic concepts (see Exercises 41–45). The following concept is used in this semantic proof: A set Δ of formulas of propositional logic is *maximally complete* if for every formula A, $A \in \Delta$ or $\neg A \in \Delta$.

Let us now use compactness Theorem I to answer our question about reducing $\Gamma \models B$ to a tautology.

Theorem 6 (Compactness theorem for propositional logic, Version II): If $\Gamma \models B$, then there is a finite $\Delta \subseteq \Gamma$ such that $\Delta \models B$.

Proof: Assume $\Gamma \models B$. Then $\Gamma \cup \{\neg B\}$ is unsatisfiable (Theorem 4). By compactness Theorem I, there is a finite subset Δ_0 of $\Gamma \cup \{\neg B\}$ that is unsatisfiable. Let $\Delta = \{A : A \in \Gamma \text{ and } A \in \Delta_0\}$. Clearly Δ is a finite subset of Γ and $\Delta \cup \{\neg B\}$ is unsatisfiable. It follows from Theorem 4 that $\Delta \models B$, as required. ∎

Corollary 2: If $\Gamma \models B$, then there is a finite $\{A_1, \ldots, A_n\} \subseteq \Gamma$ such that $(A_1 \wedge \ldots \wedge A_n) \to B$ is a tautology.

Proof: By compactness Theorem II, there is a finite $\{A_1, \ldots, A_n\} \subseteq \Gamma$ such that $\{A_1, \ldots, A_n\} \models B$. By Theorem 3, $(A_1 \wedge \ldots \wedge A_n) \to B$ is a tautology. ∎

——————————*Exercises for Section 2.2*——————————

1. List all possible 1-ary truth functions.
2. Count the number of 2-ary truth functions and describe two that are distinct from the truth functions $H_\vee$, $H_\wedge$, $H_\to$, and $H_\leftrightarrow$.
3. Write a truth table for the exclusive *or* of English. Then express "*p* or *q* but not both" in terms of $p, q, \neg, \vee,$ and $\wedge$.
4. (*Wason selection task*) Four cards are displayed as follows:

$$\boxed{A} \qquad \boxed{B} \qquad \boxed{2} \qquad \boxed{3}$$

Each card has a number on one side and a letter on the other side. Specify which cards must be turned over to establish the truth of the following rule: *If a card has a vowel on one side, then the card has an even number on the other side.* (Wason, P. C. and P. N. Johnson-Laird, *Psychology of Reasoning: Structure and Content*. Cambridge, MA: Harvard University Press, 1972.)
5. Suppose we assign the following truth function to the conditional:

A	B	$A \to B$
T	T	T
T	F	F
F	T	T
F	F	F

Note that this truth function simply repeats the truth value of B. Show that with this truth function for →, the argument form $p \to q \therefore \neg q \to \neg p$ is invalid. (If in the table we put "F, T" or "F, F" in the last two rows, the resulting truth function is $H_{\leftrightarrow}$ or $H_{\wedge}$, respectively.)

6. (*Characterizations of* →) Show that each of the following is a tautology.
   (1) $(A \to B) \leftrightarrow (\neg A \vee B)$;   (3) $(A \to B) \leftrightarrow [\neg A \vee (A \wedge B)]$.
   (2) $(A \to B) \leftrightarrow \neg(A \wedge \neg B)$;

7. (*Paradoxes of material implication*) Show that each of the following is a tautology.
   (1) $A \to (B \to A)$ [a true proposition A is implied by any proposition B];
   (2) $\neg A \to (A \to B)$ [a false proposition A implies any proposition B];
   (3) $(A \to B) \vee (B \to A)$ [of any two propositions, one implies the other].

8. Let ϕ be a truth assignment such that $\phi(p_1) = T$, $\phi(p_2) = F$, $\phi(p_3) = F$. Find $\phi(p_1 \to (\neg p_2 \wedge p_3))$.

9. Prove the following equivalences.
   (1) $\phi(\neg A) = H_{\neg}(\phi(A))$ if and only if ϕ satisfies TA¬;
   (2) $\phi(A \vee B) = H_{\vee}(\phi(A), \phi(B))$ if and only if ϕ satisfies TA∨;
   (3) $\phi(A \wedge B) = H_{\wedge}(\phi(A), \phi(B))$ if and only if ϕ satisfies TA∧;
   (4) $\phi(A \to B) = H_{\to}(\phi(A), \phi(B))$ if and only if ϕ satisfies TA→;
   (5) $\phi(A \leftrightarrow B) = H_{\leftrightarrow}(\phi(A), \phi(B))$ if and only if ϕ satisfies TA↔.

10. Prove that each of the following is a tautology.
    (1) $A \to A$;
    (2) $\neg(A \wedge \neg A)$;
    (3) $[(A \to B) \to A] \to A$ (Peirce's law);
    (4) $[(A \wedge B) \to C] \leftrightarrow [A \to (B \to C)]$ (law of exportation);
    (5) $(A \to B) \to [(A \to \neg B) \to \neg A]$.

11. For each of the following, decide if the formula is a tautology.
    (1) $(\neg A \wedge A) \to B$;   (6) $(A \to B) \vee (B \to C)$;
    (2) $[A \leftrightarrow (B \leftrightarrow C)] \leftrightarrow [(A \leftrightarrow B) \leftrightarrow C]$;   (7) $\neg(A \vee B) \to \neg A$;
    (3) $[(A \to B) \wedge \neg A] \to \neg B$;   (8) $(A \to B) \vee (\neg A \to B)$;
    (4) $[(A \wedge B) \to C] \to [(A \to B) \to C]$;   (9) $\neg\neg A \to A$.
    (5) $[(A \to B) \to C] \to [(A \wedge B) \to C]$;

12. (*Axioms and rules of inference for the formal system P in Section 3.1*) Prove the following:
    (1) $\neg A \vee A$ is a tautology (law of excluded middle);
    (2) $A \vee (B \vee C) \models (A \vee B) \vee C$ (associative rule);
    (3) $A \models B \vee A$ (expansion rule);
    (4) $A \vee A \models A$ (contraction rule);
    (5) $\{A \vee B, \neg A \vee C\} \models B \vee C$ (cut rule).

13. (*Axioms of Russell and Whitehead for* ¬ *and* ∨) Prove that PM1–PM4 are tautologies (see p. 104).

14. (*Axioms of Rosser for* ¬ *and* ∧) Prove that R1–R3 are tautologies (see p. 104).

15. (*Axioms of Frege for* ¬ *and* →) Prove that LF1–LF3 are tautologies (see p. 107).

16. (*Axioms of Lukasiewicz for* ¬ *and* →) Prove that LUK1–LUK3 are tautologies (see Exercise 13 of Section 3.5).

17. Let A be a formula whose only connective is ¬. Can A be a tautology? Justify your answer.

18. Let A be a formula whose only connective is ∨. Can A be a tautology? Justify your answer.

19. Let A be a formula whose only connective is →. Can A be a tautology? Justify your answer.

20. Let A be the formula $[(p_1 \to p_2) \wedge \neg p_1] \to \neg p_2$. Find a truth assignment that satisfies A and one that does not.

21. Use a Cantor diagonal argument to show that the number of truth assignments is uncountable.

22. Let A and B be formulas. Show that $\{A, \neg A\} \models B$.

23. Let Γ and Δ be sets of formulas. Prove the following:
    (1) If $\Gamma \vDash A$ and $\Gamma \subseteq \Delta$, then $\Delta \vDash A$;
    (2) $\Gamma \cup \{A\} \vDash B$ if and only if $\Gamma \vDash A \to B$;
    (3) If $\Gamma \cup \{\neg A\} \vDash B$ and $\Gamma \cup \{\neg A\} \vDash \neg B$, then $\Gamma \vDash A$;
    (4) If $\Gamma \cup \{A\} \vDash C$ and $\Gamma \cup \{B\} \vDash C$, then $\Gamma \cup \{A \vee B\} \vDash C$.

24. Let A and B be formulas. Show that $A \leftrightarrow B$ is a tautology if and only if $\{A\} \vDash B$ and $\{B\} \vDash A$.

25. Prove that A is a tautology if and only if $\neg A$ is unsatisfiable.

26. Let $\Gamma \vDash A$ and $\Delta \vDash \neg A$. Show that $\Gamma \cup \Delta$ is unsatisfiable.

27. Show that $\Gamma = \{p_n \vee p_{n+1}, \neg p_n \vee \neg p_{n+1} : n \in \mathbb{N}^+\}$ is satisfiable.

28. Complete the proof of Theorem 1.

29. Prove Theorem 3.

30. Show that each of the following argument forms is valid by showing that $\{A_1, \dots, A_n\} \vDash B$.
    (1) $A \to B$, $A \therefore B$ [modus ponens];
    (2) $A \to B$, $\neg B \therefore \neg A$ [modus tollens];
    (3) $A \to B \therefore \neg B \to \neg A$ [contrapositive];
    (4) $A \to B$, $B \to C \therefore A \to C$ [hypothetical syllogism];
    (5) $A \vee B$, $\neg A \therefore B$ [disjunctive syllogism];
    (6) $B \to C \therefore (A \wedge B) \to C$ [strengthening the antecedent];
    (7) $A \to B \therefore A \to (B \vee C)$ [weakening the consequent];
    (8) $A \to B$, $\neg A \to B \therefore B$;
    (9) $A \to B$, $A \to \neg B \therefore \neg A$.

31. Show that each of the argument forms in Exercise 9 in Section 3.2 is valid by showing that $\{A_1, \dots, A_n\} \vDash B$.

32. Show that each of the following argument forms is invalid.
    (1) $p \to q$, $q \therefore p$ [fallacy of affirming the consequent];
    (2) $p \to q$, $\neg p \therefore \neg q$ [fallacy of denying the antecedent].

33. Refer to Exercise 3 in Section 2.1. Translate each argument into the language of propositional logic and classify the resulting argument form as valid or invalid.

34. Refer to Exercise 8 in Section 3.2. Translate each argument into the language of propositional logic and show that the argument form $A_1, \dots, A_n, \therefore B$ is valid.

35. These puzzles are similar to some that appear in Smullyan's book *Forever Undecided*. The inhabitants of the planet GP are either green or purple. Green people always tell the truth and purple people never tell the truth. Consider two people named X and Y.
    (1) X says: *I am purple and Y is purple*. What color is X? What color is Y?
    (2) X says: *I am purple or Y is green*. What color is X? What color is Y?
    (3) X says: *If I am green, then Y is green*. What color is X? What color is Y?
    (4) X says: *I am green if and only if Y is green*. What color is Y?

36. Let Γ be a set of formulas that is unsatisfiable. Show that there is a finite subset of Γ that is unsatisfiable.

37. Derive compactness theorem I from compactness theorem II. (*Hint:* If Γ is unsatisfiable, then $\Gamma \vDash A \wedge \neg A$.)

38. Prove the converse of compactness theorem I.

39. Prove the converse of compactness theorem II.

40. A set Γ of formulas is called a *tautology* if for every truth assignment ϕ, there is at least one formula $A \in \Gamma$ such that ϕ satisfies A. (*Note:* In the case in which Γ is a single formula $\{A\}$, we obtain the statement that A is a tautology.) Prove that if Γ is a tautology, then there is a finite subset $\Delta \subseteq \Gamma$ such that Δ is a tautology.

41. Let Γ be finitely satisfiable and let A be a formula. Prove that $\Gamma \cup \{A\}$ or $\Gamma \cup \{\neg A\}$ is finitely satisfiable.

42. Let $\Gamma_0, \Gamma_1, \Gamma_2, \dots$ be sets of formulas with $\Gamma_0 \subseteq \Gamma_1 \subseteq \Gamma_2 \dots$ (increasing) and each Γ_n finitely satisfiable. Prove that $\Delta = \bigcup_{n \in \mathbb{N}} \Gamma_n$ is finitely satisfiable. (*Hint:* See Exercise 6 in Section 1.4).

43. (*Lindenbaum's construction*) Let Γ be finitely satisfiable. Show that there is a set Δ of formulas such that $\Gamma \subseteq \Delta$ and Δ is both finitely satisfiable and maximally complete ($A \in \Delta$ or $\neg A \in \Delta$ for every formula A). (*Hints:* See the preceding two exercises. Let $A_0, A_1, A_2, \ldots$ be all formulas (see Corollary 1 in Section 1.5). Construct a sequence $\Gamma_0, \Gamma_1, \Gamma_2, \ldots$ of sets of formulas with $\Gamma_0 \subseteq \Gamma_1 \subseteq \Gamma_2 \ldots$ as follows:

$$\Gamma_0 = \Gamma; \quad \Gamma_{n+1} = \begin{cases} \Gamma_n \cup \{A_n\} & \Gamma_n \cup \{A_n\} \text{ finitely satisfiable;} \\ \Gamma_n \cup \{\neg A_n\} & \Gamma_n \cup \{A_n\} \text{ not finitely satisfiable.} \end{cases}$$

Show that each Γ_n is finitely satisfiable; let $\Delta = \bigcup_{n \in \mathbb{N}} \Gamma_n$.

44. Let Δ be finitely satisfiable and maximally complete. Prove the following for all formulas A and B:
   (1) $A \in \Delta$ or $\neg A \in \Delta$ but not both;
   (2) $A \vee B \in \Delta$ if and only if $A \in \Delta$ or $B \in \Delta$;
   (3) $A \wedge B \in \Delta$ if and only if $A \in \Delta$ and $B \in \Delta$;
   (4) $A \to B \in \Delta$ if and only if $\neg A \in \Delta$ or $B \in \Delta$;
   (5) $A \leftrightarrow B \in \Delta$ if and only if $\{A, B\} \subseteq \Delta$ or $\{\neg A, \neg B\} \subseteq \Delta$.

45. (*Semantic proof of compactness Theorem I*) Let Γ be finitely satisfiable. Show that Γ is satisfiable. (*Hint:* By Lindenbaum's construction (Exercise 43), there is a set Δ of formulas such that $\Gamma \subseteq \Delta$ and Δ is both finitely satisfiable and maximally complete. Define $\phi: \text{PROP} \to \{T, F\}$ by $\phi(p_n) = T \Leftrightarrow p_n \in \Delta$. Use the preceding exercise and induction on formulas to show that for every formula A, $\phi(A) = T \Leftrightarrow A \in \Delta$.)

2.3

Adequate Sets of Connectives

Most of the results in this section are motivated by two questions: (1) Can we omit one or more of the connectives $\neg$, $\vee$, $\wedge$, $\to$, and $\leftrightarrow$ without diminishing the expressive power of the language of propositional logic? More precisely, can some of the truth functions $H_{\neg}$, $H_{\vee}$, $H_{\wedge}$, $H_{\to}$, and $H_{\leftrightarrow}$ be calculated in terms of the remaining functions? (2) Do we need additional connectives? In other words, is there some n-ary truth function H that cannot be calculated in terms of the five functions $H_{\neg}$, $H_{\vee}$, $H_{\wedge}$, $H_{\to}$, and $H_{\leftrightarrow}$?

Definition 1: Two formulas A and B are said to be *logically equivalent*, written $A \simeq B$, if the formula $A \leftrightarrow B$ is a tautology.

Logical equivalence plays a key role in settling the two questions (1) and (2). We begin our study of this important property with a lemma that gives us a variety of techniques for proving that two formulas are logically equivalent (left as an exercise for the reader).

Lemma 1: Let A and B be formulas. The following are equivalent:
   (1) $A \simeq B$ (i.e., A and B are logically equivalent);
   (2) $\phi(A) = \phi(B)$ for every truth assignment ϕ;
   (3) $\phi(A) = T$ if and only if $\phi(B) = T$ for every truth assignment ϕ;
   (4) $\phi(A) = F$ if and only if $\phi(B) = F$ for every truth assignment ϕ.

Example 1:

We use Lemma 1(3) to show that $A \wedge B \simeq \neg(\neg A \vee \neg B)$.

$$\phi(A \wedge B) = T \Leftrightarrow \phi(A) = T \text{ and } \phi(B) = T \qquad\qquad \text{TA}\wedge$$
$$\Leftrightarrow \phi(\neg A) = F \text{ and } \phi(\neg B) = F \qquad\qquad \text{TA}\neg$$
$$\Leftrightarrow \phi(\neg A \vee \neg B) = F \qquad\qquad \text{TA}\vee$$
$$\Leftrightarrow \phi(\neg(\neg A \vee \neg B)) = T \qquad\qquad \text{TA}\neg \qquad \square$$

Informally speaking, logically equivalent formulas say the same thing. Consequently, if two formulas A and B are logically equivalent, one may replace A with B in a third formula. More precisely, the following result holds.

Theorem 1 (Replacement theorem, semantic version): Let $A \simeq B$, let C be a formula in which A occurs, and let D be a formula obtained from C by replacing one or more of the occurrences of A with B. Then $C \simeq D$.

Proof: We give an informal proof; an alternate proof using induction on formulas is outlined in the exercises. Let ϕ be a truth assignment; we are required to show that $\phi(C) = \phi(D)$. Since A and B are logically equivalent, $\phi(A) = \phi(B)$. Suppose we now calculate $\phi(C)$ and $\phi(D)$. The formulas C and D are alike except that some (but not necessarily all) occurrences of A have been replaced with B. But $\phi(A) = \phi(B)$, so in the calculation of $\phi(C)$ and $\phi(D)$ we obtain $\phi(C) = \phi(D)$, as required. ■

Lemma 2 (Tools for conversion to $\neg$, $\vee$): The following logical equivalences hold:

LE1 $A \rightarrow B \simeq \neg A \vee B$;   **LE3** $A \leftrightarrow B \simeq (A \rightarrow B) \wedge (B \rightarrow A)$;

LE2 $A \wedge B \simeq \neg(\neg A \vee \neg B)$;   **LE4** $A \simeq \neg\neg A$.

Proof: See Example 1 for a proof of LE2. We use Lemma 1(4) to prove LE1, and leave LE3 and LE4 as exercises for the reader.

$$\phi(A \rightarrow B) = F \Leftrightarrow \phi(A) = T \text{ and } \phi(B) = F \qquad\qquad \text{TA}\rightarrow$$
$$\Leftrightarrow \phi(\neg A) = F \text{ and } \phi(B) = F \qquad\qquad \text{TA}\neg$$
$$\Leftrightarrow \phi(\neg A \vee B) = F \qquad\qquad \text{TA}\vee. \qquad ■$$

Example 2:

We find a formula logically equivalent to $p \leftrightarrow q$ whose only connectives are $\neg$ and $\vee$.

(1) $p \leftrightarrow q$	Given
(2) $(p \rightarrow q) \wedge (q \rightarrow p)$	LE3
(3) $(\neg p \vee q) \wedge (\neg q \vee p)$	Replacement Theorem, LE1
(4) $\neg[\neg(\neg p \vee q) \vee \neg(\neg q \vee p)]$	LE2 $\qquad\qquad \square$

Example 2 suggests that every formula is logically equivalent to a formula whose connectives are from $\{\neg, \vee\}$. Informally, this means that we can omit $\wedge$, $\rightarrow$, and $\leftrightarrow$ from our language without diminishing its expressive power.

Theorem 2 (Conversion to $\neg, \vee$): For any formula A, there is a formula B such that

(1) A and B are logically equivalent;
(2) A and B have the same propositional variables;
(3) the connectives in B are from $\{\neg, \vee\}$.

Proof: Let A be a formula. The proof is by induction on the number of connectives in A. The result obviously holds for propositional variables. Suppose A is $B \wedge C$. The formulas B and C have fewer connectives than A, hence by the induction hypothesis there are formulas B^* and C^* with these properties: $B \simeq B^*$; $C \simeq C^*$; B and B^* have the same propositional variables; C and C^* have the same propositional variables; the only connectives that occur in B^* and C^* are from $\{\neg, \vee\}$. We now have

(1) $B \wedge C$ Formula A
(2) $\neg(\neg B \vee \neg C)$ LE2
(3) $\neg(\neg B^* \vee \neg C^*)$ Replacement Theorem (twice)

The formula $\neg(\neg B^* \vee \neg C^*)$ is logically equivalent to A, has the same propositional variables as A, and its connectives are from $\{\neg, \vee\}$. We leave the remaining cases to the reader (negation and disjunction require little work, and the arguments for the conditional and biconditional are similar to that given for the case in which A is $B \wedge C$). ∎

The key idea of this proof is summarized as follows: If A has an occurrence of $\wedge$, say $(B \wedge C)$, replace it with the logically equivalent formula $\neg(\neg B \vee \neg C)$; if A has an occurrence of $\rightarrow$, say $(B \rightarrow C)$, replace it with $(\neg B \vee C)$; if A has an occurrence of $\leftrightarrow$, say $(B \leftrightarrow C)$, replace it with $(B \rightarrow C) \wedge (C \rightarrow B)$ and then with $\neg[\neg(\neg B \vee C) \vee \neg(\neg C \vee B)]$. Continuing in this way, we eventually eliminate all occurrences of $\wedge$, $\rightarrow$, and $\leftrightarrow$. Occurrences of $\neg\neg A$ can be simplified to A using $A \simeq \neg\neg A$. In a similar fashion, every formula of propositional logic is logically equivalent to a formula whose connectives are from $\{\neg, \wedge\}$ or from $\{\neg, \rightarrow\}$. Thus we have answered the first question at the beginning of this section. Let us now turn to the second question.

If we are given a formula A, we know how to construct a truth table for A. In other words, A determines a truth function which we denote by H_A; if A has n distinct propositional variables, the truth function H_A is n-ary. Let us illustrate this idea with an example.

Example 3:

Let A be the formula $p \wedge \neg q$; we describe H_A, the 2-ary truth function determined by A.

p	q	$\neg q$	$p \wedge \neg q$
T	T	F	F
T	F	T	T
F	T	F	F
F	F	T	F

$H_A(T, F) = T;$

$H_A(T, T) = H_A(F, T) = H_A(F, F) = F.$ □

Let us give an informal description of H_A before defining it precisely. Suppose A has propositional variables $p_1, \ldots, p_n$; in this case $H_A: \{T, F\}^n \to \{T, F\}$ is calculated by

$$H_A(x_1, \ldots, x_n) = \text{truth value of } A \text{ when } p_1, \ldots, p_n \text{ are assigned}$$
$$\text{truth values } x_1, \ldots, x_n \text{ respectively.}$$

A precise definition of H_A in terms of truth assignments is as follows.

Definition 2: Let A be a formula with propositional variables $p_1, \ldots, p_n$. The truth function *determined by A* is the n-ary function $H_A: \{T, F\}^n \to \{T, F\}$ defined by

$$H_A(x_1, \ldots, x_n) = \phi(A),$$

where ϕ is any truth assignment such that $\phi(p_k) = x_k$ for $1 \le k \le n$.

This definition of H_A is well defined; by Theorem 1 of Section 2.2, $\phi(A) = \psi(A)$ for any two truth assignments ϕ and ψ that agree on all the propositional variables that occur in A. We will now prove a result about the functions H_A that will be needed later in this section.

Lemma 3: Let A and B be formulas, each with propositional variables $p_1, \ldots, p_n$. Then $A \approx B$ if and only if $H_A = H_B$.

Proof: Assume $A \approx B$. Given $\langle x_1, \ldots, x_n \rangle \in \{T, F\}^n$, we are required to show that $H_A(x_1, \ldots, x_n) = H_B(x_1, \ldots, x_n)$. Let ϕ be a truth assignment such that $\phi(p_k) = x_k$ for $1 \le k \le n$. Then $H_A(x_1, \ldots, x_n) = \phi(A)$ and $H_B(x_1, \ldots, x_n) = \phi(B)$. But $A \approx B$, and hence $\phi(A) = \phi(B)$; this gives $H_A(x_1, \ldots, x_n) = H_B(x_1, \ldots, x_n)$, as required. We leave the proof of the other direction as an exercise. ∎

To summarize our discussion thus far: Given a formula A with propositional variables $p_1, \ldots, p_n$, we can find the n-ary function H_A. But now we want to go in the opposite direction: Given an n-ary truth function H, find a formula A with propositional variables $p_1, \ldots, p_n$ such that $H = H_A$. Let us give an example before proving the general result.

Example 4:

The truth table of a 3-ary truth function H follows. We find a formula A with propositional variables p, q, r and connectives $\neg, \vee, \wedge$ such that the truth function determined by A is H: that is, $H = H_A$.

p	q	r	
T	T	T	F
T	T	F	T *
T	F	T	F
T	F	F	F
F	T	T	T *
F	T	F	F
F	F	T	F
F	F	F	F

There are two 3-tuples at which the value of H is T, namely $\langle T, T, F \rangle$ and $\langle F, T, T \rangle$. Now consider the two formulas $p \wedge q \wedge \neg r$ and $\neg p \wedge q \wedge r$. The first formula has truth value T for the assignment $\langle T, T, F \rangle$ and truth value F otherwise; the second formula has truth value T for the assignment $\langle F, T, T \rangle$ and truth value F otherwise. Now let A be the disjunction of these two formulas:

$$(p \wedge q \wedge \neg r) \vee (\neg p \wedge q \wedge r).$$

A has truth value T for the two assignments $\langle T, T, F \rangle$ and $\langle F, T, T \rangle$ and truth value F for every other assignment. Thus we have found a formula A such that $H = H_A$.   □

Example 4 suggests that every truth function is determined by a formula that just uses the connectives $\neg, \vee, \wedge$. In other words, we do not need additional connectives. We now prepare the way to prove this fact.

Definition 3: A *literal* is either a propositional variable or the negation of a propositional variable (i.e., p or $\neg p$). We often use L to denote literals. A formula A is in *disjunctive normal form* (DNF) if it has the form

$$C_1 \vee \ldots \vee C_k \qquad (k = 1 \text{ allowed})$$

where each C_i $(1 \leq i \leq k)$ is a conjunction of literals. Thus each C_i has the form $L_1 \wedge \ldots \wedge L_{n_i}$ $(n_i = 1$ allowed), where each L_j $(1 \leq j \leq n_i)$ is a literal.

For example, the two formulas p and $p \wedge \neg q$ are in DNF, as is the formula $(p \wedge q \wedge \neg r) \vee (\neg p \wedge q \wedge r)$ of Example 4. On the other hand, $\neg(p \vee q)$ is not in DNF. We now show that every n-ary truth function H is determined by a formula in DNF. The proof generalizes the construction in Example 4. We begin with a lemma that is the first step of this construction.

Lemma 4: Let $\langle x_1, \ldots, x_n \rangle \in \{T, F\}^n$. Then there is a formula C such that

(1) C is of the form $L_1 \wedge \ldots \wedge L_n$, where each L_i is either p_i or $\neg p_i$;
(2) for any truth assignment ϕ, $\phi(C) = T \Leftrightarrow \phi(p_1) = x_1, \ldots, \phi(p_n) = x_n$.

Proof: For $1 \le i \le n$, let L_i be the literal defined by

$$L_i = \begin{cases} p_i & \text{if } x_i = T; \\ \neg p_i & \text{if } x_i = F. \end{cases} \qquad \blacksquare$$

Theorem 3 (Disjunctive normal form): Let H be an n-ary truth function. Then there is a formula A such that (1) A is in disjunctive normal form; (2) the propositional variables in A are $p_1, \ldots, p_n$; (3) $H_A = H$.

Proof: We use the following notation: $\mathbf{x}$ denotes an element of $\{T, F\}^n$; in other words, $\mathbf{x}$ is an n-tuple whose coordinates are T or F. First consider the special case in which every value of H is F (i.e., $H(\mathbf{x}) = F$ for all $\mathbf{x} \in \{T, F\}^n$); the required formula is $(p_1 \wedge \neg p_1) \vee \ldots \vee (p_n \wedge \neg p_n)$. Now let $\mathbf{x}_1, \ldots, \mathbf{x}_k$ be the elements of $\{T, F\}^n$ for which H has value T. For $1 \le j \le k$, apply Lemma 4 to $\mathbf{x}_j$ to obtain a formula C_j with these properties: (1_j) C_j is of the form $L_{1,j} \wedge \ldots \wedge L_{n,j}$, where each $L_{i,j}$ $(1 \le i \le n)$ is either p_i or $\neg p_i$; (2_j) for any truth assignment ϕ, $\phi(C_j) = T \Leftrightarrow \langle \phi(p_1), \ldots, \phi(p_n) \rangle = \mathbf{x}_j$. Let A be the formula $C_1 \vee \ldots \vee C_k$. By construction, A has propositional variables $p_1, \ldots, p_n$ and is in DNF. It remains to prove that $H_A = H$. Let $\mathbf{x} = \langle x_1, \ldots, x_n \rangle \in \{T, F\}^n$; we are required to prove that $H_A(x_1, \ldots, x_n) = H(x_1, \ldots, x_n)$. Let ϕ be a truth assignment such that $\phi(p_1) = x_1, \ldots, \phi(p_n) = x_n$; note that $\langle \phi(p_1), \ldots, \phi(p_n) \rangle = \mathbf{x}$. Now $H_A(x_1, \ldots, x_n) = \phi(A)$ by definition of H_A; hence it suffices to show that

$$H(x_1, \ldots, x_n) = \phi(A). \qquad (*)$$

We consider two cases. First suppose $H(x_1, \ldots, x_n) = T$. Then $\mathbf{x} = \mathbf{x}_j$ for some $j, 1 \le j \le k$. By (2_j), $\phi(C_j) = T$ and hence $\phi(A) = T$. This proves $(*)$ for this case. Next, suppose $H(x_1, \ldots, x_n) = F$. Then $\mathbf{x} \ne \mathbf{x}_j$ for $1 \le j \le k$; hence by (2_j) we have $\phi(C_j) = F$ for $1 \le j \le k$. But A is $C_1 \vee \ldots \vee C_k$; hence $\phi(A) = F$, as required. $\blacksquare$

Semantic algorithm for conversion to DNF: From the proof of Theorem 3 (or from Example 4), we see that there is an algorithm that, given a formula A with propositional variables $p_1, \ldots, p_n$, finds a formula B with propositional variables $p_1, \ldots, p_n$ that is in DNF and is logically equivalent to A. (1) Write out the truth table of A. If the truth value of A is always F, the required formula is $(\neg p_1 \wedge p_1) \vee \ldots \vee (\neg p_n \wedge p_n)$. (2) Otherwise, locate the rows in which A has truth value T. For each such row, form a conjunction as follows: All propositional variables among $p_1, \ldots, p_n$ with truth value T together with the negation of all propositional variables among $p_1, \ldots, p_n$ with truth value F. (3) The required formula is the disjunction of the formulas obtained in step (2).

Theorem 3 answers Question 2 at the beginning of this section: We do *not* need additional connectives; in fact, we see that $\neg$, $\vee$, and $\wedge$ suffice. This result suggests the following definition.

Definition 4: A set $\mathbf{C}$ of connectives is *adequate* if for every $n \geq 1$ and every n-ary truth function H, there is a formula A with propositional variables $p_1, \ldots, p_n$ such that (1) $H_A = H$; (2) the connectives that occur in A are from the set $\mathbf{C}$.

By the disjunctive normal form theorem, $\{\neg, \wedge, \vee\}$ is an adequate set of connectives. One can combine this result with Theorem 2 to show that $\{\neg, \vee\}$ is an adequate set of connectives. Other adequate sets of connectives are $\{\neg, \wedge\}$ and $\{\neg, \rightarrow\}$. It is also possible to define a single 2-ary connective that is adequate (see Exercise 15). The proof that $\{\neg, \vee\}$ is an adequate set of connectives is an easy consequence of the following theorem.

Theorem 4: Let $\mathbf{C}$ be a set of connectives. The following are equivalent: (1) $\mathbf{C}$ is adequate; (2) for every $n \geq 1$ and every formula A with propositional variables $p_1, \ldots, p_n$, there is a formula B with propositional variables $p_1, \ldots, p_n$ such that $A \simeq B$ and the connectives in B are from the set $\mathbf{C}$.

Proof: We prove (2) $\Rightarrow$ (1) and leave (1) $\Rightarrow$ (2) to the reader. Let $n \geq 1$ and let H be an n-ary truth function; we are required to find a formula B with propositional variables $p_1, \ldots, p_n$ whose connectives are from the set $\mathbf{C}$ and such that $H_B = H$. By Theorem 3, there is a formula A in DNF with propositional variables $p_1, \ldots, p_n$ such that $H = H_A$. Now apply assumption (2) to the formula A to obtain a formula B with propositional variables $p_1, \ldots, p_n$ such that $A \simeq B$ and the connectives in B are from the set $\mathbf{C}$. Since $A \simeq B$, $H_A = H_B$ (Lemma 3). But $H = H_A$; hence $H = H_B$ and B is the required formula. ∎

Theorem 5 (Adequacy of $\neg, \vee$***):*** The set $\{\neg, \vee\}$ is an adequate set of connectives.

Proof: It suffices to verify (2) of Theorem 4. Let $n \geq 1$ and let A be a formula with propositional variables $p_1, \ldots, p_n$. By Theorem 2 (conversion to $\neg, \vee$), there is a formula B with propositional variables $p_1, \ldots, p_n$ such that $A \simeq B$ and the connectives in B are from $\{\neg, \vee\}$. ∎

────────────── *Exercises for Section 2.3* ──────────────

1. (*Logical equivalence is an equivalence relation*) Prove the following:
   (1) $A \simeq A$;   (3) If $A \simeq B$ and $B \simeq C$, then $A \simeq C$.
   (2) If $A \simeq B$, then $B \simeq A$;
2. Prove Lemma 1.
3. Complete the proof of Lemma 2 by proving LE3 and LE4.
4. (*Conversion to* $\neg, \vee$) For each of the following formulas, find a logically equivalent formula whose connectives are from $\{\neg, \vee\}$; use LE1–LE4 and the replacement theorem.
   (1) $(p \wedge q) \rightarrow r$;   (2) $p \rightarrow (q \leftrightarrow r)$;   (3) $p \wedge (q \leftrightarrow r)$.

5. (*Conversion to* ¬, ∧) The following logical equivalences can be used to convert any formula A into a logically equivalent formula whose connectives are from $\{¬, ∧\}$.

   I. $A ∨ B ≃ ¬(¬A ∧ ¬B)$;   III. $A ↔ B ≃ (A → B) ∧ (B → A)$;
   II. $A → B ≃ ¬(A ∧ ¬B)$;   IV. $¬¬A ≃ A$.

   (1) Use LE1–LE4 and the replacement theorem to justify I and II.
   (2) For each of the following formulas, find a logically equivalent formula whose connectives are from $\{¬, ∧\}$; use I–IV and the replacement theorem. (a) $p ↔ q$; (b) $p → (q ∨ r)$; (c) $p → (q → r)$.

6. (*Conversion to* ¬, →) The following logical equivalences can be used to convert any formula A into a logically equivalent formula whose connectives are from $\{¬, →\}$.

   I. $A ∨ B ≃ ¬A → B$;   III. $A ↔ B ≃ (A → B) ∧ (B → A)$;
   II. $A ∧ B ≃ ¬(A → ¬B)$;   IV. $¬¬A ≃ A$.

   (1) Use LE1–LE4 and the replacement theorem to justify I and II.
   (2) For each of the following formulas, find a logically equivalent formula whose connectives are from $\{¬, →\}$; use I–IV and the replacement theorem. (a) $p ↔ q$; (b) $p ∧ (q ∨ r)$; (c) $p ∨ (q ∧ r)$.

7. (*Laws of negation*) Prove the following:
   (1) $¬(A ∧ B) ≃ (¬A ∨ ¬B)$ (de Morgan);   (3) $¬(A → B) ≃ (A ∧ ¬B)$;
   (2) $¬(A ∨ B) ≃ (¬A ∧ ¬B)$ (de Morgan);   (4) $¬(A ↔ B) ≃ (¬A ↔ B)$.

8. (*Relations among connectives*) Prove the following:
   (1) $A ∨ B ≃ [(A → B) → B]$ (∨ can be expressed in terms of →);
   (2) $A ∧ B ≃ [A ↔ (A → B)]$ (∧ can be expressed in terms of → and ↔);
   (3) $A → B ≃ [A ↔ (A ∧ B)]$ (→ can be expressed in terms of ∧ and ↔).

9. (*Lemma for the replacement theorem*) Let $A ≃ B$ and $C ≃ D$. Prove the following:
   (1) $¬A ≃ ¬B$;   (4) $A → C ≃ B → D$;
   (2) $A ∨ C ≃ B ∨ D$;   (5) $A ↔ C ≃ B ↔ D$.
   (3) $A ∧ C ≃ B ∧ D$;

10. (*Replacement theorem*) Prove the replacement theorem using induction on the number of connectives. Thus, let $A ≃ B$ and let $S(n)$ say: If a formula with n connectives has occurrences of A and a new formula is constructed by replacing one or more of the occurrences of A with B, then the new formula is logically equivalent to the original formula. Prove that $S(n)$ is true for all $n ≥ 0$. (*Hints:* First consider the special case in which the given formula is all of A; use the results in the preceding exercise.)

11. Complete the proof of Theorem 2 that every formula is logically equivalent to a formula that uses just the connectives ¬ and ∨. Consider the cases in which A is of the form $¬B$, $B ∨ C$, $B → C$, or $B ↔ C$.

12. Complete the proof of Lemma 3.

13. Prove that $\{¬, ∧\}$ is an adequate set of connectives.

14. Prove that $\{¬, →\}$ is an adequate set of connectives.

15. (NOR *and* NAND) The connectives NOR (denoted ↓) and NAND (denoted |) are defined by truth tables as follows:

A	B	$A ↓ B$
T	T	F
T	F	F
F	T	F
F	F	T

A	B	$A \mid B$
T	T	F
T	F	T
F	T	T
F	F	T

   (1) Write $¬p$ in terms of ↓.
   (2) Write $p ∨ q$ in terms of ↓.
   (3) Show that $\{↓\}$ is an adequate set of connectives. (*Hint:* $\{¬, ∨\}$ is adequate.)

(4) Write $p \to q$ in terms of $\downarrow$.

(5) Write $\neg p$ in terms of $|$.

(6) Write $p \land q$ in terms of $|$.

(7) Show that $\{|\}$ is an adequate set of connectives. (*Hint:* $\{\neg, \land\}$ is adequate.)

(8) Write $p \to q$ in terms of $|$.

16. Let $*$ be a binary connective such that $\{*\}$ is adequate. Prove that $H_*(T, T) = F$ and $H_*(F, F) = T$. (*Hint:* Suppose $H_*(T, T) = T$. Let FOR($*$) be all formulas whose only connective is $*$. Show that $H_A(T) = T$ for each $A \in$ FOR($*$) with propositional variable p_1. Conclude that there is no $A \in$ FOR($*$) with propositional variable p_1 such that $H_A = H_\neg$. *Note:* ϕ: PROP $\to \{T, F\}$ extends by $\phi(A * B) = H_*(\phi(A), \phi(B))$.)

17. Show that the only adequate binary connectives are NOR and NAND. (*Hint:* By the preceding exercise, the only possibilities for a binary connective that by itself is adequate are the following:

A	B	$A *_1 B$	$A *_2 B$	$A *_3 B$	$A *_4 B$
T	T	F	F	F	F
T	F	F	T	T	F
F	T	T	F	T	F
F	F	T	T	T	T

Now $*_3$ and $*_4$ are NAND and NOR, so it remains to show that neither $\{*_1\}$ nor $\{*_2\}$ is adequate. Note that $*_1$ reverses the truth value of A and $*_2$ reverses the truth value of B. Let FOR ($*_1$) be all formulas whose only connective is $*_1$. Show that $H_A(T) \neq H_A(F)$ for each $A \in$ FOR($*_1$) with propositional variable p_1. Conclude that there is no $A \in$ FOR($*_1$) with propositional variable p_1 such that $H_A = H$, where $H(T) = H(F)$.)

18. Show that $\{\land, \lor, \to\}$ is not an adequate set of connectives. (*Hint:* Let FOR($\land, \lor, \to$) be all formulas with connectives $\land$, $\lor$, and $\to$. Show that $H_A(T) = T$ for every $A \in$ FOR($\land$, $\lor, \to$) with propositional variable p_1. Conclude that there is no $A \in$ FOR($\land, \lor, \to$) with propositional variable p_1 such that $H_A = H_\neg$.

19. Show that $\{\neg, \leftrightarrow\}$ is not an adequate set of connectives. (*Hint:* Let $A \in$ FOR($\neg, \leftrightarrow$) with propositional variables p_1 and p_2; show that the truth table for A has either all T's, all F's, or two T's and two F's.)

20. (*Algebra of logical equivalence*) Prove the following:

(1) $\neg(A \land B) \approx \neg A \lor \neg B$
(2) $\neg(A \lor B) \approx \neg A \land \neg B$ } (de Morgan);

(3) $A \lor (B \lor C) \approx (A \lor B) \lor C$
(4) $A \land (B \land C) \approx (A \land B) \land C$ } (associative);

(5) $A \lor (B \land C) \approx (A \lor B) \land (A \lor C)$
(6) $A \land (B \lor C) \approx (A \land B) \lor (A \land C)$ } (distributive);

(7) $A \lor B \approx B \lor A$
(8) $A \land B \approx B \land A$ } (commutative);

(9) $A \lor A \approx A$
(10) $A \land A \approx A$ } (idempotent);

(11) $A \lor (B \land \neg B) \approx A$
(12) $A \land (B \lor \neg B) \approx A$ } (simplification);

(13) $A \lor (A \land B) \approx A$
(14) $A \land (A \lor B) \approx A$ } (absorption).

21. (*Syntactic proofs of logical equivalence*) Use the following tools to prove the logical equivalences that follow. LE1–LE4, replacement theorem, algebra of logical equivalence (preceding exercise).

(1) $A \to B \approx \neg B \to \neg A$;
(2) $A \to (B \to C) \approx (A \land B) \to C$;
(3) $A \leftrightarrow B \approx (A \land B) \lor (\neg A \land \neg B)$;

(4) $\neg(A \leftrightarrow B) \approx (A \land \neg B) \lor (\neg A \land B)$;
(5) $\neg(A \leftrightarrow B) \approx A \leftrightarrow \neg B$.

22. (*Semantic algorithm for conversion to* DNF) For each formula below, find a logically equivalent formula in DNF. Use the algorithm described in the text or use Example 4 as a model.

(1) $p \rightarrow (q \rightarrow p)$;   (5) $p \rightarrow (q \leftrightarrow r)$;
(2) $\neg(p \rightarrow p)$;   (6) $\neg[p \rightarrow (q \vee (p \wedge r))]$;
(3) $\neg[p \rightarrow (\neg q \wedge r)]$;   (7) $\neg[(p \rightarrow q) \vee (p \wedge r)]$.
(4) $p \leftrightarrow q$;

23. (*Syntactic algorithm for conversion to* DNF) Refer to Exercise 20. Use the following algorithm to convert the formulas in the preceding exercise to DNF:
(1) Eliminate $\rightarrow$ and $\leftrightarrow$ using LE1 and LE3.
(2) Push negation signs toward the propositional variables using de Morgan's laws.
(3) Eliminate $\neg\neg$ using $\neg\neg A \simeq A$.
(4) Use the distributive laws

$$A \wedge (B \vee C) \simeq (A \wedge B) \vee (A \wedge C) \quad \text{and} \quad (A \vee B) \wedge C \simeq (A \wedge C) \vee (B \wedge C).$$

24. A formula A is in *conjunctive normal form* (CNF) if it has the form $D_1 \wedge \ldots \wedge D_k$ ($k = 1$ allowed), where each D_i ($1 \le i \le k$) is a disjunction of literals. For example, $p, p \vee \neg q$, and $(p \vee q \vee \neg r) \wedge (\neg q \vee r) \wedge \neg r$ are in CNF. Suppose we have an algorithm that converts a formula to DNF. Construct another algorithm that converts a formula to CNF (i.e., given A, the algorithm finds a formula B logically equivalent to A and in CNF).

25. (*Semantic algorithm for conversion to* CNF) Refer to the preceding exercise for the definition of CNF. We describe an algorithm based on truth tables that converts a formula A with propositional variables $p_1, \ldots, p_n$ into a formula B with propositional variables $p_1, \ldots, p_n$ such that B is in CNF and is logically equivalent to A.
(1) Write out the truth table of A. If A is a tautology, the required formula in CNF is $(\neg p_1 \vee p_1) \wedge \ldots \wedge (\neg p_n \vee p_n)$.
(2) Otherwise, locate the rows in which A has truth value F. For each such row, form a disjunction as follows: All propositional variables among $p_1, \ldots, p_n$ with truth value F together with the negation of all propositional variables among $p_1, \ldots, p_n$ with truth value T.
(3) The required formula is the conjunction of the formulas obtained in step (2). Use this algorithm to convert the following formulas to CNF:

(1) $\neg(p \rightarrow (q \rightarrow p))$;   (5) $\neg(p \rightarrow (q \leftrightarrow r))$;
(2) $p \rightarrow p$;   (6) $p \rightarrow [q \vee (p \wedge r)]$;
(3) $p \rightarrow (\neg q \wedge r)$;   (7) $(p \rightarrow q) \vee (p \wedge r)$.
(4) $\neg(p \leftrightarrow q)$;

26. (*Syntactic algorithm for conversion to* CNF) Use the following algorithm to convert the formulas in the preceding exercise to CNF:
(1) Eliminate $\rightarrow$ and $\leftrightarrow$ using LE1 and LE3.
(2) Push negation signs toward the propositional variables using de Morgan's laws.
(3) Eliminate $\neg\neg$ using $\neg\neg A \simeq A$.
(4) Use the distributive laws

$$A \vee (B \wedge C) \simeq (A \vee B) \wedge (A \vee C) \quad \text{and} \quad (A \wedge B) \vee C \simeq (A \vee C) \wedge (B \vee C).$$

27. Let A be a formula in CNF, say $A = D_1 \wedge \ldots \wedge D_k$. Show that A is a tautology if and only if each disjunct $D_i, 1 \le i \le k$, has a pair of complementary literals ($p, \neg p$ are complementary literals).

28. Let A be a formula in DNF, say $A = C_1 \vee \ldots \vee C_k$. Show that A is unsatisfiable if and only if each conjunct $C_i, 1 \le i \le k$, has a pair of complementary literals ($p, \neg p$ are complementary literals).

29. (*Negation normal form*) The set of formulas in *negation normal form* (NNF) is defined inductively as follows: **F1**: p and $\neg p$ are in NNF for every $p \in$ PROP; **F2**: If A and B are in NNF, so are $A \wedge B$ and $A \vee B$. **F3**: Every formula in NNF is obtained by a finite number of applications of F1 and F2. The only connectives that occur in formulas in

NNF are $\neg, \vee,$ and $\wedge$; moreover, the only occurrences of $\neg$ are before propositional variables. For example, $p \vee \neg q$ is in NNF but $\neg(p \vee \neg q)$ is not. But $\neg(p \vee \neg q)$ is logically equivalent to $\neg p \wedge q$, which is in NNF. Show that every formula is logically equivalent to a formula in NNF.

30. (*Syntactic algorithm for conversion to* NNF) For each of the following formulas, use the first three steps of the algorithm in Exercise 23 to convert the formula to NNF; then convert the formula to both DNF and CNF.

(1) $p \wedge (\neg q \vee r)$;   (3) $p \vee [q \wedge \neg(r \wedge s)]$.
(2) $\neg[(p \wedge q) \vee (r \vee s)]$;

31. (*Duality*) Let FOR $(\neg, \vee, \wedge)$ be all formulas whose only connectives are $\neg, \vee,$ and $\wedge$. For each formula $A \in$ FOR $(\neg, \vee, \wedge)$, let A^* be the formula obtained from A by interchanging $\vee$ and $\wedge$ and replacing each $p \in$ PROP with $\neg p$; for example, if A is $(\neg p \vee q) \wedge \neg r$, then A^* is $(\neg\neg p \wedge \neg q) \vee \neg\neg r$. A precise inductive definition of A^* proceeds as follows: $p^* = \neg p$; $(\neg A)^* = \neg A^*$; $(A \vee B)^* = A^* \wedge B^*$; $(A \wedge B)^* = A^* \vee B^*$. Show that A and $\neg A^*$ are logically equivalent.

32. This problem outlines an inductive proof of the adequacy of $\{\neg, \vee, \wedge\}$.

(1) List the four 1-ary truth functions and, for each such function H, find a formula A that uses just $p_1, \neg, \vee,$ and $\wedge$ such that $H = H_A$.

(2) Let $n \geq 1$, assume the result for n-ary functions, and let H be an $n + 1$-ary truth function. Define two n-ary truth functions H_T and H_F by $H_T(x_1, \ldots, x_n) = H(x_1, \ldots, x_n, T)$ and $H_F(x_1, \ldots, x_n) = H(x_1, \ldots, x_n, F)$. Apply the induction hypothesis to H_T and H_F to obtain formulas B and C with propositional variables $p_1, \ldots, p_n$ and connectives $\neg, \vee,$ and $\wedge$ such that $H_T = H_B$ and $H_F = H_C$. Let A be the formula $(B \wedge p_{n+1}) \vee (C \wedge \neg p_{n+1})$ and show that $H = H_A$.

33. Alter the language of propositional logic by adding a symbol $\bot$, whose intuitive meaning is *always false*. The formulas of this new language are defined inductively as follows: **F1**: Each $p \in$ PROP is a formula and $\bot$ is a formula. **F2**: If A and B are formulas, so are $\neg A$, $(A \vee B)$, $(A \wedge B)$, $(A \to B)$, and $(A \leftrightarrow B)$. **F3**: Every formula is obtained by a finite number of applications of F1 and F2. The definition of a truth assignment is modified as follows: A truth assignment is a function ϕ: PROP $\cup \{\bot\} \to \{T, F\}$ such that $\phi(\bot) = F$. Such a truth assignment extends to all formulas in the usual way.

(1) Show that $\neg A \approx (A \to \bot)$ and $A \vee B \approx ((A \to \bot) \to B)$.

(2) Explain why $\{\to\}$ is an adequate set of connectives for this language.

34. (FOR/$\approx$ *is a Boolean algebra*) Let FOR be the set of all formulas of propositional logic. By Exercise 1, logical equivalence $\approx$ is an equivalence relation on FOR.

(1) Prove the following:

(a) If $A \approx C$ and $B \approx D$, then $(A \vee B) \approx (C \vee D)$ and $(A \wedge B) \approx (C \wedge D)$;
(b) if $A \approx B$, then $\neg A \approx \neg B$;
(c) for any two formulas A and B, $\neg A \vee A \approx \neg B \vee B$ and $\neg A \wedge A \approx \neg B \wedge B$.

(2) Define operations $+, \times,$ and $'$ on FOR/$\approx\ = \{[A]: A \in$ FOR$\}$ and elements 0 and 1 of FOR/$\approx$ as follows:

$$[A] + [B] = [A \vee B]; \qquad [A]' = [\neg A]; \qquad [\neg A \vee A] = 1;$$
$$[A] \times [B] = [A \wedge B]; \qquad\qquad\qquad\quad [\neg A \wedge A] = 0.$$

By (1), these definitions are independent of the choice of representative elements chosen. Show that $\langle$FOR/$\approx, +, \times, ', 0, 1\rangle$ is a Boolean algebra (defined in Section 1.6).

3

Propositional Logic

$\boxed{I}$ n this chapter, we introduce a formal system P whose intended interpretation is propositional logic. What motivates this approach to propositional logic? Suppose $A_1, \ldots, A_k \therefore B$ is a valid argument form and we wish to verify this fact. A purely mechanical way to do this is to use truth tables and show that $(A_1 \wedge \ldots \wedge A_k) \to B$ is a tautology (see Section 2.2). A difficulty with this procedure is that truth tables grow exponentially; if n propositional variables occur, the truth table has 2^n rows. Is there a more efficient way to proceed? Perhaps we can show that $A_1, \ldots, A_k \therefore B$ is valid by deriving it from simpler argument forms known to be valid. A disadvantage here is that we would need to keep track of a long list of valid argument forms. A more mathematical approach is to select a short list of valid argument forms and then to show that all valid argument forms can be derived from these; in other words, to axiomatize propositional logic. Since the truths of propositional logic are tautologies, the theorems of this axiom system should also be tautologies.

Although the verification of argument forms is the original motivation for this approach, it will not be our primary concern. Once a formal system P for propositional logic is introduced, we automatically have the concept of a *proof in* P. We will then study the relationship between *provability* and *truthhood*. In Section 3.2, we will prove the soundness theorem, which states that every theorem of the formal system P is a tautology.

Sections 3.3 and 3.4 are devoted to proving the converse of the soundness theorem. This result, called the *adequacy theorem,* states that every tautology is a theorem of P. This is considerably harder to prove than the soundness theorem. These two fundamental results are used to show that the formal system P is consistent and decidable. We are also able to establish the satisfying result that an argument form $A_1, \ldots, A_k \therefore B$ is valid if and only if B is a theorem of the formal system P when the formulas $A_1, \ldots, A_k$ are used as additional axioms.

To summarize, we gain insight into propositional logic by putting it into the framework of a formal system and then studying properties of that formal system: consistency, adequacy of the axioms and rules, decidability. Our main objective is to prove results *about* the formal system P rather than to prove results *within* the formal system P. In other words, we are more interested in *metalogic*—the study of properties of formal systems whose intended inter-pretation is logic—than in traditional logic and the classification of argument forms.

3.1

The Formal System P

The formal system P for propositional logic is described in three steps, as follows:

Language: The symbols of the language are

$p_1, p_2, p_3 \ldots$	(an infinite list of propositional variables);
$\neg, \vee$	(two symbols whose intended interpretation is *not*, and *or*);
(,)	(left and right parentheses for punctuation).

The formulas of P are defined inductively according to these rules:

F1 Each propositional variable is a formula.

F2 If A and B are formulas, so are $\neg A$ and $(A \vee B)$.

F3 Every formula is obtained by a finite number of applications of the two rules F1 and F2.

Axioms: For each formula A, $(\neg A \vee A)$ is an axiom, called a *propositional axiom.*

Rules of inference: Let A, B, C be formulas. There are four rules of inference:

$$\text{Associative Rule } \frac{(A \vee (B \vee C))}{((A \vee B) \vee C)} \qquad \text{Contraction Rule } \frac{(A \vee A)}{A}$$

$$\text{Expansion Rule } \frac{A}{(B \vee A)} \qquad \text{Cut Rule } \frac{(A \vee B), (\neg A \vee C)}{(B \vee C)}$$

The letters A, B, C, D, and E are used to denote formulas, and Γ and Δ denote sets of formulas. In writing formulas of P, we use the conventions given in Section 2.1 on omission of outer parentheses and occasional use of brackets. As is required of all formal systems, the language of P has the property that there is a mechanical method of deciding whether an arbitrary expression is a formula.

Note that the number of axioms is infinite, but there is just *one* axiom scheme, namely $\neg A \vee A$. For example, the formulas $\neg\neg p_3 \vee \neg p_3$ and $\neg(p_1 \vee p_2) \vee (p_1 \vee p_2)$ are specific axioms of P; note also that $\neg\neg(B \vee C) \vee \neg(B \vee C)$ is an axiom of P. It is clear that the formal system P is *axioma-tized*; this means that there is an algorithm that decides (YES or NO in a finite number of steps) whether an arbitrary formula of P is an axiom of P.

In order to keep the list of axioms and rules of inference as short and simple as possible, the language of P has just two logical connectives: $\neg$ and $\vee$.

From the discussion in Section 2.3, this is an adequate set of connectives. The remaining connectives, →, ∧, and ↔, are defined symbols. The conventions on omission of outer parentheses and occasional use of brackets apply to formulas that contain defined symbols.

Definition 1: Let A and B be formulas of P. We introduce new symbols →, ∧, and ↔ into the language of P as follows: $A \to B$ means $\neg A \lor B$; $A \land B$ means $\neg(\neg A \lor \neg B)$; $A \leftrightarrow B$ means $(A \to B) \land (B \to A)$.

Example 1:

Let us write the formula $A \leftrightarrow \neg B$ in the language of P.

(1) $A \leftrightarrow \neg B$	GIVEN
(2) $(A \to \neg B) \land (\neg B \to A)$	DEF ↔
(3) $(\neg A \lor \neg B) \land (\neg\neg B \land A)$	DEF → (twice)
(4) $\neg[\neg(\neg A \lor \neg B) \lor \neg(\neg\neg B \lor A)]$	DEF ∧ □

Now that the formal system P is available, the two concepts *proof in* P and *theorem of* P are automatically defined; recall the notation $\vdash_P A$ from Section 1.3. Let us now define the following more general notion of a proof, in which we are allowed to use a set Γ of formulas of P as "additional axioms."

Definition 2: Let Γ be a set of formulas of P. A *proof in* P *using* Γ (or a *proof with assumptions* Γ) is a finite sequence $A_1, \ldots, A_n$ of formulas of P such that for $1 \le k \le n$, one of the following three conditions is satisfied:

(1) A_k is an axiom;

(2) $A_k \in \Gamma$;

(3) $k > 1$ and A_k is the conclusion of a rule of inference whose hypotheses are among the formulas $A_1, \ldots, A_{k-1}$.

In this case, we also say that $A_1, \ldots, A_n$ is a *proof of* A_n in P *using* Γ and write $\Gamma \vdash_P A_n$ or, more simply, $\Gamma \vdash A_n$ (i.e., omit the subscript P). We also say that A_n is a *theorem of* Γ.

Comments Let $A_1, \ldots, A_n$ be a proof in P using Γ.

(1) For $1 \le k < n$, $A_1, \ldots, A_k$ is also a proof in P using Γ and $\Gamma \vdash_P A_k$.

(2) The first formula A_1 is an axiom, or $A_1 \in \Gamma$.

(3) If in Definition 2 we omit the second condition $A_k \in \Gamma$, we say that $A_1, \ldots, A_n$ is a *proof in* P and that A_n is a *theorem of* P, written $\vdash_P A_n$, or more simply $\vdash A_n$.

A proof in P, or a proof in P using Γ, is often referred to as a *formal* proof. When writing formal proofs, abbreviations are used as follows:

- AXIOM for propositional axiom
- ASSOC for associative rule
- EXP for expansion rule

- CTN for contraction rule
- CUT for cut rule

Derived Rules of Inference

It is convenient to add to the list of four rules of inference. The additional rules are derived from propositional axioms and from the four official rules of inference; it follows that we can use these new rules in formal proofs. A derived rule of inference is written schematically as

$$\frac{A_1, \ldots, A_n}{B},$$

and is understood as follows: If $A_1, \ldots, A_n$ are theorems of P, then B is a theorem of P (more generally, if $\Gamma \vdash A_1, \ldots, \Gamma \vdash A_n$, then $\Gamma \vdash B$). We begin with three rules: the *commutative rule*, the *new expansion rule*, and the *new associative rule*. After stating each rule, we will give its derivation.

Commutative Rule (CM): $\dfrac{A \vee B}{B \vee A}$

Derivation:

(1) $A \vee B$ HYP
(2) $\neg A \vee A$ AXIOM
(3) $B \vee A$ CUT ■

These three lines show that if we can obtain the formula $A \vee B$ as a theorem, then we can also obtain the formula $B \vee A$ as a theorem. If we were challenged about the legitimacy of using the commutative rule in a formal proof, we could respond by saying that the rule can always be replaced by lines (2) and (3) above. In formal proofs, an application of the commutative rule is denoted CM.

Strictly speaking, the three lines that justify the commutative rule are not a formal proof in P, but rather a *proof scheme*. If in the three lines above we replaced each occurrence of A with a specific formula of P, say p_1, and likewise replaced each occurrence of B with a specific formula, say $\neg p_2$, then the resulting sequence of formulas would be an official proof in P of the formula $\neg p_2 \vee p_1$ using the formula $p_1 \vee \neg p_2$. Proof schemes such as this are as close as we will come to writing out formal proofs in P. Henceforth, we will not distinguish between formal proofs in P and proof schemes but will refer to either as a formal proof, or simply a proof.

New Expansion Rule: $\dfrac{A}{A \vee B}$

Derivation:

(1) A	HYP
(2) $B \lor A$	EXP
(3) $A \lor B$	CM ■

New Associative Rule: $\dfrac{(A \lor B) \lor C}{A \lor (B \lor C)}$

Derivation:

(1) $(A \lor B) \lor C$	HYP
(2) $C \lor (A \lor B)$	CM
(3) $(C \lor A) \lor B$	ASSOC
(4) $B \lor (C \lor A)$	CM
(5) $(B \lor C) \lor A$	ASSOC
(6) $A \lor (B \lor C)$	CM ■

In formal proofs we will not distinguish between the official expansion rule and the *new* expansion rule; EXP will denote an application of either rule. The same applies to the official associative rule and the *new* associative rule: ASSOC will denote an application of either rule. Finally, in light of the commutative rule, we will not be fussy about the order of the disjuncts when using the cut rule. A similar comment applies to the derived rule of inference, disjunctive syllogism, which can be viewed as a special case of the cut rule.

Modus Ponens (MP): $\dfrac{A, A \to B}{B}$

Derivation:

(1) A	HYP
(2) $A \to B$	HYP
(3) $\neg A \lor B$	DEF $\to$
(4) $A \lor B$	EXP (1)
(5) $B \lor B$	CUT
(6) B	CTN ■

$\neg\neg$ *Rules* $(\neg\neg$ *RULE):* $\dfrac{A}{\neg\neg A}$; $\dfrac{\neg\neg A}{A}$; $\dfrac{A \lor B}{\neg\neg A \lor B}$; $\dfrac{\neg\neg A \lor B}{A \lor B}$.

Derivation of First $\neg\neg$ Rule:

(1) A	HYP
(2) $\neg\neg A \vee \neg A$	AXIOM
(3) $\neg\neg A \vee A$	EXP (1)
(4) $\neg\neg A \vee \neg\neg A$	CUT (2, 3)
(5) $\neg\neg A$	CTN ∎

The proofs of the other three $\neg\neg$ rules are left to the reader. There are six additional derived rules of inference, which are quite useful. We justify one of these (the join rule) and leave the others as an exercise for the reader.

Disjunctive Syllogism (DS) $\dfrac{A \vee B, \neg A}{B}$

Hypothetical Syllogism (HS) $\dfrac{A \rightarrow B, B \rightarrow C}{A \rightarrow C}$

Modus Tollens (MT) $\dfrac{A \rightarrow B, \neg B}{\neg A}$

Contrapositive (CP) $\dfrac{A \rightarrow B}{\neg B \rightarrow \neg A}$

Separation Rules (SEP) $\dfrac{A \wedge B}{A} , \dfrac{A \wedge B}{B}$

Join Rule (JOIN) $\dfrac{A, B}{A \wedge B}$

Derivation of Join Rule:

(1) A	HYP
(2) B	HYP
(3) $\neg(\neg A \vee \neg B) \vee (\neg A \vee \neg B)$	AXIOM
(4) $[\neg(\neg A \vee \neg B) \vee \neg A] \vee \neg B$	ASSOC
(5) $\neg\neg B$	$\neg\neg$ RULE (2)
(6) $\neg(\neg A \vee \neg B) \vee \neg A$	DS (4, 5)
(7) $\neg\neg A$	$\neg\neg$ RULE (1)
(8) $\neg(\neg A \vee \neg B)$	DS (6, 7)
(9) $A \wedge B$	DEF $\wedge$ ∎

Comment on strategy In the previous derivation, our goal was line (8), where $A \wedge B$ is expressed in terms of $\neg$ and $\vee$. There are no rules for deriving a formula of the form $\neg C$, so the only way to obtain $\neg C$ is to introduce an axiom of the form $\neg C \vee C$; this is the justification for line (3).

We end this section with an example of a theorem of P (more precisely, a theorem *scheme*). The reader is advised to glance at the *end* of the proof to see the motivation for the steps at its beginning. Specifically, the goal is line (4), which is $(A \wedge B) \to B$ expressed in terms of $\neg$ and $\vee$; this motivates the choice of the axiom in line (1).

Example 2:

Let A and B be formulas of P. Then $\vdash (A \wedge B) \to B$.

(1) $\neg B \vee B$	AXIOM
(2) $\neg A \vee (\neg B \vee B)$	EXP
(3) $(\neg A \vee \neg B) \vee B$	ASSOC
(4) $\neg\neg(\neg A \vee \neg B) \vee B$	$\neg\neg$ RULE
(5) $\neg(A \wedge B) \vee B$	DEF $\wedge$
(6) $(A \wedge B) \to B$	DEF $\to$

□

─────────── *Exercises for Section 3.1* ───────────

1. Justify these versions of the $\neg\neg$ rule:

   (1) $\dfrac{\neg\neg A}{A}$,   **(2)** $\dfrac{A \vee B}{\neg\neg A \vee B}$,   **(3)** $\dfrac{\neg\neg A \vee B}{A \vee B}$

2. Justify these derived rules of inference:
   (1) Disjunctive syllogism   **(4)** Contrapositive
   (2) Modus tollens   **(5)** Separation rules
   (3) Hypothetical syllogism

3. Show that $\vdash A \to (B \to (A \wedge B))$. Then use this result to justify the join rule.

4. Show that $\vdash (A \wedge B) \to A$. Then use this result and MP to justify a separation rule.

5. Justify these derived rules of inference:

   (1) $\dfrac{A \vee B}{A \vee (C \vee B)}$ (This is not the expansion rule.)

   (2) $\dfrac{A \vee (B \vee B)}{A \vee B}$ (This is not the contraction rule.)

   (3) $\dfrac{\neg(A \vee B)}{\neg(A \vee \neg B)}$ (This is not one of the $\neg\neg$ rules.)

   (4) $\dfrac{A \vee B}{\neg A \to B}$ (This is not the definition of $\to$.)

   (5) $\dfrac{A \vee B}{\neg(\neg A \wedge \neg B)}$ (This is not the definition of $\wedge$.)

6. Prove the following:
   (1) $\vdash A \to (A \vee B)$.   **(3)** $\vdash A \to \neg(\neg A \wedge B)$.
   (2) $\vdash \neg(A \wedge \neg A)$ **(4)** $\vdash (A \to B) \vee (B \to A)$.

7. Prove the following:
   (1) $\{A, \neg B\} \vdash \neg(\neg A \vee B)$.   **(4)** $\{A \to (B \wedge C)\} \vdash A \to B$.
   (2) $\{\neg(A \vee B)\} \vdash \neg A$.   **(5)** $\{(A \vee B) \to C\} \vdash A \to C$.
   (3) $\{A \to B, C \vee A\} \vdash C \vee B$.   **(6)** $\{A \to C\} \vdash (A \wedge B) \to C$.

8. Let Γ and Δ be sets of formulas of P. Prove the following:
   (1) If $\Gamma \subseteq \Delta$ and $\Gamma \vdash A$, then $\Delta \vdash A$.
   (2) If $\Gamma \cup \{A\} \vdash B$ and $\Gamma \vdash A$, then $\Gamma \vdash B$.
   (3) If $\Gamma \vdash (A \rightarrow B)$, then $\Gamma \cup \{A\} \vdash B$.
   (4) If $\{A_1, \ldots, A_n\} \vdash B$ and $\vdash A_1, \ldots, \vdash A_n$, then $\vdash B$.
   (5) If $\Delta \vdash A$ and $\Gamma \vdash B$ for every $B \in \Delta$, then $\Gamma \vdash A$.

9. Let Γ be a set of formulas of P. Prove the following:
   (1) If $\Gamma \vdash A$, then there is a finite $\Delta \subseteq \Gamma$ such that $\Delta \vdash A$.
   (2) Let $\{\Gamma_n : n \in \mathbb{N}\}$ be a collection of formulas of P with $\Gamma_0 \subseteq \Gamma_1 \subseteq \Gamma_2 \ldots$ (i.e., increasing) and let $\Delta = \bigcup_{n \in \mathbb{N}} \Gamma_n$. If $\Delta \vdash A$, then there is some $n \in \mathbb{N}$ such that $\Gamma_n \vdash A$.

10. (*Independence of contraction rule; Shoenfield*) This problem outlines a proof that there are theorems of P that cannot be proved without the contraction rule. (1) Give a formal proof that $\vdash \neg\neg(\neg p \vee p)$; do not use derived rules of inference. (2) Show that $\neg\neg(\neg p \vee p)$ does not have a proof without the contraction rule. (*Hint:* Define $f : \text{FOR} \rightarrow \{0, 1\}$ by $f(p) = 0$; $f(\neg A) = 1$; $f(A \vee B) = 0$. Calculate $f(\neg\neg(\neg p \vee p))$.)

11. Show that every proof that uses the contraction rule also uses the expansion rule. (*Hint:* Define $f : \text{FOR} \rightarrow \{0, 1\}$ by $f(p) = 0$; $f(\neg A) \neq f(A)$; $f(A \vee B) = 1 \Leftrightarrow f(A) = f(B)$; then prove (1) $f(\neg A \vee A) = 0$; (2) if $f(A \vee (B \vee C)) = 0$, then $f((A \vee B) \vee C) = 0$; (3) if $f(A \vee B) = f(\neg A \vee C) = 0$, then $f(B \vee C) = 0$; (4) $f(A \vee A) = 1$.)

12. (*Independence of expansion rule; Shoenfield*) This problem outlines a proof that there are theorems of P that cannot be proved without the expansion rule. (1) Give a formal proof that $\vdash p \vee (\neg p \vee p)$; do not use derived rules of inference. (2) Show that $p \vee (\neg p \vee p)$ does not have a proof without the expansion rule. (*Hint:* Use the function in Exercise 11.) (*Note:* Independence of the expansion rule can also be obtained from Exercises 10 and 11.)

13. (*Independence of cut rule; Shoenfield*) This problem outlines a proof that there are theorems of P that cannot be proved without the cut rule. (1) Give a formal proof that $\vdash \neg\neg(\neg p \vee p)$; do not use derived rules of inference. (2) Show that $\neg\neg(\neg p \vee p)$ does not have a proof without the cut rule. (*Hint:* Calculate $f(\neg\neg(\neg p \vee p))$, where $f : \text{FOR} \rightarrow \{0, 1\}$ is defined by $f(p) = 0$; $f(A \vee B) = \min\{f(A), f(B)\}$;

$$f(\neg A) = \begin{cases} 0 & A \text{ is } \neg p \vee p \\ 0 & A \text{ is not } \neg p \vee p \text{ and } f(A) = 1 \\ 1 & A \text{ is not } \neg p \vee p \text{ and } f(A) = 0. \end{cases}$$

14. (*Independence of associative rule; Shoenfield*) This problem outlines a proof that there are theorems of P that cannot be proved without the associative rule. (1) Give a formal proof that $\vdash \neg(\neg(\neg p \vee p) \vee \neg(\neg p \vee p))$; do not use derived rules of inference. (2) Show that this formula does not have a proof without the associative rule. (*Hint:* Define $f : \text{FOR} \rightarrow \mathbb{Z}$ by $f(p) = 0$; $f(\neg A) = 1 - f(A)$; $f(A \vee B) = f(A) \times f(B) \times [1 - f(A) - f(B)]$.)

────────── 3.2 ──────────

Soundness Theorem

In this section, we will prove the fundamental result that every theorem of P is a tautology. We will begin with a quick review of semantic concepts as they apply to the language of P (see Section 2.2). A *truth assignment* is a function $\phi : \text{PROP} \rightarrow \{T, F\}$; by induction on the number of occurrences of $\neg$ and $\vee$ in a formula, ϕ assigns a truth value T or F to every formula of P such that TA¬ and TA∨ hold. Since $\wedge$, $\rightarrow$, and $\leftrightarrow$ are defined symbols, the conditions TA→, TA∧, and TA↔ in Section 2.2 are derived from TA¬ and TA∨. For example,

$$\phi(A \rightarrow B) = F \Leftrightarrow \phi(\neg A \vee B) = F \qquad \text{DEF} \rightarrow$$
$$\Leftrightarrow \phi(\neg A) = F \text{ and } \phi(B) = F \qquad \text{TA}\vee$$
$$\Leftrightarrow \phi(A) = T \text{ and } \phi(B) = F \qquad \text{TA}\neg \quad \blacksquare$$

Definition 1: A formula A of P is a *tautology* if $\phi(A) = T$ for every truth assignment ϕ. *Notation:* $\models A$.

We emphasize that the definition of a tautology depends on the language of P, but not on the axioms or rules of inference of P. Specifically, it depends on the meaning of the symbols $\neg$ and $\vee$ as given by TA¬ and TA∨.

We now turn to the problem of showing that every theorem of P is a tautology. The proof is by induction on theorems (see Section 1.3); the property Q is that of being a tautology. Thus we want to show that every axiom of P is a tautology and that for every rule of inference of P, if all hypotheses are tautologies, then the conclusion is a tautology. The technical details are contained in the following more general result:

Lemma 1 (Satisfiability properties for P): Let ϕ be an arbitrary truth assignment.

(1) ϕ satisfies $\neg A \vee A$;

(2) if ϕ satisfies $A \vee (B \vee C)$, then ϕ satisfies $(A \vee B) \vee C$;

(3) if ϕ satisfies A, then ϕ satisfies $B \vee A$;

(4) if ϕ satisfies $A \vee A$, then ϕ satisfies A;

(5) if ϕ satisfies both $A \vee B$ and $\neg A \vee C$, then ϕ satisfies $B \vee C$.

Proof: We will prove (5) and leave (1)–(4) as exercises for the reader. Let $\phi(A \vee B) = T$ and $\phi(\neg A \vee C) = T$. From these two conditions and TA∨, we have

(a) $\phi(A) = T$ or $\phi(B) = T$;   *(b)* $\phi(\neg A) = T$ or $\phi(C) = T$.

Now consider two cases. If $\phi(A) = F$, then by (a) we must have $\phi(B) = T$. From $\phi(B) = T$ and TA∨, it easily follows that $\phi(B \vee C) = T$. Now suppose that $\phi(A) = T$. By (b) and TA¬, we have $\phi(C) = T$, and again $\phi(B \vee C) = T$ as required. $\blacksquare$

Lemma 1, on satisfiability, can be summarized as follows: Let ϕ be any truth assignment. Then (1) ϕ satisfies every axiom of P, and (2) for each of the four rules of inference of P, if ϕ satisfies each hypothesis of the rule, then ϕ satisfies the conclusion of the rule. This gives the following result (the details are left to the reader; a more general version will be proved later in this section).

Theorem 1 (Soundness theorem for P, special case): Every theorem of the formal system P is a tautology; in other words, if $\vdash A$, then $\models A$.

The soundness theorem is well named; it states that the axioms and rules of inference of P are sound in the sense that the theorems of P are always "true."

The soundness theorem is often used to show that a formula is *not* a theorem of P. For example, $p_1 \lor \neg p_2$ is not a tautology; it follows from the soundness theorem that $p_1 \lor \neg p_2$ cannot be a theorem of P. We will now use the soundness theorem to establish some important properties of the formal system P.

Definition 2: P is *consistent* if there is no formula A such that both $\vdash A$ and $\vdash \neg A$; P is *complete* if for every formula A of P, either $\vdash A$ or $\vdash \neg A$; P is *decidable* if there is an algorithm that, given an arbitrary formula A, decides whether $\vdash_P A$.

We now ask: Is P consistent? Is P complete? Is P decidable? The first two questions are easily settled by the soundness theorem:

Corollary 1: P is consistent.

Proof: Suppose not. Then there is some formula A of P such that both $\vdash A$ and $\vdash \neg A$. By the soundness theorem, both A and $\neg A$ are tautologies. Let ϕ be a truth assignment for P. Since A and $\neg A$ are tautologies, $\phi(A) = T$ and $\phi(\neg A) = T$. This contradicts TA¬. ■

Corollary 2: P is not complete.

Proof: Consider the simplest possible formula, namely p_1. Neither p_1 nor $\neg p_1$ is a theorem of P. To see that p_1 is not a theorem, let ϕ be any truth assignment such that $\phi(p_1) = F$. Then ϕ does not satisfy p_1, so p_1 cannot be a tautology. By the soundness theorem, p_1 cannot be a theorem. A similar argument shows that $\neg p_1$ is also not a theorem of P. ■

We have answered the first two questions using the soundness theorem. There remains the question of decidability. It turns out that P *is* decidable, but to prove this we also need the converse of the soundness theorem.

Theorem 2 (Adequacy theorem for P, special case): Every tautology is a theorem of the formal system P; in other words, if $\models A$, then $\vdash A$.

The adequacy theorem is also well named; it states that the axioms and rules of inference of P are adequate to prove all tautologies as theorems of P. The adequacy theorem is considerably more difficult to prove than the soundness theorem, and a major portion of this chapter is devoted to its proof (see Sections 3.3 and 3.4), but it seems appropriate at this point to assume this result and use it to show that P is decidable.

Corollary 3: The formal system P is decidable.

Proof: We write the required algorithm.

(1) Input a formula A.

(2) Use truth tables to decide whether A is a tautology.

(3) If *Yes*, print YES and halt (A is a theorem of P by the adequacy theorem).

(4) If *No*, print NO and halt (A is a not a theorem of P by the soundness theorem). ∎

We now generalize the soundness theorem. Recall from Section 2.2 the following generalization of a tautology: A is a *tautological consequence of* Γ, written $\Gamma \vDash A$, if every truth assignment that satisfies Γ also satisfies A.

Theorem 3 (Soundness theorem for P): If $\Gamma \vdash A$, then $\Gamma \vDash A$.

Proof: We will return to first principles and use complete induction to prove the result. Let $S(n)$ say: If there is a proof of A using Γ in n steps, then $\Gamma \vDash A$. We show that $S(n)$ holds for all $n \geq 1$. We then have: If $\Gamma \vdash A$, then there is a proof of A using Γ in n steps for some $n \geq 1$, so by $S(n)$ we have $\Gamma \vDash A$ as required.

Beginning step $S(1)$ states: If there is a proof of A using Γ in one step, then $\Gamma \vDash A$. Since the proof has just one step, either A is an axiom or $A \in \Gamma$. Let ϕ be a truth assignment that satisfies Γ. If A is an axiom, then ϕ satisfies A by Lemma 1(1); if $A \in \Gamma$, then ϕ satisfies A, since ϕ satisfies Γ.

Induction step Let $n \geq 1$ and assume that $S(1), \ldots, S(n)$ hold. Let $A_1, \ldots, A_n, A_{n+1}$, with $A_{n+1} = A$, be a proof of A using Γ in $n + 1$ steps and let ϕ be a truth assignment that satisfies Γ; we are required to show that ϕ satisfies A. Now if A is an axiom, or if $A \in \Gamma$, proceed as in the beginning step. Otherwise, A is the conclusion of a rule of inference whose hypotheses are among $A_1, \ldots, A_n$. By the induction hypotheses $S(1), \ldots, S(n)$, we have $\Gamma \vDash A_1, \ldots, \Gamma \vDash A_n$; hence ϕ satisfies $A_1, \ldots, A_n$. It easily follows from Lemma 1 (2), (3), (4), and (5) that ϕ satisfies A as required. ∎

Recall that an argument form $A_1, \ldots, A_n \therefore B$ is valid if B is a tautological consequence of $\{A_1, \ldots, A_n\}$—that is, if $\{A_1, \ldots, A_n\} \vDash B$ (see Section 2.2). The following is an immediate consequence of the soundness theorem:

Corollary 4:   If $\{A_1, \ldots, A_n\} \vdash B$,   then   the   argument   form
$A_1, \ldots, A_n \therefore B$ is valid.

Example 1:

We illustrate Corollary 4 by showing that the following argument is valid:
If you drink coffee, then you will get nervous. If you drink beer, then you
will get sleepy. You drink coffee or beer. Therefore you will get nervous or
sleepy.
We will be somewhat informal in translating the argument into the
language of propositional logic:

$$C \rightarrow N, B \rightarrow S, C \vee B \therefore N \vee S.$$

By Corollary 4, it suffices to show that $\{C \rightarrow N, B \rightarrow S, C \vee B\} \vdash N \vee S$.

(1) $C \rightarrow N$	HYP	
(2) $\neg C \vee N$	DEF $\rightarrow$	
(3) $B \rightarrow S$	HYP	
(4) $\neg B \vee S$	DEF $\rightarrow$	
(5) $C \vee B$	HYP	
(6) $B \vee N$	CUT (2, 5)	
(7) $N \vee S$	CUT (4, 6)	$\square$

——————— Exercises for Section 3.2 ———————

1. Use TA¬ and TA∨ and the definitions of →, ∧, and ↔ to prove TA →, TA∧, and
   TA↔.
2. Complete the proof of Lemma 1.
3. For each of the following, show that the formula is a tautology by showing that it is a
   theorem of P:
   (1) $(A \rightarrow B) \vee (B \rightarrow C)$.   (4) $\neg\neg A \rightarrow A$.
   (2) $\neg(A \vee B) \rightarrow \neg A$.   (5) $A \rightarrow (B \rightarrow A)$.
   (3) $(A \rightarrow B) \vee (\neg A \rightarrow B)$.   (6) $\neg A \rightarrow (A \rightarrow B)$.
4. Let A be a theorem of P. Prove the following:
   (1) There is at least one occurrence of $\neg$ in A.
   (2) There is at least one occurrence of $\vee$ in A.
5. Let $A_1, \ldots, A_n$ be theorems of P and let $\{A_1, \ldots, A_n\} \models B$. Show that B is a tautology.
6. Let $\Gamma = \{p_1, \neg p_2 \vee p_3\}$. Show that there is a no proof in P of p_3 using Γ.
7. Let $\Gamma = \{p_n : n \geq 2\}$. Prove the following:
   (1) There is·no proof of p_1 in P using Γ.
   (2) There is no proof of $\neg p_1$ in P using Γ.
8. Use the corollary of the soundness theorem to show that each of the following
   arguments is valid:
   (1) If the Cardinals or the Pirates win the pennant, then the Mets will be unhappy. If the
       Mets are unhappy, then they will hire a new manager. The Cardinals win the
       pennant. Therefore the Mets will hire a new manager.
   (2) If n is even, then 2 divides n. If n is odd, then 2 does not divide n. Therefore n
       cannot be both even and odd.
   (3) If the function f is differentiable, then the function f^2 is differentiable. If f^2 is

differentiable, then it is continuous. But f^2 is not continuous. Therefore f is not differentiable.

(4) If Fred takes easy courses or studies hard, then he will graduate and get a good job. Fred does not get a good job. Therefore Fred did not take easy courses.

(5) If f is continuous at c and g is continuous at c, then $f + g$ is continuous at c. But $f + g$ is not continuous at c. Therefore f is not continuous at c or g is not continuous at c.

9. Use the corollary of the soundness theorem to show that the following argument forms are valid:
  (1) $A \rightarrow B, C \rightarrow D, A \lor C \therefore B \lor D$.
  (2) $A \rightarrow B, C \rightarrow D, \neg(B \lor D) \therefore \neg(A \lor C)$. (Destructive dilemma.)
  (3) $A \rightarrow B, C \rightarrow D, A \land C \therefore B \land D$.
  (4) $A \rightarrow B, C \rightarrow D, \neg(B \land D) \therefore \neg(A \land C)$.
  (5) $A \rightarrow B, C \rightarrow D, \neg B \land \neg D \therefore \neg A \land \neg C$.
  (6) $A \rightarrow B, C \rightarrow D, \neg B \lor \neg D \therefore \neg A \lor \neg C$.
  (7) $(A \lor B) \rightarrow (C \land D), \neg C \therefore \neg A$.
  (8) $A \rightarrow B, C \rightarrow D, (B \lor D) \rightarrow E, \neg E \therefore \neg(A \lor C)$.

10. Use the corollary of the soundness theorem to show that the following argument forms are valid:
  (1) $A \lor B, C \lor D, \neg A \lor \neg C, \neg B \lor \neg D, B \lor C \therefore \neg A \land (B \land (C \land \neg D))$.
  (2) $A \lor (B \lor C), A \lor \neg B, \neg A \lor D, C \lor \neg D \therefore C$.
  (3) $A \lor (B \lor C), \neg A \lor B, A \lor \neg C, \neg B \lor C \therefore A \land (B \land C)$.
  (4) $A \lor B, B \rightarrow (A \land C) \therefore A$.

11. (*Basis of independence proofs*) Let $\Gamma \cup \{\neg A\}$ be satisfiable. Show that A is not a theorem of Γ.

12. (*Consistency*) A set Γ of formulas is *consistent* if there is no formula A such that both $\Gamma \vdash A$ and $\Gamma \vdash \neg A$. Prove the following:
  (1) Γ is consistent if and only if there is at least one formula that is not a theorem of Γ.
  (2) If Γ is satisfiable, then Γ is consistent.
  (3) Γ is consistent if and only if every finite subset of Γ is consistent.

13. (*Completeness*) A set Δ of formulas is *complete* if for every formula A, either $\Delta \vdash A$ or $\Delta \vdash \neg A$. Prove the following:
  (1) If Δ is complete and $\Delta \vdash A \lor B$, then $\Delta \vdash A$ or $\Delta \vdash B$.
  (2) If Δ is complete and ϕ satisfies Δ, then for every formula A, $\phi(A) = T$ if and only if $\Delta \vdash A$.

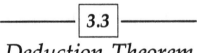

Deduction Theorem

The adequacy theorem is considerably more difficult to prove than the soundness theorem, and for its proof we need a number of intermediate results. This section is devoted to one such result, namely the deduction theorem. We emphasize that the deduction theorem has practical applications in addition to being a key theoretical result.

Theorem 1 (Deduction): If $\Gamma \cup \{A\} \vdash B$, then $\Gamma \vdash A \rightarrow B$. In particular, for the special case $\Gamma = \varnothing$, we have: If $\{A\} \vdash B$, then $\vdash A \rightarrow B$.

The deduction theorem states (for the case $\Gamma = \varnothing$): If we can write out a formal proof of the formula B using the formula A as an additional hypothesis, then there is a formal proof of the formula $A \rightarrow B$ without A as an additional hypothesis. The deduction theorem reflects the way we proceed in proving a theorem in mathematics: Given the statement "If +++, then ***," we prove *** using +++ as an additional hypothesis. Before proving the deduction theorem, let us give some examples illustrating its use in formal proofs.

Example 1:

We show that $\{A \rightarrow C, B \rightarrow C\} \vdash \neg C \rightarrow (\neg A \wedge \neg B)$.

(1) $A \rightarrow C$ HYP (formula of Γ)
(2) $B \rightarrow C$ HYP (formula of Γ)
(3) $\neg C$ HYP (deduction theorem)
(4) $\neg A$ MT $(1, 3)$
(5) $\neg B$ MT $(2, 3)$
(6) $\neg A \wedge \neg B$ JOIN

In summary, we have shown that $\{A \rightarrow C, \ B \rightarrow C, \ \neg C\} \vdash \neg A \wedge \neg B$; from the deduction theorem, we obtain $\{A \rightarrow C, B \rightarrow C\} \vdash \neg C \rightarrow (\neg A \wedge \neg B)$ as required.   □

Example 2:

The deduction theorem can be used more than once in a formal proof. We illustrate this by showing that $\vdash [(A \wedge B) \rightarrow C] \leftrightarrow [A \rightarrow (B \rightarrow C)]$. First, by the definition of $\leftrightarrow$, it suffices to show that

       (a) $\vdash [(A \wedge B) \rightarrow C] \rightarrow [A \rightarrow (B \rightarrow C)]$

and

       (b) $\vdash [A \rightarrow (B \rightarrow C)] \rightarrow [(A \wedge B) \rightarrow C]$,

and then use the join rule. We prove (a) and leave (b) as an exercise for the reader

(1) $(A \wedge B) \rightarrow C$ HYP
(2) A HYP
(3) B HYP
(4) $A \wedge B$ JOIN $(2, 3)$
(5) C MP $(1, 4)$

In summary, $\{(A \wedge B) \rightarrow C, A, B\} \vdash C$. We now have:

$\{(A \wedge B) \rightarrow C, A\} \vdash B \rightarrow C$	DEDUCTION THEOREM
$\{(A \wedge B) \rightarrow C\} \vdash A \rightarrow (B \rightarrow C)$	DEDUCTION THEOREM
$\vdash [(A \wedge B) \rightarrow C] \rightarrow [A \rightarrow (B \rightarrow C)]$	DEDUCTION THEOREM

       □

We now turn to the proof of the deduction theorem. The first step is to obtain certain generalized rules of inference (generalized in the sense that if we ignore the formula A, we obtain the original rule of inference). We have listed these rules roughly in the order of difficulty of their derivation.

Generalized Expansion Rule
(GEN EXP)

$$\frac{A \vee B}{A \vee (C \vee B)}$$

Generalized Contraction Rule
(GEN CTN)

$$\frac{A \vee (B \vee B)}{A \vee B}$$

Generalized Cut Rule
(GEN CUT)

$$\frac{A \vee (B \vee C), A \vee (\neg B \vee D)}{A \vee (C \vee D)}$$

Generalized Associative Rule
(GEN ASSOC)

$$\frac{A \vee (B \vee (C \vee D))}{A \vee ((B \vee C) \vee D)}$$

Derivation of generalized expansion rule

(1) $A \vee B$	HYP
(2) $B \vee A$	CM
(3) $C \vee (B \vee A)$	EXP
(4) $(C \vee B) \vee A$	ASSOC
(5) $A \vee (C \vee B)$	CM ■

Derivation of generalized contraction rule: Our strategy is to obtain the formula $(A \vee B) \vee (A \vee B)$ and then use the contraction rule.

(1) $A \vee (B \vee B)$	HYP
(2) $(A \vee B) \vee B$	ASSOC
(3) $B \vee (A \vee B)$	CM
(4) $A \vee (B \vee (A \vee B))$	EXP
(5) $(A \vee B) \vee (A \vee B)$	ASSOC
(6) $A \vee B$	CTN ■

Derivation of generalized cut rule: Our strategy is to obtain $B \vee ((C \vee D) \vee A)$ and $\neg B \vee ((C \vee D) \vee A)$. We then eliminate B by the cut rule, use the contraction rule to obtain $(C \vee D) \vee A$, and finally use the commutative rule to obtain $A \vee (C \vee D)$ as required. Given this hint, the reader may want to write his or her own proof rather than read the one below.

(1)	$A \vee (B \vee C)$	HYP
(2)	$(A \vee B) \vee C$	ASSOC
(3)	$((A \vee B) \vee C) \vee D$	EXP (add D)
(4)	$(A \vee B) \vee (C \vee D)$	ASSOC
(5)	$(C \vee D) \vee (A \vee B)$	CM
(6)	$((C \vee D) \vee A) \vee B$	ASSOC
(7)	$B \vee ((C \vee D) \vee A)$	CM
(8)	$A \vee (\neg B \vee D)$	HYP
(9)	$(A \vee \neg B) \vee D$	ASSOC
(10)	$D \vee (A \vee \neg B)$	CM
(11)	$C \vee (D \vee (A \vee \neg B))$	EXP (add C)
(12)	$(C \vee D) \vee (A \vee \neg B)$	ASSOC
(13)	$((C \vee D) \vee A) \vee \neg B$	ASSOC
(14)	$\neg B \vee ((C \vee D) \vee A)$	CM
(15)	$((C \vee D) \vee A) \vee ((C \vee D) \vee A)$	CUT (note 7, 14)
(16)	$(C \vee D) \vee A$	CTN
(17)	$A \vee (C \vee D)$	CM ∎

Derivation of generalized associative rule: The strategy is to use the expansion rule to obtain the theorem

$$[A \vee ((B \vee C) \vee D)] \vee [A \vee ((B \vee C) \vee D))]$$

and then to use the contraction rule. Given this hint, the reader may want to write his or her own proof rather than read the following.

(1)	$A \vee (B \vee (C \vee D))$	HYP
(2)	$(A \vee B) \vee (C \vee D)$	ASSOC
(3)	$(C \vee D) \vee (A \vee B)$	CM
(4)	$C \vee (D \vee (A \vee B))$	ASSOC
(5)	$B \vee (C \vee (D \vee (A \vee B)))$	EXP (add B)
(6)	$(B \vee C) \vee (D \vee (A \vee B))$	ASSOC
(7)	$((B \vee C) \vee D) \vee (A \vee B)$	ASSOC
(8)	$A \vee [((B \vee C) \vee D) \vee (A \vee B)]$	EXP (add A)
(9)	$[A \vee ((B \vee C) \vee D)] \vee (A \vee B)$	ASSOC

To simplify notation, let E denote the formula $[A \lor ((B \lor C) \lor D)]$.

(9) $E \lor (A \lor B)$		
(10) $(E \lor A) \lor B$	ASSOC	
(11) $((E \lor A) \lor B) \lor C$	EXP (add C)	
(12) $(E \lor A) \lor (B \lor C)$	ASSOC	
(13) $((E \lor A) \lor (B \lor C)) \lor D$	EXP (add D)	
(14) $(E \lor A) \lor ((B \lor C) \lor D)$	ASSOC	
(15) $E \lor (A \lor ((B \lor C) \lor D))$	ASSOC	
(16) $A \lor ((B \lor C) \lor D)$	CTN ■	

Proof of the deduction theorem: We want to prove that if $\Gamma \cup \{A\} \vdash B$, then $\Gamma \vdash A \to B$. The proof is by induction on the number of steps in the proof of B using $\Gamma \cup \{A\}$. Let $S(n)$ say: If a formula, say B, has a proof using $\Gamma \cup \{A\}$ in n steps, then $\Gamma \vdash A \to B$. (In this statement, A is fixed but B varies.) We will prove that $S(n)$ holds for all $n \geq 1$. Now $\Gamma \cup \{A\} \vdash B$ implies that there is a proof of B using $\Gamma \cup \{A\}$ in n steps for some $n \geq 1$, so by $S(n)$ we have $\Gamma \vdash A \to B$ as required.

Beginning step $S(1)$ states: If a formula, say B, has a proof using $\Gamma \cup \{A\}$ in one step, then $\Gamma \vdash A \to B$. Now in the case of a one-step proof of B using $\Gamma \cup \{A\}$, either B is an axiom, $B \in \Gamma$, or B is A. Suppose B is an axiom or $B \in \Gamma$. Then we have

(1) B	AXIOM or $B \in \Gamma$	
(2) $\neg A \lor B$	EXP	
(3) $A \to B$	DEF $\to$	

and so $\Gamma \vdash A \to B$ as required. Now suppose B is A, in which case we need to show that $\Gamma \vdash A \to A$. Now $A \to A$ is the same as $\neg A \lor A$; since this is an axiom of P, it is certainly a theorem of Γ.

Induction step Now let $n \geq 1$ and assume that $S(1), \ldots, S(n)$ are all true. Let $A_1, \ldots, A_n, A_{n+1}$, with $A_{n+1} = B$, be a proof of B using $\Gamma \cup \{A\}$ in $n + 1$ steps; we are required to show that $\Gamma \vdash A \to B$. If B is an axiom, if $B \in \Gamma$, or if B is A, proceed as in the beginning step. So assume that B is the conclusion of a rule of inference whose hypotheses are among $A_1, \ldots, A_n$.

Now there are four rules of inference, and each possibility must be considered; we discuss the cut rule and leave the associative, expansion, and contraction rules as exercises for the reader.

Suppose B is obtained using the cut rule. Then B is of the form $C \lor D$ and the proof of B using $\Gamma \cup \{A\}$ looks something like this:

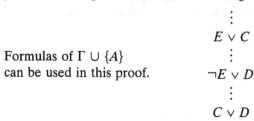

We are required to show that $\Gamma \vdash A \to (C \lor D)$. Now the formulas $E \lor C$ and $\neg E \lor D$ have proofs using $\Gamma \cup \{A\}$ in fewer than $n + 1$ steps; from the induction hypotheses $S(1), \ldots, S(n)$, it follows that $\Gamma \vdash A \to (E \lor C)$ and $\Gamma \vdash A \to (\neg E \lor D)$. We now have

(1) $A \to (E \lor C)$ THM of Γ (induction hypothesis)

(2) $\neg A \lor (E \lor C)$ DEF $\to$

(3) $A \to (\neg E \lor D)$ THM of Γ (induction hypothesis)

(4) $\neg A \lor (\neg E \lor D$ DEF $\to$

(5) $\neg A \lor (C \lor D)$ GEN CUT (2, 4)

(6) $A \to (C \lor D)$ DEF $\to$ ∎

Exercises for Section 3.3

1. Prove the converse of the deduction theorem.
2. Complete the proof of the deduction theorem by discussing the case in which B is obtained using
   (1) the expansion rule (use generalized expansion rule);
   (2) the contraction rule (use generalized contraction rule);
   (3) the associative rule (use generalized associative rule).
3. Justify these derived rules of inference:
   (1) $\dfrac{A \lor (B \lor C)}{A \lor (C \lor B)}$ (generalized commutative rule)
   (2) $\dfrac{A \lor ((B \lor C) \lor D)}{A \lor (B \lor (C \lor D))}$
   (3) $\dfrac{\neg(A \lor B)}{\neg(B \lor A)}$ (not the commutative rule)
   (4) $\dfrac{\neg(A \lor B) \lor C}{\neg(B \lor A) \lor C}$ (not the generalized commutative rule)
   (5) $\dfrac{\neg(A \lor (B \lor C))}{\neg((A \lor B) \lor C)}$ (not the associative rule)
4. (*Axioms of Russell and Whitehead for* $\neg$ *and* $\lor$) Prove that PM1–PM4 are theorems of the formal system P (see p. 104).
5. (*Axioms of Rosser for* $\neg$ *and* $\land$) Prove that R1–R3 are theorems of the formal system P (see p. 104).
6. (*Axioms of Frege for* $\neg$ *and* $\to$) Prove that LF1–LF3 are theorems of the formal system P (see p. 107).
7. (*Axioms of Lukasiewicz for* $\neg$ *and* $\to$) Prove that LUK1–LUK3 are theorems of the formal system P (see Exercise 13 in Section 3.5).
8. (*Laws of negation*) Show that each formula of Exercise 7 in Section 2.3 is a theorem of the formal system P (replace $\approx$ with $\leftrightarrow$).
9. (*Relations among connectives*) Show that each formula of Exercise 8 in Section 2.3 is a theorem of the formal system P (replace $\approx$ with $\leftrightarrow$).
10. Each of the following is a tautology; verify this by showing that each is a theorem of P.
    (1) $\vdash (A \to B) \to [(\neg A \to B) \to B]$.
    (2) $\vdash [A \to (B \to C)] \to [B \to (A \to C)]$.
    (3) $\vdash [(A \land B) \to C] \leftrightarrow [A \to (B \to C)]$.
    (4) $\vdash [A \land (\neg B \lor B)] \leftrightarrow A$.
    (5) $\vdash [A \lor (B \land \neg B)] \leftrightarrow A$.
    (6) $\vdash [(A \to B) \land (C \to D)] \to [(A \land C) \to (B \land D)]$.
    (7) $\vdash [(A \to B) \land (C \to D)] \to [(A \lor C) \to (B \lor D)]$.
    (8) $\vdash [(A \land B) \to \neg B] \to (A \to \neg B)$.

11. Give formal proofs for the following:
    (1) $\vdash [(A \lor B) \land (A \lor C)] \to [A \lor (B \land C)]$.
    (2) $\vdash [(A \to B) \to A] \to A$ (Peirce's law).
    (3) $\vdash (A \leftrightarrow B) \to [(A \land B) \lor (\neg A \land \neg B)]$.

12. Prove the following:
    (1) $\{A \to (B \lor C), (A \land B) \to D, (A \land C) \to E\} \vdash (A \land \neg E) \to D$.
    (2) $\{(A \lor B) \lor C, D \to \neg A, E \to \neg C\} \vdash (D \land E) \to B$.

13. (*Proof by cases*) Use the deduction theorem to justify the following proof procedure: If $\Gamma \cup \{A\} \vdash C$ and $\Gamma \cup \{B\} \vdash C$, then $\Gamma \cup \{A \lor B\} \vdash C$ and also $\Gamma \vdash (A \lor B) \to C$. Then prove the following:
    (1) $\vdash [A \lor (B \land C)] \to [(A \lor B) \land (A \lor C)]$.
    (2) $\vdash [(A \land B) \lor (A \land C)] \to [A \land (B \lor C)]$.
    (3) $\vdash [(A \to B) \lor (A \to C)] \to [A \to (B \lor C)]$.
    (4) $\vdash [(A \land B) \lor (\neg A \land \neg B)] \to (A \leftrightarrow B)$.
    (5) $\{A, B \lor C\} \vdash (A \land B) \lor (A \land C)$.

14. (*Proof by contradiction*) Use the deduction theorem to justify the following proof procedure: If $\Gamma \cup \{\neg A\} \vdash B$ and $\Gamma \cup \{\neg A\} \vdash \neg B$, then $\Gamma \vdash A$. Then prove the following:
    (1) $\{(A \lor B) \to (C \land D), D \to [(B \lor C) \to \neg A]\} \vdash \neg A$.
    (2) $\{\neg A \to (B \land C), (C \lor D) \to E, E \to \neg B\} \vdash A$.

15. Justify the following proof procedure: If $\Gamma \cup \{A\} \vdash B$ and $\Gamma \cup \{\neg A\} \vdash B$, then $\Gamma \vdash B$.

16. Show that $\Gamma \cup \{A\} \vdash \neg B$ if and only if $\Gamma \cup \{B\} \vdash \neg A$.

17. Let Γ be a set of formulas that is decidable. Show that $\Gamma \cup \{C\}$ is also decidable.

18. Two formulas A and B are *provably equivalent* if $\vdash A \leftrightarrow B$. Prove the following:
    (1) $\vdash A \leftrightarrow B$ if and only if $\vdash A \to B$ and $\vdash B \to A$.
    (2) Provable equivalence is an equivalence relation (i.e., $\vdash A \leftrightarrow A$; if $\vdash A \leftrightarrow B$, then $\vdash B \leftrightarrow A$; if $\vdash A \leftrightarrow B$ and $\vdash B \leftrightarrow C$, then $\vdash A \leftrightarrow C$).
    (3 $\vdash A \leftrightarrow B$ if and only if $\vdash \neg A \leftrightarrow \neg B$.
    (4) If $\vdash A \leftrightarrow B$ and $\vdash C \leftrightarrow D$, then $\vdash (A \lor C) \leftrightarrow (B \lor D)$.

19. (*Replacement theorem for* P) Let A and B be provably equivalent (i.e., $\vdash A \leftrightarrow B$), let C be a formula in which A occurs, and let C^* be obtained from C by replacing one or more of the occurrences of A with B. Prove that $\vdash C \leftrightarrow C^*$. (*Hints:* First consider the special case in which C is all of A; then use induction on the number of connectives in C and results from Exercise 18.

20. Consider the formula $\neg(p_1 \lor \neg p_2)$ with the truth table shown below. Show the following:
    (1) $\{p_1, p_2\} \vdash \neg\neg(p_1 \lor \neg p_2)$.
    (2) $\{p_1, \neg p_2\} \vdash \neg\neg(p_1 \lor \neg p_2)$.
    (3) $\{\neg p_1, p_2\} \vdash \neg(p_1 \lor \neg p_2)$.
    (4) $\{\neg p_1, \neg p_2\} \vdash \neg\neg(p_1 \lor \neg p_2)$.

p_1	p_2	$\neg(p_1 \lor \neg p_2)$
T	T	F
T	F	F
F	T	T
F	F	F

21. (*Kalmar's proof of the adequacy theorem*) This problem outlines a proof of the adequacy theorem in the following form: If A is a tautology, then $\vdash A$. For each truth assignment ϕ and each formula A, define A^ϕ by $A^\phi = A$ if $\phi(A) = T$ and $A^\phi = \neg A$ if $\phi(A) = F$. Prove the following:
    (1) If the propositional variables in A are among $p_1, \ldots, p_n$ then $\{p_1^\phi, \ldots, p_n^\phi\} \vdash A^\phi$. (*Hint:* Use induction on the number of connectives in A. This result is illustrated in Exercise 20 for the formula $\neg(p_1 \lor \neg p_2)$.)
    (2) If A is a tautology with propositional variables among $p_1, \ldots, p_n$, then $\{p_1^\phi, \ldots, p_{n-1}^\phi\} \vdash A$. (*Hint:* Let ψ agree with ϕ everywhere except that $\phi(p_n) \neq \psi(p_n)$. By (1) applied to ϕ and ψ, $\{p_1^\phi, \ldots, p_{n-1}^\phi, p_n\} \vdash A$ and $\{p_1^\phi, \ldots, p_{n-1}^\phi, \neg p_n\} \vdash A$ (note that $A^\phi = A^\psi = A$, since A is a tautology); use the deduction theorem to show $\{p_1^\phi, \ldots, p_{n-1}^\phi\} \vdash A$.)
    (3) If A is a tautology with propositional variables among $p_1, \ldots, p_n$, then $\{p_1^\phi, \ldots, p_{n-2}^\phi\} \vdash A$.
    (4) If A is a tautology, then $\vdash A$.

$\boxed{3.4}$

Model Existence Theorem and Adequacy Theorem

In this section, we will prove the converse of the soundness theorem, namely the adequacy theorem. A special case of this important result states that every tautology is a theorem of P. Informally, this means that the axioms and rules of inference of P are adequate to prove that all tautologies are theorems of the formal system P. The general version of the adequacy theorem states that if $\Gamma \models A$, then $\Gamma \vdash A$. In the process of proving this result, we will systematically study properties of sets of formulas of P. The first property we will study is that of consistency. Recall that P *is* consistent; on the other hand, it is certainly possible to have both $\Gamma \vdash A$ and $\Gamma \vdash \neg A$; for example, if $\Gamma = \{p_1, \neg p_1\}$, then both $\Gamma \vdash p_1$ and $\Gamma \vdash \neg p_1$ hold.

Definition 1: Γ is *consistent* if there is no formula A such that both $\Gamma \vdash A$ and $\Gamma \vdash \neg A$. If Γ is not consistent, we say that Γ is *inconsistent*. Thus Γ is inconsistent if there is at least one formula A such that both $\Gamma \vdash A$ and $\Gamma \vdash \neg A$. For example, $\Gamma = \{p_1, \neg p_1\}$ is inconsistent.

The next theorem, a consequence of the soundness theorem, gives a procedure for establishing the consistency of a set of formulas.

Theorem 1 (Consistency theorem for P): If Γ is satisfiable, then Γ is consistent.

Proof: Let ϕ be a truth assignment that satisfies Γ. Suppose, however, that Γ is not consistent. Then there is some formula A such that both $\Gamma \vdash A$ and $\Gamma \vdash \neg A$. By the soundness theorem, $\Gamma \models A$ and $\Gamma \models \neg A$. But ϕ satisfies Γ, so $\phi(A) = T$ and $\phi(\neg A) = T$. This contradicts TA¬. ∎

Example 1:

The set $\Gamma = \{p_n : n \in \mathbb{N}^+\}$ is consistent, since the truth assignment ϕ defined by $\phi(p_n) = T$ for all $n \in \mathbb{N}^+$ satisfies Γ. □

The converse of the consistency theorem (Γ consistent $\Rightarrow$ Γ satisfiable) is also true, but is much more difficult to prove. This converse is called the *model existence theorem*, because a truth assignment that satisfies Γ is often called a model of Γ. The adequacy theorem follows rather easily from the model existence theorem, so we will now direct our attention toward proving this important result. We begin with several results about consistency.

Lemma 1: If $\Delta \vdash A$, then there is a finite $\Delta_0 \subseteq \Delta$ such that $\Delta_0 \vdash A$.

Proof: Fix a proof of A using Δ, and let $\Delta_0 = \{B : B \in \Delta$ and B is used in this proof of $A\}$. Since a proof has a finite number of steps, the set Δ_0 is finite. Moreover, $\Delta_0 \subseteq \Delta$ and $\Delta_0 \vdash A$ as required. ∎

Theorem 2: If every finite subset of Δ is consistent, then Δ is consistent.

Proof: Suppose Δ is inconsistent. Then there is a formula A such that both $\Delta \vdash A$ and $\Delta \vdash \neg A$. By Lemma 1, there are finite subsets Δ_1 and Δ_2 of Δ such that $\Delta_1 \vdash A$ and $\Delta_2 \vdash \neg A$. Let $\Delta_0 = \Delta_1 \cup \Delta_2$. Then Δ_0 is a finite subset of Δ such that $\Delta_0 \vdash A$ and $\Delta_0 \vdash \neg A$; hence Δ_0 is inconsistent, a contradiction of the assumption that every finite subset of Δ is consistent. ∎

Theorem 3 (Characterization of consistency): Let Γ be a set of formulas of P.

(1) Γ is consistent if and only if there is at least one formula that is not a theorem of Γ.

(2) Γ is inconsistent if and only if every formula is a theorem of Γ.

Proof: First of all, (1) and (2) are equivalent, so we need only prove (2). Assume Γ is inconsistent. Then there is at least one formula A such that both $\Gamma \vdash A$ and $\Gamma \vdash \neg A$. Let B be an arbitrary formula of P; the following steps show that $\Gamma \vdash B$:

(1) A THM of Γ
(2) $\neg A$ THM of Γ
(3) $A \vee B$ EXP (1)
(4) B DS (2, 3)

We leave the proof of the other direction of (2) to the reader. ∎

Suppose that Γ is a set of formulas and that A is not a theorem of Γ. By Theorem 3, Γ is consistent. Moreover, Theorem 4 states that we can add $\neg A$ to

Γ and still have a consistent set of formulas. The deduction theorem is used to prove this important result (and this is the only time it is used).

Theorem 4 (Extension): If A is not a theorem of Γ, then $Γ \cup \{\neg A\}$ is consistent.

Proof: Suppose $Γ \cup \{\neg A\}$ is inconsistent. Then every formula is a theorem of $Γ \cup \{\neg A\}$ (see Theorem 3); in particular, A is a theorem of $Γ \cup \{\neg A\}$, and we have $Γ \cup \{\neg A\} \vdash A$. By the deduction theorem, $Γ \vdash \neg A \rightarrow A$. We now obtain a contradiction by showing that A is a theorem of Γ:

(1)	$\neg A \rightarrow A$	THM of Γ
(2)	$\neg\neg A \vee A$	DEF →
(3)	$\neg A \vee A$	AXIOM
(4)	$A \vee A$	CUT (2, 3)
(5)	A	CTN ∎

We need yet another property of sets of formulas before we can give the promised proof of the model existence theorem.

Definition 2: A set Γ of formulas of P is *complete* if for every formula A of P, either $Γ \vdash A$ or $Γ \vdash \neg A$. The special case $Γ = \varnothing$ gives the definition that P is complete (in Section 3.2 we used the soundness theorem to show that P is *not* complete).

Lemma 2: Let Δ be complete. Then $Δ \vdash A \vee B$ if and only if $Δ \vdash A$ or $Δ \vdash B$.

Proof: One direction is easy and does not require the hypothesis that Δ is complete: Suppose that either A or B is a theorem of Δ, say $Δ \vdash A$; by the expansion rule, $Δ \vdash A \vee B$. Now suppose that $Δ \vdash A \vee B$. If $Δ \vdash A$, there is nothing to prove, so assume that A is not a theorem of Δ. Since Δ is complete, $Δ \vdash \neg A$; by disjunctive syllogism, $Δ \vdash B$. ∎

Example 2:

The set $Γ = \{p_n : n \geq 2\}$ is *not* complete. To prove this, we show that neither p_1 nor $\neg p_1$ is a theorem of Γ. Suppose $Γ \vdash \neg p_1$; by the soundness theorem, $Γ \models \neg p_1$. Define a truth assignment $\phi : \text{PROP} \rightarrow \{T, F\}$ by $\phi(p_n) = T$ for all $n \in \mathbb{N}^+$. Clearly ϕ satisfies Γ, and since $Γ \models \neg p_1$, we have $\phi(\neg p_1) = T$. This contradicts the fact that $\phi(p_1) = T$. By a similar argument, p_1 is also not a theorem of Γ. □

A set of formulas Γ may be consistent but not complete. But by carefully adding formulas to Γ, one can construct a set Δ with $Γ \subseteq Δ$ such that Δ is both consistent *and* complete; this is called Lindenbaum's theorem. To illustrate: $\Gamma = \{p_n : n \geq 2\}$ is not complete (see Example 2); on the other hand, $\Delta = \{p_n : n \in \mathbb{N}^+\}$, the set obtained from Γ by adding the formula p_1 to Γ, *is* complete (see the exercises).

Theorem 5 (Lindenbaum): Let Γ be consistent. Then there is a set Δ of formulas such that (1) $\Gamma \subseteq \Delta$; (2) Δ is consistent; and (3) Δ is complete.

Proof: Let $A_0, A_1, A_2, \ldots$ be a list of all formulas of P (a formal language with a countable number of symbols has a countable number of formulas; see Section 1.5). By induction, construct a sequence $\Gamma_0, \Gamma_1, \Gamma_2, \ldots$ of sets of formulas of P as follows:

$$\Gamma_0 = \Gamma; \qquad \Gamma_{n+1} = \begin{cases} \Gamma_n & \Gamma_n \vdash A_n; \\ \Gamma_n \cup \{\neg A_n\} & A_n \text{ not a theorem of } \Gamma_n. \end{cases}$$

The sequence is obviously increasing; that is, $\Gamma_0 \subseteq \Gamma_1 \subseteq \Gamma_2 \ldots$. Let Δ be the union of $\Gamma_0, \Gamma_1, \Gamma_2, \ldots$; that is, $\Delta = \bigcup_{n \in \mathbb{N}} \Gamma_n$. Clearly $\Gamma \subseteq \Delta$, so it remains to prove that Δ is consistent and complete.

For all $n \in \mathbb{N}$, Γ_n is consistent The proof is by induction. For $n = 0$, $\Gamma_0 = \Gamma$, and Γ is consistent by hypothesis. Now let $n \geq 0$ and assume as an induction hypothesis that Γ_n is consistent. We consider two cases: If $\Gamma_n \vdash A_n$, then $\Gamma_{n+1} = \Gamma_n$ and the consistency of Γ_{n+1} is an immediate consequence of the consistency of Γ_n. If A_n is not a theorem of Γ_n, then $\Gamma_{n+1} = \Gamma_n \cup \{\neg A_n\}$ and Γ_{n+1} is consistent by the extension theorem.

Δ is consistent We know that each Γ_n is consistent. Suppose, however, that $\Delta = \bigcup_{n \in \mathbb{N}} \Gamma_n$ is not consistent. Then there is some finite $\Delta_0 \subseteq \Delta$ such that Δ_0 is not consistent (Theorem 2). We now have

$\Delta_0 \subseteq \bigcup_{n \in \mathbb{N}} \Gamma_n$ with $\Gamma_0 \subseteq \Gamma_1 \subseteq \Gamma_2 \ldots$ (increasing);
Δ_0 is finite;

it follows that there is some $n \in \mathbb{N}$ such that $\Delta_0 \subseteq \Gamma_n$ (see Exercise 6 in Section 1.4). Since Δ_0 is inconsistent, Γ_n is inconsistent, a contradiction of the previous result.

Δ is complete Let A be a formula of P. Since $A_0, A_1, A_2, \ldots$ is a list of all formulas of P, there is some $n \in \mathbb{N}$ such that $A = A_n$. We are required to show that $\Delta \vdash A_n$ or $\Delta \vdash \neg A_n$. We consider two cases: If $\Gamma_n \vdash A_n$, then $\Delta \vdash A_n$ follows immediately from the fact that $\Gamma_n \subseteq \Delta$. Suppose A_n is not a theorem of Γ_n. Then $\neg A_n \in \Gamma_{n+1}$, and from this $\Delta \vdash \neg A_n$ easily follows. ∎

Theorem 6 (Model existence theorem for P): If Γ is consistent, then Γ is satisfiable.

Proof: By Lindenbaum's theorem, there is a set Δ of formulas such that $\Gamma \subseteq \Delta$ and Δ is both consistent and complete. We will construct a truth assignment that satisfies every theorem of Δ; since $\Gamma \subseteq \Delta$, this truth assignment will also satisfy Γ. Define $\phi : \text{PROP} \to \{T, F\}$ by $\phi(p_n) = T \Leftrightarrow \Delta \vdash p_n$; recall

that ϕ extends to a function $\phi: \text{FOR} \rightarrow \{T, F\}$ and that ϕ satisfies TA¬ and
TA∨. The proof is complete if we can show that for every formula A,

$$\phi(A) = T \Leftrightarrow \Delta \vdash A. \tag{*}$$

The proof of (∗) is by induction on the number of connectives in the formula
A. The result is obvious in the case in which A is a propositional variable.

Negation Suppose A is $\neg B$. The formula B has fewer connectives than A,
hence by the induction hypothesis, $\phi(B) = T \Leftrightarrow \Delta \vdash B$.

$$\phi(A) = T \Leftrightarrow \phi(\neg B) = T \qquad\qquad A \text{ is } \neg B$$
$$\Leftrightarrow \phi(B) = F \qquad\qquad\qquad \text{TA¬}$$
$$\Leftrightarrow B \text{ not a theorem of } \Delta \qquad \text{Induction hypothesis}$$
$$\Leftrightarrow \Delta \vdash \neg B \qquad\qquad\qquad \Delta \text{ consistent and complete}$$
$$\Leftrightarrow \Delta \vdash A \qquad\qquad\qquad A \text{ is } \neg B$$

We elaborate on the equivalence of the third and fourth lines. If B is not a
theorem of Δ, then $\Delta \vdash \neg B$ by the completeness of Δ. If $\Delta \vdash \neg B$, then B is not
a theorem of Δ, by the consistency of Δ.

Disjunction Suppose A is $B \vee C$. Both B and C have fewer connectives
than A; hence by the induction hypothesis, $\phi(B) = T \Leftrightarrow \Delta \vdash B$ and $\phi(C) = T \Leftrightarrow \Delta \vdash C$.

$$\phi(A) = T \Leftrightarrow \phi(B \vee C) = T \qquad\qquad A \text{ is } B \vee C$$
$$\Leftrightarrow \phi(B) = T \text{ or } \phi(C) = T \qquad \text{TA∨}$$
$$\Leftrightarrow \Delta \vdash B \text{ or } \Delta \vdash C \qquad\qquad \text{Induction hypothesis}$$
$$\Leftrightarrow \Delta \vdash B \vee C \qquad\qquad\qquad \text{Lemma 2 } (\Delta \text{ complete})$$
$$\Leftrightarrow \Delta \vdash A \qquad\qquad\qquad\qquad A \text{ is } B \vee C \blacksquare$$

Theorem 7 (Adequacy theorem for P): If $\Gamma \models A$, then $\Gamma \vdash A$.

Proof: Suppose A is not a theorem of Γ. Then $\Gamma \cup \{\neg A\}$ is consistent
by the extension theorem. By the model existence theorem, there is a
truth assignment ϕ that satisfies $\Gamma \cup \{\neg A\}$; in other words, ϕ satisfies Γ
and $\phi(\neg A) = T$. But $\Gamma \models A$; hence we obtain $\phi(A) = T$. This contradicts
$\phi(\neg A) = T$. $\blacksquare$

Corollary 1 (Bernays, Post): Every tautology is a theorem of P.

Corollary 2: If $A_1, \ldots, A_n \therefore B$ is a valid argument form, then
$\{A_1, \ldots, A_n\} \vdash B$.

We have completed our goal of showing that a formula is a theorem of P if
and only if it is a tautology. In other words, the formal system P enables us to
capture the concept of a tautology syntactically. It is worthwhile to step back
and summarize some of the results we have obtained in carrying out this
program.

The word *complete* has a technical meaning: Every formula or its negation
is a theorem. There is, however, an informal, nontechnical meaning of

complete that is used in logic and applies to any formal system F. Quite generally, a formal system F is *complete* if everything we want to be a theorem is a theorem. For the formal system P, we want tautologies to be theorems, so in this nontechnical sense P is complete. With this terminology in mind, we will summarize the main results of this chapter in two theorems:

Completeness Theorem for P (Version I): Γ is consistent if and only if Γ is satisfiable.

Completeness Theorem for P (Version II): $\Gamma \vdash A$ if and only if $\Gamma \models A$.

Version I combines the consistency theorem and the model existence theorem; version II combines the soundness theorem and the adequacy theorem. The reader should keep these two fundamental results in mind. Each states the equivalence of a syntactic concept and a semantic concept.

In Section 2.2, we stated two versions of the compactness theorem; at that time, compactness theorem II was derived from compactness theorem I, but the proof of I was postponed until later. We can now easily obtain this result from completeness theorem I.

Theorem 8 (Compactness theorem for propositional logic, Version I): If Γ is finitely satisfiable, then Γ is satisfiable.

Proof: Suppose Γ is not satisfiable. By the model existence theorem, Γ is not consistent. By Theorem 2, there is a finite $\Delta \subseteq \Gamma$ such that Δ is not consistent. By the consistency theorem, Δ is unsatisfiable. This contradicts the finite satisfiability of Γ. ∎

─────────── *Exercises for Section 3.4* ───────────

1. For each of the following, decide whether Γ is consistent or inconsistent. To show that Γ is consistent, find a truth assignment that satisfies Γ. To show that Γ is inconsistent, either show that Γ is unsatisfiable or find some formula A such that both $\Gamma \vdash A$ and $\Gamma \vdash \neg A$.
   (1) $\Gamma = \{p_1, p_2, \neg(p_2 \vee p_3)\}$.
   (2) $\Gamma = \{p_1 \vee \neg p_2, \neg(p_1 \vee p_2)\}$.
   (3) $\Gamma = \{p_1 \vee \neg p_2, \neg(p_1 \vee p_2), p_1 \rightarrow p_2\}$.
   (4) $\Gamma = \{\neg(p_1 \vee p_2), \neg p_2 \vee p_3, (p_2 \vee p_3) \rightarrow p_4\}$.
2. Let Γ and Δ be such that $\Gamma \models A$ and $\Delta \models \neg A$. Show that $\Gamma \cup \Delta$ is not consistent.
3. Prove the converse of the extension theorem.
4. Prove the extension theorem without using Theorem 3.
5. Let Γ be the set consisting of all formulas of P of the form $\neg A \vee B$. Is Γ consistent?
6. Let $A_1, \ldots, A_n$ be tautologies and let $\{A_1, \ldots, A_n\} \models B$. Show that $\vdash B$.
7. Show that $\Gamma \vdash A$ if and only if $\Gamma \cup \{\neg A\}$ is unsatisfiable.
8. Let Γ be a consistent set of formulas. Prove the following:
   (1) If $\Gamma \vdash A \vee B$, then $\Gamma \cup \{A\}$ is consistent or $\Gamma \cup \{B\}$ is consistent.
   (2) If $\Gamma \vdash \neg(A \vee B)$, then $\Gamma \cup \{\neg A, \neg B\}$ is consistent.

9. Suppose that A is not a theorem of P and that $\vdash A \vee B$. Show that $\{\neg A, B\}$ is consistent.

10. Let Γ be inconsistent. Show that Γ is complete.

11. Assume that $\Gamma \vdash A$ or $\Gamma \vdash \neg A$ and likewise $\Gamma \vdash B$ or $\Gamma \vdash \neg B$. Show that $\Gamma \vdash \neg A \vee B$ or $\Gamma \vdash \neg(\neg A \vee B)$.

12. Let Δ be a set of formulas of P such that for all $n \in \mathbb{N}^+$, $\Delta \vdash p_n$ or $\Delta \vdash \neg p_n$. Show that Δ is complete.

13. (*Characterization of completeness*) Show that the following are equivalent:
    (1) Δ is complete.
    (2) If $\Delta \cup \{A\}$ is consistent, then $\Delta \vdash A$.

14. (*Characterization of maximal completeness*) Let Δ be a consistent set of formulas of P. Show that the following are equivalent:
    (1) Δ is maximally complete ($A \in \Delta$ or $\neg A \in \Delta$ for every formula A).
    (2) Δ is maximally consistent (if $\Delta \cup \{A\}$ is consistent, then $A \in \Delta$).

15. Show that the following are equivalent:
    (1) Γ is consistent and complete.
    (2) There is exactly one truth assignment that satisfies Γ.

16. Show that PROP is consistent and complete (see Exercise 15).

17. Let Δ be a finite set of formulas that is consistent. Show that Δ cannot be complete.

18. Show that the following are equivalent:
    (1) Γ is decidable (there is an algorithm that decides $\Gamma \vdash A$).
    (2) There is an algorithm that, given a formula A, decides if $\Gamma \cup \{A\}$ is consistent.

19. Use the soundness theorem and the adequacy theorem to prove the deduction theorem.

20. Use the extension theorem to prove the deduction theorem.

21. Use the adequacy theorem to prove the model existence theorem.

22. Use completeness theorem II to prove verison II of the compactness theorem: *If $\Gamma \vDash A$, then there is a finite $\Delta \subseteq \Gamma$ such that $\Delta \vDash A$.*

23. (*Application of the compactness theorem*) Assume compactness theorem I and the following special case of the adequacy theorem: If $\vDash A$, then $\vdash A$. Prove the adequacy theorem.

24. Let Γ be consistent. Show that $\Gamma \cup \{A\}$ is consistent or $\Gamma \cup \{\neg A\}$ is consistent.

25. (*Variation of Lindenbaum's construction*) Let $A_0, A_1, A_2, \ldots$ be all formulas of P and let Γ be a consistent set of formulas. Construct a sequence $\Gamma_0, \Gamma_1, \Gamma_2, \ldots$ of sets of formulas of P as follows:

$$\Gamma_0 = \Gamma; \qquad \Gamma_{n+1} = \begin{cases} \Gamma_n \cup \{A_n\} & \Gamma_n \cup \{A_n\} \text{ consistent;} \\ \Gamma_n \cup \{\neg A_n\} & \Gamma_n \cup \{A_n\} \text{ not consistent.} \end{cases}$$

Let $\Delta = \bigcup_{n \in \mathbb{N}} \Gamma_n$. Prove that Δ is consistent and maximally complete.

26. (*Truth sets*) A set Δ of formulas of $\{\neg, \vee\}$ is a *truth set* if it satisfies the following for all formulas A and B: (1) $A \in \Delta$ or $\neg A \in \Delta$ but not both; (2) $A \vee B \in \Delta$ if and only if $A \in \Delta$ or $B \in \Delta$. Let Δ be a truth set. Show that there is a truth assignment ϕ such that for every formula A, $\phi(A) = T \Leftrightarrow A \in \Delta$. Thus every truth set is satisfiable, and hence is consistent. (*Hint*: Define ϕ: PROP $\rightarrow \{T, F\}$ by $\phi(p_n) = T \Leftrightarrow p_n \in \Delta$.)

27. (*Model existence theorem and compactness theorem*) This exercise gives a unified approach to the model existence theorem (Γ consistent $\Rightarrow$ Γ satisfiable) and the compactness theorem (Γ finitely satisfiable $\Rightarrow$ Γ satisfiable). Let Q be a property of sets of formulas of $\{\neg, \vee\}$ (e.g., Q = consistent or Q = finitely satisfiable). Consider the following conditions on Q:

   H (hereditary): If $\Gamma \subseteq \Delta$ and Δ has property Q, then Γ has property Q.
   C (compact): If every finite subset of Δ has property Q, then Δ has property Q.
   E (extension): If Γ has property Q and A is any formula, then $\Gamma \cup \{A\}$ or $\Gamma \cup \{\neg A\}$ has property Q.
   I (incompatible): If Δ has property Q and A, B are formulas, then $\{A, \neg A\}$,

$\{\neg A, \neg B, A \lor B\}$, $\{A, \neg(A \lor B)\}$, and $\{B, \neg(A \lor B)\}$ are not subsets of Δ. Prove the following:

(1) Q satisfies the four conditions H, C, E, and I for Q = consistent and Q = finitely satisfiable.

(2) Let Q satisfy H, C, and E and let Γ have property Q. Then there is a set Δ of formulas such that $\Gamma \subseteq \Delta$, Δ has property Q, and Δ is maximally complete. (*Hint:* Exercise 25.)

(3) Let Q satisfy I, and let Δ have property Q and be maximally complete. Then Δ is a truth set (see Exercise 26), and hence is satisfiable.

(4) The model existence theorem and the compactness theorem hold.

28. (*Truth sets = truth assignments*) Refer to Exercise 26, where the following result is proved: *If Γ is a truth set, then there is a truth assignment ϕ such that $\Gamma = \{A: \phi(A) = T\}$.* Prove the following:

(1) Let ϕ be a truth assignment and let $\Gamma_\phi = \{A: \phi(A) = T\}$; then Γ_ϕ is a truth set.

(2) A is a tautology if and only if A belongs to every truth set.

(3) A is satisfiable if and only if A belongs to at least one truth set.

29. (*Hintikka sets*) Let Δ be a set of formulas of $\{\neg, \lor\}$ satisfying the following conditions:

(1) There is no $p \in \text{PROP}$ such that $\{p, \neg p\} \subseteq \Delta$.

(2) If $\neg\neg A \in \Delta$, then $A \in \Delta$.

(3) If $(A \lor B) \in \Delta$, then $A \in \Delta$ or $B \in \Delta$.

(4) If $\neg(A \lor B) \in \Delta$, then $\neg A \in \Delta$ and $\neg B \in \Delta$.

Δ is called a *Hintikka* set. For example, $\{\neg\neg p_1, p_1\}$ is a Hintikka set. Show that every Hintikka set Δ is satisfiable. (*Hint:* define $\phi: \text{PROP} \to \{T, F\}$ by $\phi(p_n) = T \Leftrightarrow p_n \in \Delta$, and then prove that the following holds for every formula A: If $A \in \Delta$, then $\phi(A) = T$. Note that all formulas of $\{\neg, \lor\}$ are of the form $p, \neg p, \neg\neg B, B \lor C$, or $\neg(B \lor C)$.)

30. (*Turing's theorem*) Let Γ be axiomatized and complete. Show that Γ is decidable. (See Section 5.5 for definitions and hints.)

31. Answer the following questions about PROP and justify each answer: (1) Is PROP consistent? (2) Is PROP complete? (3) Is PROP axiomatized? (4) Is PROP decidable?

3.5

Hilbert-Style Proof Systems for Propositional Logic

The formal system P is a syntactic approach to propositional logic, with *not* and *or* as basic connectives. It is not surprising that there are other formal systems with the same language $\{\neg, \lor\}$ as P, but with different axioms and rules of inference. There are also formal systems that use connectives other than $\neg$ and $\lor$. Before we proceed further, let us summarize the requirements for a suitable formal system for propositonal logic: The set of connectives is adequate (e.g., $\{\neg, \lor\}$, $\{\neg, \to\}$, or $\{\neg, \land\}$); the soundness and adequacy theorems hold. In this section, we will describe a number of different formal systems for propositional logic; in each case, we will want to verify that these two properties hold. To simplify the discussion, we will emphasize the special cases of the soundness and adequacy theorems rather than the more general versions $\Gamma \vdash A \Rightarrow \Gamma \vDash A$ and $\Gamma \vDash A \Rightarrow \Gamma \vdash A$. (Exercise 23 in Section 3.4 suggests a proof that the adequacy theorem follows from the compactness theorem and the special case of the adequacy theorem.)

For each system, there is just one rule of inference, namely *modus ponens*

(MP): $A, A \rightarrow B / B$. Thus the emphasis will be on stating the axioms of each formal system. All of the formal systems discussed in this section are examples of *Hilbert-style proof systems*. Although they are not a precisely defined concept, such proof systems have the following characteristics: They are easy to describe (and hence suitable for coding, as required for a proof of Gödel's incompleteness theorem); they have many axioms and few rules of inference; there is smooth development of the metalogic (soundness theorem and adequacy theorem); they are not designed for ease of writing out proofs within the system; their axioms are easily modified to give nonclassical logics, such as intuitionistic logic.

The Formal System PM: This formal system for propositional logic was used by Russell and Whitehead in *Principia Mathematica* and later by Hilbert and Ackermann in *Principles of Mathematical Logic*. The language is $\{\neg, \vee\}$, and there are four axiom schemes, as follows:

> *PM1* $(A \vee A) \rightarrow A$.
> *PM2* $A \rightarrow (A \vee B)$.
> *PM3* $(A \rightarrow B) \rightarrow [(C \vee A) \rightarrow (C \vee B)]$.
> *PM4* $(A \vee B) \rightarrow (B \vee A)$.

As usual, $A \rightarrow B$ means $\neg A \vee B$. Thus axiom PM1 in the original language of PM is $\neg(A \vee A) \vee A$, and MP is $\neg A \vee B, A / B$. In Exercise 1, the reader is asked to prove that P and PM have the same theorems. Since the soundness theorem and the adequacy theorem hold for P, they also hold for PM. Two other versions of propositional logic with language $\{\neg, \vee\}$ are discussed in the exercises.

The Formal System R: This formal system for propositional logic, due to Rosser, has language $\{\neg, \wedge\}$ and three axiom schemes as follows:

> *R1* $A \rightarrow (A \wedge A)$.
> *R2* $(A \wedge B) \rightarrow A$.
> *R3* $(A \rightarrow B) \rightarrow [\neg(B \wedge C) \rightarrow \neg(C \wedge A)]$.

The formula $A \rightarrow B$ means $\neg(A \wedge \neg B)$. Exercise 4, outlines a complete development of propositional logic for the formal system *R*.

The Formal System L: The formal system L for propositional logic has language $\{\neg, \rightarrow\}$ and four axiom schemes that are carefully chosen so that the deduction and adequacy theorems are easy to prove:

> *L1* $A \rightarrow (B \rightarrow A)$.
> *L2* $[A \rightarrow (B \rightarrow C)] \rightarrow [(A \rightarrow B) \rightarrow (A \rightarrow C)]$.
> *L3* $\neg A \rightarrow (A \rightarrow B)$.
> *L4* $(\neg A \rightarrow A) \rightarrow A$.

The semantics of L proceed as expected: A truth assignment is a function $\phi: \text{PROP} \rightarrow \{T, F\}$; ϕ has a unique extension to FOR $(\neg, \rightarrow)$, such that TA$\neg$ and TA$\rightarrow$ hold; a formula A of L is a *tautology* if $\phi(A) = T$ for every truth assignment ϕ. It is a straightforward exercise to check that each axiom of L is a tautology (use TA¬ and TA→). Moreover, if both A and $A \to B$ are tautologies, then so is B. It follows by induction on theorems that every theorem of L is a tautology. In summary, we have the following.

Theorem 1 (Soundness theorem for L): If $\vdash_L A$, then A is a tautology.

Theorem 2 (Consistency theorem for L): L is consistent.

The techniques for working in L and related systems are quite different from those used in P. Roughly speaking, the key is to use clever substitutions. We will illustrate with a proof of the important result $\vdash_L A \to A$. Note for future reference that L1 and L2 are the only axioms used in the proof.

Lemma 1: $\vdash_L A \to A$.

Proof: First, a word about strategy. If we replace C with A in L2, we obtain $[(A \to (B \to A)] \to [(A \to B) \to (A \to A)]$. In order to get $A \to A$ by two applications of MP, we need to select B so that both $A \to (B \to A)$ and $A \to B$ are axioms; the choice $B = A \to A$ works.

(1) $[A \to ((A \to A) \to A)] \to$	
$[(A \to (A \to A)) \to (A \to A)]$	AXIOM L2
(2) $A \to ((A \to A) \to A)$	AXIOM L1
(3) $(A \to (A \to A)) \to (A \to A)$	MP
(4) $A \to (A \to A)$	AXIOM L1
(5) $A \to A$	MP ∎

Theorem 3 (Deduction theorem for L): If $\Gamma \cup \{A\} \vdash_L B$, then $\Gamma \vdash_L A \to B$.

Proof: We outline the main ideas; a precise proof proceeds by induction on the number of steps in the proof of B using $\Gamma \cup \{A\}$. (For details, see Section 3.3.) Assume $\Gamma \cup \{A\} \vdash_L B$; we are required to show that $\Gamma \vdash_L A \to B$. (Henceforth, we will omit the subscript L.) There are three cases:

- B *is an axiom of* L *or* $B \in \Gamma$. In this case, it is obvious that $\Gamma \vdash B$. By L1, we have $\vdash B \to (A \to B)$, and MP gives $\Gamma \vdash A \to B$ as required.
- B *is* A. In this case, we are required to show that $\Gamma \vdash A \to A$; see Lemma 1.
- B *is obtained from* C *and* $C \to B$ *by an application of* MP. First note

that $\Gamma \cup \{A\} \vdash C$ and $\Gamma \cup \{A\} \vdash B \to C$. Moreover, the formulas C and $C \to B$ come at an earlier stage in the proof of B using $\Gamma \cup \{A\}$; hence, by induction hypothesis, $\Gamma \vdash A \to C$ and $\Gamma \vdash A \to (C \to B)$. Now, by L2, we have

$$\vdash [A \to (C \to B)] \to [(A \to C) \to (A \to B)],$$

and two applications of MP give $\Gamma \vdash A \to B$ as required. ∎

Comment From this proof, we draw the following conclusion: The deduction theorem holds for any formal system for propositional logic that has the connective $\to$, that has MP as its only rule of inference, and such that $\vdash A \to (B \to A)$ and $\{A \to B, A \to (B \to C)\} \vdash A \to C$.

Lemma 2 (Characterization of consistency): Γ is consistent if and only if there is at least one formula of L that is not a theorem of Γ; Γ is inconsistent if and only if every formula of L is a theorem of Γ.

Proof: Assume Γ is inconsistent. Then there is at least one formula A such that $\Gamma \vdash A$ and $\Gamma \vdash \neg A$. Let B be an arbitrary formula of L. By L3, $\vdash \neg A \to (A \to B)$; MP twice gives $\Gamma \vdash B$. ∎

Theorem 4 (Extension theorem for L): If A is not a theorem of Γ, then $\Gamma \cup \{\neg A\}$ is consistent.

Proof: Suppose $\Gamma \cup \{\neg A\}$ is inconsistent. Then $\Gamma \cup \{\neg A\} \vdash A$; by the deduction theorem, $\Gamma \vdash \neg A \to A$. By axiom L4, we have $\vdash (\neg A \to A) \to A$; MP then gives $\Gamma \vdash A$, a contradiction. ∎

Theorem 5 (Model existence theorem for L): If Γ is consistent, then Γ is satisfiable.

Proof: By Lindenbaum's theorem (Theorem 5 in Section 3.4), there is a set Δ of formulas such that $\Gamma \subseteq \Delta$ and Δ is both consistent and complete. It suffices to construct a truth assignment ϕ that satisfies every theorem of Δ. Define $\phi \colon \text{PROP} \to \{T, F\}$ by $\phi(p_n) = T \Leftrightarrow \Delta \vdash p_n$; ϕ has a unique extension $\phi \colon \text{FOR}(\neg, \to) \to \{T, F\}$ such that TA¬ and TA→ hold. It suffices to show that $\phi(A) = T \Leftrightarrow \Delta \vdash A$ for every formula A. The proof is by induction on the number of connectives in the formula A. For a propositional variable, the result is obvious; for the case in which A is $\neg B$, see the proof of Theorem 6 in Section 3.4. Finally, suppose A is $B \to C$. By the induction hypothesis, $\phi(B) = T \Leftrightarrow \Delta \vdash B$ and $\phi(C) = T \Leftrightarrow \Delta \vdash C$.

$\phi(B \to C) = T$	
$\Leftrightarrow \phi(B) = F$ or $\phi(C) = T$	TA→
$\Leftrightarrow B$ not a theorem of Δ or $\Delta \vdash C$	Induction hypothesis
$\Leftrightarrow \Delta \vdash \neg B$ or $\Delta \vdash C$	Δ consistent and complete
$\Leftrightarrow \Delta \vdash B \to C$	

We will justify the last equivalence. Assume $\Delta \vdash \neg B$; from axiom L3, we have

$\vdash \neg B \to (B \to C)$, and from MP we obtain $\Delta \vdash B \to C$ as required. Assume $\Delta \vdash C$; from axiom L1, we have $\vdash C \to (B \to C)$, and by MP we again obtain $\Delta \vdash B \to C$. Now suppose $\Delta \vdash B \to C$. If $\Delta \vdash \neg B$, then there is nothing to prove. If $\neg B$ is not a theorem of Δ, then by completeness, $\Delta \vdash B$. We now have $\Delta \vdash B$ and $\vdash B \to C$; MP then gives $\Delta \vdash C$ as required. ∎

Theorem 6 (Adequacy theorem for L): Every tautology is a theorem of L.

Proof: Exercise for the reader (see Theorem 7 of Section 3.4). ∎

The Formal System LF: This formal system for propositional logic, due to Frege, has language $\{\neg, \to\}$ and three axiom schemes as follows:

LF1 $A \to (B \to A)$.
LF2 $[A \to (B \to C)] \to [(A \to B) \to (A \to C)]$.
LF3 $(\neg B \to \neg A) \to (A \to B)$.

LF had advantages over L: There are fewer axioms, and axiom LF3 (which states the familiar contrapositive property) has more intuitive appeal than L3 and L4 of L. Note that LF1 and LF2 are the same as L1 and L2 of L; since the only rule of inference is MP, the deduction theorem holds in LF (see our earlier comment). We have already proved that the soundness and adequacy theorms hold for L. To see that these two results also hold for LF, it suffices to show that L and LF have the same theorems.

Theorem 7: The formal systems L and LF have the same theorems.

Proof that every theorem of LF is a theorem of L: Consider the three axioms of LF. The first two, LF1 and LF2, are also axioms of L, and hence theorems of L. The third axiom, LF3, is obviously a tautology of the language of L; it follows from the adequacy theorem for L that LF3 is a theorem of L. In summary, each axiom of LF is a theorem of L. Moreover, the two systems have the same rule of inference. It follows by induction on theorems that every theorem of LF is a theorem of L.

Proof that every theorem of L is a theorem of LF: It suffices to show that each axiom of L is a theorem of LF. This is obvious for L1 and L2, so it remains to prove that $\vdash_{LF} \neg A \to (A \to B)$ and $\vdash_{LF} (\neg A \to A) \to A$. These are proved in Lemmas 3 and 5. (*Reminder:* The deduction theorem holds for LF.) ∎

Lemma 3: $\vdash_{LF} \neg A \to (A \to B)$.

Proof: By the deduction theorem, it suffices to show that $\{\neg A\} \vdash_{LF} A \to B$.

(1) $\neg A$ HYP
(2) $\neg A \to (\neg B \to \neg A)$ AXIOM LF1

(3) $\neg B \to \neg A$ MP
(4) $(\neg B \to \neg A) \to (A \to B)$ AXIOM LF3
(5) $A \to B$ MP ∎

Lemma 4: $\vdash_{LF} (\neg A \to A) \to (\neg A \to B)$.

Proof: By the deduction theorem (twice), it suffices to show that $\{\neg A \to A, \neg A\} \vdash_{LF} B$.

(1) $\neg A \to A$ HYP
(2) $\neg A$ HYP
(3) A MP
(4) $\neg A \to (A \to B)$ LEMMA 3
(5) B MP (twice) ∎

Comment Informally, Lemma 4 states that if $\neg A$ implies A, then $\neg A$ implies any formula.

Lemma 5: $\vdash_{LF} (\neg A \to A) \to A$.

Proof: By the deduction theorem, it suffices to show that $\{\neg A \to A\} \vdash_{LF} A$. Let B be a formula that is a theorem of LF (for example, an axiom of LF).

(1) $\neg A \to A$ HYP
(2) $(\neg A \to A) \to (\neg A \to \neg B)$ LEMMA 4
(3) $\neg A \to \neg B$ MP
(4) $(\neg A \to \neg B) \to (B \to A)$ AXIOM LF3
(5) $B \to A$ MP
(6) B THM of LF
(7) A MP ∎

Note: In the proof of Theorem 7, we used the adequacy theorem of L to show that $\vdash_L (\neg B \to \neg A) \to (A \to B)$. Just for fun, let's give a formal derivation of this result; the deduction theorem is used twice.

(1) $\neg B \to \neg A$ HYP
(2) A HYP
(3) $A \to (\neg A \to B)$ AXIOM L3 and DT
(4) $\neg A \to B$ MP
(5) $\neg B \to B$ HS (1, 4)
(6) $(\neg B \to B) \to B$ AXIOM L4
(7) B MP

To justify line (3), use the axiom $\neg A \to (A \to B)$ and the deduction theorem. In line (5), we use HS (= hypothetical syllogism) as a derived rule of inference; this is easily verified using the deduction theorem. ∎

—————————*Intuitionistic Propositional Logic*—————————

The laws of logic used in mathematics were developed by Euclid and others to reason about finite mathematical objects. We shall refer to this logic as *classical* logic. The scope of mathematics was greatly expanded by Cantor's introduction of set theory and infinite sets. However, despite this addition to mathematics, the logic used to reason about mathematical objects remained the same.

In 1908 the Dutch mathematician L. E. J. Brouwer wrote a paper entitled "The Untrustworthiness of the Principles of Logic." In this paper, Brouwer challenges the assumption that classical logic is universally valid and independent of the subject matter under discussion. In particular, Brouwer felt that though classical logic applies to finite sets, it does not necessarily apply to infinite sets. Thus Brouwer and his followers developed a system of logic, called *intuitionistic logic,* that applies specifically to mathematics and that emphasizes the idea of a *constructive proof.* The following result illustrates this point.

Theorem 8: There exist irrational numbers a and b such that a^b is rational.

Proof: Either $\sqrt{2}^{\sqrt{2}}$ is rational or $\sqrt{2}^{\sqrt{2}}$ is not rational. In the first case, take $a = \sqrt{2}$ and $b = \sqrt{2}$; in the second case, take $a = \sqrt{2}^{\sqrt{2}}$ and $b = \sqrt{2}$. ∎

Although we have proved the existence of a and b, we do not know what a and b are. Brouwer's objection to this proof is that we have no construction that tells us how to find a and b. Note that the proof uses the law of the excluded middle, $\neg A \vee A$. Neither $\neg A \vee A$ nor $\neg\neg A \to A$ is a theorem of intuitionistic logic. On the other hand, every theorem of intuitionistic logic is a theorem of classical logic. In intuitionistic logic, the connectives $\neg$, $\vee$, $\wedge$, and $\to$ are independent of one another. This contrasts with classical logic, in which $(\neg A \vee B) \leftrightarrow (A \to B)$, $(A \wedge B) \leftrightarrow \neg(\neg A \vee \neg B)$, and so on. The semantics of intuitionistic logic is considerably more difficult than that of classical logic; in particular, it cannot be described with a finite number of truth values.

We will now give a formal system J for intuitionistic propositional logic with language $\{\neg, \vee, \wedge, \to\}$. The only rule of inference is MP (modus ponens), and there are ten axiom schemes as follows:

J1 $A \to (B \to A)$.

J2 $[A \to (B \to C)] \to [(A \to B) \to (A \to C)]$.

J3 $(A \wedge B) \to A$.

J4 $(A \wedge B) \to B$.

J5 $A \to (B \to (A \wedge B))$.

J6 $A \to (A \vee B)$.

J7 $A \to (B \vee A)$.

J8 $(A \to C) \to [(B \to C) \to ((A \vee B) \to C)]$.

J9 $(A \to B) \to [(A \to \neg B) \to \neg A]$.

J10 $\neg A \to (A \to B)$.

As in classical logic, $A \leftrightarrow B$ is defined as $(A \to B) \wedge (B \to A)$. This list of axioms has the property that if J10 is replaced with $\neg\neg A \to A$, we obtain classical propositional logic; in other words, we obtain a formal system for propositional logic, with language $\{\neg, \vee, \wedge, \to\}$, in which the soundness and adequacy theorems hold.

——————Exercises for Section 3.5——————

1. (*The formal system* PM *of Russell and Whitehead*) This exercise outlines a proof that P and PM have the same theorems (and that hence the soundness and adequacy theorems hold for PM). The axioms of PM are:

   PM1 $(A \vee A) \to A$.   PM3 $(A \to B) \to [(C \vee A) \to (C \vee B)]$.

   PM2 $A \to (A \vee B)$.   PM4 $(A \vee B) \to (B \vee A)$.

   (1) Show that every axiom of PM is a theorem of P and that modus ponens is a derived rule of inference of P. Conclude that every theorem of PM is a theorem of P.
   (2) Justify (in order!) the following derived rules of inference for PM:
       (a) $\{A \to B, B \to C\} \vdash_{\text{PM}} A \to C$.
       (b) $\{A \to C\} \vdash_{\text{PM}} (A \vee B) \to (C \vee B)$.
       (c) $\{A \to C, B \to C\} \vdash_{\text{PM}} (A \vee B) \to C$.
       (d) $\vdash_{\text{PM}} [A \vee (B \vee C)] \to [(A \vee B) \vee C]$.   (*Hint:* Show $\vdash_{\text{PM}} A \to [(A \vee B) \vee C]$, $\vdash_{\text{PM}} B \to [(A \vee B) \vee C]$, and $\vdash_{\text{PM}} C \to [(A \vee B) \vee C]$; then use (c).)
   (3) Show that $\vdash_{\text{PM}} \neg A \vee A$ and that each rule of inference of P is a derived rule of inference of PM; conclude that each theorem of P is a theorem of PM.
   (4) Give a direct proof that the deduction theorem holds for PM by proving (a) $\vdash_{\text{PM}} A \to (B \to A)$ and (b) $\{A \to B, A \to (B \to C)\} \vdash_{\text{PM}} A \to C$.

2. (*Variation of* PM, *due to Bernays*) Let PB be the formal system with language $\{\neg, \vee\}$, rule of inference MP, and these axiom schemes:

   PB1 $(A \vee A) \to A$.   PB3 $(A \to B) \to [(C \vee A) \to (C \vee B)]$.

   PB2 $A \to (A \vee B)$.   PB4 $(A \vee (B \vee C)) \to (B \vee (A \vee C))$.

   This problem outlines a proof that PM and PB have the same theorems (and that hence the soundness and adequacy theorems hold for PB).
   (1) Show that every theorem of PB is a theorem of PM.
   (2) Prove the following in PB:
       (a) $\vdash_{\text{PB}} (A \vee B) \to [A \vee (B \vee A)]$.
       (b) $\{A \to B, B \to C\} \vdash_{\text{PB}} A \to C$.
       (c) $\vdash_{\text{PB}} (A \vee B) \to (B \vee A)$.
   (3) Show that every theorem of PM is a theorem of PB.

3. (*Another variation of* PM) Let PV be the formal system with language $\{\neg, \vee\}$, rule of inference MP, and these axiom schemes:

   PV1 $(A \vee A) \to A$.   PV3 $(A \to B) \to [(C \vee A) \to (B \vee C)]$.

   PV2 $A \to (A \vee B)$.

   This exercise outlines a proof that PM and PV have the same theorems.
   (1) Show that every theorem of PV is a theorem of PM.
   (2) Prove the following in PV:
       (a) $\vdash_{\text{PV}} A \vee \neg A$, $\vdash_{\text{PV}} A \to \neg\neg A$, $\vdash_{\text{PV}} \neg\neg A \to A$.
       (b) $\vdash_{\text{PV}} (A \to B) \to (\neg B \to \neg A)$.
       (c) $\{A \to B, B \to C\} \vdash_{\text{PV}} A \to \neg\neg C$.

(d) $\vdash_{PV} (A \lor \neg\neg A) \to \neg\neg A$. (*Hint:* First obtain $\vdash_{PV} (A \lor \neg\neg A) \to (A \lor A)$.)

(e) $\vdash_{PV} \neg\neg A \lor \neg A$.

(f) $\vdash_{PV} A \to A$.

(g) $\vdash_{PV} (A \lor B) \to (B \lor A)$.

(h) $\{A \to B, B \to C\} \vdash_{PV} A \to C$.

(i) $\{A \to B\} \vdash_{PV} ((C \to A) \to (C \to B))$.

(j) $\vdash_{PV} [A \to (B \lor C)] \to [A \to (C \lor B)]$.

(k) $\vdash_{PV} (A \to B) \to [(C \lor A) \to (C \lor B)]$. (*Hint:* (h), (j).)

(3) Show that every theorem of PM is a theorem of PV.

4. (*The system* R *of Rosser with language* $\{\neg, \land\}$) This exercise outlines a proof that the soundness and adequacy theorems hold for the formal system R, due to Rosser. The axioms are

R1 $A \to (A \land A)$.   R3 $(A \to B) \to [\neg(B \land C) \to \neg(C \land A)]$.

R2 $(A \land B) \to A$.

Moreover, $A \to B$ means $\neg(A \land \neg B)$. Every truth assignment ϕ has a unique extension $\bar{\phi} : \text{FOR } (\neg, \land) \to \{T, F\}$, such that TA$\neg$ and TA$\land$ hold.

(1) Show that every theorem of R is a tautology and that R is consistent.

(2) Prove the following in R:

(a) $\vdash \neg(\neg A \land A)$, $\vdash \neg\neg A \to A$, $\vdash A \to \neg\neg A$.

(b) $\vdash (A \to B) \to (\neg B \to \neg A)$.

(c) $\{A \to B, B \to C, C \to D\} \vdash A \to D$.

(d) $\{\neg A \to \neg B\} \vdash B \to A$.

(e) $\{A \to B\} \vdash (C \land A) \to (B \land C)$.

(f) $\{A \to B, A \to C\} \vdash A \to (C \land B)$. (*Hint:* (c), (e).)

(g) $\vdash \neg\neg A \to (A \land A)$.

(h) $\vdash A \to A$. (*Hint:* (c).)

(i) $\{A \to B, B \to C\} \vdash A \to C$.

(j) $\vdash (A \land B) \to (B \land A)$. (*Hint:* (d).)

(k) $\vdash (A \land B) \to B$.

(l) $\vdash A \to (B \to A)$. (*Hint:* Start with R3.)

(m) $\vdash [\neg C \land (A \land B)] \to [A \land \neg\neg(B \land \neg C)]$. (*Hint:* (f).)

(n) $\{A \to (B \to C)\} \vdash (A \land B) \to C$. (*Hint:* R3, (b), (m).)

(o) $\{A \to B, A \to (B \to C)\} \vdash A \to C$. (*Hint:* (e), (n).)

(3) Prove the deduction theorem for R.

(4) Prove the following in R:

(a) $\vdash \neg A \to (A \to B)$. (*Hint:* DT, (d), (l).)

(b) $\vdash (\neg A \to A) \to A$. (*Hint:* (d).)

(c) $\vdash A \to (B \to (A \land B))$. (*Hint:* DT, (e), (l).)

(5) Prove the adequacy theorem for R.

5. Let P* have language $\{\neg, \lor\}$ and let Δ be consistent and complete. Assume that P* has the following derived rules of inference: $A/B \lor A$; $A/A \lor B$; $A \lor B, \neg A/B$. Show that Δ is satisfiable.

6. Let L* have language $\{\neg, \to\}$ and rule of inference MP. Assume the deduction theorem holds for L*. Show the following in L*:

(1) $\vdash A \to (B \to A)$.   (2) $\vdash [A \to (B \to C)] \to [(A \to B) \to (A \to C)]$.

7. (*Three variations of* L) Recall the axioms of L:

L1 $A \to (B \to A)$.

L2 $[A \to (B \to C)] \to [(A \to B) \to (A \to C)]$.

L3 $\neg A \to (A \to B)$.

L4 $(\neg A \to A) \to A$.

In this exercise, we consider three variations of L. For each variation LV, show that L and LV have the same theorems.

Axioms for variation 1 of L: LV1 $A \rightarrow (B \rightarrow A)$

LV2 $[A \rightarrow (B \rightarrow C)] \rightarrow [(A \rightarrow B) \rightarrow (A \rightarrow C)]$

LV3 $(\neg B \rightarrow \neg A) \rightarrow [(\neg B \rightarrow A) \rightarrow B]$.

Axioms for variation 2 of L: LV1 $A \rightarrow (B \rightarrow A)$

LV2 $[A \rightarrow (B \rightarrow C)] \rightarrow [(A \rightarrow B) \rightarrow (A \rightarrow C)]$

LV3 $(A \rightarrow B) \rightarrow [(A \rightarrow \neg B) \rightarrow \neg A]$

LV4 $\neg\neg A \rightarrow A$.

Axioms for variation 3 of L: LV1 $A \rightarrow (B \rightarrow A)$

LV2 $[A \rightarrow (B \rightarrow C)] \rightarrow [(A \rightarrow B) \rightarrow (A \rightarrow C)]$

LV3 $(A \rightarrow B) \rightarrow (\neg B \rightarrow \neg A)$

LV4 $\neg\neg A \rightarrow A$

LV5 $(A \rightarrow \neg A) \rightarrow \neg A$.

Note that for each variation LV, the first two axioms are L1 and L2; it follows that the deduction theorem holds for LV and that $\vdash_{LV} A \rightarrow A$. Recall that the adequacy theorem holds for L.

8. (*Hilbert's axioms for propositional logic*) Let HIL be the formal system with language $\{\neg, \rightarrow\}$, rule of inference MP, and these axioms:

HIL1 $A \rightarrow (B \rightarrow A)$.

HIL2 $[A \rightarrow (A \rightarrow B)] \rightarrow (A \rightarrow B)$.

HIL3 $[A \rightarrow (B \rightarrow C)] \rightarrow [B \rightarrow (A \rightarrow C)]$.

HIL4 $(B \rightarrow C) \rightarrow [(A \rightarrow B) \rightarrow (A \rightarrow C)]$.

HIL5 $A \rightarrow (\neg A \rightarrow B)$.

HIL6 $(A \rightarrow B) \rightarrow [(\neg A \rightarrow B) \rightarrow B]$.

(1) Show that the deduction theorem holds for HIL.

(2) Show that L and HIL have the same theorems.

9. (*Variation of L*) In this exercise, we give a variation LV of L and outline a direct proof of the adequacy theorem for LV. The axioms of LV are:

LV1 $A \rightarrow (B \rightarrow A)$.

LV2 $[A \rightarrow (B \rightarrow C)] \rightarrow [(A \rightarrow B) \rightarrow (A \rightarrow C)]$.

LV3 $\neg A \rightarrow (A \rightarrow B)$.

LV4 $(A \rightarrow \neg A) \rightarrow \neg A$.

LV5 $\neg\neg A \rightarrow A$.

Note that axioms LV1 and LV2 are the same as L1 and L2 and so the deduction theorem holds for LV. Prove the following:

(1) If $\neg A$ is not a theorem of Γ, then $\Gamma \cup \{A\}$ is consistent.

(2) If A is not a theorem of Γ, then $\Gamma \cup \{\neg A\}$ is consistent.

(3) The adequacy theorem holds for LV.

10. (*Variation of LF*) Show that LF and LFV have the same theorems, where LFV has axioms as follows:

LFV1 $A \rightarrow (B \rightarrow A)$.

LFV2 $[A \rightarrow (B \rightarrow C)] \rightarrow [(A \rightarrow B) \rightarrow (A \rightarrow C)]$.

LFV3 $(A \rightarrow B) \rightarrow (\neg B \rightarrow \neg A)$.

LFV4 $A \rightarrow \neg\neg A$.

LFV5 $\neg\neg A \rightarrow A$.

Note that LFV is obtained from LF by replacing LF3 with LFV3–LFV5.

11. (*Deduction theorem in sharper focus*) In this exercise, we consider fragments K and I of propositional logic, in which $\rightarrow$ is the only connective and MP the only rule of

inference; the main goal is to show that the deduction theorem holds for each system. It suffices to prove that $\vdash A \rightarrow (B \rightarrow A)$ and $\{A \rightarrow B, A \rightarrow (B \rightarrow C)\} \vdash A \rightarrow C$.

(1) The following steps outline a proof that the deduction theorem holds for the formal system K with these axioms:

K1 $A \rightarrow (B \rightarrow A)$.
K2 $(A \rightarrow B) \rightarrow [(B \rightarrow C) \rightarrow (A \rightarrow C)]$.
K3 $[A \rightarrow (A \rightarrow B)] \rightarrow (A \rightarrow B)$.

(a) Prove the following in K (DT not yet available!):

- $\vdash_K A \rightarrow A$.
- $\{A \rightarrow B, B \rightarrow C\} \vdash_K A \rightarrow C$.
- $\vdash_K A \rightarrow [(A \rightarrow B) \rightarrow ((A \rightarrow B) \rightarrow B)]$.
- $\vdash_K A \rightarrow [(A \rightarrow B) \rightarrow B]$.
- $\vdash_K [((B \rightarrow C) \rightarrow C) \rightarrow (A \rightarrow C)] \rightarrow [B \rightarrow (A \rightarrow C)]$.
- $\vdash_K (A \rightarrow (B \rightarrow C)) \rightarrow (B \rightarrow (A \rightarrow C))$.

(b) Show that the deduction theorem holds for K.

(2) Let I ($=$ *pure implicational calculus*) be the formal system with axioms

I1 $A \rightarrow (B \rightarrow A)$.
I2 $(A \rightarrow B) \rightarrow [(B \rightarrow C) \rightarrow (A \rightarrow C)]$.
I3 $((A \rightarrow B) \rightarrow A) \rightarrow A$ (Peirce's law).

Note that I is obtained from K by replacing axiom K3 with I3. The following steps outline a proof that every theorem of K is a theorem of I and that the deduction theorem holds for I.

(a) Prove the following in I (HS is available in I):

- $\vdash_I (A \rightarrow (A \rightarrow B)) \rightarrow [((A \rightarrow B) \rightarrow A) \rightarrow (A \rightarrow B)]$.
- $\vdash_I [A \rightarrow (A \rightarrow B)] \rightarrow (A \rightarrow B)$.

(b) Show that every theorem of K is a theorem of I.
(c) Show that the deduction theorem holds for I.
(d) Use the matrix below to show that axiom I3 of I is not a theorem of K. (*Hint:* If $A = 2$ and $B = 1$, then

$$((A \rightarrow B) \rightarrow A) \rightarrow A = 2.)$$

A \ B	0	1	2
0	0	1	2
1	0	0	0
2	0	1	0

12. (*Pure implicational calculus*) Let I ($=$ *implication*) be the formal system with language $\{\rightarrow\}$, rule of inference MP, and axioms

I1 $A \rightarrow (B \rightarrow A)$.
I2 $[A \rightarrow (B \rightarrow C)] \rightarrow [(A \rightarrow B) \rightarrow (A \rightarrow C)]$.
I3 $[(A \rightarrow B) \rightarrow A] \rightarrow A$ (Peirce's law).

This exercise outlines a proof that every tautology of the language $\{\rightarrow\}$ is a theorem of I. By axioms I1 and I2, the deduction theorem holds for I; also note that HS holds for I. A set Γ of formulas of $\{\rightarrow\}$ is *consistent* if there is at least one formula that is not a theorem of Γ, and is *complete* if for every formula A, $\Gamma \vdash A$ or $\Gamma \vdash A \rightarrow B$ for every formula B. Prove the following.

(1) If $\Delta \vdash A \rightarrow E$ and $\Delta \vdash (A \rightarrow B) \rightarrow E$, then $\Delta \vdash E$. (*Hint:* First obtain $\Delta \vdash (E \rightarrow B) \rightarrow E$ (use the deduction theorem).)

(2) (*Extension*) If $A \rightarrow E$ is not a theorem of Γ, then E is not a theorem of $\Gamma \cup \{A\}$ (so $\Gamma \cup \{A\}$ is consistent).

(3) (*Lindenbaum*) If Γ is consistent, then there is a set Δ of formulas with $\Gamma \subseteq \Delta$ such that Δ is consistent and complete. (*Hint:* Let $A_0, A_1, \ldots$ be all formulas of $\{\rightarrow\}$ and let E be a formula that is not a theorem of Γ. Take $\Delta = \bigcup_{n \in \mathbb{N}} \Gamma_n$, where $\Gamma_0 \subseteq \Gamma_1 \subseteq \ldots$ are constructed as follows:

$$\Gamma_0 = \Gamma; \qquad \Gamma_{n+1} = \begin{cases} \Gamma_n \cup \{A_n\} & A_n \rightarrow E \text{ is not a theorem of } \Gamma_n; \\ \Gamma_n & A_n \rightarrow E \text{ is a theorem of } \Gamma_n. \end{cases}$$

To prove the consistency of Δ, show that E is not a theorem of Δ; use (1) to prove the completeness of Δ.)

(4) (*Model existence*) Let Γ be consistent. Then there is a formula B of $\{\rightarrow\}$ and a truth assignment ϕ such that (a) $\phi(B) = F$; (b) if $A \in \Gamma$, then $\phi(A) = T$. (*Hint:* Let $\Gamma \subseteq \Delta$ with Δ consistent and complete, and define $\phi: \text{PROP} \rightarrow \{T, F\}$ by $\phi(p_n) = T \Leftrightarrow \Delta \vdash p_n$.)

(5) If A is not a theorem of I, then A is not a theorem of $\{A \rightarrow A_n : n \in \mathbb{N}\}$, where $\{A_n : n \in \mathbb{N}\}$ is the set of all formulas of $\{\rightarrow\}$. (*Hint:* First prove that for all $n \geq 0$, A is not a theorem of $\{A \rightarrow A_k : 0 \leq k \leq n\}$.)

(6) (*Adequacy*) If A is a tautology of $\{\rightarrow\}$, then A is a theorem of I. (*Hint:* Suppose A is not a theorem of I; use (5) and (4).)

13. (*The system* LUK *of Lukasiewicz* ("*woo-ka-SHEH-vitch*")) This exercise outlines a proof that L and LUK have the same theorems. The language for LUK is $\{\neg, \rightarrow\}$, the rule of inference is MP, and the axioms are

LUK1 $(A \rightarrow B) \rightarrow [(B \rightarrow C) \rightarrow (A \rightarrow C)]$.

LUK2 $(\neg A \rightarrow A) \rightarrow A$.

LUK3 $A \rightarrow (\neg A \rightarrow B)$.

(1) Show that every theorem of LUK is a theorem of L.

(2) Prove the following in LUK:

  (a) $\{A \rightarrow B, B \rightarrow C\} \vdash_{\text{LUK}} A \rightarrow C$.

  (b) $\vdash_{\text{LUK}} [(\neg A \rightarrow B) \rightarrow (\neg B \rightarrow B)] \rightarrow [A \rightarrow (\neg B \rightarrow B)]$.

  (c) $\vdash_{\text{LUK}} (\neg B \rightarrow \neg A) \rightarrow [A \rightarrow (\neg B \rightarrow B)]$.

  (d) $\vdash_{\text{LUK}} (\neg B \rightarrow \neg A) \rightarrow [((\neg B \rightarrow B) \rightarrow B) \rightarrow (A \rightarrow B)]$.

  (e) $\vdash_{\text{LUK}} B \rightarrow [((\neg B \rightarrow B) \rightarrow B) \rightarrow (A \rightarrow B)]$.

  (f) $\{B\} \vdash_{\text{LUK}} A \rightarrow B$.

  (g) $\vdash_{\text{LUK}} A \rightarrow [(\neg B \rightarrow B) \rightarrow B]$.

  (h) $\vdash_{\text{LUK}} [((\neg B \rightarrow B) \rightarrow B) \rightarrow (A \rightarrow B)] \rightarrow [\neg(A \rightarrow B) \rightarrow (A \rightarrow B)]$.

  (i) $\vdash_{\text{LUK}} [((\neg B \rightarrow B) \rightarrow B) \rightarrow (A \rightarrow B)] \rightarrow (A \rightarrow B)$.

  (j) $\vdash_{\text{LUK}} B \rightarrow (A \rightarrow B)$. (*Hint:* (e), (i).)

  (k) $\vdash_{\text{LUK}} (\neg B \rightarrow \neg A) \rightarrow (A \rightarrow B)$. (*Hint:* (d), (i).)

  (l) $\vdash_{\text{LUK}} \neg A \rightarrow (A \rightarrow B)$.

  (m) $\vdash_{\text{LUK}} ((A \rightarrow B) \rightarrow A) \rightarrow (\neg A \rightarrow A)$.

  (n) $\vdash_{\text{LUK}} ((A \rightarrow B) \rightarrow A) \rightarrow A$.

(3) Show that the deduction theorem holds for LUK. (*Hint:* Exercise 11.)

(4) Show that every theorem of L is a theorem of LUK.

14. (*Intuitionistic propositional logic for* $\{\neg, \rightarrow\}$) We describe a formal system KA, due to Kanger, that gives the theorems of intuitionistic propositional logic for $\{\neg, \rightarrow\}$. The axioms of KA are:

KA1 $A \rightarrow (B \rightarrow A)$.

KA2 $[A \rightarrow (B \rightarrow C)] \rightarrow [(A \rightarrow B) \rightarrow (A \rightarrow C)]$.

KA3 $(A \rightarrow \neg B) \rightarrow (B \rightarrow \neg A)$.

KA4 $\neg A \rightarrow (\neg A \rightarrow A)$.

Now introduce three variations KAV of KA as follows:

Axioms for variation 1 of KA: **KAV1** $A \to (B \to A)$.

                              KAV2 $[A \to (B \to C)] \to [(A \to B) \to (A \to C)]$.

                              KAV3 $\neg A \to (A \to B)$.

                              KAV4 $(A \to \neg A) \to \neg A$.

Axioms for variation 2 of KA: **KAV1** $A \to (B \to A)$.

                              KAV2 $[A \to (B \to C)] \to [(A \to B) \to (A \to C)]$.

                              KAV3 $(A \to B) \to [(A \to \neg B) \to \neg A]$.

                              KAV4 $\neg\neg A \to (\neg A \to A)$.

Axioms for variation 3 of KA: **KAV1** $A \to (B \to A)$.

                              KAV2 $[A \to (B \to C)] \to [(A \to B) \to (A \to C)]$.

                              KAV3 $(A \to \neg B) \to (B \to \neg A)$.

                              KAV4 $\neg(A \to A) \to A$.

(1) Show that the four systems have the same theorems. (*Hints:* The DT is available in all systems; see Lemmas 3–5.)

(2) Show that every theorem of KA is a theorem of L. (*Hint:* Use the adequacy theorem for L.)

(3) Show that we obtain classical logic by adding $\neg\neg A \to A$ (i.e., the resulting formal system has the same theorems as L). (*Hint:* Work with variation 2. It suffices to show that $\{\neg A \to A\} \vdash A$.)

(4) Show that we obtain classical logic by adding $((A \to \neg A) \to A) \to A$, a special case of Peirce's law. (*Hint:* Work with variation 1. It suffices to show that $\{\neg A \to A\} \vdash A$.)

15. (*Formal system for propositional logic with language* $\{\neg, \vee, \wedge, \to\}$) Let H denote the formal system for propositional logic with language $\{\neg, \vee, \wedge, \to\}$ and axioms as follows:

H1 $A \to (B \to A)$.

H2 $[A \to (B \to C)] \to [(A \to B) \to (A \to C)]$.

H3 $(A \wedge B) \to A$.

H4 $(A \wedge B) \to B$.

H5 $(A \to (B \to (A \wedge B)))$.

H6 $A \to (A \vee B)$.

H7 $A \to (B \vee A)$.

H8 $(A \to C) \to [(B \to C) \to ((A \vee B) \to C)]$.

H9 $(A \to \neg B) \to (B \to \neg A)$.

H10 $\neg(A \to A) \to B$.

H11 $A \vee \neg A$.

Note that axioms H1 and H2 are L1 and L2. If we omit H11, we obtain intuitionistic propositional logic; if we omit H10 and H11, we obtain minimal propositional logic.

(1) Show the following in H. (*Hint:* See Exercise 14, on KA).

     (a) $\vdash \neg A \to (A \to B)$.   (b) $\vdash (\neg A \to A) \to A$.

(2) Show that the adequacy theorem holds for H.

(3) Let H* be the formal system obtained by omitting H11 from H. Show that H* and J have the same theorems.

16. (*Law of excluded middle not a theorem of intuitionistic logic*) This exercise outlines a proof that $\neg A \vee A$ and $\neg\neg A \to A$ are not theorems of J. Introduce an $n + 1$-valued logic as follows: A truth assignment is a function $\phi: \text{PROP} \to \{0, 1, \ldots, n\}$; such an

assignment extends to all formulas FOR $(\neg, \vee, \wedge, \rightarrow)$ of J according to these rules:

$$\phi(A \vee B) = \min\{\phi(A), \phi(B)\}. \qquad \phi(A \wedge B) = \max\{\phi(A), \phi(B)\}.$$

$$\phi(\neg A) = \begin{cases} 0 & \phi(A) = n; \\ n & \phi(A) < n. \end{cases} \qquad \phi(A \rightarrow B) = \begin{cases} 0 & \phi(A) \geq \phi(B); \\ \phi(B) & \phi(A) < \phi(B). \end{cases}$$

Show that $\phi(A) = 0$ for every theorem of J and that $\phi(\neg A \vee A) = 0$, $\phi(\neg\neg A \rightarrow A) = 0$ need not hold for $n \geq 2$.

17. (*No finite semantics for intuitionistic logic*) This problem outlines a proof that the semantics of intuitionistic logic cannot be captured with a finite number of truth values.

   (1) Recall that $A \leftrightarrow B$ means $(A \rightarrow B) \wedge (B \rightarrow A)$. For $n \geq 2$, let D_n denote the formula

$$(p_1 \leftrightarrow p_2) \vee (p_1 \leftrightarrow p_3) \vee \ldots \vee (p_1 \leftrightarrow p_n)$$
$$\vee \, (p_2 \leftrightarrow p_3) \vee \ldots \vee (p_2 \leftrightarrow p_n)$$
$$\vdots$$
$$\vee \, (p_{n-1} \leftrightarrow p_n).$$

   For example, D_3 is the formula $(p_1 \leftrightarrow p_2) \vee (p_1 \leftrightarrow p_3) \vee (p_2 \leftrightarrow p_3)$. Show that D_n is not a theorem of J. (*Hint:* Use the $n + 1$-valued logic in Exercise 16.) Also show that for $n \geq 3$, D_n is a tautology of classical logic.

   (2) Consider the language with connectives $\neg, \vee, \wedge, \rightarrow$. A *matrix* for this language is a 6-tuple, $M = \langle S, S_0, H_\neg, H_\vee, H_\wedge, H_\rightarrow \rangle$, where S is a nonempty set whose elements are called *truth values*, S_0 is a subset of S whose elements are called *designated values*, and $H_\vee, H_\wedge, H_\rightarrow$, and $H_\neg$ are truth functions for $\vee, \wedge, \rightarrow$, and $\neg$. A *truth assignment* for M is a function ϕ: PROP $\rightarrow S$. Such an assignment extends to FOR $(\neg, \vee, \wedge, \rightarrow)$ in the usual way; for example, $\phi(A \vee B) = H_\vee(\phi(A), \phi(B))$, $\phi(A \wedge B) = H_\wedge(\phi(A), \phi(B))$, and so on. Show that there is *no* matrix M with S finite such that for every formula A,

$$\vdash_J A \Leftrightarrow \phi(A) \in S_0 \quad \text{for every truth assignment } \phi \text{ for } M.$$

   (*Hint:* Assume that M exists and that S, the set of all truth values, has n elements. The formula D_{n+1} is not a theorem of J, so there is a truth assignment ϕ for M such that $\phi(D_{n+1}) \notin S_0$. Choose j, k with $1 \leq j < k \leq n + 1$ such that $\phi(p_j) = \phi(p_k)$. Let E_{n+1} be obtained from D_{n+1} by replacing $(p_j \leftrightarrow p_k)$ with $(p_k \leftrightarrow p_k)$. Argue that $\phi(D_{n+1}) = \phi(E_{n+1})$ and that $\phi(E_{n+1}) \in S_0$.)

18. (*Failure of extension theorem*) This problem outlines a proof that the extension theorem does not hold for J.

   (1) Prove the following in J; the deduction theorem is available.
      (a) $\vdash (A \rightarrow B) \rightarrow (\neg B \rightarrow \neg A)$;   (b) $\vdash \neg(A \vee B) \rightarrow (\neg A \wedge \neg B)$.

   (2) Give an example of a formula A that is not a theorem of J and such that $\{\neg A\}$ is inconsistent.

   (3) Prove the following version of the extension theorem for J: If $\neg A$ is not a theorem of Γ, then $\Gamma \cup \{A\}$ is consistent.

19. (*Classical versus intuitionistic logic*) Let H be the formal system obtained from J by replacing axiom J10 with $\neg\neg A \rightarrow A$.

   (1) Show that every theorem of J is a theorem of H.

   (2) Show that H is a suitable formal system for propositional logic with language $\{\neg, \vee, \wedge, \rightarrow\}$ by showing that the soundness and adequacy theorems hold for H. One significant change is in the proof of the model existence theorem. For this, we must consider the cases in which A is $\neg B$, $B \vee C$, $B \wedge C$, and $B \rightarrow C$.

   (3) Show that for any formula A, $\vdash_H A \Leftrightarrow \vdash_J \neg\neg A$. (*Hint:* Prove the following in J.)
      (a) $\vdash_J A \rightarrow \neg\neg A$.
      (b) $\vdash_J (A \rightarrow B) \rightarrow (\neg B \rightarrow \neg A)$.
      (c) $\vdash_J \neg(\neg\neg A \rightarrow A) \rightarrow \neg A$, $\vdash_J \neg(\neg\neg A \rightarrow A) \rightarrow \neg\neg A$.
      (d) $\vdash_J \neg\neg(\neg\neg A \rightarrow A)$.
      (e) $\vdash_J (A \rightarrow \neg B) \rightarrow (B \rightarrow \neg A)$.

(f) $\vdash_J (A \to \neg\neg B) \to (\neg\neg A \to \neg\neg B)$. (*Hint:* (b), (e).)

(g) $\{\neg\neg A, \neg\neg (A \to B)\} \vdash_J \neg\neg B$. (*Hint:* First show $\vdash_J \neg\neg A \to ((A \to B) \to \neg\neg B)$.).

(4) Show that for any formula A, $\vdash_H \neg A \Leftrightarrow \vdash_J \neg A$.

(5) Let A be any formula whose only connectives are $\neg$ and $\wedge$. Show that $\vdash_H A \Leftrightarrow \vdash_J A$.

20. (*Theorems of intuitionistic logic*) Show that the following are theorems of J. (*Hint:* See part (3) of the previous problem and Exercise 14, on KA.)

(1) $\vdash_J \neg\neg(\neg A \vee A)$.

(2) $\vdash_J \neg\neg\neg A \to \neg A$.

(3) $\vdash_J (\neg A \vee A) \to (\neg\neg A \to A)$.

(4) $\vdash_J \neg(A \wedge \neg A)$.

(5) $\vdash_J (\neg A \vee B) \to (A \to B)$.

(6) $\vdash_J (\neg A \vee \neg B) \to \neg(A \wedge B)$.

(7) $\vdash_J (A \vee B) \to \neg(\neg A \wedge \neg B)$.

(8) $\vdash_J (\neg A \wedge \neg B) \to \neg(A \vee B)$.

(9) $\vdash_J (A \vee B) \to (\neg A \to B)$.

(10) $\vdash_J (A \to B) \to \neg(A \wedge \neg B)$.

21. (*Nontheorems of intuitionistic logic*) Show that the following are not theorems of J. (*Hint:* Use the fact that $\neg A \vee A$ and $\neg\neg A \to A$ are not theorems of J, or use the method of Exercise 16.)

(1) $(\neg B \to \neg A) \to (A \to B)$.

(2) $(\neg A \to B) \to (\neg B \to A)$.

(3) $(A \to B) \to (\neg A \vee B)$.

(4) $\neg(A \wedge \neg B) \to (A \to B)$.

(5) $(\neg A \to B) \to [(\neg A \to \neg B) \to A]$.

(6) $\neg(\neg A \wedge \neg B) \to (A \vee B)$.

(7) $\neg(\neg A \vee \neg B) \to (A \wedge B)$.

22. (*Heyting's formal system for intuitionistic propositional logic*) HEY is the formal system with language $\{\neg, \vee, \wedge, \to\}$ and these axioms:

HEY1 $A \to (B \to A)$.

HEY2 $[(A \to B) \wedge (B \to C)] \to (A \to C)$.

HEY3 $A \to (A \wedge A)$.

HEY4 $(A \wedge B) \to (B \wedge A)$.

HEY5 $(A \to B) \to [(A \wedge C) \to (B \wedge C)]$.

HEY6 $[A \wedge (A \to B)] \to B$.

HEY7 $A \to (A \vee B)$.

HEY8 $(A \vee B) \to (B \vee A)$.

HEY9 $[(A \to C) \wedge (B \to C)] \to [(A \vee B) \to C]$.

HEY10 $[(A \to B) \wedge (A \to \neg B)] \to \neg A$.

HEY11 $\neg A \to (A \to B)$.

Prove that J and HEY have the same theorems. Begin by proving the following in HEY. (*Note:* DT holds for HEY by axiom HEY1 and (6).)

(1) $\{A, B\} \vdash_{HEY} A \wedge B$.

(2) $\{A \to B, B \to C\} \vdash_{HEY} A \to C$.

(3) $\{A \to (B \to C)\} \vdash_{HEY} (A \wedge B) \to [(B \to C) \wedge B]$.

(4) $\{A \to (B \to C)\} \vdash_{HEY} (A \wedge B) \to C$.

(5) $\{A \to B\} \vdash_{HEY} A \to (B \wedge A)$.

(6) $\{A \to B, A \to (B \to C)\} \vdash_{HEY} A \to C$.

(7) $\vdash_{HEY} (B \wedge A) \to [(A \to B) \wedge A]$.

(8) $\vdash_{HEY} (B \wedge A) \to B$.

3.6

Gentzen-Style Proof Systems for Propositional Logic

The fundamental computational problem of propositional logic is $\{A_1, \ldots, A_n\} \models B$. By the completeness theorem for P, $\{A_1, \ldots, A_n\} \models B$ if and only if $\{A_1, \ldots, A_n\} \vdash B$; thus we can verify that B is a tautological consequence of $\{A_1, \ldots, A_n\}$ by writing a proof of B in the formal system P using $\{A_1, \ldots, A_n\}$ as additional axioms. But if B is not a tautological consequence of $\{A_1, \ldots, A_n\}$, we cannot expect to discover this fact by looking for a formal proof. In this respect, the method of truth tables has an advantage: It is a decision method for $\{A_1, \ldots, A_n\} \models B$. But the method of truth tables has at least two disadvantages: exponential growth and lack of elegance by comparison with the formal system approach. Thus it is desirable to find a procedure that combines the best aspects of truth tables and formal systems. In other words, we seek a computational procedure that applies to $\{A_1, \ldots, A_n\} \models B$ and has these characteristics: It is similar in spirit to writing a proof in a formal system; it gives a YES or NO answer in a finite number of steps.

In this section we describe *Gentzen-style proof systems*, or *sequent calculi*; these are proof procedures that have the two characteristics stated above. Gentzen-style proof systems reverse many of the properties of Hilbert-style proof systems: They are somewhat awkward to describe; they make it easy to write formal proofs; they have few axioms and many rules. Perhaps the major advantage of Gentzen-style proof systems is the fact that they provide a decision method for $\{A_1, \ldots, A_n\} \models B$. For this reason, Gentzen-style proof systems are of interest to theoretical computer scientists who do research in automated theorem proving.

—Semantic Trees and the Sequent Calculus G_L —

We will develop a proof system that we denote by G_L; G is for Gentzen and L denotes the fact that the rules are based on the left-hand rules of the original Gentzen sequent calculus. We approach G_L in two steps: First we proceed informally, describing the notion of a *semantic tree*, or a *truth tree*. One should perhaps think of a semantic tree as a variation of a truth table that is written in the spirit of a formal proof. These ideas were introduced and refined by a number of logicians including Beth, Hintikka, Schütte, and Smullyan. Semantic trees give us a practical procedure for deciding $\{A_1, \ldots, A_n\} \models B$. We then introduce the sequent calculus G_L; here we give a precise definition of a formal proof and prove soundness and adequacy theorems. Informally speaking, proofs in G_L are semantic trees turned upside down.

The method of semantic trees is based on the following result from Section 2.2.

Lemma 1: $\{A_1, \ldots, A_n\} \models B$ if and only if $\{A_1, \ldots, A_n, \neg B\}$ is unsatisfiable. In particular, B is a tautology if and only if $\neg B$ is unsatisfiable.

Lemma 1 allows us to replace the problem $\{A_1, \ldots, A_n\} \models B$ with the problem *Is* $\{A_1, \ldots, A_n, \neg B\}$ *unsatisfiable?* For this reason, the semantic tree method is often referred to as a *refutation procedure*.

Recall that a *literal* is a formula of the form p or $\neg p$, where p is a propositional variable. For a finite set Γ of formulas that consists solely of literals, we can easily tell by inspection whether Γ is unsatisfiable. For example, $\Gamma = \{p, \neg q, r, q\}$ is unsatisfiable, whereas $\Delta = \{p, \neg q, r, \neg s\}$ is satisfiable. We record the precise result as a lemma.

Lemma 2: Suppose every formula in Γ is a literal. Then Γ is unsatisfiable if and only if there is a propositional variable p such that $\{p, \neg p\} \subseteq \Gamma$.

Now suppose we are given the problem $\{A_1, \ldots, A_n\} \models B$. Form the set $\Gamma = \{A_1, \ldots, A_n, \neg B\}$; by Lemma 1, it suffices to decide whether Γ is unsatisfiable. The basic idea of semantic trees is to break down the formulas in Γ into subformulas, eventually obtaining $\Delta_1, \ldots, \Delta_k$, where for $1 \leq j \leq k$, Δ_j is a finite set of formulas and the following hold: (1) The only formulas in Δ_j are literals (or at least we can tell by inspection whether Δ_j is satisfiable or unsatisfiable); (2) if $\Delta_1, \ldots, \Delta_k$ are all unsatisfiable, then Γ is unsatisfiable; (3) if one of $\Delta_1, \ldots, \Delta_k$ is satisfiable, then Γ is satisfiable.

Now let us turn to the rules for decomposing the formulas in $\{A_1, \ldots, A_n, \neg B\}$. First of all, to keep the list of rules as short as possible, we choose an adequate set of connectives $\{\neg, \vee\}$ and define the remaining connectives in terms of these. Now for both (2) and (3) to hold, the rules used in going from Γ to $\Delta_1, \ldots, \Delta_k$ must preserve unsatisfiability *in both directions*. We will now state two lemmas that provide the key to formulating the required rules.

Lemma 3: All formulas of the language $\{\neg, \vee\}$ are of the form $p, \neg p, \neg\neg A, A \vee B,$ or $\neg(A \vee B)$.

Proof: Let A be a formula that uses only $\neg$ and $\vee$, and assume A is not a propositional variable. Then A is of the form $\neg B$ or $B \vee C$. If A is of the form $B \vee C$, there is nothing to prove. Suppose A is $\neg B$. The formula B is a propositional variable p or is of the form $\neg C$ or $C \vee D$. It follows that A is of the form $\neg p, \neg\neg C,$ or $\neg(C \vee D)$, as required. ∎

Notation Throughout this section, Γ and Δ will denote *finite* sets of formulas of propositional logic. If Γ is a set of formulas and $B_1, \ldots, B_n$ are formulas, we will often write $\Gamma, B_1, \ldots, B_n$ instead of the proper $\Gamma \cup \{B_1, \ldots, B_n\}$. In addition, we will often omit braces $\{,\}$ when working with finite sets of formulas.

Lemma 4: The following hold:

(1) $\Gamma, \neg\neg A$ is unsatisfiable if and only if Γ, A is unsatisfiable.

(2) $\Gamma, A \vee B$ is unsatisfiable if and only if Γ, A is unsatisfiable and Γ, B is unsatisfiable.

(3) $\Gamma, \neg(A \vee B)$ is unsatisfiable if and only if $\Gamma, \neg A, \neg B$ is unsatisfiable.

We leave the proof of Lemma 4 as an exercise. However, note that part

(2) can be stated as follows: $\Gamma, A \vee B$ is satisfiable if and only if Γ, A is satisfiable or Γ, B is satisfiable.

Decomposition rules for G_L:

$$\neg\neg\text{-RULE} \quad \frac{\Gamma, \neg\neg A}{\Gamma, A} \qquad\qquad \vee\text{-RULE} \quad \frac{\Gamma, A \vee B}{\Gamma, A \quad \Gamma, B}$$

$$\neg\vee\text{-RULE} \quad \frac{\Gamma, \neg(A \vee B)}{\Gamma, \neg A, \neg B}$$

The $\neg\neg$-rule simplifies $\Gamma, \neg\neg A$ to Γ, A; the $\neg\vee$-rule simplifies $\Gamma, \neg(A \vee B)$ to $\Gamma, \neg A, \neg B$; the $\vee$-rule allows us to replace $\Gamma, A \vee B$ by the two simpler sets of formulas Γ, A and Γ, B. Note that for each rule, unsatisfiability is preserved *in both directions*.

Example 1:

We verify that $\{p \to q, p \to r\} \models p \to (q \wedge r)$. By Lemma 1, it suffices to show that $\{p \to q, p \to r, \neg[p \to (q \wedge r)]\}$ is unsatisfiable. First, replace occurrences of $\wedge$ and $\to$ to obtain $\{\neg p \vee q, \neg p \vee r, \neg[\neg p \vee \neg(\neg q \vee \neg r)]\}$. We now use the decomposition rules to construct the following semantic tree:

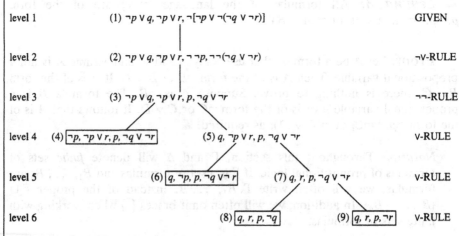

level 1	(1) $\neg p \vee q, \neg p \vee r, \neg[\neg p \vee \neg(\neg q \vee \neg r)]$	GIVEN
level 2	(2) $\neg p \vee q, \neg p \vee r, \neg\neg p, \neg\neg(\neg q \vee \neg r)$	$\neg\vee$-RULE
level 3	(3) $\neg p \vee q, \neg p \vee r, p, \neg q \vee \neg r$	$\neg\neg$-RULE
level 4	(4) $\boxed{\neg p, \neg p \vee r, p, \neg q \vee \neg r}$ (5) $q, \neg p \vee r, p, \neg q \vee \neg r$	$\vee$-RULE
level 5	(6) $\boxed{q, \neg p, p, \neg q \vee \neg r}$ (7) $q, r, p, \neg q \vee \neg r$	$\vee$-RULE
level 6	(8) $\boxed{q, r, p, \neg q}$ (9) $\boxed{q, r, p, \neg r}$	$\vee$-RULE

In the proof above, we have drawn a box around each set of formulas that contains both p and $\neg p$ for some propositional variable p. Each such set of formulas is obviously unsatisfiable. By repeated applications of Lemma 4, we see that $\{p \to q, p \to r, \neg[p \to (q \wedge r)]\}$ is unsatisfiable. First of all, sets 8 and 9 in level 6 are unsatisfiable; hence set 7 in level 5 is unsatisfiable. Now set 6 in level 5 is also unsatisfiable, so set 5 in level 4 is also unsatisfiable. Continuing in this way, we eventually see that set 1 is unsatisfiable, as required.   □

What happens if we apply these ideas to a situation in which B is not a tautological consequence of $\{A_1, \ldots, A_n\}$?

Example 2:

We use the semantic tree method to decide whether

$$\{p \rightarrow (q \vee r), q \rightarrow s, p\} \models s. \tag{*}$$

Form the set $\{p \rightarrow (q \vee r), q \rightarrow s, p, \neg s\}$, and then eliminate occurrences of $\rightarrow$ to obtain $\{\neg p \vee (q \vee r), \neg q \vee s, p, \neg s\}$. Now apply the decomposition rules to construct the following semantic tree:

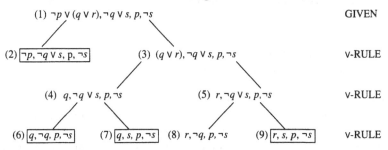

	GIVEN
(1) $\neg p \vee (q \vee r), \neg q \vee s, p, \neg s$	
(2) $\boxed{\neg p, \neg q \vee s, p, \neg s}$ · · · (3) $(q \vee r), \neg q \vee s, p, \neg s$	$\vee$-RULE
(4) $q, \neg q \vee s, p, \neg s$ · · · (5) $r, \neg q \vee s, p, \neg s$	$\vee$-RULE
(6) $\boxed{q, \neg q, p, \neg s}$ (7) $\boxed{q, s, p, \neg s}$ (8) $r, \neg q, p, \neg s$ (9) $\boxed{r, s, p, \neg s}$	$\vee$-RULE

We use Lemma 4 to show that $\{\neg p \vee (q \vee r), \neg q \vee s, p, \neg s\}$ *is* satisfiable, so (*) does *not* hold. Set 8 is satisfiable; hence set 5 is satisfiable; since set 5 is satisfiable, set 3 is satisfiable; finally, since set 3 is satisfiable, set 1 is satisfiable. Moreover, set 8—that is, $\{r, \neg q, p, \neg s\}$—tells us how to construct a truth assignment ϕ that satisfies $\{p \rightarrow (q \vee r), q \rightarrow s, p, \neg s\}$: $\phi(p) = T$, $\phi(q) = F$, $\phi(r) = T$, and $\phi(s) = F$. □

We now describe the sequent calculus G_L. Strictly speaking, G_L is not a formal system: We work with finite sets of formulas rather than formulas; we also "prove" unsatisfiability, so G_L is a refutation procedure. Nevertheless, G_L is like a formal system in spirit, and we will use the terminology of formal systems in describing G_L.

The Sequent Calculus G_L: An axiom of G_L is any finite set Γ of formulas of $\{\neg, \vee\}$ such that $\{p, \neg p\} \subseteq \Gamma$ for some propositional variable p. The rules of G_L are the decomposition rules turned upside down:

$$\neg\neg\text{-RULE}\ \frac{\Gamma, A}{\Gamma, \neg\neg A} \qquad \neg\vee\text{-RULE}\ \frac{\Gamma, \neg A, \neg B}{\Gamma, \neg(A \vee B)}$$

$$\vee\text{-RULE}\ \frac{\Gamma, A \quad \Gamma, B}{\Gamma, A \vee B}$$

Definition 1: Let Γ be a finite set of formulas of $\{\neg, \vee\}$. A *proof of* Γ in G_L is a finite sequence $\Gamma_1, \ldots, \Gamma_n$ of finite sets of formulas of $\{\neg, \vee\}$ with $\Gamma_n = \Gamma$ such that for $1 \leq k \leq n$, one of the following holds: (1) Γ_k is an axiom of G_L. (2) $k > 1$ and Γ_k is the conclusion of a rule of inference of G_L whose hypotheses are among $\Gamma_1, \ldots, \Gamma_{k-1}$. We write $G_L \vdash \Gamma$ to denote the fact that there is a proof of Γ in G_L. For an example of a proof in G_L, read Example 1, starting at the bottom, at level 6.

Theorem 1 (Soundness theorem for G_L): If $G_L \vdash \Gamma$, then Γ is unsatisfiable.

Proof: Assume $G_L \vdash \Gamma$. Now each axiom of G_L is unsatisfiable, and each rule of inference of G_L preserves unsatisfiability; it follows by induction on theorems that Γ is unsatisfiable. The details are left to the reader. ∎

Corollary 1: If $G_L \vdash \{A_1, \ldots, A_n, \neg B\}$, then $\{A_1, \ldots, A_n\} \models B$; in particular, if $G_L \vdash \neg B$, then B is a tautology.

Now let us turn to the proof of the converse of Theorem 1. The following idea is used: the *degree of* Γ, denoted $\deg \Gamma$, is the natural number calculated as follows: $+1$ for each occurrence of $\neg$ in a formula of Γ; $+2$ for each occurrence of $\vee$ in a formula of Γ. For example, $\deg\{p, q, r\} = 0$ and $\deg\{\neg p, p \vee (\neg q \vee r)\} = 6$. Note that if Γ is unsatisfiable, then $\deg \Gamma \geq 1$.

Theorem 2 (Adequacy theorem for G_L): If Γ is unsatisfiable, then $G_L \vdash \Gamma$.

Proof: First consider the special case in which each formula in Γ is a literal (p or $\neg p$). Since Γ is unsatisfiable, there is a propositional variable p such that $\{p, \neg p\} \subseteq \Gamma$. Thus Γ is an axiom of G_L; hence $G_L \vdash \Gamma$ as required. Now let $S(n)$ say: If Γ is unsatisfiable and $\deg \Gamma = n$, then $G_L \vdash \Gamma$. We use complete induction to show that $S(n)$ holds for all $n \geq 1$.

Beginning step Let Γ be unsatisfiable with $\deg \Gamma = 1$. Then every formula in Γ is a literal; hence the special case applies and $G_L \vdash \Gamma$.

Induction step Let $n \geq 1$ and assume $S(1), \ldots, S(n)$ hold. Let Γ be unsatisfiable with $\deg \Gamma = n + 1$. If every formula in Γ is a literal, the special case applies and we are finished. Thus we may assume that there is a formula $A \in \Gamma$ such that A is not a literal. We will now consider three cases:

- A is $\neg\neg B$. Write Γ in the form $\Gamma_0, \neg\neg B$ and let $\Delta = \Gamma_0, B$. Since Γ is unsatisfiable, Δ is unsatisfiable (Lemma 4(1)). Now $\deg \Delta = n - 1$; hence the induction hypothesis applies and $G_L \vdash \Gamma_0, B$. By the $\neg\neg$-rule, $G_L \vdash \Gamma_0, \neg\neg B$ as required.

- A is $B \vee C$. Write Γ in the form $\Gamma_0, B \vee C$ and let $\Delta_B = \Gamma_0, B$ and $\Delta_C = \Gamma_0, C$. Since Γ is unsatisfiable, both Δ_B and Δ_C are unsatisfiable (Lemma 4(2)). Now $\deg \Delta_B \leq n - 1$ and $\deg \Delta_C \leq n - 1$; hence the induction hypothesis applies and $G_L \vdash \Gamma_0, B$ and $G_L \vdash \Gamma_0, C$. By the $\vee$-rule, $G_L \vdash \Gamma_0, B \vee C$ as required.

- A is $\neg(B \vee C)$. Write Γ in the form $\Gamma_0, \neg(B \vee C)$ and let $\Delta = \Gamma_0, \neg B, \neg C$. Since Γ is unsatisfiable, Δ is unsatisfiable (Lemma 4(3)). Now $\deg \Delta = n$; hence the induction hypothesis applies and $G_L \vdash \Gamma_0, \neg B, \neg C$. By the $\neg\vee$-rule, $G_L \vdash \Gamma_0, \neg(B \vee C)$. ∎

The Sequent Calculus G_R

In this section, we develop another proof system for propositional logic, which we denote by G_R. Again, G is for Gentzen; R refers to the fact that the right-hand rules of the original Gentzen sequent calculus are used. In a certain sense, G_R is the dual of G_L, and we will let the discussion of G_L motivate G_R; results will be stated without proofs. Recall that the sequent calculus G_L is a *refutation* procedure: We prove Γ *unsatisfiable*. But for G_R, the emphasis is on proving that Γ is a *tautology*. Here is the required definition.

Definition 2: A set Γ of formulas is a *tautology* if for every truth assignment ϕ, there exists at least one formula $A \in \Gamma$ such that $\phi(A) = T$. Note that whenever Γ consists of a single formula, say $\Gamma = \{A\}$, the definition reduces to the statement that A is a tautology.

Lemma 5: Suppose that every formula in Γ is a literal. Then Γ is a tautology if and only if there is a propositional variable p such that $\{p, \neg p\} \subseteq \Gamma$.

Lemma 6: $\{A_1, \ldots, A_n\} \models B$ if and only if $\{\neg A_1, \ldots, \neg A_n, B\}$ is a tautology.

Lemma 7: The following hold:
(1) $\Gamma, \neg\neg A$ is a tautology if and only if Γ, A is a tautology.
(2) $\Gamma, A \vee B$ is a tautology if and only if Γ, A, B is a tautology.
(3) $\Gamma, \neg(A \vee B)$ is a tautology if and only if $\Gamma, \neg A$ is a tautology and $\Gamma, \neg B$ is a tautology.

Decomposition Rules for G_R

$$\neg\neg\text{-RULE} \quad \frac{\Gamma, \neg\neg A}{\Gamma, A} \qquad\qquad \neg\vee\text{-RULE} \quad \frac{\Gamma, \neg(A \vee B)}{\Gamma, \neg A \qquad \Gamma, \neg B}$$

$$\vee\text{-RULE} \quad \frac{\Gamma, A \vee B}{\Gamma, A, B}$$

Computationally, we proceed as follows: Given $\{A_1, \ldots, A_n\} \models B$, form the set $\{\neg A_1, \ldots, \neg A_n, B\}$ and then apply the decomposition rules to decide whether $\{\neg A_1, \ldots, \neg A_n, B\}$ is a tautology; note that the decomposition rules preserve this property in both directions. The sequent calculus G_R is officially described as follows.

The Sequent Calculus G_R: An axiom of G_R is any finite set Γ of formulas such that $\{p, \neg p\} \subseteq \Gamma$ for some propositional variable p. The rules of inference are the decomposition rules turned upside down.

Theorem 3 (Soundness and adequacy theorem for G_R): $G_R \vdash \Gamma$ if and only if Γ is a tautology.

Corollary 2: If $G_R \vdash \{\neg A_1, \ldots, \neg A_n, B\}$, then $\{A_1, \ldots, A_n\} \models B$. In particular, if $G_R \vdash B$, then B is a tautology.

—————— The Sequent Calculus G ——————

We will now describe a slight variation of the original Gentzen sequent calculus. In Hilbert-style proof systems, we work with formulas; in the sequent calculi G_L and G_R, we work with finite sets of formulas. In his original sequent calculus, Gentzen worked with *pairs* of finite sets of formulas. In G_L and G_R, the negation $\neg$ is handled either two at a time or in conjunction with $\vee$; moreover, the success of the method depends on the insight that all formulas either are literals or are of the form $\neg\neg A$, $A \vee B$, or $\neg(A \vee B)$. If we work with pairs of finite sets of formulas, the rules for $\neg$ are simplified and are independent of the other connectives. As in the presentation of G_R, we will rely on the discussion of semantic trees and G_L to motivate the development of G. Results will be stated without proof and left as exercises for the reader. In G, we include $\wedge$ among the connectives of the language.

In G_L, we prove Γ is *unsatisfiable*; in G_R, we prove Γ is a *tautology*. In the sequent calculus G, we prove that for a pair Γ, Δ of finite sets of formulas, Δ is a *tautological consequence* of Γ. Here is the required definition.

Definition 3: Let Γ and Δ be sets of formulas. We say that Δ is a *tautological consequence* of Γ if the following holds: Whenever a truth assignment ϕ satisfies Γ, there is at least one formula $B \in \Delta$ such that $\phi(B) = T$. We denote this by writing $\Gamma \models \Delta$. Note that whenever Δ consists of a single formula, say $\Delta = \{B\}$, the definition reduces to the statement that B is a tautological consequence of Γ. Special cases: $\Gamma = \varnothing$ (write $\models \Delta$; Δ is a tautology); $\Delta = \varnothing$ (write $\Gamma \models$; Γ is unsatisfiable).

Lemma 8: Suppose that the formulas of Γ and Δ are propositional variables. Then $\Gamma \models \Delta$ if and only if there is a propositional variable p such that $p \in \Gamma$ and $p \in \Delta$.

Lemma 9: Let $A_1, \ldots, A_n, B_1, \ldots, B_k$ be formulas. The following are equivalent: (1) $\{A_1, \ldots, A_n\} \models \{B_1, \ldots, B_k\}$; (2) $\{\neg A_1, \ldots, \neg A_n, B_1, \ldots, B_k\}$ is a tautology; (3) $\{A_1, \ldots, A_n, \neg B_1, \ldots, \neg B_k\}$ is unsatisfiable. (4) $(A_1 \wedge \ldots \wedge A_n) \to (B_1 \vee \ldots \vee B_k)$ is a tautology.

Lemma 10: The following hold:
(1) $\Gamma, \neg A \models \Delta$ if and only if $\Gamma \models \Delta, A$.
(2) $\Gamma \models \Delta, \neg A$ if and only if $\Gamma, A \models \Delta$.
(3) $\Gamma \models \Delta, A \vee B$ if and only if $\Gamma \models \Delta, A, B$.
(4) $\Gamma, A \vee B \models \Delta$ if and only if $\Gamma, A \models \Delta$ and $\Gamma, B \models \Delta$.
(5) $\Gamma \models \Delta, A \wedge B$ if and only if $\Gamma \models \Delta, A$ and $\Gamma \models \Delta, B$.
(6) $\Gamma, A \wedge B \models \Delta$ if and only if $\Gamma, A, B \models \Delta$.

Decomposition Rules for G:

$\neg$-RULES $\quad \dfrac{\Gamma, \neg A \models \Delta}{\Gamma \models \Delta, A} \qquad\qquad\qquad \dfrac{\Gamma \models \Delta, \neg A}{\Gamma, A \models \Delta}$

$\vee$-RULES $\quad \dfrac{\Gamma \models \Delta, A \vee B}{\Gamma \models \Delta, A, B} \qquad\qquad \dfrac{\Gamma, A \vee B \models \Delta}{\Gamma, A \models \Delta \quad \Gamma, B \models \Delta}$

$\wedge$-RULES $\quad \dfrac{\Gamma \models \Delta, A \wedge B}{\Gamma \models \Delta, A \quad \Gamma \models \Delta, B} \qquad \dfrac{\Gamma, A \wedge B \models \Delta}{\Gamma, A, B \models \Delta}$

Note that by Lemma 10, the decomposition rules preserve tautological consequence in both directions.

Example 3:

We write a semantic tree to verify that $p \wedge q, \neg(p \wedge r) \models \neg r$.

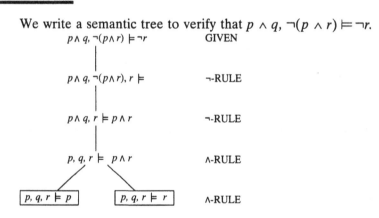

Repeated applications of Lemma 10 show that $p \wedge q, \neg(p \wedge r) \models \neg r$, as required. $\qquad\qquad\qquad\qquad\qquad\qquad\qquad\qquad\qquad\qquad$ □

Now let us turn to the formal development of G.

Definition 4: A *Gentzen sequent* (or just *sequent*) is an ordered pair $\langle \Gamma, \Delta \rangle$, where Γ and Δ are finite sets of formulas. It is traditional to replace the ordered-pair notation with the more convenient $\Gamma \Rightarrow \Delta$; $\Rightarrow$ is called the *sequent symbol*; $\Gamma = \varnothing$ or $\Delta = \varnothing$ is allowed. This is the basic syntactic concept of G.

The Sequent Calculus G: An axiom of G is any sequent $\Gamma \Rightarrow \Delta$ such that there is some propositional variable p with $p \in \Gamma$ and $p \in \Delta$. The rules of inference of G are the decomposition rules turned upside down and with $\models$ replaced by $\Rightarrow$. Note that if $\Gamma \Rightarrow \Delta$ is an axiom, then Δ is a tautological consequence of Γ. Moreover, for each rule, tautological consequence is preserved in both directions.

Definition 5: Let $\Gamma \Rightarrow \Delta$ be a sequent. A *proof of* $\Gamma \Rightarrow \Delta$ *in* G is a finite sequence $\Gamma_1 \Rightarrow \Delta_1, \ldots, \Gamma_n \Rightarrow \Delta_n$ of sequents with $\Gamma_n \Rightarrow \Delta_n$ the given sequent $\Gamma \Rightarrow \Delta$ and such that for $1 \leq k \leq n$, one of the following holds: (1) $\Gamma_k \Rightarrow \Delta_k$ is an axiom of G; (2) $k > 1$ and $\Gamma_k \Rightarrow \Delta_k$ is the conclusion of a rule of inference of G whose hypotheses are among $\Gamma_1 \Rightarrow \Delta_1, \ldots, \Gamma_{k-1} \Rightarrow \Delta_{k-1}$. We write $G \vdash \Gamma \Rightarrow \Delta$ to denote the fact that there is a proof of $\Gamma \Rightarrow \Delta$ in G.

Turn Example 3 upside down and replace $\models$ with $\Rightarrow$ to obtain an example of a proof in G. Here is an official proof in G; it is much easier to follow by starting at the bottom.

Example 4:

We show $G \vdash p \vee (q \wedge r) \Rightarrow (p \vee q) \wedge (p \vee r)$.

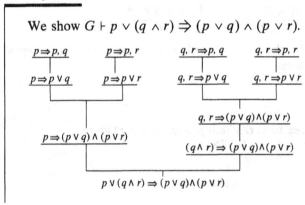

Theorem 4 (Soundness and adequacy theorem for G): $G \vdash \Gamma \Rightarrow \Delta$ if and only if $\Gamma \models \Delta$.

Corollary 3: If $G \vdash \{A_1, \ldots, A_n\} \Rightarrow B$, then $\{A_1, \ldots, A_n\} \models B$. In particular, if $G \vdash \emptyset \Rightarrow B$, then B is a tautology.

——————Exercises for Section 3.6——————

1. Prove Lemma 4.
2. (*Axioms of Frege for* $\neg$ *and* $\rightarrow$) Use the decomposition rules of G_L to show that each of LF1–LF3 is a tautology (see p. 107).
3. For each formula below, decide whether the formula is or is not a tautology. Use the decomposition rules of G_L.
   (1) $(\neg p \rightarrow p) \rightarrow p$.
   (2) $[(p \wedge q) \rightarrow r] \rightarrow [(p \rightarrow q) \rightarrow r]$.
   (3) $[(p \rightarrow q) \rightarrow p] \rightarrow p$.
   (4) $(p \rightarrow q) \rightarrow (q \rightarrow p)$.
   (5) $[(p \rightarrow q) \wedge \neg p] \rightarrow \neg q$.
   (6) $(p \rightarrow q) \vee (\neg p \rightarrow q)$.
4. (*Additional rules for* G_L, *decomposition form*) In this exercise, we justify the following decomposition rules for $\wedge$ and $\rightarrow$:

$$\wedge\text{-RULE} \quad \frac{\Gamma, A \wedge B}{\Gamma, A, B} \qquad \neg\wedge\text{-RULE} \quad \frac{\Gamma, \neg(A \wedge B)}{\Gamma, \neg A \quad \Gamma, \neg B}$$

$$\rightarrow\text{-RULE} \quad \frac{\Gamma, A \rightarrow B}{\Gamma, \neg A \quad \Gamma, B} \qquad \neg\rightarrow\text{-RULE} \quad \frac{\Gamma, \neg(A \rightarrow B)}{\Gamma, A, \neg B}$$

Prove the following.

(1) $\Gamma, A \wedge B$ is unsatisfiable if and only if Γ, A, B is unsatisfiable.

(2) $\Gamma, \neg(A \wedge B)$ is unsatisfiable if and only if $\Gamma, \neg A$ is unsatisfiable and $\Gamma, \neg B$ is unsatisfiable.

(3) $\Gamma, A \rightarrow B$ is unsatisfiable if and only if $\Gamma, \neg A$ is unsatisfiable and Γ, B is unsatisfiable.

(4) $\Gamma, \neg(A \rightarrow B)$ is unsatisfiable if and only if $\Gamma, A, \neg B$ is unsatisfiable.

5. For each of the following, verify $\{A_1, \ldots, A_n\} \models B$ by showing that $\{A_1, \ldots, A_n, \neg B\}$ is unsatisfiable. Use the decomposition rules of G_L.

   (1) $\{q \rightarrow r\} \models (p \wedge q) \rightarrow r$.

   (2) $\{p \rightarrow q\} \models p \rightarrow (q \vee r)$.

   (3) $\{(p \wedge q) \vee (p \wedge r)\} \models p \wedge (q \vee r)$.

   (4) $\{p \wedge (q \vee r)\} \models (p \wedge q) \vee (p \wedge r)$.

   (5) $\{p \rightarrow q, r \rightarrow s, p \wedge r\} \models q \wedge s$.

   (6) $\{p \rightarrow q, r \rightarrow s, p \vee r\} \models q \vee s$.

6. For each of the following, decide $\{A_1, \ldots, A_n\} \models B$ by deciding whether $\{A_1, \ldots, A_n, \neg B\}$ is unsatisfiable. If it is satisfiable, find a truth assignment ϕ that satisfies $\{A_1, \ldots, A_n, \neg B\}$. Use the decomposition rules of G_L.

   (1) $\{p \rightarrow (q \wedge r), \neg(\neg p \vee r)\} \models p \wedge \neg q$.

   (2) $\{\neg(p \wedge q), \neg p\} \models q$.

   (3) $\{(p \vee q) \rightarrow (r \wedge s), \neg r\} \models \neg p$.

   (4) $\{\neg p \vee \neg q, \neg p \vee q, p \vee \neg q\} \models p \vee q$.

7. Prove Lemma 6.

8. Prove Lemma 7.

9. Prove the soundness theorem for G_R.

10. Prove the adequacy theorem for G_R.

11. (*Axioms of Frege for $\neg$ and $\rightarrow$*) Use the decomposition rules of G_R to show that each of LF1–LF3 is a tautology (see p. 107).

12. For each formula in Exercise 3, decide whether the formula is a tautology. Use the decomposition rules of G_R.

13. (*Additional rules for G_R, decomposition form*) In this exercise, we justify the following decomposition rules for $\wedge$ and $\rightarrow$:

$$\wedge\text{-RULE} \quad \frac{\Gamma, A \wedge B}{\Gamma, A \quad \Gamma, B} \qquad\qquad \neg\wedge\text{-RULE} \quad \frac{\Gamma, \neg(A \wedge B)}{\Gamma, \neg A, \neg B}$$

$$\rightarrow\text{-RULE} \quad \frac{\Gamma, A \rightarrow B}{\Gamma, \neg A, B} \qquad\qquad \neg\rightarrow\text{-RULE} \quad \frac{\Gamma, \neg(A \rightarrow B)}{\Gamma, A \quad \Gamma, \neg B}$$

Prove the following:

   (1) $\Gamma, A \wedge B$ is a tautology if and only if Γ, A is a tautology and Γ, B is a tautology.

   (2) $\Gamma, \neg(A \wedge B)$ is a tautology if and only if $\Gamma, \neg A, \neg B$ is a tautology.

   (3) $\Gamma, A \rightarrow B$ is a tautology if and only if $\Gamma, \neg A, B$ is a tautology.

   (4) $\Gamma, \neg(A \rightarrow B)$ is a tautology if and only if Γ, A is a tautology and $\Gamma, \neg B$ is a tautology.

14. Refer to Exercise 5. Verify $\{A_1, \ldots, A_n\} \models B$ by showing that $\{\neg A_1, \ldots, \neg A_n, B\}$ is a tautology. Use the decomposition rules of G_R.

15. Refer to Exercise 6. Decide $\{A_1, \ldots, A_n\} \models B$ by deciding whether $\{\neg A_1, \ldots, \neg A_n, B\}$ is a tautology. If it is not a tautology, find a truth assignment ϕ with $\phi(\neg A_1) = \ldots = \phi(\neg A_n) = \phi(B) = F$. Use the decomposition rules of G_R.

16. Prove Lemma 9.

17. Prove Lemma 10.

18. Prove the soundness theorem for G: If $G \vdash \Gamma \Rightarrow \Delta$, then $\Gamma \models \Delta$.

19. Prove the adequacy theorem for G: If $\Gamma \models \Delta$, then $G \vdash \Gamma \Rightarrow \Delta$.

20. (*Axioms of Frege for $\neg$ and $\rightarrow$*) Use the decomposition rules of G to show that each of LF1–LF3 is a tautology (see p. 107).

21. For each formula in Exercise 3, decide whether the formula is or is not a tautology. Use the decomposition rules of G.

22. (*Additional rules for G, decomposition form*) In this exercise, we justify the following decomposition rules for →:

$$\rightarrow\text{-RULES} \quad \frac{\Gamma, A \rightarrow B \vDash \Delta}{\Gamma \vDash \Delta, A \quad \Gamma, B \vDash \Delta} \qquad \frac{\Gamma \vDash \Delta, A \rightarrow B}{\Gamma, A \vDash \Delta, B}$$

Prove the following.

(1) $\Gamma, A \rightarrow B \vDash \Delta$ if and only if $\Gamma \vDash \Delta, A$ and $\Gamma, B \vDash \Delta$.

(2) $\Gamma \vDash \Delta, A \rightarrow B$ if and only if $\Gamma, A \vDash \Delta, B$.

23. Refer to Exercise 5. Verify $\{A_1, \ldots, A_n\} \vDash B$ using the decomposition rules of G.

24. Refer to Exercise 6. In each case, use the decomposition rules of G to decide whether $\{A_1, \ldots, A_n\} \vDash B$. If not, find a truth assignment ϕ such that $\phi(A_1) = \ldots = \phi(A_n) = T$ and $\phi(B) = F$.

First-Order Languages

irst-order logic is often described as the *logic of quantifiers*. Somewhat more precisely, first-order logic is obtained by adding the logical quantifiers *for all* (denoted ∀) and *there exists* (denoted ∃) to propositional logic. To see the need for these additional concepts, consider the following argument.

All great logicians are intellectually gifted.

Gödel is a great logician.

∴ Gödel is intellectually gifted.

From the point of view of propositional logic, this argument has the form $p, q \therefore r$; by the methods of Chapter 2, this is not a valid argument form. Nevertheless, the given argument is valid. Thus we see that propositional logic is not strong enough to verify the validity of certain arguments and that additional concepts are needed.

In this chapter, we will study first-order languages. These languages are more expressive than the language of propositional logic, and their additional expressive power will enable us to translate an argument such as the one above so that its validity can be verified. To anticipate, the argument about Gödel has the form

$\forall x(L(x) \to I(x))$ $(g = \text{Gödel})$

$L(g)$ $(L(x) = x$ is a great logician$)$

$\therefore I(g)$ $(I(x) = x$ is intellectually gifted$)$

Formal systems for first-order logic will be described in Chapter 5. By writing formal proofs within these systems, we can demonstrate the validity of argument forms that involve quantifiers and connectives. Thus, this chapter is a study of the languages of the formal systems we will consider in Chapter 5. Alternatively, this chapter is a semantic approach to first-order logic.

| **4.1** |

A Language for Arithmetic

In this section, we will describe a specific first-order language that we denote by L_{NN}. The intended interpretation of L_{NN} is arithmetic—that is, the study of the set $\mathbb{N}$ of natural numbers under addition and multiplication. This language is important for the following reason: In Chapter 7, we will outline a proof of Gödel's incompleteness theorem; Gödel's theorem is stated in terms of a formal system whose language is L_{NN}.

To motivate our description of L_{NN}, consider these statements of arithmetic:

- Four is even.
- Five is prime.
- Every natural number is either even or odd.

First let us restate each sentence in a more mathematical style (the third sentence is expressed in three somewhat different ways).

(1) There is a natural number a such that $4 = 2 \times a$.

(2) There do not exist natural numbers a and b such that $5 = a \times b$, $1 < a$, and $1 < b$.

(3) For every natural number a, there is a natural number b such that $a = 2 \times b$ or $a = 2 \times b + 1$.

(4) For every natural number a, there is a natural number b such that $a = 2 \times b$ or there is a natural number b such that $a = 2 \times b + 1$.

(5) For every natural number a, there is a natural number b such that $a = 2 \times b$ or there is a natural number c such that $a = 2 \times c + 1$.

From (1)–(5) we see that, in addition to symbols for the logical connectives, we need symbols for the logical quantifiers *there exists* and *for all*; for arithmetical concepts such as *addition, multiplication, equality,* and *less than*; and for variables that range over the set $\mathbb{N}$ of natural numbers. We also need to refer to the natural numbers zero, one, two, and so on. Rather than have a symbol for each natural number, we instead have a symbol for zero, namely 0, and a symbol S, whose intended interpretation is the successor function (defined by $f(x) = x + 1$). Thus $S0$, $SS0$, $SSS0$, and so on mean one, two, three, and so on.

We are now ready to describe the first-order language L_{NN}. The description is divided into two steps: symbols and formulas.

Symbols of L_{NN}: The symbols of L_{NN} (and of all first-order languages) are divided into two groups:

Logical symbols

$x_1, x_2, x_3, \ldots$	(infinite list)
$\neg, \vee, \wedge, \rightarrow, \leftrightarrow$	(logical connectives)
$\forall, \exists$	(quantifiers)
$(,)$	(left and right parentheses)
$=$	(symbol for equality)

The symbols $x_1, x_2, x_3, \ldots$ are called *individual variables*. For the time being, we will think of these as variables that range over the set $\mathbb{N}$ of natural numbers. The other new symbols are named and interpreted as follows:

Symbol	Name	Interpretation
∀	Universal quantifier	For all
∃	Existential quantifier	There exists
=	Equality (2-ary relation symbol)	Equals

Nonlogical symbols There are five nonlogical symbols for L_{NN}: $0, S, +, \times$, and $<$. The type and meaning of each symbol is given in the table below.

Symbol	Type	Intended interpretation
0	Constant symbol	Zero
S	1-ary function symbol	Successor function on $\mathbb{N}$
+	2-ary function symbol	Addition on $\mathbb{N}$
×	2-ary function symbol	Multiplication on $\mathbb{N}$
<	2-ary relation symbol	Less-than relation on $\mathbb{N}$

The distinction between logical and nonlogical symbols is roughly this: Logical symbols are common to all first-order languages and always have the same interpretation or use; nonlogical symbols differ from language to language and have a variety of meanings.

Formulas of L_{NN}: Before defining the formulas of L_{NN}, we must define the *terms* of L_{NN}. Informally, the terms of L_{NN} are the expressions that can be interpreted as natural numbers. The terms of L_{NN} are defined inductively as follows:

T1 The constant symbol 0 is a term and each individual variable x_n is a term.

T2 If s and t are terms, so are St, $(s + t)$, and $(s \times t)$.

T3 Every term is obtained by a finite number of applications of T1 and T2.

For example, x_5, $SSS0$, $(x_1 + Sx_2)$, and $((x_3 \times S0) + SS0)$ are terms of L_{NN}.

Informally, the formulas of L_{NN} are the expressions that assert some property of natural numbers. The formulas of L_{NN} are defined inductively as follows:

F1 If s and t are terms, then $(s = t)$ and $(s < t)$ are formulas (called *atomic* formulas).

F2 If A and B are formulas, so are $\neg A$, $(A \vee B)$, $(A \wedge B)$, $(A \rightarrow B)$, and $(A \leftrightarrow B)$.

F3 If A is a formula and x_n is an individual variable, then $\forall x_n A$ and $\exists x_n A$ are formulas.

F4 Every formula is obtained by a finite number of applications of F1, F2, and F3.

The language L_{NN}, and in fact all first-order languages, has the important property that there is an algorithm that, given an arbitrary expression, decides whether the expression is a formula. The letters A, B, C, D, and E are used to denote formulas, and conventions about parentheses in propositional logic carry over to first-order languages. In addition, parentheses are often omitted in the terms $(s + t)$ and $(s \times t)$ when no ambiguity is likely to occur; for example, we write $S0 + S0 = SS0$ rather than $(S0 + S0) = SS0$. With these conventions in mind, let us translate the English sentences (1)–(5), given earlier in this section, into the formal language L_{NN} (the ideas in Section 4.3 can be used to show that (3), (4), and (5) are logically equivalent and hence "say the same thing").

(1) $\exists x_1(SSSS0 = SS0 \times x_1)$.
(2) $\neg\exists x_1\exists x_2[(SSSSS0 = x_1 \times x_2) \wedge (S0 < x_1) \wedge (S0 < x_2)]$.
(3) $\forall x_1\exists x_2[(x_1 = SS0 \times x_2) \vee (x_1 = (SS0 \times x_2) + S0)]$.
(4) $\forall x_1[\exists x_2(x_1 = SS0 \times x_2) \vee \exists x_2(x_1 = (SS0 \times x_2) + S0)]$.
(5) $\forall x_1[\exists x_2(x_1 = SS0 \times x_2) \vee \exists x_3(x_1 = (SS0 \times x_3) + S0)]$.

Example 1 ($\forall x = \neg\exists x\neg$):

The formula $\forall x_1(x_1 + S0 = S0 + x_1)$ can be written without the universal quantifier $\forall$. Informally, the formula says that every natural number commutes with 1 under addition. In other words, it is not the case that there is some natural number that does not commute with 1 under addition. Thus we can also write $\neg\exists x_1\neg(x_1 + S0 = S0 + x_1)$.   □

Example 2 ($\exists x = \neg\forall x\neg$):

The formula $\exists x_1(SSSS0 = SS0 \times x_1)$ can be written without the existential quantifier $\exists$. Informally, the formula says that 4 is even; that is, there is some natural number a such that $4 = 2 \times a$. In other words, it is not the case that for every natural number a, $4 \neq 2 \times a$. Thus we can also write $\neg\forall x_1\neg(SSSS0 = SS0 \times x_1)$.   □

We will conclude this section with some informal comments about free and bound variables. Although these ideas are discussed in the context of the language L_{NN}, there is no great loss of generality in doing so since our discussion easily carries over to all first-order languages. Also, these ideas are discussed precisely and with full generality in Section 4.4.

Definition 1: Consider an occurrence of $\forall x_n$ or $\exists x_n$ in a formula A. The *scope* of $\forall x_n$ or $\exists x_n$ is the subformula B of A to which the quantifier applies.

Example 3:

Consider the following formulas:

(1) $\forall x_1 \exists x_2 (x_1 < x_2)$.

(2) $\forall x_1 [\exists x_2 (x_1 = SS0 \times x_2) \vee \exists x_3 (x_1 = (SS0 \times x_3) + S0)]$.

In (1), the scope of $\forall x_1$ is $\exists x_2 (x_1 < x_2)$ and the scope of $\exists x_2$ is $(x_1 < x_2)$. In (2), the scope of $\exists x_2$ is $(x_1 = SS0 \times x_2)$.   □

Definition 2: Consider an occurrence of x_n in A. This occurrence is *bound* if it satisfies one of the following: (a) x_n immediately follows $\forall$ or $\exists$; (b) x_n is in the scope of $\forall x_n$ or $\exists x_n$ (i.e., there is a subformula of A of the form $\forall x_n B$ or $\exists x_n B$ and x_n occurs in B). Otherwise, this occurrence of x_n in A is *free*. Thus each occurrence of x_n in A is either free or bound, but not both.

Example 4:

There are three occurrences of x_1 in the formula

$$(SSSS0 = SS0 \times x_1) \rightarrow \exists x_1 (SSSS0 = SS0 \times x_1).$$

The first occurrence is free; the second occurrence is bound, since it satisfies (a); the third occurrence is bound, since it satisfies (b).   □

Definition 3: The variable x_n is *bound* in A if there is at least one bound occurrence of x_n in A, and is *free* in A if there is at least one free occurrence of x_n in A. Thus a variable can be both free and bound in a formula. A formula with no free variables is called a *closed* formula, or a *sentence*. The result of translating an English statement into a first-order language such as L_{NN} will always be a formula with no free variables—that is, a sentence.

Example 5:

The variable x_1 is both bound and free in the formula

$$(SSSS0 = SS0 \times x_1) \rightarrow \exists x_1 (SSSS0 = SS0 \times x_1).$$

Free occurrence   Bound occurrences
of x_1 of x_1 □

What is the role of bound variables? Consider the definite integral $\int_0^1 x^2 \, dx$. If we replace x with y to obtain $\int_0^1 y^2 \, dy$, the value of the integral is unchanged (1/3 in either case). In mathematics, this is an example of what is known as a *dummy variable*. A bound occurrence of the variable x_n is like a dummy variable. Intuitively speaking, if we replace bound occurrences of x_n with a new variable, the meaning of the formula is unchanged. The precise circumstances under which bound occurrences of a variable can be replaced by another variable will be given in Sections 4.4 and 5.4. For now, we will illustrate with an example.

Example 6:

In $\forall x_1[\exists x_2(x_1 = SS0 \times x_2) \vee \exists x_2(x_1 = (SS0 \times x_2) + S0)]$, we can replace bound occurrences of x_2 with x_3 to obtain either of the following:

$$\forall x_1[\exists x_2(x_1 = SS0 \times x_2) \vee \exists x_3(x_1 = (SS0 \times x_3) + S0)],$$
$$\forall x_1[\exists x_3(x_1 = SS0 \times x_3) \vee \exists x_3(x_1 = (SS0 \times x_3) + S0)].$$ □

The role of free variables is that of *substitution*. If the variable x_n occurs free in the formula A, and t is some term, then we can replace all free occurrences of x_n with t, provided that the new formula "says the same thing about t that the original formula says about x_n." The important idea of substitution is discussed in detail in Section 4.4 and will be used in Chapter 5 when we introduce formal systems for first-order logic.

Acknowledgment A number of the ideas in this section are inspired by Chapter 8 ("*Typographical Number Theory*") of Douglas R. Hofstadter's *Gödel, Escher, Bach* (New York: Basic Books, 1979).

────────── *Exercises for Section 4.1* ──────────

1. For the language L_{NN}, classify each of the following expressions as a term, an atomic formula, a nonatomic formula, or none of these.
   (1) $\forall x_1 \forall x_2(x_1 + x_2 = x_2 + x_1)$.   (3) $0 + x_1$.
   (2) $SS0 + SS0 = SSSS0$.   (4) $(x_1 = 0) \times SS0$.
2. Write the following sentences in the language L_{NN} (these are the nonlogical axioms of formal arithmetic; see Section 7.1).
   (1) For every natural number a, the successor of a is not zero.
   (2) For all natural numbers a and b, if a and b have the same successor, then a and b are equal.
   (3) For every natural number a, the sum of a and zero is equal to a.
   (4) For all natural numbers a and b, the sum of a and the successor of b equals the successor of the sum of a and b.
   (5) For every natural number a, the product of a and zero is equal to zero.
   (6) For all natural numbers a and b, the product of a and the successor of b is equal to the product of a and b, plus a.
   (7) There is no natural number less than zero.
   (8) For all natural numbers a and b, if a is less than the successor of b, then a is less than b or a is equal to b.
   (9) For all natural numbers a and b, a is less than b, a is equal to b, or b is less than a.
3. Express the following statements in the language L_{NN}:
   (1) For all natural numbers a and b, if a and b are even, then the product of a and b is even.
   (2) For every natural number a greater than 1, there is a prime p that divides a.
   (3) There is no natural number a greater than 1 that equals its own square.
   (4) The equation $x^2 + y^2 = z^2$ has a solution in positive integers.
   (5) The equation $x^3 + y^3 = z^3$ has no solution in positive integers.
   (6) The number 5 is not a perfect square.
   (7) If a natural number is the sum of three consecutive cubes, then the number cannot be a perfect square.

(8) There are infinitely many odd numbers.

(9) Multiplication on ℕ is associative.

4. The division algorithm (DA) states that for all natural numbers a and b with $b \neq 0$, there exist unique natural numbers q and r such that $a = bq + r$ and $r < b$. Express the DA in L_{NN}.

5. Goldbach's conjecture (1742) states that every even number greater than 2 is the sum of two primes. State this conjecture in the language L_{NN}.

6. Interpret the following sentences of L_{NN}. (This exercise is adapted from Hofstadter's *Gödel, Escher, Bach* 1979, 212.)

   (1) $\forall x_1 \, \exists x_2 (SS0 \times x_2 = x_1)$.
   (3) $\exists x_1 \, \exists x_2 (SS0 \times x_2 = x_1)$.
   (2) $\neg \forall x_1 \, \exists x_2 (SS0 \times x_2 = x_1)$.
   (4) $\exists x_1 \, \forall x_2 \neg (SS0 \times x_2 = x_1)$.

7. For each formula below and for each occurrence of x_1 or x_2 in the formula, state whether the occurrence is free or bound. In addition, identify the formulas that are sentences (= no free variables).

   (1) $\forall x_1 \, \forall x_2 [(x_1 < x_2) \vee (x_2 < x_1) \vee (x_1 = x_2)]$.
   (2) $\forall x_1 \, \exists x_2 (x_1 < x_2) \vee [(0 < x_1) \vee (0 = x_2)]$.
   (3) $\forall x_1 [(x_1 = x_2) \vee \exists x_2 ((x_2 < x_1) \vee (x_2 < x_1))]$.
   (4) $(x_1 = SSSS0) \rightarrow [\exists x_2 (x_1 = SS0 \times x_2) \leftrightarrow \exists x_2 (SSSS0 = SS0 \times x_2)]$.

$\boxed{4.2}$

First-Order Languages, Interpretations, and Models

—————— First-Order Languages ——————

The language L_{NN} is a *particular* first-order language, whose intended interpretation is elementary arithmetic. We will now describe a *general* first-order language L; the description is divided into two steps.

Symbols of L: There are two types of symbols for a first-order language L.

Logical symbols

$x_1, x_2, x_3, \dots$	(infinite list)
$\neg \ \vee$	(logical connectives)
$\forall$	(quantifier)
( ) ,	(punctuation)
$=$	(symbol for equality)

The symbols $x_1, x_2, x_3, \dots$ are called *individual variables*; informally, they represent the elements of some nonempty set. The letters x, y, and z are used to denote individual variables. The symbols $\neg$, $\vee$, $\forall$, and $=$ are interpreted as *not, or, for all,* and *equals,* respectively. Note that we have omitted the connectives $\wedge$, $\rightarrow$, and $\leftrightarrow$ and the existential quantifier $\exists$; these will be defined symbols. There is one new symbol for punctuation, the comma; its use will soon become apparent.

Nonlogical symbols The nonlogical symbols fall into three categories:

- Constant symbols
- n-ary function symbols ($n = 1, 2, \ldots$)
- n-ary relation symbols ($n = 1, 2, \ldots$)

We will illustrate with the language L_{NN}: 0 is a constant symbol; S is a 1-ary function symbol; $+$ and $\times$ are 2-ary function symbols; $<$ is a 2-ary relation symbol. The constant, function, and relation symbols vary from language to language. In particular, a first-order language may have no constant symbols, no function symbols, and no relation symbols (there is always the logical symbol $=$, whose interpretation is the 2-ary relation of equality). On the other hand, we assume that the number of symbols of each kind is countable. In most cases, the total number is finite; for example, L_{NN} has five nonlogical symbols.

Formulas of L: First we will define the terms. Informally, the terms of a first-order language are the expressions that represent the objects under discussion. The terms are defined inductively as follows:

T1 Each individual variable x_n and each constant symbol c is a term.

T2 If F is an n-ary function symbol and $t_1, \ldots, t_n$ are terms, then $F(t_1, \ldots, t_n)$ is a term (this is where the comma symbol is used).

T3 Every term is obtained by a finite number of applications of T1 and T2.

The letters $s, t, u,$ and v, possibly with subscripts, are used to denote terms. As in the case of L_{NN}, we often omit parentheses in terms to improve readability. For a binary function symbol $\circ$, we often write $s \circ t$ or $(s \circ t)$ rather than the precise $\circ(s, t)$. Now that terms are available, the formulas are defined inductively as follows:

F1 If s and t are terms, $(s = t)$ is a formula. If R is an n-ary relation symbol and $t_1, \ldots, t_n$ are terms, then $R(t_1, \ldots, t_n)$ is a formula. (These are called *atomic* formulas.)

F2 If A and B are formulas and x_n is a variable, then $\neg A$, $(A \vee B)$, and $\forall x_n A$ are formulas.

F3 Every formula is obtained by a finite number of applications of F1 and F2.

The letter L is used to denote an arbitrary first-order language; $A, B, C, D,$ and E denote formulas. Conventions about parentheses for the language L_{NN} carry over to all first-order languages. In the case of a 2-ary relation symbol R, we often write sRt or (sRt) rather than the precise $R(s, t)$. Defined symbols are as follows: $A \to B$ means $\neg A \vee B$; $A \wedge B$ means $\neg(\neg A \vee \neg B)$; $A \leftrightarrow B$ means $(A \to B) \wedge (B \to A)$; $\exists x_n A$ means $\neg\forall x_n \neg A$. We freely use defined symbols when translating English sentences into a first-order language.

All first-order languages have the same logical symbols; the difference between languages lies in the choice of the nonlogical symbols. Once the nonlogical symbols have been specified, the rules for obtaining the terms and formulas are known and are precisely described. In summary:

To describe a first-order language L, it suffices to list the nonlogical symbols of L.

We will now give several examples of first-order languages.

The Language L_G for Group Theory: A *group* is an ordered triple $\langle G, e, \circ \rangle$ that satisfies the axioms G1–G3 (see Section 1.6). To describe a language L_G for groups, it suffices to list its nonlogical symbols. There are two:

e (constant symbol); $\circ$ (2-ary function symbol).

The axioms G1–G3, written in L_G, are

G1 $\forall x_1 \forall x_2 \forall x_3 [x_1 \circ (x_2 \circ x_3) = (x_1 \circ x_2) \circ x_3]$.
G2 $\forall x_1 (x_1 \circ e = x_1)$.
G3 $\forall x_1 \exists x_2 (x_1 \circ x_2 = e)$.

The Language L_{ST} for Set Theory: There are two primitive concepts in set theory: *set* and *is an element of*. Thus the language L_{ST} for set theory has just one nonlogical symbol:

$\in$ (2-ary relation symbol).

We write $x \in y$ or $(x \in y)$ rather than the proper $\in (x, y)$. Informally, variables range over all sets, and $x \in y$ states that the set x is an element of the set y. See Section 6.4 for a list of the axioms of Zermelo-Frankel set theory written in the language L_{ST}. For example, the existence of the empty set is expressed in L_{ST} by the sentence $\exists x_1 \forall x_2 \neg (x_2 \in x_1)$.

Example 1 (Russell's paradox):

Russell's paradox arises when we construct the set A consisting of all sets b such that b is not an element of itself (see Section 1.4). The existence of A can be asserted in L_{ST} as follows:

$$\exists x_1 \forall x_2 [(x_2 \in x_1) \leftrightarrow \neg (x_2 \in x_2)].$$

In the next section, we will show that the *negation* of this sentence is logically valid.   □

The Language of Equality L_{EQ}: This language has *no* nonlogical symbols! Nevertheless, we can express certain ideas in L_{EQ} because the 2-ary logical symbol $=$ is available. For example, the following sentence of L_{EQ} asserts the existence of exactly two individuals:

$$\exists x_1 \exists x_2 [\neg (x_1 = x_2) \wedge \forall x_3 ((x_3 = x_1) \vee (x_3 = x_2))].$$

The Language L_S ***for Syllogistic Logic:*** A syllogism is a classical type of argument that was first studied by Aristotle, the founder of logic. Consider the following four kinds of sentences:

- All *A* are *B*.   • Some *A* are *B*.
- No *A* are *B*.   • Some *A* are not *B*.

A syllogism is an argument consisting of three such sentences: The first two are the hypotheses; the third is the conclusion. Each sentence has two different letters; three letters (say *A*, *B*, and *C*) are used altogether, and each letter occurs exactly twice. A first-order language L_S for syllogisms has an infinite number of 1-ary relation symbols, say $P_1, P_2, P_3, \ldots$. (There are no constant symbols, no function symbols, and no *n*-ary relation symbols for $n > 1$; in fact, this is a case in which the logical symbol = is not used.) We illustrate by translating the following syllogism into L_S (we use *A*, *B*, and *C* instead of the proper $P_1, P_2, P_3, \ldots$):

All *A* are *B*.	$\forall x[A(x) \rightarrow B(x)]$.
Some *C* are not *B*.	$\exists x[C(x) \wedge \neg B(x)]$.
$\therefore$ Some *C* are not *A*.	$\therefore \exists x[C(x) \wedge \neg A(x)]$.

The second sentence, $\exists x[C(x) \wedge \neg B(x)]$, can also be written as $\neg \forall x[C(x) \rightarrow B(x)]$; that is, *not all C are B*; in the exercises, we will ask the reader to justify this, using methods from Chapter 2.

The Language L_R ***for Relations:*** We now introduce a language that is suitable for expressing not only syllogisms but also arguments that involve individuals and *n*-ary relations for $n > 1$. The language L_R has no function symbols. On the other hand, there are an infinite number of symbols of the other types. Specifically, we have

Constant symbols (infinite list)
n-ary relation symbols (infinite list, each $n \geq 1$)

Since the language L_R is used for expressing arguments in ordinary discourse, we want considerable freedom in choosing constant and relation symbols; thus we have not specified the precise nonlogical symbols of L_R.

Example 2:

We translate the following argument into L_R:

Fred admires Monet. Whoever admires Monet admires Matisse. Fred admires only great painters. $\therefore$ Matisse is a great painter.

Use relation and constant symbols as follows: $A(x, y) = x$ admires y; $G(x) = x$ is a great painter; f = Fred; m = Monet; t = Matisse. We then have

$$A(f, m), \ \forall x[A(x, m) \rightarrow A(x, t)], \ \forall x[A(f, x) \rightarrow G(x)], \ \therefore G(t). \qquad \square$$

——————Interpretations and Models——————

The language L_{NN} was introduced with a specific interpretation in mind. Variables range over $\mathbb{N}$; 0, S, $+$, $\times$, $<$ are interpreted as *zero, successor function, addition, multiplication*, and *less than*, respectively. We will now give the precise definition of an interpretation of a first-order language. According to this definition, it is possible for a language to have many different interpretations; in particular, the language L_{NN} has interpretations other than the intended interpretation.

Definition 1: Let L be a first-order language. An *interpretation* I *of* L consists of

(1) a non-empty set D_I (called the *domain* of I);

(2) for each constant symbol c of L, a specific element c_I of D_I;

(3) for each n-ary function symbol F of L, an n-ary operation F_I on D_I;

(4) for each n-ary relation symbol R of L, an n-ary relation R_I on D_I.

The set D_I is also referred to as the *universe* of the interpretation; the variables $x_1, x_2, \ldots,$ range over D_I. We emphasize that D_I is nonempty and that for each constant symbol c, each n-ary function symbol F, and each n-ary relation symbol R, we have

$$c_I \in D_I, \quad F_I : D_I^n \to D_I, \quad \text{and} \quad R_I \subseteq D_I^n.$$

Example 3:

We give an interpretation I of the language L_G of groups such that the formulas G1–G3 are all true in I:

$$D_I = \mathbb{Z} \quad \text{and} \quad e_I = 0.$$
$$a \circ_I b = a + b \ (\circ_I \text{ is ordinary addition on } \mathbb{Z}).$$

In other words, the set $\mathbb{Z}$ of integers under addition with identity 0 is a group. This interpretation can also be written as $\langle \mathbb{Z}, 0, + \rangle$.   □

Example 4:

We give another interpretation I of the language L_G:

$$D_I = \mathbb{N} \quad \text{and} \quad e_I = 1.$$
$$a \circ_I b = a \times b \ (\circ_I \text{ is multiplication on } \mathbb{N}).$$

For this interpretation, G3 is false while G1 and G2 are true. In other words, the set $\mathbb{N}$ of natural numbers under multiplication does not satisfy the requirement for right inverses and thus is not a group. □

Examples 3 and 4 illustrate two important features of first-order languages and their interpretations: A first-order language L has many different interpretations, and a sentence of L may be true in one interpretation of L and false in another.

The Standard Interpretation $\mathcal{N}$ of L_{NN}: The language L_{NN} was introduced with a specific interpretation in mind; henceforth, we will refer to this as the *standard interpretation* of L_{NN} and denote it by $\mathcal{N}$. This important interpretation is described succinctly by

$$\mathcal{N} = \langle \mathbb{N}, 0_{\mathbb{N}}, S_{\mathbb{N}}, +_{\mathbb{N}}, \times_{\mathbb{N}}, <_{\mathbb{N}} \rangle;$$

notation is explained as follows:

The domain of the standard interpretation $\mathcal{N}$ is $\mathbb{N}$.

The symbol 0 is interpreted as the natural number zero.

The symbol S is interpreted as the successor function on $\mathbb{N}$.

The symbol + is interpreted as addition of natural numbers.

The symbol $\times$ is interpreted as multiplication of natural numbers.

The symbol $<$ is interpreted as the less-than relation on $\mathbb{N}$.

Example 5:

Let L be the first-order language with constant symbol c, 2-ary function symbol F, and 2-ary relation symbol R. We will give an interpretation of L in which the following sentence is true and one in which it is false:

$$\forall x_1 \forall x_2 [R(x_1, x_2) \rightarrow R(F(c, x_1), F(c, x_2))].$$

True $D_I = \mathbb{N}$. $F_I(a, b) = $ sum of a and b.

   $c_I = 1$. $R_I = \{\langle a, b \rangle : a, b \in \mathbb{N} \text{ and } a \text{ is less than } b\}$.

Informally, we have $\forall x_1 \forall x_2 [(x_1 < x_2) \rightarrow (1 + x_1 < 1 + x_2)]$, and this is a true statement about the natural numbers.

False $D_I = \mathbb{Z}$. $F_I(a, b) = $ product of a and b.

   $c_I = -1$. $R_I = \{\langle a, b \rangle : a, b \in \mathbb{Z} \text{ and } a \text{ is greater than } b\}$.

Informally, we have $\forall x_1 \forall x_2 [(x_1 > x_2) \rightarrow (-x_1 > -x_2)]$, and this is a false statement about the integers. □

In Examples 3–5, we have asserted that certain sentences are true or false in a given interpretation. In doing so, we relied on intuition and common sense to decide the proper truth value, and for the remainder of this section we will continue to take this approach. In the next section, we will define precisely "the sentence A is true in the interpretation I."

Definition 2: Let Γ be a set of formulas of L. A *model* of Γ is an interpretation of L with the following additional property: Every formula of Γ is true in the interpretation.

The letters M and N are used to denote models. For any first-order language L, the interpretations of L are the same as the models of $\Gamma = \varnothing$. In the most important cases, the formulas in Γ will actually be sentences. We emphasize that not every set Γ of sentences has a model.

Example 6:

Let $\Gamma = \{G1, G2, G3\}$, where G1–G3 are the axioms for a group written in L_G. The models of Γ are precisely the groups; in particular, Example 3 is an interpretation of L_G that is a model of Γ. On the other hand, Example 4 is an interpretation of L_G that is *not* a model of Γ.   □

Example 7:

Let L be the language with 2-ary relation symbol R, and let Γ be the following set of sentences of L:

(1) $\forall x_1 R(x_1, x_1)$.
(2) $\forall x_1 \forall x_2 [R(x_1, x_2) \rightarrow R(x_2, x_1)]$.
(3) $\exists x_1 \exists x_2 \exists x_3 [R(x_1, x_2) \wedge R(x_2, x_3) \wedge \neg R(x_1, x_3)]$.
(4) $\forall x_1 \exists x_2 [\neg(x_1 = x_2) \wedge R(x_1, x_2)]$.

We will construct a model M of Γ. By (1) and (2), the relation R_M is reflexive and symmetric; by (3), R_M is not transitive. By (4), for each $a \in D_M$, there exists $b \in D_M$, $b \neq a$, such that $\langle a, b \rangle \in R_M$. The following works:

$$D_M = \{1, 2, 3\}.$$
$$R_M = \{\langle 1, 1 \rangle, \langle 2, 2 \rangle, \langle 3, 3 \rangle, \langle 1, 2 \rangle, \langle 2, 1 \rangle, \langle 2, 3 \rangle, \langle 3, 2 \rangle\}.$$ □

———————————*Exercises for Section 4.2*———————————

1. Let L be a first-order language with no function symbols and three constant symbols *a, b, c.* What are the terms of L?
2. Let L be a first-order language with one 1-ary function symbol F and one constant symbol *c*. What are the variable-free terms of L (i.e., the terms with no variables)?
3. For the language L_G, classify each expression below as a term, an atomic formula, a nonatomic formula, or none of these.
   (1) $(x_1 \circ (e \circ e))$.
   (2) $\exists x_1 (e \circ x_1)$.
   (3) $\forall x_1 \forall x_2 \exists x_3 (x_1 \circ x_3 = x_2)$.
   (4) $e \circ x_1 = x_1 \circ (e \circ x_1)$.

4. Suppose in the language L_G for groups we omit the constant symbol e, keeping only the nonlogical symbol $\circ$. Write axioms for a group in this new language. Specifically, replace axioms G2 and G3 with a single axiom that does not use the constant symbol e.

5. What are the atomic formulas of L_{ST}?

6. Translate the following sentences of L_{ST}:
   (1) $\forall x_1 \forall x_2 \exists x_3 \forall x_4 [(x_4 \in x_3) \leftrightarrow ((x_4 \in x_1) \vee (x_4 \in x_2))]$.
   (2) $\forall x_1 \forall x_2 \exists x_3 \forall x_4 [(x_4 \in x_3) \leftrightarrow ((x_4 \in x_1) \wedge (x_4 \in x_2))]$.
   (3) $\forall x_1 \forall x_2 \exists x_3 \forall x_4 [(x_4 \in x_3) \leftrightarrow ((x_4 = x_1) \vee (x_4 = x_2))]$.

7. State the axiom of extensionality in L_{ST}: For all sets a and b, if a and b have the same elements, then a and b are equal.

8. State the power set axiom in L_{ST}: For every set a, there is a set b whose elements are the subsets of a.

9. State the axiom of regularity in L_{ST}: For every nonempty set a, there is an element b of a such that a and b are disjoint.

10. Express the following in the language of equality L_{EQ}:
    (1) There are at least two individuals.
    (2) There are at most two individuals.
    (3) There are exactly three individuals.
    (4) For any two individuals, there is another individual distinct from both of them.

11. Let L have a 1-ary relation symbol R. Express the following in L:
    (1) Exactly one individual has property R.
    (2) All but two individuals have property R.

12. (*Induction on terms*) Use complete induction to prove the following:
    Theorem: To prove that every term of L has property Q, it suffices to show that each variable and each constant symbol has property Q, and that if F is an n-ary function symbol and $t_1, \ldots, t_n$ are terms, each with property Q, then $F(t_1, \ldots, t_n)$ has property Q.

13. Look up the definition of a vector space in a linear algebra book. Describe a language L_{VS} for vector spaces and write the axioms for a vector space in the language L_{VS}. (*Hint:* Assume that the individual variables range over vectors and scalars. To distinguish between the two, include among the nonlogical symbols two 1-ary relation symbols V and S, interpreted as follows: $V(x) = x$ is a vector; $S(x) = x$ is a scalar.)

14. Use the definition of $\exists$ in terms of $\forall$ and various logical equivalences and replacements in Chapter 2 to explain why the two formulas $\exists x(C(x) \wedge \neg B(x))$ and $\neg \forall x(C(x) \rightarrow B(x))$ "say the same thing."

15. Use the definition of $\exists$ in terms of $\forall$ and various logical equivalences and replacements discussed in Chapter 2 to explain why the two formulas $\neg \exists x(B(x) \wedge C(x))$ and $\forall x(B(x) \rightarrow \neg C(x))$ "say the same thing."

16. (*The language L_{OF} for ordered fields*) Let L_{OF} have nonlogical symbols as follows: $0, 1$ (constant symbols); $+, \times$ (2-ary function symbols); $<$ (2-ary relation symbol). An interpretation of L_{OF} is the real number system, described as follows:

$$\mathscr{R} = \langle \mathbb{R}, 0_R, 1_R, +_R, \times_R, <_R \rangle.$$

In other words, the domain of the interpretation is $\mathbb{R}$; 0 and 1 are interpreted as the real numbers zero and one; $+$ and $\times$ are interpreted as addition and multiplication on $\mathbb{R}$; and $<$ is interpreted as the less-than relation on $\mathbb{R}$. Express the following in L_{OF}:
    (1) Every positive real number has a square root.
    (2) Every polynomial $ax + b$ has a real root.
    (3) There is a polynomial $ax^2 + bx + c$ that has no real root.

17. A *densely ordered set without first or last element* is an ordered pair $\langle D, < \rangle$, where D is a nonempty set and $<$ is a binary relation on D (which we express as *less than*) such that the following hold for all $a, b \in D$:

DO1 a is not less than a.

DO2 $<$ is a transitive relation on D.

DO3 *a* is less than *b*, *b* is less than *a*, or *a* and *b* are the same.

DO4 If *a* is less than *b*, there exists *c* ∈ *D* such that *a* is less than *c* and *c* is less than *b*.

DO5 *D* has no first element.

DO6 *D* has no last element.

(1) Describe a language L_{DO} for densely ordered sets and express DO1–DO6 in L_{DO}. (*Hint:* see Section 6.3.)
(2) Is ℚ with the usual less-than relation a model of DO1–DO6?
(3) Is ℤ with the usual less-than relation a model of DO1–DO6?
18. Write each of the following sentences in the language L_S. ($H(x) = x$ is human; $S(x) = x$ is selfish.)

(1) All humans are selfish.   (3) Some humans are selfish.
(2) No humans are selfish.   (4) Some humans are not selfish.
19. Consider the following two sentences:

(a) Some *A* are *B*.   (b) No *A* are *B*.

(1) Write (a) without using ∀.   (3) Write (b) without using ∀.
(2) Write (a) without using ∃.   (4) Write (b) without using ∃.
20. Express these syllogisms in the language L_S and in each case decide whether the syllogism is valid or invalid. (To decide validity, use Venn diagrams, as discussed in Section 1.4, or use your intuition.) These examples appear in Carroll's *Symbolic Logic* (New York: Dover, 1958).

(1) All wasps are unfriendly. No puppies are unfriendly. ∴ Puppies are not wasps.
(2) All these bonbons are chocolate creams. All these bonbons are delicious. ∴ All these chocolate creams are delicious.
(3) No wheelbarrows are comfortable vehicles. No uncomfortable vehicles are popular. ∴ No wheelbarrows are popular.
(4) Some healthy people are fat. No unhealthy people are strong. ∴ Some fat people are not strong.
21. In each of the following, the hypotheses of an argument are given. Write these sentences in the language L_S and draw the correct conclusion of each argument. These exercises appear in Carroll's *Symbolic Logic* (New York: Dover, 1958).

(1) Babies are illogical. Nobody is despised who can manage a crocodile. Illogical people are despised. ∴? ($B(x) = x$ is a baby; $I(x) = x$ is illogical; $D(x) = x$ is despised; $C(x) = x$ can manage a crocodile.)
(2) No one takes the *Times* unless he is well educated. No hedgehogs can read. Those who cannot read are not well educated. ∴? ($T(x) = x$ takes the *Times*; $W(x) = x$ is well educated; $H(x) = x$ is a hedgehog; $R(x) = x$ can read.)
(3) Everyone who is not a lunatic can do logic. No lunatics are fit to serve on a jury. None of your sons can do logic. ∴? ($L(x) = x$ is a lunatic; $Lo(x) = x$ can do logic; $J(x) = x$ is fit to serve on a jury; $S(x) = x$ is a son of yours.)
22. Let L_R have a 2-ary relation symbol *L*, interpreted as $L(x, y) = x$ loves *y*. Express the following in L_R:

(1) Everyone loves someone.
(2) Everyone is loved by someone.
(3) Someone loves everyone.
(4) There is someone who is loved by everyone.
(5) Everyone loves everyone.
(6) Someone loves someone.
(7) Everyone loves at least two people.
23. Express the following argument in the language L_R: There is a man whom all men despise. ∴ There is at least one man who despises himself. (*Hint:* $M(x) = x$ is a man; $D(x, y) = x$ despises *y*.)

24. Refer to Exercise 4 in Section 5.2; express each argument in the language L_R.
25. Refer to Exercise 5 in Section 5.2; express each argument in the language L_R.
26. For each sentence below, give an interpretation of L in which the sentence is true and one in which it is false.
    (1) $\forall x_1 R(x_1, x_1)$ (i.e., R is reflexive).
    (2) $\forall x_1 \forall x_2 [R(x_1, x_2) \rightarrow R(x_2, x_1)]$ (i.e., R is symmetric).
    (3) $\forall x_1 \forall x_2 [R(x_1, x_2) \rightarrow (R(x_2, x_1) \rightarrow (x_1 = x_2))]$ (i.e., R is antisymmetric).
    (4) $\forall x_1 \forall x_2 \forall x_3 [R(x_1, x_2) \rightarrow (R(x_2, x_3) \rightarrow R(x_1, x_3))]$ (i.e., R is transitive).
    (5) $\forall x_1 [R(c, x_1) \rightarrow R(c, F(c, x_1))]$.
    (6) $\forall x_1 [\neg(x_1 = c) \rightarrow \exists x_2 R(x_1, x_2)]$.
    (7) $\forall x_1 \forall x_2 [R(x_1, x_2) \rightarrow R(F(c, x_2), F(c, x_1))]$.
    (8) $\forall x_1 \forall x_2 [R(x_1, x_2) \rightarrow R(F(x_2, x_1), F(x_1, x_2))]$.
    (9) $\forall x_1 \exists x_2 R(x_1, x_2) \rightarrow \exists x_1 \forall x_2 R(x_1, x_2)$.
27. Let L have 2-ary relation symbol R and let x, y, z be distinct variables.
    (1) Find a finite model of $\Gamma = \{\forall x \exists y R(x, y), \forall x \neg R(x, x)\}$.
    (2) Find a model of $\Delta = \Gamma \cup \{\forall x \forall y \forall z [(R(x, y) \wedge R(y, z)) \rightarrow R(x, z)]\}$.
28. Let L have 2-ary relation symbol R. For each set $\Gamma = \{S1, S2, S3, S4\}$ of sentences of L, find (if possible!) a model of Γ.
    (1) S1 $\forall x_1 R(x_1, x_1)$.
        S2 $\forall x_1 \forall x_2 \forall x_3 [(R(x_1, x_2) \wedge R(x_2, x_3)) \rightarrow R(x_1, x_3)]$.
        S3 $\exists x_1 \exists x_2 [R(x_1, x_2) \wedge \neg R(x_2, x_1)]$.
        S4 $\forall x_1 \exists x_2 [\neg(x_1 = x_2) \wedge R(x_1, x_2)]$.
    (2) S1 $\forall x_1 \forall x_2 \forall x_3 [(R(x_1, x_2) \wedge R(x_2, x_3)) \rightarrow R(x_1, x_3)]$.
        S2 $\forall x_1 \forall x_2 [R(x_1, x_2) \rightarrow R(x_2, x_1)]$.
        S3 $\exists x_1 \neg R(x_1, x_1)$.
        S4 $\exists x_1 \exists x_2 [\neg(x_1 = x_2) \wedge R(x_1, x_2)]$.
    (3) S1 $\forall x_1 \forall x_2 [R(x_1, x_2) \rightarrow R(x_2, x_1)]$.
        S2 $\exists x_1 \exists x_2 \exists x_3 [\neg(x_1 = x_2) \wedge \neg(x_1 = x_3) \wedge \neg(x_2 = x_3)]$.
        S3 $\forall x_1 \forall x_2 \forall x_3 [(R(x_1, x_2) \wedge R(x_1, x_3)) \rightarrow (x_2 = x_3)]$.
        S4 $\forall x_1 \exists x_2 [\neg(x_1 = x_2) \wedge R(x_1, x_2)]$.
29. Let $\Gamma = \{\forall x_1 \neg(F(x_1) = c), \forall x_1 \forall x_2 ((F(x_1) = F(x_2)) \rightarrow (x_1 = x_2))\}$, where L has constant symbol c and 1-ary function symbol F.
    (1) Find an infinite model of Γ (i.e., D_M is an infinite set).
    (2) Does Γ have a finite model?
30. Refer to Exercise 11 in Section 5.2. Find a model of $\Gamma = \{W1, \ldots, W7\}$.
31. Refer to Exercise 12 in Section 5.2. Find a model of $\Gamma = \{EG1, \ldots, EG7\}$.
32. (*Chess*) We introduce a first-order language L_C to discuss the game of chess. Assume the game is played on a board whose squares are green and yellow. The language L_C has relation symbols with intended interpretation as follows:

- $Y(x) = x$ is on a yellow square.
- $G(x) = x$ is on a green square.
- $W(x) = x$ is a white piece.
- $B(x) = x$ is a black piece.
- $K(x) = x$ is a knight.

- $P(x) = x$ is a pawn.
- $Q(x) = x$ is a queen.
- $S(x) = x$ is a bishop.
- $A(x, y) = x$ attacks y.

(1) Translate the following sentences of L_C (x, y, z are distinct variables):
    (a) $\exists x \exists y [K(x) \wedge W(x) \wedge P(y) \wedge B(y) \wedge A(x, y)]$.
    (b) $\exists x [K(x) \wedge W(x) \wedge \exists y (P(y) \wedge B(y) \wedge A(x, y))]$.
    (c) $\exists x [Q(x) \wedge W(x) \wedge \forall y ((P(y) \wedge B(y)) \rightarrow A(x, y))]$.
    (d) $\exists x \forall y [Q(x) \wedge W(x) \wedge ((P(y) \wedge B(y)) \rightarrow A(x, y))]$.
    (e) $\exists x [Q(x) \wedge W(x) \wedge \forall y ((Q(y) \wedge W(y)) \rightarrow (x = y))]$.
    (f) $\exists x \exists y [W(x) \wedge K(x) \wedge W(y) \wedge K(y) \wedge \forall z ((W(z) \wedge K(z)) \rightarrow ((z = x) \wedge (z = y)))]$.

(2) Write the following in L_C:

    (*a*) A white queen is attacking a black pawn.

    (*b*) There do not exist three white knights.

    (*c*) There is exactly one white queen and exactly one black queen.

    (*d*) There are at least two white knights and one white queen.

    (*e*) There is just one white queen, and it is on a yellow square.

    (*f*) No white bishop on a green square can attack any black piece on a yellow square.

    (*g*) If a white knight is on a green square and is attacking a black piece, then the black piece is on a yellow square.

4.3

Tarski's Definition of Truth

Let A be a sentence of L and let I be an interpretation of L. In this section, we will define precisely the following statement: *A is true in the interpretation* I. Note that we do not speak of a sentence of L as being true; rather, we speak of a sentence as being true in an interpretation I of L. The definition we give is due to Tarski and plays a fundamental role in the semantics of first-order languages. The following notation is used:

    VAR = individual variables $x_1, x_2, \ldots$

    TRM = terms of L

    FOR = formulas of L

Let I be an interpretation of L. Our ultimate goal is to assign a truth value T or F to every sentence of L. The first step is to introduce the concept of an *assignment in* I. By considering all possible assignments in I, we in effect consider all possible ways of assigning elements of the domain of the interpretation to the individual variables.

Definition 1: Let I be an interpretation of L. An *assignment in* I is a function ϕ from the set of individual variables into the domain of I; in other words,

$$\phi: \text{VAR} \to D_I.$$

We use the Greek letters ϕ and ψ to denote assignments. For example, consider the language L_{NN}; the function $\phi: \text{VAR} \to \mathbb{N}$ defined by $\phi(x_n) = 2n$ is an assignment in the standard interpretation of L_{NN}; informally, ϕ interprets the nth variable x_n as the natural number $2n$.

Extend ϕ to Terms: Given an assignment ϕ in I, we extend ϕ so that it assigns an element of D_I to each term of L; in other words, $\phi:\text{TRM} \to D_I$. This extension, *which we continue to denote by ϕ,* is defined by induction on the number of symbols in the term.

    CONST SYM For each constant symbol c of L, $\phi(c) = c_I$.

    FNC SYM If F is an n-ary function symbol of L and $t_1, \ldots, t_n$ are terms of L, then $\phi(F(t_1, \ldots, t_n)) = F_I(\phi(t_1), \ldots, \phi(t_n))$.

We emphasize that *once we know $\phi(x_n)$ for each individual variable x_n, the value of $\phi(t)$ for any term t of L is automatically determined.*

Example 1:

Consider the language L_{NN}; let 0^n be defined inductively by $0^0 = 0$, $0^{n+1} = S0^n$; for example, $0^3 = SSS0$. Let $\phi: VAR \to \mathbb{N}$ be an assignment in the standard interpretation of L_{NN}. We will use induction to show that $\phi(0^n) = n$ for all $n \in \mathbb{N}$; informally speaking, in the standard interpretation ($\mathcal{N}$), the term 0^n is interpreted as the natural number n.

Beginning step $\phi(0) = 0_\mathbb{N}$ CONST SYM

           $= $ zero   The symbol 0 is interpreted as the number zero in $\mathcal{N}$.

Induction step Let $n \geq 0$ and assume that $\phi(0^n) = n$.

$\phi(0^{n+1}) = \phi(S0^n)$ Definition: 0^{n+1} is $S0^n$

      $= S_\mathbb{N}(\phi(0^n))$ FNC SYM

      $= S_\mathbb{N}(n)$ Induction hypothesis: $\phi(0^n) = n$

      $= n + 1$ $S_\mathbb{N}$ is the successor function in $\mathcal{N}$.   □

Example 2:

Let ϕ be an assignment in the standard interpretation of L_{NN} such that $\phi(x_1) = 3$ and $\phi(x_2) = 2$. We find $\phi(((S0 + x_1) \times x_2))$. Informally, it is clear that $\phi(((S0 + x_1) \times x_2)) = 8$. Formally, we proceed as follows:

      $\phi(S0 + x_1) = +_\mathbb{N}(\phi(S0), \phi(x_1))$ FNC SYM

             $= +_\mathbb{N}(1, 3)$ $\phi(S0) = 1, \phi(x_1) = 3$

             $= 4$ $+_\mathbb{N}$ is addition in $\mathcal{N}$

$\phi(((S0 + x_1) \times x_2))$

      $= \times_\mathbb{N}(\phi(S0 + x_1), \phi(x_2))$ FNC SYM

      $= \times_\mathbb{N}(4, 2)$ Calculation above, $\phi(x_2) = 2$

      $= 8$ $\times_\mathbb{N}$ is multiplication in $\mathcal{N}$.   □

Extend ϕ to Formulas: Given an assignment ϕ in L, we further extend ϕ so that it assigns a truth value T or F to each formula A of the language L; in other words, $\phi: FOR \to \{T, F\}$. This extension, *which we continue to denote by ϕ*, is unique and is defined by induction on the number of occurrences of ¬, ∨, and ∀ in the formula. The most difficult case is to assign a truth value to $\forall x A$

under the assumption that we have assigned a truth value T or F to A for every assignment in I. To pave the way for this part of the extension, we introduce the following definition.

Definition 2: Two assignments ϕ and ψ in I are x_n-*variants* if $\phi(x_k) = \psi(x_k)$ for all $k \neq n$. Thus, x_n-variants ϕ and ψ agree everywhere except possibly at x_n, where they may or may not differ. Note that every assignment ϕ is a x_n-variant of itself.

We now extend an assignment $\phi: \text{VAR} \to D_I$ in I so that it assigns a truth value T or F to each formula A of L. There are four cases (note that the definition of $\phi(s = t)$ forces us to interpret the symbol = as equality).

Atomic Suppose A is an atomic formula. There are two possibilities for A: $(s = t)$ or $R(t_1, \ldots, t_n)$. We have

$$\phi(s = t) \text{ is } \begin{cases} T & \text{if } \phi(s) \text{ and } \phi(t) \text{ are the same element of } D_I; \\ F & \text{if } \phi(s) \text{ and } \phi(t) \text{ are different elements of } D_I. \end{cases}$$

$$\phi(R(t_1, \ldots, t_n)) = \begin{cases} T & \langle \phi(t_1), \ldots, \phi(t_n) \rangle \in R_I; \\ F & \langle \phi(t_1), \ldots, \phi(t_n) \rangle \notin R_I. \end{cases}$$

Negation If A is $\neg B$, $\phi(A) = \begin{cases} T & \phi(B) = F; \\ F & \phi(B) = T. \end{cases}$

Disjunction If A is $B \vee C$, $\phi(A) = \begin{cases} T & \phi(B) = T \text{ or } \phi(C) = T; \\ F & \phi(B) = F \text{ and } \phi(C) = F. \end{cases}$

Quantifier If A is $\forall x_n B$,

$$\phi(A) = \begin{cases} T & \text{if } \psi(B) = T \text{ for every assignment } \psi \text{ in I that} \\ & \text{is an } x_n\text{-variant of } \phi; \\ F & \text{if } \psi(B) = F \text{ for at least one assignment} \\ & \psi \text{ in I that is an } x_n\text{-variant of } \phi. \end{cases}$$

The definition $\phi(\forall x_n B) = T$ can be explained intuitively as follows: $\phi: \text{VAR} \to D_I$ assigns a value to each of the variables $x_1, x_2, \ldots$; the formula $\forall x_n B$ is true for this assignment if and only if the formula B is true for every possible way of assigning a value to x_n (for $x_k \neq x_n$, the value assigned to x_k remains the same).

Summary: An assignment $\phi: \text{VAR} \to D_I$ in I assigns an element of D_I to every term of L and a truth value T or F to every formula of L so that these conditions hold.

CONST SYM	For each constant symbol c of L, $\phi(c) = c_I$.
FNC SYM	If F is an n-ary function symbol of L and $t_1, \ldots, t_n$ are terms of L, then $\phi(F(t_1, \ldots, t_n)) = F_I(\phi(t_1), \ldots, \phi(t_n))$.
TA(=)	$\phi(s = t)$ is $T \Leftrightarrow \phi(s)$ and $\phi(t)$ are the same element of D_I.
TA(R)	$\phi(R(t_1, \ldots, t_n)) = T \Leftrightarrow \langle \phi(t_1), \ldots, \phi(t_n) \rangle \in R_I$.
TA¬	$\phi(A) \neq \phi(\neg A)$.
TA∨	$\phi(A \vee B) = T \Leftrightarrow \phi(A) = T \text{ or } \phi(B) = T$.

TA∀ $\phi(\forall x_n A) = T \Leftrightarrow \psi(A) = T$ for every ψ that is an x_n-variant of ϕ.

$\phi(\forall x_n A) = F \Leftrightarrow \psi(A) = F$ for at least one ψ that is an x_n-variant of ϕ.

As in propositional logic, we can derive the conditions TA→, TA∧, and TA↔ (see Sections 2.2 and 3.2). In addition, there is the following useful result on ∃:

Lemma 1: The condition **TA∃** holds for any assignment ϕ in I:

$\phi(\exists x_n A) = T \Leftrightarrow \psi(A) = T$ for at least one ψ that is an x_n-variant of ϕ.

$\phi(\exists x_n A) = F \Leftrightarrow \psi(A) = F$ for every ψ that is an x_n-variant of ϕ.

Proof: We prove the first version of TA∃.

$$\phi(\exists x_n A) = T \Leftrightarrow \phi(\neg\forall x_n \neg A) = T$$
$$\Leftrightarrow \phi(\forall x_n \neg A) = F$$
$$\Leftrightarrow \psi(\neg A) = F \text{ for at least one } \psi \text{ that is an } x_n\text{-variant of } \phi$$
$$\Leftrightarrow \psi(A) = T \text{ for at least one } \psi \text{ that is an } x_n\text{-variant of } \phi. \qquad \blacksquare$$

Definition 3: An assignment ϕ in I *satisfies A* if $\phi(A) = T$. We also say that *A* is *satisfiable* if there is at least one interpretation I of L and at least one assignment ϕ in I such that ϕ satisfies A (i.e., $\phi(A) = T$).

Example 3:

Let I be an interpretation of a first-order language L.

(1) Every assignment in I satisfies the formula $x_1 = x_1$.

(2) There is an assignment in I that satisfies the formula $x_1 = x_2$.

(3) Every assignment in I satisfies the sentence $\forall x_1 \exists x_2 (x_1 = x_2)$.

(4) If D_I has at least two distinct elements, then there is an assignment in I that does not satisfy $x_1 = x_2$.

Proof of (1): Let ϕ be an arbitrary assignment in I. By TA(=), $\phi(x_1 = x_1)$ is T, provided that $\phi(x_1)$ and $\phi(x_1)$ are the same element of D_I; this is obvious.

Proof of (2): Let ϕ be any assignment in I such that $\phi(x_1)$ and $\phi(x_2)$ are the same element of D_I; it follows immediately from TA(=) that ϕ satisfies the formula $x_1 = x_2$.

Proof of (3): We are required to prove that $\phi(\forall x_1 \exists x_2 (x_1 = x_2)) = T$ for an arbitrary assignment ϕ in I. By TA$\forall$, it suffices to prove that $\psi(\exists x_2(x_1 = x_2)) = T$, where ψ is any assignment in I that is an x_1-variant of ϕ. For this, TA$\exists$ requires us to find an assignment ψ' in I that is an x_2-variant of ψ and such that $\psi'(x_1 = x_2)$ is T. Define ψ' by

$$\psi'(x_k) = \begin{cases} \psi(x_k) & k \neq 2; \\ \psi(x_1) & k = 2. \end{cases}$$

Clearly, ψ' is an x_2-variant of ψ, and since $\psi'(x_1)$ and $\psi'(x_2)$ are the same element of D_1, $\psi'(x_1 = x_2)$ is T as required.

Proof of (4): Let $a, b \in D_1$ with $a \neq b$; let ϕ be any assignment in I such that $\phi(x_1) = a$ and $\phi(x_2) = b$. Then $\phi(x_1)$ and $\phi(x_2)$ are distinct elements of D_1; hence $\phi(x_1 = x_2)$ is F and ϕ does not satisfy the formula $x_1 = x_2$, as required. □

Definition 4 (Tarski's definition of truth): The formula A is *true in the interpretation* I if every assignment in I satisfies A (i.e., if $\phi(A) = T$ for every assignment ϕ in I). We also say that A is *false in the interpretation* I if no assignment in I satisfies A (i.e., if $\phi(A) = F$ for every assignment ϕ in I).

Comments

- A formula cannot be both true and false in a given interpretation.
- A formula A is true in an interpretation I if and only if $\neg A$ is false in I.
- A formula with free variables may be neither true nor false in a given interpretation.
- A sentence is either true or false in a given interpretation. This is a nontrivial result, which we will prove in the next section.

Definition 5: A formula A of L is *logically valid* if it is true in every interpretation I of L; in other words, A is logically valid if $\phi(A) = T$ for every possible interpretation I of L and every possible assignment ϕ in I.

Logical validity plays the same role in first-order logic that tautological plays in propositional logic. For example, we say that two formulas A and B of a first-order language L are *logically equivalent* if $A \leftrightarrow B$ is logically valid.

Example 4:

The formulas $x_1 = x_1$ and $\forall x_1 \exists x_2(x_1 = x_2)$ are logically valid; the formula $x_1 = x_2$ is not logically valid. □

Example 5:

The formula $\forall x_n A \rightarrow A$ is logically valid.

Proof: Let I be any interpretation of L and let ϕ be any assignment in I;

we are required to show that ϕ satisfies $\forall x_n A \to A$. Suppose not; that is, suppose that $\phi(\forall x_n A \to A) = F$. By TA$\to$, we have $\phi(\forall x_n A) = T$ and $\phi(A) = F$. Now ϕ is an x_n-variant of itself, so if we apply TA$\forall$ to $\phi(\forall x_n A) = T$, we obtain $\phi(A) = T$. This contradicts $\phi(A) = F$. $\square$

Example 6:

The formula $R(x_1, c) \to \exists x_2 R(x_1, x_2)$ is logically valid.

Proof: Let I be any interpretation of L and let ϕ be any assignment in I. Suppose ϕ does not satisfy the given formula. From TA$\to$, we obtain

$$\textbf{(1)} \ \phi(R(x_1, c)) = T; \qquad \textbf{(2)} \ \phi(\exists x_2 R(x_1, x_2)) = F.$$

From (1) we obtain $\langle \phi(x_1), c_I \rangle \in R_I$ (use TA(R), then CONST SYM). Define $\psi: \text{VAR} \to D_I$ by $\psi(x_k) = \phi(x_k)$ for $k \neq 2$ and $\psi(x_2) = c_I$. Clearly, ψ is an x_2-variant of ϕ; TA$\exists$ applied to (2) gives $\psi(R(x_1, x_2)) = F$, or $\langle \psi(x_1), \psi(x_2) \rangle \notin R_I$. But $\psi(x_1) = \phi(x_1)$ and $\psi(x_2) = c_I$; hence $\langle \phi(x_1), c_I \rangle \notin R_I$. This contradicts the earlier result $\langle \phi(x_1), c_I \rangle \in R_I$. $\square$

Example 7:

The formula $\exists x_1 \forall x_2 A \to \forall x_2 \exists x_1 A$ is logically valid.

Proof: Let I be any interpretation of L and let ϕ be any assignment in I. Suppose, however, that ϕ does not satisfy the given formula. By TA$\to$, we have $\phi(\exists x_1 \forall x_2 A) = T$ and $\phi(\forall x_2 \exists x_1 A) = F$. Apply TA$\exists$ and TA$\forall$ to obtain

$$\textbf{(1)} \ \phi'(\forall x_2 A) = T \ (\phi' \text{ an } x_1\text{-variant of } \phi).$$
$$\textbf{(2)} \ \phi''(\exists x_1 A) = F \ (\phi'' \text{ an } x_2\text{-variant of } \phi).$$

The next step is to construct an assignment ψ in I that is an x_2-variant of ϕ' and at the same time is an x_1-variant of ϕ''. Let

$$\psi(x_k) = \begin{cases} \phi(x_k) & k > 2; \\ \phi'(x_1) & k = 1 \text{ (so } \psi \text{ is an } x_2\text{-variant of } \phi'); \\ \phi''(x_2) & k = 2 \text{ (so } \psi \text{ is an } x_1\text{-variant of } \phi''). \end{cases}$$

Now apply TA$\forall$ to (1) to obtain $\psi(A) = T$ and apply TA$\exists$ to (2) to obtain $\psi(A) = F$. This gives a contradiction. $\square$

Example 8 (Russell's paradox):

Let A be the formula

$$\neg \exists x_1 \forall x_2 [R(x_2, x_1) \leftrightarrow \neg R(x_2, x_2)].$$

We show that A is logically valid. Note that A is the *negation* of the formula in Example 1 of Section 4.2, except that $\in$ has been replaced by a general 2-ary relation symbol R. Let B be the formula $R(x_2, x_1) \leftrightarrow \neg R(x_2, x_2)$, so A can be written as $\neg \exists x_1 \forall x_2 B$. Let I be an interpretation of L and let ϕ be an assignment in I. Suppose $\phi(A) = F$—that is, $\phi(\neg \exists x_1 \forall x_2 B) = F$. Then $\phi(\exists x_1 \forall x_2 B) = T$; hence there is an assignment ϕ' in I that is an x_1-variant of ϕ such that $\phi'(\forall x_2 B) = T$. Define ψ by

$$\psi(x_k) = \begin{cases} \phi'(x_k) & k \neq 2; \\ \phi'(x_1) & k = 2. \end{cases}$$

Now ψ is an x_2-variant of ϕ'; hence $\psi(B) = T$. But B is $R(x_2, x_1) \leftrightarrow \neg R(x_2, x_2)$, and so, by TA$\leftrightarrow$ and TA$\neg$, we obtain $\psi(R(x_2, x_1)) \neq \psi(R(x_2, x_2))$. There are two cases to consider. Suppose $\psi(R(x_2, x_1)) = T$ and $\psi(R(x_2, x_2)) = F$; we then have (1) $\langle \psi(x_2), \psi(x_1) \rangle \in R_I$; (2) $\langle \psi(x_2), \psi(x_2) \rangle \notin R_I$. Now $\psi(x_1) = \phi'(x_1)$ and $\psi(x_2) = \phi'(x_1)$; hence $\psi(x_1) = \psi(x_2)$. Let a be this common value. From (1) and (2), we then obtain $\langle a, a \rangle \in R_I$ and $\langle a, a \rangle \notin R_I$, a contradiction. A similar contradiction occurs in the case $\psi(R(x_2, x_1)) = F$ and $\psi(R(x_2, x_2)) = T$. $\square$

Example 9:

The formula $\forall x(A \lor B) \to (A \lor \forall xB)$ is not logically valid. To prove this, it suffices to find an interpretation I and an assignment ϕ in I that does not satisfy the formula. Let A be the formula $O(x)$, where $O(x)$ means x is odd; let B be the formula $E(x)$, where $E(x)$ means x is even. The given formula then becomes

$$\forall x[O(x) \lor E(x)] \to [O(x) \lor \forall xE(x)].$$

Let $\mathbb{N}$ be the domain of the interpretation. Then $\forall x[O(x) \lor E(x)]$ is true in I and $\forall xE(x)$ is false in I. Moreover, there is an assignment ϕ in I such that $\phi(O(x)) = F$ (e.g., $\phi(x) = 2$); this assignment does not satisfy the formula as required. $\square$

Theorem 1 (Modus ponens for interpretations): Let I be an interpretation of L. If both A and $A \to B$ are true in I, then B is true in I.

Proof: Let ϕ be any assignment in I. By assumption, we have $\phi(A) = T$ and $\phi(A \to B) = T$; TA$\to$, applied to the second result, gives $\phi(A) = F$ or $\phi(B) = T$. But $\phi(A) = T$; hence $\phi(B) = T$ as required. ∎

Definition 6: Let A be a formula whose free variables are $y_1, \ldots, y_k$. The *closure* of A, denoted A', is the formula $\forall y_1 \ldots \forall y_k A$. Note that the closure A' of A has no free variables and thus is a sentence. For example, the closure of $x_1 = x_2$ is the sentence $\forall x_1 \forall x_2(x_1 = x_2)$.

Theorem 2 (Closure theorem for interpretations): Let I be an interpretation of L.

(1) A is true in I $\Leftrightarrow \forall x_n A$ is true in I.

(2) A is true in I $\Leftrightarrow$ the closure A' of A is true in I.

Proof of (1): Assume A is true in I, and let ϕ be an arbitrary assignment in I. To prove $\phi(\forall x_n A) = T$, it suffices to show that $\psi(A) = T$ for every assignment ψ in I that is an x_n-variant of ϕ. But A is true in I, so all assignments in I satisfy A. Now assume $\forall x_n A$ is true in I. By Example 5, $\forall x_n A \rightarrow A$ is logically valid and hence is true in I. Theorem 1 now applies, and A is true in I.

Proof of (2): This follows by repeated application of (1). Let $y_1, \ldots, y_k$ be the free variables in A. We have

$$A \text{ is true in } I \Leftrightarrow \forall y_k A \text{ is true in } I$$
$$\vdots$$
$$\Leftrightarrow \forall y_1 \ldots \forall y_k A \text{ is true in } I. \quad \blacksquare$$

Corollary 1: Let Γ be a set of formulas of L and let $\Gamma' = \{A' : A \in \Gamma\}$. Then Γ and Γ' have the same models.

Proof: Left as an exercise for the reader. $\blacksquare$

Example 10:

The closure of $x_1 = x_1$ is $\forall x_1 (x_1 = x_1)$; both formulas are logically valid. The closure of $x_1 = x_2$ is $\forall x_1 \forall x_2 (x_1 = x_2)$; both formulas are true in any interpretation I such that D_I has just one element; if D_I has more than one element, $x_1 = x_2$ is neither true nor false in I and $\forall x_1 \forall x_2 (x_1 = x_2)$ is false in I. $\quad \square$

─────────── *Exercises for Section 4.3* ───────────

1. For each assignment ϕ in the standard interpretation of L_{NN} and each term t or formula A of L_{NN}, find $\phi(t)$ or $\phi(A)$. First find the result informally, then use the conditions CONST SYM, FNC SYM, TA($=$), and TA(R).
   (1) $\phi(x_1) = 3$, t is $((x_1 + S0) \times (Sx_1 + SS0))$.
   (2) $\phi(x_1) = 1$, $\phi(x_2) = 6$, t is $(S(x_1 + S0) + SS(x_2 \times SS0))$.
   (3) $\phi(x_1) = 2$, A is $x_1 + S0 = SSS0$.
   (4) $\phi(x_1) = 3$, A is $x_1 + S0 = SSS0$.
   (5) $\phi(x_1) = 2$, A is $x_1 + S0 < SS0 + SS0$.
   (6) $\phi(x_1) = 3$, A is $x_1 + SS0 < SS0 + SS0$.
2. Prove the following for the standard interpretation of L_{NN}:
   (1) $\exists x_1 (SS0 + x_1 = SSSSS0)$ is true in $\mathcal{N}$.
   (2) $\forall x_1 (0 < x_1)$ is false in $\mathcal{N}$.
   (3) $SS0 + x_1 = SSSSS0$ is neither true nor false in $\mathcal{N}$.
3. Let I be an interpretation of L and let t be a variable-free term of L (a term with no variables). Show that $\phi(t) = \psi(t)$ for any two assignments ϕ and ψ in L.

4. Let I be an interpretation of L. Show that A is false in I if and only if $\exists x A$ is false in I.
5. Give an example of a formula A and an interpretation I such that the following does *not* hold: A is false in I $\Leftrightarrow A'$ is false in I.
6. (∀-*distribution and* ∃-*distribution rules*) Let $A \to B$ be true in an interpretation I. Prove the following:
   (1) $\forall x_n A \to \forall x_n B$ is true in I.
   (2) $\exists x_n A \to \exists x_n B$ is true in I.
7. Let L have 1-ary relation symbol R and let I be an interpretation of L with domain D_{I} (nonempty by definition). Prove the following:
   (1) $\forall x_1 R(x_1)$ is true in I $\Leftrightarrow a \in R_{\mathrm{I}}$ for all $a \in D_{\mathrm{I}}$ (i.e., $R_{\mathrm{I}} = D_{\mathrm{I}}$).
   (2) $\exists x_1 R(x_1)$ is true in I $\Leftrightarrow a \in R_{\mathrm{I}}$ for some $a \in D_{\mathrm{I}}$ (i.e., $R_{\mathrm{I}} \neq \varnothing$).
8. Prove that the following formulas are logically valid:
   (1) $\forall x_1 R(x_1) \to R(c)$.　　　　(5) $R(x_2) \to \exists x_1 R(x_1)$.
   (2) $\forall x_1 R(x_1) \to R(x_2)$.　　　(6) $\exists x_2 R(x_2) \to \exists x_1 R(x_1)$.
   (3) $\forall x_1 R(x_1) \to \forall x_2 R(x_2)$.　(7) $\forall x_1 R(x_1) \to \exists x_1 R(x_1)$.
   (4) $R(c) \to \exists x_1 R(x_1)$.　　　(8) $\forall x_1 P(x_1, c) \to P(x_2, c)$.
9. Show that the following formulas are logically valid:
   (1) $(x_1 = x_2) \to [R(c, F(x_1)) \to R(c, F(x_2))]$;
   (2) $\forall x_2 [R(x_2) \leftrightarrow (x_2 = x_1)] \leftrightarrow [R(x_1) \wedge \forall x_2 (R(x_2) \to (x_2 = x_1))]$.
10. Prove the following:
   (1) A is logically valid if and only if $\forall x A$ is logically valid.
   (2) A is satisfiable if and only if $\exists x A$ is satisfiable.
11. Show that $A \to \exists x_n A$ is logically valid.
12. (*Distribution of quantifiers*) Show that the following formulas are logically valid:
   (1) $\exists x_n (A \vee B) \leftrightarrow (\exists x_n A \vee \exists x_n B)$.
   (2) $\forall x_n (A \wedge B) \leftrightarrow (\forall x_n A \wedge \forall x_n B)$.
13. Show the following:
   (1) $(\forall x_n A \vee \forall x_n B) \to \forall x_n (A \vee B)$ is logically valid.
   (2) $\forall x_n (A \vee B) \to (\forall x_n A \vee \forall x_n B)$ is not logically valid.
   (3) $\exists x_n (A \wedge B) \to (\exists x_n A \wedge \exists x_n B)$ is logically valid.
   (4) $(\exists x_n A \wedge \exists x_n B) \to \exists x_n (A \wedge B)$ is not logically valid.
   (5) $(\exists x_n A \vee B) \to \exists x_n (A \vee B)$ is logically valid.
   (6) $\exists x_n (A \vee B) \to (\exists x_n A \vee B)$ is not logically valid.
   (7) $\forall x_n (A \wedge B) \to (\forall x_n A \wedge B)$ is logically valid.
   (8) $(\forall x_n A \wedge B) \to \forall x_n (A \wedge B)$ is not logically valid.
14. Prove the following:
   (1) $R(x_1) \leftrightarrow R(x_2)$ is not logically valid.
   (2) $\forall x_1 R(x_1) \leftrightarrow \forall x_2 R(x_2)$ is logically valid.
15. The sentence "All A are B" is translated as $\forall x_n (A \to B)$ or $\neg \exists x_n (A \wedge \neg B)$. Similarly, "No A are B" is translated as $\neg \exists x_n (A \wedge B)$ or $\forall x_n (A \to \neg B)$. Show that the following are logically valid:
   (1) $\forall x_n (A \to B) \leftrightarrow \neg \exists x_n (A \wedge \neg B)$.　　(2) $\neg \exists x_n (A \wedge B) \leftrightarrow \forall x_n (A \to \neg B)$.
16. (*Reduction of syllogisms to set theory*) Let I be an interpretation of L. Prove the following:
   (1) $\forall x_1 [Q(x_1) \to R(x_1)]$ is true in I if and only if $Q_{\mathrm{I}} \subseteq R_{\mathrm{I}}$.
   (2) $\exists x_1 [Q(x_1) \wedge R(x_1)]$ is true in I if and only if $Q_{\mathrm{I}} \cap R_{\mathrm{I}} \neq \varnothing$.
   (3) $\exists x_1 [Q(x_1) \wedge \neg R(x_1)]$ is true in I if and only if $Q_{\mathrm{I}} \cap (D_{\mathrm{I}} - R_{\mathrm{I}}) = \varnothing$.
   (4) $\forall x_1 [Q(x_1) \to \neg R(x_1)]$ is true in I if and only if $Q_{\mathrm{I}} \cap R_{\mathrm{I}} = \varnothing$.
17. Let L_{R} be the language of relations and let $A_1, \ldots, A_k, B$ be sentences of L. The argument form $A_1, \ldots, A_k \therefore B$ is *valid* if B is true in every interpretation in which $A_1, \ldots, A_k$ are true. Use this definition to prove the validity of the following argument forms:
   (1) $\forall x_1 [L(x_1) \to I(x_1)]$, $L(g) \therefore I(g)$.
   (2) $\forall x_1 [A(x_1) \to B(x_1)]$, $\forall x_1 [B(x_1) \to C(x_1)] \therefore \forall x_1 [A(x_1) \to C(x_1)]$.
   (3) $\forall x_1 [A(x_1) \to B(x_1)]$, $\exists x_1 [C(x_1) \wedge A(x_1)] \therefore \exists x_1 [C(x_1) \wedge B(x_1)]$.

18. Show that the following formulas are logically valid:
    (1) $\forall x_1 \exists x_2[P(x_1) \to P(x_2)]$.
    (2) $\exists x_1[P(x_1) \to Q(x_2)] \to \exists x_1[P(x_1) \to Q(x_1)]$.

19. (*Properties of equality*) Prove that the following formulas are logically valid:
    (1) $(x_1 = x_2) \to (x_2 = x_1)$ (symmetric property).
    (2) $(x_1 = x_2) \to [(x_2 = x_3) \to (x_1 = x_3)]$ (transitive property).
    (3) $(x_k = t) \to [F(x_1, \ldots, x_k, \ldots, x_n) = F(x_1, \ldots, t, \ldots, x_n)]$.
    (4) $(x_k = t) \to [R(x_1, \ldots, x_k, \ldots, x_n) \to R(x_1, \ldots, t, \ldots, x_n)]$.

20. Let L be a first-order language with formulas A, B, C, and D. Prove the following:
    (1) $A \leftrightarrow B$ is logically valid $\Leftrightarrow A \to B$ and $B \to A$ are logically valid.
    (2) $A \leftrightarrow B$ is logically valid $\Leftrightarrow \phi(A) = \phi(B)$ holds for every interpretation I of L and every assignment ϕ in I.
    (3) If $A \leftrightarrow B$ is logically valid, then $\neg A \leftrightarrow \neg B$ is logically valid.
    (4) If $A \leftrightarrow B$ and $C \leftrightarrow D$ are logically valid, then $(A \vee C) \leftrightarrow (B \vee D)$ is logically valid.
    (5) If $A \leftrightarrow B$ is logically valid, then $\forall x A \leftrightarrow \forall x B$ is logically valid.
    (6) If $A \leftrightarrow B$ is logically valid, then $\exists x A \leftrightarrow \exists x B$ is logically valid.

21. (*Replacement theorem for first-order logic, semantic version*) Let $A \leftrightarrow B$ be logically valid and let C be a formula in which A occurs. Let C* be the formula obtained from C by replacing one or more of the occurrences of A with B. Show that $C \leftrightarrow C^*$ is logically valid. (*Hint:* Use induction on the number of occurrences of $\neg$, $\vee$, and $\forall$ in C. First consider the special case in which the given formula C is the same as A; use results of Exercise 20.)

4.4

Agreement Theorem and Substitution for Free Variables

In this section, we prove two results about assignments. The first is called the *agreement theorem*; the second is the *substitution theorem for assignments*. These two theorems play a fundamental role in the semantics of first-order logic. Both are used in the proof of the soundness theorem for first-order logic. Another application of the agreement theorem is the important result that a sentence is either true or false in a given interpretation. The substitution theorem for assignments is related to the important idea of substituting a term t for the free occurrences of x in A.

Agreement Theorem

An informal definition of x *occurs free in A* has already been given (see Section 4.1). A precise inductive definition proceeds as follows.

Definition 1: The variable x *occurs free in A* (or x *is free in A*, or x *is a free variable of A*) if and only if one of the following holds:

(1) A is atomic and x occurs in A.

(2) A is $\neg B$ and x occurs free in B.

(3) A is $B \vee C$ and x occurs free in B or x occurs free in C.

(4) A is $\forall yB$ with $y \neq x$ and x occurs free in B.

Theorem 1 (Agreement): Let I be an interpretation of L.

(1) $\phi(t) = \psi(t)$ for any two assignments ϕ and ψ in I that agree on all variables in t.

(2) $\phi(A) = \psi(A)$ for any two assignments ϕ and ψ in I that agree on all free variables in A.

(3) If A is a sentence, then $\phi(A) = \psi(A)$ for any two assignments ϕ and ψ in I.

Proof of (1): Let ϕ and ψ be two assignments in I that agree on all variables that occur in t. The proof that $\phi(t) = \psi(t)$ is by induction on the number of symbols in t.

- *t is a variable or a constant symbol* If t is an individual variable x_n, then $\phi(t) = \psi(t)$, since it is given that ϕ and ψ agree on all variables in t. Suppose t is the constant symbol c. By CONST SYM, we have $\phi(t) = c_I$ and $\psi(t) = c_I$; hence $\phi(t) = \psi(t)$, as required.

- *t is a term of the form $F(t_1, \ldots, t_n)$* Now ϕ and ψ agree on all variables in t; hence ϕ and ψ certainly agree on all variables in $t_1, \ldots, t_n$. Moreover, the terms $t_1, \ldots, t_n$ have fewer symbols than t; hence the induction hypothesis applies and $\phi(t_k) = \psi(t_k)$ for $1 \leq k \leq n$. Finally,

$$
\begin{aligned}
\phi(t) &= \phi(F(t_1, \ldots, t_n)) & \\
&= F_I(\phi(t_1), \ldots, \phi(t_n)) & \text{FNC SYM} \\
&= F_I(\psi(t_1), \ldots, \psi(t_n)) & \text{Induction hypothesis} \\
&= \psi(F(t_1, \ldots, t_n)) & \text{FNC SYM} \\
&= \psi(t).
\end{aligned}
$$

Proof of (2): We are required to show that any two assignments in I that agree on all free variables in A assign the same truth value to A. The proof is by induction on the number of occurrences of $\neg$, $\vee$, and $\forall$ in A.

Atomic In this case, A is $(s = t)$ or $R(t_1, \ldots, t_n)$. Let ϕ and ψ be two assignments in I that agree on all free variables in A. By (1), $\phi(s) = \psi(s)$, $\phi(t) = \psi(t)$, and $\phi(t_k) = \psi(t_k)$ for $1 \leq k \leq n$. The following equivalences show that $\phi(s = t) = \psi(s = t)$ and $\phi(R(t_1, \ldots, t_n)) = \psi(R(t_1, \ldots, t_n))$:

$$
\begin{aligned}
\phi(s = t) \text{ is } T &\Leftrightarrow \phi(s) \text{ and } \phi(t) \text{ are the same element of } D_I & \text{TA}(=) \\
&\Leftrightarrow \psi(s) \text{ and } \psi(t) \text{ are the same element of } D_I & \text{Part (1)} \\
&\Leftrightarrow \psi(s = t) \text{ is } T & \text{TA}(=).
\end{aligned}
$$

$$
\begin{aligned}
\phi(R(t_1, \ldots, t_n)) = T &\Leftrightarrow \langle \phi(t_1), \ldots, \phi(t_n) \rangle \in R_I & \text{TA}(R) \\
&\Leftrightarrow \langle \psi(t_1), \ldots, \psi(t_n) \rangle \in R_I & \text{Part (1)} \\
&\Leftrightarrow \psi(R(t_1, \ldots, t_n)) = T & \text{TA}(R).
\end{aligned}
$$

Negation Suppose A is $\neg B$. Let ϕ and ψ be two assignments in I that agree on all free variables in A; $\phi(B) = \psi(B)$ by induction hypothesis. The following equivalence shows that $\phi(A) = \psi(A)$:

$$\begin{aligned} \phi(A) = T &\Leftrightarrow \phi(B) = F \qquad \text{TA}\neg \\ &\Leftrightarrow \psi(B) = F \qquad \text{Induction hypothesis} \\ &\Leftrightarrow \psi(A) = T \qquad \text{TA}\neg. \end{aligned}$$

Disjunction Suppose A is $B \lor C$ and let ϕ and ψ agree on all free variables in A; $\phi(B) = \psi(B)$ and $\phi(C) = \psi(C)$ by induction hypothesis. The following equivalence shows that $\phi(A) = \psi(A)$:

$$\begin{aligned} \phi(A) = T &\Leftrightarrow \phi(B) = T \text{ or } \phi(C) = T \qquad \text{TA}\lor \\ &\Leftrightarrow \psi(B) = T \text{ or } \psi(C) = T \qquad \text{Induction hypothesis} \\ &\Leftrightarrow \psi(A) = T \qquad\qquad\quad \text{TA}\lor. \end{aligned}$$

Quantifier Suppose A is $\forall x_n B$ and let ϕ and ψ agree on all free variables in A; it suffices to prove that $\phi(\forall x_n B) = T \Leftrightarrow \psi(\forall x_n B) = T$. Suppose $\phi(\forall x_n B) = T$. To prove $\psi(\forall x_n B) = T$, we prove $\psi'(B) = T$, where ψ' is an arbitrary assignment in I that is an x_n-variant of ψ. Let ϕ' be the assignment in I defined by

$$\phi'(x_k) = \begin{cases} \phi(x_k) & k \neq n; \\ \psi'(x_n) & k = n. \end{cases}$$

Clearly ϕ' is an x_n-variant of ϕ; hence from $\phi(\forall x_n B) = T$ and TA$\forall$, we obtain $\phi'(B) = T$. We now argue that ϕ' and ψ' agree on all free variables in B. If this is so, then the induction hypothesis gives $\phi'(B) = \psi'(B)$, and since $\phi'(B) = T$, we obtain $\psi'(B) = T$ as required.

Let x_k be a variable that occurs free in B; we need to show that $\phi'(x_k) = \psi'(x_k)$. For $k = n$, $\phi'(x_n) = \psi'(x_n)$ by the construction of ϕ'. Suppose $k \neq n$. Then x_k is free in A (x_k is free in B, A is $\forall x_n B$, and $k \neq n$); since ϕ and ψ agree on all free variables in A, $\phi(x_k) = \psi(x_k)$. We also have $\phi'(x_k) = \phi(x_k)$ (ϕ and ϕ' are x_n-variants and $k \neq n$) and $\psi'(x_k) = \psi(x_k)$ (ψ and ψ' are x_n-variants and $k \neq n$). Hence $\phi'(x_k) = \psi'(x_k)$ as required.

By an entirely similar argument, $\psi(\forall x_n B) = T$ implies $\phi(\forall x_n B) = T$.

Proof of (3): Assume A is a sentence; let ϕ and ψ be any two assignments in I. Since A has no free variables, ϕ and ψ automatically agree on all free variables in A; by (2), $\phi(A) = \psi(A)$ as required. ∎

Corollary 1: Let I be an interpretation of L and let A be a sentence of L.

(1) If $\phi(A) = T$ for some assignment ϕ in I, then A is true in I.
(2) If $\phi(A) = F$ for some assignment ϕ in I, then A is false in I.
(3) A is either true or false in I.

Proof: Left as an exercise for the reader. ∎

Example 1:

We will show that the sentence $\forall x_1 \forall x_2 (x_1 = x_2)$ is false in any interpretation I such that D_I has at least two distinct elements. Since $\forall x_1 \forall x_2 (x_1 = x_2)$ is a sentence, it is either true or false in I (Corollary 1). Suppose $\forall x_1 \forall x_2 (x_1 = x_2)$ is true in I. By the closure theorem for interpretations (see Section 4.3), $(x_1 = x_2)$ is also true in I. But the domain D_I has at least two elements, so there is an assignment in I that does not satisfy $(x_1 = x_2)$. This gives a contradiction; hence $\forall x_1 \forall x_2 (x_1 = x_2)$ is false in I, as claimed. $\square$

Example 2:

The formula $\exists x_n (A \vee B) \to (\exists x_n A \vee B)$ is logically valid (provided that x_n is not free in B). If not, then there is some interpretation I of L and some assignment ϕ in I such that ϕ does not satisfy the formula. By TA$\to$, we have

(1) $\phi(\exists x_n (A \vee B)) = T$;

(2) $\phi(\exists x_n A \vee B) = F$.

From (1) and TA$\exists$, we obtain $\psi(A \vee B) = T$, where ψ is an x_n-variant of ϕ; TA$\vee$ then gives $\psi(A) = T$ or $\psi(B) = T$.

Suppose $\psi(B) = T$. From (2), we have $\phi(B) = F$; hence $\phi(B) \neq \psi(B)$. On the other hand, ϕ and ψ are x_n-variants and x_n does not occur free in B (hypothesis); hence ϕ and ψ agree on all free variables in B. By the agreement theorem, $\phi(B) = \psi(B)$. This contradicts $\phi(B) \neq \psi(B)$.

Suppose $\psi(A) = T$. From (2), we have $\phi(\exists x_n A) = F$; applying TA$\exists$ to this result gives $\psi(A) = F$. This contradicts $\psi(A) = T$. $\square$

Comment 1 The requirement in Example 2 that x_n not be free in B cannot be omitted. Consider, for example, the following formula of L_{NN}:

$$\exists x_1 [(x_1 < 0) \vee (0 < x_1)] \to [\exists x_1 (x_1 < 0) \vee (0 < x_1)].$$

This formula is not true in the standard interpretation of L_{NN}: Any assignment ϕ in $\mathcal{N}$ with $\phi(x_1) = 0$ does not satisfy the formula.

Comment 2 The converse of the formula in Example 2 is also logically valid (left as an exercise for the reader; this time, the requirement that x_n not be free in B may be omitted). This result and Example 2 combine to give the following *prenex rule:* If x_n is not free in B, then $\exists x_n (A \vee B) \leftrightarrow (\exists x_n A \vee B)$ is logically valid. Informally, this means that whenever x_n is not free in B, the two formulas $\exists x_n (A \vee B)$ and $(\exists x_n A \vee B)$ "say the same thing." To illustrate, let us consider the following two sentences of L_{NN}:

(1) $\forall x_1 [\exists x_2 (Sx_2 = x_1) \vee (x_1 = 0)].$

(2) $\forall x_1 \exists x_2 [(Sx_2 = x_1) \vee (x_1 = 0)].$

Sentence (1) is perhaps the "natural way" to state in L_{NN} that every natural number is the successor of some natural number or is zero. Sentence (2) is obtained from sentence (1) by pulling the quantifier $\exists x_2$ to the front; this is allowed because the variable x_2 does not occur free in the formula $(x_1 = 0)$. Thus the sentences (1) and (2) "say the same thing." (Prenex rules and their applications are further discussed in Section 5.6.)

Add $\forall$-Rule: In Chapter 5, we will give axioms and rules of inference for first-order logic. In addition to axioms and rules for manipulating $\neg$ and $\vee$, there will be axioms and rules for introducing and eliminating the quantifier $\forall$. The rule for introducing a quantifier is called the *add $\forall$-rule,* and is stated as follows:

$$\frac{A \vee B}{A \vee \forall x_n B} \quad \text{(provided } x_n \text{ is not free in } A\text{).}$$

Note the restriction that x_n does not occur free in A. This restriction is necessary, as the following example will show. Consider these two formulas of the language L_{NN}:

$$\frac{[(x_1 < S0) \vee (x_1 = S0)] \vee (S0 < x_1)}{[(x_1 < S0) \vee (x_1 = S0)] \vee \forall x_1(S0 < x_1)}$$

Note that the restriction in the add $\forall$-rule has been violated: x_1 *does* occur free in $[(x_1 < S0) \vee (x_1 = S0)]$. In the standard interpretation of L_{NN}, the formula above the line is true but the formula below the line is neither true nor false. This is not acceptable; we require that rules of inference preserve truth.

We now use the agreement theorem to prove that the add $\forall$-rule preserves truth, provided that the restriction on the variable is satisfied. This result is a key step in the proof of the soundness theorem for first-order logic.

Theorem 2 (Add $\forall$-rule preserves truth): Let I be an interpretation of L and let x_n be a variable that does not occur free in A. If $A \vee B$ is true in I, then $A \vee \forall x_n B$ is also true in I.

Proof: Suppose there is an assignment ϕ in I such that $\phi(A \vee \forall x_n B) = F$. From TA$\vee$, we have $\phi(A) = F$ and $\phi(\forall x_n B) = F$. TA$\forall$, applied to $\phi(\forall x_n B) = F$, gives $\psi(B) = F$, where ψ is an x_n-variant of ϕ. Since ϕ and ψ are x_n-variants, they agree everywhere except possibly at x_n. But x_n does not occur free in A; hence ϕ and ψ agree on all free variables in A. By the agreement theorem, $\phi(A) = \psi(A)$. But $\phi(A) = F$; hence $\psi(A) = F$. We also have $\psi(B) = F$; hence $\psi(A \vee B) = F$. This contradicts the hypothesis that $A \vee B$ is true in I. ∎

—— *Substitution and the Substitution Theorem* ——
for Assignments

We now come to a critical topic in first-order logic: substituting a term for a free variable. This type of substitution is important in formal systems for first-order logic (see Chapter 5). First we introduce some notation:

$s_x[t]$ The term obtained from s by replacing each occurrence of x in s with t.

$A_x[t]$ The formula obtained from A by replacing each *free* occurrence of x in A with t.

For example, in the language L_{NN}, if s is $(x_1 + SSS0)$ and t is $S0$, then $s_{x_1}[t]$ is $(S0 + SSS0)$; if A is the formula $\exists x_2(x_1 = SS0 \times x_2)$ and t is the term $SSSS0$, then $A_{x_1}[t]$ is the formula $\exists x_2(SSSS0 = SS0 \times x_2)$. Note that if x does not occur free in A, no substitution takes place and $A_x[t]$ is just A. Although these descriptions of $s_x[t]$ and $A_x[t]$ are perfectly straightforward, there are occasions when a precise inductive definition is needed.

Definition 2: Let s and t be terms. The term $s_x[t]$ is defined inductively as follows:

(1) If s is the variable x, then $s_x[t]$ is t; if s is a constant symbol c or a variable $y \neq x$, then $s_x[t]$ is s (no substitution takes place).
(2) If s is $F(u, \ldots, v)$, then $s_x[t]$ is $F(u_x[t], \ldots, v_x[t])$.

Lemma 1: Let I be an interpretation of L, let s and t be terms of L, and let ϕ and ψ be two assignments in I such that

(1) ϕ and ψ are x_n-variants;
(2) $\psi(x_n) = \phi(t)$.

Then $\psi(s) = \phi(s_{x_n}[t])$.

Proof: The proof is by induction on the number of symbols in the term s.

s is a variable or a constant symbol   Suppose s is the variable x_n. Then $s_{x_n}[t]$ is t and $\psi(s) = \phi(s_{x_n}[t])$ reduces to $\psi(x_n) = \phi(t)$; this is just hypothesis (2). Suppose s is a constant symbol c. Then $s_{x_n}[t]$ is also c, and $\psi(s) = \phi(s_{x_n}[t])$ reduces to $\psi(c) = \phi(c)$; this follows from the fact that $\psi(c) = c_I$ and $\phi(c) = c_I$ (CONST SYM property). Finally, suppose s is a variable x_k with $k \neq n$. Then $s_{x_n}[t]$ is x_k and $\psi(s) = \phi(s_{x_n}[t])$ reduces to $\psi(x_k) = \phi(x_k)$; this follows from the fact that ϕ and ψ are x_n-variants and $k \neq n$.

s is a term of the form $F(u, \ldots, v)$ In this case, $s_{x_n}[t]$ is $F(u_{x_n}[t], \ldots, v_{x_n}[t])$, and by induction hypothesis we have $\psi(u) = \phi(u_{x_n}[t]), \ldots, \psi(v) = \phi(v_{x_n}[t])$. The following calculations show that $\psi(s) = \phi(s_{x_n}[t])$:

$$\psi(s) = \psi(F(u, \ldots, v))$$
$$= F_i(\psi(u), \ldots, \psi(v)) \qquad \text{FNC SYM}$$
$$= F_i(\phi(u_{x_n}[t]), \ldots, \phi(v_{x_n}[t])) \qquad \text{Induction hypothesis}$$

$$\phi(s_{x_n}[t]) = \phi(F[u_{x_n}[t], \ldots, v_{x_n}[t]])$$
$$= F_i(\phi(u_{x_n}[t]), \ldots, \phi(v_{x_n}[t])) \qquad \text{FNC SYM.} \quad \blacksquare$$

Definition 3: The formula $A_x[t]$ is defined inductively as follows:

(1) If A is $(u = v)$, then $A_x[t]$ is $(u_x[t] = v_x[t])$.
(2) If A is $R(u, \ldots, v)$, then $A_x[t]$ is $R(u_x[t], \ldots, v_x[t])$.
(3) If A is $\neg B$, then $A_x[t]$ is $\neg B_x[t]$.
(4) If A is $B \vee C$, then $A_x[t]$ is $B_x[t] \vee C_x[t]$.
(5) If A is $\forall x B$, then $A_x[t]$ is A.
(6) If A is $\forall y B$ with $y \neq x$, then $A_x[t]$ is $\forall y B_x[t]$.

Now suppose x is free in the formula A. If we replace the free occurrences of x by a term t, we obtain the formula $A_x[t]$. But we want the formula $A_x[t]$ to "say the same thing about t that the original formula A says about x." Thus there are restrictions on when this type of substitution is allowed.

Example 3:

Let A be the formula $\exists x_1(x_2 = SS0 \times x_1)$ of L_{NN}; note that x_2 occurs free in A. Informally, A says that x_2 is even (in the standard interpretation). We now replace x_2 with some terms t:

Term t	$A_{x_2}[t]$	*Interpretation of* $A_{x_2}[t]$ *in* $\mathcal{N}$
$SSSS0$	$\exists x_1(SSSS0 = SS0 \times x_1)$	The number 4 is even.
$x_3 + SS0$	$\exists x_1(x_3 + SS0 = SS0 \times x_1)$	$x_3 + SS0$ is even.
x_1	$\exists x_1(x_1 = SS0 \times x_1)$	There is some number a such that $a = 2 \times a$.

In the last substitution, the meaning of $A_{x_2}[t]$ changes; the formula no longer says something about being even. The difficulty with this last substitution is that when we replace the free variable x_2 with x_1, the new variable x_1 comes within the scope of the quantifier $\exists x_1$; this type of substitution must be avoided. □

We will now give the precise conditions under which the free occurrences of x_n in A may be replaced by a term t. Two definitions are given. The first is a good working definition, but is somewhat informal; the second is a precise inductive definition used for proving theorems and coding (see Chapter 8).

Definition 4 (Informal): The term t is *substitutable for x in A* if this condition holds: No *free* occurrence of x in A is within the scope of $\forall y$ or $\exists y$, where y is a variable that occurs in t.

Thus t is *not* substitutable for x in A precisely when there is a variable y that occurs in t and there is a free occurrence of x in A that is within the scope of $\forall y$ or $\exists y$. We can picture this situation as follows:

$$\ldots \exists y(\ldots x \ldots) \ldots \qquad \qquad \text{Formula } A$$

$$\text{free occurrence of } x \text{ in } A$$

If we now replace x with t, and if y occurs in t, then this occurrence of y will become bound; this must be avoided. There are several cases in which t is automatically substitutable for x in A: when the term t is variable-free (has no variables); when the term t is x; when the variable x has no free occurrences in A; when the formula A has no quantifiers; and when no variable in t has a bound occurrence in A.

Definition 5 (Formal): The term t is *substitutable for x in A* if and only if one of the following holds:

(1) A is atomic (there are no quantifiers in an atomic formula).

(2) A is $\neg B$ and t is substitutable for x in B.

(3) A is $B \vee C$, t is substitutable for x in B, *and* t is substitutable for x in C.

(4) A is $\forall y B$ and x is not free in A.

(5) A is $\forall y B$, t is substitutable for x in B, and y does not occur in t.

Theorem 3 (Substitution theorem for assignments): Let I be an interpretation of L and let t be substitutable for x_n in A. Suppose ϕ and ψ are two assignments in I such that

(1) ϕ and ψ are x_n-variants;

(2) $\psi(x_n) = \phi(t)$.

Then $\psi(A) = \phi(A_{x_n}[t])$.

Proof: The proof is by induction on the number of occurrences of $\neg, \vee$, and $\forall$ in the formula A.

Atomic In this case, A is of the form $(u = v)$ or $R(u, \ldots, v)$. We will discuss the second case and leave the first case to the reader. If A is

$R(u, \ldots, v)$, then $A_{x_n}[t]$ is $R(u_{x_n}[t], \ldots, v_{x_n}[t])$. Let ϕ and ψ be two assignments in I that satisfy (1) and (2). By Lemma 1, we have $\psi(u) = \phi(u_{x_n}[t]), \ldots, \psi(v) = \phi(v_{x_n}[t])$. The following calculations show that $\psi(A) = \phi(A_{x_n}[t])$:

$$\psi(A) = T \Leftrightarrow \psi(R(u, \ldots, v)) = T$$
$$\Leftrightarrow \langle \psi(u), \ldots, \psi(v) \rangle \in R_I \qquad \text{TA}(R)$$
$$\Leftrightarrow \langle \phi(u_{x_n}[t]), \ldots, \phi(v_{x_n}[t]) \rangle \in R_I \qquad \text{Lemma 1}$$

$$\phi(A_{x_n}[t]) = T \Leftrightarrow \phi(R(u_{x_n}[t], \ldots, v_{x_n}[t])) = T$$
$$\Leftrightarrow \langle \phi(u_{x_n}[t]), \ldots, \phi(v_{x_n}[t]) \rangle \in R_I \qquad \text{TA}(R).$$

Negation Suppose A is $\neg B$, and let ϕ and ψ satisfy (1) and (2). Now t is substitutable for x_n in B; hence, by induction hypothesis, we have $\psi(B) = \phi(B_{x_n}[t])$. From this it easily follows that $\psi(\neg B) = \phi(\neg B_{x_n}[t])$; that is, $\psi(A) = \phi(A_{x_n}[t])$ as required.

Disjunction Suppose A is $B \vee C$, and let ϕ and ψ satisfy (1) and (2). Note that $A_{x_n}[t]$ is $B_{x_n}[t] \vee C_{x_n}[t]$. Now t is substitutable for x_n in B and also in C; hence the induction hypothesis gives $\psi(B) = \phi(B_{x_n}[t])$ and $\psi(C) = \phi(C_{x_n}[t])$. We leave it to the reader to verify that $\psi(B \vee C) = \phi(B_{x_n}[t] \vee C_{x_n}[t])$ and thus that $\psi(A) = \phi(A_{x_n}[t])$.

Quantifier Suppose A is $\forall x_k B$, and let ϕ and ψ satisfy (1) and (2). Now t is substitutable for x_n in A; hence one of the following holds: x_n is not free in A; t is substitutable for x_n in B and x_k does not occur in t.

First, suppose x_n is not free in A. Then $A_{x_n}[t]$ is just A, and we are required to show that $\psi(A) = \phi(A)$. Now ϕ and ψ are x_n-variants and x_n does not occur free in A; hence ϕ and ψ agree on all free variables in A; by the agreement theorem, $\psi(A) = \phi(A)$ as required.

Now assume that t is substitutable for x_n in B and that x_k does not occur in t. We can also assume that $n \neq k$ (if $n = k$, then A is $\forall x_n B$ and x_n is not free in A; this case has already been covered). It suffices to show that $\psi(\forall x_k B) = T \Leftrightarrow \phi(\forall x_k B_{x_n}[t]) = T$. First suppose that $\psi(\forall x_k B) = T$. We will prove $\phi(\forall x_k B_{x_n}[t]) = T$ by showing that $\phi'(B_{x_n}[t]) = T$, where ϕ' is an arbitrary assignment in I that is an x_k-variant of ϕ. Define ψ' by

$$\psi'(x_i) = \begin{cases} \psi(x_i) & i \neq k; \\ \phi'(x_k) & i = k. \end{cases}$$

Now ψ' is an x_k-variant of ψ and $\psi(\forall x_k B) = T$; hence $\psi'(B) = T$ by TA$\forall$. We will now prove that ϕ' and ψ' satisfy the hypotheses (1) and (2) of the theorem.

(1) ψ' and ϕ' are x_n-variants Let $i \neq n$. For $i = k$, $\psi'(x_i) = \phi'(x_i)$ by construction of ψ'. For $i \neq k$ we have $\psi'(x_i) = \psi(x_i)$ (ψ' and ψ are x_k-variants and $i \neq k$), $\psi(x_i) = \phi(x_i)$ (ϕ and ψ are x_n-variants and $i \neq n$), and $\phi(x_i) = \phi'(x_i)$ (ϕ and ϕ' are x_k-variants and $i \neq k$). Thus $\psi'(x_i) = \phi'(x_i)$ as required.

(2) $\psi'(x_n) = \phi'(t)$ The two assignments ϕ and ϕ' are x_k-variants and x_k

has no occurrences in t; hence ϕ and ϕ' agree on all variables in t. By part (1) of the agreement theorem, we have $\phi(t) = \phi'(t)$. But $\psi'(x_n) = \psi(x_n)$ (ψ' and ψ are x_k-variants and $n \neq k$) and $\psi(x_n) = \phi(t)$ (hypothesis (1) of the theorem), so $\psi'(x_n) = \phi'(t)$ as required.

Since ϕ' and ψ' satisfy (1) and (2) and t is substitutable for x_n in B, the induction hypothesis applies and $\psi'(B) = \phi'(B_{x_n}[t])$. But $\psi'(B) = T$, so $\phi'(B_{x_n}[t]) = T$ as required.

Now suppose $\phi(\forall x_k B_{x_n}[t]) = T$. By an argument similar to the one just given, $\psi(\forall x_k B) = T$. ∎

Corollary 2 (Special case $\psi = \phi$): Let I be an interpretation of L and let t be substitutable for x_n in A. If ϕ is any assignment in I such that $\phi(x_n) = \phi(t)$, then $\phi(A) = \phi(A_{x_n}[t])$.

Substitution Axioms: Let t be substitutable for x_n in A. The formula

$$\forall x_n A \rightarrow A_{x_n}[t] \quad (t \text{ substitutable for } x_n \text{ in } A)$$

is called a *substitution axiom* (of first-order logic). These axioms will be used in Chapter 5, when we introduce a formal system for first-order logic. A substitution axiom captures the Aristotelian principle of logic: One may pass from the universal to the particular. More technically, a substitution axiom allows us to *drop* the universal quantifier $\forall$ and *replace* the free occurrences of x_n in A by a term t that is substitutable for x_n in A. Note that the formula $\forall x_n A \rightarrow A$ is always a substitution axiom; this is so since x_n is substitutable for x_n in any formula and $A_{x_n}[x_n]$ is just A. A major application of Theorem 3 is the result that substitution axioms are logically valid.

Example 4:

Each of the following is a substitution axiom:

(1) $\forall x_1 Q(x_1) \rightarrow Q(c)$.   (4) $\forall x_1 \exists x_2 R(x_1, x_2) \rightarrow \exists x_2 R(c, x_2)$.
(2) $\forall x_1 Q(x_1) \rightarrow Q(x_1)$.   (5) $\forall x_1 \exists x_2 R(x_1, x_2) \rightarrow \exists x_2 R(x_1, x_2)$.
(3) $\forall x_1 Q(x_1) \rightarrow Q(x_2)$.   (6) $\forall x_1 \exists x_2 R(x_1, x_2) \rightarrow \exists x_2 R(x_3, x_2)$.

On the other hand, $\forall x_1 \exists x_2 R(x_1, x_2) \rightarrow \exists x_2 R(x_2, x_2)$ is *not* a substitution axiom; the condition that x_2 is substitutable for x_1 in the formula $\exists x_2 R(x_1, x_2)$ is *not* satisfied. Note that the formula is false if we interpret R as *less than* on $\mathbb{N}$; in this case, the formula says $\forall x_1 \exists x_2 (x_1 < x_2) \rightarrow \exists x_2 (x_2 < x_2)$, which is false. □

Theorem 4 (Logical validity of substitution axioms): Assume t is substitutable for x_n in A. Then the formula $\forall x_n A \rightarrow A_{x_n}[t]$ is logically valid.

Proof: Let I be an interpretation of L and let ϕ be an assignment in I. Suppose that ϕ does not satisfy $\forall x_n A \rightarrow A_{x_n}[t]$. By TA→, we have $\phi(\forall x_n A) = T$ and $\phi(A_{x_n}[t]) = F$. Let ψ be the assignment in I defined by $\psi(x_k) = \phi(x_k)$ for $k \neq n$ and $\psi(x_n) = \phi(t)$. Clearly, ψ is an x_n-variant of ϕ, so $\psi(A) = T$ (use $\phi(\forall x_n A) = T$ and TA$\forall$). We now have the following: t is substitutable for x_n in A; $\psi(x_n) = \phi(t)$; ψ is an x_n-variant of ϕ. The substitution theorem for assignments applies, and $\psi(A) = \phi(A_{x_n}[t])$. From this equation and $\psi(A) = T$, we obtain $\phi(A_{x_n}[t]) = T$. This contradicts $\phi(A_{x_n}[t]) = F$. ∎

Equality Axioms: Let t be substitutable for x_n in A. The formula

$$(x_n = t) \rightarrow (A \leftrightarrow A_{x_n}[t]) \qquad (t \text{ substitutable for } x_n \text{ in } A)$$

is called an *equality axiom* (of first-order logic). These axioms capture Leibnitz's principle: If two things are the same, then what is true of one is true of the other.

Example 5:

The following formulas are instances of the equality axiom:

(1) $(x_1 = x_2) \rightarrow [(x_1 = x_3) \leftrightarrow (x_2 = x_3)]$.
(2) $(x_1 = x_2) \rightarrow [(x_2 = x_1) \leftrightarrow (x_2 = x_2)]$.
(3) $(x_1 = x_2) \rightarrow [(S0 < x_1) \leftrightarrow (S0 < x_2)]$.   □

Theorem 5 (Logical validity of equality axioms): If t is substitutable for x_n in A, then $(x_n = t) \rightarrow (A \leftrightarrow A_{x_n}[t])$ is logically valid.

Proof: Let I be an interpretation of L and let ϕ be an assignment in I. Suppose, however, that ϕ does not satisfy the equality axiom. By TA$\rightarrow$ and TA$\leftrightarrow$, $\phi(x_n = t)$ is T and $\phi(A) \neq \phi(A_{x_n}[t])$. We then have $\phi(x_n) = \phi(t)$; hence the substitution theorem for assignments (special case $\psi = \phi$) applies and $\phi(A) = \phi(A_{x_n}[t])$. This contradicts $\phi(A) \neq \phi(A_{x_n}[t])$. ∎

The next result will be used in Section 5.6 to prove results on the proper procedure for replacement of bound variables in a formula (also see the exercises).

Theorem 6 (Reverse substitution): Let x and y be distinct variables such that (1) y is substitutable for x in A; (2) y does not occur free in A. Then x is substitutable for y in $A_x[y]$ and $(A_x[y])_y[x]$ is A.

Outline of proof: First of all, suppose s is a term such that y does not occur in s. We can show by induction on terms that $(s_x[y])_y[x]$ is s. Now let x and y satisfy (1) and (2). The proof is by induction on formulas. For A atomic, the proof uses the result that $(s_x[y])_y[x]$ is s; the details are left to the reader. The cases in which A is $\neg B$ or A is $B \vee C$ are straightforward and are left to the reader; it remains to consider the case in which A is $\forall z B$.

First suppose x is not free in A. Then $A_x[y]$ is just A, and we are required to show that x is substitutable for y in A and $A_y[x]$ is A; both properties follow from the hypothesis that y is not free in A. Now assume x is free in A; note that $z \neq x$. By hypothesis, y is substitutable for x in $\forall z B$; it follows that y is substitutable for x in B and $z \neq y$. By the induction hypothesis, x is substitutable for y in $B_x[y]$ and $(B_x[y])_y[x]$ is B. Now $A_x[y]$ is $\forall z B_x[y]$, and we have

$$(A_x[y])_y[x] = (\forall z B_x[y])_y[x] = \forall z (B_x[y])_y[x] = \forall z B = A.$$

Finally, since x is substitutable for y in $B_x[y]$ and $z \neq x$, x is substitutable for y in $\forall z B_x[y]$. Thus x is substitutable for y in $A_x[y]$ as required. ∎

Simultaneous substitution will be used extensively in Chapter 7, when we discuss expressibility of relations. This type of substitution is described as follows: Let s be a term, A a formula, $y_1, \ldots, y_n$ distinct variables, and $t_1, \ldots, t_n$ terms.

$s_{y_1, \ldots, y_n}[t_1, \ldots, t_n]$ Term obtained from s by replacing each occurrence of y_k with t_k, $1 \le k \le n$

$A_{y_1, \ldots, y_n}[t_1, \ldots, t_n]$ Formula obtained from A by replacing each *free* occurrence of y_k with t_k, $1 \le k \le n$.

—————————*Exercises for Section 4.4*—————————

1. Use the agreement theorem to show that the following are logically valid:
   (1) $A \rightarrow \forall x_n A$ (x_n not free in A).
   (2) $\exists x_n A \rightarrow A$ (x_n not free in A).
2. (*Prenex rules*) Show that the following are logically valid:
   (1) $\neg \forall x_n A \leftrightarrow \exists x_n \neg A$.
   (2) $\neg \exists x_n A \leftrightarrow \forall x_n \neg A$.
   (3) $(\exists x_n A \vee B) \leftrightarrow \exists x_n (A \vee B)$ (x_n not free in B).
   (4) $(\forall x_n A \vee B) \leftrightarrow \forall x_n (A \vee B)$ (x_n not free in B).
   (5) $(\exists x_n A \wedge B) \leftrightarrow \exists x_n (A \wedge B)$ (x_n not free in B).
   (6) $(\forall x_n A \wedge B) \leftrightarrow \forall x_n (A \wedge B)$ (x_n not free in B).
3. Give an inductive definition of x_n *occurs bound in A*.
4. Is the following a legitimate application of the add $\forall$-rule for formulas of L_{NN}? Justify your answer.

$$\frac{\neg(x_1 = 0) \vee (x_1 = 0)}{\neg(x_1 = 0) \vee \forall x_1(x_1 = 0)}$$

5. Give an example of formulas A and B such that x does not occur free in A and such that $(A \vee B) \rightarrow (A \vee \forall x B)$ is *not* logically valid. Does this contradict the add $\forall$-rule?
6. ($\exists$-*introduction and* $\forall$-*introduction rules*) Let I be an interpretation of L and assume that $A \rightarrow B$ is true in I. Prove the following:
   (1) $\exists x_n A \rightarrow B$ is true in I (x_n not free in B).
   (2) $A \rightarrow \forall x_n B$ is true in I (x_n not free in A).
7. Prove Corollary 1.
8. Let A be the formula $\exists x_2(x_1 = x_2 \times x_2)$ of L_{NN}; note that x_1 is free in A. For each term t, write the formula $A_{x_1}[t]$ and decide whether t is substitutable for x_1 in A: (1) $SSSS0$; (2) $x_3 + SSS0$; (3) $x_1 + x_3$; (4) x_2.
9. (*Substitution theorem*) Assume t is substitutable for x_n in A. Show that $A_{x_n}[t] \rightarrow \exists x_n A$ is logically valid.
10. (*Substitution rule and existential generalization*) Let I be an interpretation of L and let t be substitutable for x_n in A. Show the following:
    (1) If A is true in I, then $A_{x_n}[t]$ is true in I.
    (2) If $A_{x_n}[t]$ is true in I, then $\exists x_n A$ is true in I.

11. Identify the formulas that are substitution axioms:
    (1) $\forall x_2 \exists x_1 R(x_2, x_1) \rightarrow \exists x_1 R(c, x_1)$.   (3) $\forall x_2 \exists x_1 R(x_2, x_1) \rightarrow \exists x_1 R(x_2, x_1)$.
    (2) $\forall x_2 \exists x_1 R(x_2, x_1) \rightarrow \exists x_1 R(x_1, x_1)$.   (4) $\forall x_2 \exists x_1 R(x_2, x_1) \rightarrow \exists x_1 R(x_3, x_1)$.

12. Let A be a formula of L_{NN} with x_n free in A. Show that $\forall x_n A$ is true in $\mathcal{N}$ (the standard interpretation of L_{NN}) if and only if $A_{x_n}[0^k]$ is true in $\mathcal{N}$ for all $k \in \mathbb{N}$.

13. Explain why $\forall x_1 \forall x_2 R(x_1, x_2) \rightarrow \forall x_2 R(x_2, x_2)$ is not a substitution axiom. Is $\forall x_1 \forall x_2 R(x_1, x_2) \rightarrow \forall x_2 R(x_2, x_2)$ logically valid? Justify your answer.

14. Let t be substitutable for x_n in A and assume that x_n does not occur in t. Prove that the following are logically valid:
    (1) $A_{x_n}[t] \leftrightarrow \exists x_n (A \wedge (x_n = t))$.
    (2) $A_{x_n}[t] \leftrightarrow \forall x_n ((x_n = t) \rightarrow A)$.

15. Explain why each of the following formulas of L_{NN} is an equality axiom (identify t, A, and $A_x[t]$).
    (1) $(x_1 = x_2) \rightarrow [(Sx_1 = Sx_3) \leftrightarrow (Sx_2 = Sx_3)]$.
    (2) $(x_2 = x_1) \rightarrow [(x_2 < x_3) \leftrightarrow (x_1 < x_3)]$.
    (3) $(x_2 = x_3) \rightarrow [(x_1 + x_2 = x_2 + x_1) \leftrightarrow (x_1 + x_3 = x_3 + x_1)]$.

16. (*Variations of the equality axioms*) Refer to Theorem 3 in Section 5.1. Show that EQ AX(1)–EQ AX(4) are logically valid.

17. Use the inductive definition of $A_x[t]$ to show that if x is not free in A, then $A_x[t]$ is A.

18. Use the inductive definition that t is substitutable for x in A to prove the following:
    (1) If t has no variables, then t is substitutable for x in A.
    (2) If t is x, then t is substitutable for x in A.
    (3) If no variable in t has a bound occurrence in A, then t is substitutable for x in A.
    (4) If x has no free occurrences in A, then t is substitutable for x in A.

19. Give an example of a formula A and variables x and y such that $(A_x[y])_y[x]$ is not the same as A.

20. Let A be the formula $x_1 = x_2$. Show that $A_{x_1, x_2}[x_2, x_3]$ is not the same as $(A_{x_1}[x_2])_{x_2}[x_3]$.

21. Let I be an interpretation of L, let A be a formula of L, and let Γ be a set of formulas of L. Let x_n and x_k be distinct variables such that x_k has no free occurrences in any formula of $\Gamma \cup \{A\}$ and such that x_k is substitutable for x_n in A. Prove that if $\Gamma \cup \{\neg \forall x_n A\}$ is satisfiable, then $\Gamma \cup \{\neg \forall x_n A, \neg A_{x_n}[x_k]\}$ is also satisfiable. (*Hint:* Agreement theorem and substitution theorem for assignments.)

22. This problem outlines a proof of the following result: Let x and y be distinct variables such that y is substitutable for x in A and y does not occur free in A; then $\forall x A \leftrightarrow \forall y A_x[y]$ is logically valid. Prove the following:
    (1) $\forall x A \rightarrow A_x[y]$ is a substitution axiom (and hence logically valid).
    (2) If $B \rightarrow C$ is logically valid and y does not occur free in B, then $B \rightarrow \forall y C$ is logically valid.
    (3) $\forall x A \rightarrow \forall y A_x[y]$ is logically valid.
    (4) $\forall y A_x[y] \rightarrow \forall x A$ is logically valid (*Hint:* Reverse substitution).
    (5) $\forall x A \leftrightarrow \forall y A_x[y]$ is logically valid.

23. (*Replacement of bound variables*) Let x and y be distinct variables and let A^* be obtained from A by replacing an occurrence of $\forall x B$ or $\exists x B$ with $\forall y B_x[y]$ or $\exists y B_x[y]$ respectively, where y is substitutable for x in B and y does not occur free in B. Show that $A \leftrightarrow A^*$ is logically valid. (*Hint:* Use Exercise 22 and the replacement theorem for first-order logic (Exercise 21 in Section 4.3).)

24. Given a formula A, a variable x, and a term t, show that there is a formula A^* such that $A \leftrightarrow A^*$ is logically valid and t is substitutable for x in A^*. (*Hint:* Use induction on the number of connectives in A. The hard case is when A is of the form $\forall y B$, where y is a variable distinct from x; use Exercise 23.) We will illustrate the result. Suppose A is the formula $\exists x_2(x_1 = SS0 \times x_2)$ of L_{NN}; the term x_2 is not substitutable for x_1 in A. Let A^* be the formula $\exists x_3(x_1 = SS0 \times x_3)$; the term x_2 *is* substitutable for x_1 in A^*.

5

First-Order Logic

$\boxed{T}$ his chapter is a study of first-order logic—the logic of connectives and quantifiers—from a syntactic point of view. Our overall goal is to introduce formal systems for first-order logic and then to show that the concept of logical validity can be captured syntactically.

As in propositional logic, there is a practical reason for the syntactic approach, namely *verification of validity*. Given sentences $A_1, \ldots, A_n, B$ of a first-order language L, we say that the argument form $A_1, \ldots, A_n \therefore B$ is *valid* if B is true in every model of $\{A_1, \ldots, A_n\}$. Although the methods of Chapter 4 can be used to verify the validity of an argument form, there is another approach: to introduce a formal system and then to prove the metatheorem that if $\{A_1, \ldots, A_n\} \vdash B$, then B is true in every model of $\{A_1, \ldots, A_n\}$. This syntactic approach provides an alternate method of establishing validity that is usually quicker and more elegant than the semantic approach.

There is also a theoretical reason for the syntactic approach. In the case of propositional logic, truth tables give a purely mechanical method for deciding whether a formula is a tautology; from this point of view, formal systems such as P are superfluous. The situation in first-order logic is quite different: A major result of twentieth-century logic states that there is a first-order language L for which there is *no* algorithm such that, given an arbitrary formula A of L, the algorithm decides whether A is logically valid. This result is due independently to Church and to Turing, and was proved around 1936. But by introducing formal systems for first-order logic, we at least have a mechanical method for generating proofs and thus *listing* all theorems (= logically valid formulas) of first-order logic.

To introduce these formal systems, we start with a first-order language L and then add axioms and rules of inference to obtain a formal system, which we denote by FO_L and call a *first-order logic*. Since FO_L depends on the language L, there are many different first-order logics (one for each language). As in propositional logic, our goal is to prove the following fundamental results for a formula A of L and a set Γ of formulas of L:

(1) A is a theorem of FO_L if and only if A is logically valid.

(2) $\Gamma \vdash A$ if and only if A is true in every model of Γ.

(3) Γ is consistent if and only if Γ has a model.

$$\boxed{5.1}$$

The Formal System FO_L

Let L be a first-order language. The formal system obtained by adding the following (logical) axioms and rules of inference to L is called a *first-order logic* (or *predicate calculus*) and is denoted by FO_L.

Axioms Let A be a formula of L.

Propositional Axioms	$\neg A \vee A$.
Substitution Axioms	$\forall x_n A \rightarrow A_{x_n}[t]$ (t substitutable for x_n in A). $\forall x_n A \rightarrow A$ (special case).
Reflexive Axiom	$\forall x_1(x_1 = x_1)$.
Equality Axioms	$(x_n = t) \rightarrow (A \leftrightarrow A_{x_n}[t])$ (t substitutable for x_n in A).

Rules of inference Let A, B, and C be formulas of L.

Associative Rule	$\dfrac{A \vee (B \vee C)}{(A \vee B) \vee C}$
Contraction Rule	$\dfrac{A \vee A}{A}$
Expansion Rule	$\dfrac{A}{B \vee A}$
Cut Rule	$\dfrac{A \vee B, \neg A \vee C}{B \vee C}$
Add $\forall$-Rule	$\dfrac{A \vee B}{A \vee \forall x_n B}$ (x_n not free in A)

We emphasize that FO_L depends on the language L. For example, if L has a constant symbol c, then $\forall x_1(x_1 = x_2) \rightarrow (c = x_2)$ is a substitution axiom of FO_L. In spite of this dependence on L, we will usually omit the subscript L and use FO to denote the first-order logic under discussion.

The notations $\vdash_{FO} A$ and $\Gamma \vdash_{FO} A$ have their usual meanings. For example, $\Gamma \vdash_{FO} A$ states that there is a proof in FO of A using Γ. We usually omit the subscript FO and write $\vdash A$ or $\Gamma \vdash A$. In formal proofs, the axioms and rules of inference are abbreviated as follows:

PROP AX	REF AX	ASSOC	CTN
SUBST AX	EQ AX	EXP	CUT
		ADD $\forall$	

───── *Substitution Theorems and Derived* ─────
Rules of Inference

Each rule of inference of the formal system P is also a rule of inference for FO; in addition, $\neg A \vee A$ is an axiom of FO. It follows that each derived rule of inference of P is also a derived rule of inference of FO. As a reminder to the reader, we have, among others:

- Commutative rule (CM)
- Modus ponens (MP)
- Modus tollens (MT)
- Hypothetical syllogism (HS)
- Contrapositive (CP)

- $\neg\neg$ rule ($\neg\neg$ RULE)
- Disjunctive syllogism (DS)
- Join rule (JOIN)
- Separation rules (SEP)

Since the commutative rule is available, we will not be concerned with the order of the disjuncts when applying the add ∨-rule.

We now obtain theorems and derived rules of inference for introducing or eliminating the quantifiers ∀ and ∃. These results are divided into two groups: the consequences of the substitution axioms and the consequences of the add ∨-rule.

Consequences of Substitution Axioms:

- *Substitution Theorems*

  $\vdash_{FO} A_{x_n}[t] \rightarrow \exists x_n A$ (t substitutable for x_n in A)

  $\vdash_{F0} A \rightarrow \exists x_n A$ (special case)

- *∀-Elimination Rule*

  $$\frac{\forall x_n A}{A_{x_n}[t]} \quad (t \text{ substitutable for } x_n \text{ in } A)$$

  $$\frac{\forall x_n A}{A} \quad (\text{special case of } ∀\text{-elimination rule})$$

- *Existential Generalization*

  $$\frac{A_{x_n}[t]}{\exists x_n A} \quad (t \text{ substitutable for } x_n \text{ in } A)$$

  $$\frac{A}{\exists x_n A} \quad (\text{special case of existential generalization})$$

In formal proofs, these theorems and rules are abbreviated by

SUBST THM   ∀-ELIM   ∃-GEN

Justification of substitution theorems The formula $A_{x_n}[t] \rightarrow \exists x_n A$ is essentially the contrapositive of a substitution axiom. More precisely

(1) $\forall x_n \neg A \rightarrow \neg A_{x_n}[t]$	SUBST AX
(2) $\neg\forall x_n \neg A \vee \neg A_{x_n}[t]$	DEF $\rightarrow$
(3) $\exists x_n A \vee \neg A_{x_n}[t]$	DEF ∃
(4) $\neg A_{x_n}[t] \vee \exists x_n A$	CM
(5) $A_{x_n}[t] \rightarrow \exists x_n A$	DEF $\rightarrow$. ∎

Justification of ∀-elimination rule

(1) $\forall x_n A$	HYP
(2) $\forall x_n A \rightarrow A_{x_n}[t]$	SUBST AX
(3) $A_{x_n}[t]$	MP. ∎

Justification of existential generalization

(1) $A_{x_n}[t]$	HYP
(2) $A_{x_n}[t] \rightarrow \exists x_n A$	SUBST THM
(3) $\exists x_n A$	MP. ∎

Example 1:

$\{\forall x(Q(x) \rightarrow R(x)), \neg R(c)\} \vdash \exists x \neg Q(x).$

(1) $\forall x(Q(x) \rightarrow R(x))$	HYP
(2) $Q(c) \rightarrow R(c)$	∀-ELIM
(3) $\neg R(c)$	HYP
(4) $\neg Q(c)$	MT
(5) $\exists x \neg Q(x)$	∃-GEN. □

Consequences of Add ∀-Rule: The following rules follow from the add ∀-rule; some justifications are left to the reader.

- *Generalization*
$$\frac{A}{\forall x_n A}$$

- *Substitution Rule*
$$\frac{A}{A_{x_n}[t]} \quad (t \text{ substitutable for } x_n \text{ in } A)$$

- *∀-Introduction Rule*
$$\frac{A \rightarrow B}{A \rightarrow \forall x_n B} \quad (x_n \text{ not free in } A)$$

- *∃-Introduction Rule*
$$\frac{A \rightarrow B}{\exists x_n A \rightarrow B} \quad (x_n \text{ not free in } B)$$

- *∀-Distribution Rule*
$$\frac{A \rightarrow B}{\forall x_n A \rightarrow \forall x_n B}$$

- *∃-Distribution Rule*
$$\frac{A \rightarrow B}{\exists x_n A \rightarrow \exists x_n B}$$

In formal proofs, these rules are abbreviated as follows:

GEN	∀-INTRO	∀-DIST
SUBST RULE	∃-INTRO	∃-DIST

Justification of generalization

(1) A	HYP
(2) $\forall x_n A \vee A$	EXP
(3) $\forall x_n A \vee \forall x_n A$	ADD ∀ (x_n not free in $\forall x_n A$)
(4) $\forall x_n A$	CTN. ∎

Justification of ∃-introduction rule

(1) $A \to B$	HYP
(2) $\neg A \vee B$	DEF $\to$
(3) $\forall x_n \neg A \vee B$	ADD $\forall$ (x_n not free in B)
(4) $\neg\neg \forall x_n \neg A \vee B$	$\neg\neg$ RULE
(5) $\neg \exists x_n A \vee B$	DEF $\exists$
(6) $\exists x_n A \to B$	DEF $\to$. ∎

Justification of ∃-distribution rule

(1) $A \to B$	HYP
(2) $B \to \exists x_n B$	SUBST THM
(3) $A \to \exists x_n B$	HS
(4) $\exists x_n A \to \exists x_n B$	∃-INTRO (x_n not free in $\exists x_n B$). ∎

We now give a number of examples illustrating the use of these rules. The first two results show that $\exists x \forall y A \to \forall y \exists x A$ and $\forall x_1 \exists x_2 (x_1 = x_2)$ are theorems of first-order logic (both formulas are logically valid; see Section 4.3).

Example 2:

$\vdash_{\text{FO}} \exists x \forall y A \to \forall y \exists x A$.

(1) $\forall y A \to A$	SUBST AX
(2) $\exists x \forall y A \to \exists x A$	∃-DIST
(3) $\exists x \forall y A \to \forall y \exists x A$	$\forall$-INTRO (y not free in $\exists x \forall y A$). □

Example 3:

$\vdash_{\text{FO}} \forall x_1 \exists x_2 (x_1 = x_2)$.

(1) $(x_1 = x_1) \to \exists x_2 (x_1 = x_2)$	SUBST THM
(2) $\forall x_1 (x_1 = x_1) \to \forall x_1 \exists x_2 (x_1 = x_2)$	$\forall$-DIST
(3) $\forall x_1 (x_1 = x_1)$	REF AX
(4) $\forall x_1 \exists x_2 (x_1 = x_2)$	MP. □

Example 4:

Let x, y, and z be distinct variables. We will show the following:

$$\{\forall x \, \forall y \, \forall z [R(x, y) \to (R(y, z) \to R(x, z))], \forall x \neg R(x, x)\}$$
$$\vdash \forall x \, \forall y [R(x, y) \to \neg R(y, x)].$$

Informally, any relation that is transitive and antireflexive is antisymmetric.

(1) $\forall x \, \forall y \, \forall z [R(x, y) \to (R(y, z) \to R(x, z))]$	HYP
(2) $\forall y \, \forall z [R(x, y) \to (R(y, z) \to R(x, z))]$	$\forall$-ELIM

(3) $\forall z[R(x, y) \rightarrow (R(y, z) \rightarrow R(x, z))]$	$\forall$-ELIM
(4) $R(x, y) \rightarrow [R(y, x) \rightarrow R(x, x)]$	$\forall$-ELIM
(5) $\neg R(x, y) \vee [\neg R(y, x) \vee R(x, x)]$	DEF $\rightarrow$ (twice)
(6) $[\neg R(x, y) \vee \neg R(y, x)] \vee R(x, x)$	ASSOC
(7) $\forall x \neg R(x, x)$	HYP
(8) $\neg R(x, x)$	$\forall$-ELIM
(9) $\neg R(x, y) \vee \neg R(y, x)$	DS (6, 8)
(10) $R(x, y) \rightarrow \neg R(y, x)$	DEF $\rightarrow$
(11) $\forall x \, \forall y [R(x, y) \rightarrow \neg R(y, x)]$	GEN (twice). $\square$

Example 5:

$\vdash_{FO} \forall x_1 R(x_1, x_1) \rightarrow \forall x_2 \, \exists x_3 R(x_2, x_3)$.

(1) $\forall x_1 R(x_1, x_1) \rightarrow R(x_2, x_2)$	SUBST AX
(2) $R(x_2, x_2) \rightarrow \exists x_3 R(x_2, x_3)$	SUBST THM
(3) $\forall x_1 R(x_1, x_1) \rightarrow \exists x_3 R(x_2, x_3)$	HS
(4) $\forall x_1 R(x_1, x_1) \rightarrow \forall x_2 \, \exists x_3 R(x_2, x_3)$	$\forall$-INTRO (x_2 not free in $\forall x_1 R(x_1, x_1)$). $\square$

Recall these ideas: A formula with no free variables is a *closed* formula, or a *sentence*; if $y_1, \ldots, y_k$ are the free variables in A, the formula $\forall y_1 \ldots \forall y_k A$ has no free variables and is called the *closure* of A; the closure of a formula A is denoted A'. In Section 4.3, we proved that a formula A is true in an interpretation if and only if its closure A' is true in that interpretation; this is called the *closure theorem for interpretations*. There is a syntactic version of this result.

Theorem 1 (Closure theorem for provability): Let A be a formula of L and let Γ be a set of formulas of L.

(1) $\Gamma \vdash A$ if and only if $\Gamma \vdash \forall x_n A$.

(2) $\Gamma \vdash A$ if and only if $\Gamma \vdash A'$.

Proof: To prove (1), use generalization and the $\forall$-elimination rule. The proof of (2) follows by repeated application of (1) as follows: Let $y_1, \ldots, y_k$ be the free variables in A. Then A' is $\forall y_1 \ldots \forall y_k A$, and we have

$$\Gamma \vdash A \Leftrightarrow \Gamma \vdash \forall y_k A$$
$$\vdots$$
$$\Leftrightarrow \Gamma \vdash \forall y_1 \ldots \forall y_{k-1} \forall y_k A. \quad \blacksquare$$

────────── *Tautology Theorem* ──────────

We now introduce an idea that is a great time-saver when we write formal proofs in first-order logic. The idea, in a nutshell, is this: Certain formulas of a first-order language L have the *form* of a tautology of propositional logic; every tautology of propositional logic has a proof in the formal system P (adequacy theorem for P); all of the axioms and rules of P are incorporated into the axioms and rules of first-order logic, FO$_L$. It follows that each formula of first-order logic that has the form of a tautology is a theorem of first-order logic. Before stating precise results, let us illustrate with an example. Consider the formula $R(x_1) \to [\exists x_2 Q(x_2) \vee R(x_1)]$. Familiar? Suppose we replace $R(x_1)$ with p and $\exists x_2 Q(x_2)$ with q to obtain $p \to (q \vee p)$, a tautology of propositional logic. It follows from the adequacy theorem that there is a proof of $p \to (q \vee p)$ in the formal system P. If in this proof we now replace each occurrence of p with $R(x_1)$ and each occurrence of q with $\exists x_2 Q(x_2)$, the result is a proof in first-order logic of the formula $R(x_1) \to [\exists x_2 Q(x_2) \vee R(x_1)]$.

To use this idea effectively, we must be familiar with a wide range of tautologies of propositional logic. Once we have a candidate such as $p \to (q \vee p)$ in hand, there are two ways to verify that it is indeed a tautology: Use truth assignments (or truth tables), or write out a formal proof of $p \to (q \vee p)$ in P (or show $\{p\} \vdash_P (q \vee p)$ and then use the deduction theorem). We will now formalize these ideas.

Definition 1: A formula A of L is a *tautology of* L if it is a substitution instance of a tautology of propositional logic. More precisely, A is a tautology of L if there is a tautology B of propositional logic with propositional variables $p_1, \ldots, p_n$ and there are formulas $C_1, \ldots, C_n$ of L such that A is the result of replacing each occurrence of p_i in B with C_i for $1 \le i \le n$.

Example 6:

> The formula $\exists x_1 R(x_1) \to [\forall x_2 Q(x_2) \to \exists x_1 R(x_1)]$ is a tautology of first-order logic. To see this, note that $p \to (q \to p)$ is a tautology of propositional logic and that the given formula can be obtained from it by replacing p with $\exists x_1 R(x_1)$ and q with $\forall x_2 Q(x_2)$. □

Theorem 2 (Tautology theorem): If A is a tautology of L, then $\vdash_{FO} A$.

Proof: Since A is a tautology of L, there is a tautology B of propositional logic with propositional variables $p_1, \ldots, p_n$ and there are formulas $C_1, \ldots, C_n$ of L such that A is the result of replacing each occurrence of p_i in B with C_i for $1 \le i \le n$. By the adequacy theorem for propositional logic, B is a therorem of the formal system P. Let $B_1, \ldots, B_k$ with $B_k = B$ be a proof of B in P. In each formula of the proof $B_1, \ldots, B_k$, replace each occurrence of p_i with C_i for

$1 \le i \le n$ (if the propositional variable p_i for $i > n$ occurs somewhere in the proof, just replace it with $(x_1 = x_1)$). Denote this new sequence of formulas by $D_1, \ldots, D_k$, and note that D_k is A. Now each axiom and each rule of inference of P is an axiom or rule of inference of FO_L; it follows that $D_1, \ldots, D_k$ is a proof of the formula A in FO_L as required. ∎

Example 7:

We list a number of tautologies of propositional logic that are often used in conjunction with the tautology theorem.

(1) $(p \leftrightarrow q) \to (p \to q)$.

(2) $(p \leftrightarrow q) \to (q \to p)$.

(3) $[p \to (q \to r)] \to [q \to (p \to r)]$.

(4) $[p \to (q \to r)] \to [(p \wedge q) \to r]$.

(5) $(p \to q) \to [(p \to r) \to (p \to (q \wedge r))]$.

(6) $(p \to r) \to [(q \to r) \to ((p \vee q) \to r)]$.

(7) $(p \to q) \to [(p \wedge r) \to (q \wedge r)]$.

(8) $(p \to q) \to [(p \vee r) \to (q \vee r)]$. □

In formal proofs, we denote an application of the tautology theorem by TT.

Example 8:

We will show that $\vdash_{FO} \exists x(A \wedge B) \to (\exists xA \wedge B)$ (x not free in B); the key is the tautology $(p \to q) \to [(p \wedge r) \to (q \wedge r)]$.

(1) $A \to \exists xA$	SUBST THM
(2) $(A \to \exists xA) \to [(A \wedge B) \to (\exists xA \wedge B)]$	TT
(3) $(A \wedge B) \to (\exists xA \wedge B)$	MP
(4) $\exists x(A \wedge B) \to (\exists xA \wedge B)$	∃-INTRO (x not free in $\exists xA \wedge B$).

(*Note:* Line 2 is obtained by replacing p with A, q with $\exists xA$, and r with B.) □

Example 9 (Russell's paradox):

In Example 8 in Section 4.3, we proved that the sentence $\neg\exists x\forall y[R(y, x) \leftrightarrow \neg R(y, y)]$ is logically valid; we will now show that it is also a theorem of first-order logic. (In reading the formal proof, begin with lines 7–4.)

(1) $\forall y[R(y, x) \leftrightarrow \neg R(y, y)] \to [R(x, x) \leftrightarrow \neg R(x, x)]$	SUBST AX
(2) $\neg[R(x, x) \leftrightarrow \neg R(x, x)] \to \neg\forall y[R(y, x) \leftrightarrow \neg R(y, y)]$	CP
(3) $\neg[R(x, x) \leftrightarrow \neg R(x, x)]$	TT $(\neg(p \leftrightarrow \neg p))$
(4) $\neg\forall y[R(y, x) \leftrightarrow \neg R(y, y)]$	MP

(5) $\forall x \neg \forall y[R(y, x) \leftrightarrow \neg R(y, y)]$ GEN

(6) $\neg\neg\forall x \neg \forall y[R(y, x) \leftrightarrow \neg R(y, y)]$ $\neg\neg$ RULE

(7) $\neg\exists x \forall y[R(y, x) \leftrightarrow \neg R(y, y)]$ DEF $\exists$.   □

Equality Axioms

The formula $(x_n = t) \rightarrow (A \leftrightarrow A_{x_n}[t])$ (t is substitutable for x_n in A) is an equality axiom. As noted in Section 4.4, these axioms capture Leibniz's law: If two things are the same, then whatever is true of one is true of the other. We now use the tautology theorem to obtain useful variations of the equality axioms. To see the need for these results, suppose we wish to show that the transitive law $(x_1 = x_2) \rightarrow [(x_2 = x_3) \rightarrow (x_1 = x_3)]$ is a theorem of first-order logic. A direct application of the equality axiom gives $(x_1 = x_2) \rightarrow [(x_1 = x_3) \leftrightarrow (x_2 = x_3)]$, but this does not quite have the form we want.

Theorem 3 (Variations of the equality axiom): Let t be substitutable for x in A. Each of the following formulas is a theorem of FO$_L$:

EQ AX(1) $(x = t) \rightarrow (A \rightarrow A_x[t])$

EQ AX(2) $(x = t) \rightarrow (A_x[t] \rightarrow A)$

EQ AX(3) $A \rightarrow ((x = t) \rightarrow A_x[t])$

EQ AX(4) $A_x[t] \rightarrow ((x = t) \rightarrow A)$

Proof: We will prove (4) and leave the other proofs as exercises for the reader. The key is the tautology $[p \rightarrow (q \leftrightarrow r)] \rightarrow [r \rightarrow (p \rightarrow q)]$.

(1) $(x = t) \rightarrow (A \leftrightarrow A_x[t])$ EQ AX

(2) $[(x = t) \rightarrow (A \leftrightarrow A_x[t])] \rightarrow [A_x[t] \rightarrow ((x = t) \rightarrow A)]$ TT

(3) $A_x[t] \rightarrow ((x = t) \rightarrow A)$ MP.

(*Note:* Line 2 is obtained by replacing p with $(x = t)$, q with A, and r with $A_x[t]$.) ∎

Corollary 1 (Symmetric and transitive laws): Let s, t, and u be terms of L.

(1) $\vdash_{F0} (s = t) \rightarrow (t = s)$;

(2) $\vdash_{F0} [(s = t) \wedge (t = u)] \rightarrow (s = u)$.

Proof of (1): The equality axiom gives $(x_1 = x_2) \rightarrow [(x_2 = x_1) \leftrightarrow (x_2 = x_2)]$, but this is not the form we want. Instead, we use EQ AX(4). Choose a variable x that does not occur in t.

(1) $(t = t) \rightarrow [(x = t) \rightarrow (t = x)]$ EQ AX(4)

(2) $\forall x_1(x_1 = x_1)$ REF AX

(3) $t = t$ $\forall$-ELIM

(4) $(x = t) \rightarrow (t = x)$ MP (3, 1)

(5) $(s = t) \rightarrow (t = s)$ SUBST RULE. ■

Proof of (2): By the tautology $[p \rightarrow (q \rightarrow r)] \rightarrow [(p \wedge q) \rightarrow r]$, it suffices to show that $\vdash_{FO} (s = t) \rightarrow [(t = u) \rightarrow (s = u)]$ and then to use MP. Choose a variable x that does not occur in t or u.

(1) $(x = t) \rightarrow [(t = u) \rightarrow (x = u)]$ EQ AX(2)

(2) $(s = t) \rightarrow [(t = u) \rightarrow (s = u)]$ SUBST RULE. ■

From Corollary 1, we obtain two useful rules of inference:

Tran Rule $\dfrac{s = t, t = u}{s = u}$ Sym Rule $\dfrac{s = t}{t = s}$

In mathematical proofs and calculations, we often take for granted certain properties of equality. For example, from $a = b$, we conclude that $a + c = b + c$. In formal proofs, this is justified by equality axioms; the following is a typical result:

Function Rule $\dfrac{s = t}{F(s, u) = F(t, u)}$

Justification Choose a variable x that does not occur in t or u.

(1) $s = t$ HYP

(2) $(x = t) \rightarrow [(F(t, u) = F(t, u)) \rightarrow (F(x, u) = F(t, u))]$ EQ AX(2)

(3) $(s = t) \rightarrow [(F(t, u) = F(t, u)) \rightarrow (F(s, u) = F(t, u))]$ SUBST RULE

(4) $(F(t, u) = F(t, u)) \rightarrow (F(s, u) = F(t, u))$ MP (1, 3)

(5) $\forall x_1(x_1 = x_1)$ REF AX

(6) $F(t, u) = F(t, u)$ $\forall$-ELIM

(7) $F(s, u) = F(t, u)$ MP (4, 6). ■

Example 10:

$\{R(b), \neg R(c)\} \vdash \neg(b = c)$.

(1) $R(b)$ HYP

(2) $\neg R(c)$ HYP

(3) $R(x) \rightarrow [(x = c) \rightarrow R(c)]$ EQ AX(3)

(4) $R(b) \rightarrow [(b = c) \rightarrow R(c)]$ SUBST RULE

(5) $(b = c) \rightarrow R(c)$ MP (1, 4)

(6) $\neg R(c) \rightarrow \neg(b = c)$ CP

(7) $\neg(b = c)$ MP (2, 6). $\square$

——————————*Exercises for Section 5.1*——————————

1. Justify the following derived rules of inference:
   (1) Substitution rule   (3) ∀-distribution rule
   (2) ∀-introduction rule

2. Show the following:
   (1) $\vdash \forall x A \rightarrow \exists x A$.
   (2) $\vdash \exists x \exists y A \rightarrow \exists y \exists x A$.
   (3) $\vdash \forall x \forall y A \rightarrow \forall y \forall x A$.
   (4) $\vdash \forall x \exists y A \rightarrow \exists x \exists y A$.
   (5) $\vdash \forall x \exists y A \rightarrow \exists y \exists x A$.
   (6) $\vdash \forall x \forall y A \rightarrow \forall x A$.
   (7) $\vdash A \rightarrow \forall x A$ (x not free in A).
   (8) $\vdash \exists x A \rightarrow A$ (x not free in A).
   (9) $\vdash \forall x (A \lor B) \rightarrow (A \lor \forall x B)$
       (x not free in A).

3. Justify the following derived rules of inference:
   (1) $\dfrac{A \rightarrow B}{\forall x A \rightarrow B}$ (3) $\dfrac{A \lor B}{A \lor \exists x B}$
   (2) $\dfrac{A \rightarrow B}{A \rightarrow \exists x B}$ (4) $\dfrac{A \lor \forall x B}{A \lor B}$

4. Show the following:
   (1) $\vdash \forall x_1 P(x_1) \rightarrow \forall x_2 P(x_2)$.
   (2) $\vdash \exists x_1 P(x_1) \rightarrow \exists x_2 P(x_2)$.
   (3) $\vdash \forall x_1 R(x_1, x_2) \rightarrow \forall x_3 R(x_3, x_2)$.
   (4) $\vdash \exists x_3 R(x_3, x_2) \rightarrow \exists x_1 R(x_1, x_2)$.
   (5) $\vdash \forall x_1 \forall x_2 R(x_1, x_2) \rightarrow R(x_3, x_4)$.
   (6) $\vdash \forall x_1 \forall x_2 R(x_1, x_2) \rightarrow R(x_2, x_1)$.
   (7) $\vdash R(x_3, x_4) \rightarrow \exists x_1 \exists x_2 R(x_1, x_2)$.
   (8) $\vdash R(x_2, x_1) \rightarrow \exists x_1 \exists x_2 R(x_1, x_2)$.
   (9) $\vdash \forall x_1 \exists x_2 R(x_1, x_2) \rightarrow \exists x_1 R(c, x_1)$.

5. Which of the following is an instance of a substitution theorem?
   (1) $R(x_1, c) \rightarrow \exists x_2 R(x_1, x_2)$.
   (2) $[Q(x_1, x_1) \leftrightarrow R(x_1, x_1)] \rightarrow \exists x_2 [Q(x_2, x_1) \leftrightarrow R(x_2, x_2)]$.
   (3) $\forall x_1 Q(x_1, c) \rightarrow \exists x_2 \forall x_1 Q(x_2, c)$.

6. Criticize this "proof" that $R(x_2) \rightarrow \forall x_2 R(x_2)$ is a theorem of first-order logic:
   (1) $\neg R(x_1) \lor R(x_1)$ PROP AX
   (2) $\neg R(x_1) \lor \forall x_2 R(x_1)$ ADD ∀ (x_2 not free in $\neg R(x_1)$)
   (3) $\neg R(x_2) \lor \forall x_2 R(x_2)$ SUBST RULE (x_2 for x_1)
   (4) $R(x_2) \rightarrow \forall x_2 R(x_2)$ DEF → .
   Then show that $R(x_2) \rightarrow \forall x_2 R(x_2)$ is not logically valid.

7. Let x and y be distinct variables. Prove the following:
   (1) $\{\forall x \, \forall y [R(x, y) \rightarrow \neg R(y, x)]\} \vdash \forall x \neg R(x, x)$.
   (2) $\vdash \exists x R(x, c) \rightarrow \exists x \exists y R(x, y)$.
   (3) $\vdash R(b, c) \rightarrow \exists x \exists y R(x, y)$.

8. Which of the following formulas are tautologies of L? Of those that are not tautologies, which are logically valid?
   (1) $R(x_1, x_2) \rightarrow [(x_1 = x_2) \rightarrow R(x_1, x_2)]$.
   (2) $R(x_1, x_2) \rightarrow [(x_1 = x_3) \rightarrow R(x_3, x_2)]$.
   (3) $[R(x_1, x_2) \land (x_1 = x_2)] \rightarrow R(x_1, x_2)$.
   (4) $[R(x_1, x_2) \land (x_1 = x_2)] \rightarrow R(x_2, x_2)$.
   (5) $R(x_1, x_3) \rightarrow \exists x_2 R(x_1, x_2)$.
   (6) $\neg R(x_1, x_2) \rightarrow [R(x_1, x_2) \rightarrow (x_1 = x_2)]$.

9. Verify the tautologies in Example 7.

10. Prove EQ AX(1)–EQ AX(3) in Theorem 3 (variations of the equality axiom).

11. Prove the following (s, t, u, and v are terms):
    (1) $\vdash_{\text{F0}} (s = t) \rightarrow [R(s, u) \rightarrow R(t, u)]$.
    (2) $\vdash_{\text{F0}} (u = v) \rightarrow [R(t, u) \rightarrow R(t, v)]$.
    (3) $\vdash_{\text{F0}} [(s = t) \land (u = v)] \rightarrow [R(s, u) \rightarrow R(t, v)]$.

(4) $\vdash_{FO} (s = t) \rightarrow [F(s, u) = F(t, u)]$.
(5) $\vdash_{FO} (u = v) \rightarrow [F(t, u) = F(t, v)]$.
(6) $\vdash_{FO} [(s = t) \wedge (u = v)] \rightarrow [F(s, u) = F(t, v)]$.

12. Show the following:
  (1) $\vdash \exists x[F(x) = c] \rightarrow \exists x(x = c)$.
  (2) $\exists x(x = c) \rightarrow \exists x[F(x) = c]$ is not logically valid.

13. Fix a first-order language L and let FO be the first-order logic with language L. Show that FO and FO* have the same theorems, where FO* is a variation of FO obtained by replacing the add ∀-rule with the following axiom scheme and rule:

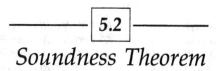

$$\forall x(A \rightarrow B) \rightarrow (A \rightarrow \forall xB) \quad (x \text{ not free in } A) \qquad GEN \ \frac{A}{\forall x_n A}.$$

14. Let A be a formula such that (1) $\vdash_{FO} A$; (2) A has no occurrences of the equality symbol $=$. Show that there is a proof of A that does not use the reflexive axiom or the equality axioms.

15. Let Γ be a set of formulas of L, let $\Gamma' = \{A' : A \in \Gamma\}$, and let B be an arbitrary formula of L. Prove that $\Gamma \vdash B$ if and only if $\Gamma' \vdash B$.

---------- 5.2 ----------

Soundness Theorem

The goal of this section is to prove the soundness theorem. The simplest version of this fundamental result states that every theorem of first-order logic is logically valid. Informally speaking, this means that the axioms and the rules of inference of first-order logic are sound in the sense that each theorem of FO_L is true in every possible interpretation of L. From this result, it easily follows that first-order logic is consistent.

Theorem 1 (Soundness theorem for first-order logic): Let Γ be a set of formulas of a first-order language L. If $\Gamma \vdash A$, then A is true in every model of Γ.

Proof: Let M be a model of Γ and let $A_1, \ldots, A_n$ be a proof of A using Γ. We will give an informal inductive argument that A is true in M (see Theorem 3 in Section 3.2 for a precise formulation of the induction argument). Consider A_k, a step in the proof of A using Γ; we will show that each A_k ($1 \le k \le n$) is true in M. If $A_k \in \Gamma$, then A_k is true in M, since M is a model of Γ. Suppose A_k is one of the following axioms: propositional, substitution, reflexive, equality. Each of these is logically valid, and therefore true in M. (This is obvious for propositional axioms and the reflexive axiom; in Section 4.4, the substitution theorem for assignments is used to show that substitution axioms and equality axioms are logically valid.) Finally, assume A_k is the conclusion of a rule of inference whose hypotheses are among $A_1, \ldots, A_{k-1}$; by induction hypothesis, $A_1, \ldots, A_{k-1}$ are true in M. If the rule of inference is the expansion, contraction, associative, or cut rule, proceed as in Chapter 3 (see

Lemma 1 in Section 3.2). Suppose the rule of inference is the add ∀-rule, in which case our proof looks like this:

$$\vdots$$

$$B \vee C$$

$$\vdots$$

$$B \vee \forall x C$$

$$\vdots$$

Now $B \vee C$ is true in M by induction hypothesis; since the add ∀-rule preserves truth (see Section 4.4), it follows that $B \vee \forall x C$ is true in M. ∎

Corollary 1: Let L be a first-order language. Then every theorem of the first-order logic FO_L is logically valid.

Proof: Assume A is a theorem of FO_L; that is, $\varnothing \vdash A$. By the soundness theorem, A is true in every model of $\varnothing$. But the models of $\varnothing$ are just the interpretations of L; thus A is true in every interpretation of L as required. ∎

Definition 1: A first-order logic FO_L is *consistent* if there is no formula A of L such that both $\vdash A$ and $\vdash \neg A$; otherwise FO_L is *inconsistent*.

Corollary 2 (Consistency of first-order logic): For any first-order language L, the first-order logic FO_L is consistent.

Proof: Suppose FO_L is inconsistent. Then there is a formula A of L such that both $\vdash A$ and $\vdash \neg A$. By Corollary 1, both A and $\neg A$ are logically valid. Let I be any interpretation of L. Then both A and $\neg A$ are true in I, and this is impossible. ∎

Suppose we are given an argument in everyday discourse that involves connectives, quantifiers, relations, and possibly individuals. There is a first-order language L, and there are sentences $A_1, \ldots, A_n, B$ of L, such that $A_1, \ldots, A_n \therefore B$ is the *form* of this argument. As in propositional logic, the given argument is *valid* if its corresponding argument form $A_1, \ldots, A_n \therefore B$ is valid. The validity of an argument form is defined as follows.

Definition 2: Let $A_1, \ldots, A_n, B$ be sentences of L. The argument form $A_1, \ldots, A_n \therefore B$ is *valid* if B is true in every model of $\{A_1, \ldots, A_n\}$.

Corollary 3 (Verification of validity): If $\{A_1, \ldots, A_n\} \vdash B$, then the argument form $A_1, \ldots, A_n \therefore B$ is valid.

The next two examples illustrate the use of Corollary 3.

Example 1:

We will show that the following argument form is valid (see Example 2 in Section 4.2): $A(f, m)$, $\forall x[A(x, m) \to A(x, t)]$, $\forall x[A(f, x) \to G(x)] \therefore G(t)$.

(1)	$A(f, m)$	HYP
(2)	$\forall x[A(x, m) \to A(x, t)]$	HYP
(3)	$A(f, m) \to A(f, t)$	$\forall$-ELIM
(4)	$A(f, t)$	MP (1, 3)
(5)	$\forall x[A(f, x) \to G(x)]$	HYP
(6)	$A(f, t) \to G(t)$	$\forall$-ELIM
(7)	$G(t)$	MP (4, 6). $\qquad\qquad \square$

Existential quantifiers can be tricky to handle in formal proofs. One basic technique is to use the $\exists$-distribution rule, as illustrated in the following example.

Example 2:

$\forall x[P(x) \to Q(x)]$, $\exists x \neg Q(x) \therefore \exists x \neg P(x)$ is a valid argument form.

(1)	$\forall x[P(x) \to Q(x)]$	HYP
(2)	$P(x) \to Q(x)$	$\forall$-ELIM
(3)	$\neg Q(x) \to \neg P(x)$	CP
(4)	$\exists x \neg Q(x) \to \exists x \neg P(x)$	$\exists$-DIST
(5)	$\exists x \neg Q(x)$	HYP
(6)	$\exists x \neg P(x)$	MP. $\qquad\qquad \square$

We will now use the soundness theorem to give a sufficient condition for the consistency of a set of formulas.

Definition 3: A set Γ of formulas of L is *consistent* if there is no formula A of L such that both $\Gamma \vdash A$ and $\Gamma \vdash \neg A$; otherwise, Γ is *inconsistent*.

Theorem 2 (Consistency theorm for first-order logic): If Γ has a model, then Γ is consistent.

Proof: Left as an exercise for the reader. ∎

Example 3:

Let $\Gamma = \{G1, G2, G3, \forall x_1(x_1 \circ x_1 = e), \exists x_1 \neg(x_1 = e)\}$, where the language is L_G and G1–G3 are the axioms for a group written in L_G. We will show that Γ is consistent by finding a model of Γ. It suffices to find a group such that every element is its own right inverse and there is at least one element of the group that is distinct from the right identity. Let $D_M = \{1, -1\}$, $e_M = 1$, and $\circ_M = $ multiplication. $\qquad\qquad \square$

─────────────── *Exercises for Section 5.2* ───────────────

1. Prove the following:
   (1) $\{\neg\exists x(A \wedge B)\} \vdash \forall x(A \rightarrow \neg B)$.
   (2) $\{\forall x(A \rightarrow \neg B)\} \vdash \neg\exists x(A \wedge B)$.
2. Show the validity of the following arguments:
   (1) All A are B. No C are B. $\therefore$ No C are A.
   (2) All A are D. All B are C. No C are D. $\therefore$ No A are B.
3. Show the validity of the following argument forms:
   (1) $R(c) \therefore \exists x[R(x) \vee Q(x)]$.
   (2) $\forall x[Q(x) \rightarrow R(x)]$, $Q(c) \therefore \exists x[Q(x) \wedge R(x)]$.
   (3) $\forall x[P(x) \rightarrow Q(x)]$, $\neg\exists x[P(x) \wedge Q(x)] \therefore \forall x \neg P(x)$.
   (4) $\forall x[P(x) \vee Q(x)]$, $\forall x[P(x) \rightarrow R(x)]$, $\exists x \neg R(x) \therefore \exists x Q(x)$.
   (5) $R(c) \rightarrow \exists x[Q(x) \wedge \neg Q(x)] \therefore \neg R(c)$.
   (6) $\exists x Q(x) \therefore \exists x[Q(x) \vee R(x)]$.
   (7) $\exists x[(P(x) \vee Q(x)] \rightarrow \exists x R(x)$, $\forall x[R(x) \rightarrow S(x)]$, $\exists x P(x) \therefore \exists x S(x)$.
4. Show the validity of the following arguments:
   (1) Babies are illogical. Nobody is despised who can manage a crocodile. Illogical people are despised. $\therefore$ Babies cannot manage crocodiles. (Lewis Carroll)
   (2) For every natural number n, if n is prime, then $n = 2$ or n is odd. 5 is prime and $5 \neq 2$. $\therefore$ 5 is odd. ($P(x) = x$ is prime; $T(x) = x$ equals 2; $O(x) = x$ is odd; $f = 5$.)
   (3) All Frenchmen drink coffee or wine. Pierre is a Frenchman who does not drink coffee. All wine drinkers are healthy and live full lives. $\therefore$ Pierre is healthy. ($F(x) = x$ is a Frenchman; $C(x) = x$ drinks coffee; $W(x) = x$ drinks wine; $H(x) = x$ is healthy; $L(x) = x$ lives a full life; p = Pierre.)
   (4) There are no pink elephants. $\therefore$ All pink elephants can fly. ($P(x) = x$ is a pink elephant; $F(x) = x$ can fly.)
   (5) Whoever drinks beer is a Mets fan. Fred is not a Mets fan. Ralph drinks beer. $\therefore$ Fred is not Ralph. ($B(x) = x$ is a beer drinker; $M(x) = x$ is a Mets fan; f = Fred; r = Ralph.)
   (6) Anyone who likes Sarah likes Jennifer. Alice does not like Jennifer. $\therefore$ Alice does not like Sarah. ($L(x, y) = x$ likes y; a = Alice; j = Jennifer; s = Sarah.)
   (7) Tom is taller than Dick. Dick is taller than Harry. "Taller than" is a transitive relation. $\therefore$ Tom is taller than Harry. ($T(x, y) = x$ is taller than y; t = Tom; d = Dick; h = Harry.)
   (8) Mark Twain is Samuel Clemens. Mark Twain wrote *Tom Sawyer*. $\therefore$ Samuel Clemens wrote *Tom Sawyer*. ($W(x, y) = x$ wrote y; m = Mark Twain; c = Samuel Clemens; t = *Tom Sawyer*.)
5. Show the validity of the following arguments:
   (1) Fred is jealous of anyone who flirts with one of his girl friends. Sue Ellen is a girl friend of Fred. Ralph flirts with Mary Lou. Whoever flirts with Mary Lou flirts with Sue Ellen. $\therefore$ Fred is jealous of Ralph. ($J(x)$ = Fred is jealous of x; $F(x, y) = x$ flirts with y; $G(x) = x$ is a girl friend of Fred; s = Sue Ellen; m = Mary Lou; r = Ralph.)
   (2) All mothers love all children. Fred is a child. Alice does not love Fred. $\therefore$ Alice is not a mother. ($L(x, y) = x$ loves y; $C(x) = x$ is a child; $M(x) = x$ is a mother; a = Alice; f = Fred.)
   (3) Pythons will squeeze anyone who steps on their tails. Anyone who is squeezed by a python suffers a broken arm. Fred does not have a broken arm. $\therefore$ Fred did not step on the tail of a python. ($P(x) = x$ is a python; $S(x, y) = x$ squeezes y; $T(x, y) = x$ steps on the tail of y; $B(x) = x$ has a broken arm; f = Fred.)
   (4) All Yankee fans hate all Mets fans. All Mets fans drink beer. Fred is a Yankee fan and Ralph is a Mets fan. $\therefore$ There is a Yankee fan who hates some beer drinker. ($Y(x) = x$ is a Yankee fan; $M(x) = x$ is a Mets fan; $B(x) = x$ drinks beer; $H(x, y) = x$ hates y; f = Fred; r = Ralph.)

6. Show the validity of the following argument forms:
   (1) $R(b)$, $\neg R(c) \therefore \exists x \exists y \neg(x = y)$.
   (2) $R(c)$, $\forall x(\neg(x = c) \rightarrow R(x)) \therefore \forall x R(x)$.
   (3) $\forall x((x = b) \vee (x = c))$, $Q(b)$, $R(c) \therefore \forall x(Q(x) \vee R(x))$.
   (4) $\neg(a = b)$, $b = c \therefore \neg(a = c)$.

7. (*Tautologies of L versus logically valid formulas of L*) Logical validity plays the same role in first-order logic that tautology plays in propositional logic. (1) Show that every tautology of L is logically valid. (2) Show that $\forall x_1(x_1 = x_1) \rightarrow (x_1 = x_1)$ is a logically valid formula of L that is not a tautology of L.

8. (*Logical consequence of Γ versus true in every model of Γ*) A is a *logical consequence* of Γ if the following holds: If I is any interpretation of L and if ϕ is any assignment in I that satisfies Γ (i.e., $\phi(B) = T$ for all $B \in \Gamma$), then ϕ also satisfies A. *Notation:* $\Gamma \vDash A$ (for $\Gamma = \varnothing$ this reduces to the definition of logical validity). Prove the following:
   (1) If $\Gamma \vDash A$, then A is true in every model of Γ.
   (2) If every formula in Γ is a sentence, and if A is true in every model of Γ, then $\Gamma \vDash A$.
   (3) The formula $\forall x_1 R(x_1)$ is true in every model of $\{R(x_1)\}$, but $\forall x_1 R(x_1)$ is not a logical consequence of $\{R(x_1)\}$.

9. In group theory, we show that the commutative law $a \circ b = b \circ a$ is not a theorem by giving an example of a group for which the commutative law fails. Prove the following formal version of this procedure: If $\Gamma \cup \{\neg A\}$ has a model, then A is not a theorem of Γ.

10. Let $\Gamma = \{G1, G2, G3\}$, where G1–G3 are the axioms for a group written in L_G. Prove the following:
   (1) $\Gamma \cup \{\forall x_1(x_1 \circ x_1 = e)\}$ is consistent.
   (2) $\Gamma \cup \{\neg \forall x_1(x_1 \circ x_1 = e)\}$ is consistent.
   (3) The sentence $\forall x_1(x_1 \circ x_1 = e)$ is not a theorem of Γ.
   (4) The sentence $\neg \forall x_1(x_1 \circ x_1 = e)$ is not a theorem of Γ.

11. This exercise is due to Hao Wang (*A Survey of Mathematical Logic* (Peking: Science Press, 1962), 15). Let L have constant symbols a, b, and c and 2-ary relation symbol R; show that $\Gamma = \{W1, \ldots, W7\}$ is consistent by finding a model of Γ.

W1 $\forall x_1[(x_1 = a) \vee (x_1 = b) \vee (x_1 = c)]$.

W2 $\neg(a = b) \wedge \neg(a = c) \wedge \neg(b = c)$.

W3 $\forall x_1 \neg R(x_1, x_1)$.

W4 $\forall x_1 \forall x_2 \forall x_3[(R(x_1, x_2) \wedge R(x_1, x_3)) \rightarrow (x_2 = x_3)]$.

W5 $\forall x_1 \forall x_2 \forall x_3[(R(x_2, x_1) \wedge R(x_3, x_1)) \rightarrow (x_2 = x_3)]$.

W6 $\forall x_1 \exists x_2 R(x_1, x_2)$.

W7 $R(a, b)$.

12. (*An elementary geometry EG*) We introduce an elementary geometry EG. There are three undefined concepts: *point*, *line*, and *is on*. The axioms are $\Gamma = \{EG1, \ldots, EG7\}$:

EG1 $\forall x_1[(P(x_1) \vee L(x_1)) \wedge \neg(P(x_1) \wedge L(x_1))]$.

EG2 $\forall x_1 \forall x_2[O(x_1, x_2) \rightarrow (P(x_1) \wedge L(x_2))]$.

EG3 $\exists x_1 \exists x_2[P(x_1) \wedge P(x_2) \wedge \neg(x_1 = x_2)]$.

EG4 $\forall x_1 \forall x_2[(P(x_1) \wedge P(x_2) \wedge \neg(x_1 = x_2)) \rightarrow \exists x_3(O(x_1, x_3) \wedge O(x_2, x_3))]$.

EG5 $\forall x_1[L(x_1) \rightarrow \exists x_2(P(x_2) \wedge \neg O(x_2, x_1))]$.

EG6 $\forall x_1[L(x_1) \rightarrow \exists x_2 \exists x_3(\neg(x_2 = x_3) \wedge O(x_2, x_1) \wedge O(x_3, x_1))]$.

EG7 $\forall x_1 \forall x_2[(L(x_1) \wedge L(x_2) \wedge \neg(x_1 = x_2)) \rightarrow \exists x_3(O(x_3, x_1) \wedge O(x_3, x_2))]$.

   (1) Find a finite model of EG. (*Hint:* $D_M = P_M \cup L_M$, where P_M is the set of points and L_M is the set of lines.)
   (2) Find an infinite model of Γ.

13. Let L_{EQ} be the first-order language of equality. Show that the first-order logic FO with language L_{EQ} is not complete (there is a sentence A of L_{EQ} such that neither A nor $\neg A$ is a theorem of FO).

14. (*Independence of axioms*) A set Γ of formulas is *independent* if there is no formula $A \in \Gamma$ such that $\Gamma - \{A\} \vdash A$.
    (1) Suppose, for all $A \in \Gamma$, that $(\Gamma - \{A\}) \cup \{\neg A\}$ has a model. Show that Γ is independent.
    (2) Show that $\Gamma = \{G1, G2, G3\}$ is independent, where G1, G2, and G3 are the axioms for a group written in L_G.

15. A formula A is *consistent with* Γ if $\Gamma \cup \{A\}$ has a model, and is *independent of* Γ if both A and $\neg A$ are consistent with Γ (i.e., $\Gamma \cup \{A\}$ and $\Gamma \cup \{\neg A\}$ each have a model).
    (1) Let A be consistent with Γ. Show that $\neg A$ is not a theorem of Γ.
    (2) Let A be the formula $\forall x(x \circ x = e)$ of L_G. Is A consistent with $\{G1, G2, G3\}$? Is A independent of $\{G1, G2, G3\}$?

5.3

The Deduction Theorem and the Equality Theorem

In this section, we will prove two results that are needed for the model existence theorem.

The Deduction Theorem

The deduction theorem is a powerful proof technique for first-order logic. But, as in the case of propositional logic, the deduction theorem also has an important theoretical role: It is a key step in the proof of the adequacy theorem. Unfortunately, the deduction theorem does not carry over verbatim from propositional logic. The following example shows why.

Example 1:

The following is a formal proof that $\{x_1 = S0\} \vdash \forall x_1(x_1 = S0)$.

(1) $x_1 = S0$ HYP
(2) $\forall x_1(x_1 = S0)$ GEN.

On the other hand, the formula $(x_1 = S0) \to \forall x_1(x_1 = S0)$ is *not* a theorem of first-order logic, since it is not true in the standard interpretation of L_{NN}. What's wrong here? In this proof of $\{x_1 = S0\} \vdash \forall x_1(x_1 = S0)$, generalization is used with the variable x_1, and x_1 occurs free in the assumption $x_1 = S0$. Moreover, generalization is a derived rule of inference obtained from the add $\forall$-rule. With these observations in mind, we will now formulate a correct version of the deduction theorem. $\square$

Theorem 1 (Deduction theorem): If $\Gamma \cup \{A\} \vdash B$, where this proof has no application of the add $\forall$-rule (or a rule derived from it) that involves a variable free in A, then $\Gamma \vdash A \to B$.

Proof: The proof is quite similar to the corresponding result for propositional logic, so many details will be skipped. Let $S(n)$ say: If some formula, say B, has a proof using $\Gamma \cup \{A\}$ in n steps, and this proof has no application of the add $\forall$-rule involving a variable that is free in A, then $\Gamma \vdash A \to B$. We use induction to prove $S(n)$ for all $n \geq 1$.

Beginning step The statement $S(1)$ is true as in the corresponding proof for propositional logic (see Section 3.3).

Induction step Let $n \geq 1$ and assume that $S(1), \ldots, S(n)$ are true. Suppose that there is a proof of B using $\Gamma \cup \{A\}$ in $n + 1$ steps, and that the proof has no application of the add $\forall$-rule involving a variable free in A. If B is an axiom, or if $B \in \Gamma$, proceed as in the beginning step. Thus we may assume that B is obtained by a rule of inference of first-order logic. There are five such rules, and each must be considered. But for four of these rules (associative, cut, expansion, and contraction), the proof proceeds as in the deduction theorem for P. Therefore, it remains to consider the case in which the add $\forall$-rule is used. In this case, B has the form $C \vee \forall x_n D$, and the proof looks something like this:

We may use formulas of $\Gamma \cup \{A\}$ in this proof.

$$(k) \quad C \vee D$$

$$(n + 1) \quad C \vee \forall x_n D \quad (x_n \text{ not free in } C).$$

The induction hypothesis $S(k)$ gives $\Gamma \vdash A \to (C \vee D)$, and we are required to show $\Gamma \vdash A \to (C \vee \forall x_n D)$. Now the proof of $C \vee \forall x_n D$ using $\Gamma \cup \{A\}$ *does* have an application of the add $\forall$-rule involving the variable x_n; hence x_n cannot be free in A (hypothesis of the deduction theorem). Also, x_n is not free in C (required for a correct application of the add $\forall$-rule). It follows that x_n is not free in $\neg A \vee C$. We now have the following proof using Γ:

(1) $A \to (C \vee D)$ — THM of Γ

(2) $\neg A \vee (C \vee D)$ — DEF $\to$

(3) $(\neg A \vee C) \vee D$ — ASSOC

(4) $(\neg A \vee C) \vee \forall x_n D$ — ADD $\forall$ (x_n not free in $\neg A \vee C$)

(5) $\neg A \vee (C \vee \forall x_n D)$ — ASSOC

(6) $A \to (C \vee \forall x_n D)$ — DEF $\to$. ∎

Note We will list the derived rules of inference that are obtained from the add $\forall$-rule. Thus whenever one of these rules is used in a proof using the deduction theorem, the variable cannot appear free in the formula A.

GEN	∀-INTRO	∀-DIST
SUBST RULE	∃-INTRO	∃-DIST

There is one situation in which the restriction in the deduction theorem is automatically satisfied.

Corollary 1 (Special case of the deduction theorem): Let A be a sentence of L. If $\Gamma \cup \{A\} \vdash B$, then $\Gamma \vdash A \to B$.

Example 2 (Prenex rule):

We use the deduction theorem to show that

$$\vdash_{\text{FO}} \forall x(A \lor B) \leftrightarrow (\forall xA \lor B) \quad (x \text{ not free in } B).$$

Since the given formula is of the form $C \leftrightarrow D$, it suffices to show $\vdash C \to D$ and $\vdash D \to C$ separately and then to use JOIN and the definition of $\leftrightarrow$ to obtain $\vdash C \leftrightarrow D$.

Proof that: $\vdash_{\text{FO}} \forall x(A \lor B) \to (\forall xA \lor B)$

(1) $\forall x(A \lor B)$	HYP
(2) $A \lor B$	∀-ELIM
(3) $\forall xA \lor B$	ADD ∀ (x not free in B).

We have shown that $\{\forall x(A \lor B)\} \vdash \forall xA \lor B$; although the proof uses the add ∀-rule with the variable x, x is *not* free in $\forall x(A \lor B)$. Thus the deduction theorem applies and $\vdash_{\text{FO}} \forall x(A \lor B) \to (\forall xA \lor B)$ as required.

Proof that: $\vdash_{\text{FO}} (\forall xA \lor B) \to \forall x(A \lor B)$

(1) $\forall xA \lor B$	HYP
(2) $\forall xA \to A$	SUBST AX
(3) $\neg \forall xA \lor A$	DEF $\to$
(4) $A \lor B$	CUT $(1, 3)$
(5) $\forall x(A \lor B)$	GEN.

We have shown that $\{\forall xA \lor B\} \vdash \forall x(A \lor B)$; although the proof uses generalization on the variable x, x is *not* free in $\forall xA \lor B$ (since, by assumption, x is not free in B). Therefore, the deduction theorem applies and $\vdash_{\text{FO}} (\forall xA \lor B) \to \forall x(A \lor B)$ as required. $\quad\square$

We now introduce the ∃-elimination rule, a technique for handling existential quantifiers; ∃-ELIM is used to denote an application of this rule.

Theorem 2 (∃-elimination rule): Assume x does not occur free in B. To prove either

(1) $\Gamma \vdash \exists x A \to B$ or   (2) $\Gamma \cup \{\exists x A\} \vdash B$,

proceed as follows: Drop the quantifier $\exists x$ and prove $\Gamma \cup \{A\} \vdash B$, where this proof has no application of the add ∀-rule (or a rule derived from it) that involves a variable free in A.

Proof: By the deduction theorem, $\Gamma \vdash A \to B$. Since x is not free in B, the ∃-introduction rule applies and $\Gamma \vdash \exists x A \to B$; this proves (1). To obtain (2), use (1) and MP. ■

Example 3 (Prenex rule):

$\vdash \exists x(A \wedge B) \leftrightarrow (\exists x A \wedge B)$ (x not free in B).

Proof that: $\vdash \exists x(A \wedge B) \to (\exists x A \wedge B)$. We use ∃-ELIM.

(1) $A \wedge B$	∃-ELIM
(2) A	SEP
(3) $A \to \exists x A$	SUBST THM
(4) $\exists x A$	MP
(5) B	SEP (1)
(6) $\exists x A \wedge B$	JOIN.

We have proved $\{A \wedge B\} \vdash \exists x A \wedge B$ with no application of the add ∀-rule (or a rule derived from it). Moreover, the variable x does not occur free in $\exists x A \wedge B$. Hence Theorem 2(1) applies, and we have $\vdash \exists x(A \wedge B) \to (\exists x A \wedge B)$ as required.

Proof that: $\vdash (\exists x A \wedge B) \to \exists x(A \wedge B)$

(1) $\exists x A \wedge B$	HYP
(2) $\exists x A$	SEP
(3) B	SEP
(4) $B \to (A \to (A \wedge B))$	TT ($p \to (q \to (p \wedge q))$)
(5) $A \to (A \wedge B)$	MP
(6) $\exists x A \to \exists x(A \wedge B)$	∃-DIST
(7) $\exists x(A \wedge B)$	MP (2, 6).

We have proved $\{\exists x A \wedge B\} \vdash \exists x(A \wedge B)$; although the proof uses the ∃-distribution rule with the variable x, x is *not* free in $\exists x A \wedge B$. Thus the deduction theorem applies, and $\vdash (\exists x A \wedge B) \to \exists x(A \wedge B)$ as required. □

Example 4:

We verify the validity of the following syllogism: There is no rational number whose square is two; there is a real number whose square is two. ∴ There is a real number that is not a rational number. The required argument form is:

$$\forall x[Q(x) \to \neg S(x)] \qquad Q(x) = x \text{ is rational.}$$
$$\exists x[R(x) \wedge S(x)] \qquad S(x) = \text{square of } x \text{ is } 2.$$
$$\therefore \exists x[R(x) \wedge \neg Q(x)] \qquad R(x) = x \text{ is real.}$$

We use ∃-ELIM to write a formal proof for

$$\{\forall x[Q(x) \to \neg S(x)], \exists x[R(x) \wedge S(x)]\} \vdash \exists x[R(x) \wedge \neg Q(x)]. \qquad (*)$$

(1) $R(x) \wedge S(x)$ ∃-ELIM
(2) $\forall x[Q(x) \to \neg S(x)]$ HYP
(3) $Q(x) \to \neg S(x)$ ∀-ELIM
(4) $S(x)$ SEP (1)
(5) $\neg\neg S(x)$ $\neg\neg$ RULE
(6) $\neg Q(x)$ MT (3, 5)
(7) $R(x)$ SEP (1)
(8) $R(x) \wedge \neg Q(x)$ JOIN
(9) $\exists x[R(x) \wedge \neg Q(x)]$ ∃-GEN.

This proof has no application of the add ∀-rule (or a rule derived from it), and the variable x does not occur free in $\exists x[R(x) \wedge \neg Q(x)]$. Hence Theorem 2(2) applies and (*) holds as required.   □

────────────── *The Equality Theorem* ──────────────

In this section, we will obtain consequences of the equality axioms. Corollary 2, which follows, is used in the proof of the model existence theorem.

Lemma 1: $\vdash (y = z) \to (u = u_y[z])$, where u is a term and y and z are variables.

Proof: We have:

(1) $(u_y[z] = u_y[z]) \to [(y = z) \to (u = u_y[z])]$ EQ AX(4)
(2) $\forall x_1(x_1 = x_1)$ REF AX
(3) $u_y[z] = u_y[z]$ ∀-ELIM
(4) $(y = z) \to (u = u_y[z])$ MP(1, 3). ∎

Theorem 3 (Equality theorem): Let u be a term, let A be a formula, and let $y_1, \ldots, y_n$ be distinct variables. Assume $z_1, \ldots, z_n$ are variables such that for $1 \leq k \leq n$, z_k is substitutable for y_k in A. Then

(1) $\vdash [(y_1 = z_1) \wedge \ldots \wedge (y_n = z_n)] \rightarrow (u = u_{y_1, \ldots, y_n}[z_1, \ldots, z_n])$.

(2) $\vdash [(y_1 = z_1) \wedge \ldots \wedge (y_n = z_n)] \rightarrow (A \leftrightarrow A_{y_1, \ldots, y_n}[z_1, \ldots, z_n])$.

Proof: The proof of (1) is by induction on the number of variables $y_1, \ldots, y_n$. The case $n = 1$ follows from Lemma 1. Now assume that $n \geq 1$ and that the result holds for n. The induction hypothesis, applied to the term $u_{y_1}[z_1]$ and the variables $y_2, \ldots, y_{n+1}, z_2, \ldots, z_{n+1}$, gives

$$\vdash [(y_2 = z_2) \wedge \ldots \wedge (y_{n+1} = z_{n+1})]$$
$$\rightarrow (u_{y_1}[z_1] = u_{y_1, y_2, \ldots, y_{n+1}}[z_1, z_2, \ldots, z_{n+1}]).$$

Let v denote $u_{y_1, y_2, \ldots, y_{n+1}}[z_1, z_2, \ldots, z_{n+1}]$. With this simpler notation, the above result becomes $\vdash [(y_2 = z_2) \wedge \ldots \wedge (y_{n+1} = z_{n+1})] \rightarrow (u_{y_1}[z_1] = v)$. By Lemma 1, we have $\vdash (y_1 = z_1) \rightarrow (u = u_{y_1}[z_1])$. These two results and the tautology theorem give

$$\vdash [(y_1 = z_1) \wedge \ldots \wedge (y_{n+1} = z_{n+1})] \rightarrow [(u = u_{y_1}[z_1]) \wedge (u_{y_1}[z_1] = v)].$$

(Idea: Use the tautology $(p \rightarrow q) \rightarrow [(r \rightarrow s) \rightarrow ((p \wedge r) \rightarrow (q \wedge s))]$ and MP twice.) Now $\vdash [(u = u_{y_1}[z_1]) \wedge (u_{y_1}[z_1] = v)] \rightarrow (u = v)$ (transitive property of equality), and by HS we obtain (1) as required.

The proof of (2) is similar to that of (1); it proceeds by induction on the number of variables $y_1, \ldots, y_n$. Note that for the beginning step $n = 1$, (2) is an instance of an equality axiom. The details are left to the reader. ∎

Corollary 2: Let $s_1, \ldots, s_n, t_1, \ldots, t_n$ be terms, let F be an n-ary function symbol, and let R be an n-ary relation symbol.

(1) $\vdash [(s_1 = t_1) \wedge \ldots \wedge (s_n = t_n)] \rightarrow [F(s_1, \ldots, s_n) = F(t_1, \ldots, t_n)]$.

(2) $\vdash [(s_1 = t_1) \wedge \ldots \wedge (s_n = t_n)] \rightarrow [R(s_1, \ldots, s_n) \leftrightarrow R(t_1, \ldots, t_n)]$.

Proof: Choose distinct variables $y_1, \ldots, y_n, z_1, \ldots, z_n$ that do not occur in any of the terms. By (1) of the equality theorem, with u the term $F(y_1, \ldots, y_n)$:

$$\vdash [(y_1 = z_1) \wedge \ldots \wedge (y_n = z_n)] \rightarrow [F(y_1, \ldots, y_n) = F(z_1, \ldots, z_n)].$$

Now apply the substitution rule to this $2n$ times: Replace y_1 with s_1, z_1 with $t_1, \ldots, y_n$ with s_n, and z_n with t_n to obtain (1). The proof of (2) is similar: This time, use part (2) of the equality theorem with the formula $R(y_1, \ldots, y_n)$. The details are left to the reader. ∎

We end this section with two rules of inference that are immediate consequences of Corollary 2:

Function Rule
$$\frac{(s_1 = t_1), \ldots, (s_n = t_n)}{F(s_1, \ldots, s_n) = F(t_1, \ldots, t_n)}$$

Relation Rule
$$\frac{(s_1 = t_1), \ldots, (s_n = t_n), R(s_1, \ldots, s_n)}{R(t_1, \ldots, t_n)}$$

—————————Exercises for Section 5.3—————————

1. Show the following: (1) $\{x_1 = x_2\} \vdash x_1 = x_3$; (2) $(x_1 = x_2) \rightarrow (x_1 = x_3)$ is not logically valid. Why does this not contradict the deduction theorem?

2. Write formal proofs for the following; the tautology and deduction theorems are available.

   (1) $\vdash (\forall xA \lor \forall xB) \rightarrow \forall x(A \lor B)$.
   (2) $\vdash \exists x(A \lor B) \leftrightarrow (\exists xA \lor \exists xB)$.
   (3) $\vdash \forall x(A \land B) \leftrightarrow (\forall xA \land \forall xB)$.
   (4) $\vdash \exists x(A \land B) \rightarrow (\exists xA \land \exists xB)$.

3. Show the following:

   (1) $\vdash \forall x(A \lor B) \rightarrow (\forall xA \lor \exists xB)$.
   (2) $\vdash (\forall xA \lor \exists xB) \rightarrow \exists x(A \lor B)$.
   (3) $\vdash (\forall xA \land \exists xB) \rightarrow \exists x(A \land B)$.
   (4) $\vdash \forall x(A \rightarrow B) \rightarrow (\forall xA \rightarrow \forall xB)$.
   (5) $\vdash \forall x(A \rightarrow B) \rightarrow (\exists xA \rightarrow \exists xB)$.
   (6) $\vdash \exists x(A \rightarrow B) \rightarrow (\forall xA \rightarrow \exists xB)$.
   (7) $\vdash (\exists xA \rightarrow \forall xB) \rightarrow \forall x(A \rightarrow B)$.
   (8) $\vdash (\forall xA \rightarrow B) \rightarrow \exists x(A \rightarrow B)$.
   (9) $\vdash \forall x(A \rightarrow B) \leftrightarrow \neg\exists x(A \land \neg B)$.

4. (*Prenex rules*) Refer to Exercise 2 in Section 4.4. Show that each prenex rule is a theorem of first-order logic.

5. (*Existential quantifiers in formal proofs*) Use ∃-ELIM to show that the following are valid:

   (1) $\exists xP(x)$, $\forall x[P(x) \rightarrow Q(x)] \therefore \exists x[P(x) \land Q(x)]$.
   (2) $\forall x[P(x) \rightarrow (Q(x) \lor R(x))]$, $\exists x[P(x) \land \neg Q(x)] \therefore \exists x[P(x) \land R(x)]$.
   (3) $\neg\exists x[Q(x) \land R(x)]$, $\exists x[P(x) \land R(x)] \therefore \exists x[P(x) \land \neg Q(x)]$.
   (4) $\forall x[P(x) \rightarrow Q(x)]$, $\forall x[P(x) \rightarrow R(x)]$, $\exists xP(x) \therefore \exists x[Q(x) \land R(x)]$.
   (5) $\forall x[P(x) \rightarrow Q(x)]$, $\exists x[P(x) \lor R(x)] \therefore \exists x[Q(x) \lor R(x)]$.
   (6) $\forall x[P(x) \land Q(x)]$, $\forall x[P(x) \rightarrow R(x)]$, $\exists x[Q(x) \rightarrow S(x)] \therefore \exists x[R(x) \land S(x)]$.

6. (*Proof by cases*) Justify the following proof procedure: If $\Gamma \cup \{A\} \vdash C$ and $\Gamma \cup \{B\} \vdash C$, and these two proofs have no application of the add ∀-rule (or a rule derived from it) involving a variable that occurs free in A or B respectively, then $\Gamma \vdash (A \lor B) \rightarrow C$. Then use proof by cases to show that $\vdash (\exists xA \lor \exists xB) \rightarrow \exists x(A \lor B)$.

7. Show valid: $\forall x\exists yR(x, y)$, $\forall x\forall y(R(x, y) \rightarrow R(y, x))$, $\forall x \,\forall y \,\forall z[R(x, y) \rightarrow (R(y, z) \rightarrow R(x, z))] \therefore \forall xR(x, x)$.

8. Show valid: No A are B. All B are C. There is something with property B. $\therefore$ Some C are not A.

9. Show that the following arguments are valid:

   (1) All red-headed people are hard workers. All hard workers are happy. There are some rich people who are not happy. $\therefore$ There are some rich people who are not red-headed. ($R(x) = x$ is red-headed; $W(x) = x$ is a hard worker; $H(x) = x$ is happy; $I(x) = x$ is rich.)
   (2) All ellipses are curves. Fred draws an ellipse. $\therefore$ Someone draws a curve. ($E(x) = x$ is an ellipse; $C(x) = x$ is a curve; $D(x, y) = x$ draws y; $f = $ Fred.)
   (3) There is a man whom all men despise. $\therefore$ There is at least one man who despises himself. ($M(x) = x$ is a man; $D(x, y) = x$ despises y.)

10. Assume t is substitutable for x in A and that x does not occur in t. Prove the following:

    (1) $\vdash A_x[t] \leftrightarrow \forall x[(x = t) \rightarrow A)]$.
    (2) $\vdash A_x[t] \leftrightarrow \exists x[(x = t) \land A)]$.

11. Show valid: $a = b$, $c = d$, $\neg(b = d) \therefore \neg(a = c)$.

12. Let Γ be a set of sentences that is inconsistent and let B be an arbitrary formula of L. Show that there exist $A_1, \ldots, A_n \in \Gamma$ such that $(A_1 \land \ldots \land A_n) \rightarrow B$ is logically valid.

13. Consider the following syllogism: All A are B. Some C are not B. $\therefore$ Some C are not A.

    (1) Write the appropriate argument form.
    (2) Prove the argument form valid by writing a formal proof. Use the tautology theorem, but not the deduction theorem or the ∃-elimination rule.
    (3) Prove the argument form valid by writing a formal proof. This time, use the ∃-elimination rule.
    (4) Give a semantic proof of validity (i.e., use the definition of a valid argument form to prove validity).

$$\boxed{5.4}$$

The Model Existence Theorem

In this section, we will prove the model existence theorem: If Γ is consistent, then Γ has a model. As in the case of propositional logic, this is a key step in the proof of the adequacy theorem. Given a consistent set Γ of formulas of L, how will we construct a model of Γ? Roughly speaking, from the symbols of L! Somewhat more precisely, we construct an interpretation of L called the *canonical interpretation of* Γ. The domain of this interpretation consists of equivalence classes of variable-free terms of L; constant, function, and relation symbols are then given a rather natural interpretation. Under suitable conditions on Γ, the canonical interpretation of Γ is indeed a model of Γ.

Description of the canonical interpretation CI_Γ *of* Γ: Let Γ be a set of formulas of a language L that has at least one constant symbol. Let VFT be the set of all variable-free terms of L; since we are assuming that L has at least one constant symbol, VFT $\neq \emptyset$. Define a 2-ary relation $\sim$ on VFT by $s \sim t \Leftrightarrow \Gamma \vdash s = t$; note that $\sim$ depends on Γ. The following properties hold for all variable-free terms s, t, and u of L:

ER1 $t \sim t$.   **ER3** If $s \sim t$ and $t \sim u$, then $s \sim u$.

ER2 If $s \sim t$, then $t \sim s$.

To see this, use the definition of $\sim$ to rewrite the properties as follows: ER1, $\Gamma \vdash t = t$; ER2, if $\Gamma \vdash s = t$, then $\Gamma \vdash t = s$; ER3, if $\Gamma \vdash s = t$ and $\Gamma \vdash t = u$, then $\Gamma \vdash s = u$. We now see that ER1 follows from the reflexive axiom and that ER2 and ER3 follow from equality axioms (see Corollary 1 in Section 5.1). Since $\sim$ satisfies ER1–ER3, $\sim$ is an equivalence relation on VFT. For each $t \in$ VFT, let $[t] = \{s: s \in \text{VFT and } s \sim t\}$; $[t]$ is an equivalence class of $\sim$. By Theorem 3 in Section 1.4, $s \sim t \Leftrightarrow [s] = [t]$.

Although the canonical interpretation CI_Γ depends on Γ, we usually drop the subscript Γ to avoid excessive notation. The *domain* of CI is the collection of all equivalence classes of variable-free terms; in other words,

$$D_{\text{CI}} = \{[t]: t \in \text{VFT}\}, \quad \text{or} \quad D_{\text{CI}} = \text{VFT}/\!\sim.$$

Constant, function, and relation symbols are interpreted as follows:

CI(c) For each constant symbol c, $c_{\text{CI}} = [c]$.

CI(F) For each n-ary function symbol F, $F_{\text{CI}}: D^n_{\text{CI}} \to D_{\text{CI}}$ is defined by $F_{\text{CI}}(\langle [t_1], \ldots, [t_n] \rangle) = [F(t_1, \ldots, t_n)]$.

CI(R) For each n-ary relation symbol R, $R_{\text{CI}} \subseteq D^n_{\text{CI}}$ is defined by $\langle [t_1], \ldots, [t_n] \rangle \in R_{\text{CI}} \Leftrightarrow \Gamma \vdash R(t_1, \ldots, t_n)$.

Note the dependence of the relation R_{CI} on Γ. It is necessary to check that

F_{CI} and R_{CI} are well defined (i.e., independent of the choice of representative elements). For this, it suffices to show that if $\Gamma \vdash s_1 = t_1, \ldots, \Gamma \vdash s_n = t_n$, then $\Gamma \vdash F(s_1, \ldots, s_n) = F(t_1, \ldots, t_n)$ and $\Gamma \vdash R(s_1, \ldots, s_n) \Leftrightarrow \Gamma \vdash R(t_1, \ldots, t_n)$. These two results follow easily from Corollary 2 of the equality theorem (see Section 5.3).

Example 1:

Let Γ be all sentences of L_{NN} that are true in the standard interpretation $\mathcal{N}$ of L_{NN}. The following holds for any sentence A of L_{NN}: $\Gamma \vdash A$ if and only if A is true in $\mathcal{N}$ (use the soundness theorem). Examples of variable-free terms of L_{NN} are 0, $S0$, $SS0$, and $S0 + S0$. Moreover, $(S0 + S0) = SS0$ is true in $\mathcal{N}$; hence $\Gamma \vdash (S0 + S0) = SS0$. From this, it follows that $(S0 + S0) \sim SS0$ (definition of $\sim$); hence $[S0 + S0] = [SS0]$. Also note that $[S0] +_{CI} [S0] = [S0 + S0] = [SS0]$. Finally, $0 = S0$ is not true in the standard interpretation; hence $0 = S0$ is not a theorem of Γ. Thus $0 \sim S0$ fails, and $[0] \neq [S0]$. $\square$

Example 2:

Let $\Gamma = \{G1, G2, G3\}$, where G1–G3 are the axioms of a group expressed in the language L_G. The canonical interpretation of Γ is easily described. There is exactly one constant symbol, e, and the variable-free terms of L_G are e, $e \circ e$, $e \circ (e \circ e)$, $(e \circ e) \circ e$, and so on. But it is intuitively clear that if t is any one of these terms, then $\Gamma \vdash t = e$, and so $t \sim e$. Thus all variable-free terms are in $[e]$, and we have: $D_{CI} = \{[e]\}$; $e_{CI} = [e]$; $\circ_{CI}([e], [e]) = [e]$. In other words, the canonical interpretation of Γ is (up to isomorphism) the group with a single element. $\square$

Lemma 1 (Terms): Let Γ be a set of formulas of a language L with at least one constant symbol and let $\phi: VAR \rightarrow D_{CI}$ be an assignment in the canonical interpretation CI of Γ. Then $\phi(t) = [t]$ for every variable-free term t of L.

Proof: By induction on the number of symbols in t. First, suppose t is a constant symbol c. Then $\phi(c) = c_{CI} = [c]$ (use CONST SYM and CI(c) respectively). Now suppose t is a term of the form $F(t_1, \ldots, t_n)$. In this case, $t_1, \ldots, t_n$ are variable-free terms with fewer symbols than t; hence, by induction hypothesis, $\phi(t_k) = [t_k]$ for $1 \leq k \leq n$. We have:

$$
\begin{aligned}
\phi(t) &= \phi(F(t_1, \ldots, t_n)) \\
&= F_{CI}(\phi(t_1), \ldots, \phi(t_n)) &&\text{FNC SYM} \\
&= F_{CI}([t_1], \ldots, [t_n]) &&\text{Induction hypothesis} \\
&= [F(t_1, \ldots, t_n)] &&\text{CI}(F) \\
&= [t]. \quad \blacksquare
\end{aligned}
$$

Comment Now that Lemma 1 is available, we will no longer need the two conditions CI(c) and CI(F).

Lemma 2 (Atomic sentences): Let Γ be a set of formulas of a language L with at least one constant symbol. Then for any atomic sentence A of L,

$$\Gamma \vdash A \Leftrightarrow A \text{ is true in the canonical interpretation of } \Gamma.$$

Proof: There are two cases: Either A is $(s = t)$ or A is $R(t_1, \ldots, t_n)$. Let ϕ be an assignment in the canonical interpretation of Γ. Since A is a sentence, we have: A true in CI $\Leftrightarrow \phi(A) = T$. By Lemma 1, $\phi(s) = [s]$, $\phi(t) = [t]$, and $\phi(t_k) = [t_k]$ for $1 \le k \le n$. The following two lists of equivalent statements prove Lemma 2:

(1) $(s = t)$ true in CI
(2) $\phi(s = t)$ is T
(3) $\phi(s)$ and $\phi(t)$ are the same element of D_{CI} TA($=$)
(4) $[s]$ and $[t]$ are the same element of D_{CI} Lemma 1
(5) $[s] = [t]$ Definition of equality
(6) $s \sim t$
(7) $\Gamma \vdash s = t$ Definition of $\sim$.

(1) $R(t_1, \ldots, t_n)$ true in CI
(2) $\phi(R(t_1, \ldots, t_n)) = T$
(3) $\langle \phi(t_1), \ldots, \phi(t_n) \rangle \in R_{CI}$ TA(R)
(4) $\langle [t_1], \ldots, [t_n] \rangle \in R_{CI}$ Lemma 1
(5) $\Gamma \vdash R(t_1, \ldots, t_n)$ CI(R). ∎

Comment Now that Lemma 2 is available, we will no longer need the condition CI(R).

We want to prove that the conclusion of Lemma 2 extends to all sentences of L (under the assumption that Γ is consistent). But to do this, we need two additional assumptions about Γ.

Definition 1: Let Γ be a set of formulas of L.

(1) Γ is *complete* if for each sentence A of L, either $\Gamma \vdash A$ or $\Gamma \vdash \neg A$.
(2) Γ has the *Henkin property* if for each formula B with exactly one free variable x, if $\Gamma \vdash \neg \forall x B$, then there is a variable-free term t such that $\Gamma \vdash \neg B_x[t]$. We call t a *witness* to $\neg \forall x B$.

The following example shows why we need the Henkin property.

Example 3:

Let L have constant symbol c and let $\Gamma = \{\neg \forall x(x = c)\}$. Any interpretation of L whose domain has at least two distinct elements is a model of Γ;

hence Γ is consistent. Now let us describe the canonical interpretation CI_Γ of Γ: $D_{CI} = \{[c]\}$; $c_{CI} = [c]$. It is not difficult to see that $\neg\forall x(x = c)$ is false in CI_Γ; hence CI_Γ is not a model of Γ. Note, however, that Γ does not have the Henkin property.   □

Theorem 1 (Henkin): Let L be a first-order language with at least one constant symbol and let Γ be a set of formulas of L that is consistent, complete, and Henkin. Then the canonical interpretation CI_Γ of Γ is a model of Γ.

Proof: We first prove that for any sentence A of L, $\Gamma \vdash A \Leftrightarrow A$ true in CI. The proof is by induction on the number of occurrences of $\neg$, $\vee$, and $\forall$ in A. The case in which A is atomic follows from Lemma 2.

Negation Suppose A is $\neg B$; by induction hypothesis, $\Gamma \vdash B \Leftrightarrow B$ true in CI. We have:

> A true in CI
>> $\Leftrightarrow B$ false in CI
>> $\Leftrightarrow B$ not a theorem of Γ Induction hypothesis
>> $\Leftrightarrow \Gamma \vdash \neg B$ Γ complete ($\Rightarrow$) and consistent ($\Leftarrow$)
>> $\Leftrightarrow \Gamma \vdash A$.

Disjunction Left as an exercise for the reader (see Theorem 6 in Section 3.4).

Quantifier Suppose A is $\forall xB$. First, assume x is not free in B. Then B is a sentence, the induction hypothesis applies to B, and we have

> A true in CI $\Leftrightarrow B$ true in CI   Closure theorem (interpretations)
>> $\Leftrightarrow \Gamma \vdash B$ Induction hypothesis
>> $\Leftrightarrow \Gamma \vdash \forall xB$ Closure theorem (provability)
>> $\Leftrightarrow \Gamma \vdash A$.

Now assume x is free in B. In this case, B is not a sentence, so the induction hypothesis does not apply to B. We are required to show that $\forall xB$ is true in CI $\Leftrightarrow \Gamma \vdash \forall xB$. First, assume $\forall xB$ true in CI. Since Γ is complete and $\forall xB$ is a sentence, either $\Gamma \vdash \forall xB$ or $\Gamma \vdash \neg\forall xB$. Assume $\Gamma \vdash \neg\forall xB$; under this assumption, we will prove that there is a variable-free term t of L such that (1) $\Gamma \vdash B_x[t]$ and (2) $\Gamma \vdash \neg B_x[t]$. But (1) and (2) together contradict the consistency of Γ, so completeness of Γ gives $\Gamma \vdash \forall xB$ as required. We still need to prove (1) and (2) assuming $\Gamma \vdash \neg\forall xB$. Since Γ has the Henkin property and $\Gamma \vdash \neg\forall xB$, there is a variable-free term t such that $\Gamma \vdash \neg B_x[t]$; this proves (2). The formula $\forall xB \to B_x[t]$ is a substitution axiom, and hence is true in any interpretation. In addition, $\forall xB$ is true in CI, so by modus ponens for interpretations, $B_x[t]$ is true in CI. Now $B_x[t]$ *is* a sentence; hence the induction hypothesis applies and $\Gamma \vdash B_x[t]$. This proves (1).

Now assume $\Gamma \vdash \forall xB$; it suffices to show B true in CI (closure theorem for interpretations). Let $\phi: VAR \to VFT/\sim$ be an arbitrary assignment in CI; we are required to show $\phi(B) = T$. There is a variable-free term t such that $\phi(x) = [t]$. By Lemma 1, we have $\phi(t) = [t]$; hence $\phi(x) = \phi(t)$. By the substitution theorem for assignments (special case $\phi = \psi$), $\phi(B) = \phi(B_x[t])$.

So to prove $\phi(B) = T$, it suffices to show that $B_x[t]$ is true in CI. Consider this proof in Γ:

  (1) $\forall xB$ THM of Γ
  (2) $\forall xB \rightarrow B_x[t]$ SUBST AX
  (3) $B_x[t]$ MP.

In summary, $B_x[t]$ is a sentence and $\Gamma \vdash B_x[t]$; by the induction hypothesis, $B_x[t]$ is true in CI.

As a consequence of the above, we have: If $\Gamma \vdash A$, then A is true in CI (A a sentence of L). It remains to show that CI is a model of Γ. Let $A \in \Gamma$. Then $\Gamma \vdash A$; hence $\Gamma \vdash A'$. By the result just proved, A' is true in CI. So A is true in CI as required (closure theorem for interpretations). ■

Henkin's theorem is a special case of the model existence theorem: It holds for Γ consistent, complete, and Henkin. To obtain the model existence theorem from this special case, we need to show that for Γ consistent, there is a set Δ of formulas such that $\Gamma \subseteq \Delta$ and Δ is consistent, complete, and Henkin. The construction of Δ is a modification of the proof of Lindenbaum's theorem for propositional logic. First we need some additional results, two of which are familiar.

Lemma 3 (Characterization of consistency): Γ is consistent if and only if there is at least one formula of L that is not a theorem of Γ; Γ is inconsistent if and only if every formula of L is a theorem of Γ.

Proof: See the proof in Section 3.4 of the corresponding result for propositional logic. ■

Theorem 2 (Extension): Let A be a sentence of L that is not a theorem of Γ. Then $\Gamma \cup \{\neg A\}$ is consistent.

Proof: Left as an exercise for the reader (see Theorem 4 in Section 3.4).

Theorem 3 (Theorem on constants): Let c be a constant symbol of L that does not occur in A or in any formula of Γ. If $\Gamma \vdash A_x[c]$, then $\Gamma \vdash \forall xA$.

Proof: Let $A_1, \ldots, A_n$ (with $A_n = A_x[c]$) be a proof of $A_x[c]$ using Γ. Choose a new variable y that does not occur in A or in the proof $A_1, \ldots, A_n$. For $1 \leq k \leq n$, let B_k be the formula obtained from A_k by replacing each occurrence of c with y. Now A_n is $A_x[c]$ and c does not occur in A; hence $B_n = (A_x[c])_c[y]) = A_x[y]$. We will now make several observations, which together show that $B_1, \ldots, B_n$ is a proof using Γ. Let $1 \leq k \leq n$. (1) If $A_k \in \Gamma$, then $B_k = A_k$ (c does not occur in any formula in Γ), so $B_k \in \Gamma$. (2) If A_k is a logical axiom, then B_k is an axiom of the same kind. (3) If $k > 1$ and A_k is the conclusion of a rule of inference whose hypotheses are among $A_1, \ldots, A_{k-1}$, then B_k is the conclusion of the same rule with hypotheses among $B_1, \ldots, B_{k-1}$. In summary, $B_1, \ldots, B_n$ is a proof using Γ and B_n is

$A_x[y]$; hence $\Gamma \vdash A_x[y]$. Note that by reverse substitution (see Section 4.4), x is substitutable for y in $A_x[y]$ and $(A_x[y])_y[x] = A$. Finally, $\Gamma \vdash \forall x A$ as follows:

(1) $A_x[y]$ THM of Γ

(2) $\forall y A_x[y]$ GEN

(3) $\forall y A_x[y] \rightarrow A$ SUBST AX

(4) A MP

(5) $\forall x A$ GEN. ∎

Corollary 1: Let Γ be a consistent set of formulas of L and let A be a formula with exactly one free variable x such that $\Gamma \vdash \neg \forall x A$. Suppose c is a constant symbol of L that does not occur in A or in any formula of Γ. Then $\Gamma \cup \{\neg A_x[c]\}$ is consistent.

Proof: Suppose $\Gamma \cup \{\neg A_x[c]\}$ is not consistent. Then every formula is a theorem of $\Gamma \cup \{\neg A_x[c]\}$. In particular, $\Gamma \cup \{\neg A_x[c]\} \vdash A_x[c]$; hence $\Gamma \vdash \neg A_x[c] \rightarrow A_x[c]$ (deduction theorem). Now consider the following proof in Γ:

(1) $\neg A_x[c] \rightarrow A_x[c]$ THM of Γ

(2) $\neg\neg A_x[c] \vee A_x[c]$ DEF $\rightarrow$

(3) $A_x[c] \vee A_x[c]$ $\neg\neg$ RULE

(4) $A_x[c]$ CTN.

The theorem on constants applies and $\Gamma \vdash \forall x A$. We now have $\Gamma \vdash \forall x A$ and $\Gamma \vdash \neg \forall x A$, a contradiction of the consistency of Γ. ∎

Theorem 4 (Lindenbaum-Henkin): Let Γ be consistent and let $c_1, c_2, \ldots$ be an infinite list of constant symbols of L that do not occur in any of the formulas of Γ. Then there is a set Δ of formulas of L such that $\Gamma \subseteq \Delta$ and Δ is consistent, complete, and Henkin.

Proof: Let $A_0, A_1, A_2, \ldots$ be a list of all sentences of L. By induction, construct a sequence $\Gamma_0, \Gamma_1, \Gamma_2, \ldots$ of sets of formulas of L as follows:

$\Gamma_0 = \Gamma$;

$$\Gamma_{n+1} = \begin{cases} \Gamma_n & \Gamma_n \vdash A_n \text{ and } A_n \text{ is } not \text{ of the form } \neg\forall x A, \\ & \text{where } A \text{ has exactly one free variable } x. \\ \Gamma_n \cup \{\neg A_x[c]\} & \Gamma_n \vdash A_n \text{ and } A_n \text{ is of the form } \neg\forall x A, \text{ where} \\ & A \text{ has exactly one free variable } x. \\ \Gamma_n \cup \{\neg A_n\} & A_n \text{ is not a theorem of } \Gamma_n. \end{cases}$$

Note In the case in which Γ_{n+1} is defined by $\Gamma_{n+1} = \Gamma_n \cup \{\neg A_x[c]\}$, the constant symbol c is the first symbol in the list $c_1, c_2, \ldots$ that does not occur in any of the formulas of Γ_n or in A_n. Such a constant symbol c exists: None of the constant symbols $c_1, c_2, \ldots$ occur in the formulas of Γ, and only a finite number of formulas are added to Γ to obtain Γ_n. By choosing c in this way, we guarantee that the hypotheses of Corollary 1 are satisfied.

Each Γ_n *is consistent* The proof is by induction. Obviously, Γ_0 is consistent. Assume that Γ_n is consistent. There are three cases, based on the construction of Γ_{n+1}: If $\Gamma_{n+1} = \Gamma_n$, then Γ_{n+1} is consistent by the induction hypothesis that Γ_n is consistent. If $\Gamma_{n+1} = \Gamma_n \cup \{\neg A_x[c]\}$, then Γ_{n+1} is consistent by Corollary 1. Finally, if $\Gamma_{n+1} = \Gamma_n \cup \{\neg A_n\}$, then Γ_{n+1} is consistent by the extension theorem.

Now let $\Delta = \bigcup_{n \in \mathbb{N}} \Gamma_n$. Clearly $\Gamma \subseteq \Delta$, so it remains to prove that Δ is consistent, complete, and Henkin. The proof that Δ is consistent and complete proceeds as in the case for propositional logic; for details, see the proof of Lindenbaum's theorem in Section 3.4.

Δ *is Henkin* Let A be a formula with exactly one free variable x such that $\Delta \vdash \neg \forall x A$; it suffices to show that there is a constant symbol c such that $\Delta \vdash \neg A_x[c]$. Since $\neg \forall x A$ is a sentence, there exists $n \in \mathbb{N}$ such that $A_n = \neg \forall x A$. Assume, for a moment, that we have proved that $\Gamma_n \vdash \neg \forall x A$; it then follows by the construction of Γ_{n+1} that $\neg A_x[c] \in \Gamma_{n+1}$, hence $\Delta \vdash \neg A_x[c]$ as required. It remains to prove $\Gamma_n \vdash \neg \forall x A$. Suppose, by way of contradiction, that $\neg \forall x A$ is not a theorem of Γ_n. By the construction of Γ_{n+1} we have $\neg \neg \forall x A \in \Gamma_{n+1}$, hence $\Delta \vdash \neg \neg \forall x A$. But we also have $\Delta \vdash \neg \forall x A$, a contradiction of the consistency of Δ. ■

Definition 2: Let Γ be a set of formulas of L and let L^+ be the language obtained from L by adding new constant symbols to L (perhaps an infinite number, say $c_1, c_2, \ldots$). In this situation, we may regard Γ as a set of formulas of L^+; we denote this by writing Γ^+, and we call Γ^+ an *extension of Γ by constants*.

Every theorem of Γ is a theorem of Γ^+, but the converse is false; Γ^+ has additional theorems, since its language has additional constant symbols. For example, if c is a new constant symbol, then $\forall x(x = x) \rightarrow (c = c)$ is a substitution axiom of L^+ and hence is a theorem of Γ^+. But $\forall x(x = x) \rightarrow (c = c)$ is not even a formula of L; hence it cannot be a theorem of Γ. However, we do have the following result.

Theorem 5 (Extension by constants): Let Γ be a set of formulas of L and let Γ^+ be an extension of Γ by constants.

(1) For every formula A of L, $\Gamma \vdash A$ if and only if $\Gamma^+ \vdash A$.

(2) Γ is consistent if and only if Γ^+ is consistent.

Proof of (1): If $\Gamma \vdash A$, then obviously $\Gamma^+ \vdash A$. Let L^+ denote the language obtained from L by adding the new constant symbols. Suppose $\Gamma^+ \vdash A$, and let $A_1, \ldots, A_n$ be formulas of L^+ such that $A_n = A$ and $A_1, \ldots, A_n$ is a proof using Γ^+. We emphasize that these are formulas of L^+ but not necessarily of L, and hence we cannot immediately say that $\Gamma \vdash A$. But we can convert this into a proof of A using Γ, and each formula of this new proof is a formula of L.

Let the new constant symbols that occur in the proof $A_1, \ldots, A_n$ be $c_1, \ldots, c_k$ (these are constant symbols of L^+ but not of L). Choose k new variables $y_1, \ldots, y_k$ that do not occur anywhere at all in the proof $A_1, \ldots, A_n$. For $1 \leq j \leq n$, let B_j be the formula of L obtained from A_j by replacing those occurrences of $c_1, \ldots, c_k$ in A_j with $y_1, \ldots, y_k$ respectively. Note that B_n is just A_n, since A is a formula of L. As in the proof of the theorem on constants, $B_1, \ldots, B_n$ is a set of formulas of L that is a proof of A using Γ.

Proof of (2): It is clear that if Γ^+ is consistent, then Γ is consistent. Suppose, then, that Γ is consistent, and let us show that Γ^+ is consistent. If not, then every formula of L^+ is a theorem of Γ^+; in particular, $\Gamma^+ \vdash \neg(x = x)$. But $\neg(x = x)$ is a formula of L; hence by (1) we have $\Gamma \vdash \neg(x = x)$. But obviously $\Gamma \vdash (x = x)$, so we have contradicted the consistency of Γ. ∎

Theorem 6 (Model existence theorem): If Γ is consistent, then Γ has a model.

Proof: Assume Γ is consistent. We would like to apply the Lindenbaum-Henkin theorem and assert that $\Gamma \subseteq \Delta$, where Δ is consistent, complete, and Henkin. But to do this, we must arrange to have a countable number of constant symbols that do not occur in any of the formulas of Γ.

Let L^+ be the language obtained from L by adding an infinite list of new constant symbols $c_1, c_2, \ldots$ to L. We then obtain Γ^+, an extension of Γ by constants. Since Γ is consistent, Γ^+ is consistent. Apply the Lindenbaum-Henkin theorem to Γ^+: There is a set Δ of formulas of L^+ such that $\Gamma \subseteq \Delta$ and Δ is consistent, complete, and Henkin. By Henkin's theorem, the canonical interpretation of Δ is a model of Δ. Now $\Gamma \subseteq \Delta$; hence if we ignore the interpretation that is given to the constant symbols $c_1, c_2, \ldots$, we obtain a model of Γ as required. ∎

Summary: We will now summarize the three major steps in the proof of the model existence theorem. Let Γ be a consistent set of formulas of L.

(1) Obtain a new language L^+ by adding a countable number of new constant symbols $c_1, c_2, \ldots$ to L; Γ is still consistent.

(2) By the Henkin-Lindenbaum construction, there is an extension Δ of Γ that is consistent, complete, and Henkin.

(3) By Henkin's theorem, the canonical interpretation of Δ is a model of Δ. If we ignore the interpretation given to the new constant symbols $c_1, c_2, \ldots$, the canonical interpretation of Δ is a model of Γ.

Once the model existence theorem is available, we can easily obtain the adequacy theorem; this will be done in the next section. Other consequences of the model existence theorem (together with the consistency theorem—see Section 5.2) are the Löwenheim-Skolem theorem and the compactness theorem; proofs and applications of these two important theorems will be given in Section 6.2. For now, we will simply state these results and leave the proofs as an exercise for the reader. (Both the Löwenheim-Skolem theorem and the compactness theorem are stated entirely in terms of semantic concepts; in the exercises, we will outline a semantic proof of the compactness theorem.)

Theorem 7 (Löwenheim-Skolem): If Γ has a model, then Γ has a countable model.

Theorem 8 (Compactness theorem for first-order logic, Version I): If every finite subset of Γ has a model, then Γ has a model.

Theorem 9 (Compactness theorem for first-order logic, Version II): If A is true in every model of Γ, then there is a finite $\Delta \subseteq \Gamma$ such that A is true in every model of Δ.

──────────*Exercises for Section 5.4*──────────

1. Let Γ be the set of all sentences of L_{NN} that are true in the standard interpretation of L_{NN}. The following are variable-free terms of L_{NN}. Which belong to the same equivalence class in the canonical interpretation of Γ?
   (1) $S0$ (3) $SS0 + S0$ (5) $SSS0$
   (2) $S0 + S0$ (4) $S0 \times S0$ (6) $SS0 + (0 \times SSSS0)$
2. Refer to the axioms of NN (= formal arithmetic) in Section 7.1. The language of NN is L_{NN}; assume that the standard interpretation of L_{NN} is a model of NN. This problem shows that the formula A in the extension theorem must be a sentence. Consider the formula $(x_1 < S0)$ of L_{NN}.
   (1) Explain why $(x_1 < S0)$ is not a theorem of NN.
   (2) Let $\Gamma = \{NN1, \ldots, NN9\} \cup \{\neg(x_1 < S0)\}$. Show that Γ is inconsistent.
3. Prove Lemma 3 (characterization of consistency).
4. Prove Theorem 2 (extension theorem).
5. Let Γ_1 and Γ_2 be sets of formulas of L with these properties: $\Gamma_1 \subseteq \Gamma_2$; Γ_1 is complete; Γ_2 has a model. Show that Γ_1 and Γ_2 have the same theorems.
6. Let $\Gamma = \{\neg(b = c), \forall x_1 \neg R(x_1, x_1), \forall x_1[R(x_1, b) \vee R(x_1, c)], \forall x_1[F(x_1) = b]\}$.
   (1) Show that Γ is consistent by finding a model of Γ.
   (2) Describe the canonical interpretation CI_Γ of Γ. In other words, describe D_{CI}, b_{CI}, c_{CI}, F_{CI}, and R_{CI}, and justify your results.
   (3) Is the canonical interpretation CI_Γ of Γ a model of Γ?
7. Let $\Gamma = \{\neg(b = c), \forall x_1 R(x_1, x_1), \forall x_1 \forall x_2 \forall x_3[(R(x_1, x_2) \wedge R(x_1, x_3)) \rightarrow (x_2 = x_3)], \forall x_1[(F(x_1) = b) \vee (F(x_1) = c)], \forall x_1 \neg[F(x_1) = x_1]\}$.
   (1) Show that Γ is consistent by finding a model of Γ.
   (2) Describe the canonical interpretation CI_Γ of Γ. In other words, describe D_{CI}, b_{CI}, c_{CI}, F_{CI}, and R_{CI}, and justify your results.
   (3) Is the canonical interpretation CI_Γ of Γ a model of Γ?
8. Let $\Gamma = \{\neg(b = c), \neg\forall x[(x = b) \vee (x = c)]\}$.
   (1) Find a model of Γ.
   (2) Describe the canonical interpretation of Γ.
   (3) Is CI_Γ a model of Γ? Is Γ Henkin?
9. Let $\Gamma = \{\neg(b = c), \neg\forall x(x = b), \neg\forall x(x = c)\}$.
   (1) Find a model of Γ.
   (2) Describe the canonical interpretation of Γ.
   (3) Is CI_Γ a model of Γ? Is Γ Henkin?
10. (*Theorem on constants*) Refer to the axioms of NN in Section 7.1.
    (1) Show that NN $\vdash \neg\forall x_1(x_1 = 0)$.
    (2) Show that $\Gamma \cup \{\neg(0 = 0)\}$ is not consistent, where $\Gamma = \{NN1, \ldots, NN9\}$. Why does this not contradict the corollary of the theorem on constants?
11. Let Γ be a set of formulas of L. Show that Γ is complete if and only if there is no sentence A of L such that both $\Gamma \cup \{A\}$ and $\Gamma \cup \{\neg A\}$ are consistent.

12. (*Canonical interpretation of formal arithmetic*) Refer to Section 7.1 for a description of the formal system NN with language L_{NN}. Prove the following. (*Hint:* Assume the results of Theorem 1 in Section 7.1 on basic properties of the formal system NN).
    (1) For each variable-free term t of L_{NN}, there exists $n \geq 0$ such that NN $\vdash t = 0^n$.
    (2) $D_{CI} = \{[0^n]: n \in \mathbb{N}\}$.
    (3) $S_{CI}([0^n]) = [0^{n+1}]$.
    (4) $+_{CI}([0^m], [0^n]) = [0^{m+n}]$.
    (5) $\times_{CI}([0^m], [0^n]) = [0^{m \times n}]$.
    (6) $<_{CI} = \{\langle [0^m], [0^n] \rangle : m$ is less than $n\}$.
13. (*Henkin's theorem for a first-order language without equality*) Let L be a first-order language without the equality symbol $=$. Assume, however, that L has at least one relation symbol and a constant symbol. In this case, the proof of Henkin's theorem is simplified. Let Δ be a set of sentences of L such that Δ is consistent, complete, and Henkin. Show that the following interpretation I of L is a model of Δ: $D_I = $ VFT; $c_I = c$; $F_I: D_I^n \rightarrow D_I$ is defined by $F_I(t_1, \ldots, t_n) = F(t_1, \ldots, t_n)$; $R_I \subseteq D_I^n$ is defined by $\langle t_1, \ldots, t_n \rangle \in R_I \Leftrightarrow \Gamma \vdash R(t_1, \ldots, t_n)$.
14. (*First-order truth set*) Let L be a first-order language with at least one constant symbol. Let Δ be a set of sentences of L satisfying these conditions: (1) For every sentence A of L, $A \in \Delta$ or $\neg A \in \Delta$ but not both; (2) for all sentences A and B of L, $A \vee B \in \Delta$ if and only if $A \in \Delta$ or $B \in \Delta$; (3) for every sentence A of L, $A \in \Delta$ if and only if $\forall x A \in \Delta$; (4) if A has exactly one free variable x, then $\neg \forall x A \in \Delta$ if and only if there is a variable-free term t of L such that $\neg A_x[t] \in \Delta$; (5) for variable-free terms s, t, and u of L, $(t = t) \in \Delta$; if $(s = t) \in \Delta$, then $(t = s) \in \Delta$; if $(s = t) \in \Delta$ and $(t = u) \in \Delta$, then $(s = u) \in \Delta$; (6) if $s_1, \ldots, s_n, t_1, \ldots, t_n$ are variable-free terms of L with $(s_1 = t_1) \in \Delta, \ldots, (s_n = t_n) \in \Delta$, then $F(s_1, \ldots, s_n) = F(t_1, \ldots, t_n) \in \Delta$ and $R(s_1, \ldots, s_n) \in \Delta \Leftrightarrow R(t_1, \ldots, t_n) \in \Delta$. Show that Δ has a model. (*Hint:* $D_I = $ VFT$/\sim$, where $s \sim t \Leftrightarrow (s = t) \in \Delta$; prove that for every sentence A of L, $A \in \Delta \Leftrightarrow A$ true in I.)
15. (*Semantic proof of the compactness theorem*) We will outline a proof of the compactness theorem that uses only semantic concepts. A set Γ of sentences is said to have the *finite model property* (FMP) if every finite subset of Γ has a model. With this terminology, the compactness theorem states: If Γ has the FMP, then Γ has a model. Prove the following:
    (1) Let Γ have the FMP and let A be a sentence. Then $\Gamma \cup \{A\}$ or $\Gamma \cup \{\neg A\}$ has FMP.
    (2) Let $\Gamma_0, \Gamma_1, \Gamma_2, \ldots$ be sets of sentences with $\Gamma_0 \subseteq \Gamma_1 \subseteq \Gamma_2 \ldots$ and each Γ_n has FMP. Then $\Delta = \bigcup_{n \in \mathbb{N}} \Gamma_n$ has the FMP.
    (3) Let A be a formula with exactly one free variable x. Let c be a constant symbol that does not occur in A or in any sentence of Γ. If $\Gamma \cup \{\neg \forall x A\}$ has the FMP, then $\Gamma \cup \{\neg \forall x A, \neg A_x[c]\}$ has the FMP. (*Hint:* Let $\Delta \subseteq \Gamma$ be finite and let M be a model of $\Delta \cup \{\neg \forall x A\}$. Give a (possibly new) interpretation of c to obtain a model M* of $\Delta \cup \{\neg \forall x A, \neg A_x[c]\}$; use the substitution theorem for assignments.)
    (4) Let A be a formula with exactly one free variable x. Let t be a variable-free term of L. Then $\{\forall x A, \neg A_x[t]\}$ has no model.
    (5) (*Lindenbaum-Henkin*) Let Γ have the FMP and let $c_1, c_2, \ldots$ be an infinite sequence of constant symbols that do not occur in any sentence of Γ. Then there is a set Δ of sentences such that $\Gamma \subseteq \Delta$ and (a) Δ has FMP; (b) Δ is maximally complete (for every sentence A, $A \in \Delta$ or $\neg A \in \Delta$); (c) if A has exactly one free variable x, then $\neg \forall x A \in \Delta$ if and only if there is a variable-free term t of L such that $\neg A_x[t] \in \Delta$. (*Hint:* Mimic the proof of the Lindenbaum-Henkin theorem.)
    (6) Let Δ be a set of sentences satisfying (a), (b), and (c). Then Δ is a truth set (see Exercise 14), and hence Δ has a model.
    (7) If Γ has the FMP, then Γ has a model.
16. (*First-order Hintikka set*) Let L be a first-order language with at least one constant symbol. Let Δ be a set of sentences of L satisfying these conditions: (1) If $\neg \neg A \in \Delta$, then $A \in \Delta$; (2) if $A \vee B \in \Delta$, then $A \in \Delta$ or $B \in \Delta$; (3) if $\neg (A \vee B) \in \Delta$, then $\neg A \in \Delta$ and $\neg B \in \Delta$; (4) if $\forall x A \in \Delta$, then $A_x[t] \in \Delta$ for every variable-free term t of L; (5) if $\neg \forall x A \in \Delta$, then there is a variable-free term t such that $\neg A_x[t] \in \Delta$; (6) if s and t

are variable-free terms of L, then $\{s = t, \neg(s = t)\} \not\subseteq \Delta$; (7) for any n-ary relation symbol R and any variable-free terms $t_1, \ldots, t_n$ of L, $\{R(t_1, \ldots, t_n), \neg R(t_1, \ldots, t_n)\} \not\subseteq \Delta$; (8) for variable-free terms s, t, and u of L, $(t = t) \in \Delta$; if $(s = t) \in \Delta$, then $(t = s) \in \Delta$; if $(s = t) \in \Delta$ and $(t = u) \in \Delta$, then $(s = u) \in \Delta$; (9) if $s_1, \ldots, s_n, t_1, \ldots, t_n$ are variable-free terms of L with $(s_1 = t_1) \in \Delta, \ldots, (s_n = t_n) \in \Delta$, then $F(s_1, \ldots, s_n) = F(t_1, \ldots, t_n) \in \Delta$ and $R(s_1, \ldots, s_n) \in \Delta \Leftrightarrow R(t_1, \ldots, t_n) \in \Delta$. Show that Δ has a model. (*Hint:* $D_I = VFT/{\sim}$, where $s \sim t \Leftrightarrow (s = t) \in \Delta$; show that for every sentence A of L, $A \in \Delta \Rightarrow A$ true in I (every sentence of L is one of the following: atomic, $\neg A$ with A atomic, $\neg\neg A$, $A \lor B$, $\neg(A \lor B)$, $\forall x A$, $\neg \forall x A$).)

17. Prove the Löwenheim-Skolem theorem.

18. Prove compactness theorem I. (*Note:* See Section 6.2 for further exercises on the compactness theorem.)

19. Let Γ be a set of sentences of L and let A be a formula with exactly one free variable x. Prove the following:

    (1) $\Gamma \cup \{\forall x A\}$ has a model if and only if for every variable-free term t of L, $\Gamma \cup \{\forall x A, A_x[t]\}$ has a model.

    (2) $\Gamma \cup \{\neg \forall x A\}$ has a model if and only if $\Gamma \cup \{\neg A_x[c]\}$ has a model, where c is a constant symbol that does not occur in A or in any sentence in Γ.

    (*Note:* Since we are working exclusively with sentences, we can replace *model* with *satisfiable*; see Exercise 8 in Section 5.2.)

20. Let A be a sentence and let Γ be a set of sentences of L. Prove that A is true in every model of Γ if and only if $\Gamma \cup \{\neg A\}$ has no model.

21. (*Semantic trees for first-order logic*) Refer to Exercises 19 and 20 and to the description of G_L (left-hand Gentzen sequent calculus) in Section 3.6. To simplify our discussion, we work with a first-order language *without* equality. There are two rules for the quantifier $\forall$, which we write in decomposition form as follows:

$\forall$-*Rule* $\dfrac{\Gamma, \forall x A}{\Gamma, \forall x A, A_x[t]}$ (t any variable-free term)

$\neg\forall$-*Rule* $\dfrac{\Gamma, \neg\forall x A}{\Gamma, \neg A_x[c]}$ (c a new constant symbol not in Γ or A).

Example: We will show that $\exists x(R(x) \to Q(c)) \to (\forall x R(x) \to Q(c))$ is logically valid by showing that $\neg[\neg\exists x(\neg R(x) \lor Q(c)) \lor (\neg\forall x R(x) \lor Q(c))]$ has no model.

$\neg[\forall x \neg(\neg R(x) \lor Q(c)) \lor (\neg\forall x R(x) \lor Q(c))]$	GIVEN
$\neg\forall x \neg(\neg R(x) \lor Q(c)), \neg(\neg\forall x R(x) \lor Q(c))$	$\neg\lor$-RULE
$\neg\forall x \neg(\neg R(x) \lor Q(c)), \forall x R(x), \neg Q(c)$	$\neg\lor$-RULE, $\neg\neg$-RULE
$\neg R(d) \lor Q(c), \forall x R(x), \neg Q(c)$	$\neg\forall$-RULE, $\neg\neg$-RULE
$\neg R(d), \forall x R(x), \neg Q(c)$ $\boxed{Q(c), \forall x R(x), \neg Q(c)}$	$\lor$-RULE
$\boxed{\neg R(d), \forall x R(x), R(d), \neg Q(c)}$	$\forall$-RULE.

Example: We will show that $\exists y \forall x[R(x) \to R(y)]$ is logically valid by showing that $\forall y \neg\forall x[\neg R(x) \lor R(y)]$ has no model.

(1) $\forall y \neg\forall x[\neg R(x) \lor R(y)]$ GIVEN

(2) $\forall y \neg\forall x[\neg R(x) \lor R(y)], \neg\forall x[\neg R(x) \lor R(c)]$ $\forall$-RULE

(3) $\forall y \neg\forall x[\neg R(x) \lor R(y)], \neg[\neg R(d) \lor R(c)]$ $\neg\forall$-RULE

(4) $\forall y \neg\forall x[\neg R(x) \lor R(y)], R(d), \neg R(c)$ $\neg\lor$-RULE, $\neg\neg$-RULE

(5) $\forall y \neg \forall x[\neg R(x) \vee R(y)], \neg \forall x[\neg R(x) \vee R(d)], R(d), \neg R(c)$ ∀-RULE
(6) $\forall y \neg \forall x[\neg R(x) \vee R(y)], \neg[\neg R(e) \vee R(d)], R(d), \neg R(c)$ ¬∀-RULE
(7) $\boxed{\forall y \neg \forall x[\neg R(x) \vee R(y)], R(e), \neg R(d), R(d), \neg R(c)}$ ¬∨-RULE, ¬¬-RULE.

(*Note:* If we turn these two examples of semantic trees upside down, we obtain formal proofs in the first-order version of the Gentzen sequent calculus G_L.)

(1) For each sentence A, show that A is logically valid by showing that $\neg A$ has no model; use semantic trees as above.

   (a) $\forall x R(x) \rightarrow \exists x R(x)$.
   (b) $\forall x R(x) \rightarrow \forall y R(y)$.
   (c) $\exists x \forall y R(x, y) \rightarrow \forall y \exists x R(x, y)$.
   (d) $\exists x[P(x) \wedge Q(x)] \rightarrow [\exists x P(x) \wedge \exists x Q(x)]$.
   (e) $[\forall x P(x) \vee \forall x Q(x)] \rightarrow \forall x[P(x) \vee Q(x)]$.
   (f) $\exists x[P(x) \vee Q(x)] \rightarrow [\exists x P(x) \vee \exists x Q(x)]$.
   (g) $[\exists x P(x) \vee \exists x Q(x)] \rightarrow \exists x[P(x) \vee Q(x)]$.
   (h) $\forall x \exists y R(x, y) \rightarrow \forall x \exists y \exists z[R(x, y) \wedge R(y, z)]$.

(2) For each argument form $A_1, \ldots, A_n \therefore B$, prove validity by showing that $\{A_1, \ldots, A_n, \neg B\}$ has no model; use semantic trees.

   (a) $\forall x(P(x) \rightarrow Q(x)), \forall x(R(x) \rightarrow \neg Q(x)) \therefore \forall x(R(x) \rightarrow \neg P(x))$.
   (b) $\forall x(P(x) \rightarrow Q(x)), \exists x(R(x) \wedge \neg Q(x)) \therefore \exists x(R(x) \wedge \neg P(x))$.
   (c) $\exists x \neg Q(x), \forall x(Q(x) \vee R(x)) \therefore \exists x R(x)$.
   (d) $\exists x(P(x) \rightarrow Q(x)), \forall x \neg Q(x) \therefore \exists x \neg P(x)$.

22. (*Soundness theorem for first-order sequent calculus* G_L). Let G_L be the first-order version of the Gentzen sequent calculus (assume a first-order language L without equality). The rules are: ¬¬-RULE, ∨-RULE, ¬∨-RULE, ∀-RULE, and ¬∀-RULE. Show that if Γ is a set of sentences of L and $G_L \vdash \Gamma$, then Γ has no model. (*Hint:* See Section 3.6 and Exercise 19 of this section. The ∀-RULE and the ¬∀-RULE are the decomposition rules in the previous exercise, turned upside down. (*Note:* Hintikka sets play an important role in the proof of the adequacy theorem for G_L; see, for example, Fitting, *First-Order Logic and Automated Theorem Proving* (New York: Springer-Verlag 1990) 115 or Bell and Machover *A Course in Mathematical Logic* (Amsterdam: North-Holland 1977) 91.

5.5

Gödel's Completeness Theorems; Decidability and Listability

In this section, we will use the model existence theorem to prove the adequacy theorem. We will then summarize the major theorems of first-order logic in two statements known as Gödel's completeness theorems. An application of these results is the existence of an algorithm that lists all logically valid formulas of a first-order language L.

Theorem 1 (Adequacy theorem for first-order logic): If A is true in every model of Γ, then $\Gamma \vdash A$.

Proof: Suppose A is not a theorem of Γ. Then A' ($=$ closure of A) is also not a theorem of Γ (closure theorem for provability). By the extension theorem, $\Gamma \cup \{\neg A'\}$ is consistent. By the model existence theorem, $\Gamma \cup \{\neg A'\}$ has a model; call it M. Note that M is a model of Γ and that $\neg A'$ is true in M.

Now A, being true in every model of Γ, is true in M. By the closure theorem for interpretations, A' is true in M. Thus we have both A' and $\neg A'$ true in M, a contradiction. ■

Corollary 1: Every logically valid formula of L is a theorem of the first-order logic FO_L.

Corollary 2 (Provability of valid argument forms): In first-order logic, if $A_1, \ldots, A_n \therefore B$ is a valid argument form, then $\{A_1, \ldots, A_n\} \vdash B$.

As in propositional logic, we can summarize the major metalogical results of first-order logic in two theorems. The first combines the consistency theorem and the model existence theorem; the second combines the soundness theorem and the adequacy theorem. We remind the reader that the word *complete* is used in the nontechnical sense of "everything we want to be a theorem is a theorem."

Gödel's Completeness Theorem I: Γ is consistent if and only if Γ has a model.

Gödel's Completeness Theorem II: $\Gamma \vdash A$ if and only if A is true in every model of Γ.

Two special cases of II are: A is logically valid if and only if $\vdash_{\text{FO}} A$; $A_1, \ldots, A_n \therefore B$ is a valid argument form if and only if $\{A_1, \ldots, A_n\} \vdash_{\text{FO}} B$.

Let us illustrate the significance of the adequacy theorem with an example from group theory. Consider the following theorem and proof, taken from a hypothetical modern algebra text:

Theorem: Cancellation on the right holds in any group.

Proof: Let H be a group and let $a, b, c \in H$. We show: If $ba = ca$, then $b = c$. Let 1 be the identity element of H and let $d \in H$ be such that $ad = 1$. Then

$ba = ca$	(given)
$(ba)d = (ca)d$	(multiply on right by d)
$b(ad) = c(ad)$	(associative law holds in H)
$b1 = c1$	(replace ad by 1)
$b = c$	($b1 = b, c1 = c$ since 1 is identity) ■

Now let A be the following sentence of the language L_G of groups:

$$\forall x_1 \forall x_2 \forall x_3 [(x_1 \circ x_3 = x_2 \circ x_3) \rightarrow (x_1 = x_2)].$$

The mathematical proof given above may be regarded as verification that A is true in every model of $\Gamma = \{G1, G2, G3\}$, where G1–G3 are the axioms of a group, written in L_G. It follows from the adequacy theorem that $\Gamma \vdash A$. In other words, first-order logic has sufficient axioms and rules of inference to *write out a formal proof of A*; it is understood, of course, that in this formal proof we can use the sentences G1–G3 as additional axioms.

This discussion can be summarized as follows: Suppose we have a sentence *A* and a set Γ of sentences of some first-order language L. If there is a traditional mathematical proof of *A* using sentences in Γ as nonlogical axioms, then in first-order logic there is a formal proof of *A* using Γ.

Historical note At the 1928 meeting of the International Congress of Mathematicians, David Hilbert raised two important questions that have greatly influenced the development of logic in the twentieth century. (For now, our interest is in the first question; the second question will be discussed in Sections 6.1 and 7.1). **Q1**: Is first-order logic complete (in the non-technical sense)? More precisely, does every logically valid formula have a formal proof in first-order logic? **Q2**: Is arithmetic complete? Within a period of just three years, Kurt Gödel, the greatest logician of the twentieth century, solved both of these important problems. By completeness theorem II, first-order logic is indeed complete.

────────── *Decidability and Listability* ──────────

The tautology problem for propositional logic is solvable; in other words, there is an algorithm that decides whether an arbitrary formula of propositional logic is a tautology (use truth tables). There is a corresponding problem for first-order logic.

Definition 1: Let L be a first-order language. The *validity problem for* L asks whether there is an algorithm that, given an arbitrary formula *A* of L, decides whether *A* is logically valid. If such an algorithm exists, we say that the validity problem for L is *solvable*; otherwise, the validity problem for L is *unsolvable*.

Hilbert often emphasized the importance of the validity problem or, as it is often called, the *Entscheidungsproblem* (decision problem). By Gödel's completeness theorem II, the validity problem for L reduces to a problem about the formal system FO_L.

Definition 2: Let L be a first-order language. The formal system FO_L is *decidable* if there is an algorithm that, given an arbitrary formula *A* of L, decides whether $\vdash_{F0} A$. If there is no such algorithm, we say that FO_L is *undecidable*.

Theorem 2: The validity problem for L is solvable if and only if the formal system FO_L is decidable.

Proof: Let *A* be an arbitrary formula of L. By Gödel's completeness theorem II, *A* is logically valid $\Leftrightarrow \vdash_{F0} A$. It follows that an algorithm that decides logical validity also decides theoremhood, and vice versa. ∎

The solution of the validity problem depends on the language L, but in most cases the problem is unsolvable. This was first proved by Church and, independently, by Turing. At the end of this section we will state precise results, but for now we simply record this important result.

Theorem 3 (Church-Turing, 1936): There is a first-order language L for which the validity problem is unsolvable.

Although the validity problem for L may not be solvable, the next-best thing is true: There is a mechanical way of listing the logically valid formulas of L. To do this, it suffices to mechanically list the theorems of FO_L. The basic idea is quite simple: *Mechanically print out all proofs.* We now prepare the way to prove this precisely, and in the more general setting of a set Γ of formulas of L. In our subsequent discussion, the following notation is used: FOR_L = set of all formulas of L.

To mechanically print out proofs using Γ, we need an algorithm for deciding whether $A_1, \ldots, A_n$ is a proof using Γ; for this, we need an algorithm for recognizing nonlogical axioms, that is, formulas in Γ.

Definition 3: A set $\Gamma \subseteq FOR_L$ is *axiomatized* if there is an algorithm that, given $A \in FOR_L$, decides whether $A \in \Gamma$.

For example, if Γ is finite, then Γ is certainly axiomatized. If $\Gamma \subseteq FOR_L$ is axiomatized, we can regard Γ as a formal system: The language is L; the rules are those of first-order logic; the axioms are the logical axioms of first-order logic *plus* the formulas of Γ (called *nonlogical* axioms).

Theorem 4 (The algorithm CHKPRF): Let $\Gamma \subseteq FOR_L$ be axiomatized. Then there is an algorithm CHKPRF (= check proof) that, given an arbitrary finite sequence of formulas $A_1, \ldots, A_n$ of L, decides whether $A_1, \ldots, A_n$ is a proof using Γ.

Proof: We write the required algorithm CHKPRF.

(1) Input $A_1, \ldots, A_n$ and set $k = 1$.
(2) Check the following: A_k is a logical axiom; $A_k \in \Gamma$ (Γ is axiomatized); $k > 1$ and A_k is the conclusion of a rule of inference with hypotheses among $A_1, \ldots, A_{k-1}$.
(3) If all three fail, print NO and halt.
(4) Is $k = n$?
(5) If *Yes*, print YES and halt; otherwise, add 1 to k and go to (2). ∎

Definition 4: Let $\Gamma \subseteq FOR_L$.

(1) Γ is *decidable* if there is an algorithm that decides whether $\Gamma \vdash A$.
(2) Γ is *listable,* or *effectively enumerable,* if there is an algorithm that lists all theorems of Γ and only theorems of Γ.

In other words, we require an algorithm with these properties: The output of the algorithm is a list of formulas; every formula in the output list is a theorem of Γ; every theorem of Γ occurs at least once in the output list. Since the number of theorems of Γ is infinite, the required algorithm for listability continues forever.

Before studying decidability and listability in more detail, let us make the following convention: *All languages have a finite number of nonlogical symbols.* Though this assumption is by no means an absolute requirement, it does simplify proofs; moreover, all of the important languages encountered later in this text satisfy this requirement.

Our first result states that decidability implies listability. To prove this precisely, it is convenient to introduce some notation. For each $n \geq 1$, let $\text{FOR}_L(n)$ be the set of all formulas of L with $\leq n$ symbols and with variables among $x_1, \ldots, x_n$. For example, $(x_1 = x_1) \in \text{FOR}_L(5)$ and $(x_1 < x_9) \in \text{FOR}_L(9)$. Note that $\text{FOR}_L(n) \subseteq \text{FOR}_L(n + 1)$ for all $n \in \mathbb{N}^+$.

Lemma 1: For each $n \geq 1$, $\text{FOR}_L(n)$ is finite and there is a mechanical method of listing all formulas in $\text{FOR}_L(n)$. Moreover, $\text{FOR}_L = \bigcup_{n \in \mathbb{N}^+} \text{FOR}_L(n)$.

Proof: For each $n \geq 1$, the set $\text{FOR}_L(n)$ is finite, since the number of nonlogical symbols of L is finite. One way to mechanically list all formulas in $\text{FOR}_L(n)$ is simply to write down all possible expressions with $\leq n$ symbols, using only variables among $x_1, \ldots, x_n$, and then select all expressions that are formulas. To prove $\text{FOR}_L = \bigcup_{n \in \mathbb{N}^+} \text{FOR}_L(n)$, suppose $A \in \text{FOR}_L$; then $A \in \text{FOR}_L(n)$, where $n = \max \{$number of symbols in A, largest subscript k of a variable x_k in $A\}$. ■

Theorem 5: If Γ is decidable, then Γ is listable.

Proof: Let Φ be an algorithm that decides $\Gamma \vdash A$; we are required to write an algorithm that lists all theorems of Γ and only theorems of Γ.

(1) Let $n = 1$.

(2) Make a list of all formulas in $\text{FOR}_L(n)$; there are only a finite number. For each formula $A \in \text{FOR}_L(n)$, use the algorithm Φ to decide whether $\Gamma \vdash A$. If *Yes*, print the formula A.

(3) Add 1 to n and go to step 2. ■

There is an alternative approach to listability that further clarifies its relationship with decidability (see also Section 1.7).

Definition 5: A set $\Gamma \subseteq \text{FOR}_L$ is *semidecidable* if there is an algorithm such that for each input $A \in \text{FOR}_L$, (1) if $\Gamma \vdash A$, the algorithm halts in a finite number of steps and prints YES; (2) if A is not a theorem of Γ, the algorithm continues forever.

Theorem 6 (Characterization of listability): Γ is listable if and only if Γ is semidecidable.

We omit the proof (left as an exercise for the reader) because we will consistently work with the property of listability. But in subsequent theorems of this section, the reader may want to replace listability with semidecidability and prove the resulting theorem; note, for example, that Theorem 5 is obvious with this replacement.

Now suppose Γ is axiomatized. Although Γ may not be decidable, it is listable. The basic idea of the proof is straightforward: Print out one-line proofs, then print out two-line proofs, and so on. There are, however, some technical problems with this procedure. First, suppose we consider proofs with just one line; we have, among others, $(x_1 = x_1) \vee \neg(x_1 = x_1)$, $(x_2 = x_2) \vee \neg(x_2 = x_2)$, and so on. In other words, there are an infinite number of one-line proofs! We can avoid this difficulty by specifying that for n-line proofs, we consider only those with variables among $x_1, \ldots, x_n$. But even with this restriction, there are still an infinite number of one-line proofs — for example, $(x_1 = x_1) \vee \neg(x_1 = x_1)$, $\neg(x_1 = x_1) \vee \neg\neg(x_1 = x_1)$, $\neg\neg(x_1 = x_1) \vee \neg\neg\neg(x_1 = x_1)$, and so on. We can avoid this difficulty by specifying that for n-line proofs, each formula in the proof must have at most n symbols. With these observations in mind, we prove the following result.

Theorem 7: If $\Gamma \subseteq \text{FOR}_L$ is axiomatized, then Γ is listable.

Proof: Since Γ is axiomatized, there is an algorithm CHKPRF that decides whether a sequence $A_1, \ldots, A_n$ is a proof. We are required to write an algorithm that lists all theorems of Γ and only theorems of Γ.

(1) Set $n = 1$.

(2) Construct all possible finite sequences $A_1, \ldots, A_k$ of formulas of L with these two properties: $k \leq n$; each of the formulas $A_1, \ldots, A_k$ is in $\text{FOR}_L(n)$. The number of such sequences is finite.

(3) For each such sequence, say $A_1, \ldots, A_k$, use CHKPRF to decide whether $A_1, \ldots, A_k$ is a proof using Γ. If so, print the formula A_k.

(4) Add 1 to n and go to step 2.

It is clear that the output of this algorithm consists of formulas, each of which is a theorem of Γ. To complete the proof, we need to check that if $\Gamma \vdash A$, then A is an output. Suppose $\Gamma \vdash A$, and let $A_1, \ldots, A_k$ ($A_k = A$) be a proof of A using Γ. Choose n so large that $k \leq n$ and each of the formulas $A_1, \ldots, A_k$ is in $\text{FOR}_L(n)$. Then A is listed at stage n of the algorithm as required. ∎

Corollary 3: There is an algorithm that lists all logically valid formulas of L and only logically valid formulas of L.

Proof: By completeness theorem II, the set of logically valid formulas of L is the same as the set of theorems of FO_L. Now use Theorem 7 with $\Gamma = \varnothing$. ∎

In his famous paper "On computable numbers, with an application to the Entscheidungsproblem" (Proceedings of the London Mathematical Society, Ser 2, 42 (1936), 259), Turing makes the following important observation.

Theorem 8 (Turing): If Γ is axiomatized and complete, then Γ is decidable.

Proof: If Γ is inconsistent, then all formulas are theorems of Γ and the required algorithm is obvious: Print YES for every input A. Hence we may assume that Γ is consistent. By the closure theorem for provability, $\Gamma \vdash A \Leftrightarrow \Gamma \vdash A'$ for any $A \in \text{FOR}_L$; hence it suffices to write an algorithm that decides whether an arbitrary sentence of L is a theorem of Γ. Since Γ is axiomatized, the algorithm CHKPRF is available.

(1) Input a sentence A of L and set $n = 1$.

(2) Construct all possible finite sequences $A_1, \ldots, A_k$ of formulas of L having these three properties: $k \leq n$; each of the formulas $A_1, \ldots, A_k$ is in $\text{FOR}_L(n)$; $A_1, \ldots, A_k$ is a proof using Γ (CHKPRF is used here). The number of such proofs is finite.

(3) Among these proofs, is there a proof of A?

(4) If *Yes*, print YES (A is a theorem of Γ) and halt.

(5) Among these proofs, is there a proof of $\neg A$?

(6) If *Yes*, print NO (A not a theorem of Γ by consistency of Γ) and halt.

(7) Add 1 to n and go to (2).

It remains to show that the algorithm does not continue forever. It is here that completeness is used. Let A be a sentence of L (i.e., an input). Since Γ is complete, $\Gamma \vdash A$ or $\Gamma \vdash \neg A$. Suppose $\Gamma \vdash A$, and let $A_1, \ldots, A_k$ $(A_k = A)$ be a proof of A using Γ. Choose n so large that $k \leq n$ and each of the formulas $A_1, \ldots, A_k$ is in $\text{FOR}_L(n)$. Then at stage n, the algorithm discovers this proof of A, prints YES, and halts. A similar argument holds in the case in which $\Gamma \vdash \neg A$ (at some stage n the algorithm discovers a proof of $\neg A$, prints NO, and halts). ■

Comment The alert reader will notice that in the proof of Turing's theorem, there are two possible algorithms for decidability. We have proved the existence of such an algorithm, although we may not know which algorithm is the correct one.

What happens if we start with Γ decidable and then extend it, by adding either new nonlogical axioms or new constant symbols?

Theorem 9 (Finite extension): Let $\Gamma \subseteq \text{FOR}_L$ and let Δ be obtained from Γ by adding a finite number of sentences of L. If Γ is decidable, then Δ is decidable.

Proof: Let $A_1, \ldots, A_n$ be the finite number of sentences added to Γ to obtain Δ; in other words, $\Delta = \Gamma \cup \{A_1, \ldots, A_n\}$. By the deduction theorem and MP, the following holds for every formula C of L:

$$\Delta \vdash C \Leftrightarrow \Gamma \vdash (A_1 \wedge \ldots \wedge A_n) \rightarrow C.$$

The following algorithm establishes the decidability of Δ:

(1) Input a formula C of L.

(2) Does $\Gamma \vdash (A_1 \wedge \ldots \wedge A_n) \to C$ hold? (Γ is decidable.)

(3) If *Yes*, then C is theorem of Δ; print YES and halt.

(4) If *No*, then C is not a theorem of Δ; print NO and halt. ∎

Corollary 4: Let Δ be a finite set of sentences of L. If Δ is undecidable, then the validity problem for L is unsolvable.

Proof: Suppose the validity problem for L is solvable. By Gödel's completeness theorem, A is logically valid $\Leftrightarrow \vdash_{FO} A$; it follows that FO_L is decidable. By Theorem 9 ($\Gamma = \varnothing$), Δ is decidable, a contradiction. ∎

The converse of Corollary 4 is false: If L is the first-order language with just one 2-ary relation symbol R, then the validity problem for L is unsolvable. But if we add the axioms DO1–DO6 for a densely ordered set without first or last elements (see Section 6.3), the resulting theory is decidable.

Theorem 10 (Extension by constants): Let $\Gamma \subseteq FOR_L$ and let Γ^+ be an extension of Γ by constants. Then Γ is decidable if and only if Γ^+ is decidable.

Proof: Let L^+ be the language obtained from L by adding new constant symbols to obtain Γ^+. First assume Γ is decidable. We claim that there is an algorithm that, given a formula A^+ of L^+, finds a formula A of L such that $\Gamma \vdash A \Leftrightarrow \Gamma^+ \vdash A^+$. Assuming this is done, the following algorithm establishes the decidability of Γ^+:

(1) Input $A^+ \in FOR_{L^+}$.

(2) Obtain $A \in FOR_L$ such that $\Gamma \vdash A \Leftrightarrow \Gamma^+ \vdash A^+$.

(3) Use the algorithm for Γ to decide whether $\Gamma \vdash A$ (Γ is decidable).

(4) If *Yes*, then $\Gamma^+ \vdash A^+$; print YES and halt.

(5) If *No*, then A^+ not a theorem of Γ^+; print NO and halt.

We will now give the mechanical construction of A from A^+ so that $\Gamma \vdash A \Leftrightarrow \Gamma^+ \vdash A^+$. Given $A^+ \in FOR_{L^+}$, let $c_1, \ldots, c_k$ be the new constant symbols that occur in A^+ (these are symbols of L^+ but not of L) and let $y_1, \ldots, y_k$ be the first k variables in the list $x_1, x_2, \ldots$ that do not occur in A^+. The formula A is obtained from A^+ by replacing each occurrence of c_i with y_i, $1 \leq i \leq k$. In other words, we have $A = A^+_{c_1, \ldots, c_k}[y_1, \ldots, y_k]$. Note also that $A_{y_1, \ldots, y_k}[c_1, \ldots, c_k] = A^+$. If $\Gamma \vdash A$, then $\Gamma^+ \vdash A^+$ follows by k applications of the substitution rule; if $\Gamma^+ \vdash A^+$, then $\Gamma \vdash A$ follows by k applications of the theorem on constants (Theorem 3 in Section 5.4).

Now assume Γ^+ is decidable. By the theorem on extension by constants (Theorem 5 in Section 5.4), we have $\Gamma \vdash A \Leftrightarrow \Gamma^+ \vdash A$ for any formula A of L. One can now write an algorithm similar to the one given above and show that Γ is decidable as required. ∎

We will end this section with a precise delineation between solvable and unsolvable instances of the validity problem. For more detail and references, see Monk 1976, and Boolos and Jeffrey 1989. (*Note:* Theorem 13(2), below, is an exercise in Section 9.6.)

Theorem 11 (Löwenheim): If the only nonlogical symbols of L are 1-ary relation symbols (= allowed), then the validity problem for L is solvable.

Theorem 12 (Ehrenfeucht): If the only nonlogical symbol of L is a single 1-ary function symbol, then the validity problem for L is solvable.

Theorem 13 (Trachtenbrot): The validity problem for L is unsolvable if the language L satisfies any one of the following conditions:

(1) L has at least one n-ary relation symbol for $n > 1$ (= may be omitted).

(2) L has at least one n-ary function symbol for $n > 1$.

(3) L has at least two 1-ary function symbols.

────────── *Exercises for Section 5.5* ──────────

1. Let Γ and Δ be sets of sentences of L. Show that Γ and Δ have the same theorems if and only if Γ and Δ have the same models.
2. Use Gödel's completeness theorem II to prove compactness theorem II (see Section 5.4).
3. Let $\Gamma \subseteq FOR_L$ be listable, and assume there is also an algorithm that lists all formulas of L that are not theorems of Γ. Show that Γ is decidable.
4. (*Due to Craig*) Prove the following:
   (1) Let $\Gamma = \{A_n : n \in \mathbb{N}\}$ and $\Delta = \{B_n : n \in \mathbb{N}\}$, where $B_n = A_0 \wedge \ldots \wedge A_n$ for $n \in \mathbb{N}$. Then for every formula C of L, $\Gamma \vdash C \Leftrightarrow \Delta \vdash C$.
   (2) If there is an algorithm that lists all formulas in Γ, then there is a set Δ of formulas such that (a) Δ is axiomatized; (b) for every formula C of L, $\Gamma \vdash C \Leftrightarrow \Delta \vdash C$.
5. Generalize Theorem 7 as follows: If there is an algorithm that lists all formulas of Γ, then Γ is listable. (*Hint:* See Exercise 4(2).)
6. Generalize Theorem 8 as follows: If there is an algorithm that lists all formulas of Γ, and Γ is complete, then Γ is decidable. (*Hint:* See Exercise 4(2).)
7. (*Listable = semidecidable*) Prove Theorem 5.
8. (*Satisfiability problem*) The *satisfiability problem for* L asks whether there is an algorithm that decides whether an arbitrary formula of L is satisfiable. Prove that the validity problem for L is solvable if and only if the satisfiability problem for L is solvable. (*Hint: A* is logically valid if and only if $\neg A$ is not satisfiable.)
9. Let L be a first-order language and let L^+ be obtained from L by adding new constant symbols to L. Show that the validity problem for L is solvable if and only if the validity problem for L^+ is solvable.
10. Let $\Gamma \subseteq FOR_L$ be axiomatized, and let Δ be the set of all formulas of L that are true in every model of Γ. Prove that there is an algorithm that lists all formulas of Δ and only formulas of Δ.

5.6

Replacement Theorem and Prenex Form

In this section, we will treat additional topics in first-order logic. The results obtained are of the form $\vdash_{FO} A \leftrightarrow A^*$, where A and A^* are related in some way. The approach is syntactic; in other words, in each case we give a formal proof of $\vdash_{FO} A \leftrightarrow A^*$. But by Gödel's completeness theorem II, we could also establish these results semantically by showing that $A \leftrightarrow A^*$ is logically valid. Theorems 1–3 of this section are used in the proof of Gödel's incompleteness theorem (see Section 7.3).

———— The Replacement Theorem ————

Two formulas B and C of L are *provably equivalent using* Γ if $\Gamma \vdash B \leftrightarrow C$. With this terminology, the replacement theorem states: If B and C are provably equivalent, and A^* is obtained from A by replacing one or more occurrences of B with C, then A and A^* are provably equivalent. We begin with basic properties of provably equivalent formulas.

Lemma 1 *The following hold:*

(1) If $\Gamma \vdash A \leftrightarrow B$, then $\Gamma \vdash \neg A \leftrightarrow \neg B$.
(2) If $\Gamma \vdash A \leftrightarrow B$ and $\Gamma \vdash C \leftrightarrow D$, then $\Gamma \vdash (A \vee C) \leftrightarrow (B \vee D)$.
(3) If $\Gamma \vdash A \leftrightarrow B$, then $\Gamma \vdash \forall x A \leftrightarrow \forall x B$.

Theorem 1 (Replacement theorem for first-order logic, syntactic version): Let $\Gamma \vdash B \leftrightarrow C$, let A be a formula in which B occurs, and let A^* be obtained from A by replacing one or more of the occurrences of B with C. Then $\Gamma \vdash A \leftrightarrow A^*$.

Proof: We first consider the special case in which B is all of A. Then $A^* = C$, and we are required to show $\Gamma \vdash B \leftrightarrow C$; this holds by hypothesis. Henceforth, we will assume that B is not all of A. We now proceed by induction on the number of occurrences of $\neg$, $\vee$, and $\forall$ in A. If A is an atomic formula, we are in the special case.

Negation Suppose A is $\neg D$. Then A^* is $\neg D^*$, where D^* is obtained from D by replacing one or more occurrences of B with C. By induction hypothesis, $\Gamma \vdash D \leftrightarrow D^*$; by Lemma 1(1), $\Gamma \vdash \neg D \leftrightarrow \neg D^*$.

Disjunction Suppose A is $D \vee E$. Then A^* is $D^* \vee E^*$, where D^* either is D or is obtained from D by replacing one or more occurrences of B with C, and likewise for E^*. By induction hypothesis, $\Gamma \vdash D \leftrightarrow D^*$ and $\Gamma \vdash E \leftrightarrow E^*$; by Lemma 1(2), $\Gamma \vdash (D \vee E) \leftrightarrow (D^* \vee E^*)$.

Quantifier Suppose A is $\forall x D$. Then A^* is $\forall x D^*$, where D^* is obtained from D by replacing one or more occurrences of B with C. By induction hypothesis, $\Gamma \vdash D \leftrightarrow D^*$; by Lemma 1(3), $\Gamma \vdash \forall x D \leftrightarrow \forall x D^*$. ∎

Corollary 1: Let $\Gamma \vdash B \leftrightarrow C$ and let $A*$ be obtained from A by replacing one or more occurrences of B with C. If $\Gamma \vdash A$, then $\Gamma \vdash A*$.

In formal proofs, we will not distinguish between the replacement theorem and Corollary 1, but will refer to both as RT.

Example 1 (Prenex rules):

$\vdash \forall x \neg A \leftrightarrow \neg \exists x A$ and $\vdash \exists x \neg A \leftrightarrow \neg \forall x A$.

(1)	$\forall x \neg A \leftrightarrow \neg \neg \forall x \neg A$	TT $(p \leftrightarrow \neg \neg p)$
(2)	$\forall x \neg A \leftrightarrow \neg \exists x A$	DEF $\exists$

(1)	$\exists x \neg A \leftrightarrow \exists x \neg A$	TT $(p \leftrightarrow p)$
(2)	$\exists x \neg A \leftrightarrow \neg \forall x \neg \neg A$	DEF $\exists$
(3)	$\neg \neg A \leftrightarrow A$	TT $(\neg \neg p \leftrightarrow p)$
(4)	$\exists x \neg A \leftrightarrow \neg \forall x A$	RT $(2, 3)$.

□

The next result gives conditions under which occurrences of a bound variable can be replaced with a new variable.

Theorem 2 (Change of bound variables (CBV)): Let x and y be distinct variables such that y is substitutable for x in B and y does not occur free in B. Let $A*$ be obtained from A by replacing one or more occurrences of $\forall x B$ or $\exists x B$ in A with $\forall y B_x[y]$ or $\exists y B_x[y]$ respectively. Then (1) $\vdash A \leftrightarrow A*$; (2) if $\vdash A$, then $\vdash A*$.

Proof: Part (2) follows easily from (1). To prove (1), it suffices to show $\vdash \forall x B \leftrightarrow \forall y B_x[y]$ and $\vdash \exists x B \leftrightarrow \exists y B_x[y]$ and then use the replacement theorem. We prove the required result for $\forall$ and leave $\exists$ to the reader. By the theorem on reverse substitution (see Section 4.4), the two hypotheses guarantee that x is substitutable for y in $B_x[y]$ and that $(B_x[y])_y[x]$ is B.

(1)	$\forall x B \rightarrow B_x[y]$	SUBST AX
(2)	$\forall x B \rightarrow \forall y B_x[y]$	$\forall$-INTRO (y not free in $\forall x B$)
(3)	$\forall y B_x[y] \rightarrow B$	SUBST AX $((B_x[y])_y[x]$ is $B)$
(4)	$\forall y B_x[y] \rightarrow \forall x B$	$\forall$-INTRO (x not free in $\forall y B x[y]$)
(5)	$(\forall x B \rightarrow \forall y B_x[y]) \wedge (\forall y B_x[y] \rightarrow \forall x B)$	JOIN $(2, 4)$
(6)	$\forall x B \leftrightarrow \forall y B_x[y]$	DEF $\leftrightarrow$. ∎

The CBV theorem is often used in the following situation: The term t is not substitutable for x in A, but by changing bound variables, we can find a formula $A*$ such that $\vdash A \leftrightarrow A*$ and t *is* substitutable for x in $A*$. For example, $x_2 + S0$ is not substitutable for x_1 in $\exists x_2(x_1 = SS0 \times x_2)$, but if we replace the bound occurrences of x_2 with x_3, we obtain $\exists x_3(x_1 = SS0 \times x_3)$, and $x_2 + S0$ is substitutable for x_1 in this formula.

Example 2:

We will show that $\vdash \exists x_1 \forall x_2 R(x_1, x_2) \rightarrow \exists x_2 \forall x_1 R(x_2, x_1)$.

(1) $\exists x_1 \forall x_2 R(x_1, x_2) \rightarrow \exists x_1 \forall x_2 R(x_1, x_2)$ TT $(p \rightarrow p)$

(2) $\exists x_1 \forall x_2 R(x_1, x_2) \rightarrow \exists x_3 \forall x_2 R(x_3, x_2)$ CBV

(3) $\exists x_1 \forall x_2 R(x_1, x_2) \rightarrow \exists x_3 \forall x_1 R(x_3, x_1)$ CBV

(4) $\exists x_1 \forall x_2 R(x_1, x_2) \rightarrow \exists x_2 \forall x_1 R(x_2, x_1)$ CBV.   □

Lemma 2: Let x be a variable and let t be a variable-free term. For any formula A of L, $\vdash A_x[t] \leftrightarrow \exists x((x = t) \wedge A)$.

Proof that: $\vdash A_x[t] \rightarrow \exists x((x = t) \wedge A)$

(1) $((t = t) \wedge A_x[t]) \rightarrow \exists x((x = t) \wedge A)$ SUBST THM

(2) $\forall x_1(x_1 = x_1)$ REF AX

(3) $t = t$ ∀-ELIM

(4) $(t = t) \rightarrow [A_x[t] \rightarrow ((t = t) \wedge A_x[t])]$ TT $(p \rightarrow [q \rightarrow (p \wedge q)])$

(5) $A_x[t] \rightarrow ((t = t) \wedge A_x[t])$ MP

(6) $A_x[t] \rightarrow \exists x((x = t) \wedge A)$ HS (5, 1).

Proof that: $\vdash \exists x((x = t) \wedge A) \rightarrow A_x[t]$
We use the tautology $[p \rightarrow (q \rightarrow r)] \rightarrow [(p \wedge q) \rightarrow r]$.

(1) $(x = t) \rightarrow (A \rightarrow A_x[t])$ EQ AX(1)

(2) $[(x = t) \rightarrow (A \rightarrow A_x[t])] \rightarrow [((x = t) \wedge A) \rightarrow A_x[t]]$ TT

(3) $((x = t) \wedge A) \rightarrow A_x[t]$ MP

(4) $\exists x((x = t) \wedge A) \rightarrow A_x[t]$ ∃-INTRO.   ∎

Theorem 3: Let $y_1, \ldots, y_k$ be distinct variables and let $t_1, \ldots, t_k$ be variable-free terms. For any formula A of L,

$$\vdash A_{y_1, \ldots, y_k}[t_1, \ldots, t_k] \leftrightarrow \exists y_1 \ldots \exists y_k[(y_1 = t_1) \wedge \ldots \wedge (y_k = t_k) \wedge A].$$

Proof: By induction on the number of variables $y_1, \ldots, y_k$. For $k = 1$, the result follows from Lemma 2. Let $k \geq 1$ and assume that the result holds for k. To simplify notation, let B denote $A_{y_1, y_2, \ldots, y_{k+1}}[t_1, t_2, \ldots, t_{k+1}]$. The main steps are outlined as follows:

- The induction hypothesis with $A_{y_{k+1}}[t_{k+1}]$, $y_1, \ldots, y_k$, $t_1, \ldots, t_k$ gives

$$\vdash B \leftrightarrow \exists y_1 \ldots \exists y_k((y_1 = t_1) \wedge \ldots \wedge (y_k = t_k) \wedge A_{y_{k+1}}[t_{k+1}]). \quad (*)$$

- Lemma 2 with A, y_{k+1}, t_{k+1} gives $\vdash A_{y_{k+1}}[t_{k+1}] \leftrightarrow \exists y_{k+1}((y_{k+1} = t_{k+1}) \wedge A)$.

- By the replacement theorem, we can replace $A_{y_{k+1}}[t_{k+1}]$ in $(*)$ with $\exists y_{k+1}((y_{k+1} = t_{k+1}) \wedge A)$ to obtain

$$\vdash B \leftrightarrow \exists y_1 \ldots \exists y_k[(y_1 = t_1) \wedge \ldots \wedge (y_k = t_k)$$
$$\wedge \exists y_{k+1}((y_{k+1} = t_{k+1}) \wedge A)]. \quad (**)$$

- By k applications of the prenex rule $\vdash \exists x(C \wedge D) \leftrightarrow (C \wedge \exists xD)$ (x not free in C) and the replacement theorem, the quantifier $\exists y_{k+1}$ in (**) can be moved to the front of [...], giving the desired result. (The prenex rule is proved in Example 3 of Section 5.3; more precisely, see PR13, later in this section.) ■

--------------------- *Prenex Form* ---------------------

We will begin with one more example of a prenex rule.

Example 3 (Prenex rule):

$\vdash \exists x(A \vee B) \leftrightarrow (\exists xA \vee B)$ (x not free in B).

Proof that: $\vdash \exists x(A \vee B) \to (\exists xA \vee B)$

(1) $A \to \exists xA$	SUBST THM
(2) $(A \to \exists xA) \to [(A \vee B)$ $\to (\exists xA \vee B)]$	TT $((p \to q) \to [(p \vee r) \to (q \vee r)])$
(3) $(A \vee B) \to (\exists xA \vee B)$	MP
(4) $\exists x(A \vee B) \to (\exists xA \vee B)$	$\exists$-INTRO (x not free in $\exists xA \vee B$). ■

Proof that: $\vdash (\exists xA \vee B) \to \exists x(A \vee B)$

The basic idea is to prove $\vdash \exists xA \to \exists x(A \vee B)$ and $\vdash B \to \exists x(A \vee B)$ and then to use the tautology $(p \to r) \to [(q \to r) \to ((p \vee q) \to r)]$.

(1) $A \to (A \vee B)$	TT $(p \to (p \vee q))$
(2) $\exists xA \to \exists x(A \vee B)$	$\exists$-DIST
(3) $B \to (A \vee B)$	TT $(q \to (p \vee q))$
(4) $(A \vee B) \to \exists x(A \vee B)$	SUBST THM
(5) $B \to \exists x(A \vee B)$	HS

Replace $p \to r$ with line (2) and $q \to r$ with line (5) to obtain

(6) $(\exists xA \to \exists x(A \vee B)) \to [(B \to \exists x(A \vee B))$ $\to ((\exists xA \vee B) \to \exists x(A \vee B))]$	
(7) $(B \to \exists x(A \vee B)) \to ((\exists xA \vee B) \to \exists x(A \vee B))$	MP $(2, 6)$
(8) $(\exists xA \vee B) \to \exists x(A \vee B)$	MP $(5, 7)$. ■

Formulas of first-order logic are often classified in terms of the number of quantifiers in the formula and the order in which the quantifiers occur in the formula. For example, the formula $\forall x \exists yA$, which has alternating quantifiers, is considered more complicated than the formula $\forall x \forall y \forall zA$, in which all quantifiers are the same. In order to perform this classification precisely, and to compare formulas directly, it is necessary to put formulas in what is called prenex form. The basic idea is to write the formula in such a way that all quantifiers occur at the beginning of the formula.

Definition 1: A formula A is in *prenex form* if it has no quantifiers or has the form $Q_1 y_1 \ldots Q_n y_n B$, where each Q_k, $1 \le k \le n$, is a quantifier (either $\forall$ or $\exists$), B contains no quantifiers, and the variables $y_1, \ldots, y_n$ are distinct. For example, $x_1 = x_2$ and $\forall x_4 \exists x_1 (x_1 = x_4)$ are in prenex form, but $\forall x_1 R(x_1) \to \exists x_2 R(x_2)$ is not.

Every formula can be put into prenex form. More precisely, for each formula A, there is a formula A^* such that A^* is in prenex form and $\vdash A \leftrightarrow A^*$. Before proving this result, we will state fundamental prenex rules and then illustrate the rules with an example.

Theorem 4 (Prenex rules): Let B and C be formulas of L.

PR1 $\vdash_{F0} \neg \forall x B \leftrightarrow \exists x \neg B$.

PR2 $\vdash_{F0} \neg \exists x B \leftrightarrow \forall x \neg B$.

PR3 $\vdash_{F0} (\forall x B \vee C) \leftrightarrow \forall x (B \vee C)$ (x not free in C).

PR4 $\vdash_{F0} (\exists x B \vee C) \leftrightarrow \exists x (B \vee C)$ (x not free in C).

PR5 $\vdash_{F0} (B \vee \forall x C) \leftrightarrow \forall x (B \vee C)$ (x not free in B).

PR6 $\vdash_{F0} (B \vee \exists x C) \leftrightarrow \exists x (B \vee C)$ (x not free in B).

Proof: Most of these results have already been proved: PR1, PR2, and PR4 in Examples 1 and 3 of this section; PR3 in Example 2 of Section 5.3. PR5 and PR6 follow from PR3 and PR4 respectively. ∎

Example 4:

We will put the formula $\exists x \forall y R(x, y) \to \forall y \exists x R(x, y)$ in prenex form. The tools used are the prenex rules, the replacement theorem (RT), and the theorem on change of bound variables (CBV). The letters $x, y, z,$ and w denote distinct variables. We write a sequence of formulas, each provably equivalent to the next:

(1) $\exists x \forall y R(x, y) \to \forall y \exists x R(x, y)$ GIVEN

(2) $\neg \exists x \forall y R(x, y) \vee \forall y \exists x R(x, y)$ DEF $\to$

(3) $\forall x \neg \forall y R(x, y) \vee \forall y \exists x R(x, y)$ PR2, RT

(4) $\forall x \exists y \neg R(x, y) \vee \forall y \exists x R(x, y)$ PR1, RT

(5) $\forall x [\exists y \neg R(x, y) \vee \forall y \exists x R(x, y)]$ PR3

(6) $\forall x \exists y [\neg R(x, y) \vee \forall y \exists x R(x, y)]$ PR4, RT

(7) $\forall x \exists y [\neg R(x, y) \vee \forall z \exists x R(x, z)]$ CBV (replace y with z)

(8) $\forall x \exists y \forall z [\neg R(x, y) \vee \exists x R(x, z)]$ PR5, RT

(9) $\forall x \exists y \forall z [\neg R(x, y) \vee \exists w R(w, z)]$ CBV (replace x with w)

(10) $\forall x \exists y \forall z \exists w [\neg R(x, y) \vee R(w, z)]$ PR6, RT.   □

We now turn to the proof that every formula is provably equivalent to a formula in prenex form. As in Example 4, the tools used are the prenex rules, the replacement theorem, and the theorem on change of bound variables. The following notation is used: Q denotes a quantifier, either $\forall$ or $\exists$; Q' denotes the quantifier opposite Q. In discussing prenex from, we will treat $\exists$ as a symbol of the original language rather than as a defined symbol.

Theorem 5 (Prenex form): Let A be a formula. Then there is a formula A^* in prenex form such that $\vdash_{FO} A \leftrightarrow A^*$.

Proof: The proof is by induction on the number of occurrences of $\neg$, $\vee$, $\forall$, and $\exists$ in A. Each atomic formula is already in prenex form.

Negation Suppose A is $\neg B$. By the induction hypothesis applied to B, there is a formula B^* in prenex form, say $Qy_1 \ldots Qy_k C$, such that $\vdash B \leftrightarrow Qy_1 \ldots Qy_k C$. By Lemma 1(1), $\vdash \neg B \leftrightarrow \neg Qy_1 \ldots Qy_k C$. By k applications of the prenex rules PR1 and PR2 and the replacement theorem, we have $\vdash \neg B \leftrightarrow Q'y_1 \ldots Q'y_k \neg C$ as required.

Disjunction Suppose A is $B \vee C$. By the induction hypothesis, $\vdash B \leftrightarrow B^*$ and $\vdash C \leftrightarrow C^*$, where B^* has prenex form $Qy_1 \ldots Qy_k D$ and C^* has prenex form $Qz_1 \ldots Qz_n E$. By the theorem on change of bound variables, we may assume that $y_1, \ldots, y_k, z_1, \ldots, z_n$ are distinct from one another, that $y_1, \ldots, y_k$ are distinct from any free variables in E, and that $z_1, \ldots, z_n$ are distinct from any free variables in D. By Lemma 1(2),

$$\vdash (B \vee C) \leftrightarrow [(Qy_1 \ldots Qy_k D) \vee (Qz_1 \ldots Qz_n E)].$$

Finally, by k applications of PR3 and PR4 and the replacement theorem, followed by n applications of PR5 and PR6 and the replacement theorem, we have $\vdash (B \vee C) \leftrightarrow [Qy_1 \ldots Qy_k Qz_1 \ldots Qz_n (D \vee E)]$ as required.

Quantifier Suppose A is $\forall x B$. By the induction hypothesis applied to B, there is a formula B^* in prenex form, say $Qy_1 \ldots Qy_k C$, such that $\vdash B \leftrightarrow Qy_1 \ldots Qy_k C$. By the theorem on change of bound variables, we may assume that $y_1, \ldots, y_k$ are distinct from x. By Lemma 1(3) we have $\vdash \forall x B \leftrightarrow \forall x Qy_1 \ldots Qy_k C$ as required. The case in which A is $\exists x B$ is similar, and is left to the reader. ∎

Additional Prenex Rules: It is convenient to have prenex rules for $\wedge$ and $\rightarrow$, thereby avoiding the elimination of these defined symbols. Proofs of these additional rules are outlined in the exercises.

PR7 $\vdash_{FO} (\exists x B \rightarrow C) \leftrightarrow \forall x (B \rightarrow C)$ (x not free in C).
PR8 $\vdash_{FO} (\forall x B \rightarrow C) \leftrightarrow \exists x (B \rightarrow C)$ (x not free in C).
PR9 $\vdash_{FO} (B \rightarrow \forall x C) \leftrightarrow \forall x (B \rightarrow C)$ (x not free in B).
PR10 $\vdash_{FO} (B \rightarrow \exists x C) \leftrightarrow \exists x (B \rightarrow C)$ (x not free in B).
PR11 $\vdash_{FO} (\exists x B \wedge C) \leftrightarrow \exists x (B \wedge C)$ (x not free in C)
PR12 $\vdash_{FO} (\forall x B \wedge C) \leftrightarrow \forall x (B \wedge C)$ (x not free in C).
PR13 $\vdash_{FO} (B \wedge \exists x C) \leftrightarrow \exists x (B \wedge C)$ (x not free in B).
PR14 $\vdash_{FO} (B \wedge \forall x C) \leftrightarrow \forall x (B \wedge C)$ (x not free in B).

Skolemization

We will end this section with a brief and somewhat informal discussion of a method of eliminating existential quantifiers in a sentence. This procedure is called *Skolemization* and is very useful in automated theorem proving. Though Skolemization does not preserve logical equivalence, the following property does hold: The original sentence A is true in some model if and only if the Skolemization of A is true in some model.

The basic idea is illustrated by the following example: Let M be a model of the sentence $\forall x_1 \exists x_2 R(x_1, x_2)$ and let F be a 1-ary function symbol. Since $\forall x_1 \exists x_2 R(x_1, x_2)$ is true in M, it follows that for each $m \in D_M$, there exists $m' \in D_M$ such that $\langle m, m' \rangle \in R_M$. Let $f: D_M \rightarrow D_M$ be the function defined by $f(m) = m'$. Now let N denote the model obtained from M by interpreting the function symbol F as this function f; in other words, $F_N: D_N \rightarrow D_N$ is defined by $F_N(m) = f(m)$ for all $m \in D$. In this case, the sentence $\forall x_1 R(x_1, F(x_1))$ is true in N. This new sentence is called the *Skolemization of* $\forall x_1 \exists x_2 R(x_1, x_2)$ and is obtained from $\forall x_1 \exists x_2 R(x_1, x_2)$ by omitting $\exists x_2$ and replacing x_2 in $R(x_1, x_2)$ with $F(x_1)$. This special case of Skolemization is summarized in the following theorem.

Theorem 6: Let Γ be a set of sentences and let F be a 1-ary function symbol that does not occur in any sentence in Γ. Then $\Gamma \cup \{\forall x_1 \exists x_2 R(x_1, x_2)\}$ has a model if and only if $\Gamma \cup \{\forall x_1 R(x_1, F(x_1))\}$ has a model.

Proof: We first show that $\vdash \forall x_1 R(x_1, F(x_1)) \rightarrow \forall x_1 \exists x_2 R(x_1, x_2)$.

(1) $R(x_1, F(x_1)) \rightarrow \exists x_2 R(x_1, x_2)$ SUBST THM
(2) $\forall x_1 R(x_1, F(x_1)) \rightarrow \forall x_1 \exists x_2 R(x_1, x_2)$ $\forall$-DIST.

From this result, it follows easily that any model of $\Gamma \cup \{\forall x_1 R(x_1, F(x_1))\}$ is also a model of $\Gamma \cup \{\forall x_1 \exists x_2 R(x_1, x_2)\}$.

Now assume M is a model of $\Gamma \cup \{\forall x_1 \exists x_2 R(x_1, x_2)\}$; we will construct a model of $\Gamma \cup \{\forall x_1 R(x_1, F(x_1))\}$ by giving a (possibly new) interpretation to the function symbol F. Let $m \in D_M$. Choose an assignment ϕ_m in M such that $\phi_m(x_1) = m$. Now $\phi_m(\exists x_2 R(x_1, x_2)) = T$; hence there is an assignment ψ_m in M that is a 2-variant of ϕ_m such that $\psi_m(R(x_1, x_2)) = T$; that is, $\langle \psi_m(x_1), \psi_m(x_2) \rangle \in R_M$. Let $\psi_m(x_2) = m'$, so that we obtain $\langle m, m' \rangle \in R_M$. Let N be the interpretation obtained from M by interpretating the function symbol F as follows: $F_N(m) = m'$. We emphasize that this is the only difference between the two interpretations M and N; in particular, $D_M = D_N$ and $R_M = R_N$. We will prove that N is a model of $\Gamma \cup \{\forall x_1 R(x_1, F(x_1))\}$. Now M is a model of Γ, and the symbol F does not occur in any sentence in Γ; hence N is also a model of Γ. Thus it remains to prove that $\forall x_1 R(x_1, F(x_1))$ is true in N. By the closure theorem for interpretations, it suffices to show that $R(x_1, F(x_1))$ is true in N. Let ϕ be an arbitrary assignment in N, and let $\phi(x_1) = m$. Then

$$\phi(R(x_1, F(x_1)) = T \Leftrightarrow \langle \phi(x_1), F_N(\phi(x_1)) \rangle \in R_N$$
$$\Leftrightarrow \langle m, F_N(m) \rangle \in R_N \Leftrightarrow \langle m, m' \rangle \in R_N,$$

so $\phi(R(x_1, F(x_1)) = T$ as required. ∎

Let us describe, somewhat informally, the procedure for obtaining the Skolemization of a sentence that is written in terms of ¬, ∨, ∀, and ∃. Working from left to right, do the following for each subformula of A of the form $\exists z[\dots z \dots]$: If $\exists z[\dots z \dots]$ is not within the scope of a universally quantified variable, eliminate $\exists z$ and replace each occurrence of z in $[\dots z \dots]$ with c, a new constant symbol (called a *Skolem constant*). Otherwise, assume $\exists z[\dots z \dots]$ is within the scope of the universally quantified variables $y_1, \dots, y_k$; eliminate $\exists z$ and replace each occurrence of z in $[\dots z \dots]$ with the term $F(y_1, \dots, y_k)$, where F is a new function symbol (called a *Skolem function*). We illustrate Skolemization with a number of examples. For a more detailed discussion, see Fitting, *First-Order Logic and Automated Theorem Proving* (New York: Springer-Verlag, 1990).

Sentence A	Skolemization of A
$\exists x R(x)$.	$R(c)$.
$\exists x \forall y \exists z R(x, y, z)$.	$\forall y R(c, y, F(y))$.
$\forall x \exists y \exists z R(x, y, z)$.	$\forall x R(x, F(x), G(x))$.
$\forall x \forall y \exists z R(x, y, z)$.	$\forall x \forall y R(x, y, F(x, y))$.
$\forall x[\forall y \exists z Q(x, y, z) \vee \exists z R(x, z)]$.	$\forall x[\forall y Q(x, y, F(x, y)) \vee R(x, G(x))]$.
$\forall x \exists y[R(x, y) \vee \exists z Q(x, y, z)]$.	$\forall x[R(x, F(x)) \vee Q(x, F(x), G(x))]$.

——————— *Exercises for Section 5.6* ———————

1. Show that provable equivalence is an equivalence relation.
2. Prove Lemma 1.
3. (*Completion of proof of theorem* 2) Prove $\vdash \exists x A \leftrightarrow \exists y A_x[y]$ (y is substitutable for x in A and y does not occur free in A).
4. (*Additional prenex rules*) Use PR1–PR6 to prove the additional prenex rules PR7–PR14; the replacement theorem is available.
5. (*Completion of proof of Theorem* 5) Let $\vdash A \leftrightarrow B$; show that $\vdash \exists x A \leftrightarrow \exists x B$. Then complete the proof of the prenex form theorem.
6. For each formula below, find a formula in prenex form that is provably equivalent to it. (*Note:* x, y, and z denote distinct variables.)
   (1) $\forall x R(x) \rightarrow \exists y P(x, y)$.
   (2) $\exists x P(x, y) \wedge [\forall y Q(y) \vee \neg \exists y R(y)]$.
   (3) $[\forall x P(x, y) \vee \exists y Q(y)] \rightarrow \forall x R(x)$.
   (4) $\forall x \forall y Q(x, y, z) \vee \forall y \forall z R(x, y, z)$.
   (5) $\neg[\forall x \forall y R(x, y) \wedge \exists x \exists y R(x, y)]$.
7. Use prenex rules to show the following (x and y denote distinct variables).
   (1) $\vdash \forall x \exists y[R(x) \rightarrow R(y)]$.
   (2) $\vdash \exists x \forall y[R(x) \rightarrow R(y)]$.
   (3) $\{\forall x \exists y[Q(x) \rightarrow R(x, y)], \exists x Q(x)\} \vdash \exists x \exists y Q(x, y)$.
8. Let A and B be two formulas satisfying these conditions: If $\vdash A$, then $\vdash B$; if $\vdash B$, then $\vdash A$. Can you conclude that $\vdash A \leftrightarrow B$? Justify your answer.
9. Find the Skolemization of the following sentences:
   (1) $\forall x[\exists y \forall z R(x, y, z) \vee \exists u Q(u)]$.
   (3) $\exists x \forall y[Q(x, y) \vee \exists z R(x, y, z)]$.
   (2) $\forall x \exists y[\exists z Q(x, y, z) \vee \forall u R(x, u, y)]$.
   (4) $\forall x[\exists y Q(y) \vee \forall z \exists u R(x, z, u)]$.
10. Refer to the examples of sentences A and their Skolemization at the end of this section. Prove the following:
   (1) $\vdash R(c) \rightarrow \exists x R(x)$.
   (2) $\vdash \forall y R(c, y, F(y)) \rightarrow \exists x \forall y \exists z R(x, y, z)$.
   (3) $\vdash \forall x R(x, F(x), G(x)) \rightarrow \forall x \exists y \exists z R(x, y, z)$.
   (4) $\vdash \forall x \forall y R(x, y, F(x, y)) \rightarrow \forall x \forall y \exists z R(x, y, z)$.

(5) $\vdash \forall x[\forall y Q(x, y, F(x, y)) \vee R(x, G(x))] \rightarrow \forall x[\forall y \exists z Q(x, y, z) \vee \exists z R(x, z)]$.

(6) $\vdash \forall x[R(x, F(x)) \vee Q(x, F(x), G(x))] \rightarrow \forall x \exists y[R(x, y) \vee \exists z Q(x, y, z)]$.

11. Let Γ be a set of sentences and let c be a constant symbol that does not occur in any sentence in Γ. Show that $\Gamma \cup \{\exists x_1 R(x_1)\}$ has a model if and only if $\Gamma \cup \{R(c)\}$ has a model.

6

Mathematics and Logic

I

n Chapters 2 through 5, we have used mathematics to study logic. A quick review: We began with propositional logic, then added quantifiers to obtain first-order logic; precise mathematical definitions of all concepts were given, and in each case the goal was to prove certain metalogical results such as the soundness and adequacy theorems. The theme of this chapter, and in fact of the remainder of the book, is the interplay between mathematics and logic.

6.1

First-Order Theories and Hilbert's Program

First-Order Theories

Group theory, arithmetic, and set theory are fundamental, important branches of mathematics. The axioms G1–G3 for a group and the Peano postulates P1–P7 for arithmetic are stated in Section 1.6; the axioms ZF1–ZF9 of Zermelo-Frankel set theory are discussed in Section 6.4. In arithmetic, the focus is on questions such as the following: Are there an infinite number of primes? Is every even number greater than 2 the sum of two primes?

But we are interested in questions that are quite different in spirit from these. For example: Is there a simple list of axioms for arithmetic such that all true statements of arithmetic, and only true statements, can be derived from these axioms? This question obviously involves logic. In particular, we need a formal language L in which to express statements of arithmetic, a precise definition of a mathematical proof, and a set of formulas of L as a candidate for the simple list of axioms.

Definition 1: A *first-order theory* (*theory* for short) consists of a first-order language L together with a set Γ of sentences of L. The sentences in Γ are called the *nonlogical axioms* of the theory. To describe a first-order theory, it suffices (1) to describe the language L, and (2) to state the nonlogical axioms of the theory (i.e., the sentences in Γ).

We have been studying sets of formulas throughout Chapters 2 through 5, so one might ask the reason for this new terminology. The idea is that a first-order language L and a set Γ of sentences of L are introduced to formalize a *mathematical theory* such as group theory, number theory, or set theory. Also, note the emphasis on sentences rather than formulas; this corresponds to the fact that in translating the axioms of a mathematical theory into a first-order language L, the resulting formulas are actually sentences.

We are especially interested in theories Γ that are *axiomatized* (there is an algorithm that decides whether $A \in \Gamma$). If Γ is finite, then Γ is certainly axiomatized, in which case we say that Γ is *finitely axiomatized*. Recall the following results from Section 5.5: (1) If Γ is axiomatized, then Γ is listable; (2) if Γ is axiomatized and complete, then Γ is decidable (Turing). We will now give a number of examples of first-order theories.

The Theory of Groups: This theory is denoted G. The language is L_G and the nonlogical axioms are $\Gamma = \{G1, G2, G3\}$, where

> **G1** $\forall x_1 \forall x_2 \forall x_3 [x_1 \circ (x_2 \circ x_3) = (x_1 \circ x_2) \circ x_3]$.
> **G2** $\forall x_1 (x_1 \circ e = x_1)$.
> **G3** $\forall x_1 \exists x_2 (x_1 \circ x_2 = e)$.

Group theory is finitely axiomatized and consistent (every group is a model of Γ). In the next section, we will answer the following questions about group theory:

> **Q1** $\Gamma = \{G1, G2, G3\}$ is a finite set of sentences of L_G whose models are precisely the groups. Is there a set Δ of sentences of L_G whose models are precisely the infinite groups? Can Δ be finite?
>
> **Q2** Is there a set Δ of sentences of L_G whose models are precisely the finite groups?

Peano Arithmetic: This theory is denoted PA. The language L_{PA} of Peano arithmetic is the language L_{NN} with the 2-ary relation symbol $<$ omitted; in other words, the nonlogical symbols are $0, S, +,$ and $\times$. The nonlogical axioms of PA are the closures of these formulas:

> **PA1** $\neg(Sx_1 = 0)$.
> **PA2** $(Sx_1 = Sx_2) \rightarrow (x_1 = x_2)$.
> **PA3** $[A_x[0] \wedge \forall x(A \rightarrow A_x[Sx])] \rightarrow \forall x A$.
> **PA4** $x_1 + 0 = x_1$.
> **PA5** $x_1 + Sx_2 = S(x_1 + x_2)$.
> **PA6** $x_1 \times 0 = 0$.
> **PA7** $x_1 \times Sx_2 = (x_1 \times x_2) + x_1$.

The axioms for PA are obviously based on the Peano postulates P1–P7 (see Section 1.6). Axiom PA3 holds for any formula A of L_{PA}, and is the first-order version of induction. PA is not finitely axiomatized, since PA3 is actually an axiom scheme rather than a single axiom. But PA is axiomatized, and hence listable.

Let us compare axiom scheme PA3 with Peano postulate P3 on induction. There is a difficulty in stating P3 in the language L_{PA}. In the standard interpretation, variables range over *elements* of $\mathbb{N}$ but not over *subsets* of $\mathbb{N}$. So, though we can say *for every element a of* $\mathbb{N}, \ldots$, we cannot say *for every subset A of* $\mathbb{N}, \ldots$. But PA3 does allow us to state induction for the subsets of $\mathbb{N}$ that are defined by some formula A of L_{PA}.

Is PA consistent? This is a somewhat delicate question and will be discussed later in this section. For now, let us take the point of view that the standard interpretation $\mathcal{N} = \langle \mathbb{N}, 0_{\mathbb{N}}, S_{\mathbb{N}}, +_{\mathbb{N}}, \times_{\mathbb{N}} \rangle$ of L_{PA} (ignore the interpretation of $<$) is a model of PA and hence that PA is consistent. We will often refer to $\mathcal{N}$ as the *standard model* of PA.

Let us illustrate the use of the nonlogical axioms of PA with an example of a formal proof in PA. To facilitate this task, we will list certain frequently used derived rules of inference of first-order logic; all are consequences of the equality axioms (see Section 5.1).

Tran Rule
$$\frac{s = t, t = u}{s = u}$$

Sym Rule
$$\frac{s = t}{t = s}$$

Function Rules
$$\frac{t = u}{St = Su}$$

$$\frac{s = t}{s + u = t + u} \quad \frac{s = t}{u + s = u + t}$$

Example 1:

We will show that $PA \vdash \forall x_1 (0 \times x_1 = 0)$. By axiom PA3, it suffices to show that $PA \vdash 0 \times 0 = 0$ and $PA \vdash [(0 \times x_1) = 0] \rightarrow [(0 \times Sx_1) = 0]$.

(1) $\forall x_1 (x_1 \times 0 = 0)$ PA6
(2) $0 \times 0 = 0$ $\forall$-ELIM

We use the deduction theorem to prove the second part.

(1) $(0 \times x_1) = 0$ HYP
(2) $\forall x_1 \forall x_2 [x_1 \times Sx_2 = (x_1 \times x_2) + x_1]$ PA7
(3) $\forall x_2 [0 \times Sx_2 = (0 \times x_2) + 0]$ $\forall$-ELIM
(4) $0 \times Sx_1 = (0 \times x_1) + 0$ $\forall$-ELIM
(5) $(0 \times x_1) + 0 = 0 + 0$ FUNCTION RULE (1)
(6) $0 \times Sx_1 = 0 + 0$ TRAN RULE
(7) $\forall x_1 (x_1 + 0 = x_1)$ PA4
(8) $0 + 0 = 0$ $\forall$-ELIM
(9) $0 \times Sx_1 = 0$ TRAN RULE (6, 8).   □

Have we included enough axioms in PA? In other words, does PA $\vdash A$ hold for every sentence A that is true in the standard interpretation? Similar questions, which we will answer in Chapter 7, are:

Q3 Is PA complete? (*No,* by Gödel's incompleteness theorem.)

Q4 Is PA decidable? (*No,* by Church's theorem and Church's thesis.)

Comment on $<$ We have not included $<$ among the nonlogical symbols of the language of PA. If $<$ were included, then we would want to add the closures of these formulas as additional nonlogical axioms:

PA8 $\neg(x_1 < 0)$.

PA9 $(x_1 < Sx_2) \leftrightarrow [(x_1 < x_2) \vee (x_1 = x_2)]$.

PA10 $(x_1 < x_2) \vee (x_1 = x_2) \vee (x_1 < x_2)$.

An alternate approach would be to introduce $<$ as a defined symbol as follows: $s < t$ means $\exists x(s + Sx = t)$, where x is the first variable that does not occur in either s or t. With this approach, PA8–PA10 are theorems of PA.

We will illustrate these ideas by proving PA8, below; formal proofs of PA9 and PA10 will be outlined in the exercises. In Chapter 7, we will replace PA with a finitely axiomatized theory NN, whose language is L_{NN} and whose nonlogical axioms are PA1–PA10, with the induction scheme PA3 omitted and PA9 simplified. The theory NN may be regarded as a slimmed-down version of PA that is still strong enough to serve as the basis for Gödel's incompleteness theorem and the related theorems of Church and Tarski.

Example 2:

We will show that PA $\vdash \forall x_1 \neg(x_1 < 0)$. By the definition of $<$, it suffices to show that PA $\vdash \forall x_1 \neg \exists x_2(x_1 + Sx_2 = 0)$. We will show PA $\vdash \neg(x_1 + Sx_2 = 0)$, and then use generalization twice.

(1) $\forall x_1 \neg(Sx_1 = 0)$	PA1
(2) $\neg(S(x_1 + x_2) = 0)$	$\forall$-ELIM
(3) $\forall x_1 \forall x_2[x_1 + Sx_2 = S(x_1 + x_2)]$	PA5
(4) $x_1 + Sx_2 = S(x_1 + x_2)$	$\forall$-ELIM (twice)
(5) $(x_3 = S(x_1 + x_2))$ $\rightarrow [\neg(S(x_1 + x_2) = 0) \rightarrow \neg(x_3 = 0)]$	EQ AX(2)
(6) $(x_1 + Sx_2 = S(x_1 + x_2))$ $\rightarrow [\neg(S(x_1 + x_2) = 0) \rightarrow \neg(x_1 + Sx_2 = 0)]$	SUBST RULE
(7) $\neg(S(x_1 + x_2) = 0) \rightarrow \neg(x_1 + Sx_2 = 0)$	MP(4, 6)
(8) $\neg(x_1 + Sx_2 = 0)$	MP(2, 7). □

The easiest way to describe a first-order theory is to give an explicit list of nonlogical axioms; this was done in the case of PA and group theory. An advantage of this method is that it always leads to an axiomatized theory. There is, however, a model-theoretic technique for describing theories.

Complete Arithmetic: The language of complete arithmetic is L_{PA}; the nonlogical axioms are

$$\text{Th}(\mathcal{N}) = \{A: A \text{ is a sentence of } L_{PA} \text{ and } A \text{ is true in}$$
$$\text{the standard interpretation } \mathcal{N} \text{ of } L_{PA}\}.$$

We will use Th $(\mathcal{N})$ ($= theory of \mathcal{N}$) to denote complete arithmetic. How useful is complete arithmetic? Is it cheating to declare the axioms of a theory to be the true sentences when we may not know which sentences are true? Is the sentence of L_{PA} that asserts Goldbach's conjecture a nonlogical axiom of Th $(\mathcal{N})$? We will not answer these questions now but instead will study properties of Th $(\mathcal{N})$ and its relationship to PA.

Theorem 1: The following hold for complete arithmetic Th $(\mathcal{N})$:

(1) The standard interpretation $\mathcal{N}$ is a model of Th $(\mathcal{N})$.

(2) Th $(\mathcal{N})$ is consistent and complete.

(3) For every sentence A of L_{PA}, Th $(\mathcal{N}) \vdash A$ if and only if $A \in$ Th $(\mathcal{N})$.

Proof: The standard interpretation is a model of Th $(\mathcal{N})$, since Th $(\mathcal{N})$ is *defined* as the set of sentences that are true in $\mathcal{N}$. This proves (1). Th $(\mathcal{N})$ is consistent because it has a model, namely $\mathcal{N}$. To see that Th $(\mathcal{N})$ is complete, let A be a sentence of L_{PA}. If A is true in $\mathcal{N}$, then $A \in$ Th $(\mathcal{N})$, so Th $(\mathcal{N}) \vdash A$. If A is false in $\mathcal{N}$, then $\neg A$ is true in $\mathcal{N}$, so $\neg A \in$ Th $(\mathcal{N})$; hence Th $(\mathcal{N}) \vdash \neg A$. This proves (2). Finally, let A be a sentence of L_{PA} such that Th $(\mathcal{N}) \vdash A$. Since $\mathcal{N}$ is a model of Th $(\mathcal{N})$, A is true in $\mathcal{N}$ (soundness theorem), so $A \in$ Th $(\mathcal{N})$ as required. ■

Theorem 2: Assume that the standard interpretation is a model of PA. Then every theorem of PA is a theorem of complete arithmetic.

Proof: Let PA $\vdash A$. Then PA $\vdash A'$ as well ($A' =$ closure of A), and by the soundness theorem A' is true in $\mathcal{N}$. Hence $A' \in$ Th $(\mathcal{N})$, so A' is certainly a theorem of Th $(\mathcal{N})$. By the closure theorem for provability, Th $(\mathcal{N}) \vdash A$ as required. ■

It is instructive to compare complete arithmetic and Peano arithmetic. Every theorem of PA is a theorem of complete arithmetic; complete arithmetic has many desirable properties, including consistency and completeness. Peano arithmetic is axiomatized; what about complete arithmetic?

Definition 2: A theory Γ with language L is *axiomatizable* if there is a set Δ of sentences of L such that (1) Δ is an axiomatized theory, and (2) Γ and Δ have the same theorems.

Note that an axiomatized theory is automatically axiomatizable. Here is the idea behind this definition: We are given a theory Γ with language L, but there is no obvious algorithm for deciding whether $A \in \Gamma$ for an arbitrary sentence A of L; perhaps we can replace Γ with Δ, where Δ is obviously axiomatized and Γ and Δ have the same theorems.

We now ask additional questions about complete arithmetic and PA; all have been answered by Gödel and Tarski (see Chapter 7).

> *Q5* Is complete arithmetic axiomatizable? (*No*, by Gödel.)
>
> *Q6* Do PA and complete arithmetic have the same theorems? (*No*, by Gödel.)
>
> *Q7* Is complete arithmetic listable? (*No*, by Tarski.)

The following table summarizes our discussion of PA and complete arithmetic (assume that the standard interpretation is a model of PA):

Property	Peano Arithmetic	Complete Arithmetic
Axiomatizable	Yes	No
Consistent	Yes	Yes
Complete	No	Yes
Decidable	No	No
Listable	Yes	No

We ask a final question about PA and complete arithmetic (Skolem has shown that the answer is *yes* in both cases; see Section 6.2).

> *Q8* Is there a model of PA that is *not* isomorphic to $\mathcal{N}$, the standard model of PA? What about complete arithmetic?

The Theory of Fields: This theory is denoted F. The language L_F has four nonlogical symbols: 0, 1 (constant symbols); $+$, $\times$ (2-ary function symbols). The nonlogical axioms of F are the closures of the following formulas of L_F (these are the field axioms written in the first-order language L_F):

F1 $x_1 + (x_2 + x_3) = (x_1 + x_2) + x_3$.
F2 $x_1 + x_2 = x_2 + x_1$.
F3 $x_1 + 0 = x_1$.
F4 $\exists x_2(x_1 + x_2 = 0)$.
F5 $x_1 \times (x_2 \times x_3) = (x_1 \times x_2) \times x_3$.
F6 $x_1 \times x_2 = x_2 \times x_1$.
F7 $x_1 \times 1 = x_1$.
F8 $\neg(x_1 = 0) \rightarrow \exists x_2(x_1 \times x_2 = 1)$.
F9 $\neg(0 = 1)$.
F10 $x_1 \times (x_2 + x_3) = (x_1 \times x_2) + (x_1 \times x_3)$.

The Theory of Ordered Fields: This theory is denoted OF. The language L_{OF} is obtained from L_F by adding a 2-ary relation symbol $<$. The nonlogical axioms are the axioms F1–F10 of the theory of fields, together with the closures of these formulas of L_{OF}:

O1 $\neg(x_1 < x_1)$.

O2 $[(x_1 < x_2) \wedge (x_2 < x_3)] \rightarrow (x_1 < x_3)$.

O3 $(x_1 < x_2) \vee (x_2 < x_1) \vee (x_1 = x_2)$.

O4 $(x_1 < x_2) \rightarrow [(x_1 + x_3) < (x_2 + x_3)]$.

O5 $[(x_1 < x_2) \wedge (0 < x_3)] \rightarrow [(x_1 \times x_3) < (x_2 \times x_3)]$.

$\langle \mathbb{Q}, 0_\mathbb{Q}, 1_\mathbb{Q}, +_\mathbb{Q}, \times_\mathbb{Q}, <_\mathbb{Q} \rangle$ and $\langle \mathbb{R}, 0_\mathbb{R}, 1_\mathbb{R}, +_\mathbb{R}, \times_\mathbb{R}, <_\mathbb{R} \rangle$ are models of OF; $\langle \mathbb{C}, 0_\mathbb{C}, 1_\mathbb{C}, +_\mathbb{C}, \times_\mathbb{C} \rangle$ ($\mathbb{C}$ is the set of complex numbers) is a model of F but not of OF; in other words, $\langle \mathbb{C}, 0_\mathbb{C}, 1_\mathbb{C}, +_\mathbb{C}, \times_\mathbb{C} \rangle$ is a field but not an ordered field (one cannot define a 2-ary relation R on $\mathbb{C}$ such that O1–O5 also hold).

We will use the theory of ordered fields to describe the real number system. The *least upper bound property* (LUBP) states: Every nonempty set that has an upper bound has a least upper bound. The real number system is often defined as *the unique complete ordered field*. The basis of this definition is the following fundamental theorem from analysis.

Theorem 3 (Existence and uniqueness of $\mathbb{R}$): There is a model of OF that satisfies the least upper bound property. Moreover, this model is unique.

Theorem 3 is proved as follows. Assume the existence of the natural number system; basic concepts of set theory are then used to construct a set $\mathbb{R}$, elements $0_\mathbb{R}, 1_\mathbb{R} \in \mathbb{R}$, 2-ary operations $+_\mathbb{R}, \times_\mathbb{R}$ on $\mathbb{R}$, and a 2-ary relation $<_\mathbb{R}$ on $\mathbb{R}$ such that $\mathscr{R} = \langle \mathbb{R}, 0_\mathbb{R}, 1_\mathbb{R}, +_\mathbb{R}, \times_\mathbb{R}, <_\mathbb{R} \rangle$ is an ordered field that also satisfies the LUBP. This complete ordered field, called the *real number system* and denoted by $\mathscr{R}$, is unique in the following sense: If $\langle D_M, 0_M, 1_M, +_M, \times_M, <_M \rangle$ is a model of OF that also satisfies the LUBP, then there is a one-to-one function h from $\mathbb{R}$ onto D_M such that for all $a, b \in \mathbb{R}$, (1) $h(a +_\mathbb{R} b) = h(a) +_M h(b)$; (2) $h(a \times_\mathbb{R} b) = h(a) \times_M h(b)$; (3) $a <_\mathbb{R} b \Leftrightarrow h(a) <_M h(b)$. The function h is called an *isomorphism,* and $\mathscr{R}$ and $\langle D_M, 0_M, 1_M, +_M, \times_M, <_M \rangle$ are said to be *isomorphic.*

We have not stated the LUBP in the language L_{OF}; in the next section, we will show that there is no set Γ of sentences of L_{OF} whose only model (up to isomorphism) is the real number system.

The Theory of an interpretation I: This is a generalization of complete arithmetic. Given an interpretation I of a first-order language L, the *theory of* I, denoted Th (I), is the set of sentences true in I; in other words,

$$\text{Th (I)} = \{A : A \text{ is a sentence of L and } A \text{ is true in I}\}.$$

Basic properties of Th (I) are summarized as follows (proof similar to that of Theorem 1).

Theorem 4: Let I be an interpretation of L. (1) I is a model of Th (I); (2) Th (I) is consistent and complete; (3) for every sentence A of L, Th (I) ⊢ A ⇔ A ∈ Th (I).

Example 3:

> Consider the language L_{OF} of ordered fields and the interpretation $\mathscr{R} = \langle \mathbb{R}, 0_\mathbb{R}, 1_\mathbb{R}, +_\mathbb{R}, \times_\mathbb{R}, <_\mathbb{R} \rangle$ of L_{OF}; $\mathscr{R}$ is a model of Th ($\mathscr{R}$) and Th ($\mathscr{R}$) is consistent and complete; Th ($\mathscr{R}$) is called the *theory of real numbers* and is decidable (due to Tarski). □

─────────────── *Hilbert's Program* ───────────────

We will first discuss the distinction between *axiom systems* and *axiomatized theories*. The first known example of an axiom system is Euclid's set of axioms for plane geometry. Euclid's treatment of geometry is widely admired for its use of explicit axioms and step-by-step proofs. Other examples of axiom systems include the Peano postulates for the natural numbers, the axioms for a group, the axioms for an ordered field, and the Zermelo–Frankel axioms for set theory.

Axiom systems are *classical* or *modern*. A classical axiom system attempts to uniquely characterize a mathematical structure. Peano's axioms P1–P3 for arithmetic, Euclid's axioms for plane geometry, and the Zermelo–Frankel axioms for set theory are all examples of classical axiom systems.

A modern axiom system is introduced with the goal of studying a collection of mathematical structures that have certain properties in common. Examples include the axioms for a group and the axioms for a field. Thus in working with the axioms for a group, we are studying the properties that are common to a wide range of mathematical structures such as the integers under addition, the positive rationals under multiplication, {1, −1} under multiplication, and so on.

Then what is the difference between an axiom system and an axiomatized theory? We quote Frege, the great German mathematician and philosopher who was the first to write down axioms for first-order logic and who, together with Boole, deserves to be called the founder of modern logic.

> It cannot be demanded that everything be proved, because that is impossible; but we can require that all propositions used without proof be expressly declared to be so.... Furthermore I demand—and in this I go beyond Euclid—that all of the methods of inference used must be specified in advance.

> *Grundgestze der Arithmetik (The Basic Laws of Arithmetic)*, 1893

In other words, an axiomatized theory goes *beyond* an axiom system in these respects: The axioms are written in a first-order language; the principles of logic that are allowed in constructing proofs are those of first-order logic. In an axiomatized theory, we specify *both* the mathematical axioms *and* the principles of logic; there are *no hidden assumptions*.

We have noted on earlier occasions that a formal system is the appropriate framework for answering questions about consistency and completeness. An axiomatized theory Γ with language L is a formal system: The language is L; the rules are those of first-order logic; the axioms are the logical axioms of first-order logic *plus* the sentences of Γ. An axiomatized theory is the *formal* counterpart of an axiomatized branch of mathematics. Oversimplifying somewhat,

Axiom system + first-order logic = axiomatized theory.

Axiomatized theories play an important role in what is known as *Hilbert's program*. This is not a precisely defined concept, and we will interpret it rather broadly. During the period 1900–1930, Hilbert outlined the following areas of research in logic and in the foundations of mathematics:

(1) Give axiom systems for the various branches of mathematics.

(2) Prove the existence of a common core of logical axioms and rules of inference that suffice for all mathematical reasoning.

(3) Develop *proof theory*.

(4) Give an absolute consistency proof for arithmetic.

(5) Find a simple list of axioms for arithmetic and prove the resulting axiomatized theory to be consistent and complete.

(6) Justify the use of infinite sets in mathematics.

(7) Prove the consistency of mathematics.

We have already discussed axiom systems and their relationship to axiomatized theories. We now discuss (2)–(5) and save (6) and (7) for Section 6.4. Part (2), which asks for a description of the logic used by mathematicians, is answered by Gödel's completeness theorem in the following sense: Suppose we have an axiom system and a mathematical proof of a statement Φ. Suppose further that this mathematical theory has a first-order formulation with language L and nonlogical axioms Γ. Let A be a sentence of L that expresses Φ. Our mathematical proof of Φ shows that A is true in every model of Γ; hence, by Gödel's completeness theorem, we have $\Gamma \vdash A$. In summary, first-order logic suffices for mathematical reasoning.

Proof theory is a new branch of mathematics, in which the concept of a mathematical proof is the principal object of study. One of the main goals of proof theory is to establish the consistency of fundamental branches of mathematics such as arithmetic and set theory. Proof theory is Hilbert's approach to the foundations of mathematics, and is often referred to as *metamathematics*. A byproduct of proof theory is that of *verification*: A proposed mathematical proof in a subject such as group theory can be translated (in theory if not in practice) into a formal proof and then mechanically checked for correctness.

In the year 1900, Hilbert gave a talk at the International Congress of Mathematicians. In his address, he proposed a list of 23 problems that would challenge mathematicians of the twentieth century. One of the problems on his list is the following.

Hilbert's Second Problem: Prove the consistency of the real number system.

Why was Hilbert interested in proving the consistency of the real number system? In 1899, Hilbert wrote a book on geometry, entitled *Grundlagen der Geometrie*. In this book, Hilbert updated Euclid's axioms for geometry and corrected certain flaws (for example, the lack of a clear distinction between definitions and axioms). But in addition, Hilbert's work was highly original and thoroughly modern. Euclid's treatment of geometry is based on Aristotle's approach to axiom systems, namely the assumption that the axioms are "obviously true" statements about the concepts being studied; Hilbert's approach, on the other hand, is modern in the sense that the axioms give meaning to the undefined concepts. For Hilbert, we must be able to replace *point, line, plane* with *table, chair, beer mug* and still proceed to prove theorems without adjustments. (See Smorynski, "Hilbert's Programme", *CWT Quarterly* 1 (1988), for a further discussion of these ideas.)

Another modern characteristic of Hilbert's book is his emphasis on proving the consistency of his axiom system. Well-known mathematicians before Hilbert's time, such as Pasch, Peano, and Frege, had emphasized the desirability of proving axiom systems consistent, and Hilbert shared this point of view. There are at least two good reasons for consistency proofs. First of all, the axioms for a branch of mathematics are selected with the goal of proving all true statements about the subject under discussion; there is, however, the danger that one may select too many axioms, and so a consistency proof is in order. Second, not all axiom systems are obviously consistent; although the consistency of the axioms for Euclidean geometry may seem self-evident, what about axiom systems for non-Euclidean geometries?

Hilbert obtained a relative consistency proof for his axioms for Euclidean geometry; he proved that if the real number system is consistent, then so are his axioms for plane geometry. Research by Cantor, Dedekind, Weierstrass, and Peano shows that, beginning with the natural number system and using basic concepts of set theory, we can construct the integers, the rationals, and the reals. From this it is reasonable to conclude that the consistency of the natural number system implies the consistency of the real number system. Thus we have two relative consistency results: If the real number system is consistent, then Euclidean geometry is consistent; if arithmetic is consistent, then the real number system is consistent.

Hilbert realized that not all consistency proofs could be relative consistency proofs and that eventually one must give an absolute consistency proof for some fundamental branch of mathematics. For this, Hilbert focused on arithmetic, surely the most fundamental of all the branches of mathematics.

Many mathematicians of Hilbert's generation did not feel the need for a consistency proof of arithmetic; indeed, for them the consistency of arithmetic was self-evident. Earlier in this section, we argued that Peano arithmetic is consistent because the standard interpretation is a model of the axioms

PA1–PA7. But this consistency proof is by no means a rigorous argument; in particular, we did not systematically verify the truth of the axioms PA1–PA7, but rather stated that they are self-evident properties of natural numbers. (*Note:* The situation with regard to the axioms of group theory is quite different. First of all, we can give examples of finite groups, even of a group with one element, and moreover it is possible to write out a table that interprets the composition ∘. We can then verify in a mechanical way and in a finite number of steps that each of the group axioms is true.) To emphasize the need for a proof of the consistency of arithmetic, Hilbert states, in his essay "On the Infinite," (In From Frege to Gödel, 367–92).

> What we have twice experienced, once with the paradoxes of the infinitesimal calculus and once with the paradoxes of set theory, will not be experienced a third time, nor ever again.

In short, Hilbert wanted to avoid "unpleasant surprises in the future" (Benacerraf and Putnam (eds.) 1983).

The standard method of proving consistency is to give a model. Hilbert observes in his essay "On the Infinite" (1925) that the laws of physics point to a discrete rather than a continuous world, to a universe that is unbounded but finite; in other words, that neither the infinitely small nor the infinitely large exists in nature. Since any model of arithmetic is necessarily infinite, the possibility of finding a physical model of arithmetic is ruled out.

Thus Hilbert was faced with the difficult problem of proving consistency without the use of a model. His strategy for an absolute consistency proof of arithmetic uses his proof theory as follows: (1) Introduce an axiomatized theory for arithmetic; from our earlier discussion, the obvious choice is PA. (2) Give a finitist proof that the sentence $\neg(0 = 0)$ is *not* a theorem of PA. This proof would be a combinatorial argument that there is no finite sequence $A_1, \ldots, A_n$ of formulas of L_{PA} that is a formal proof of $\neg(0 = 0)$. (3) Since a theory is consistent if it has at least one formula that is not a theorem, PA is consistent.

Hilbert claimed that this procedure would show that arithmetic is consistent, because if there were an inconsistency in arithmetic, then presumably this inconsistent argument could be translated into a formal proof in PA, thereby showing that PA is not consistent.

Hilbert's interest in completeness is explained by his belief that axiom systems are the proper approach to studying the various branches of mathematics. In particular, for branches such as arithmetic, Euclidean geometry, and set theory, we should be able to give a simple, consistent list of axioms such that every sentence that is true can be derived from the axioms. Related to this is Hilbert's *formalist* approach to mathematics. According to this view, mathematics is the study of axiom systems. Primitive concepts such as *set, is an element of, natural number, zero,* and *successor* need not be given any meaning or interpretation. Moreover, there is a mechanical way to obtain theorems: by writing down finite sequences of formulas according to certain rules. But to avoid the conclusion that mathematics is a meaningless game, Hilbert wanted classical axiom systems to be complete in the sense that every sentence that is true in the intended interpretation can be derived from the axioms.

The part of Hilbert's program that asks for proofs of the consistency and completeness of arithmetic received a devastating blow in 1931. Remarkable theorems of Kurt Gödel suggest that it is extremely unlikely that Hilbert's requirement of a finitist proof of the consistency of arithmetic can be achieved. Moreover, an axiomatized theory such as PA cannot be both consistent and complete. We state the results here; for a further discussion, see Chapter 7.

Theorem 5 (Gödel's incompleteness theorem I, PA version):
Every axiomatized consistent extension of PA is incomplete.

Theorem 6 (Gödel's incompleteness theorem II, PA version):
One cannot prove the consistency of PA by methods that can be formalized in PA.

────────── *Exercises for Section 6.1* ──────────

1. Show that group theory is not complete. (*Hint:* Let A be the sentence $\exists x_1 [\neg(x_1 = e) \wedge (x_1 \circ x_1 = e)]$; show that neither A nor $\neg A$ is a theorem of group theory.)

2. Let $\Gamma = \{G1, G2, G3\}$ be the nonlogical axioms of group theory, and let Δ be all sentences of L_G that are true in every model of Γ. Show that the theory Δ is finitely axiomatizable.

3. Prove the following in PA without PA3:
   (1) PA $\vdash S0 \times S0 = S0$.
   (2) PA $\vdash S0 + SS0 = SS0 + S0$.
   (3) PA $\vdash S0 \times SS0 = SS0$.
   (4) PA $\vdash \forall x_1[(x_1 + S0) = Sx_1]$.

4. Use PA3 to prove the following in PA:
   (1) PA $\vdash \forall x_1 \neg(Sx_1 = x_1)$. (*Hint:* PA $\vdash \neg(S0 = 0)$, PA $\vdash \neg(Sx_1 = x_1) \rightarrow \neg(SSx_1 = Sx_1)$.)
   (2) PA $\vdash \forall x_1(0 + x_1 = x_1)$.
   (3) PA $\vdash \forall x_1(x_1 \times S0 = x_1)$.

5. Suppose that PA $\vdash A_x[0]$ and PA $\vdash A_x[Sx]$. Show that PA $\vdash \forall x A$.

6. Prove PA $\vdash \forall x_1[(x_1 = 0) \vee \exists x_2(x_1 = Sx_2)]$.

7. (*Properties of addition and multiplication*) Prove the following:
   (1) PA $\vdash \forall x_1 \forall x_2(x_1 + Sx_2 = Sx_1 + x_2)$ (PA3 with x_2).
   (2) PA $\vdash \forall x_1 \forall x_2(x_1 + x_2 = x_2 + x_1)$ (PA3 with x_1).
   (3) PA $\vdash \forall x_1 \forall x_2 \forall x_3[x_1 + (x_2 + x_3) = (x_1 + x_2) + x_3]$ (PA3 with x_3).
   (4) PA $\vdash \forall x_1 \forall x_2(Sx_1 \times x_2 = x_1 \times x_2 + x_2)$ (PA3 with x_2).
   (5) PA $\vdash \forall x_1 \forall x_2(x_1 \times x_2 = x_2 \times x_1)$ (PA3 with x_1).

8. (*Proof of PA9 from PA1–PA7*). Prove PA9 by proving the following:
   PA $\vdash \exists x_3(x_1 + Sx_3 = Sx_2) \leftrightarrow [\exists x_3(x_1 + Sx_3 = x_2) \vee (x_1 = x_2)]$. *Hints:*
   (1) PA $\vdash (x_1 + x_3 = x_2) \leftrightarrow (x_1 + Sx_3 = Sx_2)$. (PA5, PA2).
   (2) PA $\vdash (x_1 + 0 = x_2) \rightarrow \exists x_3(x_1 + x_3 = x_2)$.
   (3) PA $\vdash (x_1 = x_2) \rightarrow \exists x_3(x_1 + Sx_3 = Sx_2)$ ((1), (2).)
   (4) PA $\vdash (x_1 + Sx_3 = x_2) \rightarrow \exists x_3(x_1 + x_3 = x_2)$.
   (5) PA $\vdash \exists x_3(x_1 + Sx_3 = x_2) \rightarrow \exists x_3(x_1 + Sx_3 = Sx_2)$.
   (6) PA $\vdash [\exists x_3(x_1 + Sx_3 = x_2) \vee (x_1 = x_2)] \rightarrow \exists x_3(x_1 + Sx_3 = Sx_2)$. ((3), (5).)
   (7) PA $\vdash (x_1 + x_3 = x_2) \rightarrow [\exists x_3(x_1 + Sx_3 = x_2) \vee (x_1 + 0 = x_2)]$. (PA3 with x_3.)
   (8) PA $\vdash \exists x_3(x_1 + Sx_3 = Sx_2) \rightarrow [\exists x_3(x_1 + Sx_3 = x_2) \vee (x_1 = x_2)]$.

9. (*Proof of PA10 from PA1–PA9*) Prove PA $\vdash (x_1 < x_2) \vee (x_1 = x_2) \vee (x_2 < x_1)$. *Hint:* prove the following and then use PA3 with x_1.
   (1) PA $\vdash \forall x_1[(x_1 = 0) \vee (0 < x_1)]$.
   (2) PA $\vdash \forall x_1 \forall x_2[(x_1 < x_2) \rightarrow (Sx_1 < Sx_2)]$. (PA3 with x_2, PA8, PA9.)
   (3) PA $\vdash \forall x_1 \forall x_2[(x_1 < x_2) \rightarrow ((Sx_1 < x_2) \vee (Sx_1 = x_2))]$. ((2), PA9.)

10. Prove Theorem 4 on properties of Th (I).

11. Let Γ be a theory with language L and let M be a model of Γ. Prove the following:
  (1) If Γ ⊢ *A*, then Th (M) ⊢ *A*.
  (2) Γ is complete if and only if Γ and Th (M) have the same theorems.
12. We describe a field F_2 with exactly two elements 0, 1. Addition and multiplication are defined as follows:

+	0	1		×	0	1
0	0	1		0	0	0
1	1	0		1	0	1

   F_2 is a field. Show that F_2 is not an ordered field.
13. $\mathscr{C} = \langle \mathbb{C}, 0_{\mathrm{C}}, 1_{\mathrm{C}}, +_{\mathrm{C}}, \times_{\mathrm{C}} \rangle$ is a field. Show that $\mathscr{C}$ is not an ordered field.

6.2

The Löwenheim-Skolem Theorem and the Compactness Theorem

In this section, the model existence theorem will be used to prove two key results in model theory: the Löwenheim-Skolem theorem and the compactness theorem. We will then give applications of these two fundamental results, many of which reveal shortcomings in the expressive power of first-order logic. This will lead us to a discussion of extensions of first-order logic.

—— The Löwenheim-Skolem Theorem ——

Definition 1: An interpretation I of L is *finite* if its domain D_I is a finite set, *countable* if D_I is a countable set, and *uncountable* if D_I is an uncountable set.

Theorem 1 (Löwenheim-Skolem): Let Γ be a first-order theory. If Γ has a model, then Γ has a countable model.

Proof: Since Γ has a model, Γ is consistent. Now recall the model existence theorem: Γ consistent ⇒ Γ has a model. In the proof of this result, we obtain an extension Δ of Γ that is consistent, complete, and Henkin; the canonical interpretation of Δ is then used to construct a model of Γ. Now the domain of a canonical interpretation consists of equivalence classes of variable-free terms, and this is a countable set. Thus Γ has a countable model as required. ■

Here is a striking application of the Löwenheim-Skolem theorem: The real number system $\mathscr{R} = \langle \mathbb{R}, 0_{\mathbb{R}}, 1_{\mathbb{R}}, +_{\mathbb{R}}, \times_{\mathbb{R}}, <_{\mathbb{R}} \rangle$ is a model of the theory of ordered fields. Can we add additional sentences of L_{OF} to this theory so that $\mathcal{R}$ is the only model?

Corollary 1: There is no set Γ of sentences of L_{OF} whose only model (up to isomorphism) is the real number system.

Proof: Let Γ be a set of sentences of L_{OF} such that $\mathcal{R} = \langle \mathbb{R}, 0_{\mathbb{R}}, 1_{\mathbb{R}}, +_{\mathbb{R}}, \times_{\mathbb{R}}, <_{\mathbb{R}} \rangle$ is a model of Γ. By the Löwenheim-Skolem theorem, Γ has a countable model M. Clearly M is not isomorphic to $\mathcal{R}$, since the domain of M is a countable set whereas $\mathbb{R}$ is an uncountable set. ∎

Why does Corollary 1 not contradict the uniqueness of the real number system (see Theorem 3 in Section 6.1)? The difficulty, as one might expect, is that the LUBP cannot be expressed in L_{OF}; informally speaking, in first-order logic we can quantify over the elements of the domain of an interpretation, but not over the subsets of that domain.

---------- *The Compactness Theorem* ----------

Theorem 2 (Compactness theorem): If every finite subset of Γ has a model, then Γ has a model.

Proof: Suppose Γ has no model. By the model existence theorem, Γ is not consistent. Let A be a formula of L such that both $\Gamma \vdash A$ and $\Gamma \vdash \neg A$. Since proofs have only a finite number of steps, there is a finite $\Delta \subseteq \Gamma$ such that $\Delta \vdash A$ and $\Delta \vdash \neg A$ (see Lemma 1 in Section 3.4). It follows that Δ is not consistent; hence Δ has no model (consistency theorem). This contradicts the hypothesis that every finite subset of Γ has a model. ∎

We will give a number of applications of this powerful theorem, but first let us prove the following lemma on capturing the infinite models of a theory.

Lemma 1: There is a set Σ of sentences of L_{EQ} such that every model of Σ is infinite.

Proof: For each $n \geq 1$, there is a sentence σ_n of L_{EQ} such that every model of σ_n has at least n elements; for example, σ_3 is

$$\exists x_1 \exists x_2 \exists x_3 [\neg(x_1 = x_2) \wedge \neg(x_1 = x_3) \wedge \neg(x_2 = x_3)].$$

The required set of sentences is $\Sigma = \{\sigma_n : n \geq 1\}$. ∎

We emphasize that the set Σ is infinite. Now consider the theory of groups with language L_G and nonlogical axioms $\Gamma = \{G1, G2, G3\}$; this theory has both finite and infinite models. Moreover, we can use $\Sigma = \{\sigma_n : n \geq 1\}$ to isolate the infinite models of Γ: The models of $\Gamma \cup \Sigma$ are precisely the infinite groups. We call $\Gamma \cup \Sigma$ with language L_G the *theory of infinite groups*.

Let us summarize the situation with respect to group theory: $\Gamma = \{G1, G2, G3\}$ is a finite set of sentences of L_G whose models are precisely the groups; $\Gamma \cup \Sigma$ is an infinite set of sentences of L_G whose models are precisely the infinite groups. Is there a set Δ of sentences of L_G whose models are precisely the finite groups? Is the theory of infinite groups finitely axiomatizable? Both questions are answered negatively by the following consequence of the compactness theorem.

Theorem 3: If Γ has arbitrarily large finite models, then Γ has an infinite model.

Proof: A precise statement of the hypothesis is: For all $n \in \mathbb{N}^+$, there exists $k \geq n$ such that Γ has a model with k elements. Let $\Delta = \Gamma \cup \Sigma$, where $\Sigma = \{\sigma_n : n \geq 1\}$ is the set of sentences in Lemma 1. Note that the models of Δ are the infinite models of Γ.

We will prove that every finite subset of Δ has a model. Fix $n \in \mathbb{N}^+$; it suffices to show that $\Gamma \cup \{\sigma_i : 1 \leq i \leq n\}$ has a model. Now Γ has arbitrarily large finite models; hence there is a model M of Γ with k elements, where $k \geq n$. Let us show that M is also a model of $\{\sigma_i : 1 \leq i \leq n\}$. Consider σ_i, where $1 \leq i \leq n$. Informally, σ_i states that there are at least i elements; since the domain of M has k elements and $k \geq n \geq i$, σ_i is true in M. This completes the proof that every finite subset of Δ has a model.

By the compactness theorem, $\Delta = \Gamma \cup \Sigma$ has a model M. Clearly M is a model of Γ; moreover, M is infinite, since it is also a model of Σ. Thus Γ has an infinite model as required. ∎

Corollary 2: There is no set of sentences of L_G whose models are precisely the finite groups.

Proof: Suppose there is a set Δ of sentences of L_G whose models are the finite groups. Now for each $n \geq 1$, there is a group H_n with 2^n elements (see Exercise 10); this shows that Δ has arbitrarily large finite models. By Theorem 3, Δ has an infinite model. But this contradicts the assumption that the models of Δ are the finite groups. ∎

Corollary 3: There is no finite set of sentences of L_G whose models are precisely the infinite groups.

Proof: Suppose there is a finite set $\{A_1, \ldots, A_n\}$ of sentences of L_G whose models are the infinite groups. Let A be the sentence $A_1 \wedge \ldots \wedge A_n$. We show that $\Delta = \{G1, G2, G3, \neg A\}$ has arbitrarily large finite models. For each $n \geq 1$, there is a group H_n with 2^n elements (see Exercise 10). Since H_n is a group, it is a model of $\{G1, G2, G3\}$. But H_n cannot be a model of A, because then H_n would be infinite. Thus $\neg A$ is true in H_n and hence H_n is a model of Δ.

Since Δ has arbitrarily large finite models, Δ has an infinite model $M = \langle D_M, e_M, \circ_M \rangle$. Now $\{G1, G2, G3\} \subseteq \Delta$, so M must be a group. Thus M is an infinite group; hence A is true in M. This contradicts the fact that M is a model of $\neg A$. ∎

We now combine the Löwenheim-Skolem theorem and the compactness theorem to answer the following questions about Peano arithmetic and complete arithmetic: Do the axioms of PA uniquely characterize arithmetic? Is $\mathcal{N}$ the only countable model of complete arithmetic?

Definition 2: A model M of complete arithmetic Th $(\mathcal{N})$ is said to be *nonstandard* if it is *not* isomorphic to $\mathcal{N}$.

Skolem proved the existence of nonstandard models of complete arithmetic. The proof we give uses the following fact: If $\langle D_M, 0_M, S_M \rangle$ and $\langle \mathbb{N}, 0_\mathbb{N}, S_\mathbb{N} \rangle$ are isomorphic, then Peano postulate P3 on induction holds in $\langle D_M, 0_M, S_M \rangle$ (see Exercise 6 in Section 1.6).

Theorem 4 (Skolem, 1934, first version): There is a countable nonstandard model of complete arithmetic.

Proof: It suffices to construct a countable model M of Th $(\mathcal{N})$ for which induction fails. Let L^+ be the language obtained from L_{PA} by adding a new constant symbol c and let

$$\Delta = \text{Th}(\mathcal{N}) \cup \{\neg(0^n = c): n \geq 0\}.$$

We assert that every finite subset of Δ has a model. To prove this, it suffices to show that for $n \geq 0$, Th $(\mathcal{N}) \cup \{\neg(0^n = c): 0 \leq k \leq n\}$ has a model. This is easy: Take the standard model $\mathcal{N}$ and interpret c as $n + 1$ (recall that 0^k is interpreted as the natural number k). By the compactness theorem, there is a model M of Δ. By the Löwenheim-Skolem theorem, we can assume that M is countable. In summary, M is a countable model of Th $(\mathcal{N})$ and the sentences $\{\neg(0^n = c): n \geq 0\}$ are all true in M.

It remains to prove that $M = \langle D_M, 0_M, S_M \rangle$ does not satisfy induction. Let $A = \{S_M^n(0_M): n \in \mathbb{N}\}$, where $S_M^n(0_M) \in D_M$ is defined inductively by $S_M^0(0_M) = 0_M$ and $S_M^{n+1}(0_M) = S_M(S^n(0_M))$. Clearly, $0_M \in A$ and $S_M(a) \in A$ whenever $a \in A$; in other words, $A \subseteq D_M$ satisfies the hypotheses of induction. But $A \neq D_M$. To prove this, consider $c_M \in D_M$. Suppose $c_M \in A$, say $c_M = S_M^n(0_M)$. Now $S_M^n(0_M) = (0^n)_M$ (left as an exercise for the reader; see Example 1 in Section 4.3); hence $c_M = (0^n)_M$. But $\neg(0^n = c)$ is true in M; hence $c_M \neq (0^n)_M$—a contradiction, as required. ∎

Let us put Skolem's theorem in a somewhat different light to further emphasize shortcomings of the expressive power of first-order languages.

Theorem 5 (Skolem, second version): There is no set Γ of sentences of L_{PA} (or L_{NN}) whose only model (up to isomorphism) is the standard model.

Proof: Let Γ be a set of sentences of L_{PA} such that $\mathcal{N}$ is a model of Γ; note that $\Gamma \subseteq \text{Th}\,(\mathcal{N})$. It suffices to produce a model of Γ not isomorphic to $\mathcal{N}$. By Theorem 4, there is a model M of $\text{Th}\,(\mathcal{N})$ that is not isomorphic to $\mathcal{N}$. But $\Gamma \subseteq \text{Th}\,(\mathcal{N})$, so M is the required model of Γ. ■

Note The language for Theorems 4 and 5 need only have the constant symbol 0 and the 1-ary function symbol S.

According to Theorem 5, Peano arithmetic PA has a countable model that is not isomorphic to $\mathcal{N}$. Why does this not contradict Dedekind's result that the Peano postulates P1–P3 uniquely characterize the natural number system (see Section 1.6)? The reason, briefly, is that axiom PA3 of Peano arithmetic does not completely capture the induction axiom P3. Note that PA3 applies to a countable number of properties (= formulas), whereas P3 applies to uncountably many subsets of $\mathbb{N}$.

Until now, we have always assumed that the number of nonlogical symbols in a first-order language is countable. There is, however, no reason why a language cannot have uncountably many symbols. Moreover, both the model existence theorem and the compactness theorem hold for uncountable languages. Let us assume the compactness theorem for uncountable languages and use it to prove an upward version of the Löwenheim-Skolem theorem.

Theorem 6 (Upward Löwenheim-Skolem theorem, special case): If Γ has an infinite model, then Γ has an uncountable model.

Proof: Let M be an infinite model of Γ, let $\{c_x : x \in \mathbb{R}\}$ be distinct new constant symbols (one for each real number x), and let L^+ be the language obtained from L by adding these new constant symbols to L; note that L^+ is an uncountable language. In this proof, we use the fact (not proved by us) that the compactness theorem applies to uncountable languages. Let $\Delta = \Gamma \cup \{\neg(c_x = c_y)): x, y \in \mathbb{R}$ and $x \neq y\}$. We will prove that $\Gamma \cup \Delta_0$ has a model, where Δ_0 is a finite subset of $\{\neg(c_x = c_y): x, y \in \mathbb{R}$ and $x \neq y\}$. Let $\{c_{x_1}, \ldots, c_{x_n}\}$ be the new constant symbols that occur in the formulas in Δ_0. Choose n distinct elements $a_1, \ldots, a_n$ of D_M (use the fact that D_M is infinite). Then M is a model of $\Gamma \cup \Delta_0$, provided that the new constant symbols are interpreted as follows: $(c_{x_1})_M = a_1, \ldots, (c_{x_n})_M = a_n;\ (c_x) = a_1$ for $x \neq x_k$, $1 \leq k \leq n$.

By the compactness theorem, Δ has a model N. This model N of Δ is also a model of Γ, and since the sentences $\{\neg(c_x = c_y): x, y \in \mathbb{R}$ and $x \neq y\}$ are all true in N, D_N has a subset equipotent to $\mathbb{R}$, and hence is uncountable as required. ■

From this result and Theorem 3, we obtain another version of the upward Löwenheim-Skolem theorem.

Theorem 7 (Upward Löwenheim-Skolem theorem, second version): If Γ has arbitrarily large finite models, then Γ has an uncountable model.

——*Isomorphisms and Elementary Equivalence*——

The notion of an isomorphism has already been used to express the uniqueness of the natural number system and the real number system (see Sections 1.6 and 6.1). Let us now define the general notion of isomorphic interpretations.

Definition 3: Two interpretations I and J of L are *isomorphic*, denoted $I \approx J$, if there is a one-to-one and onto function $h: D_I \to D_J$ such that

(1) $h(c_I) = c_J$ (each constant symbol c);

(2) $h(F_I(a_1, \ldots, a_n)) = F_J(h(a_1), \ldots, h(a_n))$ (each n-ary function symbol F and all $a_1, \ldots, a_n \in D_I$);

(3) $\langle a_1, \ldots, a_n \rangle \in R_I \Leftrightarrow \langle h(a_1), \ldots, h(a_n) \rangle \in R_J$ (each n-ary relation symbol R and all $a_1, \ldots, a_n \in D_I$). A one-to-one function from D_I onto D_J that satisfies (1)–(3) is called an *isomorphism from* I *to* J. Informally speaking, isomorphic interpretations of L are mathematically the same.

Example 1:

$\langle \mathbb{Z}, 0, + \rangle$ and $\langle 2\mathbb{Z}, 0, + \rangle$ ($2\mathbb{Z} = \{2a : a \in \mathbb{Z}\}$) are isomorphic models of the theory of groups; the required isomorphism is the function $h: \mathbb{Z} \to 2\mathbb{Z}$ defined by $h(a) = 2a$.   □

Definition 4: A theory Γ is *categorical* if any two models of Γ are isomorphic, and is *ω-categorical* if any two countably infinite models of Γ are isomorphic.

By Skolem's theorem, Peano arithmetic and complete arithmetic are not ω-categorical. On the other hand, the theory of densely ordered sets without first or last elements is ω-categorical (see Section 6.3). The next result tells us that no "interesting" first-order theory is categorical.

Theorem 8: If Γ is a first-order theory with an infinite model, then Γ is not categorical.

Proof: By the Löwenheim-Skolem theorem, Γ has a countable model M. By the upward Löwenheim-Skolem theorem, there is an uncountable model N of Γ. Since D_M is countable and D_N is uncountable, M and N cannot be isomorphic. Thus Γ is not categorical as required. ∎

We now generalize the isomorphism property.

Definition 5: Two interpretations I and J of L are *elementarily equivalent*, written I ≡ J, if for every sentence A of L, A is true in I if and only if A is true in J; in other words, Th (I) = Th (J).

The next lemma shows how to construct elementarily equivalent models.

Lemma 2: Let I be an interpretation of L. If M is a model of Th (I), then M ≡ I (i.e., M and I are elementarily equivalent).

Proof: Let A be a sentence of L; we are required to show that A is true in I if and only if A is true in M. First assume A is true in I. Then $A \in$ Th (I), and since M is a model of Th (I), A is true in M. Now assume A is true in M. Suppose, however, that A is false in I. Then $\neg A \in$ Th (I), and since M is a model of Th (I), $\neg A$ is true in M, a contradiction. ∎

To see how the lemma applies, consider a nonstandard model M of complete arithmetic (Skolem's theorem). Since M is a model of Th ($\mathcal{N}$), M and $\mathcal{N}$ are elementarily equivalent. This means that a nonstandard model of complete arithmetic, though not isomorphic to $\mathcal{N}$, does have the same set of true sentences as $\mathcal{N}$.

The isomorphism property of interpretations is stronger than that of elementary equivalence. The proof of the following theorem is outlined in the exercises.

Theorem 9: If I and J are isomorphic interpretations of L, then I and J are elementarily equivalent.

We will now illustrate these ideas with a fundamental result in nonstandard analysis. The following notation is used in the theory of fields: $n \cdot 1$ ($n \in \mathbb{N}^+$) denotes the term $1 + \cdots + 1$ (n occurrences of the symbol 1). For example, $3 \cdot 1$ is the term $1 + 1 + 1$.

Definition 6: An ordered field $\langle D_F, 0_F, 1_F, +_F, \times_F, <_F \rangle$ is *Archimedean* if for all $a \in D_F$, there exists $n \in \mathbb{N}^+$ such that $a <_F 1_F +_F \ldots +_F 1_F$ (1_F added to itself n times).

The field $\mathbb{Q}$ of rational numbers and the field $\mathbb{R}$ of real numbers are examples of Archimedean ordered fields. If two ordered fields are isomorphic and one of them has the Archimedean order property, then so does the other (left as an exercise for the reader). We now use the compactness theorem to prove the existence of an ordered field that is not Archimedean.

Theorem 10: There is an ordered field that is elementarily equivalent to $\mathcal{R}$ but is not Archimedean (and hence is not isomorphic to $\mathcal{R}$).

Proof: Let L^+ be the language obtained from L_{OF} by adding a new constant symbol c and let $\Delta = $ Th ($\mathcal{R}$) $\cup \{n \cdot 1 < c : n \in \mathbb{N}^+\}$. Now $\mathcal{R}$ is a model of every finite subset of Δ, so the compactness theorem applies and Δ has a model, say M. Since M is a model of Th ($\mathcal{R}$), M ≡ $\mathcal{R}$ (Lemma 2); we leave it to the reader to check that M is not Archimedean, as required. ∎

First-Order Properties

We will now explore further the expressive power (or lack thereof) of first-order languages.

Definition 7: Let L be a first-order language. A property Q of interpretations of L is said to be a *first-order property of* L if there is a set Γ of sentences of L such that for every interpretation I of L,

I is a model of $\Gamma \Leftrightarrow$ I has property Q.

The set Γ of sentences of L may be infinite; if Γ is finite, we say that Q is a *strong* first-order property. Note that if Γ is finite, say $\Gamma = \{A_1, \ldots, A_n\}$, then Γ can be replaced by a single sentence, namely $A_1 \wedge \ldots \wedge A_n$. Thus Q is a strong first-order property if there is a sentence A of L such that for every interpretation I of L, A is true in I $\Leftrightarrow$ I has property Q.

Example 2:

> Consider the language L_G of groups.
>
> (1) The property of being a group is a strong first-order property of L_G; take $\Gamma = \{G1, G2, G3\}$.
>
> (2) The property of being an infinite group is a first-order property of L_G; take $\Gamma = \{G1, G2, G3\} \cup \Sigma$.
>
> (3) The property of being an infinite group is not a strong first-order property of L_G (see Corollary 3).
>
> (4) The property of being a finite group is not a first-order property of L_G (see Corollary 2). □

Note The notion of a first-order property is often treated with a different terminology, as follows: A class $\mathfrak{J}$ of interpretations of L ($\mathfrak{J}$ is *not* a set) is *elementary* if there is a sentence A of L such that for every interpretation I of L, I $\in \mathfrak{J} \Leftrightarrow A$ is true in I; $\mathfrak{J}$ is Δ-*elementary* if there is a set Γ of sentences of L such that for every interpretation I of L, I $\in \mathfrak{J} \Leftrightarrow$ I is a model of Γ.

The following lemma is often used to show that a property Q is *not* first-order.

Lemma 3: Let Q be a property of interpretations of L. Suppose there are interpretations I and J of L such that (1) I $\equiv$ J (i.e., I and J are elementarily equivalent), and (2) I has property Q but J does not. Then Q is not a first-order property.

Proof: Suppose there is a set Γ of sentences of L such that for every interpretation I′ of L, I′ is a model of $\Gamma \Leftrightarrow$ I′ has property Q. Since J does not have property Q, there is a sentence $A \in \Gamma$ such that A is false in J. But I $\equiv$ J, so A is also false in I. Now I does have property Q, so I is a model of Γ; in particular, A is true in I. This is a contradiction. ■

Example 3:

Induction is not a first-order property of any language L with constant symbol 0 and 1-ary function symbol S. To see this, it suffices to find interpretations I and J of L such that (1) I $\equiv$ J and (2) I satisfies induction but J does not. $\langle \mathbb{N}, 0_\mathbb{N}, S_\mathbb{N} \rangle$ satisfies induction, but the countable non-standard model of arithmetic constructed in Theorem 4 does not. $\square$

Example 4:

The Archimedean order property is not first-order. By Theorem 10, there is an ordered field M such that M $\equiv \mathscr{R}$ but M is not Archimedean. $\square$

Up to this section, the results we have obtained in first-order logic fall into one of two categories: theorems about syntactic concepts such as provability or consistency (e.g., the deduction theorem and the extension theorem); theorems about the interplay between syntactic and semantic concepts (e.g., the soundness and model existence theorems). In contrast, the compactness theorem and the two Löwenheim-Skolem theorems are stated exclusively in terms of semantic concepts, namely models. Moreover, there are purely semantic proofs of these results. *Model theory* is often defined as the study of the relationship between sentences of a first-order language and its interpretations of the language. The compactness theorem and the two Löwenheim-Skolem theorems are the cornerstones of model theory.

Extensions of First-Order Logic

We have seen that first-order logic has certain shortcomings with respect to its expressive power. For example, the notion of finiteness cannot be captured with a first-order language. We can extend the expressive power of a first-order language by adding new logical symbols to that language, and we will give several examples of this. There is, however, a price to be paid: The resulting logic fails to have one or more of the important metalogical properties enjoyed by first-order logic. In particular, in all cases either the compactness theorem or the Löwenheim-Skolem theorem fails. This is no accident; a famous uniqueness result of Lindström states that first-order logic is the *only* logic that is an expressive as first-order logic and for which both the compactness theorem and the Löwenheim-Skolem theorem hold.

Incidentally, the term *first-order* refers to the fact that the individual variables $x_1, x_2, \ldots$ range over the elements of the domain of the interpretation; in particular, $x_1, x_2, \ldots$ do not range over subsets of the domain or over functions defined on the domain of the interpretation. We obtain a *second-order logic* by introducing new variables to range over relations and/or functions on the domain of the interpretation.

We will now describe several extensions of first-order logic. In each case, we start with a first-order language L and add certain logical symbols to obtain a language that is more expressive than L. We obtain a new logic by giving definitions of satisfaction and truth for this new language. In most cases, these definitions are straightforward modifications of those for first-order languages. The compactness theorem fails for all of the logics that we will now describe.

Weak Second-Order Logic L_{WII}: We will introduce a second-order language in which finiteness is captured by a single sentence. Given a first-order language L, let L_{WII} be the language obtained from L by adding an infinite number of new variable symbols $Z_1, Z_2, \ldots$. An interpretation of L_{WII} is simply an interpretation of L in which the new variables $Z_1, Z_2, \ldots$ range over the *finite* subsets of the domain of the interpretation. The formulas of L_{WII} are the formulas of L together with formulas obtained by applying these two rules: (1) For each term t of L and each variable Z_n, $Z_n(t)$ is an atomic formula of L_{WII} (intuitively, $Z_n(t)$ asserts that t is an element of the finite set Z_n); (2) for each formula A of L_{WII} and each variable Z_n, $\forall Z_n A$ and $\exists Z_n A$ are formulas of L_{WII}.

Consider the sentence $\exists Z \forall x Z(x)$ of L_{WII} (we have omitted subscripts for the variables x and Z). Informally, $\exists Z \forall x Z(x)$ states that there is a finite subset of the domain of the interpretation such that all elements of the domain belong to that set; in other words, the domain of the interpretation is finite.

The compactness theorem does not hold for weak second-order logic. To see this, suppose we start with the language L_{EQ} of equality, and let L_{WII} denote the corresponding weak second-order logic. Then $\Gamma = \Sigma \cup \{\exists Z \forall x Z(x)\}$ is a set of sentences of L_{WII} such that every finite subset of Γ has a model but Γ has no model.

Monadic Second-Order Logic L_{II}: Given a first-order language L, let L_{II} be the language obtained from L by adding an infinite number of new variable symbols $Z_1, Z_2, \ldots$. Syntactically, L_{II} is the same as L_{WII}; the difference lies in the way the new variable symbols are interpreted. This time, $Z_1, Z_2, \ldots$ range over *all* subsets of the domain of the interpretation. As an example, suppose we start with the first-order language L_{PA}. The Peano postulates P1–P3 are easily stated in L_{II}; for P3 we have

$$\forall Z[(Z(0) \wedge \forall x(Z(x) \rightarrow Z(Sx))) \rightarrow \forall x Z(x)].$$

Both the compactness theorem and the Löwenheim-Skolem theorem fail for monadic second-order logic.

Extended First-Order Logic $L_{\omega_1 \omega}$: This logic is obtained by adding a new logical symbol $\bigvee$ to L; informally, $\bigvee$ is an infinite disjunction. Given a first-order language L, the formulas of $L_{\omega_1 \omega}$ are the formulas of L together with formulas obtained by applying the following rule: If $\{A_n : n \in \mathbb{N}\}$ is an infinite collection of formulas of $L_{\omega_1 \omega}$, then $\bigvee A_n$ is a formula of $L_{\omega_1 \omega}$. The Löwenheim-Skolem theorem holds for $L_{\omega_1 \omega}$, but both the compactness theorem and the upward Löwenheim-Skolem theorem fail.

Example 5:

Suppose we start with the language L_{PA}. For each $n \in \mathbb{N}$, let A_n be the formula $(x = 0^n)$. Then $\forall x \bigvee A_n$ is a sentence of $L_{\omega_1 \omega}$ that can be written informally as

$$\forall x[(x = 0) \vee (x = 0^1) \vee (x = 0^2) \vee \ldots].$$

This sentence asserts that the domain D of any interpretation I of $L_{\omega_1 \omega}$ is the set $\{(0^n)_I : n \in \mathbb{N}\}$. Thus any model of $\forall x \bigvee A_n$ is countable; hence the upward Löwenheim-Skolem theorem does not hold for $L_{\omega_1 \omega}$. $\square$

Example 6:

Suppose we start with the language L_{EQ} of equality. Finiteness is captured by the following sentence of $L_{\omega_1 \omega}$ (and hence the compactness theorem fails for $L_{\omega_1 \omega}$):

$$\bigvee \exists x_1 \ldots \exists x_n \forall x_{n+1}[(x_1 = x_{n+1}) \vee \ldots (x_n = x_{n+1})].$$ $\square$

Extended First-Order Logic L_Q: In this logic we add a new logical quantifier symbol Q; informally, $Qx \ldots$ means that there exist infinitely many x for which $\ldots$ holds. The formulas of L_Q are the formulas of L together with formulas obtained by applying the following rule: If A is a formula of L_Q and x is a variable, then QxA is a formula of L_Q. The compactness theorem does not hold for the extended first-order logic L_Q. Suppose we start with the language L_{EQ} of pure equality. Then $\neg Qx(x = x)$ is a sentence of L_Q whose models are precisely the nonempty finite sets. Now let $\Gamma = \Sigma \cup \{\neg Qx(x = x)\}$; every finite subset of Γ has a model, but Γ itself has no model.

——————*Exercises for Section 6.2*——————

1. Let Γ be a set of formulas of L with no model. Show that there is a finite subset of Γ that has no model.
2. (*Compactness theorem*) Compactness theorem II states: If A is true in every model of Γ, then there is a finite $\Delta \subseteq \Gamma$ such that A is true in every model of Δ.
   (1) Use the compactness theorem to prove compactness theorem II.
   (2) Use compactness theorem II to prove the compactness theorem.
   (3) Prove the converse of the compactness theorem.
   (4) Prove the converse of compactness theorem II.
3. Let Γ and Δ be sets of sentences of a first-order language L such that for every interpretation I of L, I is a model of Γ if and only if I is not a model of Δ. Show that there are finite sets $\Gamma_0 \subseteq \Gamma$ and $\Delta_0 \subseteq \Delta$ such that for every sentence A, if $\Gamma \vdash A$, then $\Gamma_0 \vdash A$; if $\Delta \vdash A$, then $\Delta_0 \vdash A$. (*Hint:* $\Gamma \cup \Delta$ has no model.)
4. Let A be a sentence and Γ a set of sentences of L such that A is true in every infinite model of Γ. Show that there is a positive integer k such that A is true in every finite model of Γ whose domain has $\geq k$ elements.
5. Let L_{EQ} be the language of pure equality. An interpretation of L_{EQ} is simply a nonempty set. Prove the following:
   (1) There is a set of sentences of L_{EQ} whose models are precisely the infinite sets (i.e., *infinite* is a first-order property).
   (2) There is no set of sentences of L_{EQ} whose models are precisely the nonempty finite sets (i.e., *finite* is not a first-order property).

(3) There is no finite set of sentences of L_{EQ} whose models are precisely the infinite sets (i.e., *infinite* is not a strong first-order property).

(4) There is no set of sentences of L_{EQ} whose models are precisely the uncountable sets (i.e., *uncountable* is not a strong first-order property).

(5) There is no set of sentences of L_{EQ} whose models are precisely the nonempty countable sets (i.e., *countable* is not a strong first-order property).

6. (*Joint consistency and interpolation*) This exercise outlines a proof of special cases of the Robinson joint consistency theorem (JCT) and the Craig interpolation theorem (CIT). Let Γ and Δ be sets of sentences of L.

   (1) Assume that $\Gamma \cup \Delta$ has no model. Show that there is a sentence A such that (a) A is true in every model of Γ, and (b) $\neg A$ is true in every model of Δ.

   (2) Assume that $\Gamma \cup \Delta$ is inconsistent. Show that there is a sentence A such that $\Gamma \vdash A$ and $\Delta \vdash \neg A$. This is a special case of JCT.

   (3) Let A and C be sentences such that $\Gamma \cup \Delta \vdash A \rightarrow C$. Show that there is a sentence B such that $\Gamma \vdash A \rightarrow B$ and $\Delta \vdash B \rightarrow C$. This is a special case of CIT. (*Hint:* Let $\Gamma^* = \Gamma \cup \{A\}$ and $\Delta^* = \Delta \cup \{\neg C\}$, and use (2).)

   (4) Assume (3) and prove (2).

7. Let $\Gamma = \{\forall x \neg R(x, x), \forall x \forall y \forall z[(R(x, y) \wedge R(y, z)) \rightarrow R(x, z)], \forall x \exists y R(x, y)\}$. Show that every model of Γ is infinite.

8. Show that Γ in Exercise 29 in Section 4.2 does not have a finite model.

9. Let Γ be a theory with language L, and assume that Γ has arbitrarily large finite models. Prove the following:

   (1) There is an infinite set Δ of sentences of L such that for every interpretation I of L, I is a model of Δ if and only if I is an infinite model of Γ.

   (2) There is no set Δ of sentences of L such that for every interpretation I of L, I is a model of Δ if and only if I is a finite model of Γ.

   (3) There is no finite set Δ of sentences of L such that for every interpretation I of L, I is a model of Δ if and only if I is an infinite model of Γ.

10. Given a group $\langle G, e, \circ \rangle$ and $n \in \mathbb{N}^+$, define a set H_n and a binary operation $*$ on H_n by

$$H_n = \{\langle g_1, \ldots, g_n \rangle : g_1, \ldots, g_n \in G\};$$
$$\langle g_1, \ldots, g_n \rangle * \langle h_1, \ldots, h_n \rangle = \langle g_1 \circ h_1, \ldots, g_n \circ h_n \rangle.$$

   Show that H_n with operation $*$ is a group. Note that if G has 2 elements, then H_n has 2^n elements.

11. Refer to the proof of Skolem's theorem. Show that $S_M^n(0_M) = (0^n)_M$.

12. (*Basic properties of isomorphisms*) Let I_1, I_2, I_3 be interpretations of L. Prove the following: (1) $I_1 \simeq I_1$; (2) if $I_1 \simeq I_2$ and $I_2 \simeq I_3$, then $I_1 \simeq I_3$; (3) if $I_1 \simeq I_2$, then $I_2 \simeq I_1$.

13. Let I be an interpretation of L and let $h: D_1 \rightarrow X$ (X a set) be one-to-one and onto. Prove that there is an interpretation J of L such that $D_J = X$ and h is an isomorphism from I to J.

14. (*Isomorphic $\Rightarrow$ elementarily equivalent*) Let h be an isomorphism from I to J; this exercise outlines a proof that $I \equiv J$. The following notation is used: Given an assignment $\phi: VAR \rightarrow D_1$ in I, there is a natural assignment ϕ_J in J, namely the composition $h \circ \phi$. In summary,

$$\phi: VAR \rightarrow D_I; \ h: D_I \rightarrow D_J \text{ (isomorphism)}; \ \phi_J: VAR \rightarrow D_J, \text{ where } \phi_J = h \circ \phi.$$

   Recall that ϕ and ϕ_J have extensions $\phi: TRM \rightarrow D_I$ and $\phi: FOR \rightarrow \{T, F\}$, and $\phi_J: TRM \rightarrow D_J$ and $\phi_J: FOR \rightarrow \{T, F\}$. Prove the following:

   (1) For any assignment ϕ in I, $\phi_J(t) = h(\phi(t))$.

   (2) For any assignment ϕ in I, $\phi_J(A) = \phi(A)$.

   (3) For every sentence A of L, A is true in I if and only if A is true in J.

15. Prove the converse of Lemma 2: If $M \equiv I$, then M is a model of Th (I).

16. (*Löwenheim-Skolem theorem*) Let Γ have a model M. Prove that there is a countable model of Γ that is elementarily equivalent to M. (*Hint:* apply the LS theorem to Th (M).)

17. Let Γ have an infinite model. Prove that there are models M and N of Γ such that M is uncountable, N is countable, and M $\equiv$ N.

18. Prove that there is an uncountable model of Th $(\mathcal{R})$ that is elementarily equivalent to $\mathcal{R}$ but not isomorphic to $\mathcal{R}$.

19. Give an example of a theory that is complete but not ω-categorical.

20. (*Fields of characteristic p and of characteristic 0*) Refer to the theory of fields in Section 6.1. A field F has *characteristic p* (p a prime) if p is the smallest positive integer such that $1_F +_F \cdots +_F 1_F = 0_F$ (p copies of 1_F). If there is no such prime p, F has *characteristic 0*. The field of real numbers has characteristic 0; for every prime p there is a field F_p of characteristic p. The property of being a field of characteristic p is a strong first-order property of L_F; the required finite set of sentences is

$$\{F1, \ldots, F10\} \cup \{p \cdot 1 = 0\} \cup \{\neg(k \cdot 1 = 0): 1 \le k < p\}.$$

Prove the following:

(1) The property of being a field of characteristic 0 is a first-order property.

(2) The property of being a field of characteristic 0 is not a strong first-order property. (Thus the theory of fields of characteristic 0 is not finitely axiomatized.) (*Hint:* Suppose there is a sentence A of L_F such that for every interpretation I of L_F, A is true in I $\Leftrightarrow$ I is a field of characteristic 0. Let

$$\Delta = \{F1, \ldots, F10, \neg A\} \cup \{\neg(n \cdot 1 = 0): n \ge 1\}.$$

Use the compactness theorem to prove that Δ has a model.)

21. Let $M = \langle D_M, 0_M, 1_M, +_M, \times_M, <_M \rangle$ and $N = \langle D_N, 0_N, 1_N, +_N, \times_N, <_N \rangle$ be isomorphic models of the theory of ordered fields, with M Archimedean. Prove that N is Archimedean.

22. (*Subinterpretations*) Let M and N be interpretations of L; N is called a *subinterpretation* of M, written N $\subseteq$ M, if the following conditions hold: (a) $D_N \subseteq D_M$; (b) $c_M = c_N$ (each constant symbol c); (c) $F_M(a_1, \ldots, a_n) = F_N(a_1, \ldots, a_n)$ (each function symbol F and all $a_1, \ldots, a_n \in D_N$); (d) $R_M(a_1, \ldots, a_n) \Leftrightarrow R_N(a_1, \ldots, a_n)$ (each relation symbol R and all $a_1, \ldots, a_n \in D_N$). Prove the following (M$_1$, M$_2$, M$_3$ interpretations of L):

(1) $M_1 \subseteq M_1$.

(2) If $M_1 \subseteq M_2$ and $M_2 \subseteq M_1$, then $M_1 = M_2$.

(3) If $M_1 \subseteq M_2$ and $M_2 \subseteq M_3$, then $M_1 \subseteq M_3$.

(4) $\langle \mathbb{N}, 0_N, +_N \rangle \subseteq \langle \mathbb{Z}, 0_Z, +_Z \rangle$ (both interpretations of L_G).

(5) $\langle \mathbb{Q}, 0_Q, 1_Q, +_Q, \times_Q, <_Q \rangle \subseteq \langle \mathbb{R}, 0_R, 1_R, +_R, \times_R, <_R \rangle$ (both interpretations of L_{OF}).

23. Let M and N be interpretations of L with N $\subseteq$ M (see Exercise 22). This exercise outlines a proof that if A is a quantifier-free formula of L that is true in M, then A is also true in N. Let $\phi: \text{VAR} \to D_N$ be an assignment in N; recall that ϕ has extensions $\phi: \text{TRM} \to D_N$ and $\phi: \text{FOR} \to \{T, F\}$. Since $D_N \subseteq D_M$, ϕ is also an assignment in M; in this case, we write ϕ_M. Note that ϕ_M (regarded as an assignment in M) has extensions $\phi_M: \text{TRM} \to D_M$ and $\phi_M: \text{FOR} \to \{T, F\}$. Prove the following:

(1) For every term t of L, $\phi(t) = \phi_M(t)$.

(2) For every quantifier-free formula A of L, $\phi(A) = \phi_M(A)$.

(3) Every quantifier-free formula true in M is also true in N.

24. (*There is no set Γ of universal sentences of L_G whose models are the groups*) A sentence A is *universal* if it has the form $\forall y_1 \ldots \forall y_k B$, where B is quantifier-free. The axioms G1 and G2 of group theory are universal, but G3 is not. If we expand the language L_G by adding a 1-ary function symbol $^{-1}$ (intended interpretation: x^{-1} = right inverse of x), then $\{G1, G2, \forall x_1(x_1 \circ x_1^{-1} = e)\}$ is a set of universal sentences of this expanded language whose models are the groups. Prove the following:

(1) Let Γ be a set of universal sentences of L, let M be a model of Γ, and let N $\subseteq$ M (see Exercises 22 and 23). Prove that N is a model of Γ. (*Hints:* Exercise 23; closure theorem for interpretations.)

(2) For the language L_G, $\langle \mathbb{N}, 0_N, +_N \rangle \subseteq \langle \mathbb{Z}, 0_Z, +_Z \rangle$ but $\langle \mathbb{N}, 0_N, +_N \rangle$ is not a group.

(3) Show that there is no set Γ of universal sentences of L_G whose models are precisely the groups.

25. (*Well-ordering property is not first-order*) Let L be a language with 2-ary relation symbol $\leq$. This exercise outlines a proof that WO (see Section 1.6) is not a first-order property of L. An interpretation of L is $\langle \mathbb{N}, \leq_\mathbb{N} \rangle$, where $\leq_\mathbb{N}$ is the usual order on $\mathbb{N}$; $\langle \mathbb{N}, \leq_\mathbb{N} \rangle$ satisfies WO. It suffices to find an interpretation of L that is elementarily equivalent to $\langle \mathbb{N}, \leq_\mathbb{N} \rangle$ but does not satisfy WO (Lemma 3). Let L^+ be the language obtained from L by adding distinct new constant symbols $\{c_n : n \in \mathbb{N}\}$ and let

$$\Delta = \text{Th}(\langle \mathbb{N}, \leq_\mathbb{N} \rangle) \cup \{(c_{n+1} \leq c_n) \wedge \neg(c_{n+1} = c_n) : n \in \mathbb{N}\}.$$

Use the compactness theorem to show that Δ has a model M. Since M is a model of Th $(\langle \mathbb{N}, \leq_\mathbb{N} \rangle)$, $M \equiv \langle \mathbb{N}, \leq_\mathbb{N} \rangle$. Show that $\{(c_n)_M : n \in \mathbb{N}\} \subseteq D_M$ has no smallest element.

26. Give an example of a set Γ of sentences of L_{II} such that every finite subset of Γ has a model but Γ has no model. (*Hint:* Let L be the language with function symbol S and constant symbols 0 and c. Take the Peano postulates P1–P3 together with the sentences $\{\neg(c = 0^n) : n \in \mathbb{N}\}$.)

27. Give an example of a set Γ of sentences of $L_{\omega_1 \omega}$ such that every finite subset of Γ has a model but Γ has no model. (*Hint:* Let L be the language L_{EQ}; for each $n \geq 1$, let A_n be a sentence of L_{EQ} that states that there are at most n individuals. Consider $\Sigma \cup \{\bigvee A_n\}$.)

28. Let L be the first-order language with function symbol S and constant symbol 0.
    (1) Give a set of sentences of L_{II} whose only model is the standard model $\langle \mathbb{N}, 0_\mathbb{N}, S_\mathbb{N} \rangle$.
    (2) Give a set of sentences of $L_{\omega_1 \omega}$ whose only model is the standard model $\langle \mathbb{N}, 0_\mathbb{N}, S_\mathbb{N} \rangle$.
    Conclude that the upward Löwenheim-Skolem theorem fails for L_{II} and $L_{\omega_1 \omega}$.

29. State the Archimedean order property in $L_{\omega_1 \omega}$ (L is the language L_F of fields).

30. (*Well-ordering property in L_{II}*) Let L be a first-order language with a 2-ary relation symbol $\leq$. Express WO (see Section 1.6) in the language L_{II}.

6.3

Decidable Theories

Both first-order logic and Peano arithmetic are undecidable (for precise statements of these results, see Sections 5.5 and 7.1). But one should not immediately conclude from these two examples that all interesting first-order theories are undecidable. In this section, we will prove the positive result that the theory of densely ordered sets without first or last element is decidable. We will begin with a description of DO.

The First-Order Theory DO (Densely ordered sets without first or last element): The language L_{DO} has exactly one binary relation symbol, $<$. The nonlogical axioms are the closures of these formulas:

DO1 $\neg(x_1 < x_1)$.
DO2 $[(x_1 < x_2) \wedge (x_2 < x_3)] \rightarrow (x_1 < x_3)$.
DO3 $(x_1 < x_2) \vee (x_2 < x_1) \vee (x_1 = x_2)$.
DO4 $(x_1 < x_2) \rightarrow \exists x_3[(x_1 < x_3) \wedge (x_3 < x_2)]$.
DO5 $\exists x_2(x_2 < x_1)$.
DO6 $\exists x_2(x_1 < x_2)$.

DO1–DO3 state that $<_I$ is a linear order; DO4 is the density property (between any two elements there is another element); DO5 and DO6 state that there is no first or last element. It is obvious that DO is finitely axiomatized. DO has no finite models; $\langle \mathbb{Q}, <_\mathbb{Q} \rangle$ is a countable model of DO (the domain of this model is $\mathbb{Q}$ and $<_\mathbb{Q}$ is the less-than relation on $\mathbb{Q}$); $\langle \mathbb{R}, <_\mathbb{R} \rangle$ is an uncountable model of DO; DO is consistent.

We will give two proofs that DO is decidable. The first uses Turing's result: *If Γ is axiomatized and complete, then Γ is decidable.* This theorem reduces the problem of proving DO decidable to that of proving DO complete, so let us begin by taking a closer look at complete theories.

Theorem 1: Let Γ be a theory with language L. The following are equivalent:

(1) Γ is complete.
(2) Any two models of Γ are elementarily equivalent.
(3) Any two countable models of Γ are elementarily equivalent.

Proof of (1) $\Rightarrow$ (2): Assume Γ is complete and let M and N be models of Γ. Suppose the sentence A is true in M; we are required to show that A is also true in N. By completeness, $\Gamma \vdash A$ or $\Gamma \vdash \neg A$. Suppose $\Gamma \vdash \neg A$. Since M is a model of Γ, $\neg A$ is true in M (soundness theorem). This contradicts the assumption that A is true in M, so we have $\Gamma \vdash A$. Since N is a model of Γ, A is true in N as requried. By a similar argument, every sentence that is true in N is also true in M.

Proof of (3) $\Rightarrow$ (1): Assume that any two countable models of Γ are elementarily equivalent. Suppose, however, that Γ is not complete. Then there is a sentence A of L such that neither A nor $\neg A$ is a theorem of Γ. Since A is not a theorem of Γ, the extension theorem applies and $\Gamma \cup \{\neg A\}$ is consistent. By the model existence theorem, there is a model M of $\Gamma \cup \{\neg A\}$; by the Löwenheim-Skolem theorem, we may assume that M is countable. In summary, M is a countable model of Γ and $\neg A$ is true in M. By a similar argument, with A replaced by $\neg A$, there is a countable model N of Γ such that A is true in N. Now the same sentences are true in M and N (hypothesis); since $\neg A$ is true in M, $\neg A$ is true in N. This contradicts A true in N. ∎

Theorem 2 (Lós-Vaught, special case): Let Γ be a theory such that any two countable models of Γ are isomorphic. Then Γ is complete.

Proof: Use Theorem 1 and Theorem 9 in Section 6.2. ∎

Corollary 1: Let Γ be an axiomatized theory such that any two countable models of Γ are isomorphic. Then Γ is decidable.

Proof: Use Theorem 2 and Turing's result that every axiomatized complete theory is decidable. ∎

By the Lós-Vaught theorem, DO is complete provided that any two countable models of DO are isomorphic. This is a well-known result, due to Cantor.

Theorem 3 (Cantor): Any two countable dense linearly ordered sets without first or last elements are isomorphic.

Proof: Let $\langle X, <_X \rangle$ and $\langle Y, <_Y \rangle$ be two countable dense linearly ordered sets without first or last elements. We are required to construct a one-to-one function $h: X \rightarrow Y$ from X onto Y such that for all $a, b \in X$, $a <_X b$ if and only if $h(a) <_Y h(b)$. Let

$$P = \{f: f \text{ is an isomorphism from a finite subset}$$

$$\text{of } X \text{ onto a finite subset of } Y\}.$$

For example, suppose $a, b \in X$ with $a <_X b$ and $c, d \in Y$ with $c <_Y d$. Define $f: \{a, b\} \rightarrow \{c, d\}$ by $f(a) = c$ and $f(b) = d$; then $f \in P$. (*Notation:* For $f, g \in P$, $f \subseteq g$ means that $\text{dom} f \subseteq \text{dom} g$ and that for all $a \in \text{dom} f$, $f(a) = g(a)$; in other words, g is an *extension* of f.) We now show that P has the *back-and-forth* property:

I Given $f \in P$ and $a \in X$, there exists $g \in P$ such that $f \subseteq g$ and $a \in \text{dom} g$.

II Given $f \in P$ and $b \in Y$, there exists $g \in P$ such that $f \subseteq g$ and $b \in \text{ran} g$.

We will prove II and leave I to the reader. Let $f \in P$ and $b \in Y$ be given. Suppose $\text{dom} f = \{a_1, \ldots, a_n\}$, where the elements are written in increasing order $a_1 <_X \ldots <_X a_n$, and $\text{ran} f = \{b_1, \ldots, b_n\}$, where again the elements are written in increasing order $b_1 <_Y \ldots <_Y b_n$. Since f is an isomorphism, $f(a_k) = b_k$ for $1 \le k \le n$. We now consider four cases. (1) If $b \in \text{ran} f$, take $g = f$. (2) Suppose $b <_Y b_1$. Choose $a \in X$ such that $a <_X a_1$, and define $g: \text{dom} f \cup \{a\} \rightarrow \text{ran} f \cup \{b\}$ by $g(a) = b$ and $g(a_k) = b_k$ for $1 \le k \le n$. Note that we can always choose $a <_X a_1$, since X has no first element. Clearly, $f \subseteq g$ and $b \in \text{ran} g$. (3) Suppose $b_n <_Y b$. Proceed as in (2), except that we now choose $a \in X$ such that $a_n <_X a$; this is possible since x has no last element. (4) Suppose $b_{k-1} <_Y b <_Y b_k$ for some k with $2 \le k \le n$. Choose $a \in X$ such that $a_{k-1} <_X a <_X a_k$ (possible since X is dense) and define g as in (2).

The back-and-forth properties are used to obtain an isomorphism h from X onto Y. Since X and Y are countable, $X = \{a_n: n \ge 1\}$ and $Y = \{b_n: n \ge 1\}$. We construct a sequence $f_1 \subseteq f_2 \subseteq f_3 \ldots$ in P such that for all $n \ge 1$, $\{a_1, \ldots, a_n\} \subseteq \text{dom} f_n$ and $\{b_1, \ldots, b_n\} \subseteq \text{ran} f_n$. To begin, let f_1 be the function with domain $\{a_1\}$ and $f_1(a_1) = b_1$. Next, given $f_1 \in P$ and $a_2 \in X$, use I to obtain $g \in P$ such that $f_1 \subseteq g$ and $a_2 \in \text{dom} g$; then apply II with $g \in P$ and $b_2 \in Y$ to obtain $f_2 \in P$ such that $g \subseteq f_2$ and $b_2 \in \text{ran} f_2$. At this stage, we have $f_1 \subseteq f_2$, $\{a_1, a_2\} \subseteq \text{dom} f_2$, and $\{b_1, b_2\} \subseteq \text{ran} f_2$. Continue this process; in the general case, we construct f_{n+1} from f_n in the same way as f_2 is obtained from f_1. Finally, define $h: X \rightarrow Y$ by

$$h(a) = f_n(a), \quad \text{where} \quad a \in \text{dom} f_n.$$

To complete the proof of Cantor's theorem, we must prove (1)–(4) below; we will prove (3) and leave the others as an exercise for the reader.

(1) h is well-defined (if $a \in \operatorname{dom} f_n$ and $a \in \operatorname{dom} f_m$, then $f_n(a) = f_m(a)$).

(2) $\operatorname{dom} h = X$ (for each $a \in X$, there exists $n \geq 1$ such that $a \in \operatorname{dom} f_n$).

(3) h is one-to-one and onto.

(4) For all $a, a' \in X$, $a <_X a'$ if and only if $h(a) <_Y h(a')$.

Proof of (3): To prove h one-to-one, let $h(a) = h(a')$. By the definition of h, there exist $m, n \in \mathbb{N}^+$ such that $h(a) = f_m(a)$ and $h(a') = f_n(a')$. From $h(a) = h(a')$, we obtain $f_m(a) = f_n(a')$. Suppose $m \leq n$. Then $f_m \subseteq f_n$; hence $f_m(a) = f_n(a)$, so we have $f_n(a) = f_n(a')$. But f_n is an isomorphism, and hence is one-to-one, and $a = a'$ as required. To prove h onto, let $b_n \in Y$. By construction of f_n, we have $b_n \in \operatorname{ran} f_n$; hence there exists $a \in \operatorname{dom} f_n$ such that $f_n(a) = b_n$. So $h(a) = b_n$ as required. ∎

Corollary 2: DO is ω-categorical.

Proof: Let $M = \langle D_M, <_M \rangle$ and $N = \langle D_N, <_N \rangle$ be countably infinite models of DO. Then M and N are countable dense linearly ordered sets without first or last elements; hence, by Cantor's theorem, M and N are isomorphic as required. ∎

Theorem 4: DO is complete and decidable.

Proof: By Corollary 2 and the Lós-Vaught theorem, DO is complete. Since DO is also axiomatized, it is decidable. ∎

We have proved that DO is decidable without actually producing a decision method (we can, of course, print out all proofs). We will now use the method of *elimination of quantifiers* to give a direct proof that DO is decidable. This method of proof, though it is longer than the preceding proof, has the virtue of producing a decision method and moreover does not depend on Cantor's theorem or on the Lós-Vaught theorem.

We will make some assumptions about the language L_{DO}. First of all, we will treat $\wedge$ and $\exists$ as original symbols of the language rather than as defined symbols. Also, in writing a disjunction $B_1 \vee \ldots \vee B_n$, we will not be concerned about the placement of parentheses among the disjuncts or the order of the disjuncts $B_1, \ldots, B_n$. Similar comments apply to a conjunction $B_1 \wedge \ldots \wedge B_n$.

Definition 1: A formula A is *quantifier-free* if it has no quantifiers; it is *elementary* if it is of the form $B_1 \vee \ldots \vee B_k$, where each B_i $(1 \leq i \leq k)$ is a conjunction of atomic formulas.

Note that a disjunction of elementary formulas is again an elementary formula, and that an elementary formula has no occurrences of $\neg$. Every elementary formula is quantifier-free; on the other hand, $(x < y) \vee [\neg(x =$

$y) \wedge (y < x)]$ is a quantifier-free formula of L_{DO} that is not elementary. The next three lemmas give algorithms for simplifying formulas in the theory DO.

Lemma 1 (Elimination of ¬): For each quantifier-free formula A of L_{DO}, there is an elementary formula B of L_{DO} with the same variables as A such that $DO \vdash A \leftrightarrow B$.

Proof: We first consider the special case in which A is of the form $B_1 \wedge \ldots \wedge B_k$, where each B_i is an atomic formula or is the negation of an atomic formula. The basic idea is to use

(1) $DO \vdash \neg(x = y) \leftrightarrow [(x < y) \vee (y < x)]$
(2) $DO \vdash \neg(x < y) \leftrightarrow [(x = y) \vee (y < x)]$

to eliminate ¬ (see Exercise 4). We proceed somewhat informally. Suppose B_1 is $\neg(x = y)$. By (1) and the replacement theorem, we have

$$DO \vdash A \leftrightarrow [((x < y) \vee (y < x)) \wedge B_2 \wedge \ldots \wedge B_k].$$

By the tautology $[(p \vee q) \wedge r] \leftrightarrow [(p \wedge r) \vee (q \wedge r)]$, we then obtain

$$DO \vdash A \leftrightarrow [((x < y) \wedge B_2 \wedge \ldots \wedge B_k) \vee ((y < x) \wedge B_2 \wedge \ldots \wedge B_k)].$$

Repeat this process with $B_2, \ldots, B_k$, in each case using (1) or (2) to eliminate ¬ whenever necessary. We eventually obtain an elementary formula B with the same variables as A such that $DO \vdash A \leftrightarrow B$.

We now turn to the general case. Let A be a quantifier-free formula. Then there is a formula C in disjunctive normal form such that $\vdash A \leftrightarrow C$; here C is of the form $C_1 \vee \ldots \vee C_n$, where each C_i $(1 \leq i \leq n)$ is a conjunction of formulas, each of which is atomic or is the negation of an atomic formula. The method of constructing C is discussed in Section 2.3; to obtain the first-order version, we treat atomic formulas of A as propositional variables.

At this point, we have $\vdash A \leftrightarrow (C_1 \vee \ldots \vee C_n)$. In addition, the special case applies to each C_i. For $1 \leq i \leq n$, let B_i be an elementary formula such that $DO \vdash C_i \leftrightarrow B_i$. By the replacement theorem, $DO \vdash A \leftrightarrow (B_1 \vee \ldots \vee B_n)$, and moreover the required elementary formula for A is $B_1 \vee \ldots \vee B_n$. ∎

Lemma 2 (Elimination of ∃): Let A be a quantifier-free formula of L_{DO} and let x be a variable distinct from x_1. Then there is a quantifier-free formula A^* of L_{DO} such that

(1) $DO \vdash \exists x A \leftrightarrow A^*$.
(2) The variable x does not occur in A^*.
(3) {variables in A^*} $\subseteq$ {variables in A} $\cup \{x_1\}$.

Proof: We first prove the lemma for the special case in which A has the form $B_1 \wedge \ldots \wedge B_n$, where each B_i is atomic. The possible atomic formulas are $x = x$, $x < x$, $x = y$, $y = x$, $x < y$, $y < x$, $y = z$, and $y < z$, where y and z are variables distinct from x but not necessarily distinct from one another. We will use the following results of first-order logic (in (a), y is substitutable for x in B):

(a) $\vdash \exists x((x = y) \wedge B) \leftrightarrow B_x[y]; \quad \vdash \exists x((y = x) \wedge B) \leftrightarrow B_x[y].$

(b) $\vdash \exists x((x = x) \wedge B) \leftrightarrow \exists x B.$

(c) $\vdash \exists x(B \wedge C) \leftrightarrow (B \wedge \exists x C)$, where x does not occur free in B.

We will now consider various cases.

- $x = y \ or \ y = x$ First, suppose A is $x = y$. In first-order logic, we have $\vdash \exists x(x = y) \leftrightarrow (x_1 = x_1)$; hence the required formula A^* is $x_1 = x_1$. Now suppose A has the form $(x = y) \wedge B$ (more precisely, $(x = y) \wedge B_2 \wedge \ldots \wedge B_n$). In this case, (a) shows that the required formula A^* is $B_x[y]$. Similar arguments apply to $y = x$.

- $x < x$ First, suppose A is $(x < x)$. Now DO $\vdash \exists x(x < x) \leftrightarrow \neg(x_1 = x_1)$, so in this case the required formula A^* is $\neg(x_1 = x_1)$. If A has the form $(x < x) \wedge B$, then DO $\vdash \exists x((x < x) \wedge B) \leftrightarrow \neg(x_1 = x_1)$, and again the required formula A^* is $\neg(x_1 = x_1)$.

- $x = x$ First, suppose A is $x = x$. Now $\vdash \exists x(x = x) \leftrightarrow (x_1 = x_1)$; hence the required formula A^* in this case is $x_1 = x_1$. Suppose A has the form $(x = x) \wedge B$. By (b), we have $\vdash \exists x((x = x) \wedge B) \leftrightarrow \exists x B$, so if we find a formula B^* for B, then the required formula for A is also B^*. (This shows that we can assume that $x = x$ does not occur in A.)

- $y = z \ or \ y < z$ Suppose A is $y = z$. Since $\vdash \exists x(y = z) \leftrightarrow (y = z)$, the required formula is $y = z$. Suppose A has the form $(y = z) \wedge B$. By (c), we have $\vdash \exists x A \leftrightarrow (y = z) \wedge \exists x B$, so if we find a formula B^* for B, then the required formula for A is $(y = z) \wedge B^*$. Similar arguments apply to $y < z$.

- *Remaining cases* Three possibilities for A remain. These are

  (1) $(x < y_1) \wedge \ldots \wedge (x < y_n)$.

  (2) $(y_1 < x) \wedge \ldots \wedge (y_n < x)$.

  (3) $(y_1 < x) \wedge \ldots \wedge (y_k < x) \wedge (x < z_1) \wedge \ldots \wedge (x < z_j)$.

  From DO5, we obtain DO $\vdash \exists x[(x < y_1) \wedge \ldots \wedge (x < y_n)]$; hence the required formula A^* in case (1) is $x_1 = x_1$. Likewise, the required formula in case (2) is $x_1 = x_1$; DO6 is used here. Finally, with the help of DO4, we can show that

  $$\text{DO} \vdash \exists x[(y_1 < x) \wedge \ldots \wedge (y_k < x) \wedge (x < z_1) \wedge \ldots \wedge (x < z_j)] \leftrightarrow A^*,$$

  where A^* is $(y_1 < z_1) \wedge \ldots \wedge (y_1 < z_j) \wedge \ldots \wedge (y_k < z_1) \wedge \ldots \wedge (y_k < z_j)$. This covers all cases for A, and the proof of the special case is complete.

We now turn to the general case in which A is a quantifier-free formula. By Lemma 1, we can assume that A is elementary. Thus A has the form $A_1 \vee \ldots \vee A_k$, where each A_i is a conjunction of atomic formulas. By first-order logic, we have $\vdash \exists x A \leftrightarrow (\exists x A_1 \vee \ldots \vee \exists x A_k)$. The special case applies to each of the formulas $A_1, \ldots, A_k$; hence there are quantifier-free formulas $A_1^*, \ldots, A_k^*$ satisfying (1)–(3) of the lemma. In particular, DO $\vdash \exists x A_i \leftrightarrow A_i^*$ for $1 \leq i \leq k$. By the replacement theorem, we obtain

DO $\vdash \exists x A \leftrightarrow (A_1^* \vee \ldots \vee A_k^*)$ and $A_1^* \vee \ldots \vee A_k^*$ is the required formula for A. ∎

Lemma 3 (Elimination of ∀): Let A be a quantifier-free formula of L_{DO} and let x be a variable distinct from x_1. Then there is a quantifier-free formula A^* of L_{DO} such that

(1) DO $\vdash \forall x A \leftrightarrow A^*$.

(2) The variable x does not occur in A^*.

(3) {variables in A^*} $\subseteq$ {variables in A} $\cup \{x_1\}$.

Proof: Apply Lemma 2 to $\neg A$; there is a formula A^* such that DO $\vdash \exists x \neg A \leftrightarrow A^*$ and (2) and (3) hold. From this, we obtain DO $\vdash \forall x A \leftrightarrow \neg A^*$; hence the required formula for $\forall x A$ is $\neg A^*$. ∎

Theorem 5: DO is decidable.

Proof: We first observe that the proofs of Lemmas 1 through 3 describe algorithms for constructing the formulas said to exist in the conclusion. We construct an algorithm that, given a formula A of L_{DO}, decides whether DO $\vdash A$. By the closure theorem for provability, we can assume that the given formula A is a sentence. By the results in Section 5.6, on prenex form,

$$\vdash A \leftrightarrow Qy_1 \ldots Qy_k B,$$

where B is a quantifier-free formula. The algorithm for prenex form is such that the only variables in B are $y_1, \ldots, y_k$ (we are starting with a sentence A). Moreover, by the theorem on change of bound variables, we can assume that x_1 is distinct from $y_1, \ldots, y_k$.

Apply Lemma 2 or 3 to $Qy_k B$. There is a quantifier-free formula A_k^* such that DO $\vdash Qy_k B \leftrightarrow A_k^*$ and the variables in A_k^* are among $y_1, \ldots, y_{k-1}, x_1$. By the replacement theorem, DO $\vdash A \leftrightarrow Qy_1 \ldots Qy_{k-1} A_k^*$. Repeat this process with $Qy_{k-1} A_k^*$; after a total of k steps, we obtain a quantifier-free formula A^* such that DO $\vdash A \leftrightarrow A^*$ and the only variable in A^* is x_1. Now A^* is a quantifier-free formula, so by Lemma 1 we can assume that A^* is elementary. Thus A^* is of the form $C_1 \vee \ldots \vee C_k$, where each C_i is a conjunction, and each conjunct is either $x_1 = x_1$ or $x_1 < x_1$. Now DO $\vdash A$ if and only if DO $\vdash A^*$; hence it suffices to describe a decision method for formulas of the form of A^*. We have

DO $\vdash A^* \Leftrightarrow$ for some i, $1 \leq i \leq k$, each atomic formula in C_i is $x_1 = x_1$.

To justify this, assume DO $\vdash A^*$. Now $\langle \mathbb{Q}, <_{\mathbb{Q}} \rangle$ is a model of DO; hence the soundness theorem applies and A^* is true in $\langle \mathbb{Q}, <_{\mathbb{Q}} \rangle$. Let $\phi : \text{VAR} \to \mathbb{Q}$ be an assignment for this interpretation. Since $\phi(A^*) = T$, there is some i, $1 \leq i \leq k$, such that $\phi(C_i) = T$. Let $C_i = D_1 \wedge \ldots \wedge D_n$. Clearly, $\phi(D_j) = T$ for $1 \leq j \leq n$. But $\phi(x_1 < x_1) = F$, so no D_j is $x_1 < x_1$; in other words, each atomic formula in C_i must be $x_1 = x_1$ as required. We leave the easy proof of the other direction to the reader. ∎

By Church's theorem, any consistent extension of PA is undecidable (see Section 7.1). But certain reducts of arithmetic are decidable. In particular, if we simplify by omitting multiplication, or multiplication and addition, we obtain decidable theories. We will state, without proofs, the precise results.

Successor Arithmetic SA: The language L_{SA} has constant symbol 0 and 1-ary function symbol S. The standard interpretation of L_{SA} is $\langle \mathbb{N}, 0_{\mathbb{N}}, S_{\mathbb{N}} \rangle$. Successor arithmetic is defined semantically as follows: SA is the set of all sentences of L_{SA} that are true in $\langle \mathbb{N}, 0_{\mathbb{N}}, S_{\mathbb{N}} \rangle$ (i.e., SA is Th$(\langle \mathbb{N}, O_{\mathbb{N}}, S_{\mathbb{N}} \rangle)$). Exercises 7 and 8 outline a proof that successor arithmetic is decidable.

Presburger Arithmetic PrA: The language L_{PrA} has constant symbol 0, 1-ary function symbol S, and 2-ary function symbol $+$; the standard interpretation of L_{PrA} is $\langle \mathbb{N}, 0_{\mathbb{N}}, S_{\mathbb{N}}, +_{\mathbb{N}} \rangle$. PrA is defined semantically as follows: PrA is the set of sentences of L_{PrA} that are true in $\langle \mathbb{N}, 0_{\mathbb{N}}, S_{\mathbb{N}}, +_{\mathbb{N}} \rangle$. Presburger arithmetic is decidable (see Enderton 1972 or Cohen 1966).

We will end this section with a short list of decidable theories. For more information and further examples, see Monk, *Mathematical Logic* 1976 and the survey article "Decidable Theories" by Rubin in *Handbook of Mathematical Logic* 1977.

Decidable Theories:

(1) Presburger Arithmetic (Presburger, 1929)

(2) Theory DO (Langford, 1927)

(3) Theory of real numbers, denoted Th$(\mathscr{R})$ (Tarski, 1949)

(4) Theory of complex numbers, denoted Th$(\mathscr{C})$ (Tarski, 1949)

(5) Theory of abelian groups (Szmielew, 1949)

—————————Exercises for Section 6.3—————————

1. Show that $\langle \mathbb{Q}, <_{\mathbb{Q}} \rangle$ and $\langle \mathbb{R}, <_{\mathbb{R}} \rangle$ are models of DO that are elementarily equivalent but not isomorphic.
2. Give an example of a theory that is (1) complete but not ω-categorical; (2) complete and ω-categorical but not categorical.
3. Complete the proof of Cantor's theorem by proving (1), (2), and (4).
4. The theory LO of linearly ordered sets has language L_{DO} and nonlogical axioms DO1–DO3. Give formal proofs for the following:
   (1) LO $\vdash \forall x_1 \forall x_2 \neg[(x_1 < x_2) \wedge (x_2 < x_1)]$.
   (2) LO $\vdash \neg(x = y) \leftrightarrow [(x < y) \vee (y < x)]$.
   (3) LO $\vdash \neg(x < y) \leftrightarrow [(x = y) \vee (y < x)]$.
5. Give formal proofs for the following (y, z distinct from x):
   (1) $\vdash \exists x(x = y) \leftrightarrow (x_1 = x_1)$.
   (2) $\vdash \exists x(y = z) \leftrightarrow (y = z)$.
   (3) $\vdash \exists x((x = x) \wedge B) \leftrightarrow \exists x B$.
   (4) $\vdash \exists x((x = y) \wedge B) \leftrightarrow B_x[y]$. ($y$ substitutable for x in B)
6. Give formal proofs for the following:
   (1) DO $\vdash \exists x((x < x) \wedge B) \leftrightarrow \neg(x_1 = x_1)$.

(2) $DO \vdash \exists x[(y < x) \wedge (z < x)] \leftrightarrow (x_1 = x_1)$.

(3) $DO \vdash \exists x[(y_1 < x) \wedge (y_2 < x) \wedge (x < z_1) \wedge (x < z_2)]$
$\leftrightarrow [(y_1 < z_1) \wedge (y_1 < z_2) \wedge (y_2 < z_1) \wedge (y_2 < z_2)]$.

7. This exercise outlines the first step of the proof that successor arithmetic SA is decidable. Let A be a quantifier-free formula and x a variable. Show that there is a quantifier-free formula A^* such that $SA \vdash \exists xA \leftrightarrow A^*$, x does not occur in A^*, and the variables in A^* are among the variables in A. *Hints:* Assume that A is in DNF, say $A_1 \vee \ldots \vee A_n$. Since $\vdash \exists xA \leftrightarrow \exists xA_1 \vee \ldots \vee \exists xA_n$, it suffices to consider the case in which A is $B_1 \wedge \ldots \wedge B_k$, where each B_i is atomic or is the negation of an atomic formula. *Notation:* $y^1 = y$, $y^{k+1} = S(y^k)$, where y is a variable, possibly x; 0^k and y^k are the terms of SA. Prove the following:

(1) We can assume that x occurs at least once in each B_i. (*Hint:* If x does not occur in B, $\vdash \exists x(B \wedge C) \leftrightarrow (B \wedge \exists xC)$.)

(2) We can assume that x occurs exactly once in each B_i. (*Hint:* $(x^k = x^n)$ can be replaced with $0 = 0$ if $k = n$, and with $\neg(0 = 0)$ if $k \neq n$.)

(3) We can assume that the only atomic formulas that occur in A are of the form $(x^k = t)$. (*Note:* t, u denote terms of the form 0^n or y^n ($y \neq x$).)

(4) If each B_i is of the form $\neg(x^k = t)$, then A^* is $(0 = 0)$.

(5) Suppose B_1 is $(x^k = t)$. Let C_1 be $\neg(t = 0) \wedge \ldots \wedge \neg(t = 0^{k-1})$; this eliminates x from B_1. Fix $i \geq 2$ and suppose B_i is $x^m = u$ or $\neg(x^m = u)$; $(x^m = u) \leftrightarrow (x^{m+k} = u^k)$ is true in $\langle \mathbb{N}, 0_\mathbb{N}, S_\mathbb{N} \rangle$, so we can replace $x^m = u$ in B_i with $x^{m+k} = u^k$, and B_i becomes $x^{m+k} = u^k$ or $\neg(x^{m+k} = u^k)$. The formula $(x^k = t) \to [(x^{m+k} = u^k) \leftrightarrow (t^m = u^k)]$ is logically valid. Let C_i be the result of replacing $x^{m+k} = u^k$ in B_i with $t^m = u^k$; this eliminates x from B_i. Then $SA \vdash \exists xA \leftrightarrow C_1 \vee \ldots \vee C_k$, and A^* is $C_1 \vee \ldots \vee C_k$.

8. (*Decidability of successor arithmetic*) This exercise outlines a proof that SA is decidable. Prove the following:

(1) Given a quantifier-free formula A, there is a quantifier-free formula A^* such that $SA \vdash \exists xA \leftrightarrow A^*$, x does not occur in A^*, and the variables in A^* are among the variables in A. (*Hint:* Exercise 7.)

(2) Part (1) holds when $SA \vdash \exists xA \leftrightarrow A^*$ is replaced with $SA \vdash \forall xA \leftrightarrow A^*$.

(3) There is an algorithm for deciding whether a sentence of L_{SA} is a theorem of SA.

6.4

Zermelo-Frankel Set Theory

Plane geometry is the study of points and lines; arithmetic is the study of the natural numbers. Each of these important branches of mathematics can be described by a list of axioms. First Euclid, then Hilbert, gave axioms for Euclidean geometry; Dedekind and Peano gave axioms for arithmetic.

Set theory is the study of sets, with a special emphasis on infinite sets; indeed, set theory is often described as the mathematical study of infinity. Set theory is the creation of Georg Cantor (1845–1918), and to Cantor we are indebted for such fundamental results as these: $\mathbb{N} \approx \mathbb{Q}$; $P(\mathbb{N})$ is uncountable. However, Cantor did not axiomatize his theory of sets; we shall refer to intuitive, nonaxiomatic set theory as *Cantorian set theory*. Hilbert, in his essay "On the Infinite" (1925), had this to say about Cantor's work: "This theory is, I think, the finest product of mathematical genius and one of the supreme achievements of purely intellectual human activity."

By about 1900, it was known that certain constructions in Cantorian set theory lead to a contradiction; the best-known of these is Russell's paradox. Mathematicians realized that Cantorian set theory was too loosely described and that a careful axiomization was called for. A list of axioms was given by Zermelo in 1908, with additions and clarifications by Frankel and Skolem in 1922 and by von Neumann in 1925. Today, the mathematical theory described by these axioms is called *Zermelo-Frankel set theory*, or *ZF set theory* for short. We emphasize that ZF set theory is a classical axiom system in the sense that its purpose is to describe Cantorian set theory. In this respect, it is similar to Euclid's axioms, which describe plane geometry, and the Peano postulates, which describe arithmetic.

Let us take a closer look at the source of Russell's paradox. The following construction has intuitive appeal.

Axiom of (Unrestricted) Comprehension: Given a property Q of sets, there is a set consisting of all sets that have property Q; in other words, $\{x: x \text{ has property } Q\}$ is a set.

This axiom was first formalized by Frege; Cantor also used comprehension in his work with sets. With this axiom, little else is needed, and we can construct a wide variety of sets such as these:

Null set: $\varnothing = \{x: x \neq x\}$.

Pair: $\{x, y\} = \{z: z = x \vee z = y\}$.

Intersection: $x \cap y = \{z: z \in x \wedge z \in y\}$.

Union: $\bigcup x = \{z: \exists y(z \in y \wedge y \in x)\}$.

Power set: $P(x) = \{y: y \subseteq x\}$.

Unfortunately, the axiom of comprehension leads to a contradiction, as noted by Russell in 1902. Let $A = \{x: x \notin x\}$; both $A \in A$ and $A \notin A$ lead to a contradiction. Speaking informally, the difficulty with A is that it is "too large." Indeed, if one accepts the intuitively appealing fact that $x \notin x$ always holds, then A actually consists of all sets. One way to avoid the construction of such "large sets" is to restrict comprehension so that it only applies to sets that have already been constructed. But to compensate for this restriction on the power of comprehension, we must add additional axioms for the construction of new sets.

We proceed as follows: First we summarize the axioms of ZF set theory in a short, concise list. We then discuss each axiom in detail, with an emphasis on the purpose of that axiom. Finally, we state the axioms in the language L_{ST} of set theory. This gives rise to a formal system, which we denote ZF and whose intended interpretation, or intuitive content, is Cantorian set theory.

In our axiomatic treatment of set theory, *set* and *is an element of* (or *membership*) are the undefined concepts; informally speaking, the Zermelo-Frankel axioms give meaning to these undefined concepts. We use the letters x, y, z, u, v, a, and b to denote sets. To facilitate the statement of the power set axiom, we introduce a defined concept as follows: y is a *subset* of x, written $y \subseteq x$, if every set that is an element of y is also an element of x (i.e., $y \subseteq x \Leftrightarrow$ for every set $z, z \in y \Rightarrow z \in x$).

Zermelo-Frankel Axioms for Cantorian Set Theory:

ZF1 (*Extensionality*) If two sets x and y have the same elements, then $x = y$.

ZF2 (*Separation*) If x is a set and Q is a property of sets, then $\{z : z \in x \land z$ has property $Q\}$ is a set.

ZF3 (*Null set*) There is a set $\varnothing$ that has no elements.

ZF4 (*Pair*) For any two sets x and y, there is a set $\{x, y\}$ whose only elements are x and y.

ZF5 (*Union*) For any set x, there is a set $\bigcup x$ consisting of all sets that are the elements of the elements of x.

ZF6 (*Infinity*) There is an infinite set.

ZF7 (*Power set*) For any set x, there is a set $P(x)$ whose elements are the subsets of x.

ZF8 (*Replacement*) Let x be a set and let F be a functional (if a, u, v are sets and $F(a) = u$ and $F(a) = v$, then $u = v$). Then $\{F(y) : y \in x\}$ is a set.

ZF9 (*Regularity*) For any nonempty set x, there is a set $y \in x$ such that x and y have no elements in common.

We emphasize that the axioms of pairing, union, power, and replacement assert the existence of sets $\{x, y\}$, $\bigcup x$, $P(x)$, and $\{F(y): y \in x\}$ such that for every set z,

$$z \in \{x, y\} \Leftrightarrow z = x \lor z = y.$$
$$z \in \bigcup x \Leftrightarrow \text{there exists } y \in x \text{ such that } z \in y.$$
$$z \in P(x) \Leftrightarrow z \subseteq x.$$
$$z \in \{F(y): y \in x\} \Leftrightarrow \text{there exists } y \in x \text{ such that } z = F(y).$$

The Zermelo-Frankel axioms fall into one of three categories, as follows: They guarantee the existence of specific sets (null set, infinity); they construct a new set from a given set or sets (separation, pair, union, power, replacement); or they limit the scope of sets (extensionality, regularity). Moreover, the axioms are selected on the basis of these criteria: Each axiom is an intuitively obvious true statement about sets (regularity is a possible exception); there are enough axioms to construct the sets commonly used in mathematics and thus to provide a foundation for mathematics; the axioms avoid known contradictions such as Russell's paradox.

Zermelo realized that sets that lead to a contradiction are "large" sets such as those in Russell's paradox. His idea for avoiding these "large" sets is to limit the scope of the axioms so that only "small" sets are constructed. In particular, the following principle is used in the construction of sets: *Sets are constructed in stages.* Thus we begin with the simplest set, namely the empty set, and then use basic constructions such as pairing, union, and power set to obtain new sets. This process continues indefinitely. If all goes well, we should be able to construct all the sets we need and at the same time to avoid the construction of "large" sets that lead to a contradiction. For a further discussion of this point of view, see J. R. Shoenfield, "The Axioms of Set Theory", in the *Handbook of Mathematical Logic*, Jon Baraise, ed. (Amsterdam: North Holland, 1977).

We will now turn to a discussion of the axioms themselves. Axiom ZF1, the axiom of extensionality, is used to establish the precise relationship between the equality relation = and the membership relation ∈. The following formula of L_{ST} is logically valid (here x, y, z denote distinct individual variables):

$$(x = y) \rightarrow \forall z[z \in x \leftrightarrow z \in y].$$

Informally, this formula states that two sets that are equal have the same elements. Extensionality is the *converse* of this statement, namely

$$\forall z[z \in x \leftrightarrow z \in y] \rightarrow (x = y).$$

This converse, which states that two sets having the same elements are equal, is justified as follows: Suppose two sets x and y have the same elements. Since membership between sets is the only relation available, we cannot expect to distinguish between the two sets x and y, so we declare the two sets to be the same. We note that extensionality can be stated as follows: If $x \subseteq y$ and $y \subseteq x$, then $x = y$. Extensionality is a limiting axiom; it does not enable us to construct new sets, but rather limits the variety of sets that we can construct. Extensionality is used in the following ways: (1) Other axiom(s) of ZF set theory prove the *existence* of a set x; extensionality is used to prove the *uniqueness* of that set. (2) Two sets, such as $x \cap (y \cup z)$ and $(x \cap y) \cup (x \cap z)$, are constructed; extensionality is then used to prove that they are the same set. Thus, while the descriptions of these two sets are different (i.e., have different *intensions*), they have the same elements (i.e., have the same *extension*). The axiom of extensionality allows us to disregard differences between the descriptions of a set and instead focus exclusively on the elements in the set.

Axiom ZF2, the axiom of separation, is a restricted version of comprehension; with this axiom we can separate off, from a given set x, a subset y consisting of those elements of x that have some property Q. The somewhat vague concept of a property of sets can be made precise by stating the axiom of separation in the language L_{ST} of set theory. Note the distinction between comprehension and separation: Comprehension can yield a set that is "too large," whereas separation begins with a set x and selects a subset y of x. We illustrate the use of separation by proving the existence of $x \cap y$ (the *intersection of x and y*) and $x - y$ (the *difference of x and y*).

Lemma 1: Given sets x and y, there exist sets $x \cap y$ and $x - y$ such that for every set z,

(1) $z \in x \cap y \Leftrightarrow z \in x \wedge z \in y$.
(2) $z \in x - y \Leftrightarrow z \in x \wedge z \notin y$.

Proof: Both sets exist by separation; for example, $x \cap y = \{z : z \in x \wedge z \in y\}$; in other words, $x \cap y$ consists of all sets $z \in x$ that satisfy the property $z \in y$. ∎

We cannot duplicate Russell's paradox with separation. To see this, let x be a set, and suppose we use separation to construct a set y by $y =$

$\{z: z \in x \wedge z \notin z\}$. Consider two cases: First, suppose $y \in y$. Then $y \in x$ and $y \notin y$, a contradiction (as in the case of Russell's paradox). Next, suppose $y \notin y$. Then either $y \notin x$ or $y \in y$. Now $y \in y$ leads to a contradiction, and we are simply left with the conclusion that $y \notin x$.

With separation, we can use Russell's construction to our advantage and prove the following result.

Lemma 2: There is no universal set.

Proof: Suppose there is a set V such that $x \in V$ for every set x. Let $y = \{x: x \in V \wedge x \notin x\}$. Under the assumption that V is a set, separation applies and y is a set. As usual, consider two cases. Suppose $y \in y$. Then $y \in V$ and $y \notin y$, a contradiction. Next, suppose $y \notin y$. Then $y \notin V$ or $y \in y$. Now y is a set (by separation), and every set is an element of V, so $y \in V$. It follows that $y \in y$, a contradiction. Thus V is not a set, as required. ∎

To construct new sets, we need an old set to begin with. Axiom ZF3, the null set axiom, asserts the existence of a set that has no elements; this set is denoted $\varnothing$. We could replace the null set axiom with an axiom that simply states that there exists at least one set. For suppose there is a set x. By separation, we obtain the empty set by $\varnothing = \{y: y \in x \wedge y \neq y\}$. The axiom of extensionality can be used to prove the uniqueness of $\varnothing$.

Axiom ZF4, the axiom of pairing, enables us to construct a variety of sets with one or two elements. First of all, we have the following construction for one-element sets.

Lemma 3: For every set x, there is a set $\{x\}$ such that for every set z, $z \in \{x\} \Leftrightarrow z = x$.

Proof: Apply the pairing axiom to x, x: There is a set $\{x, x\}$ such that for all z,

$$z \in \{x, x\} \Leftrightarrow z = x \vee z = x.$$

Thus $\{x, x\}$, which we shorten to $\{x\}$, is the required set. ∎

The set $\{x\}$ is called the *singleton of x*. Here are some examples of one-element and two-element sets: $\{\varnothing\}$, $\{\{\varnothing\}\}$, $\{\{\{\varnothing\}\}\}, \ldots$; $\{\varnothing, \{\varnothing\}\}$, $\{\{\varnothing\}, \{\varnothing, \{\varnothing\}\}\}$. Another application of the pairing axiom is the existence of ordered pairs.

Definition 1: Let x and y be sets. The *ordered pair of x and y*, denoted $\langle x, y \rangle$, is defined by $\langle x, y \rangle = \{\{x\}, \{x, y\}\}$.

Several applications of pairing prove the existence of the set $\langle x, y \rangle$. This clever definition allows us to prove the next lemma, which provides the fundamental property we require of an ordered pair (a proof is outlined in the exercises).

Lemma 4: If $\langle a, b \rangle = \langle x, y \rangle$, then $a = x$ and $b = y$.

All of the sets constructed so far have at most two elements. Axiom ZF5, the union axiom, will enable us to construct larger sets. This axiom can be described informally as follows: Let x be a set with elements $y, u, v, \ldots$; the elements of $\bigcup x$ are precisely the sets that are the elements of $y, u, v, \ldots$. More precisely, $z \in \bigcup x \Leftrightarrow$ there exists $y \in x$ such that $z \in y$. As an application of union, we construct $x \cup y$, the *union of x and y*, and prove the existence of sets with n elements.

Lemma 5: Given sets x and y, there is a set $x \cup y$ such that for every set z,

$$z \in x \cup y \Leftrightarrow z \in x \vee z \in y.$$

Proof: Let $x \cup y = \bigcup \{x, y\}$; this set exists by the union and pairing axioms. For every set z,

$$z \in x \cup y \Leftrightarrow z \in \bigcup \{x, y\}$$
$$\Leftrightarrow \text{there exists } u \in \{x, y\} \text{ such that } z \in u$$
$$\Leftrightarrow (u = x \vee u = y) \wedge z \in u$$
$$\Leftrightarrow z \in x \vee z \in y. \quad \blacksquare$$

Lemma 6: Let $x_1, \ldots, x_n$ be sets. Then there is a set $\{x_1, \ldots, x_n\}$ such that for every set z,

$$z \in \{x_1, \ldots, x_n\} \Leftrightarrow (z = x_1) \vee \ldots \vee (z = x_n).$$

Proof: By induction. For $n = 1$, the result holds by Lemma 3. Let $n \geq 1$, and assume that the result holds for n sets. Given $n + 1$ sets $x_1, \ldots, x_n, x_{n+1}$, let $\{x_1, \ldots, x_n, x_{n+1}\} = \{x_1, \ldots, x_n\} \cup \{x_{n+1}\}$. This set exists by Lemma 5. Moreover, for any set z,

$$z \in \{x_1, \ldots, x_n, x_{n+1}\} \Leftrightarrow z \in \{x_1, \ldots, x_n\} \cup \{x_{n+1}\}$$
$$\Leftrightarrow z \in \{x_1, \ldots, x_n\} \vee z \in \{x_{n+1}\}$$
$$\Leftrightarrow (z = x_1) \vee \ldots \vee (z = x_n) \vee (z = x_{n+1}). \quad \blacksquare$$

Eventually, we would like to show that the natural numbers can be constructed from the Zermelo-Frankel axioms. More precisely, we want to construct sets that we can identify with the natural numbers and such that the Peano postulates P1–P3 hold. As a guiding principle, let us agree that the set identified with natural number n should have exactly n elements. With this in mind, we introduce definitions as follows:

(0) The natural number 0 is the set $\varnothing$.

(1) The natural number 1 is the set $\{0\}$.

(2) The natural number 2 is the set $\{0, 1\}$.

(3) The natural number 3 is the set $\{0, 1, 2\}$.

There is an operator S (called successor) that allows us to systematically continue this list. An informal description of S is as follows: Given a set x, $S(x)$ is the set whose elements are x together with the elements of x. For example, S applied to the set 3 $(=\{0, 1, 2\})$ gives $S(3) = \{0, 1, 2, 3\}$. With the operator S available, we can summarize the construction of the natural numbers $0, 1, 2, \ldots$ as follows: $0 = \varnothing$; $n + 1 = S(n)$.

Lemma 7: For every set x, there is a set $S(x)$ such that for every set z,

$$z \in S(x) \Leftrightarrow z \in x \lor z = x.$$

Proof: Let $S(x) = x \cup \{x\}$; $S(x)$ exists by pairing and union. For any set z,

$$z \in S(x) \Leftrightarrow z \in x \cup \{x\} \Leftrightarrow z \in x \lor z \in \{x\} \Leftrightarrow z \in x \lor z = x. \quad \blacksquare$$

To summarize: Starting with the empty set, and using union and pairing, we have been able to construct a wide variety of finite sets, including the natural numbers $0, 1, 2, \ldots$. However, these axioms do not enable us to construct an infinite set. In particular, we have no axiom that allows us to collect together the natural numbers $0, 1, 2, \ldots$ into a single (infinite) set. It is here that axiom ZF6, the axiom of infinity, comes in. In its present form this axiom is rather vague, so the first order of business is to restate it.

Definition 2: A set x is called a *successor set* if it satisfies these two conditions: (1) $\varnothing \in x$; (2) if $y \in x$, then $S(y) \in x$.

We now state ZF6 as follows: *There exists a successor set.* This axiom is absolutely fundamental; from the remaining axioms of ZF set theory, we cannot prove the existence of infinite sets. Since set theory is the mathematical study of infinity, this axiom, or its equivalent, is essential.

Let us use separation and infinity to construct the set ω of natural numbers. By the axiom of infinity, there is a successor set z. Thus $0 \in z$; if $n \in z$, then $n + 1 \in z$; or, informally, every natural number belongs to z. Unfortunately, this successor set z may include sets other than the natural numbers. To overcome this difficulty, we construct the "smallest" successor set and then *define* the set of natural numbers to be this set.

Lemma 8: There is a unique set ω such that (1) ω is a successor set; (2) ω is a subset of every successor set.

Proof: By the axiom of infinity, there is a successor set, say z. Let

$$\omega = \{y : y \in z \land y \in b \text{ for every successor set } b\}.$$

This set exists by separation applied to z. We prove the uniqueness of ω and leave the proof of (1) and (2) as exercises for the reader. Assume ω satisfies (1) and (2), and let ω' be a set that satisfies (1)' ω' is a successor set, and (2)' ω' is a subset of every successor set. To prove that $\omega = \omega'$, it suffices to show that $\omega \subseteq \omega'$ and $\omega' \subseteq \omega$. By (1) and (2)', $\omega' \subseteq \omega$; by (1)' and (2), $\omega \subseteq \omega'$. ■

In axiomatic set theory, it is customary to denote the set of natural numbers by ω rather than $\mathbb{N}$; we will do this for the remainder of this section. Later we will prove that the set ω and the operator S satisfy the Peano postulates P1–P3 (see Section 1.6).

We now turn to ZF7, the power set axiom. A major application of this axiom is to construct arbitrarily large infinite sets. Once the infinite set ω is available, we can construct larger infinite sets $P(\omega)$, $P(P(\omega))$, and so on (by a Cantor diagonal argument, x is never equipotent to $P(x)$; see Section 1.4). The power set axiom can also be used to construct the Cartesian product of two sets. But first let us prove the following lemma.

Lemma 9: Let $a, b, x,$ and y be sets with $a \in x$ and $b \in y$. Then $\langle a, b \rangle \in P(P(x \cup y))$.

Proof: First note that $u \in P(v) \Leftrightarrow u \subseteq v$. With this observation in mind, we are required to show that $\{\{a\}, \{a, b\}\} \subseteq P(x \cup y)$; that is, (1) $\{a\} \in P(x \cup y)$; and (2) $\{a, b\} \in P(x \cup y)$. To prove (2), it suffices to show that $\{a, b\} \subseteq x \cup y$; this is an easy consequence of $a \in x$ and $b \in y$. ■

Definition 3: Let x and y be sets. The *Cartesian product of x and y,* denoted $x \times y$, is defined by

$$x \times y = \{z : z \in P(P(x \cup y)) \wedge \exists a \exists b [a \in x \wedge b \in y \wedge (z = \langle a, b \rangle)]\}.$$

Note that the axiom of separation is used with the set $P(P(x \cup y))$.

The axioms we have discussed thus far were stated by Zermelo in 1908. But Zermelo overlooked an important construction that is often used in advanced set theory, and in 1922 Frankel and Skolem independently added ZF8, the axiom of replacement, to the axioms for set theory. Informally, this axiom states that if we start with a set x, and then replace every $y \in x$ with a (possibly new) set $F(y)$, we obtain a set $\{F(y) : y \in x\}$ whose elements are the sets $F(y)$.

The axiom of replacement appears in naive set theory in the form of indexed collections of sets, say $\{x_t: t \in T\}$; see Section 1.4. Here is the original application of replacement.

Example 1:

> The axiom of infinity allows us to collect the sets $0, 1, 2, \ldots$ into a single set. Suppose, however, we form the sets $\varnothing$, $P(\varnothing)$, $P(P(\varnothing)), \ldots$; Zermelo's axioms are not sufficient to construct the set consisting of these sets. But this can be done with replacement, as follows: Let F be defined by $F(\varnothing) = \varnothing$ and $F(n + 1) = P(F(n))$ for all $n \in \omega$; then $\{F(n): n \in \omega\}$ is a set whose elements are the sets $\varnothing$, $P(\varnothing)$, $P(P(\varnothing)), \ldots$.   □

The last axiom of Zermelo-Frankel set theory is ZF9, the axiom of regularity. This axiom was not added until 1925, and is due to von Neumann. The axiom of regularity is a technical, nonobvious axiom, that can be stated as follows: $\forall x[(x \neq \varnothing) \rightarrow \exists y(y \in x \wedge x \cap y = \varnothing)]$. Like extensionality, regularity does not enable us to construct new sets, but rather limits the sets that we can construct.

Let us formulate the principle that sets are constructed in stages:

(*) To construct set x, we must first construct every set y such that $y \in x$.

This principle can be used to show that the axiom of regularity is "true." For suppose we have a nonempty set x. By (*), there is some $y \in x$ such that y is constructed before (or at least as early as) any other element of x. It follows that $x \cap y = \varnothing$. On the contrary, suppose there exists $z \in x \cap y$. Since $z \in y$, z is constructed before y; but $z \in x$, and hence y is constructed at least as early as z, which leads to a contradiction.

Let us give an application of regularity. Suppose two sets x and y are related by $y \in x$. By (*), y is constructed *before* x. This observation would seem to rule out $x \in x$, and also $x \in y$ and $y \in x$ simultaneously. Can we prove this?

Lemma 10: Assume the axiom of regularity.

(1) There is no set x such that $x \in x$.

(2) There are no sets x and y such that $x \in y$ and $y \in x$.

Proof: First note that (1) follows from (2). Suppose there exist sets x and y such that $x \in y$ and $y \in x$. Apply the axiom of regularity to $\{x, y\}$: There is a set $z \in \{x, y\}$ such that $z \cap \{x, y\} = \varnothing$. Now $z \in \{x, y\}$ implies $z = x$ or $z = y$; as a matter of notation, assume $z = x$. We then have $x \cap \{x, y\} = \varnothing$; this contradicts $y \in x \cap \{x, y\}$. ■

Later we will make the claim that all of the objects of classical mathematics (e.g., rational, real, and complex number systems) can be constructed from the axioms of ZF set theory. As a first step in this direction, let us construct the natural number system from these axioms. Specifically, we will show that ω

and the successor operator S satisfy the Peano postulates P1–P3. First note that $\varnothing \in \omega$; for all $x \in \omega$, $S(x) \in \omega$.

Theorem 1 (*Existence of natural number system*): The set ω and the operator S satisfy the following conditions:

P1 For all $x \in \omega$, $S(x) \neq \varnothing$.

P2 If $S(x) = S(y)$, then $x = y$.

P3 Let $x \subseteq \omega$ satisfy the following: $\varnothing \in x$; if $y \in x$, then $S(y) \in x$. Then $x = \omega$.

Proof: P1 holds, since $x \in S(x)$. P2 states that S is one-to-one. Let $S(x) = S(y)$; in proving $x = y$, we use the consequence of the axiom of regularity that $x \in y$ and $y \in x$ is impossible. (This use of regularity is not essential and could be avoided with some extra work.) From $S(x) = S(y)$, we obtain $x \cup \{x\} = y \cup \{y\}$. Now $x \in x \cup \{x\}$; hence $x \in y \cup \{y\}$, so $x \in y$ or $x = y$. If $x = y$, we are finished. Assume, then, that $x \in y$. By a similar argument, we have $y \in x$ or $y = x$. But we are in the case $x \in y$, so $y \in x$ is impossible. Hence $y = x$ as required.

To prove P3, let $x \subseteq \omega$ satisfy these properties: $\varnothing \in x$; if $y \in x$, then $S(y) \in x$ (note that x is a successor set). To prove $x = \omega$, it suffices to show that $x \subseteq \omega$ and $\omega \subseteq x$ (extensionality). We are given that $x \subseteq \omega$, and $\omega \subseteq x$ follows from the fact that ω is a subset of every successor set (see Lemma 8(2)). ∎

The axiom system of ZF set theory differs from other axiom systems in this respect: A formal language is required for a precise statement of some of its axioms (specifically, separation and replacement). Skolem first made this observation in 1922. Let us now state all of the axioms of ZF set theory in the language L_{ST}. We use $x, y, z, u, v, w, p_1, \ldots, p_k$ for variables (distinct) and write $x \in y$ and $x = y$ for $(x \in y)$ and $(x = y)$ respectively.

ZF1 (*Extensionality*) $\forall x \forall y [\forall z (z \in x \leftrightarrow z \in y) \rightarrow x = y]$.

ZF2 (*Separation*) $\forall x \forall p_1 \ldots \forall p_k \exists y \forall u [u \in y \leftrightarrow (u \in x \wedge A)]$.

ZF3 (*Null set*) $\exists x \forall y \neg (y \in x)$.

ZF4 (*Pair*) $\forall x \forall y \exists z \forall u [u \in z \leftrightarrow (u = x \vee u = y)]$.

ZF5 (*Union*) $\forall x \exists y \forall z [z \in y \leftrightarrow \exists u (u \in x \wedge z \in u)]$.

ZF6 (*Infinity*) $\exists x [\varnothing \in x \wedge \forall y (y \in x \rightarrow S(y) \in x)]$.

ZF7 (*Power set*) $\forall x \exists y \forall z [z \in y \leftrightarrow \forall u (u \in z \rightarrow u \in x)]$.

ZF8 (*Replacement*) $\forall p_1 \ldots \forall p_k [\forall u \forall v \forall w ((A \wedge A_v[w]) \rightarrow v = w)$
$\rightarrow \forall x \exists y \forall v (v \in y \leftrightarrow \exists u (u \in x \wedge A))]$.

ZF9 (*Regularity*) $\forall x [\exists y (y \in x) \rightarrow \exists y (y \in x \wedge \neg \exists z (z \in x \wedge z \in y))]$.

Notes In separation, A is a formula with free variables $u, p_1, \ldots, p_k$; $p_1, \ldots, p_k$ are often referred to as *parameters*. In replacement, A is a formula with free variables $u, v, p_1, \ldots, p_k$ and w is substitutable for v in A. An informal statement of replacement is as follows: Assume the formula A is *functional* (for each u there is at most one v such that A);

then for any set x there is a set y that consists of all sets v for which there is a set u such that $u \in x$ and A. In infinity, we can replace the defined symbols $\varnothing$ and S as follows: $\varnothing \in x$ with $\exists y[y \in x \wedge \neg \exists z(z \in y)]$; $S(y) \in x$ with $\exists z[z \in x \wedge \forall u(u \in z \leftrightarrow (u \in y \vee u = y))]$.

If we add first-order logic to the axioms ZF1–ZF9, we obtain an axiomatized theory, or a formal system, that we denote ZF. Thus the language of the formal system ZF is L_{ST} and the nonlogical axioms are all instances of the sentences ZF1–ZF9. Henceforth we will not be overly concerned with the distinction between ZF set theory (an axiom system for Cantorian set theory) and the formal system ZF.

We now ask four questions about the list of axioms ZF1–ZF9. **Q1.** Are there enough axioms to construct the classical objects of mathematics such as the integers, the rationals, the reals, and the complex numbers? **Q2.** Are there enough axioms to construct all of the sets used in modern mathematics? **Q3.** Is ZF complete? **Q4.** Are the axioms consistent?

It is generally agreed that the axioms of Zermelo-Frankel set theory are indeed sufficient to construct the classical objects of mathematics, so Q1 gets a *yes* answer. On the other hand, the answer to Q2 is *no*; still another axiom is needed if we want to prove certain results of modern mathematics such as this one: *A countable union of countable sets is countable.* This additional axiom is somewhat controversial because of its nonconstructive nature: It asserts the existence of a set but does not tell us how to construct the set.

Axiom of Choice: Let x be a nonempty set such that (1) $y \neq \varnothing$ for all $y \in x$; (2) if $y, z \in x$ and $y \neq z$, then $y \cap z = \varnothing$. Then there is a set b such that for all $y \in x$, $b \cap y$ has exactly one element.

We use AC to denote the axiom of choice. Here is a less formal statement of the axiom: If $\{x_t : t \in T\}$ is a pairwise disjoint collection of nonempty sets, then there exists a set b that "chooses" exactly one set from each of the sets x_t. This axiom has many equivalent forms, one of which is the following (verification is left as an exercise for the reader).

Axiom of Choice (Alternate version): Let $\{x_t : t \in T\}$ be a non-empty collection of nonempty sets. Then there is a function $f: T \to \bigcup_{t \in T} x_t$ such that $f(t) \in x_t$ for all $t \in T$.

While AC seems innocent enough, it has many equivalent forms, some not so "obviously true." Example: *For every nonempty set X, there is a relation $\leq$ on X that is a well-ordering.* This is a striking result in light of the fact that no one has ever succeeded in constructing an explicit well-ordering on the set $\mathbb{R}$ of real numbers.

The situation, in a nutshell, is this: We need AC if we want to prove certain important results in modern mathematics; AC could possibly be the source of a contradiction, since some of its consequences seem paradoxical. Fortunately, Gödel has rescued us from this dilemma by proving that AC cannot lead to a contradiction unless there is already a contradiction in set theory. Gödel's theorem also shows that we cannot derive ¬AC from the axioms of ZF. To state Gödel's result precisely, we introduce notation as follows: ZFC is the formal system with language L_{ST} and axioms ZF1–ZF9 plus AC. Thus ZFC = ZF + AC.

Theorem 2 (Gödel, 1938): If ZF is consistent, then ZFC is consistent.

Corollary 1: If ZF is consistent, then ¬AC is not a theorem of ZF.

Proof: Assume ZF is consistent. By Theorem 2, ZFC is also consistent. Suppose ZF ⊢ ¬AC. Then certainly ZFC ⊢ ¬AC, and moreover ZFC ⊢ AC. This contradicts the consistency of ZFC. ■

Is AC really a new axiom? Or is it possible that we can prove AC from the axioms of ZF? Cohen has proved that this cannot be done.

Theorem 3 (Cohen, 1963): If ZF is consistent, then ZF + ¬AC is consistent.

Corollary 2: If ZF is consistent, then AC is not a theorem of ZF.

Let us now return to the question of the sufficiency of the axioms for modern mathematics. The mathematical theory ZFC does indeed seem adequate to construct all the sets of modern mathematics. Some evidence for this claim is that virtually all modern textbooks in analysis, topology, and algebra use the axioms of ZFC as the foundation for the subject at hand. The claim that ZFC is adequate for the construction of all sets used in modern mathematics is not susceptible to proof, but rather is based on empirical evidence. In connection with this assertion, let us formulate the following somewhat broader claim.

Hilbert's Thesis: All provable mathematical statements are expressible in the language L_{ST} of set theory and have a formal proof in ZF (or in ZFC).

Isolating the axioms of Zermelo-Frankel set theory is a remarkable and outstanding achievement. The formal system ZFC is a clear, precise answer to this question: What logical axioms, rules of inference, and nonlogical axioms suffice to derive the theorems of modern mathematics? In addition, ZFC shows that first-order logic is sufficient for all mathematical reasoning.

The formal system ZFC is obviously axiomatized. It is not, however, finitely axiomatized (see separation and replacement). But by introducing the additional primitive concept of a *class*, we can give a finite set of axioms for Cantorian set theory; this mathematical theory is called NBG (after von Neumann, Bernays, and Gödel) and is described in E. Mendelson, *Introduction to Mathematical Logic* 1987.

What about the completeness of ZFC? It turns out that there are many natural questions that cannot be settled on the basis of the axioms of ZFC. The best-known, and oldest, is called the *continuum hypothesis*, denoted CH and stated as follows: *For every infinite $A \subseteq \mathbb{R}$, either $A \approx \mathbb{N}$ or $A \approx \mathbb{R}$.* At the same time that Gödel and Cohen proved their results about the axiom of choice, they also proved the following.

Theorem 4 (Gödel): If ZFC is consistent, then ZFC + CH is consistent.

Theorem 5 (Cohen): If ZFC is consistent, then ZFC + ¬CH is consistent.

By Gödel's result, we cannot prove the negation of CH from ZFC (unless ZFC is inconsistent, in which case everything is provable!). Cohen's result says that we cannot prove CH from ZFC. The significance of these important results is this: A mathematician who accepts Hilbert's thesis that all mathematical reasoning can be formalized in ZFC knows that it is a waste of time to look for a mathematical proof of either CH or ¬CH. A more productive area of research is to assume additional axioms for ZFC and try to derive either CH or ¬CH. Set theorists are still looking for a statement about sets that is "obviously true" and that implies either CH or ¬CH. Proofs of the results of Gödel and Cohen are given in Jech 1978 and Kunen 1980.

We have already observed that neither CH nor ¬CH is a theorem of ZFC, so ZFC is not complete. Moreover, set-theoretic topology provides us with a host of interesting, natural sentences A of L_{ST} such that neither A nor ¬A is a theorem of ZFC. But can we perhaps add a finite number of axioms (or axiom schemes) to ZFC, say ZFC$^+$, such that ZFC$^+$ is consistent and complete? Gödel has proved that this is not possible; see Section 7.1 for a further discussion of this profound result and its impact on the axiomatic method.

Theorem 6 (Gödel's incompleteness theorem I, ZFC version): Let ZFC$^+$ be a consistent axiomatized extension of ZFC. Then ZFC$^+$ is not complete.

Now let us consider the question of the consistency of ZFC. Hilbert, starting around 1900 and continuing throughout his mathematical career, was very much interested in consistency. We have already discussed some of the reasons for his interest in the consistency of geometry and arithmetic (see Section 6.1), so we emphasize here his interest in the consistency of ZFC.

When Cantor first began his study of infinite sets in the 1870s, he was criticized by several colleagues, who claimed that infinite sets do not exist and hence are not appropriate for mathematical study. Perhaps the best-known mathematician of Cantor's generation to adopt this position was Kronker, a number theorist who strongly emphasized the constructive nature of mathematics. (He is credited with the statement "God made the integers; all the rest is the work of man.")

This opposition to infinite sets gained support when, around 1900, contradictions in Cantorian set theory began to surface. The most notable critics of Cantor's work were Brouwer and Poincaré. Poincaré stated: "Actual infinity does not exist. What we call infinite is only the endless possibility of creating new objects no matter how many objects exist already." His attitude is perhaps summed up by a remark he made in 1908: "Later generations will regard set theory as a disease from which one has recovered." A more reasoned attack on Cantor's work is Brouwer's position: The logic used in mathematics was developed by Euclid and the early Greeks to reason about finite sets; this logic does not necessarily apply to infinite sets (see the discussion on intuitionism in Section 3.5).

But Hilbert strongly opposed the suggestion that infinite sets should be eliminated from mathematics. In his essay, "On the Infinite," he states, "No one shall drive us out of the paradise that Cantor has created for us." The paradoxes in set theory even raised the possibility that mathematics itself is inconsistent. Thus Hilbert was motivated to consider the following two questions: Is mathematics consistent? Is it legitimate to use infinite sets in mathematics? To answer these questions in the affirmative, Hilbert isolated the following fundamental problem: *Prove the consistency of the formal system* ZFC.

Suppose this is done. We can then claim that mathematics is indeed consistent. For suppose mathematics is inconsistent. Then there is a mathematical proof of some obviously false statement such as $0 \neq 0$. Since ZFC formalizes all mathematical reasoning (Hilbert's thesis), one has ZFC $\vdash$ $\neg(\varnothing = \varnothing)$. This contradicts the consistency of ZFC. We can also argue that it is legitimate to use infinite sets in mathematics; this is so because ZFC axiomatizes the theory of infinite sets, and ZFC is consistent.

Hilbert's required proof of the consistency of ZFC should be syntactic in nature; it should be a direct combinatorial argument that there is no finite sequence of formulas of L_{ST} that is a proof in ZFC of $\neg(\varnothing = \varnothing)$. The proof should use only "finitist" methods. Though this is not a precisely defined concept, Hilbert specifically wanted to avoid the use of infinite sets in the proof. Remember that one of his goals was to justify the use of infinite sets in mathematics, and he wished to avoid the criticism of circularity. To summarize, Hilbert wanted an absolute consistency proof of ZFC, a proof without recourse to a (necessarily infinite) model. The idea behind Hilbert's strategy is nicely summarized by Cohen in *Set Theory and the Continuum Hypothesis* (1966, 3) (bracketed phrase added): "In this way questions concerning infinite sets are replaced by questions concerning the combinatorial possibilities of a certain formal game [proofs in ZFC]."

Now I will tell you some things you may not want to discuss with your nonmathematical friends. Hilbert's hope for a proof of the consistency of mathematics was shattered by the following remarkable theorem of Gödel. (We will discuss and explain this theorem in more detail in Chapter 7.)

Theorem 7 (Gödel's incompleteness theorem II, ZFC version):
One cannot prove the consistency of ZFC by methods formalizable in ZFC.

Somewhat more precisely, Gödel proved the following: There is a sentence CON (ZFC) of L_{ST} whose intended interpretation is "ZFC is consistent"; if ZFC is consistent, then CON (ZFC) is not a theorem of ZFC.

Let us outline the argument that Gödel's incompleteness theorem II shows that there is no mathematical proof of the consistency of mathematics. Suppose there is such a proof. Now, Hilbert's thesis states that all provable mathematical statements are expressible in the language L_{ST} of set theory and that each has a formal proof in ZFC. Thus we have a formal proof of the consistency of ZFC in ZFC, a contradiction of Gödel's incompleteness theorem II. The distinguished mathematician Hermann Weyl, on hearing about Gödel's work, asserted that mathematics is surely consistent, so God exists, but our inability to prove it consistent shows that the Devil exists.

It would seem that the only way to know the consistency of ZFC is to assume it as a new "axiom." Does the assumption that ZFC is consistent lead to a contradiction? Let us consider the following example, known as *Skolem's paradox*. Suppose ZFC is consistent. By the Löwenheim-Skolem theorem, there is a countable model of ZFC, which we denote by $\langle D_M, \in_M \rangle$. Let us assume that $\in_M$ is the usual membership relation and that D_M is a transitive set: If $a \in D_M$, then $a \subseteq D_M$ (see the Mostowski collapsing theorem, 88 in Jech). Since $\langle D_M, \in_M \rangle$ is a model of ZFC, we can use the axiom of infinity, the power axiom, and a Cantor diagonal argument to prove the existence of a set $a \in D_M$ that is uncountable (relative to $\langle D_M, \in_M \rangle$). We now have (1) a is uncountable (relative to $\langle D_M, \in_M \rangle$); (2) $a \subseteq D_M$ and D_M is countable, and hence a is countable. Are these statements contradictory? The set a is indeed countable, but the model $\langle D_M, \in_M \rangle$ "thinks" that a is uncountable for the following reason: There is no $f \in D_M$ such that f is a one-to-one function from a into the natural numbers.

We end with a brief summary of the situation with respect to the consistency of ZFC. (1) Gödel's incompleteness theorem II rules out a formal proof. (2) For a Platonist, the question of consistency is a nonissue. There is a set-theoretic universe independent of the minds of mathematicians; moreover, the axioms of ZFC are obviously true statements about this universe. Thus ZFC is consistent. (3) For the research mathematician not working in the foundations of mathematics, the following attitude is appropriate: The axioms of Zermelo-Frankel set theory have been known for more than 60 years, and no contradictions have surfaced. Moreover, if a contradiction is discovered, the problem can almost surely be corrected without serious effect on most existing mathematics.

The following books, and papers, are recommended for further reading: Bell and Machover 1977; van Dalen, Doets, and Swart 1978; Jech, Kunen, and Roitman 1990, and Shoenfield 1993.

─────────── *Exercises for Section 6.4* ───────────

1. Derive the reflexive axiom $\forall x(x = x)$ from the axiom of extensionality.
2. Prove: If $\langle a, b \rangle = \langle x, y \rangle$, then $a = x$ and $b = y$. (*Hint:* Consider two cases: (1) $\{a\} = \{x, y\}$; (2) $\{a\} = \{x\}$.
3. Prove parts (1) and (2) of Lemma 8.
4. Show that there is no set x such that $P(x) \subseteq x$. (*Hint:* Russell's paradox.)
5. (*A non-standard model*) Consider the following interpretation of L_{ST}: $D_1 = \mathbb{R}$; $\in_1 =$ less-than relation on $\mathbb{R}$. Which of these axioms of ZF are true in this interpretation?
   (1) $\forall x \forall y [\forall z(z \in x \leftrightarrow z \in y) \rightarrow x = y]$.
   (2) $\exists x \forall y \neg (y \in x)$.
   (3) $\forall x \forall y \exists z \forall u [u \in z \leftrightarrow (u = x \lor u = y)]$.
   (4) $\forall x \exists y \forall z [z \in y \leftrightarrow \exists u(z \in u \land u \in x)]$.
6. Introduce axioms for set theory as follows. *Weak pair:* If x and y are sets, there is a set z such that $x \in z$ and $y \in z$. *Weak union:* If x is a set, there is a set z such that if $y \in u$

and $u \in x$, then $y \in z$. *Weak power*: If x is a set, there is a set z such that if $y \subseteq x$, then $y \in z$. Prove the following:

(1) Separation + weak pair $\Rightarrow$ pair.
(2) Separation + weak union $\Rightarrow$ union.
(3) Separation + weak power $\Rightarrow$ power.

7. Show that the pairing axiom can be derived from the null, power, and replacement axioms. (*Hints:* First show $\varnothing$, $P(\varnothing) \in P(P(\varnothing))$, and $\varnothing \neq P(\varnothing)$. Given sets x and y, let F be defined by $F(\varnothing) = x$ and $F(z) = y$ for all $z \neq \varnothing$. Show that $u \in \{F(z): z \in P(P(\varnothing))\} \Leftrightarrow u = x \vee u = y$.)

8. Show that separation can be obtained from the null and replacement axioms. (*Hint:* Let x be a set and let Q be a property of sets; we are required to prove the existence of the set $\{y: y \in x \wedge Q(y)\}$. If there is no $y \in x$ with $Q(y)$, use the null set axiom. Otherwise, let $y_0 \in x$ with $Q(y_0)$, and define F by $F(y) = y$ if $Q(y)$ and $F(y) = y_0$ otherwise. Apply replacement to x and F.)

9. (*Axiom of replacement*) The requirement that the formula A in replacement be functional is essential. Suppose A is $u \subseteq v$ (more precisely, $\forall t (t \in u \to t \in v)$). The axiom of replacement, with x the set $\{\varnothing\}$ and no parameters, states: If $\forall u \forall v \forall w ((u \subseteq v \wedge u \subseteq w) \to v = w)$, then $\exists y \forall v [v \in y \leftrightarrow \exists u (u \in \{\varnothing\} \wedge u \subseteq v)]$.
   (1) Explain why $\forall u \forall v \forall w ((u \subseteq v \wedge u \subseteq w) \to v = w)$ is false.
   (2) Explain why $\exists y \forall v [v \in y \leftrightarrow \exists u (u \in \{\varnothing\} \wedge u \subseteq v)]$ asserts the existence of a set y such that every set is an element of y (and so y is a universal set, which is impossible).

10. Let A be a sentence and let Δ be a set of sentences of L_{ST}. Prove equivalent: (1) $\Delta \cup \{A\}$ is consistent; (2) $\neg A$ is not a theorem of Δ.

11. Prove that $n + 1 = \{0, \ldots, n\}$.

12. Let $z \in \omega$. Prove that $z = \varnothing$ or $z = S(y)$ for some $y \in \omega$. (*Hint:* Use Theorem 1(3) to show that $x = \omega$, where

$$x = \{z: z \in \omega \wedge (z = \varnothing \vee z = S(y) \text{ for some } y \in \omega)\}.)$$

13. (*Characterizations of transitivity*) A set x is *transitive* if it satisfies the following property: $z \in y$ and $y \in x$ imply $z \in x$. Prove that the following are equivalent:
   (1) x is transitive.   **(3)** For all $y \in x$, $x \cap y = y$.
   (2) For all $y \in x$, $y \subseteq x$.   **(4)** For all $y \subseteq x$, $\bigcup y \subseteq x$.

14. (*Properties of transitive sets*) Let x be a set. Prove the following:
   (1) If x is transitive, then $S(x)$ is transitive.
   (2) If x is transitive, then $P(x)$ is transitive.
   (3) If x is transitive and $y \in z_1, z_1 \in z_2, \ldots, z_{n-1} \in z_n, z_n \in x$, then $y \in x$.
   (4) If every $y \in x$ is transitive, then $\bigcup x$ is transitive.
   (5) If x is transitive and $y \in x$, then $\bigcup y \subseteq x$.
   (6) ω is transitive. (*Hint:* Let $x = \{y: y \in \omega \wedge y \subseteq \omega\}$; use Theorem 1(3) to show that $x = \omega$; also see Exercise 13.)

15. (*Transitive closure of x*) The *transitive closure* of x, denoted TC(x), is constructed as follows: $x_0 = x$; $x_{n+1} = \bigcup x_n$; TC$(x) = \bigcup \{x_n: n \in \omega\}$. The axioms of union, replacement, and infinity are used to prove the existence of TC(x). Prove the following:
   (1) $x \subseteq$ TC(x).
   (2) TC(x) is transitive.
   (3) If $x \subseteq y$ and y is transitive, then TC$(x) \subseteq y$.
   (4) x is transitive if and only if TC$(x) = x$.
   (*Hint* for (3): First show that $x_n \subseteq y$ for all $n \in \omega$; see Exercise 13.)

16. (*Ordinals*) A set α is an *ordinal* if it satisfies these two conditions: α is transitive; for all $x \in \alpha$, x is transitive. (*Example:* $\{\varnothing, \{\varnothing\}, \{\{\varnothing\}\}\}$ is transitive but is not an ordinal; $\{\varnothing, \{\varnothing\}, \{\varnothing, \{\varnothing\}\}\}$ is an ordinal.) Prove the following:
   (1) $\varnothing$ is an ordinal.
   (2) If α is an ordinal and $x \in \alpha$, then x is an ordinal.

(3) If α is an ordinal, then $S(\alpha)$ is an ordinal.

(4) If every $y \in x$ is an ordinal, then $\bigcup x$ is an ordinal.

(5) ω is an ordinal. (*Hint:* Let $x = \{y: y \in \omega \wedge y$ is transitive$\}$; use Theorem 1(3) to show that $x = \omega$; also see Exercise 14.)

17. Let ON $= \{\alpha: \alpha$ is an ordinal$\}$, TR $= \{x: x$ is a transitive set$\}$.

   (1) Show that ON is not a set. (*Hint:* No set satisfies $x \in x$).

   (2) Show that TR is not a set.

18. Let α be a set with these two properties: α is transitive; for all $x, y \in \alpha$, $x \in y$ or $y \in x$ or $x = y$. Prove that α is an ordinal (see Exercise 16 for the definition). (*Hint:* Use the axiom of regularity.)

19. (*Applications of the axiom of regularity*) Use regularity to prove the following:

   (1) There are no sets $x_1, \ldots, x_n$ such that $x_1 \in x_2, \ldots, x_{n-1} \in x_n, x_n \in x_1$.

   (2) There are no sets $x_0, x_1, x_2, \ldots$ such that $x_{n+1} \in x_n$ for all $n \in \omega$.

   (3) $x \neq \{x\}$.   (6) $x \neq \langle x, y \rangle$.

   (4) $x \neq S(x)$.   (7) $y \neq \langle x, y \rangle$.

   (5) $x \neq P(x)$.

20. Let I $= \{x: x$ is an infinite set$\}$. Since there are an infinite number of infinite sets (e.g., $\omega, P(\omega), P(P(\omega)), \ldots$), I itself is infinite, so I $\in$ I. Does this contradict regularity? Explain.

21. Use the axiom of regularity to explain why we cannot have a set x as follows: $x = \{\emptyset, \{\emptyset, \{\emptyset, \ldots\}\}\}$. (See Barwise and Moss, 1991).

22. Assume ZF is consistent. Explain why we cannot prove the consistency of ZF in ZFC.

23. (*Equivalent form of the* AC) Prove that AC is equivalent to the following: **AC*** Let $\{x_t: t \in T\}$ be a nonempty collection of nonempty sets. Then there exists a function $f: T \to \bigcup_{t \in T} x_t$ such that $f(t) \in x_t$ for all $t \in T$. (*Hint for AC $\Rightarrow$ AC**: Apply AC to $\{\{t\} \times x_t: t \in T\}$.)

24. (*Well-ordering theorem* $\Rightarrow$ AC) Prove AC using the fact that every nonempty set can be well-ordered.

25. (*AC holds for a finite number of sets*) Let $\{x_k: 1 \le k \le n\}$ be a pairwise disjoint collection of nonempty sets. Prove, without AC, that there is a set b such that for $1 \le k \le n$, $b \cap x_k$ has exactly one element. (*Hint:* Use induction.)

26. (*Russell*) Russell has used the following example to illustrate AC: Let $X = \{x_n: n \in \omega\}$ be a pairwise disjoint collection of nonempty sets.

   (1) Suppose each x_n consists of a pair of shoes (left and right). Explain how, without AC, we can describe a procedure for finding a set b such that for all $n \in \omega$, b contains exactly one shoe from each pair.

   (2) Now suppose each x_n consists of a pair of socks (same color!). Explain why we need AC to obtain a set b such that for all $n \in \omega$, b contains exactly one sock from each pair.

7

Incompleteness, Undecidability, and Indefinability

These two systems [Principia Mathematica *and Zermelo-Frankel set theory*] *are so comprehensive that in them all methods of proof today used in mathematics are formalized—that is, reduced to a few axioms and rules of inference. One might therefore conjecture that these axioms and rules of inference are sufficient to decide any mathematical question that can at all be formally expressed in these systems. It will be shown below that this is not the case, that on the contrary there are in the two systems mentioned relatively simple problems in the theory of integers that cannot be decided on the basis of the axioms.*

—Kurt Gödel, 1931

$\boxed{T}$ he main focus of this chapter is a remarkable result known as Gödel's incompleteness theorem. According to John von Neumann, one of the greatest mathematicians of the twentieth century, "Kurt Gödel's achievement in modern logic is singular and monumental—indeed it is more than a monument, it is a landmark which will remain visible far in space and time." (Remarks made on March 14, 1951, at the Einstein Award ceremonies.) Gödel's theorem tells us that it is impossible to select a consistent, easily recognizable list of axioms for arithmetic and then to prove all true statements of arithmetic from these axioms. Gödel's incompleteness theorem is one of the deepest and most important theorems in twentieth-century mathematics, and the techniques developed by Gödel to prove this result are fundamental in modern mathematical logic.

In this chapter, we will outline the main ideas of Gödel's incompleteness theorem, omitting some of the technical details. We will also prove two closely related results, due to Church and Tarski. Roughly speaking, proofs in this chapter are simplified by replacing the technical definition *recursive* with the informal concept of an algorithm; that the two approaches are the same is Church's thesis. We also note that Church's thesis allows for a simplification in the statement of these theorems. In Chapter 8, we will give precise statements of the theorems of Gödel, Church, and Tarski, along with detailed proofs that certain relations are recursive or RE.

7.1

Overview of the Theorems of Gödel, Church, and Tarski

In Chapter 1, we raised the following problem: Find a consistent, easily recognizable list of axioms for arithmetic such that all true statements of arithmetic can be derived from these axioms. To clarify the meaning of arithmetic, we introduce a formal system that we denote NN and call *formal arithmetic.* The language of NN is L_{NN}, the rules are the rules of first-order logic, and the axioms are the (logical) axioms of first-order logic *plus* the closures of these formulas (the *nonlogical* axioms):

NN1 $\neg(Sx_1 = 0)$.

NN2 $(Sx_1 = Sx_2) \rightarrow (x_1 = x_2)$.

NN3 $x_1 + 0 = x_1$.

NN4 $x_1 + Sx_2 = S(x_1 + x_2)$.

NN5 $x_1 \times 0 = 0$.

NN6 $x_1 \times Sx_2 = (x_1 \times x_2) + x_1$.

NN7 $\neg(x_1 < 0)$.

NN8 $(x_1 < Sx_2) \rightarrow [(x_1 < x_2) \vee (x_1 = x_2)]$.

NN9 $(x_1 < x_2) \vee (x_1 = x_2) \vee (x_2 < x_1)$.

Is NN consistent? It certainly seems reasonable to assert that the standard interpretation is a model of NN and so NN is indeed consistent. But there are difficulties with this claim (see Section 6.1), so whenever we need the fact that $\mathcal{N}$ is a model of NN, or the fact that NN is consistent, we will explicitly state it as a hypothesis.

The nonlogical axioms NN1–NN9 are really quite weak; there are many true statements of arithmetic that cannot be proved from these axioms. For example, we can show that $NN \vdash 0 \times 0 = 0$, $NN \vdash 0 \times S0 = 0$, and so on, but the more general result $\forall x_1(0 \times x_1 = 0)$ is not a theorem of NN (see Exercise 6). NN, unlike Peano arithmetic PA (see Section 6.1), does not have axioms for induction.

So why are the particular axioms NN1–NN9 chosen? A quick answer is that they suffice to prove the result that every recursive relation is expressible (see Section 7.3). This means nothing at this point; a more informative answer is that NN1–NN9 are chosen so that the following result holds. (For $n \geq 0$, the term 0^n is defined inductively by $0^0 = 0$ and $0^{n+1} = S0^n$; for example, 0^2 is $SS0$. In the standard interpretation, the term 0^n is interpreted as the natural number n.)

Theorem 1 (Basic properties of the formal system NN): The following hold for all $a, b \in \mathbb{N}$:

(1) If a equals b, then $NN \vdash (0^a = 0^b)$.

(2) If a does not equal b, then $NN \vdash \neg(0^a = 0^b)$.

(3) If a is less than b, then $NN \vdash (0^a < 0^b)$.

(4) If a is not less than b, then $NN \vdash \neg(0^a < 0^b)$.

(5) $NN \vdash (0^a + 0^b) = 0^{a+b}$.

(6) $NN \vdash (0^a \times 0^b) = 0^{a \times b}$.

(7) $NN \vdash \forall x[(x < 0^a) \vee (x = 0^a) \vee (0^a < x)]$.

(8) If $NN \vdash A_x[0], \ldots, NN \vdash A_x[0^{a-1}]$, then $NN \vdash (x < 0^a) \rightarrow A$.

Comment 1 It is often stated that Gödel's incompleteness theorem holds for any formal system that contains "enough arithmetic." Although Theorem 1 gives a fairly accurate description of "enough arithmetic," perhaps the best way to appreciate the meaning of this phrase is to read Section 7.3; this is where the axioms NN1–NN9 are explicitly used.

Comment 2 We will eventually prove all parts of Theorem 1; this will require a certain technical proficiency in writing formal proofs in NN. We could avoid these formal proofs by replacing the axioms NN1–NN9 with (1)–(8) of Theorem 1. However, one nice feature of NN is that it is finitely axiomatized; this is lost if we declare (1)–(8) to be the nonlogical axioms. The fact that NN has a finite number of axioms is used in Section 7.5 to prove that the validity problem for L_{NN} is unsolvable.

Comment 3 We could also replace NN with the more natural theory Peano arithmetic. But PA is actually a stronger theory than we need, and the fewer assumptions we make about NN, the broader the scope of Gödel's theorem. Also note that PA is not finitely axiomatized. Therefore, we will proceed with NN.

The following notation is used: FOR_{NN} = set of formulas of L_{NN}. Since NN is quite weak, we must add additional axioms to NN if there is to be any hope of proving all true statements of arithmetic.

Definition 1: A set $\Gamma \subseteq FOR_{NN}$ is an *extension* of NN if $\{NN1, \ldots, NN9\} \subseteq \Gamma$.

In other words, a set Γ of formulas of L_{NN} is an extension of NN if the nine sentences NN1–NN9 are among the formulas in Γ. Note that if Γ is an extension of NN and $NN \vdash A$, then $\Gamma \vdash A$. As an example of an extension of

NN, let Γ be the formulas of L_{NN} consisting of NN1–NN9 plus the commutative law $\forall x_1 \forall x_2 (x_1 + x_2 = x_2 + x_1)$.

A set $\Gamma \subseteq \text{FOR}_{NN}$ is *axiomatized* if there is an algorithm that decides whether $A \in \Gamma$ (see Section 5.5). NN (more precisely, $\{NN1, \ldots, NN9\}$) is axiomatized and, moreover, any finite extension of NN is axiomatized (even finitely axiomatized). Now let us formulate our problem precisely.

Problem (Hilbert, 1928): Find an extension Γ of NN such that (1) Γ is consistent; (2) Γ is axiomatized; (3) Γ is complete.

See Section 6.1 for a discussion of Hilbert's program and the reason for his interest in the consistency and completeness of the natural number system. In 1931, Gödel published the following theorem, showing that Hilbert's problem has no solution.

Theorem 2 (Gödel's incompleteness theorem): Every consistent axiomatized extension Γ of NN is incomplete (i.e., there is a sentence of L_{NN} such that neither it nor its negation is a theorem of Γ).

Strictly speaking, Gödel assumed a somewhat stronger hypothesis than consistency, called *ω-consistency*. In addition, the word *axiomatized* is replaced by the more precise term *recursively axiomatized* (see Section 8.7 for a precise statement).

It is instructive to draw a picture of Gödel's result. Let Γ be a consistent axiomatized extension of NN. For the sake of simplicity, assume that all formulas in Γ are sentences and that these sentences are true in the standard interpretation of L_{NN}. It follows that if A is a sentence of L_{NN} and $\Gamma \vdash A$, then A is true in $\mathcal{N}$ (soundness theorem). Moreover, each sentence of L_{NN} is either true or false in $\mathcal{N}$ (agreement theorem). So we have the following picture:

Sentences of L_{NN}

Sentences NN1–NN9

Sentences in Γ

Sentences of L_{NN} that are theorems of Γ

TRUE (in $\mathcal{N}$)   FALSE (in $\mathcal{N}$)

——— *Figure 1* ———

Gödel's incompleteness theorem tells us that no matter how we select a set Γ of axioms for arithmetic, subject only to the requirement that Γ be an extension of NN that is consistent and axiomatized, there is a sentence A of L_{NN} such that neither it nor its negation is a theorem of Γ. Now A is a sentence, so it is either true or false in $\mathcal{N}$. Suppose A is true in $\mathcal{N}$. Then A is a sentence that is true, but is not a theorem of Γ; in other words, in Figure 1, A is in the "true" box but is not in the "theorem" box. Suppose A is false in $\mathcal{N}$. Then $\neg A$ is true in $\mathcal{N}$, so $\neg A$ is a sentence that is true but is not a theorem of

Γ. In summary, in Figure 1, there is no consistent axiomatized extension Γ of NN such that

sentences that are theorems of Γ = sentences that are true in $\mathcal{N}$.

Gödel's theorem can be applied repeatedly, thereby showing that the situation is hopeless. To see this, suppose we have a consistent axiomatized extension Γ of NN. By Gödel's theorem, there is a sentence A such that neither it nor its negation is a theorem of Γ. Apparently we did not select enough nonlogical axioms for Γ. Since A is not a theorem of Γ, the extension theorem applies, and $\Gamma \cup \{\neg A\}$ is an extension of NN that is consistent and such that $\neg A$ *is* a theorem of $\Gamma \cup \{\neg A\}$. Now this extension $\Gamma \cup \{\neg A\}$ is also axiomatized, so Gödel's incompleteness theorem applies to $\Gamma \cup \{\neg A\}$ and there is a sentence B such that neither B nor $\neg B$ is a theorem of $\Gamma \cup \{\neg A\}$. This argument can be repeated indefinitely. In summary, no matter how careful we are in selecting our axioms for arithmetic, we will never succeed in obtaining a list that is consistent, axiomatized, *and* complete.

Why is Gödel's incompleteness theorem so important? Perhaps the main reason is that it reveals a shortcoming, or a limitation, in the axiomatic method. By 1930, mathematicians had great confidence in the axiomatic method: The proper way to study a branch of mathematics such as number theory or geometry is to select a simple list of axioms (statements about the subject that are "obviously true") and then to proceed to prove theorems from these axioms. An informal reading of the soundness theorem states:

Every theorem is a true statement about the subject.   (∗)

Moreover, it was expected that for a proper choice of axioms, the converse of (∗) would hold. Gödel's incompleteness theorem shows that it is not possible to achieve this goal for arithmetic, the most fundamental of all the branches of mathematics.

Although we have stated Gödel's theorem in the context of NN, the theorem holds for any formal system in which we can develop arithmetic and prove the axioms NN1–NN9 as theorems. For example, this can be done in the formal system ZF (see Section 6.4); hence we have the following version of Gödel's theorem.

Theorem 3 (Gödel's incompleteness theorem, ZF version):
Every consistent axiomatized extension Γ of ZF is incomplete (i.e., there is a sentence A of L_{ST} such that neither A nor $\neg A$ is a theorem of Γ).

Gödel's theorem has been paraphrased, or given a philosophical interpretation, in a number of different ways: (1) "Mathematical thinking is, and must be, essentially creative. This conclusion must inevitably result in at least a partial reversal of the entire axiomatic trend of the late nineteenth and early twentieth centuries, with a return to meaning and truth as being of the essence of mathematics." (Post, "Recursively enumerable sets of positive integers and their decision problems," *Bulletin Ams* 50 (1944), 295.) (2) Truth $\neq$ provability. (3) Truth cannot be captured syntactically. (4) Mind is not machine.

Two key ideas are used in the proof of Gödel's result:

- Coding (or *Gödel numbering,* in honor of Gödel)
- Expressibility

How are these two ideas used? The language L_{NN} is designed to express mathematical statements about arithmetic (e.g., primality, division algorithm, or Goldbach's conjecture). But by coding and expressibility, Gödel shows that there are formulas of L_{NN} that express logical concepts such as *axiom, proof,* and so on. In particular, there is a sentence of L_{NN} that states (in the standard interpretation): "I am not a theorem of NN." Assume there is such a sentence G of L_{NN}; we will give an informal argument that neither G nor its negation is a theorem of NN.

G is not a theorem of NN   Suppose NN $\vdash G$. By the soundness theorem, G is true in $\mathcal{N}$. But G states that it is not a theorem of NN. Thus we are in a situation in which NN $\vdash G$ and also G is not a theorem of NN; this contradicts the law of the excluded middle, $p \vee \neg p$.

$\neg G$ is not a theorem of NN   Suppose NN $\vdash \neg G$. Then $\neg G$ is true in $\mathcal{N}$, so G is false in $\mathcal{N}$. But G claims that it is not a theorem of NN, and this is false, so G *is* a theorem of NN. In summary, we have both NN $\vdash \neg G$ and NN $\vdash G$, a contradiction of the consistency of NN. ■

Note that the sentence G discusses the syntactic concept of provability ("I am not provable from the axioms of NN"). In addition, G incorporates the important notion of self-reference: G is talking about itself. In the next section, we will outline the main ideas needed to prove the existence of the sentence G.

We will now give a remarkable application of Gödel's theorem. Recall that complete arithmetic, Th $(\mathcal{N})$, is the set of all sentences of L_{NN} that are true in the standard interpretation of L_{NN}; Th $(\mathcal{N})$ is consistent and complete (see Theorem 1 in Section 6.1; in Figure 1, Th $(\mathcal{N})$ = "true" box).

Corollary 1: Assume that the standard interpretation is a model of NN. Then there is no algorithm that, given an arbitrary sentence A of L_{NN}, decides (YES or NO) whether A is true in the standard interpretation.

Proof: The hypothesis that the standard interpretation is a model of NN simply means that complete arithmetic, Th $(\mathcal{N})$, is an extension of NN. It follows from Gödel's incompleteness theorem that Th $(\mathcal{N})$ cannot be axiomatized. The argument is as follows:

Consistent + axiomatized $\Rightarrow$ *not* complete (Gödel).

But Th $(\mathcal{N})$ *is* consistent and complete; hence Th $(\mathcal{N})$ cannot be axiomatized. Since Th $(\mathcal{N})$ is not axiomatized, there is no algorithm that, given a sentence A of L_{NN}, decides whether $A \in$ Th $(\mathcal{N})$. But for any sentence A of L_{NN}, $A \in$ Th $(\mathcal{N}) \Leftrightarrow A$ is true in $\mathcal{N}$. Thus there is no algorithm that decides whether an arbitrary sentence of L_{NN} is true in $\mathcal{N}$. ■

In retrospect, the nonexistence of an algorithm for truth should come as no surprise. Indeed, it would be remarkable if there were a purely mechanical procedure for deciding the truth or falsity of unsolved problems of number theory, such as Goldbach's conjecture. In other words, discovering mathematical truth will always require creativity and ingenuity.

────────────── *Church's Theorem* ──────────────

A set $\Gamma \subseteq \text{FOR}_{\text{NN}}$ is *decidable* if there is an algorithm that decides $\Gamma \vdash A$; if no such algorithm exists, we say that Γ is *undecidable* (see Section 5.5). In 1936, Church proved the following fundamental result.

Theorem 4 (Church): Every consistent extension Γ of NN is undecidable.

Technically speaking, the conclusion should be *recursively undecidable* rather than *undecidable*; for a precise statement, see Section 8.6. The techniques developed by Gödel are used in the proof of Church's theorem. On the other hand, Church's theorem is perhaps the more fundamental result, and Gödel's theorem can be obtained from Church's theorem; this is done informally below and precisely in Section 8.7. Note that Corollary 1 can be derived from Church's theorem by taking $\Gamma = \text{Th}(\mathcal{N})$; thus Corollary 1 asserts that $\text{Th}(\mathcal{N})$ is undecidable.

Proof that Church's theorem $\Rightarrow$ Gödel's theorem: Let Γ be a consistent axiomatized extension of NN. Suppose also that Γ is complete. Now, Turing's theorem (see Theorem 8 of Section 5.5) states that

$$\text{axiomatized} + \text{complete} \Rightarrow \text{decidable};$$

hence Γ is decidable. This contradicts Church's theorem, so Γ cannot be complete, as required. ∎

Church's theorem forces us to confront the fundamental question *What is an algorithm?* Suppose $\Gamma \subseteq \text{FOR}_{\text{L}}$ is decidable. To prove this fact, it suffices to describe an algorithm and then to show that it works; this was done in Chapter 3, in connection with the formal system P for propositional logic. On the other hand, suppose Γ is undecidable. In this case, we must prove the nonexistence of an algorithm, and to do so, we need a precise definition. Algorithms and the informal notions of a decidable relation and a computable function were discussed in Section 1.7; the formal counterparts of these ideas, recursive relations and recursive functions, were discussed in Section 1.8. Church's thesis is the claim that all computable functions are recursive functions and all decidable relations are recursive relations. In Church's theorem, we prove that every consistent extension of NN is recursively undecidable; Church's thesis allows us to draw the stronger conclusion that Γ is algorithmically undecidable.

────────────── *Tarski's Theorem* ──────────────

Gödel's sentence G, which informally says "I am not a theorem of NN," is closely related to the Liar's Paradox. One version of this paradox is as follows: Let X be the sentence "I am not true." We leave it to the reader to show the following: If X is true, then X is false; if X is false, then X is true. Thus in either case we reach a contradiction. If we compare sentence G of Gödel's theorem and sentence X of the Liar's Paradox, we see that in the case of G we escape an outright contradiction because of the possibility that a sentence of L_{NN} may have the property that neither it nor its negation is a theorem of NN. This escape is not available in the case of the Liar's Paradox: Every sentence of L_{NN} is either true or false (in the standard interpretation).

Then how do we escape a contradiction in the case of the Liar's Paradox? In 1933, a few years after Gödel proved his incompleteness theorem, Tarski proved that there is no formula of L_{NN} that expresses the concept of truth. An imprecise statement of Tarski's theorem follows; a precise statement and proof will be given in Section 7.6.

Theorem 5 (Tarski): Assume that the standard interpretation is a model of NN. Then the concept of truth is not definable in L_{NN}.

Let us also mention a consequence of Tarski's theorem that generalizes Corollary 1 of this section (a proof is given in Section 7.6).

Corollary 2: Assume that the standard interpretation is a model of NN. Then Th ($\mathcal{N}$) is not listable.

——————Exercises for Section 7.1——————

1. Let Γ be an axiomatized extension of NN and let M be a model of Γ. Show that there is a sentence of L_{NN} that is true in M but is not a theorem of Γ.
2. Let Γ be a consistent axiomatized extension of NN. Show that Γ has a model that is not isomorphic to the standard interpretation $\mathcal{N}$.
3. Smullyan has written a number of books on logical puzzles that illustrate Gödel's incompleteness theorem. The following is a simplified version of one that appears in Chapter 16 of *The Lady or the Tiger?* 1982. The puzzle illustrates how it is possible for a formula to talk about itself. Consider a formal language L with just five symbols: $\neg$, P, A, [, and]. Informally, $\neg$ means *not*, P means *provable*, [and] are for punctuation, and A means *the associate of*. (If X is an expression of the language, the associate of X is the expression $X[X]$; for example, the associate of PAP is $PAP[PAP]$). The formulas of L are all expressions having one of the following forms:

   $P[X]$ (the expression X is provable).

   $\neg P[X]$ (the expression X is not provable).

   $PA[X]$ (the associate of X is provable).

   $\neg PA[X]$ (the associate of X is not provable).

   Question: What does $\neg PA[\neg PA]$ say?
4. This problem outlines a proof of the following result: If A is a sentence of L_{NN} without quantifiers, then NN $\vdash A$ or NN $\vdash \neg A$. (*Hints:* Consequences of the equality axioms as discussed in Sections 5.1 and 5.3; Theorem 1 on basic properties of NN.)
   (1) Let t be a variable-free term of L_{NN}. Show that there is a natural number a such that NN $\vdash t = 0^a$.
   (2) Let s and t be variable-free terms of L_{NN}. Prove the following:
      (a) NN $\vdash s = t$ or NN $\vdash \neg(s = t)$.
      (b) NN $\vdash s < t$ or NN $\vdash \neg(s < t)$.

(3) Let A be a sentence of L_{NN} without quantifiers. Show that $NN \vdash A$ or $NN \vdash \neg A$.

5. Assume that the standard interpretation is a model of NN. Consider a sentence of L_{NN} of the form $\exists x A$, where A has no quantifiers. Show that if $\exists x A$ is true in $\mathcal{N}$, then $NN \vdash \exists x A$. (*Hint:* Substitution theorem for assignments and Exercise 4.) Does this result also hold for a sentence of the form $\forall x A$, where A has no quantifiers?

6. This problem outlines a proof that neither $\forall x_1(0 \times x_1 = 0)$ nor $\neg \forall x_1(0 \times x_1 = 0)$ is a theorem of NN. Assume that the standard interpretation is a model of NN.

   (1) Show that $\neg \forall x_1(0 \times x_1 = 0)$ is not a theorem of NN.

   (2) Show that the following interpretation M of L_{NN} is a model of NN; assume basic facts of arithmetic.

   $D_M = \mathbb{N} \cup \{\infty\}$.   ($\infty$ is any set that is not an element of $\mathbb{N}$.)

   $$S_M(a) = \begin{cases} a + 1 & a \in \mathbb{N}; \\ \infty & a = \infty. \end{cases}$$

   $$a +_M b = \begin{cases} \text{sum of } a \text{ and } b & a, b \in \mathbb{N}; \\ \infty & a = \infty \text{ or } b = \infty. \end{cases}$$

   $$a \times_M b = \begin{cases} \text{product of } a \text{ and } b & a, b \in \mathbb{N}; \\ 0 & a = \infty \text{ and } b = 0; \\ \infty & a = \infty \text{ and } b \neq 0, \text{ or } b = \infty. \end{cases}$$

   $<_M = \{\langle a, b \rangle : a, b \in \mathbb{N} \text{ and } a \text{ is less than } b, \text{ or } a \in \mathbb{N} \text{ and } b = \infty\}$.

   (3) Show that $\forall x_1(0 \times x_1 = 0)$ is not a theorem of NN.

   (4) Show that for each $n \in \mathbb{N}$, $NN \vdash 0 \times 0^n = 0$ (use Theorem 1).

7. Let Γ be a consistent extension of NN. Show that there is no algorithm that, given an arbitrary sentence A of L_{NN}, decides (YES or NO) whether $\Gamma \cup \{\neg A\}$ is consistent.

8. By Gödel's theorem, we cannot have an extension Γ of NN that has all three of the following properties: consistent, axiomatized, complete. Give an example of an extension Γ of NN such that

   (1) Γ is consistent and axiomatized but not complete.

   (2) Γ is consistent and complete but not axiomatized.

   (3) Γ is axiomatized and complete but not consistent.

9. Let $\Delta = \{A : A \text{ is a sentence of } L_{NN} \text{ and } A \text{ is true in } \mathcal{N}\}$. Answer the following and justify your answer; assume that $\mathcal{N}$ is a model of NN. (1) Is Δ consistent? (2) Is Δ complete? (3) Is Δ axiomatized? (4) Is Δ decidable? (5) Is Δ listable?

7.2

Coding and Expressibility

——Effective Coding of the Language L_{NN} ——

We assign to each formula A of L_{NN} a natural number, called the *code of A*. This number is also referred to as the *Gödel number of A,* in honor of Gödel, the first to use this important idea in the study of formal systems. Roughly speaking, codes allow us to discuss logical concepts such as theorem, proof, and so on, in terms of relations on the set $\mathbb{N}$ of natural numbers. The process of assigning a natural number to each formula of a formal language is often called the *arithmetization of syntax.* We construct a function $g \colon \text{FOR}_{NN} \to \mathbb{N}$ satisfying two conditions Gö1 and Gö2; the value $g(A)$ is called the *code of A.*

Theorem 1 (*Effective coding of* L_NN): There is a one-to-one function g: FOR$_{\text{NN}} \to \mathbb{N}$ satisfying these two conditions:

Gö1 There is an algorithm that, given a formula A of L$_{\text{NN}}$, calculates $g(A)$.

Gö2 There is an algorithm that, given a natural number a, decides (YES or NO) whether there is a formula A such that $g(A) = a$. If the answer is YES, the algorithm finds the formula A such that $g(A) = a$.

Proof: The proof has four parts: Construct g, prove that g is one-to-one, and write the required algorithms for Gö1 and Gö2.

Construction of g: We begin by assigning an odd number to each symbol of L$_{\text{NN}}$; this function is denoted by SN, and $SN(s)$ (where s is a symbol) is called the *symbol number of s*.

$SN(() = 1$	$SN(\forall) = 9$	$SN(+) = 17$
$SN()) = 3$	$SN(=) = 11$	$SN(\times) = 19$
$SN(\neg) = 5$	$SN(0) = 13$	$SN(<) = 21$
$SN(\vee) = 7$	$SN(S) = 15$	$SN(x_n) = 21 + 2n \quad (n = 1, 2, \ldots)$

Symbol Table

Now let A be the formula $s_1 \ldots s_n$, where $s_1, \ldots, s_n$ are symbols of the language L$_{\text{NN}}$. Then

$$g(A) = p_1^{SN(s_1)} \times \ldots \times p_n^{SN(s_n)},$$

where $p_1 < \ldots < p_n$ are the first n primes in increasing order.

Example 1:

The code of a formula is obtained by generating consecutive primes and then using the symbol table. For example,

$$g(\neg(Sx_1 = 0)) = 2^5 3^1 5^{15} 7^{23} 11^{11} 13^{13} 17^3.$$

We can also go in the other direction: Given $a \in \mathbb{N}^+$, factor a, say

$$a = 2^9 3^{23} 5^1 7^{13} 11^{21} 13^{23} 17^3,$$

and then use the symbol table to recover the formula whose code is a, in this case $\forall x_1(0 < x_1)$. $\qquad\square$

We note several facts about symbol numbers and codes:

(1) Every symbol number is odd; conversely, for every odd number a, there is a unique symbol s such that $SN(s) = a$.

(2) Given a finite sequence $a_1, \ldots, a_n$ of odd numbers, we can use the symbol table to find the symbols $s_1, \ldots, s_n$ such that $SN(s_k) = a_k$ for $1 \le k \le n$. We can also mechanically check if $s_1 \ldots s_n$ (an expression in the language L$_{\text{NN}}$) is actually a formula of L$_{\text{NN}}$.

(3) The code of a formula is even and is of the form $p_1^{e_1} \times \ldots \times p_n^{e_n}$, where $p_1 < \ldots < p_n$ are the first n primes in increasing order and $e_1, \ldots, e_n$ are all odd.

(4) Not every even natural number is the code of a formula.

Proof that g is one-to-one: The key to the proof is the unique factorization theorem (Theorem 10 in the appendix on number theory). Let A and B be two formulas with $g(A) = g(B)$; we are required to prove that $A = B$. Let $A = s_1 \ldots s_n$ and $B = t_1 \ldots t_m$, where $s_1, \ldots, s_n, t_1, \ldots, t_m$ are symbols. From the definition of g,

$$g(A) = p_1^{SN(s_1)} \times \ldots \times p_n^{SN(s_n)} \quad (p_1 < \ldots < p_n \text{ first } n \text{ primes})$$
$$g(B) = p_1^{SN(t_1)} \times \ldots \times p_m^{SN(t_m)} \quad (p_1 < \ldots < p_m \text{ first } m \text{ primes}).$$

Since $g(A) = g(B)$, unique factorization gives $n = m$ and $SN(s_k) = SN(t_k)$ for $1 \le k \le n$. It follows that $s_k = t_k$ for $1 \le k \le n$, so $A = B$ as required.

Algorithm for Gö1:

(1) Input the formula $A = s_1 \ldots s_n$.

(2) Find the first n primes $p_1 < \ldots < p_n$.

(3) Use the symbol table to find $SN(s_1), \ldots, SN(s_n)$.

(4) Calculate $g(A) = p_1^{SN(s_1)} \times \ldots \times p_n^{SN(s_n)}$.

Algorithm for Gö2:

(1) Input $a \in \mathbb{N}$; if $a \le 1$, go to (7).

(2) Factor $a = p_1^{e_1} \times \ldots \times p_n^{e_n}$ ($p_1 < \ldots < p_n$ primes, $e_k \ge 1$ for $1 \le k \le n$).

(3) Are $p_1, \ldots, p_n$ the first n primes? If *No*, go to (7).

(4) Are $e_1, \ldots, e_n$ all odd? If *No*, go to (7).

(5) Use the symbol table to find symbols $s_1, \ldots, s_n$ such that $SN(s_k) = e_k$ for $1 \le k \le n$. Is $s_1 \ldots s_k$ a formula of L_{NN}? If *No*, go to (7).

(6) If *Yes*, print YES and $s_1 \ldots s_k$; halt.

(7) Print NO and halt. ∎

Now that the language L_{NN} is coded, we can define relations on $\mathbb{N}$ in terms of logical concepts. In particular, for each $\Gamma \subseteq L_{NN}$ we will define a 2-ary relation PRF_Γ and a 1-ary relation THM_Γ. The definition of PRF_Γ requires that we assign codes to proofs using Γ.

Definition 1: Let A be a formula and Γ a set of formulas of L_{NN}. A number $b \in \mathbb{N}$ is the *code of a proof of A using Γ* if

$$b = p_1^{g(A_1)} \times \ldots \times p_n^{g(A_n)},$$

where $p_1 < \ldots < p_n$ are the first n primes in increasing order and $A_1, \ldots, A_n$ is a finite sequence of formulas of L_{NN} that is a proof of A using Γ.

Definition 2: For each $\Gamma \subseteq \text{FOR}_{NN}$, define PRF_Γ and THM_Γ as follows:

(1) $\text{PRF}_\Gamma(a, b) \Leftrightarrow a$ is the code of a formula A of L_{NN} and b is the code of a proof of A using Γ.

(2) $\text{THM}_\Gamma(a) \Leftrightarrow a$ is the code of a formula A of L_{NN} and $\Gamma \vdash A$.

The next two theorems relate properties of Γ to properties of PRF_Γ or THM_Γ.

Theorem 2: If $\Gamma \subseteq \text{FOR}_{NN}$ is axiomatized, then PRF_Γ is a decidable relation.

Proof: Since Γ is axiomatized, there is an algorithm CHKPRF that, given a finite sequence $A_1, \ldots, A_n$ of formulas of L_{NN}, decides whether $A_1, \ldots, A_n$ is a proof using Γ (see Theorem 4 in Section 5.5). Here is the required algorithm for PRF_Γ:

(1) Input $a, b \in \mathbb{N}$.

(2) Use Gö2 to decide whether a is the code of a formula A. If *No*, go to (9).

(3) If *Yes*, use Gö2 to find the formula A such that $g(A) = a$.

(4) Factor $b = p_1^{e_1} \times \ldots \times p_n^{e_n}$ with $p_1 < \ldots < p_n$ and decide whether $p_1, \ldots, p_n$ are the first n primes in increasing order. If *No*, go to (9).

(5) If *Yes*, use Gö2 to decide whether $e_1, \ldots, e_n$ are codes of formulas. If some e_k $(1 \leq k \leq n)$ is not the code of a formula, go to (9).

(6) If $e_1, \ldots, e_n$ are all codes of formulas, use Gö2 to find $A_1, \ldots, A_n$ such that $g(A_k) = e_k$ for $1 \leq k \leq n$.

(7) Use CHKPRF to decide whether $A_1, \ldots, A_n$ is a proof using Γ. If *No*, go to (9).

(8) Is $A_n = A$? If *Yes*, print YES and halt.

(9) Print NO and halt. ■

Theorem 3: Let $\Gamma \subseteq \text{FOR}_{NN}$. Then Γ is decidable if and only if THM_Γ is a decidable relation.

We postpone the proof of Theorem 3 until Section 7.5 (for now, we will leave it as an exercise for the reader). But note that Theorem 3 allows us to restate Church's theorem as follows: If Γ is a consistent extension of NN, then THM_Γ is not a decidable relation. Thus Church's theorem provides us with a rather natural example of a relation on $\mathbb{N}$ that is not decidable, namely THM_Γ.

─────────────── *Expressible Relations* ───────────────

Relations on $\mathbb{N}$ are usually defined in terms of number-theoretic properties. Consider, for example, the divisibility relation DIV, defined as follows:

DIV $(a, b) \Leftrightarrow a \neq 0$ and there exists $c \in \mathbb{N}$ such that $b = a \times c$.

But coding makes it possible to define relations on $\mathbb{N}$ in terms of logical concepts (e.g., PRF_Γ and THM_Γ). It is intuitively clear that the formula $\neg(x_1 = 0) \wedge \exists x_3(x_2 = x_3 \times x_1)$ "talks about" the divisibility relation DIV. Is there a formula of L_{NN} that "talks about" the relation PRF_Γ?

Definition 3: An n-ary relation R on $\mathbb{N}$ is *expressible in* NN (or simply *expressible*) if there is a formula A with exactly n free variables $x_1, \ldots, x_n$ such that for all $a_1, \ldots, a_n \in \mathbb{N}$,

> **EX1** If $R(a_1, \ldots, a_n)$, then NN $\vdash A_{x_1, \ldots, x_n}[0^{a_1}, \ldots, 0^{a_n}]$.
>
> **EX2** If $\neg R(a_1, \ldots, a_n)$, then NN $\vdash \neg A_{x_1, \ldots, x_n}[0^{a_1}, \ldots, 0^{a_n}]$.

We also say that the formula A *expresses* the relation R. (*Reminder on notation:* $R(a_1, \ldots, a_n)$ means $\langle a_1, \ldots, a_n \rangle \in R$; $\neg R(a_1, \ldots, a_n)$ means $\langle a_1, \ldots, a_n \rangle \notin R$).

Expressibility is a precise syntactic way of capturing the idea that a formula "talks about" a relation on $\mathbb{N}$. The conditions EX1 and EX2 are explained as follows: Given $a_1, \ldots, a_n \in \mathbb{N}$, construct the sentence $A_{x_1, \ldots, x_n}[0^{a_1}, \ldots, 0^{a_n}]$ by replacing the free occurrences of $x_1, \ldots, x_n$ in A with the terms $0^{a_1}, \ldots, 0^{a_n}$ respectively (recall that in the standard interpretation, 0^a is interpreted as the natural number a). EX1 and EX2 now state that either $A_{x_1, \ldots, x_n}[0^{a_1}, \ldots, 0^{a_n}]$ or $\neg A_{x_1, \ldots, x_n}[0^{a_1}, \ldots, 0^{a_n}]$ is a theorem of NN, depending on whether $R(a_1, \ldots, a_n)$ or $\neg R(a_1, \ldots, a_n)$. We state, without proof, the following results on expressibility.

Theorem 4: Let Γ be a consistent extension of NN and let A be a formula with exactly n free variables $x_1, \ldots, x_n$ that expresses the n-ary relation R. Then for all $a_1, \ldots, a_n \in \mathbb{N}$, $R(a_1, \ldots, a_n) \Leftrightarrow \Gamma \vdash A_{x_1, \ldots, x_n}[0^{a_1}, \ldots, 0^{a_n}]$.

Theorem 5: Assume that the standard interpretation is a model of NN and let A be a formula with exactly n free variables $x_1, \ldots, x_n$ that expresses the n-ary relation R. Then for all $a_1, \ldots, a_n \in \mathbb{N}$, $R(a_1, \ldots, a_n) \Leftrightarrow A_{x_1, \ldots, x_n}[0^{a_1}, \ldots, 0^{a_n}]$ is true in $\mathcal{N}$.

We are going to give examples of expressible relations, but first we need parts (1) through (4) of Theorem 1 in Section 7.1, on basic properties of the formal system NN.

Theorem 6 (Equality/less-than in NN): The following hold for all $a, b \in \mathbb{N}$:

> **(1)** If a equals b, then NN $\vdash 0^a = 0^b$.
>
> **(2)** If a does not equal b, then NN $\vdash \neg(0^a = 0^b)$.
>
> **(3)** If a is less than b, then NN $\vdash 0^a < 0^b$.
>
> **(4)** If a is not less than b, then NN $\vdash \neg(0^a < 0^b)$.

Proof of (1): Assume that a and b are the same. We have:

(1) $\forall x_1(x_1 = x_1)$	REF AX
(2) $0^a = 0^a$	$\forall$-ELIM
(3) $0^a = 0^b$	b is the same as a.

Proof of (2): Let $S(b)$ say: If some number, say a, is greater than b, then $\mathrm{NN} \vdash \neg(0^a = 0^b)$. We use induction to prove that $S(b)$ holds for all $b \geq 0$. (*Warning:* The symbol 0 is used in two ways in this proof, namely as a constant symbol of the language L_{NN} and as the natural number zero; we rely on the context to make it clear which use is intended.)

Beginning step $S(0)$ says: If a is a natural number that is greater than 0, then $\mathrm{NN} \vdash \neg(0^a = 0)$. Since a is greater than 0, $a - 1$ is a natural number and we have

(1) $\forall x_1 \neg(Sx_1 = 0)$	NN1
(2) $\neg(S0^{a-1} = 0)$	$\forall$-ELIM
(3) $\neg(0^a = 0)$	NOTATION ($S0^{a-1}$ is 0^a).

Induction step Let $b \geq 0$ and assume that $S(b)$ holds. $S(b + 1)$ says: If some number, say a, is greater than $b + 1$, then $\mathrm{NN} \vdash \neg(0^a = 0^{b+1})$. Since a is greater than $b + 1$, $a - 1$ is greater than b, and by the induction hypothesis $S(b)$, we have $\mathrm{NN} \vdash \neg(0^{a-1} = 0^b)$. Continuing:

(1) $\neg(0^{a-1} = 0^b)$	THM of NN
(2) $\forall x_1 \forall x_2[(Sx_1 = Sx_2) \rightarrow (x_1 = x_2)]$	NN2
(3) $(S0^{a-1} = S0^b) \rightarrow (0^{a-1} = 0^b)$	$\forall$-ELIM (twice)
(4) $(0^a = 0^{b+1}) \rightarrow (0^{a-1} = 0^b)$	NOTATION ($S0^{a-1}$ is 0^a and $S0^b$ is 0^{b+1})
(5) $\neg(0^{a-1} = 0^b) \rightarrow \neg(0^a = 0^{b+1})$	CP
(6) $\neg(0^a = 0^{b+1})$	MP(1, 5).

This completes the proof that $S(b)$ holds for all $b \geq 0$.

Now, by assumption, a and b are not equal. First suppose that a is greater than b. Then by $S(b)$, we have $\mathrm{NN} \vdash \neg(0^a = 0^b)$, as required. Next, assume that b is greater than a. Then by the result proved above, with a and b reversed, we have $\mathrm{NN} \vdash \neg(0^b = 0^a)$. Continuing:

(1) $\neg(0^b = 0^a)$	THM of NN
(2) $(0^a = 0^b) \rightarrow (0^b = 0^a)$	SYM (see Cor. 1 in Section 5.1)
(3) $\neg(0^b = 0^a) \rightarrow \neg(0^a = 0^b)$	CP
(4) $\neg(0^a = 0^b)$	MP(1, 3).

Proof of (4): Let $S(b)$ say: If $a \geq b$ (i.e., if a is not less than b), then $\mathrm{NN} \vdash \neg(0^a < 0^b)$. We will use induction to prove that $S(b)$ holds for all $b \geq 0$.

Beginning step $S(0)$ says: If $a \geq 0$, then $\mathrm{NN} \vdash \neg(0^a < 0)$. We have

(1) $\forall x_1 \neg(x_1 < 0)$	NN7
(2) $\neg(0^a < 0)$	$\forall$-ELIM.

Induction step Assume that $S(b)$ holds. $S(b + 1)$ says: If $a \geq b + 1$, then NN $\vdash \neg(0^a < 0^{b+1})$. Now $a \geq b + 1$ implies $a \geq b$ and $a \neq b$, so by induction hypothesis $S(b)$ and part (2), we have NN $\vdash \neg(0^a < 0^b)$ and NN $\vdash \neg(0^a = 0^b)$. Continuing:

(1) $\neg(0^a < 0^b)$	THM of NN
(2) $\neg(0^a = 0^b)$	THM of NN
(3) $\forall x_1 \forall x_2 [(x_1 < Sx_2) \rightarrow ((x_1 < x_2) \vee (x_1 = x_2))]$	NN8
(4) $(0^a < S0^b) \rightarrow [(0^a < 0^b) \vee (0^a = 0^b)]$	∀-ELIM (twice)
(5) $(0^a < 0^{b+1}) \rightarrow [(0^a < 0^b) \vee (0^a = 0^b)]$	NOTATION ($S0^b$ is 0^{b+1})
(6) $\neg(0^a < 0^{b+1}) \vee [(0^a < 0^b) \vee (0^a = 0^b)]$	DEF →
(7) $[\neg(0^a < 0^{b+1}) \vee (0^a < 0^b)] \vee (0^a = 0^b)$	ASSOC RULE
(8) $\neg(0^a < 0^{b+1}) \vee (0^a < 0^b)$	DS(2, 7)
(9) $\neg(0^a < 0^{b+1})$	DS(1, 8).

Proof of (3): Let a be less than b; we are required to show that NN $\vdash 0^a < 0^b$. Now b is not equal to a, and b is not less than a; hence by (2) and (4), we have NN $\vdash \neg(0^b = 0^a)$ and NN $\vdash \neg(0^b < 0^a)$. Continuing:

(1) $\neg(0^b = 0^a)$	THM of NN
(2) $\neg(0^b < 0^a)$	THM of NN
(3) $\forall x_1 \forall x_2 [(x_1 < x_2) \vee (x_1 = x_2) \vee (x_2 < x_1)]$	NN9
(4) $(0^b < 0^a) \vee (0^b = 0^a) \vee (0^a < 0^b)$	∀-ELIM (twice)
(5) $(0^b = 0^a) \vee (0^a < 0^b)$	DS(2, 4)
(6) $0^a < 0^b$	DS(1, 5). ∎

Example 2:

The 2-ary relation of *equality* on $\mathbb{N}$ is expressible. The obvious choice for a formula with two free variables x_1 and x_2 is $x_1 = x_2$. We need to check that for all $a, b \in \mathbb{N}$:

EX1 If a equals b, then NN $\vdash (0^a = 0^b)$.
EX2 If a does not equal b, then NN $\vdash \neg(0^a = 0^b)$.

These are parts (1) and (2) of Theorem 6. ☐

Example 3:

The 2-ary relation *less than* on $\mathbb{N}$ is expressible. The required formula with two free variables x_1 and x_2 is $x_1 < x_2$. We need to check that for all $a, b \in \mathbb{N}$:

EX1 If a is less than b, then NN $\vdash (0^a < 0^b)$.
EX2 If a is not less than b, then NN $\vdash \neg(0^a < 0^b)$.

These are parts (3) and (4) of Theorem 6. ☐

A simple counting argument shows that not all relations on $\mathbb{N}$ are expressible.

Theorem 7: Assume NN is consistent. Then there is a 1-ary relation on $\mathbb{N}$ that is not expressible.

Outline of Proof: The number of 1-ary relations on $\mathbb{N}$ is uncountable, but the number of formulas of L_{NN} is countable; hence there are not enough formulas to express all relations on $\mathbb{N}$ (see Exercise 8 for details). ∎

So far, the only method we have of showing that an n-ary relation is expressible is to find a formula A with exactly n free variables $x_1, \ldots, x_n$ and then to prove the two expressibility conditions EX1 and EX2. But even for simple relations such as *equality* and *less than,* this requires considerable effort; for complicated relations, this method is totally impractical. Fortunately, there is an alternate method.

Theorem 8 (Gödel's theorem on expressibility): Let R be an n-ary relation on $\mathbb{N}$. If R is recursive, then R is expressible.

This important result will be proved in the next section. Although Gödel's expressibility theorem greatly simplifies the procedure for showing that a relation is expressible, it does require technical proficiency with recursive functions and relations. But these technical details can be avoided by assuming Church's thesis—in other words, by assuming that all decidable relations are recursive. Let us outline the procedure used throughout this chapter to prove that a relation R is expressible: (1) Show that R is decidable; (2) assert that R is recursive by Church's thesis; (3) use Gödel's theorem on expressibility to conclude that R is expressible. We emphasize that this is a nonessential use of Church's thesis; in all cases, the methods described in Chapter 8 can be used to give a direct proof that the relation is recursive. We will illustrate with two examples.

Example 4:

Define R by $R(a) \Leftrightarrow a$ is even, $a \geq 4$, and a is the sum of two primes. The relation R is decidable, and hence recursive (Church's thesis); by Gödel's theorem on expressibility, R is expressible. □

Example 5:

Let Γ be an axiomatized extension of L_{NN}. We will construct a sentence CON_Γ of L_{NN} that asserts the consistency of Γ. The 2-ary relation PRF_Γ is decidable (see Theorem 2), hence PRF_Γ is expressible. Let Prf_Γ be a formula of L_{NN} with exactly two free variables x_1 and x_2 that expresses

PRF$_\Gamma$. Now the 1-ary relation FOR, defined by FOR(a) $\Leftrightarrow$ a is the code of a formula of L$_{NN}$, is decidable (use Gö2); hence there is a formula *For* of L$_{NN}$ with exactly one free variable x_1 that expresses FOR. Finally, let CON$_\Gamma$ be the sentence

$$\exists x_1[\textit{For} \wedge \neg\exists x_2 \textit{Prf}_\Gamma].$$

An informal translation of this sentence is: There is a natural number a such that (1) a is the code of a formula A and (2) there is no natural number b that is the code of a proof of A. In other words, CON$_\Gamma$ states that there is at least one formula that is not a theorem of Γ; this means that Γ is consistent. Gödel's second incompleteness theorem asserts that under certain conditions the sentence CON$_\Gamma$ is not a theorem of Γ (see Section 7.4). $\qquad\qquad\square$

───────────── *Exercises for Section 7.2* ─────────────

1. Which of the following numbers are codes? If the number is a code, find the formula that has the number as its code.
   (1) 100 $\qquad\qquad$ (3) $2^1 \times 3^{13} \times 5^{11} \times 7^{13} \times 11^3$
   (2) $2^5 \times 3^1 \times 5^{21} \times 7^3$
2. Let $\Gamma \subseteq$ FOR$_{NN}$ and define a 1-ary relation on $\mathbb{N}$ by NLA$_\Gamma(a)$ $\Leftrightarrow$ a is the code of a formula A and $A \in \Gamma$. Prove the following (use Gö1 and Gö2):
   (1) Γ is axiomatized if and only if NLA$_\Gamma$ is a decidable relation.
   (2) Γ is decidable if and only if THM$_\Gamma$ is a decidable relation.
   (3) Γ is semidecidable (= listable) if and only if THM$_\Gamma$ is a semidecidable relation.
3. Prove each relation expressible by proving it decidable:
   (1) $R(a, b, c, n) \Leftrightarrow a, b, c \in \mathbb{N}^+$, $n > 2$, and $a^n + b^n = c^n$.
   (2) $R(c, n) \Leftrightarrow c \in \mathbb{N}^+$, $n > 2$, and the equation $x^n + y^n = c^n$ has a solution in positive integers.
4. (*Expressible* $\Rightarrow$ *weakly expressible*) Prove Theorem 4.
5. Assume that NN is consistent and let R be a 1-ary relation on $\mathbb{N}$ that is expressible. Argue that R is a decidable relation.
6. (*Expressible* $\Rightarrow$ *definable*) Prove Theorem 5.
7. Let P and Q be n-ary relations on $\mathbb{N}$; see Section 8.2 (Definition 1) for the definition of $\neg P$ and $P \vee Q$. Prove the following:
   (1) If P is expressible, then $\neg P$ is expressible.
   (2) If P and Q are expressible, then $P \vee Q$ is expressible.
8. Let Q and R be distinct 1-ary expressible relations on $\mathbb{N}$, and let A and B be formulas, each with exactly one free variable x_1, that express Q and R respectively. Show that A and B are distinct formulas (assuming NN is consistent). Then use this result to prove there is a 1-ary relation that is not expressible.
9. Show that each of the following relations is decidable (and hence expressible):
   (1) LAX$_{NN}(a)$ $\Leftrightarrow$ a is the code of a logical axiom of NN.
   (2) DS(a, b, c) $\Leftrightarrow$ a is the code of a formula $\neg A$, b is the code of a formula $A \vee B$, and c is the code of B.
10. (*Variations of* CON$_\Gamma$) Let *Prf* be a formula of L$_{NN}$ with exactly two free variables x_1 and x_2 that expresses PRF$_\Gamma$. Translate the following sentences of L$_{NN}$:
    (1) $\neg\exists x_2(\textit{Prf}_{x_1}[0^c])$, where $c = 2^5 \times 3^1 \times 5^{13} \times 7^{11} \times 11^{13} \times 13^3$.
    (2) $\neg\forall x_1[\textit{For} \rightarrow \exists x_2 \textit{Prf}]$.
11. (*Another variation of* CON$_\Gamma$) Define a 2-ary relation on $\mathbb{N}$ by NEG (a, b) $\Leftrightarrow$ a is the code of a formula A and b is the code of the formula $\neg A$.
    (1) Show that NEG is decidable.
    (2) Let *Neg* be a formula with exactly two free variables x_1 and x_3 that expresses

NEG. Assume *Prf* has no occurrences of x_3 or x_4. Translate the sentence $\neg \exists x_1 \exists x_2 \exists x_3 \exists x_4 (Prf \wedge Prf_{x_1, x_2}[x_3, x_4] \wedge Neg)$.

7.3

Recursive Relation $\Rightarrow$ *Expressible Relation*

In this section, we will prove Gödel's fundamental result that every recursive relation is expressible. The proof requires no knowledge of recursive functions other than the definition and the method of proof known as *induction on recursive functions* (see Section 1.8). On the other hand, we are required to write formal proofs in NN, and this is where we see the reason for choosing the particular axioms NN1–NN9 for NN. (The precise results needed are summarized in Theorem 1 of Section 7.1; we have already proved (1)–(4).) We take a three-step approach: (1) Introduce the class of *representable* functions; (2) prove that every recursive function is representable; (3) prove that a relation R is expressible if and only if K_R is representable. We then have

$$R \text{ recursive} \Rightarrow K_R \text{ recursive (definition of } R \text{ recursive)}$$
$$\Rightarrow K_R \text{ representable (by (2))}$$
$$\Rightarrow R \text{ expressible (by (3))}.$$

Definition 1: An *n*-ary function F is *representable in* NN (or simply *representable*) if there is a formula A with exactly $n + 1$ free variables $x_1, \ldots, x_{n+1}$ such that for all $a_1, \ldots, a_n \in \mathbb{N}$,

$$\text{NN} \vdash A_{x_1, \ldots, x_n}[0^{a_1}, \ldots, 0^{a_n}] \leftrightarrow (x_{n+1} = 0^{F(a_1, \ldots, a_n)}).$$

In this case, we also say that the formula A *represents* F.

This is a precise syntactic way of asserting that a formula A "talks about" a function F. To emphasize this claim, we ask the reader to verify the following (A represents F, $a_1, \ldots, a_n \in \mathbb{N}$):

(1) $\text{NN} \vdash A_{x_1, \ldots, x_n, x_{n+1}}[0^{a_1}, \ldots, 0^{a_n}, 0^{F(a_1, \ldots, a_n)}]$.

(2) For all $b \in \mathbb{N}$ with $b \neq F(a_1, \ldots, a_n)$, $\text{NN} \vdash \neg A_{x_1, \ldots, x_n, x_{n+1}}$ $[0^{a_1}, \ldots, 0^{a_n}, 0^b]$.

Our first result states that addition and multiplication are representable. But first we need two lemmas (we will be using various derived rules of inference based on the equality axioms; see Sections 5.1 and 5.3).

Lemma 1 (Addition/multiplication in NN): Let a and b be natural numbers.

(1) $\text{NN} \vdash 0^a + 0^b = 0^{a+b}$.

(2) $\text{NN} \vdash 0^a \times 0^b = 0^{a \times b}$.

Proof of (1): In the statement of this result, the symbol $+$ is used in two different ways: In the term $0^a + 0^b$, it is a symbol of the language L_{NN}; in the term 0^{a+b}, it denotes addition of natural numbers. The context should make it clear which use is intended. The proof of (1) is by induction on b.

Beginning step For $b = 0$, we must show $NN \vdash 0^a + 0 = 0^a$.

(1) $\forall x_1(x_1 + 0 = x_1)$ NN3
(2) $0^a + 0 = 0^a$ $\forall$-ELIM.

Induction step Assume $NN \vdash 0^a + 0^b = 0^{a+b}$; we are required to show that $NN \vdash 0^a + 0^{b+1} = 0^{a+b+1}$.

(1) $\forall x_1 \forall x_2[x_1 + Sx_2 = S(x_1 + x_2)]$ NN4
(2) $0^a + S0^b = S(0^a + 0^b)$ $\forall$-ELIM (twice)
(3) $0^a + 0^{b+1} = S(0^a + 0^b)$ NOTATION ($S0^b$ is 0^{b+1})
(4) $0^a + 0^b = 0^{a+b}$ THM of NN (ind hyp)
(5) $S(0^a + 0^b) = S0^{a+b}$ FUNCTION RULE
(6) $S(0^a + 0^b) = 0^{a+b+1}$ NOTATION ($S0^{a+b}$ is 0^{a+b+1})
(7) $0^a + 0^{b+1} = 0^{a+b+1}$ TRANS RULE (3, 6).

Proof of (2): Again, the proof is by induction on b.

Beginning step For $b = 0$, we must show $NN \vdash 0^a \times 0 = 0$.

(1) $\forall x_1(x_1 \times 0 = 0)$ NN5
(2) $0^a \times 0 = 0$ $\forall$-ELIM.

Induction step Assume $NN \vdash 0^a \times 0^b = 0^{a \times b}$; we are required to show that $NN \vdash 0^a \times 0^{b+1} = 0^{a \times b + a}$.

(1) $\forall x_1 \forall x_2[x_1 \times Sx_2 = (x_1 \times x_2) + x_1]$ NN6
(2) $0^a \times S0^b = (0^a \times 0^b) + 0^a$ $\forall$-ELIM (twice)
(3) $0^a \times 0^{b+1} = (0^a \times 0^b) + 0^a$ NOTATION ($S0^b$ is 0^{b+1})
(4) $0^a \times 0^b = 0^{a \times b}$ THM of NN (ind hyp)
(5) $(0^a \times 0^b) + 0^a = 0^{a \times b} + 0^a$ FUNCTION RULE
                                         (add 0^a to (4))
(6) $0^a \times 0^{b+1} = 0^{a \times b} + 0^a$ TRAN RULE (3, 5)
(7) $0^{a \times b} + 0^a = 0^{a \times b + a}$ PART 1
(8) $0^a \times 0^{b+1} = 0^{a \times b + a}$ TRAN RULE (6, 7). ∎

Lemma 2: Let x be a variable and let s and t be variable-free terms of L_{NN}. If $NN \vdash s = t$, then $NN \vdash (x = s) \leftrightarrow (x = t)$.

Proof: Let y be a variable distinct from x.

(1) $s = t$ THM of NN
(2) $(y = t) \rightarrow [(x = y) \leftrightarrow (x = t)]$ EQ AX
(3) $(s = t) \rightarrow [(x = s) \leftrightarrow (x = t)]$ SUBST RULE (s for y)
(4) $(x = s) \leftrightarrow (x = t)$ MP(1, 3). ∎

Theorem 1: Addition and multiplication are representable.

Proof: The formula $x_3 = x_1 + x_2$ represents addition. To see this, let $a, b \in \mathbb{N}$; we are required to show $NN \vdash (x_3 = 0^a + 0^b) \leftrightarrow (x_3 = 0^{a+b})$. By Lemma 1, we have $NN \vdash 0^a + 0^b = 0^{a+b}$; now apply Lemma 2, with s the term $0^a + 0^b$ and t the term 0^{a+b}, to obtain the required result.

The formula $x_3 = x_1 \times x_2$ represents multiplication. Let $a, b \in \mathbb{N}$; by a similar argument, $NN \vdash (x_3 = 0^a \times 0^b) \leftrightarrow (x_3 = 0^{a \times b})$ as required. ∎

Theorem 2: The projection functions $U_{n,k}$ are representable ($n \geq 1, 1 \leq k \leq n$).

Proof: The formula $(x_{n+1} = x_k) \wedge (x_1 = x_1) \wedge \ldots \wedge (x_n = x_n)$ represents $U_{n,k}$. To prove this, let $a_1, \ldots, a_n \in \mathbb{N}$; we are required to show that

$$NN \vdash [(x_{n+1} = 0^{a_k}) \wedge (0^{a_1} = 0^{a_1}) \wedge \ldots \wedge (0^{a_n} = 0^{a_n})] \leftrightarrow (x_{n+1} = 0^{U_{n,k}(a_1, \ldots, a_n)}).$$

The details are left to the reader. ∎

The next theorem establishes the relationship between representability and expressibility.

Theorem 3: An n-ary relation R is expressible if and only if its characteristic function K_R is representable.

Proof: Assume R is expressible. Let A be a formula with free variables $x_1, \ldots, x_n$ such that for all $a_1, \ldots, a_n \in \mathbb{N}$,

EX1 If $R(a_1, \ldots, a_n)$, then $NN \vdash A_{x_1, \ldots, x_n}[0^{a_1}, \ldots, 0^{a_n}]$.
EX2 If $\neg R(a_1, \ldots, a_n)$, then $NN \vdash \neg A_{x_1, \ldots, x_n}[0^{a_1}, \ldots, 0^{a_n}]$.

The formula

$$[A \wedge (x_{n+1} = 0)] \vee [\neg A \wedge (x_{n+1} = S0)],$$

with free variables $x_1, \ldots, x_n, x_{n+1}$, represents K_R. To see this, fix $a_1, \ldots, a_n \in \mathbb{N}$, and denote $A_{x_1, \ldots, x_n}[0^{a_1}, \ldots, 0^{a_n}]$ by B. With this notation, EX1 and EX2 become

EX1 If $R(a_1, \ldots, a_n)$, then $NN \vdash B$.

EX2 If $\neg R(a_1, \ldots, a_n)$, then $NN \vdash \neg B$.

And, we are required to show that

$$NN \vdash [(B \wedge (x_{n+1} = 0)) \vee (\neg B \wedge (x_{n+1} = S0))]$$

$$\leftrightarrow (x_{n+1} = 0^{K_R(a_1, \ldots, a_n)}). \quad (*)$$

We consider two cases: Suppose $R(a_1, \ldots, a_n)$. By EX1, we have $\text{NN} \vdash B$. Moreover, $K_R(a_1, \ldots, a_n) = 0$, so (*) reduces to

$$\text{NN} \vdash [(B \wedge (x_{n+1} = 0)) \vee (\neg B \wedge (x_{n+1} = S0))] \leftrightarrow (x_{n+1} = 0). \qquad (**)$$

The proof of (**) assuming $\text{NN} \vdash B$ is a straightforward problem in propositional logic; it amounts to showing that if $\vdash p$, then $\vdash [(p \wedge q) \vee (\neg p \wedge r)] \leftrightarrow q$. One way to proceed is to show that $p \rightarrow ([(p \wedge q) \vee (\neg p \wedge r)] \leftrightarrow q)$ is a tautology. The details are left to the reader. The case $\neg R(a_1, \ldots, a_n)$ is similar, and is left to the reader; this time, $\text{NN} \vdash \neg B$, $K_R(a_1, \ldots, a_n) = 1$, and the tautology is $\neg p \rightarrow ([(p \wedge q) \vee (\neg p \wedge r)] \leftrightarrow r)$.

Now assume K_R is representable. Let A be a formula with exactly $n + 1$ free variables $x_1, \ldots, x_{n+1}$ such that for all $a_1, \ldots, a_n \in \mathbb{N}$,

$$\text{NN} \vdash A_{x_1, \ldots, x_n}[0^{a_1}, \ldots, 0^{a_n}] \leftrightarrow (x_{n+1} = 0^{K_R(a_1, \ldots, a_n)}). \qquad (*)$$

The formula that expresses R is

$$A_{x_{n+1}}[0].$$

To see this, let $a_1, \ldots, a_n \in \mathbb{N}$ and denote $A_{x_1, \ldots, x_n, x_{n+1}}[0^{a_1}, \ldots, 0^{a_n}, 0]$ by B. With this notation, we are required to show that

EX1 If $R(a_1, \ldots, a_n)$, then $\text{NN} \vdash B$.

EX2 If $\neg R(a_1, \ldots, a_n)$, then $\text{NN} \vdash \neg B$.

To prove EX1, assume $R(a_1, \ldots, a_n)$. Then $K_R(a_1, \ldots, a_n) = 0$, and by (*) we have $\text{NN} \vdash A_{x_1, \ldots, x_n}[0^{a_1}, \ldots, 0^{a_n}] \leftrightarrow (x_{n+1} = 0)$. By the SUBST RULE (replace x_{n+1} with 0), we obtain $\text{NN} \vdash B \leftrightarrow (0 = 0)$. But $\text{NN} \vdash 0 = 0$, so $\text{NN} \vdash B$ as required.

To prove EX2, assume $\neg R(a_1, \ldots, a_n)$. Then $K_R(a_1, \ldots, a_n) = 1$, and by (*) we have $\text{NN} \vdash A_{x_1, \ldots, x_n}[0^{a_1}, \ldots, 0^{a_n}] \leftrightarrow (x_{n+1} = S0)$. By the SUBST RULE (replace x_{n+1} with 0), we obtain $\text{NN} \vdash B \leftrightarrow (0 = S0)$. But $\text{NN} \vdash \neg(0 = S0)$ (see Theorem 6 of Section 7.2), and so $\text{NN} \vdash \neg B$ as required. ∎

Theorem 4: $K_<$ is representable.

Proof: In Section 7.2 (Example 3), we proved that the relation $<$ on $\mathbb{N}$ is expressible. It follows from Theorem 3 that its characteristic function $K_<$ is representable. ∎

To summarize, thus far we have proved that all starting functions are representable. The next step is to show that the composition of representable functions is representable. We will begin with the technical result that any set of distinct variables can be used in the definition of representability.

Lemma 3 (New variables): Let F be an n-ary representable function and let $y_1, \ldots, y_n, y_{n+1}$ be any set of $n + 1$ distinct variables. Then there is a formula A with exactly $n + 1$ free variables $y_1, \ldots, y_n, y_{n+1}$ such that for all $a_1, \ldots, a_n \in \mathbb{N}$,

$$\text{NN} \vdash A_{y_1, \ldots, y_n}[0^{a_1}, \ldots, 0^{a_n}] \leftrightarrow (y_{n+1} = 0^{F(a_1, \ldots, a_n)}). \qquad (*)$$

Proof: Since F is representable, there is a formula B with exactly $n + 1$ free variables $x_1, \ldots, x_{n+1}$ such that for all $a_1, \ldots, a_n \in \mathbb{N}$,

$$\text{NN} \vdash B_{x_1, \ldots, x_n}[0^{a_1}, \ldots, 0^{a_n}] \leftrightarrow (x_{n+1} = 0^{F(a_1, \ldots, a_n)}). \tag{**}$$

Let $z_1, \ldots, z_{n+1}$ be $n + 1$ new variables that do not occur in B and that are distinct from $y_1, \ldots, y_{n+1}$. For $1 \le k \le n + 1$, replace all bound occurrences of y_k in B with z_k; call this new formula C. By the theorem on replacement of bound variables (see Section 5.6), $\vdash B \leftrightarrow C$. The desired formula A is obtained from C by replacing each free occurrence of x_k in C with y_k, $1 \le k \le n + 1$; in other words, A is $C_{x_1, \ldots, x_n, x_{n+1}}[y_1, \ldots, y_n, y_{n+1}]$. Note that A has exactly $n + 1$ free variables $y_1, \ldots, y_n, y_{n+1}$; this is so because bound occurrences of the y's have been replaced by the z's.

Let $a_1, \ldots, a_n \in \mathbb{N}$; we are required to prove (*). To simplify notation, let $F(a_1, \ldots, a_n) = b$. We have

(1) $B_{x_1, \ldots, x_n}[0^{a_1}, \ldots, 0^{a_n}] \leftrightarrow (x_{n+1} = 0^b)$ THM of NN (see **)

(2) $B \leftrightarrow C$ THM of FO$_L$

(3) $B_{x_1, \ldots, x_n}[0^{a_1}, \ldots, 0^{a_n}]$ SUBST RULE (n times)
$\leftrightarrow C_{x_1, \ldots, x_n}[0^{a_1}, \ldots, 0^{a_n}]$

(4) $C_{x_1, \ldots, x_n}[0^{a_1}, \ldots, 0^{a_n}] \leftrightarrow (x_{n+1} = 0^b)$ (1), (3), PROP LOGIC

(5) $C_{x_1, \ldots, x_n, x_{n+1}}[0^{a_1}, \ldots, 0^{a_n}, y_{n+1}]$ SUBST RULE (y_{n+1} for x_{n+1}).
$\leftrightarrow (y_{n+1} = 0^b)$

But $C_{x_1, \ldots, x_n, x_{n+1}}[0^{a_1}, \ldots, 0^{a_n}, y_{n+1}]$ is $A_{y_1, \ldots, y_n}[0^{a_1}, \ldots, 0^{a_n}]$, so (*) holds as required. ∎

We will also need Theorem 3 of Section 5.6, as follows.

Lemma 4: Let $y_1, \ldots, y_k$ be distinct variables and let $t_1, \ldots, t_k$ be variable-free terms. For any formula B, $\vdash_{\text{FO}} \exists y_1 \ldots \exists y_k[(y_1 = t_1) \wedge \ldots \wedge (y_k = t_k) \wedge B] \leftrightarrow B_{y_1, \ldots, y_k}[t_1, \ldots, t_k]$.

Theorem 5 (Composition): The composition of representable functions is representable.

Proof: Let G be a k-ary representable function, let $H_1, \ldots, H_k$ be n-ary representable functions, and let

$$F(a_1, \ldots, a_n) = G(H_1(a_1, \ldots, a_n), \ldots, H_k(a_1, \ldots, a_n));$$

we are required to show that F is representable. The formula that represents F is motivated by the following observation:

$$F(a_1, \ldots, a_n) = c \Leftrightarrow \exists b_1 \ldots \exists b_k$$
$$[H_1(a_1, \ldots, a_n) = b_1 \wedge \ldots \wedge H_k(a_1, \ldots, a_n) = b_k \wedge G(b_1, \ldots, b_k) = c].$$

We work with the $n + k + 1$ distinct variables $x_1, \ldots, x_n, x_{n+1}, y_1, \ldots, y_k$. For $1 \le i \le k$, H_i is representable, so there is a formula A^i with exactly $n + 1$ free variables $x_1, \ldots, x_n, y_i$ such that for all $a_1, \ldots, a_n \in \mathbb{N}$,

$$\text{NN} \vdash A^i_{x_1, \ldots, x_n}[0^{a_1}, \ldots, 0^{a_n}] \leftrightarrow (y_i = 0^{H_i(a_1, \ldots, a_n)}), 1 \le i \le k. \tag{*}$$

Likewise, G is representable, so there is a formula B with exactly $k + 1$ free variables $y_1, \ldots, y_k, x_{n+1}$ such that for all $b_1, \ldots, b_k \in \mathbb{N}$,

$$\text{NN} \vdash B_{y_1, \ldots, y_k}[0^{b_1}, \ldots, 0^{b_k}] \leftrightarrow (x_{n+1} = 0^{G(b_1, \ldots, b_k)}). \qquad (**)$$

The formula

$$\exists y_1 \ldots \exists y_k (A^1 \wedge \ldots \wedge A^k \wedge B)$$

with free variables $x_1, \ldots, x_{n+1}$ represents F. To see this, let $a_1, \ldots, a_n \in \mathbb{N}$. For $1 \leq i \leq k$, denote $A^i_{x_1, \ldots, x_n}[0^{a_1}, \ldots, 0^{a_n}]$ by A_i, let $H_i(a_1, \ldots, a_n) = b_i$, and let $G(b_1, \ldots, b_k) = c$. With this notation, (*) and (**) become

(1) $\text{NN} \vdash A_i \leftrightarrow (y_i = 0^{b_i})$, $1 \leq i \leq k$;
(2) $\text{NN} \vdash B_{y_1, \ldots, y_k}[0^{b_1}, \ldots, 0^{b_k}] \leftrightarrow (x_{n+1} = 0^c)$;

and we are required to show that

$$\text{NN} \vdash \exists y_1 \ldots \exists y_k (A_1 \wedge \ldots \wedge A_k \wedge B) \leftrightarrow (x_{n+1} = 0^c). \qquad (***)$$

Now from Lemma 4, we have

$$\vdash_{\text{FO}} \exists y_1 \ldots \exists y_k [(y_1 = 0^{b_1}) \wedge \ldots \wedge (y_k = 0^{b_k}) \wedge B] \leftrightarrow B_{y_1, \ldots, y_k}[0^{b_1}, \ldots, 0^{b_k}].$$

But the replacement theorem (see Section 5.6) allows us to replace $y_i = 0^{b_i}$ with A_i, for $1 \leq i \leq k$ (see (1)) and $B_{y_1, \ldots, y_k}[0^{b_1}, \ldots, 0^{b_k}]$ with $x_{n+1} = 0^c$ (see (2)). This yields (***) as required. ∎

Finally, we need to show that the minimalization of a regular representable function is representable. First we prove two lemmas.

Lemma 5: Let A be a formula of L_{NN} such that $\text{NN} \vdash A_x[0], \ldots, \text{NN} \vdash A_x[0^{n-1}]$. Then $\text{NN} \vdash (x < 0^n) \to A$.

Proof: The proof is by induction.

Beginning step We are required to show that $\text{NN} \vdash (x < 0) \to A$.

(1) $\forall x_1 \neg (x_1 < 0)$ NN7
(2) $\neg (x < 0)$ $\forall$-ELIM
(3) $\neg (x < 0) \vee A$ EXP
(4) $(x < 0) \to A$ DEF $\to$.

Induction step Let $n \geq 0$ and assume that $\text{NN} \vdash A_x[0], \ldots, \text{NN} \vdash A_x[0^n]$. By induction hypothesis, we have $\text{NN} \vdash (x < 0^n) \to A$, and we are required to show $\text{NN} \vdash (x < 0^{n+1}) \to A$.

(1) $\forall x_1 \forall x_2 [(x_1 < Sx_2)$ NN8
     $\to ((x_1 < x_2) \vee (x_1 = x_2))]$
(2) $(x_1 < S0^n) \to ((x_1 < 0^n) \vee (x_1 = 0^n))$ $\forall$-ELIM (twice)
(3) $(x_1 < 0^{n+1}) \to ((x_1 < 0^n) \vee (x_1 = 0^n))$ NOTATION ($S0^n$ is 0^{n+1})
(4) $(x < 0^{n+1}) \to ((x < 0^n) \vee (x = 0^n))$ SUBST RULE
(5) $(x < 0^n) \to A$ THM of NN (ind hyp)
(6) $A_x[0^n]$ THM of NN (hyp)

    (7) $A_x[0^n] \to ((x = 0^n) \to A)$ EQ AX (4)

    (8) $(x = 0^n) \to A$ MP

    (9) $[(x < 0^n) \vee (x = 0^n)] \to A$ TT (5, 8; see (*))

    (10) $(x < 0^{n+1}) \to A$ HS (4, 9).

(*) $(p \to r) \to [(q \to r) \to ((p \vee q) \to r)]$ and then MP with lines (5) and (8). ∎

Lemma 6: An n-ary function F is representable if and only if there is a formula A with exactly $n + 1$ free variables $x_1, \ldots, x_{n+1}$ such that for all $a_1, \ldots, a_n \in \mathbb{N}$,

R1 $NN \vdash A_{x_1, \ldots, x_n, x_{n+1}}[0^{a_1}, \ldots, 0^{a_n}, 0^{F(a_1, \ldots, a_n)}]$.

R2 $NN \vdash A_{x_1, \ldots, x_n}[0^{a_1}, \ldots, 0^{a_n}] \to (x_{n+1} = 0^{F(a_1, \ldots, a_n)})$.

Proof: Assume there is a formula A such that R1 and R2 hold; we are required to show that for all $a_1, \ldots, a_n \in \mathbb{N}$,

$$NN \vdash A_{x_1, \ldots, x_n}[0^{a_1}, \ldots, 0^{a_n}] \leftrightarrow (x_{n+1} = 0^{F(a_1, \ldots, a_n)}).$$

By R2, it suffices to show that

$$NN \vdash (x_{n+1} = 0^{F(a_1, \ldots, a_n)}) \to A_{x_1, \ldots, x_n}[0^{a_1}, \ldots, 0^{a_n}].$$

To simplify notation, let B be the formula $A_{x_1, \ldots, x_n}[0^{a_1}, \ldots, 0^{a_n}]$. We have:

    (1) $B_{x_{n+1}}[0^{F(a_1, \ldots, a_n)}]$ THM of NN (see R1)

    (2) $B_{x_{n+1}}[0^{F(a_1, \ldots, a_n)}]$ EQ AX (4)

        $\to [(x_{n+1} = 0^{F(a_1, \ldots, a_n)}) \to B]$

    (3) $(x_{n+1} = 0^{F(a_1, \ldots, a_n)}) \to B$ MP.

We leave the other half of the proof to the reader. ∎

Theorem 6 (Minimalization): Let F be an n-ary function that is the minimalization of an $n + 1$-ary regular function G. If G is representable, then so is F.

Proof: Let $a_1, \ldots, a_n \in \mathbb{N}$. By the way in which F is defined,

$$F(a_1, \ldots, a_n) = c \Leftrightarrow [G(a_1, \ldots, a_n, c)$$
$$= 0 \wedge \forall k((k < c) \to G(a_1, \ldots, a_n, k) \neq 0)]. \qquad (*)$$

Since G is representable, there is, by Lemma 6, a formula A with exactly $n + 2$ free variables $x_1, \ldots, x_{n+2}$ such that for all $a_1, \ldots, a_n, b \in \mathbb{N}$,

R1 $NN \vdash A_{x_1, \ldots, x_{n+1}, x_{n+2}}[0^{a_1}, \ldots, 0^{a_n}, 0^b, 0^{G(a_1, \ldots, a_n, b)}]$.

R2 $NN \vdash A_{x_1, \ldots, x_{n+1}}[0^{a_1}, \ldots, 0^{a_n}, 0^b] \to (x_{n+2} = 0^{G(a_1, \ldots, a_n, b)})$.

Let y be a variable that does not occur in A; (*) suggests that the formula

$$A_{x_{n+2}}[0] \wedge \forall y((y < x_{n+1}) \to \neg A_{x_{n+1}, x_{n+2}}[y, 0])$$

with free variables $x_1, \ldots, x_{n+1}$ represents F. To verify this, let

$a_1, \ldots, a_n \in \mathbb{N}$ and denote $A_{x_1, \ldots, x_n, x_{n+2}}[0^{a_1}, \ldots, 0^{a_n}, 0]$ by B. With this notation, we are required to show that:

R1′ NN $\vdash B_{x_{n+1}}[0^{F(a_1, \ldots, a_n)}] \wedge \forall y((y < 0^{F(a_1, \ldots, a_n)}) \rightarrow \neg B_{x_{n+1}}[y])$.

R2′ NN $\vdash [B \wedge \forall y((y < x_{n+1}) \rightarrow \neg B_{x_{n+1}}[y])] \rightarrow (x_{n+1} = 0^{F(a_1, \ldots, a_n)})$.

Let $F(a_1, \ldots, a_n) = c$; then $G(a_1, \ldots, a_n, c) = 0$ and $G(a_1, \ldots, a_n, k) \neq 0$ for $k < c$. We first prove that

(a) NN $\vdash B_{x_{n+1}}[0^c]$.

(b) NN $\vdash \neg B_{x_{n+1}}[0^k]$ for $0 \leq k \leq c - 1$.

To prove (a), use R1 with $b = c$. To prove (b), use R2 with $b = k$ and the SUBST RULE (replace x_{n+2} with 0) to obtain

$$\text{NN} \vdash B_{x_{n+1}}[0^k] \rightarrow (0 = 0^{G(a_1, \ldots, a_n, k)}).$$

Now $G(a_1, \ldots, a_n, k) \neq 0$; hence NN $\vdash \neg(0 = 0^{G(a_1, \ldots, a_n, k)})$ (see Theorem 6 of Section 7.2); it follows that NN $\vdash \neg B_{x_{n+1}}[0^k]$ as required.

Now from (b) and Lemma 5, we obtain NN $\vdash (x_{n+1} < 0^c) \rightarrow \neg B$. By the SUBST RULE (replace x_{n+1} by y) and GEN, we obtain

$$\text{NN} \vdash \forall y((y < 0^c) \rightarrow \neg B_{x_{n+1}}[y]).$$

The conjunction of this result with (a) gives R1′ as required.

It remains to prove R2′. We prove that

NN $\vdash (x_{n+1} < 0^c) \vee (x_{n+1} = 0^c) \vee (0^c < x_{n+1})$;

NN $\vdash B \rightarrow \neg(x_{n+1} < 0^c)$;

NN $\vdash \forall y((y < x_{n+1}) \rightarrow \neg B_{x_{n+1}}[y]) \rightarrow \neg(0^c < x_{n+1})$;

and then use propositional logic as follows:

$$p \vee q \vee r, s \rightarrow \neg p, t \rightarrow \neg r, \therefore (s \wedge t) \rightarrow q.$$

(1) $\forall x_1 \forall x_2[(x_1 < x_2) \vee (x_1 = x_2)$	NN9
$\quad \vee (x_2 < x_1)]$	
(2) $(x_1 < 0^c) \vee (x_1 = 0^c) \vee (0^c < x_1)$	$\forall$-ELIM (twice)
(3) $(x_{n+1} < 0^c) \vee (x_{n+1} = 0^c)$	SUBST RULE (x_{n+1} for x_1)
$\quad \vee (0^c < x_{n+1})$	

(1) $(x_{n+1} < 0^c) \rightarrow \neg B$	THM of NN (Lemma 5)
(2) $\neg(x_{n+1} < 0^c) \vee \neg B$	DEF $\rightarrow$
(3) $\neg B \vee \neg(x_{n+1} < 0^c)$	CM
(4) $B \rightarrow \neg(x_{n+1} < 0^c)$	DEF $\rightarrow$.

For the last result, we use the deduction theorem.

(1) $B_{x_{n+1}}[0^c]$	THM of NN (see (a))
(2) $\forall y((y < x_{n+1}) \to \neg B_{x_{n+1}}[y])$	HYP
(3) $(0^c < x_{n+1}) \to \neg B_{x_{n+1}}[0^c]$	$\forall$-ELIM
(4) $\neg(0^c < x_{n+1})$	(1), (3), PROP LOGIC. ∎

Theorem 7 (Gödel's theorem on expressibility): Every recursive relation is an expressible relation; every recursive function is a representable function.

Proof: We first show that every recursive function is representable. The proof is by induction on recursive functions, with Q the property of being representable. We have proved that all starting functions are representable; that the composition of representable functions is representable; that the minimalization of a regular representable function is representable. Thus every recursive function is representable as required.

Now let R be a recursive relation. The following steps show that R is expressible: (1) K_R is a recursive function; (2) K_R is a representable function (proved above); (3) R is an expressible relation (Theorem 3). ∎

———————————*Exercises for Section 7.3*———————————

1. Give a direct proof (i.e., without using Gödel's theorem on expressibility) that the following functions are representable:
   (1) $S(a) = a + 1$ (successor function).
   (2) $C(a) = k$ (1-ary constant function k).
   (3) $F(a) = 2a$.
   (4) $M(a, b) = \min\{a, b\}$.
   (5) sg, defined by $sg(a) = \begin{cases} 0 & a = 0; \\ 1 & a > 0. \end{cases}$
   (*Hint:* See Theorem 1 in Section 7.1.)

2. Let F be an n-ary representable function and let A be a formula that represents F. Show that for all $a_1, \ldots, a_n \in \mathbb{N}$,
   (1) $NN \vdash A_{x_1, \ldots, x_n, x_{n+1}}[0^{a_1}, \ldots, 0^{a_n}, 0^{F(a_1, \ldots, a_n)}]$.
   (2) For all $b \in \mathbb{N}$ with $b \neq F(a_1, \ldots, a_n)$, $NN \vdash \neg A_{x_1, \ldots, x_n, x_{n+1}}[0^{a_1}, \ldots, 0^{a_n}, 0^b]$.

3. Complete the proof of Theorem 1 that multiplication is representable.

4. Complete the proof of Theorem 2 that the projection function $U_{n,k}$ is representable.

5. Complete the proof of Theorem 3 by proving (*) in the case $\neg R(a_1, \ldots, a_n)$.

6. Let F and G be distinct 1-ary representable functions, and let A and B be the corresponding formulas with free variables x_1, x_2 that show F and G representable. Show that A and B are distinct formulas (assuming NN consistent).

7. Use a counting argument to show that there is a 1-ary function that is not representable. Assume NN is consistent and use Exercise 6.

8. Complete the proof of Lemma 6 (if A represents F, then R1 and R2 hold).

9. Let R be an n-ary expressible relation and let $y_1, \ldots, y_n$ be any set of n distinct variables. Show that there is a formula A with exactly n free variables $y_1, \ldots, y_n$ such that for all $a_1, \ldots, a_n \in \mathbb{N}$.

   EX1 If $R(a_1, \ldots, a_n)$, then $NN \vdash A_{y_1, \ldots, y_n}[0^{a_1}, \ldots, 0^{a_n}]$.
   EX2 If $\neg R(a_1, \ldots, a_n)$, then $NN \vdash \neg A_{y_1, \ldots, y_n}[0^{a_1}, \ldots, 0^{a_n}]$.

10. Let F be an n-ary function. Recall that the graph of F is the $n + 1$-ary relation defined by $\mathcal{G}_F = \{\langle a_1, \ldots, a_n, b \rangle: F(a_1, \ldots, a_n) = b\}$. Show that F is representable if and only if

the graph $\mathcal{G}_F$ of F is expressible. (*Hint:* Assume that $\mathcal{G}_F$ is expressible with formula A and free variables $x_1, \ldots, x_n, x_{n+1}$; consider the formula $A \wedge \forall y((y < x_{n+1}) \to \neg A_{x_{n+1}}[y])$.)

11. In this section, we have given a long, rather difficult proof that every recursive relation is expressible. Discuss the difficulties involved in trying to prove (without Church's thesis) that every decidable relation is expressible.

$$\boxed{7.4}$$

Gödel's Incompleteness Theorems

In this section, we will prove Gödel's incompleteness theorem and Rosser's generalization of Gödel's result. We will also discuss Gödel's second incompleteness theorem, on limitations of consistency proofs in arithmetic. Gödel, in his original paper, assumed a somewhat stronger condition than consistency, so the first order of business is to define this new concept.

Definition 1: A set $\Gamma \subseteq \text{FOR}_{NN}$ is ω-*consistent* if the following condition holds for every formula A of L_{NN} with exactly one free variable x: If $\Gamma \vdash A_x[0^n]$ for all $n \in \mathbb{N}$, then the formula $\neg \forall x A$ is *not* a theorem of Γ.

This somewhat unusual condition needs some explanation. Suppose that for every $n \in \mathbb{N}$ we can prove that the substitution instance $A_x[0^n]$ of A is a theorem of Γ. Since we regard variables as ranging over the set $\mathbb{N}$ of natural numbers, it is tempting to conclude that $\forall x A$ is a theorem of Γ. There are, however, two objections to this argument. First, there is the possibility that Γ has a model whose domain is distinct from $\mathbb{N}$. Second, a rule of inference of the form

$$\vdash A_x[0^n] \quad \text{for all } n \in \mathbb{N} \therefore \vdash \forall x A$$

violates the spirit of a formal system; such a rule requires us to check an infinite number of hypotheses.

So what does ω-consistency say? If $\Gamma \vdash A_x[0^n]$ for all $n \in \mathbb{N}$, then, as a consolation prize, we can assert that the *negation* of $\forall x A$ is *not* a theorem of Γ. We will prove two results about ω-consistency. The first states that ω-consistency implies consistency. The second gives a sufficient condition for a set $\Gamma \subseteq \text{FOR}_{NN}$ to be ω-consistent; this sufficient condition shows that ω-consistency is not an unreasonable assumption.

Lemma 1: If $\Gamma \subseteq \text{FOR}_{NN}$ is ω-consistent, then Γ is consistent.

Proof: To prove that Γ is consistent, it suffices to find a formula that is not a theorem of Γ. Let A be the formula $x_1 = x_1$ and let $n \in \mathbb{N}$.

(1) $\forall x_1(x_1 = x_1)$ REF AX
(2) $0^n = 0^n$ $\forall$-ELIM.

We have shown that $\Gamma \vdash A_{x_1}[0^n]$ for all $n \in \mathbb{N}$; it follows by ω-consistency that the formula $\neg \forall x_1(x_1 = x_1)$ is *not* a theorem of Γ, as required. ∎

Lemma 2: Assume that the standard interpretation of L_{NN} is a model of $\Gamma \subseteq FOR_{NN}$. Then Γ is ω-consistent.

Proof: Let A be a formula with exactly one free variable x such that $\Gamma \vdash A_x[0^n]$ for all $n \in \mathbb{N}$. Suppose, however, that $\Gamma \vdash \neg \forall x A$. By the soundness theorem $\neg \forall x A$ is true in $\mathcal{N}$; hence $\forall x A$ is false in $\mathcal{N}$. Let $\phi: VAR \to \mathbb{N}$ be an assignment in $\mathcal{N}$ such that $\phi(A) = F$, and let $\phi(x) = n$. Now $\Gamma \vdash A_x[0^n]$; hence $A_x[0^n]$ is true in $\mathcal{N}$. On the other hand, $\phi(0^n) = n$; hence $\phi(x) = \phi(0^n)$, so by the substitution theorem for assignments, $\phi(A) = \phi(A_x[0^n])$. But $\phi(A) = F$; hence $\phi(A_x[0^n]) = F$. This contradicts the assertion that $A_x[0^n]$ is true in $\mathcal{N}$. ∎

Theorem 1 (Gödel's incompleteness theorem I): Every ω-consistent axiomatized extension Γ of NN is incomplete (i.e., there is a sentence A of L_{NN} such that neither A nor $\neg A$ is a theorem of Γ).

Proof: We begin by defining a 2-ary relation R_{Γ} on $\mathbb{N}$ that is a subtle modification of the decidable relation PRF_{Γ}. Let

$$R_{\Gamma}(a, b) \Leftrightarrow a \text{ and } b \text{ satisfy conditions R1 and R2, where}$$

R1 a is the code of a formula A with exactly one free variable x_1.

R2 b is the code of a proof of $A_{x_1}[0^a]$ using Γ.

Note that R2 incorporates the idea of self-reference: The formula $A_{x_1}[0^a]$ is obtained from A by replacing the free occurrences of x_1 with the term 0^a, where a is the code of A.

Since Γ is axiomatized, the 2-ary relation PRF_{Γ} is decidable (see Theorem 2 in Section 7.2). We will take advantage of this fact to show that R_{Γ} is also a decidable relation.

(1) Input $a, b \in \mathbb{N}$.

(2) Use Gö2 to decide whether a is the code of a formula A of L_{NN}. If *No*, print NO and halt.

(3) If *Yes*, use Gö2 to find the formula A whose code is a.

(4) Does A have exactly one free variable x_1? If *No*, print NO and halt.

(5) If *Yes*, use Gö1 to find $c = g(A_{x_1}[0^a])$.

(6) Does $PRF_{\Gamma}(c, b)$ hold? (PRF_{Γ} is decidable.)

(7) If *No*, print NO and halt; if *Yes*, print YES and halt.

Henceforth, we will drop the subscript in R_{Γ}. Since R is decidable, R is expressible. Let B be a formula with exactly two free variables x_1 and x_2 such that for all $a, b \in \mathbb{N}$,

EX1 If $R(a, b)$, then $\Gamma \vdash B_{x_1, x_2}[0^a, 0^b]$.

EX2 If $\neg R(a, b)$, then $\Gamma \vdash \neg B_{x_1, x_2}[0^a, 0^b]$.

(We are allowed to replace NN with Γ in the expressibility conditions because Γ is an extension of NN.) Let p be the code of the formula $\forall x_2 \neg B$. Note that p

is the code of a formula with exactly one free variable x_1, so p satisfies R1. Moreover, the result of replacing the free occurrences of x_1 in $\forall x_2 \neg B$ with the term 0^p is $\forall x_2 \neg B_{x_1}[0^p]$. Hence we can write

$$R(p, b) \Leftrightarrow b \text{ is the code of a proof of } \forall x_2 \neg B_{x_1}[0^p] \text{ using } \Gamma. \qquad (*)$$

Let G be the sentence $\forall x_2 \neg B_{x_1}[0^p]$. We will prove that neither G nor its negation is a theorem of Γ. Before doing this, however, let us "translate" the sentence G so that we can better understand the reason behind its construction. First, write the sentence G in the equivalent form $\neg \exists x_2 B_{x_1}[0^p]$. We will proceed informally.

(1) There is no natural number b such that $B_{x_1, x_2}[0^p, 0^b]$.

(2) There is no natural number b such that $R(p, b)$ (see EX1).

(3) There is no natural number b that is the code of a proof of $\forall x_2 \neg B_{x_1}[0^p]$ using Γ (see $(*)$).

(4) The sentence $\forall x_2 \neg B_{x_1}[0^p]$ is not a theorem of Γ.

But the sentence we are translating is $\forall x_2 \neg B_{x_1}[0^p]$; hence in (4) the sentence is actually referring to itself, and we have

(5) I am not a theorem of Γ.

The above discussion is a digression and not an official part of the proof; rather, it is included here to help you better understand Gödel's construction of G. Now let us turn to the proof that neither G nor $\neg G$ is a theorem of Γ.

Proof that $G(= \forall x_2 \neg B_{x_1}[0^p])$ ***is not a theorem of*** Γ: Suppose $\Gamma \vdash G$, and let b be the code of a proof of G using Γ. From $(*)$ it follows that $R(p, b)$; hence EX1 applies and $\Gamma \vdash B_{x_1, x_2}[0^p, 0^b]$. We will now contradict the consistency of Γ by showing that $\Gamma \vdash \neg B_{x_1, x_2}[0^p, 0^b]$.

(1) $\forall x_2 \neg B_{x_1}[0^p]$ THM of Γ

(2) $\neg B_{x_1, x_2}[0^p, 0^b]$ $\forall$-ELIM.

Proof that $\neg G(= \neg \forall x_2 \neg B_{x_1}[0^p])$ ***is not a theorem of*** Γ: The formula $\neg B_{x_1}[0^p]$ has exactly one free variable x_2. By ω-consistency, to show that $\neg \forall x_2 \neg B_{x_1}[0^p]$ is not a theorem of Γ, it suffices to show that $\Gamma \vdash \neg B_{x_1, x_2}[0^p, 0^b]$ for all $b \in \mathbb{N}$. By EX2, it suffices to prove that $\neg R(p, b)$ for all $b \in \mathbb{N}$. Suppose, for some $b \in \mathbb{N}$, that $R(p, b)$. Then b is the code of a proof of $\forall x_2 \neg B_{x_1}[0^p]$ using Γ (see $(*)$). In other words, $\Gamma \vdash G$, a contradiction of what was proved above. ∎

Comment on the use of Church's thesis In this proof of Gödel's theorem, Church's thesis is used in a nonessential way: We have proved that the relation R is decidable and then have claimed, by Church's thesis, that R is recursive (and hence expressible). By the techniques presented in Chapter 8, we can give a direct proof that the relation R is recursive, but in this case we must make the stronger assumption that Γ is recursively axiomatized. See Section 8.7 for a further discussion of these ideas.

In his groundbreaking and absolutely fundamental paper, Gödel points out that his methods also show that a proof of the consistency of arithmetic requires methods beyond those formalized in arithmetic. This observation was a blow to Hilbert's hope of proving the consistency of arithmetic and set theory using finitist methods (see the discussions in Sections 6.1 and 6.4).

This second result of Gödel requires a theory somewhat stronger than NN; specifically, the axiom scheme for induction is needed. So in our discussion of Gödel's second incompleteness theorem (Gödel II), we will work with Peano arithmetic as discussed in Section 6.1. (Technical comment: We may regard PA as an extension of NN by adding $<$ to the nonlogical symbols of PA (so L_{NN} is the language of PA) and by adding NN7–NN9 to the nonlogical axioms of PA.) Let CON_{PA} denote the following sentence of L_{NN}, constructed in Example 5 of Section 7.2:

$$\exists x_1[For \wedge \neg \exists x_2 Prf_{PA}].$$

Informally, CON_{PA} asserts that there is a formula of L_{NN} that is not a theorem of PA; in other words, CON_{PA} asserts that PA is consistent. Without actually writing out a detailed proof, Gödel asserted that the sentence CON_{PA} is not a theorem of PA. A rigorous proof of this fact requires considerable work, and for this we refer the reader to Boolos, *The Unprovability of Consistency*, (Cambridge: Cambridge University Press, 1979) (see the exercises for some of the main ideas). We will instead give an informal argument along the lines that are followed in Gödel's paper.

Theorem 2 (Gödel's incompleteness theorem II): Assume PA is consistent. Then CON_{PA} is not a theorem of PA.

Informal proof: Let G_{PA} be the sentence constructed in the proof of Gödel's incompleteness theorem I. The intuitive meaning of G_{PA} is that G_{PA} is not a theorem of PA. In the proof of Gödel I, we showed the following: *If PA is consistent, then G_{PA} is not a theorem of* PA. But this argument can actually be formalized in PA (this is where we need induction); hence PA $\vdash$ CON_{PA} → G_{PA}. Now suppose PA $\vdash$ CON_{PA}; by modus ponens, we obtain PA $\vdash$ G_{PA}. This contradicts the fact that G_{PA} is not a theorem of PA. ∎

Gentzen and others have given consistency proofs of PA, but in all cases, these proofs use methods that cannot be formalized in PA. Although we have stated Gödel II in the context of PA, the theorem holds in any formal system in which one can develop Peano arithmetic. For example, this can be done in the formal system ZF (see Section 6.4); hence we have the following version of Gödel II.

Theorem 3 (Gödel's incompleteness theorem II, ZF version): Assume ZF is consistent. Then CON_{ZF} is not a theorem of ZF.

As in the case of Gödel I, Gödel II has been paraphrased, or given a philosophical interpretation, in a number of ways:

(1) If an axiom system contains as much number theory as PA, then we cannot prove the consistency of that axiom system from the axioms of the system. (Shoenfield, *Mathematical Logic*, (Reading, MA: Addison-Wesley, 1967) 213).

(2) A proof of the consistency of arithmetic requires ideas and methods that are beyond those in arithmetic.

(3) No mathematical theory worth its salt is strong enough to prove its own consistency. (Barwise and Moss, "Hypersets" 1991.)

Suppose we proved $PA \vdash CON_{PA}$. Would this constitute a proof that PA is consistent?

- *Prior to* Gödel II, such a formal proof would not be considered evidence that PA is consistent. For if PA is actually inconsistent, then every sentence of L_{NN} is a theorem of PA, including CON_{PA}.

- *After* Gödel II, such a proof would actually show that PA is inconsistent! For if PA is consistent, then by Gödel II, the sentence CON_{PA} is not a theorem of PA.

——— Gödel-Rosser Incompleteness Theorem ———

Gödel proved his celebrated incompleteness theorem in 1931. The proof hinges on finding a sentence G that says "I am not a theorem of Γ." To prove that $\neg G$ is not a theorem of Γ, we needed to assume that Γ is ω-consistent (a condition stronger than consistency). In 1936, the logician Rosser generalized Gödel's theorem by replacing ω-consistency with consistency. Rosser's idea is to construct a sentence D (which depends upon Γ) that says (roughly): "For all $b \in \mathbb{N}$, if there is a proof of me using Γ with code b, then there is a proof of my negation using Γ with code less than b." We will now give the details of Rosser's proof. Two lemmas are used, the first of which is Lemma 5 of Section 7.3. To obtain the second lemma, we need to add an additional nonlogical axiom to NN, namely the converse of NN8. Thus, let NN^+ be the extension of NN obtained by adding the closure of the formula $[(x_1 < x_2) \vee (x_1 = x_2)] \rightarrow (x_1 < Sx_2)$.

Lemma 3: Let A be a formula such that $NN \vdash A_x[0], \ldots, NN \vdash A_x[0^{b-1}]$. Then $NN \vdash (x < 0^b) \rightarrow A$.

Lemma 4: Let $b \geq 0$. Then $NN^+ \vdash \neg(x < 0^{b+1}) \rightarrow (0^b < x)$.

Proof: We have the following formal proof in NN^+:

(1) $\forall x_1 \forall x_2([(x_1 < x_2) \vee (x_1 = x_2)]$ AX of NN^+
     $\rightarrow (x_1 < Sx_2))$

(2) $[(x_1 < 0^b) \vee (x_1 = 0^b)] \rightarrow (x_1 < S0^b)$ $\forall$-ELIM

(3) $[(x_1 < 0^b) \vee (x_1 = 0^b)] \rightarrow (x_1 < 0^{b+1})$ NOTATION: $S0^b$ is 0^{b+1}

(4) $[(x < 0^b) \vee (x = 0^b)] \rightarrow (x < 0^{b+1})$ SUBST RULE

(5) $\neg[(x < 0^b) \vee (x = 0^b)] \vee (x < 0^{b+1})$ DEF $\rightarrow$

(6) $\forall x_1 \forall x_2((x_1 < x_2) \vee [(x_1 = x_2)$ NN9
     $\vee (x_2 < x_1)])$

(7) $(x_1 < 0^b) \vee [(x_1 = 0^b) \vee (0^b < x_1)]$ $\forall$-ELIM

(8) $(x < 0^b) \vee [(x = 0^b) \vee (0^b < x)]$	SUBST RULE
(9) $[(x < 0^b) \vee (x = 0^b)] \vee (0^b < x)$	ASSOC
(10) $(x < 0^{b+1}) \vee (0^b < x)$	CUT (5, 9)
(11) $\neg\neg(x < 0^{b+1}) \vee (0^b < x)$	$\neg\neg$ RULE
(12) $\neg(x < 0^{b+1}) \to (0^b < x)$	DEF $\to$. ∎

Theorem 4 (Gödel-Rosser incompleteness theorem): Every consistent axiomatized extension of NN$^+$ is incomplete.

Proof: Let Γ be a consistent axiomatized extension of NN$^+$. Let R be the 2-ary relation defined in the proof of Gödel's theorem (in terms of R1 and R2), and define Q by

$$Q(a, b) \Leftrightarrow a \text{ and } b \text{ satisfy conditions R1 and Q2,}$$

where

Q2: b is the code of a proof of $\neg A_{x_1}[0^a]$ using Γ.

Both R and Q are decidable relations; hence both are expressible. Let B be a formula with exactly two free variables x_1 and x_2 and let C be a formula with exactly two free variables x_1 and x_3 (see Exercise 9 in Section 7.3) such that for all $a, b \in \mathbb{N}$,

EX1 If $R(a, b)$, then $\Gamma \vdash B_{x_1, x_2}[0^a, 0^b]$.
EX2 If $\neg R(a, b)$, then $\Gamma \vdash \neg B_{x_1, x_2}[0^a, 0^b]$.
EX3 If $Q(a, b)$, then $\Gamma \vdash C_{x_1, x_3}[0^a, 0^b]$.
EX4 If $\neg Q(a, b)$, then $\Gamma \vdash \neg C_{x_1, x_3}[0^a, 0^b]$.

Let p be the code of the formula $\forall x_2(B \to \exists x_3((x_3 < x_2) \wedge C)$. Note that p is the code of a formula with exactly one free variable x_1, so p satisfies R1. Moreover, the result of replacing the free occurrences of x_1 in $\forall x_2(B \to \exists x_3((x_3 < x_2) \wedge C))$ with the term 0^p is

$$\forall x_2(B_{x_1}[0^p] \to \exists x_3((x_3 < x_2) \wedge C_{x_1}[0^p])).$$

Call this sentence D. Then

$$R(p, b) \Leftrightarrow b \text{ is the code of a proof of } D \text{ using } \Gamma. \qquad (*)$$
$$Q(p, b) \Leftrightarrow b \text{ is the code of a proof of } \neg D \text{ using } \Gamma. \qquad (**)$$

We will now show that neither D nor $\neg D$ is a theorem of Γ.

Proof that D is not a theorem of Γ: Suppose $\Gamma \vdash D$, and let b be the code of a proof of D using Γ. By $(*)$ we have $R(p, b)$; hence EX1 applies and $\Gamma \vdash B_{x_1, x_2}[0^p, 0^b]$. Now consider the following proof in Γ:

(1) $B_{x_1, x_2}[0^p, 0^b]$	THM of Γ
(2) $\forall x_2(B_{x_1}[0^p] \to \exists x_3((x_3 < x_2) \wedge C_{x_1}[0^p]))$	THM of Γ
(3) $B_{x_1, x_2}[0^p, 0^b] \to \exists x_3((x_3 < 0^b) \wedge C_{x_1}[0^p])$	$\forall$-ELIM
(4) $\exists x_3((x_3 < 0^b) \wedge C_{x_1}[0^p])$	MP(1, 3).

In summary, we have $\Gamma \vdash \exists x_3((x_3 < 0^b) \wedge C_{x_1}[0^p])$ (under the assumption that D is a theorem of Γ). We will reach a contradiction by showing that $\Gamma \vdash \neg \exists x_3((x_3 < 0^b) \wedge C_{x_1}[0^p])$; that is, that $\Gamma \vdash \forall x_3 \neg((x_3 < 0^b) \wedge C_{x_1}[0^p])$.

By hypothesis, Γ is consistent. Moreover, we are assuming that $\Gamma \vdash D$; hence there is no proof of $\neg D$ using Γ. Let $n \in \mathbb{N}$ be arbitrary. By (**), we have $\neg Q(p, n)$; hence EX4 applies and $\Gamma \vdash \neg C_{x_1, x_3}[0^p, 0^n]$. In particular, we have $\Gamma \vdash \neg C_{x_1, x_3}[0^p, 0], \ldots, \Gamma \vdash \neg C_{x_1, x_3}[0^p, 0^{b-1}]$; hence Lemma 3 gives $\Gamma \vdash (x_3 < 0^b) \to \neg C_{x_1}[0^p]$. Continuing:

(1) $(x_3 < 0^b) \to \neg C_{x_1}[0^p]$ THM of Γ

(2) $\neg(x_3 < 0^b) \vee \neg C_{x_1}[0^p]$ DEF $\to$

(3) $\neg\neg(\neg(x_3 < 0^b) \vee \neg C_{x_1}[0^p])$ $\neg\neg$RULE

(4) $\neg((x_3 < 0^b) \wedge C_{x_1}[0^p])$ DEF $\wedge$

(5) $\forall x_3 \neg((x_3 < 0^b) \wedge C_{x_1}[0^p])$ GEN.

Proof that $\neg D$ is not a theorem of Γ: Suppose $\Gamma \vdash \neg D$, and let b be the code of a proof of $\neg D$ using Γ. By (**) we have $Q(p, b)$; hence EX3 applies and $\Gamma \vdash C_{x_1, x_3}[0^p, 0^b]$. We are going to prove the following:

(a) $\Gamma \vdash B_{x_1}[0^p] \to \neg(x_2 < 0^{b+1})$.

(b) $\Gamma \vdash \neg(x_2 < 0^{b+1}) \to (0^b < x_2)$.

(c) $\Gamma \vdash (0^b < x_2) \to \exists x_3((x_3 < x_2) \wedge C_{x_1}[0^p])$.

Assuming that (a) through (c) hold, HS (twice) gives

$$\Gamma \vdash B_{x_1}[0^p] \to \exists x_3((x_3 < x_2) \wedge C_{x_1}[0^p]).$$

Now apply generalization with the variable x_2 to obtain $\Gamma \vdash D$. This contradicts the consistency of Γ.

Now (b) is Lemma 4, so it remains to prove (a) and (c). Since Γ is consistent and $\Gamma \vdash \neg D$, there is no proof of D in Γ. Let $n \in \mathbb{N}$ be arbitrary. From (*), it follows that $\neg R(p, n)$; hence EX2 applies and $\Gamma \vdash \neg B_{x_1, x_2}[0^p, 0^n]$. In particular, we have $\Gamma \vdash \neg B_{x_1, x_2}[0^p, 0], \ldots, \Gamma \vdash \neg B_{x_1, x_2}[0^p, 0^b]$; hence Lemma 3 gives $\Gamma \vdash (x_2 < 0^{b+1}) \to \neg B_{x_1}[0^p])$. From this, we easily obtain (a) as required.

It remains to prove (c). Recall from above that $\Gamma \vdash C_{x_1, x_3}[0^p, 0^b]$.

(1) $C_{x_1, x_3}[0^p, 0^b]$ THM of Γ

(2) $((0^b < x_2) \wedge C_{x_1, x_3}[0^p, 0^b])$ SUBST THM
     $\to \exists x_3((x_3 < x_2) \wedge C_{x_1}[0^p])$

(3) $C_{x_1, x_3}[0^p, 0^b] \to [(0^b < x_2)$ TT(*), MP(2)
     $\to \exists x_3((x_3 < x_2) \wedge C_{x_1}[0^p])]$

(4) $(0^b < x_2) \to \exists x_3((x_3 < x_2) \wedge C_{x_1}[0^p])$ MP(1, 3)

(*) $[(p \wedge q) \to r] \to [q \to (p \to r)]$. ∎

─────────────*Exercises for Section 7.4*─────────────

1. Let Γ be a consistent axiomatized extension of NN. The sentence $G = \forall x_2 \neg B_{x_1}[0^p]$, in the proof of Gödel's incompleteness theorem, is not a theorem of Γ. By the extension theorem, $\Gamma \cup \{\neg G\}$ is consistent. Prove that $\Gamma \cup \{\neg G\}$ is *not* ω-consistent.

2. Let Γ be a consistent axiomatized extension of NN. Show that there is a formula A with exactly one free variable x such that (1) $\Gamma \vdash A_x[0^n]$ for all $n \in \mathbb{N}$; (2) $\forall x A$ is not a theorem of Γ.

3. Assume that NN is ω-consistent. Prove that every RE relation (see Section 1.8) is *weakly expressible* in NN; in other words, given an n-ary RE relation R, show that there is a formula A with exactly n free variables $x_1, \ldots, x_n$ such that for all $a_1, \ldots, a_n \in \mathbb{N}$, $R(a_1, \ldots, a_n) \Leftrightarrow \text{NN} \vdash A_{x_1, \ldots, x_n}[0^{a_1}, \ldots, 0^{a_n}]$.

4. (*Proof of Gödel's theorem I using the fixed point theorem*) In Exercise 2 in Section 7.5, we will outline a proof of the **fixed point theorem**: *If A is a formula of L_{NN} with exactly one free variable x_1, then there is a sentence B with code p such that $\text{NN} \vdash B \leftrightarrow A_{x_1}[0^p]$.* Use this result to prove Gödel's incompleteness theorem I. (*Hint:* Let A be a formula with exactly two free variables x_1 and x_2 that expresses PRF_Γ; apply the fixed point theorem to $\forall x_2 \neg A$ to obtain a sentence B with code p such that $\text{NN} \vdash B \leftrightarrow \forall x_2 \neg A_{x_1}[0^p]$. Then prove the following: (1) If Γ is consistent, then B is not a theorem of Γ. (2) If Γ is ω-consistent, then $\neg B$ is not a theorem of Γ.)

5. (*Gödel's incompleteness theorem II*) In this exercise, we give an axiomatic approach to Gödel II; for hints and more detail, the reader is referred to Smullyan, *Gödel's Incompleteness Theorems,* (Oxford: Oxford University Press, 1992), 106. Fix a formula P of L_{NN} with exactly one free variable x_1. For any formula A of L_{NN}, $P_{x_1}[0^{g(A)}]$ is the formula obtained from P by replacing the free occurrences of x_1 with the term $0^{g(A)}$; note that $g(A)$ is the code of A. We denote this new formula by $P(A)$. In summary, $P(A)$ is the formula $P_{x_1}[0^{g(A)}]$. Now let $\Gamma \subseteq \text{FOR}_{NN}$; the formula P is called a *provability assertion for* Γ if it satisfies the following conditions:

   D1 If $\Gamma \vdash A$, then $\Gamma \vdash P(A)$.

   D2 $\Gamma \vdash P(A \rightarrow B) \rightarrow [P(A) \rightarrow P(B)]$.

   D3 $\Gamma \vdash P(A) \rightarrow P(P(A))$.

   D1–D3 are called *derivability conditions*, and are due to Hilbert and Bernays, with improvements by Löb. The formula $\exists x_2 Prf_{PA}$ is a provability assertion for PA. The proof, however, is very difficult; see Chapter 2 of Boolos, *The Unprovability of Consistency* (Cambridge: Cambridge University Press, 1979). Here we assume this result and use it to prove Gödel II.

   (1) Use D1–D3 to prove the following for sentences A, B, and C of L_{NN}.

   P1 If $\Gamma \vdash A \rightarrow B$, then $\Gamma \vdash P(A) \rightarrow P(B)$.

   P2 If $\Gamma \vdash A \rightarrow (B \rightarrow C)$, then $\Gamma \vdash P(A) \rightarrow [P(B) \rightarrow P(C)]$.

   P3 If $\Gamma \vdash A \rightarrow (P(A) \rightarrow B)$, then $\Gamma \vdash P(A) \rightarrow P(B)$.

   (2) Apply the fixed point theorem to $\neg P$ (see Exercise 4) to obtain a sentence B with code p such that $\Gamma \vdash B \leftrightarrow \neg P_{x_1}[0^p]$. This can also be written as $\Gamma \vdash B \leftrightarrow \neg P(B)$. Prove the following:

   (a) If Γ is consistent, then B is not a theorem of Γ. (*Hint:* D1.)

   (b) For any formula A, $\Gamma \vdash P(B) \rightarrow P(A)$. (*Hint:* P3.)

   (c) If A is not a theorem of Γ, then $\neg P(A)$ is not a theorem of Γ.

   Gödel II from (2)(c): Suppose P is the provability assertion $\exists x_2 Prf_{PA}$ for PA, and let c be the code of $\neg(0 = 0)$. Assume PA is consistent, so $\neg(0 = 0)$ is not a theorem of PA. To prove Gödel II, it suffices to show that $\neg\exists x_2 Prf_{x_1}[0^c]$ is not a theorem of PA. This follows immediately from (2)(c). (*Note:* $\neg\exists x_2 Prf_{x_1}[0^c]$ states that $\neg(0 = 0)$ is not a theorem of PA.)

6. (*Löb's theorem* $\Rightarrow$ *Gödel II*) Refer to Exercises 4 and 5. We outline a proof of Löb's theorem and then use it to prove Gödel II.

Löb's theorem: For any sentence A, if $\Gamma \vdash P(A) \rightarrow A$, then $\Gamma \vdash A$. Apply the fixed point theorem to $P \rightarrow A$ to obtain a sentence B with code p such that $\Gamma \vdash B \leftrightarrow (P_{x_1}[0^p] \rightarrow A)$; this can also be written as $\Gamma \vdash B \leftrightarrow [P(B) \rightarrow A]$. Use this result and $\Gamma \vdash P(A) \rightarrow A$ to prove $\Gamma \vdash A$. (*Hint:* P3 and D1.)

Gödel II from Löb: Suppose P is the provability assertion $\exists x_2 Prf$ for PA, and let c be the code of $\neg(0 = 0)$. Assume PA is consistent, so $\neg(0 = 0)$ is not a theorem of PA. To prove Gödel II, it suffices to show that $\neg\exists x_2 Prf_{x_1}[0^c]$ is not a theorem of PA. Suppose PA $\vdash \neg\exists x_2 Prf_{x_1}[0^c]$; by the expansion rule, PA $\vdash \exists x_2 Prf_{x_1}[0^c] \rightarrow \neg(0 = 0)$; hence by Löb's theorem, PA $\vdash \neg(0 = 0)$, a contradiction.

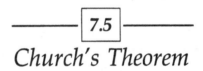

Church's Theorem

In this section, we will prove that every consistent extension Γ of NN is undecidable. The first step is to reduce the question of the decidability of Γ to that of the decidability of the 1-ary relation THM_Γ (see Section 7.2).

Lemma 1: Let $\Gamma \subseteq FOR_{NN}$. Then Γ is decidable if and only if THM_Γ is a decidable relation.

Proof: Assume Γ is decidable; the following algorithm shows that THM_Γ is decidable.

(1) Input $a \in \mathbb{N}$.
(2) Use Gö2 to decide whether a is the code of a formula. If *No*, print NO ($a \notin THM_\Gamma$) and halt.
(3) If *Yes*, use Gö2 to find the formula A such that $g(A) = a$.
(4) Is $\Gamma \vdash A$? (Γ is decidable.) If *No*, print NO and halt.
(5) If *Yes*, print YES ($a \in THM_\Gamma$) and halt.

Now assume that THM_Γ is a decidable relation. The following algorithm shows that Γ is decidable:

(1) Input a formula A of L_{NN}.
(2) Use Gö1 to find $g(A) = a$.
(3) Is $a \in THM_\Gamma$? (THM_Γ is decidable.) If *No*, print NO (A is not a theorem of Γ) and halt.
(4) If *Yes*, print YES (A is a theorem of Γ) and halt. ■

Theorem 1 allows us to prove Church's theorem in the following form: If Γ *is a consistent extension of* NN, *then* THM_Γ *is not a decidable relation.* The next lemma is a key step in the proof of both Church's theorem and Tarski's theorem.

Lemma 2: There is a 1-ary computable function DIA such that if a is the code of a formula A of L_{NN} with exactly one free variable x_1, then $DIA(a) = $ code of the formula $A_{x_1}[0^a]$.

Proof: Define DIA as follows:

$$DIA(a) = \begin{cases} \text{Code of } A_{x_1}[0^a] & a \text{ is the code of a formula } A \text{ of } L_{NN} \\ & \text{with exactly one free variable } x_1; \\ 0 & \text{otherwise.} \end{cases}$$

We prove DIA computable by writing an algorithm that computes DIA:

(1) Input $a \in \mathbb{N}$.
(2) Use Gö2 to decide whether a is the code of a formula of L_{NN}. If *No*, $DIA(a) = 0$.
(3) If *Yes*, use Gö2 to find the formula A whose code is a.
(4) Does A have exactly one free variable x_1? If *No*, $DIA(a) = 0$.
(5) If *Yes*, use Gö1 to find the code b of the formula $A_{x_1}[0^a]$ and set $DIA(a) = b$. ■

Note In Section 8.7, we will prove the existence of a recursive function DIA that has the property described in Lemma 2. In this proof, we will show DIA to be recursive by diagonalizing a certain 3-ary recursive function SUB; hence its name.

Lemma 3: Let Γ be a consistent extension of NN, let R be a 1-ary relation on $\mathbb{N}$, and let A be a formula that expresses R. Then for all $a \in \mathbb{N}$, $R(a) \Leftrightarrow \Gamma \vdash A_{x_1}[0^a]$.

Proof: Since A expresses R, it follows that for all $a \in \mathbb{N}$,

EX1 If $R(a)$, then $NN \vdash A_{x_1}[0^a]$.
EX2 If $\neg R(a)$, then $NN \vdash \neg A_{x_1}[0^a]$.

First suppose $R(a)$. By EX1, $NN \vdash A_{x_1}[0^a]$, and since Γ is an extension of NN, $\Gamma \vdash A_{x_1}[0^a]$ as required. Now assume $\Gamma \vdash A_{x_1}[0^a]$. Suppose, however, that $R(a)$ fails. Then $\neg R(a)$; hence $\Gamma \vdash \neg A_{x_1}[0^a]$ (use EX2 and the fact that Γ extends NN). This contradicts the consistency of Γ. ■

Theorem 1 (Church): Let Γ be a consistent extension of NN. Then Γ is undecidable.

Proof: It suffices to show that the 1-ary relation THM_Γ is undecidable (Lemma 1). Suppose THM_Γ is decidable. *Strategy:* the proper choice of a relation R in Lemma 3 leads to a contradiction. Define R by

$$R(a) \Leftrightarrow \neg THM_\Gamma(DIA(a)).$$

First, let us show that R is decidable (under the assumption that THM$_\Gamma$ is decidable). Here is the required algorithm:

(1) Input $a \in \mathbb{N}$.

(2) Compute DIA(a). (DIA is a computable function.)

(3) Use the algorithm for THM$_\Gamma$ to decide whether THM$_\Gamma$(DIA(a)).

(4) If *Yes*, print NO and halt.

(5) If *No*, print YES and halt.

Since R is decidable, R is expressible. Let A be a formula with exactly one free variable x_1 that expresses R. It follows from Lemma 3 that for all $a \in \mathbb{N}$,

$$\neg \text{THM}_\Gamma(\text{DIA}(a)) \Leftrightarrow \Gamma \vdash A_{x_1}[0^a].$$

Now let p be the code of the formula A; then DIA(p) is the code of the formula $A_{x_1}[0^p]$, and we have

$$\neg \text{THM}_\Gamma(\text{DIA}(p)) \Leftrightarrow \Gamma \vdash A_{x_1}[0^p].$$

The right side of this equivalence states that $A_{x_1}[0^p]$ is a theorem of Γ; the left side states that $A_{x_1}[0^p]$ is *not* a theorem of Γ. This is the required contradiction; hence THM$_\Gamma$ is not decidable, as required. ■

> **Comment on the use of Church's thesis** We can prove directly that the relation $\neg$THM$_\Gamma$(DIA(a)) is recursive under the assumption that THM$_\Gamma$ is a recursive relation. In this case, the formal conclusion of Church's theorem is that THM$_\Gamma$ is not recursive; to draw the stronger conclusion that THM$_\Gamma$ is undecidable, we must use Church's thesis. See Section 8.6 for a precise statement and proof of Church's theorem.

We will now use Church's theorem to show that the validity problem for the language L$_{NN}$ is unsolvable. This was proved by Church in 1936; a related result (but for a different language) was proved by Turing at about the same time. Precise proofs of these two results are given in Sections 8.6 and 9.2 respectively. We use the following result (Corollary 4 in Section 5.5): *Let* $\Delta \subseteq \text{FOR}_L$ *be finite; if* Δ *is undecidable, then the validity problem for L is unsolvable.*

Theorem 2 (Church, unsolvability of the validity problem for L$_{NN}$): Assume NN is consistent. Then the validity problem for L$_{NN}$ is unsolvable.

Proof: Let $\Delta = \{NN1, \ldots, NN9\}$ be the nonlogical axioms of NN. By Church's theorem, Δ is undecidable. Now Δ is obviously finite; hence the validity problem for L$_{NN}$ is unsolvable, as required. ■

────────────────**Exercises for Section 7.5**────────────────

1. (*Recursive inseparability*) Two sets (= 1-ary relations) P and Q are *recursively inseparable* if there is no recursive set R such that $P \subseteq R$ and $Q \subseteq R^c$. Define 1-ary relations THM and REF on $\mathbb{N}$ by

   THM(a) $\Leftrightarrow$ a is the code of a formula A of L_{NN} and NN $\vdash A$.
   REF(a) $\Leftrightarrow$ a is the code of a formula A of L_{NN} and NN $\vdash \neg A$.

   Assume NN is consistent. Show that THM and REF (= refutable) are recursively inseparable. (*Hint:* Suppose there is a recursive relation R such that THM $\subseteq R$ and REF $\subseteq R^c$; consider $\neg R(\text{DIA}(a))$.

2. (*Fixed point theorem*) Let A be a formula with exactly one free variable x_1. Prove that there is a sentence B with code q such that NN $\vdash B \leftrightarrow A_{x_1}[0^q]$. Intuitively, B says "I have the property expressed by A." (*Hint:* DIA is a representable function; hence there is a formula C with exactly two free variables x_1 and y such that for all $a \in \mathbb{N}$, NN $\vdash C_{x_1}[0^a] \leftrightarrow (y = 0^{\text{DIA}(a)})$. (*Note:* y is a variable that does not occur in A). Let p be the code of the formula $\forall y(C \rightarrow A_{x_1}[y])$, let B be the sentence $\forall y(C_{x_1}[0^p] \rightarrow A_{x_1}[y])$, and let q be the code of B. Note that DIA(p) $= q$, and hence

$$\text{NN} \vdash C_{x_1}[0^p] \leftrightarrow (y = 0^q). \qquad (*)$$

   Show that NN $\vdash B \leftrightarrow A_{x_1}[0^q]$ by proving the following:
   (1) NN $\vdash \forall y(C_{x_1}[0^p] \rightarrow A_{x_1}[y]) \rightarrow A_{x_1}[0^q]$. (Use (*).)
   (2) NN $\vdash A_{x_1}[0^q] \rightarrow \forall y(C_{x_1}[0^p] \rightarrow A_{x_1}[y])$. (Use (*) and EQ AX(4).))

3. Use the fixed point theorem to prove Church's theorem. (*Hint:* Assume THM$_\Gamma$ is decidable and let A be a formula with exactly one free variable x_1 that expresses $\neg$THM$_\Gamma$.

4. Use the fixed point theorem to prove that THM and REF are recursively inseparable (see Exercise 1; assume NN is consistent).

5. Use the fact that THM$_{NN}$ is not decidable but is listable to prove that the following relation is not listable: NTHM$_{NN}$(a) $\Leftrightarrow$ a is the code of a formula that is not a theorem of NN.

────────────── **7.6** ──────────────

Definability and Tarski's Theorem

Church's theorem asserts the undecidability of the 1-ary relation THM$_\Gamma$; Tarski's theorem asserts the indefinability of the 1-ary relation TR, defined as follows.

Definition 1: TR(a) $\Leftrightarrow$ a is the code of a sentence of L_{NN} that is true in the standard interpretation $\mathcal{N}$.

Definition 2: An n-ary relation R is *definable* (or *arithmetical*) if there is a formula A of L_{NN} with exactly n free variables $x_1, \ldots, x_n$ such that for all $a_1, \ldots, a_n \in \mathbb{N}$, $R(a_1, \ldots, a_n) \Leftrightarrow A_{x_1, \ldots, x_n}[0^{a_1}, \ldots, 0^{a_n}]$ is true in $\mathcal{N}$. We also say that the formula A *defines* R. A relation that is not definable is said to be *indefinable*. Finally, an n-ary function F is *definable* if its graph $\mathcal{G}_F$ is definable.

Definability, like expressibility, is a precise way of capturing the idea that a formula "talks about" a relation. We can compare the two approaches by noting that definability is semantic, whereas expressibility is syntactic. An informal reading of Tarski's theorem is that we cannot find a formula of the language L_{NN} that talks about the concept of truth. Roughly speaking, definability generalizes expressibility. More precisely, we have the following result.

Lemma 1: Assume that the standard interpretation is a model of NN. Then (1) every expressible relation is definable; (2) every representable function is definable.

Proof: Let R be an n-ary expressible relation. Then there is a formula A with exactly n free variables $x_1, \ldots, x_n$ such that for all $a_1, \ldots, a_n \in \mathbb{N}$,

EX1 If $R(a_1, \ldots, a_n)$, then $NN \vdash A_{x_1, \ldots, x_n}[0^{a_1}, \ldots, 0^{a_n}]$.
EX2 If $\neg R(a_1, \ldots, a_n)$, then $NN \vdash \neg A_{x_1, \ldots, x_n}[0^{a_1}, \ldots, 0^{a_n}]$.

This same formula defines R. To see this, let $a_1, \ldots, a_n \in \mathbb{N}$; we are required to show that

$$R(a_1, \ldots, a_n) \Leftrightarrow A_{x_1, \ldots, x_n}[0^{a_1}, \ldots, 0^{a_n}] \text{ is true in } \mathcal{N}.$$

Assume $R(a_1, \ldots, a_n)$. By EX1, we have $NN \vdash A_{x_1, \ldots, x_n}[0^{a_1}, \ldots, 0^{a_n}]$; hence by the soundness theorem, $A_{x_1, \ldots, x_n}[0^{a_1}, \ldots, 0^{a_n}]$ is true in $\mathcal{N}$. Now assume $A_{x_1, \ldots, x_n}[0^{a_1}, \ldots, 0^{a_n}]$ is true in $\mathcal{N}$. Suppose, however, that $R(a_1, \ldots, a_n)$ fails—that is, that $\neg R(a_1, \ldots, a_n)$. By EX2, we have $NN \vdash \neg A_{x_1, \ldots, x_n}[0^{a_1}, \ldots, 0^{a_n}]$; hence $\neg A_{x_1, \ldots, x_n}[0^{a_1}, \ldots, 0^{a_n}]$ is true in $\mathcal{N}$. This is a contradiction; a sentence and its negation cannot both be true in $\mathcal{N}$.

Now assume that F is an n-ary representable function; we are required to show that $\mathcal{G}_F$ is definable. By (1), it suffices to show that $\mathcal{G}_F$ is expressible. Since F is representable, there is a formula A with exactly $n + 1$ free variables $x_1, \ldots, x_{n+1}$ such that for all $a_1, \ldots, a_n \in \mathbb{N}$,

$$NN \vdash A_{x_1, \ldots, x_n}[0^{a_1}, \ldots, 0^{a_n}] \leftrightarrow (x_{n+1} = 0^{F(a_1, \ldots, a_n)}). \qquad (*)$$

This same formula expresses $\mathcal{G}_F$. To see this, let $a_1, \ldots, a_n, b \in \mathbb{N}$; we are required to show that

EX1 If $F(a_1, \ldots, a_n) = b$, then $NN \vdash A_{x_1, \ldots, x_n, x_{n+1}}[0^{a_1}, \ldots, 0^{a_n}, 0^b]$.
EX2 If $F(a_1, \ldots, a_n) \neq b$, then $NN \vdash \neg A_{x_1, \ldots, x_n, x_{n+1}}[0^{a_1}, \ldots, 0^{a_n}, 0^b]$.

We will prove EX2 and leave EX1 as an exercise for the reader. Apply the SUBST RULE to $(*)$ (replace x_{n+1} by 0^b) to obtain

$$NN \vdash A_{x_1, \ldots, x_n, x_{n+1}}[0^{a_1}, \ldots, 0^{a_n}, 0^b] \leftrightarrow (0^b = 0^{F(a_1, \ldots, a_n)}). \qquad (**)$$

Now $F(a_1, \ldots, a_n) \neq b$; hence $NN \vdash \neg(0^b = 0^{F(a_1, \ldots, a_n)})$ (see Theorem 6 in Section 7.2). It easily follows from $(**)$ that $NN \vdash \neg A_{x_1, \ldots, x_n, x_{n+1}}[0^{a_1}, \ldots, 0^{a_n}, 0^b]$ as required. ∎

The class of definable relations is very stable under logical operations. We prove the closure properties that are needed in the proof of Tarski's theorem; further results will be developed in the exercises.

Lemma 2 (Negation): If R is definable, then $\neg R$ is also definable.

Proof: Assume R is n-ary and let A be a formula with exactly n free variables $x_1, \ldots, x_n$ that defines R. Then the formula $\neg A$ defines $\neg R$. ∎

Lemma 3 (Existential quantifier): Let Q be an $n + 1$-ary definable relation. Then R is definable, where R is defined by

$$R(a_1, \ldots, a_n) \Leftrightarrow \exists b Q(a_1, \ldots, a_n, b). \qquad (*)$$

Proof: Let A be a formula with exactly $n + 1$ free variables $x_1, \ldots, x_{n+1}$ such that for all $a_1, \ldots, a_n, b \in \mathbb{N}$,

$$Q(a_1, \ldots, a_n, b) \Leftrightarrow A_{x_1, \ldots, x_n, x_{n+1}}[0^{a_1}, \ldots, 0^{a_n}, 0^b] \text{ is true in } \mathcal{N}. \qquad (**)$$

The required formula for R is $\exists x_{n+1} A$. To prove this, we must show that for all $a_1, \ldots, a_n \in \mathbb{N}$, $R(a_1, \ldots, a_n) \Leftrightarrow \exists x_{n+1} A_{x_1, \ldots, x_n}[0^{a_1}, \ldots, 0^{a_n}]$ is true in $\mathcal{N}$.

Assume $R(a_1, \ldots, a_n)$. By $(*)$, there exists $b \in \mathbb{N}$ such that $Q(a_1, \ldots, a_n, b)$. It follows from $(**)$ that $A_{x_1, \ldots, x_n, x_{n+1}}[0^{a_1}, \ldots, 0^{a_n}, 0^b]$ is true in $\mathcal{N}$. Now the formula

$$A_{x_1, \ldots, x_n, x_{n+1}}[0^{a_1}, \ldots, 0^{a_n}, 0^b] \rightarrow \exists x_{n+1} A_{x_1, \ldots, x_n}[0^{a_1}, \ldots, 0^{a_n}]$$

is a substitution theorem of first-order logic and therefore is logically valid; hence it is true in $\mathcal{N}$. By modus ponens for interpretations, $\exists x_{n+1} A_{x_1, \ldots, x_n}[0^{a_1}, \ldots, 0^{a_n}]$ is true in $\mathcal{N}$ as required.

Now assume $\exists x_{n+1} A_{x_1, \ldots, x_n}[0^{a_1}, \ldots, 0^{a_n}]$ is true in $\mathcal{N}$. To prove $R(a_1, \ldots, a_n)$, it suffices, by $(*)$, to find $b \in \mathbb{N}$ such that $Q(a_1, \ldots, a_n, b)$; by $(**)$, it suffices to find $b \in \mathbb{N}$ such that $A_{x_1, \ldots, x_n, x_{n+1}}[0^{a_1}, \ldots, 0^{a_n}, 0^b]$ is true in $\mathcal{N}$. Since $\exists x_{n+1} A_{x_1, \ldots, x_n}[0^{a_1}, \ldots, 0^{a_n}]$ is true in $\mathcal{N}$, there is an assignment $\phi: \text{VAR} \rightarrow \mathbb{N}$ in $\mathcal{N}$ such that $\phi(A_{x_1, \ldots, x_n}[0^{a_1}, \ldots, 0^{a_n}]) = T$. Let $\phi(x_{n+1}) = b$. Now $\phi(0^b) = b$ also holds; hence $\phi(x_{n+1}) = \phi(0^b)$. The substitution theorem for assignments (special case $\phi = \psi$), applied to $A_{x_1, \ldots, x_n}[0^{a_1}, \ldots, 0^{a_n}]$, gives

$$\phi(A_{x_1, \ldots, x_n}[0^{a_1}, \ldots, 0^{a_n}]) = \phi(A_{x_1, \ldots, x_n, x_{n+1}}[0^{a_1}, \ldots, 0^{a_n}, 0^b]).$$

From this it follows that $\phi(A_{x_1, \ldots, x_n, x_{n+1}}[0^{a_1}, \ldots, 0^{a_n}, 0^b]) = T$. Since this is a sentence, it is true in $\mathcal{N}$ as required. ∎

Corollary 1: Every RE relation is definable (assume that the standard interpretation is a model of NN).

Proof: Let R be an n-ary RE relation. Then there is an $n + 1$-ary recursive relation Q such that for all $a_1, \ldots, a_n \in \mathbb{N}$, $R(a_1, \ldots, a_n) \Leftrightarrow \exists b Q(a_1, \ldots, a_n, b)$. Since Q is recursive, it is expressible, and hence definable (Lemma 1). Lemma 3 now applies, and R is definable as required. ∎

Lemma 4 (Composition): Let Q be a 1-ary definable relation and let F be a 1-ary definable function. Then the 1-ary relation R defined by $R(a) \Leftrightarrow Q(F(a))$ is definable.

Proof: We have

$$R(a) \Leftrightarrow Q(F(a))$$
$$\Leftrightarrow \exists b[Q(b) \wedge F(a) = b]$$
$$\Leftrightarrow \exists b[Q(b) \wedge \mathscr{G}_F(a, b)].$$

By Lemma 3, R is definable if we can show that $Q(b) \wedge \mathscr{G}_F(a, b)$ is definable. Now Q and $\mathscr{G}_F$ are both definable; hence there is a formula A with exactly one free variable x_2 (see Exercise 5) and there is a formula B with exactly two free variables x_1 and x_2 such that for all $a, b \in \mathbb{N}$,

(1) $Q(b) \Leftrightarrow A_{x_2}[0^b]$ is true in $\mathscr{N}$;

(2) $\mathscr{G}_F(a, b) \Leftrightarrow B_{x_1, x_2}[0^a, 0^b]$ is true in $\mathscr{N}$.

Let C be the formula $A \wedge B$. Note that for all $a, b \in \mathbb{N}$, $C_{x_1, x_2}[0^a, 0^b]$ is $A_{x_2}[0^b] \wedge B_{x_1, x_2}[0^a, 0^b]$. We leave it to the reader to check that for all $a, b \in \mathbb{N}$, $Q(b) \wedge \mathscr{G}_F(a, b) \Leftrightarrow C_{x_1, x_2}[0^a, 0^b]$ is true in $\mathscr{N}$. This shows that $Q(b) \wedge \mathscr{G}_F(a, b)$ is definable as required. ■

Now we are ready to prove Tarski's theorem. The proof is similar to that of Church's theorem in Section 7.5; in particular, we will use the computable function DIA.

Theorem 1 (Tarski): Assume that the standard interpretation is a model of NN. Then TR is not definable.

Proof: Suppose TR is definable. Define a 1-ary relation R by

$$R(a) \Leftrightarrow \neg TR(DIA(a)).$$

If TR is definable, then $\neg$TR is also definable (Lemma 2). The function DIA is computable, and hence representable (see Section 7.3); by Lemma 1, DIA is definable. Lemma 4 now applies, and R is definable. Let A be a formula with exactly one free variable x_1 such that for all $a \in \mathbb{N}$,

$$\neg TR(DIA(a)) \Leftrightarrow A_{x_1}[0^a] \text{ is true in } \mathscr{N}.$$

Let p be the code of the formula A; then $DIA(p)$ is the code of the formula $A_{x_1}[0^p]$, and we have

$$\neg TR(DIA(p)) \Leftrightarrow A_{x_1}[0^p] \text{ is true in } \mathscr{N}.$$

The left side of this equivalence states that the formula $A_{x_1}[0^p]$ is not true in $\mathscr{N}$; this contradicts the right side. ■

Corollary 2: Assume that the standard interpretation is a model of NN. Then TR is not RE.

Proof: Suppose TR is RE. By Corollary 1, TR is definable, a contradiction of Tarski's theorem. ∎

See Sections 1.8 and 8.5 for a discussion of RE relations as the formal counterpart of the semidecidable relations. If we assume Church's thesis, then from Corollary 2 we can draw the stronger conclusion that TR is not listable (equivalently, not semidecidable). From this, it follows (left as an exercise for the reader) that Th($\mathcal{N}$) is not listable; in other words, for the standard interpretation of L_{NN}, *there is no algorithm that lists all true sentences, and only true sentences.*

───────────── *Exercises for Section 7.6* ─────────────

1. Prove that the following relations on $\mathbb{N}$ are definable by finding the required formula A and then proving that it works:
   (1) EVEN(a) ⇔ a is even.
   (2) DIV(a, b) ⇔ $a \neq 0$ and there exists $c \in \mathbb{N}$ such that $b = a \times c$.
   (3) COMP(a) ⇔ there exist $b, c > 1$ such that $a = b \times c$.
2. Assume TR is not listable and show that Th ($\mathcal{N}$) is not listable (work with semi-decidability if you wish).
3. Show that there is no algorithm that decides if an arbitrary sentence of L_{NN} is true in $\mathcal{N}$ (assume $\mathcal{N}$ is a model of NN).
4. (*Closure properties of definable relations*) We have already proved that the class of definable relations is closed under negation and existential quantification. Prove the following additional closure properties:
   (1) If Q and R are n-ary definable relations, then $Q \vee R$ and $Q \wedge R$ are also definable.
   (2) If the $n + 1$-ary relation Q is definable, then R is definable, where R is defined by
       $R(a_1, \ldots, a_n)$ ⇔ $\forall b Q(a_1, \ldots, a_n, b)$.
5. Let R be a 1-ary definable relation and let x be a variable. Show that there is a formula A with exactly one free variable x such that for all $a \in \mathbb{N}$, $R(a)$ ⇔ $A_x[0^a]$ is true in $\mathcal{N}$.
6. Use the fixed point theorem to prove Tarski's theorem (see Exercise 2 in Section 7.5). (*Hint:* Assume TR is definable, and let A be a formula with exactly one free variable x_1 that defines ¬TR.)
7. (*Special case of Gödel's theorem from Tarski's theorem*) Let Γ be an axiomatized extension of NN such that the standard interpretation is a model of Γ. Prove that there is a sentence of L_{NN} that is true in $\mathcal{N}$ but is not a theorem of Γ. (*Hint:* THM$_\Gamma$ ⊆ TR and THM$_\Gamma$ is listable.)
8. Assume that the standard interpretation is a model of NN. Give a direct proof, without using Tarski's theorem, that the function DIA and the relation TR cannot both be definable.
9. (*Arithmetical ⇒ definable*) An n-ary relation R is *arithmetical* if there is an $n + k$-ary expressible relation P such that for all $a_1, \ldots, a_n \in \mathbb{N}$,

$$R(a_1, \ldots, a_n) \Leftrightarrow Q b_1 \ldots Q b_k P(a_1, \ldots, a_n, b_1, \ldots, b_k).$$

(*Note:* For $1 \leq j \leq k$, Qb is either $\forall b$ (for all $b \in \mathbb{N}$) or $\exists b$ (there exists $b \in \mathbb{N}$).) We allow $k = 0$, so every expressible relation is arithmetical. (This definition is somewhat nonstandard; one usually requires P to be recursive instead of expressible; we have already proved that recursive implies expressible, and in the exercises for Section 8.7 the converse will be proved.) Assume that the standard interpretation is a model of NN. Show that every arithmetical relation is definable. (*Hint:* Induction on k; also use Lemmas 1 through 3.)

10. (*Definable* $\Rightarrow$ *arithmetical*) Assume that the standard interpretation is a model of NN. Prove the following:

    (1) Let A be a quantifier-free formula with exactly n free variables $x_1, \ldots, x_n$ and let R be an n-ary relation such that for all $a_1, \ldots, a_n \in \mathbb{N}$, $R(a_1, \ldots, a_n) \Leftrightarrow$ $A_{x_1, \ldots, x_n}[0^{a_1}, \ldots, 0^{a_n}]$ is true in $\mathcal{N}$. Then R is expressible. (*Hint:* See Exercise 4 in Section 7.1.)

    (2) Let B be a formula with exactly $n + 1$ free variables $x_1, \ldots, x_n, y$ and let Q be an $n + 1$-ary relation such that for all $a_1, \ldots, a_n, b \in \mathbb{N}$, $Q(a_1, \ldots, a_n, b) \Leftrightarrow$ $B_{x_1, \ldots, x_n, y}[0^{a_1}, \ldots, 0^{a_n}, 0^b]$ is true in $\mathcal{N}$. Then for all $a_1, \ldots, a_n \in \mathbb{N}$, $\exists b Q(a_1, \ldots, a_n, b) \Leftrightarrow \exists y B_{x_1, \ldots, x_n}[0^{a_1}, \ldots, 0^{a_n}]$ is true in $\mathcal{N}$.

    (3) If R is definable, then R is arithmetical (see Exercise 9).

8

Recursive Functions

The importance of the technical concept recursive function *derives from the overwhelming evidence that it is coextensive with the intuitive concept* effectively calculable function.

—*Emil Post* 1944, 285

T his chapter has two goals. One is a systematic study of the recursive functions, recursive relations, and RE relations; this is carried out in Sections 8.1–8.5. The other is to prove the theorems of Church, Gödel, and Tarski in the following forms: THM_Γ is not recursive; THM_Γ is RE; TR is not definable. These fundamental results were discussed in Chapter 7, but their proofs were simplified by replacing the formal concept of a recursive relation with the informal concept of a decidable relation.

8.1

Recursive Functions

Recall the following key ideas from Chapter 1:

Informal concept (Section 1.7)	*Formal counterpart (Section 1.8)*
Computable function	Recursive function
Decidable relation	Recursive relation
Semidecidable relation	RE relation

All functions are from $\mathbb{N}^n$ into $\mathbb{N}$ and all relations are on $\mathbb{N}$. In other words, we have $F: \mathbb{N}^n \to \mathbb{N}$ and $R \subseteq \mathbb{N}^n$ ($n \geq 1$). Generally speaking, F, G, and H denote functions and P, Q, and R denote relations. The reader is referred to Section 1.8 for the precise definition of a recursive function. The basic idea of the definition is this: We begin with certain starting functions ($+$, $\times$, $K_<$, $U_{n,k}$)

312

and then apply the operations of composition and minimalization to construct additional recursive functions.

In Section 1.8, we proved that each starting function is computable and that the composition and minimalization of computable functions are computable. It follows by induction on recursive functions that every recursive function is computable. The converse of this fact, that every computable function is recursive, is Church's thesis.

Showing that a function is recursive is similar to writing a proof in a formal system: Starting functions are like axioms, and the two operations of composition and minimalization are like rules of inference. But this procedure is much too cumbersome to carry out in practice, so instead we use the following, more efficient, method.

Lemma 1: Let F be a n-ary function. To show that F is recursive, it suffices to show that F satisfies one of the following three conditions:

(1) F is a starting function.

(2) F is the composition of recursive functions.

(3) F is the minimalization of a regular recursive function.

Proof: If F is a starting function, then F is certainly recursive. Suppose F is the composition of recursive functions. To simplify notation, suppose that

$$F(a_1, \ldots, a_n) = G(H(a_1, \ldots, a_n), J(a_1, \ldots, a_n)),$$

where G, H, and J are recursive. Since G is recursive, there is a finite sequence $G_1, \ldots, G_k$ of functions with $G_k = G$ and such that for $1 \le i \le k$, G_i is a starting function or $i > 1$ and G_i is obtained by composition or minimalization of functions from among $G_1, \ldots, G_{i-1}$. Likewise, for H and J, we have sequences $H_1, \ldots, H_m$ with $H_m = H$ and $J_1, \ldots, J_n$ with $J_n = J$ such that the same conditions hold. The sequence of functions

$$G_1, \ldots, G_k, H_1, \ldots, H_m, J_1, \ldots, J_n, F$$

shows that F is recursive: F is the last function in the list, and each function in the list either is a starting function or is obtained by composition or minimalization from previous functions in the list (F is the composition of G_k, H_m, and J_n).

Now suppose F is the minimalization of a regular recursive function G; in other words, $F(a_1, \ldots, a_n) = \mu x[G(a_1, \ldots, a_n, x) = 0]$. Since G is recursive, there is a finite sequence $G_1, \ldots, G_k$ of functions with $G_k = G$ and such that for $1 \le i \le k$, either G_i is a starting function or $i > 1$ and G_i is obtained by composition or minimalization of functions from among $G_1, \ldots, G_{i-1}$. The sequence of functions $G_1, \ldots, G_k, F$ shows that F is recursive. ∎

Lemma 1 is our basic technique for proving a function recursive and will be used without explicit mention. We now begin a systematic study of recursive functions. The results we obtain will fall into one of two categories: (1) recursiveness of specific functions; (2) rules or operations for constructing new recursive functions from known recursive functions. At the beginning, proofs may seem tedious and results unimpressive, but as more techniques are developed, proofs will become easier and results more substantial. The reader should constantly keep our goal in mind: to give a precise counterpart for the class of computable functions. For each theorem, the reader should verify the correctness of the statement obtained by replacing "recursive" with "computable."

Theorem 1 (Successor, constant functions): The following functions are recursive:

(1) $S(a) = a + 1$ (successor function).

(2) $C_{n,k}(a_1, \ldots, a_n) = k$ (the n-ary constant function k, where $k \in \mathbb{N}$).

Proof of (1): We will begin by showing that the function $K_<$ is regular. Let $a \in \mathbb{N}$; we are required to find $x \in \mathbb{N}$ such that $K_<(a, x) = 0$. The value $x = a + 1$ works. In fact, $a + 1$ is the *smallest* x for which $K_<(a, x) = 0$ holds. Thus we can write

$$S(a) = \mu x[K_<(a, x) = 0]$$

and S is recursive by minimalization of a regular recursive function.

Proof of (2): The proof that $C_{n,k}$ is recursive is by induction on k:

$$C_{n,0}(a_1, \ldots, a_n) = \mu x[U_{n+1,n+1}(a_1, \ldots, a_n, x) = 0]$$
$$C_{n,k+1}(a_1, \ldots, a_n) = S(C_{n,k}(a_1, \ldots, a_n)).$$

$C_{n,0}$ is recursive by minimalization of a regular recursive function; $C_{n,k+1}(a_1, \ldots, a_n)$ is recursive because it is the composition of the recursive functions S and $C_{n,k}$ ($C_{n,k}$ is recursive by the induction hypothesis). ∎

Theorem 2 (Sum/product rule): The sum and the product of two recursive functions are recursive. More precisely, if H_1 and H_2 are n-ary recursive functions, so are the n-ary functions

$$F(a_1, \ldots, a_n) = H_1(a_1, \ldots, a_n) + H_2(a_1, \ldots, a_n)$$
$$G(a_1, \ldots, a_n) = H_1(a_1, \ldots, a_n) \times H_2(a_1, \ldots, a_n).$$

Proof: Addition $+$ and multiplication $\times$ are starting functions, and hence recursive. For F, we can write

$$F(a_1, \ldots, a_n) = +(H_1(a_1, \ldots, a_n), H_2(a_1, \ldots, a_n));$$

hence F is recursive, since it is the composition of recursive functions. A similar argument holds for G. ∎

Example 1:

The 3-ary function $F(a, b, c) = 2 \times c + a \times b$ is recursive. To prove this, we begin by introducing 3-ary auxilary functions H_1 and H_2, defined by $H_1(a, b, c) = 2 \times c$ and $H_2(a, b, c) = a \times b$. By the sum rule, F is recursive provided that both H_1 and H_2 are recursive. For H_1, we have

$$H_1(a, b, c) = 2 \times c$$
$$= C_{3,2}(a, b, c) \times U_{3,3}(a, b, c).$$

Both $C_{3,2}$ and $U_{3,3}$ are recursive; hence H_1 is recursive by the product rule. A similar argument applies to H_2:

$$H_2(a, b, c) = a \times b = U_{3,1}(a, b, c) \times U_{3,2}(a, b, c). \quad \square$$

The starting function $K_<$ is a 2-ary decision function. We now introduce two useful 1-ary decision functions, *sign (sg)* and *cosign (csg)*:

$$\text{sg} (a) = \begin{cases} 0 & a = 0 \\ 1 & a \neq 0. \end{cases} \qquad \text{csg} (a) = \begin{cases} 1 & a = 0 \\ 0 & a \neq 0. \end{cases}$$

Theorem 3: The functions sg and csg are recursive.

Proof: We first show that csg is recursive and then use this result to prove that sg is recursive:

$$\text{csg} (a) = K_<(0, a) = K_<(C_{1,0}(a), U_{1,1}(a))$$
$$\text{sg} (a) = \text{csg} (\text{csg} (a)).$$

Each is the composition of recursive functions and hence is recursive. ∎

A proper application of composition requires a certain amount of fussing about with constant and projection functions. It seems clear, however, that in applying the composition rule to $F(a_1, \ldots, a_n)$, the variables $a_1, \ldots, a_n$ can be rearranged, omitted, repeated, or even replaced by constants. We will now show that this is indeed the case.

Theorem 4 (Composition, special case): Let G be a k-ary recursive function, and define an n-ary function F by

$$F(a_1, \ldots, a_n) = G(b_1, \ldots, b_k),$$

where for $1 \leq i \leq k$, $b_i \in \{a_1, \ldots, a_n\}$ or $b_i \in \mathbb{N}$. Then F is recursive.

Proof: For $1 \leq i \leq k$, let H_i be the n-ary recursive function determined as follows:

(1) If $b_i = a_j$, H_i is the projection function $U_{n,j}$ (so $H_i(a_1, \ldots, a_n) = b_i$).
(2) If $b_i \in \mathbb{N}$, H_i is the constant function C_{n,b_i} (so $H_i(a_1, \ldots, a_n) = b_i$).

Thus for $1 \leq i \leq k$, we have $b_i = H_i(a_1, \ldots, a_n)$; hence F can be written as

$F(a_1, \ldots, a_n) = G(H_1(a_1, \ldots, a_n), \ldots, H_k(a_1, \ldots, a_n))$ and F is recursive by composition as required. ∎

Example 2:

Let G be a 2-ary recursive function and let H be a 4-ary recursive function. By Theorem 4, F_1, F_2, and F_3 are all recursive:

$$F_1(a, b, c) = 3 \times b$$
$$F_2(a, b, c) = G(a, c)$$
$$F_3(a, b, c) = H(c, c, a, 6). \quad \square$$

Example 3:

Given a 2-ary recursive function G and a 3-ary recursive function H, define a 3-ary function F by

$$F(a, b, c) = G(G(3, b), H(a + b, c, 3 \times a)).$$

We will show that F is recursive. Define two auxiliary functions L_1 and L_2 by

$$L_1(a, b, c) = G(3, b) \quad \text{and} \quad L_2(a, b, c) = H(a + b, c, 3 \times a).$$

Assuming that L_1 and L_2 are recursive, F is recursive by composition as follows:

$$F(a, b, c) = G(L_1(a, b, c), L_2(a, b, c)).$$

The function L_1 is recursive by Theorem 4. To see that L_2 is recursive, introduce three auxiliary functions K_1, K_2, and K_3:

$$K_1(a, b, c) = a + b, \quad K_2(a, b, c) = c, \quad K_3(a, b, c) = 3 \times a.$$

Each of these functions is recursive by Theorem 4, so L_2 is recursive by composition as follows: $L_2(a, b, c) = H(K_1(a, b, c), K_2(a, b, c), K_3(a, b, c))$.

$\square$

Exercises for Section 8.1

1. Prove the converse of Lemma 1. In other words, prove that if F is recursive, then F is a starting function, the composition of recursive functions, or the minimalization of a regular recursive function.
2. For each $n \geq 2$, define an n-ary function A_n by $A_n(a_1, \ldots, a_n) = a_1 + \cdots + a_n$. Show that each A_n is recursive.
3. Let $F(a, b, c) = a + b + 2$. Show that F is recursive.
4. Let G and H be 2-ary recursive functions. Show that F is recursive, where F is defined by $F(a, b, c) = G(H(b, c + 2), H(G(a, 3 \times b), 5))$. Give details.
5. Let G be a 2-ary recursive function. Show that F is recursive, where F is defined by $F(a, b, c) = 8 \times c + (G(c, b) \times G(a + b, 3))$. Give details.
6. Show that each of the following functions is recursive:

(1) $F(a) = \begin{cases} 4 & a = 0 \\ a & a \neq 0. \end{cases}$

(2) $F(a) = [\sqrt{a}]$.

(3) $F(a) = [a/3]$.

(Hint for (2) and (3): Use minimalization and $K_<$. Recall that $[\ ]$ is the greatest integer function. Thus $[x]$ is the greatest integer such that $[x] \leq x$; alternatively, $[x]$ is the smallest integer such that $x < [x] + 1$.)

7. The characteristic functions for $>$, $=$, $\leq$, $\geq$, and $\neq$, namely $K_>$, $K_=$, $K_\leq$, $K_\geq$, and $K_{\neq}$, are 2-ary decision functions. For example, $K_=$ is defined by

$$K_=(a, b) = \begin{cases} 0 & a = b \\ 1 & a \neq b. \end{cases}$$

Show that each of these functions is recursive.

8. (*No 2-ary recursive function enumerates all 1-ary recursive functions*) Let G be a 2-ary function with the following property: For each 1-ary recursive function F, there exists $e \in \mathbb{N}$ (called an *index* of F) such that for all $a \in \mathbb{N}$, $F(a) = G(e, a)$. In this case, we say that G *enumerates* all 1-ary recursive functions. Show that G is *not* recursive. (*Hint:* Use a Cantor diagonal argument, assume G is recursive, and define $F: \mathbb{N} \to \mathbb{N}$ by $F(a) = G(a, a) + 1$.)

9. (*Top-down = bottom-up*) Our definition of a recursive function is a bottom-up approach, since we begin with starting functions and then construct more complicated recursive functions using composition and minimalization. Let $\mathfrak{R}^-$ be the set of all functions obtained in this way. The class of recursive functions can also be described top-down. Let FNC denote the set of all functions from $\mathbb{N}^n$ ($n \geq 1$) into $\mathbb{N}$. A subset X of FNC is said to be *recursively closed* if it satisfies these three conditions: (1) Every starting function is in X; (2) if F is the composition of $G, H_1, \ldots, H_k$ and $G, H_1, \ldots, H_k \in X$, then $F \in X$; (3) if F is the minimalization of a regular function G and $G \in X$, then $F \in X$. Let $\mathfrak{R}^+ = \cap\{X : X \subseteq \text{FNC and } X \text{ is recursively closed}\}$. Show that $\mathfrak{R}^- = \mathfrak{R}^+$.

8.2

Recursive Relations

Many functions $F: \mathbb{N}^n \to \mathbb{N}$ are naturally defined in terms of relations on $\mathbb{N}$. In this section, we will study recursive relations from the point of view of expanding our list of techniques for showing functions to be recursive. Recall this notation:

$$R(a_1, \ldots, a_n) \quad \text{for } \langle a_1, \ldots, a_n \rangle \in R$$
$$\neg R(a_1, \ldots, a_n) \quad \text{for } \langle a_1, \ldots, a_n \rangle \notin R.$$

Also, an n-ary relation R is *recursive* if its characteristic function K_R is recursive; K_R is defined by

$$K_R(a_1, \ldots, a_n) = \begin{cases} 0 & R(a_1, \ldots, a_n) \\ 1 & \neg R(a_1, \ldots, a_n). \end{cases}$$

Example 1:

The 2-ary relation $<$ is recursive, since its characteristic function $K_<$ is a starting function and hence is recursive. $\qquad \square$

Definition 1: Let P and Q be n-ary relations on $\mathbb{N}$. The n-ary relations $\neg P$, $P \vee Q$, and $P \wedge Q$ are defined by the following:

(1) $(\neg P)(a_1, \ldots, a_n) \Leftrightarrow \neg P(a_1, \ldots, a_n)$ (i.e., $\langle a_1, \ldots, a_n \rangle \notin P$).

(2) $(P \vee Q)(a_1, \ldots, a_n) \Leftrightarrow P(a_1, \ldots, a_n) \vee Q(a_1, \ldots, a_n)$.

(3) $(P \wedge Q)(a_1, \ldots, a_n) \Leftrightarrow P(a_1, \ldots, a_n) \wedge Q(a_1, \ldots, a_n)$.

We call $\neg P$ the *negation* of P, $P \vee Q$ the *disjunction* of P and Q, and $P \wedge Q$ the *conjunction* of P and Q. In set-theoretic terms, $P, Q \subseteq \mathbb{N}^n$, $\neg P$ is the complement of P in $\mathbb{N}^n$, $P \vee Q$ is the union of P and Q, and $P \wedge Q$ is the intersection of P and Q.

Theorem 1 (Negation/disjunction/conjunction rule): Let P and Q be n-ary recursive relations. Then the n-ary relations $\neg P$, $P \vee Q$, and $P \wedge Q$ are recursive.

Proof: It suffices to show that the characteristic functions $K_{\neg P}$, $K_{P \vee Q}$, and $K_{P \wedge Q}$ are recursive:

$$K_{\neg P}(a_1, \ldots, a_n) = \operatorname{csg}(K_P(a_1, \ldots, a_n)).$$
$$K_{P \vee Q}(a_1, \ldots, a_n) = K_P(a_1, \ldots, a_n) \times K_Q(a_1, \ldots, a_n).$$
$$K_{P \wedge Q}(a_1, \ldots, a_n) = \operatorname{sg}(K_P(a_1, \ldots, a_n) + K_Q(a_1, \ldots, a_n)).\ \blacksquare$$

Example 2:

> The 2-ary relation $\geq$ is recursive. This is proved by applying the negation rule to the recursive relation $<$ as follows:
>
> $$a \geq b \Leftrightarrow \neg(a < b). \quad \square$$

Theorem 2 (Composition of a relation): Let Q be a k-ary recursive relation and let $H_1, \ldots, H_k$ be n-ary recursive functions. Then the n-ary relation R defined by

$$R(a_1, \ldots, a_n) \Leftrightarrow Q(H_1(a_1, \ldots, a_n), \ldots, H_k(a_1, \ldots, a_n))$$

is recursive.

Proof: It suffices to show that the characteristic function K_R of R is recursive. Now $K_R(a_1, \ldots, a_n) = K_Q(H_1(a_1, \ldots, a_n), \ldots, H_k(a_1, \ldots, a_n))$, so K_R is recursive by composition of recursive functions. $\blacksquare$

As in the case of composition of functions, a proper application of composition of relations requires careful attention to variables and numerous applications of the projection and constant functions. The next theorem allows us to ignore these tedious details.

Theorem 3 (Composition of a relation, special case): Let Q be a k-ary recursive relation. Define an n-ary relation R by

$$R(a_1, \ldots, a_n) \Leftrightarrow Q(b_1, \ldots, b_k),$$

where for $1 \leq i \leq k$, $b_i \in \{a_1, \ldots, a_n\}$ or $b_i \in \mathbb{N}$. Then R is recursive.

Proof: The proof is similar to that of Theorem 4 in Section 8.1 and is left as an exercise for the reader. ∎

Theorem 4: The relations $>$, $\geq$, $\leq$, $=$, and $\neq$ are recursive.

Proof: We have

(1) $a > b \Leftrightarrow b < a.$ (4) $a = b \Leftrightarrow (a \leq b) \wedge (b \leq a).$

(2) $a \geq b \Leftrightarrow \neg(a < b).$ (5) $a \neq b \Leftrightarrow \neg(a = b).$

(3) $a \leq b \Leftrightarrow \neg(b < a).$

We give details for (3). The relation $R(a, b) \Leftrightarrow b < a$ is recursive by Theorem 3. The negation rule then applies, and $\neg(b < a)$ is recursive as required. ∎

Example 3:

We will show that the following 3-ary relation is recursive:

$$R(a, b, c) \Leftrightarrow [(a < b \times (c + 1)) \vee (b = 0)].$$

Introduce auxilary relations P and Q by

$$P(a, b, c) \Leftrightarrow a < b \times (c + 1), \quad Q(a, b, c) \Leftrightarrow b = 0.$$

Then $R(a, b, c) \Leftrightarrow P(a, b, c) \vee Q(a, b, c)$; hence R is recursive by the disjunction rule, provided that both P and Q are recursive. Now Q is recursive by Theorem 3 and the fact that $=$ is recursive. To check P, introduce an auxilary function H by $H(a, b, c) = b \times (c + 1)$; a standard argument shows that H is recursive. Finally,

$$P(a, b, c) \Leftrightarrow U_{3,1}(a, b, c) < H(a, b, c),$$

so P is recursive by composition of relations as required.   □

Theorem 5: Every finite subset of $\mathbb{N}$ is recursive.

Proof: Let $R \subseteq \mathbb{N}$ be finite. If $R = \varnothing$, then $R(a) \Leftrightarrow (a < a)$ shows that R is recursive. Otherwise, let $R = \{c_1, \ldots, c_n\}$; then R is recursive by

$$R(a) \Leftrightarrow (a = c_1) \vee \cdots \vee (a = c_n).$$

The use of $\cdots$ is allowed in this case because n is fixed and does not depend on a. ∎

Theorem 6 (Definition by cases): Let $G_1, \ldots, G_k$ be n-ary recursive functions and let $R_1, \ldots, R_k$ be n-ary recursive relations such that for all $a_1, \ldots, a_n \in \mathbb{N}$, exactly one of $R_1(a_1, \ldots, a_n), \ldots, R_k(a_1, \ldots, a_n)$ holds. Then F is recursive, where

$$F(a_1, \ldots, a_n) = \begin{cases} G_1(a_1, \ldots, a_n) & R_1(a_1, \ldots, a_n) \\ \vdots & \vdots \\ G_k(a_1, \ldots, a_n) & R_k(a_1, \ldots, a_n). \end{cases}$$

Proof: We use the following notation: $\mathbf{a} = \langle a_1, \ldots, a_n \rangle$; K_i is the characteristic function of the relation R_i. The function F is recursive by repeated application of the sum/product rule:

$$F(\mathbf{a}) = G_1(\mathbf{a}) \times \operatorname{csg}(K_1(\mathbf{a})) + \cdots + G_k(\mathbf{a}) \times \operatorname{csg}(K_k(\mathbf{a})). \blacksquare$$

Example 4:

Definition by cases shows that $F(a, b) = \max\{a, b\}$ is recursive:

$$F(a, b) = \begin{cases} a & a \geq b \\ b & a < b. \end{cases} \square$$

The next result broadens the scope of minimalization.

Definition 2: An $n + 1$-ary relation R is *regular* if for all $a_1, \ldots, a_n \in \mathbb{N}$, there exists $x \in \mathbb{N}$ such that $R(a_1, \ldots, a_n, x)$. Given an $n + 1$-ary regular relation R and $a_1, \ldots, a_n \in \mathbb{N}$, $\mu x R(a_1, \ldots, a_n, x)$ denotes the *smallest* x for which $R(a_1, \ldots, a_n, x)$ holds.

Theorem 7 (Minimalization of a regular relation): Let R be an $n + 1$-ary regular recursive relation. Then the n-ary function F defined by

$$F(a_1, \ldots, a_n) = \mu x R(a_1, \ldots, a_n, x)$$

is recursive.

Proof: Since R is regular and recursive, its characteristic function K_R is also regular and recursive. Moreover,

$$F(a_1, \ldots, a_n) = \mu x [K_R(a_1, \ldots, a_n, x) = 0];$$

hence F is recursive by minimalization (of a regular recursive function). $\blacksquare$

Comment The original minimalization operation, described by

$$F(a_1, \ldots, a_n) = \mu x [G(a_1, \ldots, a_n, x) = 0],$$

with G regular, is a special case of minimalization of a regular relation. To see this, define R by

$$R(a_1, \ldots, a_n, x) \Leftrightarrow G(a_1, \ldots, a_n, x) = 0,$$

and note that R is a regular relation provided that G is a regular function. We also have

$$F(a_1, \ldots, a_n) = \mu x R(a_1, \ldots, a_n, x),$$

and this is the format for minimalization of a regular relation. Unless otherwise stated, from now on *minimalization* will refer to the more general operation of minimalization of a regular relation.

We will now use the results obtained thus far to show that certain important functions in number theory are recursive.

Theorem 8 (Proper subtraction, QT, RM): The following 2-ary functions are recursive:

(1) Proper subtraction $\dotdiv$, defined by $a \dotdiv b = \begin{cases} a - b & b \le a \\ 0 & a < b. \end{cases}$

(2) $QT(a, b) = q$, the quotient when a is divided by $b \ne 0$ ($QT(a, 0) = 0$ by convention).

(3) $RM(a, b) = r$, the remainder when a is divided by $b \ne 0$ ($RM(a, 0) = 0$ by convention).

Proof of (1): First note the difference between ordinary subtraction and proper subtraction: For $a < b$, $a - b$ is negative, whereas $a \dotdiv b = 0$. Let $a, b \in \mathbb{N}$. Then

$$b \le a \Leftrightarrow \text{there is a (unique) } x \in \mathbb{N} \text{ such that } a = b + x;$$

moreover, in this case $a \dotdiv b = x$. (Note that x is also the smallest $x \in \mathbb{N}$ such that $a = b + x$.) Now introduce a 3-ary relation L by

$$L(a, b, x) \Leftrightarrow [(a = b + x) \lor (a < b)].$$

The relation L is recursive by standard arguments (disjunction rule, composition of relations). We now argue that L is also regular. Let $a, b \in \mathbb{N}$. If $a < b$, then for *any* $x \in \mathbb{N}$, we have $L(a, b, x)$; moreover, the smallest $x \in \mathbb{N}$ such that $L(a, b, x)$ is $x = 0$. On the other hand, if $b \le a$, then there exists $x \in \mathbb{N}$ such that $a = b + x$; hence we also have $L(a, b, x)$. We leave it to the reader to verify that

$$a \dotdiv b = \mu x[(a = b + x) \lor (a < b)].$$

It follows that $\dotdiv$ is recursive by minimalization of a regular recursive relation.

Proof of (2) and (3): We will now show that QT and RM are recursive. The division algorithm states that for $a, b \in \mathbb{N}$ with $b > 0$, there exist unique $q, r \in \mathbb{N}$ such that $a = bq + r$ and $0 \le r < b$. Note that $r = a \dotdiv bq$. By definition, $QT(a, b) = q$ and $RM(a, b) = r$. (For $b = 0$, we define $QT(a, 0) = RM(a, 0) = 0$.) In the proof of the division algorithm, q is the unique natural number such that $b \times q \le a < b \times (q + 1)$ (for details, see the appendix on number theory). Thus QT and RM are recursive by

$$QT(a, b) = \mu q[(a < b \times (q + 1)) \vee (b = 0)]$$
$$RM(a, b) = [a \div (b \times QT(a, b))] \times sg(b). \blacksquare$$

The class of recursive relations is closed under the logical operations of negation, disjunction, and conjunction. What about quantification? Remember that one of our goals is to prove Church's theorem, which states that the 1-ary relation THM_Γ is *not* recursive (Γ = consistent extension of NN). On the other hand, in the process of proving Gödel's incompleteness theorem, we will prove that for Γ recursively axiomatized, the 2-ary relation PRE_Γ is recursive and for all $a \in \mathbb{N}$, $THM_\Gamma(a) \Leftrightarrow \exists b PRF_\Gamma(a, b)$. This shows that the class of recursive relations is *not* closed under quantification. But, as we now show, a *bounded* quantifier applied to a recursive relation does give a recursive relation. These bounded quantifiers are defined as follows:

- $\exists x < bR(a_1, \ldots, a_n, x) \Leftrightarrow \exists x[(x < b) \wedge R(a_1, \ldots, a_n, x)]$
- $\forall x < bR(a_1, \ldots, a_n, x) \Leftrightarrow \forall x[(x < b) \rightarrow R(a_1, \ldots, a_n, x)]$.

Theorem 9 (Bounded quantifier rule, first version): Let R be an $n + 1$-ary recursive relation. Then the $n + 1$-ary relations P and Q are recursive, where

$$P(a_1, \ldots, a_n, b) \Leftrightarrow \exists x < bR(a_1, \ldots, a_n, x)$$
$$Q(a_1, \ldots, a_n, b) \Leftrightarrow \forall x < bR(a_1, \ldots, a_n, x).$$

Proof: Here is the basic idea of the proof for the relation P: Given $a_1, \ldots, a_n, b \in \mathbb{N}$, find (if possible) $x < b$ such that $R(a_1, \ldots, a_n, x)$; otherwise, let $x = b$. Then $P(a_1, \ldots, a_n, b)$ if and only if $x < b$. Here are the precise details:

$$P(a_1, \ldots, a_n, b) \Leftrightarrow \mu x[R(a_1, \ldots, a_n, x) \vee (x = b)] < b.$$

Note that the relation $[R(a_1, \ldots, a_n, x) \vee (x = b)]$ is regular and recursive.

The following sequence of steps shows that Q is also recursive.

$$\begin{aligned} Q(a_1, \ldots, a_n, b) &\Leftrightarrow \forall x < bR(a_1, \ldots, a_n, x) \\ &\Leftrightarrow \forall x[(x < b) \rightarrow R(a_1, \ldots, a_n, x)] \\ &\Leftrightarrow \neg \exists x \neg[\neg(x < b) \vee R(a_1, \ldots, a_n, x)] \\ &\Leftrightarrow \neg \exists x[(x < b) \wedge \neg R(a_1, \ldots, a_n, x)] \\ &\Leftrightarrow \neg[\exists x < b \neg R(a_1, \ldots, a_n, x)]. \end{aligned}$$

Now R is recursive, so $\neg R$ is recursive by the negation rule. By the result on bounded $\exists$, $\exists x < b \neg R(a_1, \ldots, a_n, x)$ is recursive. Finally, one more application of the negation rule shows that Q is recursive as required. $\blacksquare$

There is another version of the bounded quantifier rule. This second version follows from the first and is probably the one used more often. In any case, an application of either will be referred to as the *bounded quantifier rule*.

Theorem 10 (Bounded quantifier rule, second version): Let H

be an n-ary recursive function and let R be an $n + 1$-ary recursive relation. Then P and Q are recursive, where

$$P(a_1, \ldots, a_n) \Leftrightarrow \exists x < H(a_1, \ldots, a_n)R(a_1, \ldots, a_n, x)$$
$$Q(a_1, \ldots, a_n) \Leftrightarrow \forall x < H(a_1, \ldots, a_n)R(a_1, \ldots, a_n, x).$$

Proof: We show P recursive and leave Q to the reader. By Theorem 9, the $n + 1$-ary relation P^* defined by

$$P^*(a_1, \ldots, a_n, b) \Leftrightarrow \exists x < bR(a_1, \ldots, a_n, x)$$

is recursive. It is easy to check that

$$P(a_1, \ldots, a_n) \Leftrightarrow P^*(a_1, \ldots, a_n, H(a_1, \ldots, a_n)),$$

so P is recursive by composition as required. ∎

Example 5:

The triangular numbers are $1, 3, 6, 10, \ldots$ (that is, $1, 1 + 2, 1 + 2 + 3, \ldots$). Thus $a \in \mathbb{N}$ is a triangular number if and only if $a = n(n + 1)/2$ for some $n \geq 1$. Let TRI be defined by

$$\text{TRI}(a) \Leftrightarrow a \text{ is a triangular number.}$$

The relation TRI is recursive by the bounded quantifier rule as follows:

$$\text{TRI}(a) \Leftrightarrow \exists n < a + 1[(n \geq 1) \wedge (2 \times a = n \times (n + 1))]. \quad \square$$

Example 6:

The set COMP of composite numbers is recursive as follows:

$$\text{COMP}(a) \Leftrightarrow \exists b < a \, \exists c < a[(a = b \times c) \wedge (b > 1) \wedge (c > 1)]. \quad \square$$

At this point, it is worthwhile to step back and summarize the main results of Sections 8.1 and 8.2.

Functions S, sg, and csg are 1-ary recursive functions.
$+, \times, \div, QT$, and RM are 2-ary recursive functions.
For $n \geq 1$, $U_{n,k}$ ($1 \leq k \leq n$) and $C_{n,k}$ ($k \in \mathbb{N}$) are n-ary recursive functions.
Sum/product rule.
Composition.
Definition by cases.
Minimalization (of a regular recursive relation).

Relations $<, \leq, =, >, \geq$, and $\neq$ are 2-ary recursive relations.
Negation/disjunction/conjunction rule.
Composition of a relation.
Bounded $\exists$ and bounded $\forall$ rules.

Summary: Recursive Functions and Recursive Relations

—————————————*Exercises for Section 8.2*—————————————

1. Prove Theorem 3.
2. Prove Theorem 4, parts (1), (2), (4), and (5). (Verify details.)
3. Let P and Q be n-ary decidable relations. Show that $\neg P$, $P \vee Q$, and $P \wedge Q$ are decidable relations.
4. Let P and Q be n-ary relations on $\mathbb{N}$. Show the following:
   (1) $K_{\neg P} = 1 - K_P$.
   (2) $K_{P \vee Q} = K_P \times K_Q$.
   (3) $K_{P \wedge Q} = K_P + K_Q - (K_P \times K_Q)$.
5. Let Q be a k-ary decidable relation and let $H_1, \ldots, H_k$ be n-ary computable functions. Show that R is decidable, where R is defined by $R(a_1, \ldots, a_n) \Leftrightarrow Q(H_1(a_1, \ldots, a_n), \ldots, H_k(a_1, \ldots, a_n))$.
6. Let R be an $n + 1$-ary decidable relation that is also regular. Show that F is computable, where $F(a_1, \ldots, a_n) = \mu x R(a_1, \ldots, a_n, x)$.
7. Let H be an n-ary computable function and let R be an $n + 1$-ary decidable relation. Show that the following n-ary relations are decidable:
   (1) $P(a_1, \ldots, a_n) \Leftrightarrow \exists x < H(a_1, \ldots, a_n) R(a_1, \ldots, a_n, x)$.
   (2) $Q(a_1, \ldots, a_n) \Leftrightarrow \forall x < H(a_1, \ldots, a_n) R(a_1, \ldots, a_n, x)$.
8. Let $F: \mathbb{N} \to \mathbb{N}$ be a computable function and let $R \subseteq \mathbb{N}$ be a decidable relation. Show that $F^{-1}(R)$ is a decidable relation. (*Note:* $F^{-1}(R) = \{a : F(a) \in R\}$.)
9. Let $F: \mathbb{N} \to \mathbb{N}$ be a recursive function and let $R \subseteq \mathbb{N}$ be a recursive relation. Show that $F^{-1}(R)$ is a recursive relation.
10. Let $G_1, \ldots, G_k$ be n-ary computable functions and let $R_1, \ldots, R_k$ be n-ary decidable relations such that for all $a_1, \ldots, a_n \in \mathbb{N}$, exactly one of $R_1(a_1, \ldots, a_n), \ldots, R_k(a_1, \ldots, a_n)$ holds. Show that F is computable, where

$$F(a_1, \ldots, a_n) = \begin{cases} G_1(a_1, \ldots, a_n) & R_1(a_1, \ldots, a_n) \\ \vdots & \vdots \\ G_k(a_1, \ldots, a_n) & R_k(a_1, \ldots, a_n). \end{cases}$$

11. Let $a, b \in \mathbb{N}$. Show that $a \div (b + 1) = (a \div b) \div 1$.
12. Show that each of the following relations is recursive:
    (1) $DIV(a, b) \Leftrightarrow a \neq 0$ and a divides b.
    (2) $ODD(a) \Leftrightarrow a$ is odd.
    (3) $PRIME(a) \Leftrightarrow a$ is prime.
    (4) $GB(a) \Leftrightarrow a$ is an even number ≥ 4 and a is the sum of two primes.
13. Show that F is recursive, where $F(a) = \begin{cases} 2a & a \text{ even} \\ a + 7 & a \text{ odd}. \end{cases}$
14. Show that the following functions are recursive:
    (1) $F(a, b) = \min \{a, b\}$.
    (2) $F(a, b) = |a - b|$.
15. Let $F: \mathbb{N} \to \mathbb{N}$ be one-to-one, onto, and recursive. Show that the inverse F^{-1} of F is recursive. (*Hint:* $F^{-1}(b) = a \Leftrightarrow F(a) = b$.)
16. (*Version of definition by cases*) Let $G_1, \ldots, G_k$ be n-ary recursive functions and let $R_1, \ldots, R_{k-1}$ be n-ary recursive relations such that for all $a_1, \ldots, a_n \in \mathbb{N}$, $R_i(a_1, \ldots, a_n)$ holds for at most one i, $1 \leq i \leq k - 1$. Show that the following function is recursive:

$$F(a_1, \ldots, a_n) = \begin{cases} G_1(a_1, \ldots, a_n) & R_1(a_1, \ldots, a_n) \\ \vdots & \vdots \\ G_{k-1}(a_1, \ldots, a_n) & R_{k-1}(a_1, \ldots, a_n) \\ G_k(a_1, \ldots, a_n) & \text{otherwise.} \end{cases}$$

17. Let R be an $n + 2$-ary recursive relation and let G and H be n-ary recursive functions. Show that the following n-ary relations are recursive:
    (1) $P(a_1, \ldots, a_n) \Leftrightarrow \exists x < G(a_1, \ldots, a_n) \, \exists y < H(a_1, \ldots, a_n) R(a_1, \ldots, a_n, x, y)$.
    (2) $Q(a_1, \ldots, a_n) \Leftrightarrow \forall x < G(a_1, \ldots, a_n) \, \forall y < H(a_1, \ldots, a_n) R(a_1, \ldots, a_n, x, y)$.
18. Complete the proof of the second version of the bounded quantifier rule (Theorem 10). In other words, show that if H is an n-ary recursive function and R is an $n + 1$-ary recursive relation, then Q is recursive, where

    $$Q(a_1, \ldots, a_n, b) \Leftrightarrow \forall x < H(a_1, \ldots, a_n) R(a_1, \ldots, a_n, x).$$

19. Let $\Re$ be a class of relations on $\mathbb{N}$ that is closed under negation (i.e., if $R \in \Re$, then $\neg R \in \Re$). Prove that $\Re$ is closed under bounded $\forall$ if and only if $\Re$ is closed under bounded $\exists$.
20. (*Bounded minimalization*) Minimalization, defined by

    $$F(a_1, \ldots, a_n) = \mu x R(a_1, \ldots, a_n, x),$$

    requires that R be regular. Without this requirement, the search for $x \in \mathbb{N}$ for which $R(a_1, \ldots, a_n, x)$ holds could continue forever. With bounded minimalization, we omit the requirement that R be regular, and instead place a bound on the search for x. Prove the following:
    (1) Let R be an $n + 1$-ary recursive relation, and define $F: \mathbb{N}^{n+1} \to \mathbb{N}$ by

    $$F(a_1, \ldots, a_n, b) = \mu x \leq b R(a_1, \ldots, a_n, x).$$

    Then F is recursive. $(F(a_1, \ldots, a_n, b) = b + 1$ if there is no $x \leq b$ such that $R(a_1, \ldots, a_n, x)$.)
    (2) Let H be an n-ary recursive function, let R be an $n + 1$-ary recursive relation, and define $F: \mathbb{N}^n \to \mathbb{N}$ by

    $$F(a_1, \ldots, a_n) = \mu x \leq H(a_1, \ldots, a_n) R(a_1, \ldots, a_n, x).$$

    Then F is recursive. $(F(a_1, \ldots, a_n) = H(a_1, \ldots, a_n) + 1$ if there is no $x \leq H(a_1, \ldots, a_n)$ such that $R(a_1, \ldots, a_n, x)$.)
21. (*Bounded minimalization*) Refer to Exercise 20. Let R be an $n + 1$-ary decidable relation, and define $F: \mathbb{N}^{n+1} \to \mathbb{N}$ by

    $$F(a_1, \ldots, a_n, b) = \mu x \leq b R(a_1, \ldots, a_n, x).$$

    Show that F is computable.

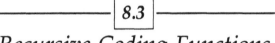

Recursive Coding Functions

In this section, we obtain recursive functions for coding n-tuples of natural numbers. This has two major applications: (1) Justify the operations of primitive recursion and course-of-values recursion, two new and powerful methods for constructing new recursive functions from known recursive functions. For example, primitive recursion allows us to show easily that the factorial function $FT(n) = n!$ and the exponential function $EXP(a, b) = a^b$ are recursive. (2) Effectively code the language L_{NN} of formal arithmetic as a key step in the proof of Church's theorem, Tarski's theorem, and Gödel's incompleteness theorem. Application (1) on primitive recursion is a continuation of our systematic study of recursive functions; (2) is an application of recursion theory to undecidability results.

In Section 7.2, we used prime numbers and exponentiation to effectively code the language L_{NN}. However, that method of coding is not available to us at this point, since we do not yet know that $EXP(a, b) = a^b$ is recursive. Instead, we will use certain important recursive functions, due to Cantor and Gödel.

——Cantor's Pairing and Projection Functions——

Cantor, in 1874, proved that $\mathbb{N} \times \mathbb{N}$ is equipotent to $\mathbb{N}$ by constructing a one-to-one function J from $\mathbb{N} \times \mathbb{N}$ onto $\mathbb{N}$. It is not difficult to find an explicit formula for J; this explicit formula shows that J is recursive. An obvious application of J is to recursively code all ordered pairs. Given an ordered pair $\langle a, b \rangle$, let $c = J(a, b)$; we call c the *code* of $\langle a, b \rangle$. Since J is onto, every $c \in \mathbb{N}$ is the code of at least one ordered pair $\langle a, b \rangle$; since J is one-to-one, distinct pairs cannot have the same code. Given a code c, we want to recursively find the pair $\langle a, b \rangle$ whose code is c. More precisely, we want 1-ary recursive functions K and L such that $K(J(a, b)) = a$ and $L(J(a, b)) = b$; K and L are often referred to as *decoding* functions.

Theorem 1 (Cantor's pairing and projection functions): There is a 2-ary recursive function J and there are 1-ary recursive functions K and L such that

(1) J is a one-to-one function from $\mathbb{N} \times \mathbb{N}$ onto $\mathbb{N}$.
(2) $J(a, b) = (a + b)(a + b + 1)/2 + a$.
(3) $K(J(a, b)) = a$ and $L(J(a, b)) = b$.
(4) $K(c) \le c$ and $L(c) \le c$.
(5) If $c > 0$, then $K(c) < c$.

Proof: We begin with a picture of $\mathbb{N} \times \mathbb{N}$ and the way J works:

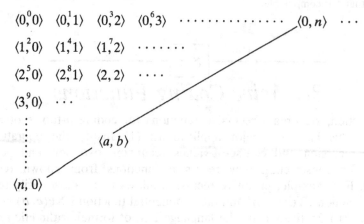

The function J assigns 0 to $\langle 0, 0 \rangle$, 1 to $\langle 0, 1 \rangle$, 2 to $\langle 1, 0 \rangle$, and proceeds down successive diagonals in this fashion. We use the following terminology: The line connecting $\langle 0, n \rangle$ and $\langle n, 0 \rangle$ is called *diagonal n*. For example, $\langle 0, 0 \rangle$ is on diagonal 0; $\langle 0, 2 \rangle$, $\langle 1, 1 \rangle$, and $\langle 2, 0 \rangle$ are on diagonal 2. We need to define J at an arbitrary $\langle a, b \rangle$ that is on diagonal n. We begin with three observations:

- $a + b = n$.
- The total number of pairs on all diagonals preceding diagonal n is $1 + 2 + \cdots + n = n(n + 1)/2$.
- Starting at $\langle 0, n \rangle$, $\langle a, b \rangle$ is the $a + 1$-element on diagonal n.

Taking into account that the first value for J is 0 rather than 1,

$$J(a, b) = \frac{n(n + 1)}{2} + a$$

$$= \frac{(a + b)(a + b + 1)}{2} + a.$$

The proof that J is one-to-one and onto is outlined in Exercises 1 and 2. J is recursive because we can write

$$J(a, b) = \text{QT}((a + b) \times (a + b + 1), 2) + a.$$

We now turn to the problem of constructing the projection functions K and L. Let $c \in \mathbb{N}$. Since J is one-to-one and onto, there is a *unique* pair $\langle a, b \rangle$ such that $J(a, b) = c$. By definition, $K(c) = a$ and $L(c) = b$. We can also describe K and L as follows:

$$K(c) = \text{unique number } a \text{ for which there exists}$$
$$b \in \mathbb{N} \text{ such that } J(a, b) = c.$$

A similar description holds for L. We are required to show that K and L are recursive and satisfy (3), (4), and (5). From $J(a, b) = c$, we have

$$\frac{(a + b)(a + b + 1)}{2} + a = c.$$

From this equation, it follows that $a \leq c$ and $b \leq c$, and that $a < c$ for $c > 0$. Since $K(c) = a$ and $L(c) = b$, it is clear that (4) and (5) hold. To see that K and L are recursive, write

$$K(c) = \mu a[\exists b < (c + 1)(J(a, b) = c)]$$
$$L(c) = \mu b[\exists a < (c + 1)(J(a, b) = c)].$$

It remains to prove (3). Let $K(J(a, b)) = a'$. By the definition of K, a' is the smallest number for which there exists b' such that $J(a', b') = J(a, b)$. But J is one-to-one, so $a = a'$, and hence $K(J(a, b)) = a$ as required. A similar argument shows that $L(J(a, b)) = b$. ∎

Example 1:

(*Collapsing quantifiers*) Let R be an $n + 2$-ary relation. We use Cantor's functions J, K, and L to show that for all $a_1, \ldots, a_n \in \mathbb{N}$,

$$\exists a\ \exists b R(a_1, \ldots, a_n, a, b) \Leftrightarrow \exists c R(a_1, \ldots, a_n, K(c), L(c)).$$

This is a very useful tool in recursion theory, and is called *collapsing a quantifier.* First assume there exist $a, b \in \mathbb{N}$ such that $R(a_1, \ldots, a_n, a, b)$. Let $c = J(a, b)$. Then $K(c) = a$ and $L(c) = b$; hence $\exists c R(a_1, \ldots, a_n, K(c), L(c))$ as required. Conversely, suppose there exists $c \in \mathbb{N}$ such that $R(a_1, \ldots, a_n, K(c), L(c))$. Let $a = K(c)$ and $b = L(c)$, then $\exists a \exists b R(a_1, \ldots, a_n, a, b)$ as required. $\qquad\square$

We will now extend Cantor's result so that it holds for all $n \geq 2$.

Theorem 2: For each $n \geq 2$, there is an n-ary recursive function J_n and there are 1-ary recursive functions $J_{n,1}, \ldots, J_{n,n}$ satisfying these conditions:

(1) J_n is a one-to-one function from $\mathbb{N}^n$ onto $\mathbb{N}$.

(2) $J_{n,k}(J_n(a_1, \ldots, a_n)) = a_k$ for $1 \leq k \leq n$.

Proof: The proof is by induction. For $n = 2$, take $J_2 = J$, $J_{2,1} = K$, and $J_{2,2} = L$. Now let $n \geq 2$ and assume that there exist recursive functions J_n and $J_{n,1}, \ldots, J_{n,n}$ satisfying (1) and (2). Define functions J_{n+1} and $J_{n+1,1}, \ldots, J_{n+1,n+1}$ by

$$J_{n+1}(a_1, \ldots, a_n, a_{n+1}) = J(J_n(a_1, \ldots, a_n), a_{n+1})$$
$$J_{n+1,k}(c) = J_{n,k}(K(c)), \quad 1 \leq k \leq n$$
$$J_{n+1,n+1}(c) = L(c).$$

It is clear that J_{n+1} and $J_{n+1,1}, \ldots, J_{n+1,n+1}$ are recursive. We leave it to the reader to check that $J_{n+1}: \mathbb{N}^{n+1} \to \mathbb{N}$ is one-to-one and onto. It remains to prove that $J_{n+1,k}(J_{n+1}(a_1, \ldots, a_n, a_{n+1})) = a_k$ for $1 \leq k \leq n + 1$. First assume $k \leq n$. Then

$$\begin{aligned} J_{n+1,k}(J_{n+1}(a_1, \ldots, a_n, a_{n+1})) &= J_{n,k}(K(J_{n+1}(a_1, \ldots, a_n, a_{n+1}))) \\ &= J_{n,k}(K(J(J_n(a_1, \ldots, a_n), a_{n+1}))) \\ &= J_{n,k}(J_n(a_1, \ldots, a_n)) \\ &= a_k. \end{aligned}$$

We leave the case $k = n + 1$ to the reader. ∎

Gödel's β-Function

Cantor's functions J, K, and L recursively code all ordered pairs. In addition, the functions J_n and $J_{n,1}, \ldots, J_{n,n}$ recursively code all n tuples, $n \geq 2$. There is, however, a drawback to this method of coding: Each of the functions J_n is onto. This means that every $c \in \mathbb{N}$ is the code of a pair, of a triple, and so on; in other words, there is the problem of overlap. We now introduce Gödel's β-function, a 2-ary recursive function that allows us to recursively code n-tuples (each $n \geq 1$) so that the following holds: If c is the code of an n-tuple $\langle a_1, \ldots, a_n \rangle$ and also the code of an m-tuple $\langle b_1, \ldots, b_m \rangle$, then $n = m$ and

$a_k = b_k$ for $1 \le k \le n$. The construction of β requires several ideas from number theory; see the appendix on number theory for background material (including a proof of Lemma 1).

Lemma 1 (Chinese remainder theorem, special case): Let $m_0, \ldots, m_n$ be positive integers that are relatively prime in pairs and let $a_0, \ldots, a_n \in \mathbb{N}$ be such that $a_k < m_k$ for $0 \le k \le n$. Then there is a number $a < m_0 \times \cdots \times m_n$ such that for $0 \le k \le n$,

$$a_k = \text{remainder when } a \text{ is divided by } m_k.$$

In our application of the Chinese remainder theorem, we will choose relatively prime numbers $m_0, \ldots, m_n$ that depend on just one number b.

Lemma 2: Let $n! \mid b$, where $b > 0$. Then the $n + 1$ numbers $1 + b$, $1 + 2b, \ldots, 1 + (n + 1)b$ are relatively prime in pairs.

Proof: Suppose we have $i, j \in \mathbb{N}$ with $1 \le i < j \le n + 1$ and a prime p such that $p \mid 1 + ib$ and $p \mid 1 + jb$. Then p divides the difference of these two numbers; that is, $p \mid b(j - i)$. Since p is a prime, $p \mid b$ or $p \mid j - i$. Suppose $p \mid b$. Now $p \mid 1 + ib$ also; hence $p \mid 1$, a contradiction. Suppose $p \mid j - i$. Now $1 \le j - i \le n$; hence $p \mid n!$ But $n! \mid b$ (hypothesis); hence $p \mid b$. But we have already shown that $p \mid b$ is impossible. ∎

Lemma 3 (Gödel): Let $a_0, \ldots, a_n \in \mathbb{N}$. Then there exist $a, b \in \mathbb{N}$ such that for $0 \le k \le n$,

$$a_k = \text{remainder when } a \text{ is divided by } 1 + b(k + 1).$$

Proof: Let $b = n! \times (\max\{a_0, \ldots, a_n\} + 1)$, and note that each $a_k < b$. Now $n! \mid b$; hence, by Lemma 2, the numbers $1 + b$, $1 + 2b, \ldots, 1 + (n + 1)b$ are relatively prime in pairs. The result now follows immediately from the Chinese remainder theorem (Lemma 1). ∎

Theorem 3 (Gödel's β-function): There is a 2-ary recursive function β satisfying these three properties:

(1) Given $a_0, \ldots, a_n \in \mathbb{N}$, there exists $c \in \mathbb{N}$ such that $\beta(c, k) = a_k$ for $0 \le k \le n$.

(2) $\beta(0, k) = 0$ for all $k \in \mathbb{N}$.

(3) If $c > 0$, then $\beta(c, k) < c$ for all $k \in \mathbb{N}$.

Proof: We proceed informally at first. Let $a_0, \ldots, a_n \in \mathbb{N}$. By Lemma 3, there exist $a, b \in \mathbb{N}$ such that for $0 \le k \le n$, a_k is the remainder when a is divided by $1 + b(k + 1)$; that is,

$$a_k = \text{RM}(a, 1 + b \times (k + 1)), \quad 0 \le k \le n.$$

We can regard the pair a, b as a code for $a_0, \ldots, a_n$. (If we know a and b, we can recover $a_0, \ldots, a_n$ from these equations.) Cantor's pairing and projection functions are now used to reduce a, b to a single code. Let $c = J(a, b)$, so that $K(c) = a$ and $L(c) = b$. We then have

$$a_k = \mathrm{RM}(K(c), 1 + L(c) \times (k + 1)), \quad 0 \le k \le n.$$

This suggests that we define β by

$$\beta(c, k) = \mathrm{RM}(K(c), 1 + L(c) \times (k + 1)).$$

It is clear that β is recursive and that (1) is satisfied.

It remains to check the technical conditions (2) and (3). Let $k \in \mathbb{N}$. By definition, $\beta(c, k)$ is the remainder when $K(c)$ is divided by $1 + L(c) \times (k + 1)$. In other words, by the division algorithm,

$$K(c) = [1 + L(c) \times (k + 1)] \times q + \beta(c, k) \quad (q = \text{quotient});$$

hence $\beta(c, k) \le K(c)$. For (2), we have $\beta(0, k) \le K(0) = 0$. To check (3), let $c > 0$ and note that $K(c) < c$ by Theorem 1(5). Then $\beta(c, k) \le K(c) < c$ as required. ■

Recursive Coding of n-tuples

Gödel's β-function is now used to recursively code n-tuples (each $n \ge 1$). We proceed informally at first. Fix $n \ge 1$ and let $a_1, \ldots, a_n \in \mathbb{N}$. Apply Gödel's β-function to $n, a_1, \ldots, a_n$; there is a number c such that

$$\beta(c, 0) = n \quad \text{and} \quad \beta(c, k) = a_k \quad \text{for} \quad 1 \le k \le n.$$

The smallest c with these two properties is the *code* of $a_1, \ldots, a_n$, and we write

$$\langle a_1, \ldots, a_n \rangle_n = c.$$

Note that β serves as a decoding function. For suppose we know that the number c is a code. Calculate $\beta(c, 0) = n$; then c is the code of an n-tuple. Now calculate $\beta(c, k)$ for $1 \le k \le n$ to obtain the entries $a_1, \ldots, a_n$ of the n-tuple.

Theorem 4 (Recursive coding functions): For each $n \ge 1$, there is an n-ary recursive function $\langle \cdots \rangle_n$ satisfying these three properties:

(1) If $c = \langle a_1, \ldots, a_n \rangle_n$, then $\beta(c, 0) = n$ and $\beta(c, k) = a_k$ for $1 \le k \le n$ (decoding).

(2) If $\langle a_1, \ldots, a_n \rangle_n = \langle b_1, \ldots, b_m \rangle_m$, then $n = m$ and $a_k = b_k$ for $1 \le k \le n$ (uniqueness).

(3) If $c = \langle a_1, \ldots, a_n \rangle_n$, then $n < c$ and $a_k < c$ for $1 \le k \le n$ (bounded).

Proof: Fix $n \ge 1$, and define $\langle \cdots \rangle_n$ by

$$\langle a_1, \ldots, a_n \rangle_n = \mu c[(\beta(c, 0) = n) \wedge (\beta(c, 1) = a_1) \wedge \cdots \wedge (\beta(c, n) = a_n)].$$

The use of $\cdots$ is allowed because n is fixed. We could also use $\forall k < n[\beta(c, k + 1) = a_{k+1}]$ in place of $(\beta(c, 1) = a_1) \wedge \cdots \wedge (\beta(c, n) = a_n)$. In either case, a standard argument shows that $\langle \cdots \rangle_n$ is recursive. Theorem 3(1), on Gödel's β-function, shows that the relation in brackets $[\cdots]$ is regular; see the infomal discussion preceding the theorem. It is obvious that (1) holds.

We will now prove (2). Suppose $\langle a_1, \ldots, a_n \rangle_n = \langle b_1, \ldots, b_m \rangle_m$, and let c be the common value. Then $\beta(c, 0) = n$ and also $\beta(c, 0) = m$; hence $n = m$ as required. By a similar argument, $a_k = b_k$ for $1 \le k \le n$.

Finally we prove (3). Let $c = \langle a_1, \ldots, a_n \rangle_n$; by (1), we have $\beta(c, 0) = n$ and $\beta(c, k) = a_k$ for $1 \le k \le n$. Thus (3) requires us to prove that $\beta(c, k) < c$ for $0 \le k \le n$. By Theorem 3(3), these inequalities hold provided that $c > 0$. Now $\beta(0, 0) = 0$ by Theorem 3(2); on the other hand, we have $\beta(c, 0) = n \ge 1$, and hence $c > 0$ as required. ∎

Let $c \in \mathbb{N}$. Is c the code of some n-tuple? If so, we can use β to find this n-tuple. What we need, then, is a recursive relation that decides if c is a code in the first place. Let CD be the 1-ary relation defined by

$$\text{CD}(c) \Leftrightarrow c \text{ is the code of some } n\text{-tuple}$$
$$\Leftrightarrow \text{ there exists } n \in \mathbb{N}^+ \text{ and } a_1, \ldots, a_n \in \mathbb{N}$$
$$\text{such that } c = \langle a_1, \ldots, a_n \rangle_n.$$

Theorem 5: The 1-ary relation CD is recursive.

Proof: We proceed informally at first. Let $c \in \mathbb{N}$. Calculate $\beta(c, 0) = n$; assuming $n \ge 1$, calculate $\beta(c, k) = a_k$ for $1 \le k \le n$. Then c is the code of the n-tuple $\langle a_1, \ldots, a_n \rangle$ provided that c is the *smallest* number such that $\beta(c, 0) = n$ and $\beta(c, k) = a_k$ for $1 \le k \le n$. Formally, we have

$$\text{CD}(c) \Leftrightarrow (\beta(c, 0) \ge 1) \wedge \forall x < c[(\beta(x, 0) \ne \beta(c, 0))$$
$$\vee \exists k < \beta(c, 0)(\beta(x, k + 1) \ne \beta(c, k + 1))]. \blacksquare$$

There is one additional result on coding that will be needed later: a 2-ary function $*$, called the *concatenation function*.

Theorem 6: There is a 2-ary recursive function $*$ with these properties:

(1) $\langle a_1, \ldots, a_n \rangle_n * \langle b_1, \ldots, b_m \rangle_m = \langle a_1, \ldots, a_n, b_1, \ldots, b_m \rangle_{n+m}.$
(2) If c is a code, then $0 * c = c * 0 = c$.

Proof: Let $x, y \in \mathbb{N}$. To motivate the definition of $*$, suppose that x and y are codes, say $x = \langle a_1, \ldots, a_n \rangle_n$ and $y = \langle b_1, \ldots, b_m \rangle_m$; in this case, our goal is to define $*$ so that $x * y = \langle a_1, \ldots, a_n, b_1, \ldots, b_m \rangle_{n+m}$. Note that $\beta(x, 0) = n$, $\beta(y, 0) = m$, $\beta(x, k) = a_k$ for $1 \le k \le n$, and $\beta(y, k) = b_k$ for $1 \le k \le m$. Apply Gödel's β-function to the sequence $n + m, a_1, \ldots, a_n, b_1, \ldots, b_m$; there is a number z such that $\beta(z, 0) = n + m$, $\beta(z, k) = a_k$ for $1 \le k \le n$, and $\beta(z, k + n) = b_k$ for $1 \le k \le m$. If z is the smallest number with these properties, then $z = \langle a_1, \ldots, a_n, b_1, \ldots, b_m \rangle_{n+m}$. Thus we define $*$ by

$$x * y = \mu z[(\beta(z, 0) = \beta(x, 0) + \beta(y, 0))$$
$$\wedge \forall k < \beta(x, 0)(\beta(z, k + 1) = \beta(x, k + 1))$$
$$\wedge \forall k < \beta(y, 0)(\beta(z, \beta(x, 0) + k + 1) = \beta(y, k + 1))].$$

The relation in brackets $[\cdots]$ is regular by Theorem 3(1), on Gödel's β-function, and is recursive by standard techniques. From the discussion, $*$ satisfies (1). The result in (2) is an easy consequence of the fact that $\beta(0, 0) = 0$. ∎

———————————— *Exercises for Section 8.3* ————————————

1. This exercise outlines a proof that J is one-to-one.
   (1) Let $a + b = c + d$ and $a < c$. Show that $J(a, b) < J(c, d)$. (This is the case in which $\langle a, b \rangle$ and $\langle c, d \rangle$ are on the same diagonal.)
   (2) Prove the inequality $(a + b)(a + b + 1)/2 + a < (a + b + 1)(a + b + 2)/2$.
   (3) Let $a + b < c + d$. Show that $J(a, b) < J(c, d)$. (This is the case in which $\langle a, b \rangle$ is on an earlier diagonal than $\langle c, d \rangle$.)
   (4) Show that J is one-to-one as follows: Let $\langle a, b \rangle \neq \langle c, d \rangle$, and show that $J(a, b) \neq J(c, d)$. Consider the two cases $a + b = c + d$ and $a + b < c + d$.
2. This exercise outlines a proof that J is onto.
   (1) Let $J(a, 0) = n$. Show that $J(0, a + 1) = n + 1$.
   (2) Let $J(a, b + 1) = n$. Show that $J(a + 1, b) = n + 1$.
   (3) Use induction to show that J is onto. Let $S(n)$ say: There is a pair $\langle a, b \rangle$ such that $J(a, b) = n$. Use (1) and (2) to prove the induction step.
3. Show that $J(K(c), L(c)) = c$.
4. Show that J_{n+1} is one-to-one and onto (see the proof of Theorem 2).
5. Show that $J_{n+1,n+1}(a_1, \ldots, a_n, a_{n+1}) = a_{n+1}$ (see the proof of Theorem 2).
6. (*Contraction of an n-ary function using* $J_{n,1}, \ldots, J_{n,n}, J_n$) Let F be an n-ary function. Define a 1-ary function F^* by $F^*(a) = F(J_{n,1}(a), \ldots, J_{n,n}(a))$.
   (1) Show that $F(a_1, \ldots, a_n) = F^*(J_n(a_1, \ldots, a_n))$.
   (2) Show that F is recursive if and only if F^* is recursive.
7. (*Contraction of an n-ary function using* β *and* $\langle \cdots \rangle_n$) Let F be an n-ary function. Define a 1-ary function F^* by $F^*(a) = F(\beta(a, 1), \ldots, \beta(a, n))$.
   (1) Show that $F(a_1, \ldots, a_n) = F^*(\langle a_1, \ldots, a_n \rangle_n)$.
   (2) Show that F is recursive if and only if F^* is recursive.
8. (*Contraction of an n-ary relation using* $J_{n,1}, \ldots, J_{n,n}, J_n$) Let R be an n-ary relation. Define a 1-ary relation R^* by $R^*(a) \Leftrightarrow R(J_{n,1}(a), \ldots, J_{n,n}(a))$.
   (1) Show that $R(a_1, \ldots, a_n) \Leftrightarrow R^*(J_n(a_1, \ldots, a_n))$.
   (2) Show that R is recursive if and only if R^* is recursive.
9. (*Contraction of an n-ary relation using* β *and* $\langle \cdots \rangle_n$) Let R be an n-ary relation. Define a 1-ary relation R^* by $R^*(a) \Leftrightarrow R(\beta(a, 1), \ldots, \beta(a, n))$.
   (1) Show that $R(a_1, \ldots, a_n) \Leftrightarrow R^*(\langle a_1, \ldots, a_n \rangle_n)$.
   (2) Show that R is recursive if and only if R^* is recursive.
10. Assume that every 1-ary computable function is recursive. Show that every n-ary computable function is recursive.
11. Show that there is a 1-ary recursive function F such that if $a = \langle a_1, \ldots, a_n \rangle_n$, then $F(a) = a_n$.
12. Show that there is a 2-ary recursive function IP ($=$ initial part) such that if $a = \langle a_1, \ldots, a_n \rangle_n$ and $1 \le k \le n$, then $\mathrm{IP}(a, k) = \langle a_1, \ldots, a_k \rangle_k$.
13. Let a, b, and c be natural numbers such that $CD(a)$, $CD(b)$, and $CD(c)$ all hold. Show that $a * (b * c) = (a * b) * c$.
14. (*Collapsing quantifiers*) Let R be an $n + 2$-ary relation. Prove the following:
   (1) $\exists a \exists b R(a_1, \ldots, a_n, a, b) \Leftrightarrow \exists c R(a_1, \ldots, a_n, \beta(c, 1), \beta(c, 2))$.
   (2) $\forall a \forall b R(a_1, \ldots, a_n, a, b) \Leftrightarrow \forall c R(a_1, \ldots, a_n, K(c), L(c))$.
   (3) $\forall a \forall b R(a_1, \ldots, a_n, a, b) \Leftrightarrow \forall c R(a_1, \ldots, a_n, \beta(c, 1), \beta(c, 2))$.

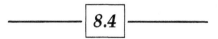

8.4

Primitive Recursion

―――――The Operation of Primitive Recursion―――――

By now we have developed a number of techniques for showing that certain functions are recursive. At this point, however, it would be quite difficult to prove that such obviously computable functions as the factorial function $FT(n) = n!$ and the exponential function $EXP(a, b) = a^b$ are recursive. To handle such functions easily, we need a new tool, called *primitive recursion*. Before giving a precise statement of this new operation, let us use the factorial function FT to illustrate the basic idea. Informally, we write

$$0! = 1$$
$$(n + 1)! = n! \times (n + 1).$$

The first equation gives the value of FT at 0; the second equation shows how to calculate the value of FT at $n + 1$ in terms of the value of FT at n. This procedure leads to a recursive function, as the following theorem states.

Theorem 1 (*Primitive recursion*): Let G be an n-ary function, let H be an $n + 2$-ary function, and define an $n + 1$-ary function F by

$$F(a_1, \ldots, a_n, 0) = G(a_1, \ldots, a_n)$$
$$F(a_1, \ldots, a_n, b + 1) = H(a_1, \ldots, a_n, b, F(a_1, \ldots, a_n, b)).$$

If both G and H are recursive, then F is recursive.

Proof: Given an n-ary function G and an $n + 2$-ary function H, there is a unique $n + 1$-ary function F satisfying these two equations. We assume this fact and concentrate on showing that F is recursive whenever both G and H are recursive. Let $a_1, \ldots, a_n, b \in \mathbb{N}$. By Gödel's β-function, there exists $c \in \mathbb{N}$ such that

(1) $\beta(c, k) = F(a_1, \ldots, a_n, k)$, $0 \leq k \leq b$.

Informally, c is a code of the calculations $F(a_1, \ldots, a_n, 0), \ldots, F(a_1, \ldots, a_n, b)$. Note that $F(a_1, \ldots, a_n, b) = \beta(c, b)$, and hence F is recursive if we can replace c with a recursive function of $a_1, \ldots, a_n, b$. To do this, we begin by eliminating F in (1):

(2) $\beta(c, 0) = G(a_1, \ldots, a_n)$.
(3) $\beta(c, k + 1) = H(a_1, \ldots, a_n, k, \beta(c, k))$, $0 \leq k \leq b - 1$.

In summary, given $a_1, \ldots, a_n, b \in \mathbb{N}$, there exists $c \in \mathbb{N}$ that satisfies (2) and (3). Moreover, any number c that satisfies (2) and (3) also satisfies (1); the proof of this fact, which proceeds by induction on $k \leq b$, is left to the reader.

We now construct an $n + 1$-ary recursive function L such that $L(a_1, \ldots, a_n, b) = c$, where c satisfies (2) and (3).

$$L(a_1, \ldots, a_n, b) = \mu c[(\beta(c, 0) = G(a_1, \ldots, a_n))$$
$$\wedge \; \forall k < b(\beta(c, k + 1) = H(a_1, \ldots, a_n, k, \beta(c, k)))].$$

Finally, $F(a_1, \ldots, a_n, b) = \beta(L(a_1, \ldots, a_n, b), b)$, and F is recursive as required. ∎

Corollary 1 (Special case of primitive recursion): Let $k \in \mathbb{N}$, let H be a 2-ary function, and define a 1-ary function F by

$$F(0) = k$$
$$F(n + 1) = H(n, F(n)).$$

If H is recursive, then F is recursive.

Proof: Left as an exercise for the reader. ∎

Example 1:

The factorial function $\text{FT}(n) = n!$ is recursive. For an informal proof, we simply write

$$0! = 1$$
$$(n + 1)! = n!(n + 1).$$

Formally, Corollary 1 requires us to find a 2-ary recursive function $H(n, x)$ such that $\text{FT}(n + 1) = H(n, \text{FT}(n))$—that is, such that $(n + 1)! = H(n, n!)$. Take $H(n, x) = (n + 1) \times x$. □

Example 2:

The exponential function $\text{EXP}(a, b) = a^b$ is recursive. This is proved informally as follows:

$$a^0 = 1$$
$$a^{b+1} = a^b \times a.$$

Formally, Theorem 1 requires us to find recursive functions G and H such that $\text{EXP}(a, 0) = G(a)$ and $\text{EXP}(a, b + 1) = H(a, b, \text{EXP}(a, b))$; in other words,

$$a^0 = G(a) \quad \text{and} \quad a^{b+1} = H(a, b, a^b).$$

Take $G(a) = 1$ and $H(a, b, x) = a \times x$. □

Example 3:

Let F be the 1-ary function defined by the following table:

a	0	1	2	3	4	5	6	7	8	9	10	$\cdots$
$F(a)$	0	1	1	2	2	2	3	3	3	3	4	$\cdots$

Thus F increases by 1 at the numbers $1, 3, 6, 10, \ldots$; these are the triangular numbers. In Example 5 of Section 8.2, we showed that the relation

$$\mathrm{TRI}(a) \Leftrightarrow a \text{ is a triangular number}$$

is a recursive relation. Hence we can write

$$F(0) = 0$$
$$F(a + 1) = F(a) + \mathrm{csg}\,(K_{\mathrm{TRI}}(a + 1)),$$

and F is recursive by primitive recursion. (*Note:* For a formal proof, the required recursive function H is $H(a, x) = x + \mathrm{csg}\,(K_{\mathrm{TRI}}(a + 1))$. □

We now use primitive recursion to justify operations for constructing recursive functions called *bounded sum* and *bounded product*.

Theorem 2 (Bounded sum/bounded product rule): Let F be an $n + 1$-ary recursive function. Then the following $n + 1$-ary functions are also recursive:

$$G(a_1, \ldots, a_n, b) = \sum_{k \leq b} F(a_1, \ldots, a_n, k)$$

$$H(a_1, \ldots, a_n, b) = \prod_{k \leq b} F(a_1, \ldots, a_n, k).$$

Proof: We will show that the bounded product H is recursive and will leave the bounded sum G to the reader. Let $a_1, \ldots, a_n \in \mathbb{N}$; informally,

$$H(a_1, \ldots, a_n, 0) = F(a_1, \ldots, a_n, 0)$$
$$H(a_1, \ldots, a_n, b + 1) = H(a_1, \ldots, a_n, b) \times F(a_1, \ldots, a_n, b + 1).$$

Formally, we need an $n + 2$-ary recursive function H_1 such that $H(a_1, \ldots, a_n, b + 1) = H_1(a_1, \ldots, a_n, b, H(a_1, \ldots, a_n, b))$—that is,

$$H(a_1, \ldots, a_n, b) \times F(a_1, \ldots, a_n, b + 1) = H_1(a_1, \ldots, a_n, b, H(a_1, \ldots, a_n, b)).$$

Take $H_1(a_1, \ldots, a_n, b, x) = x \times F(a_1, \ldots, a_n, b + 1)$. ∎

Example 4:

Let $\pi(n)$ = number of primes $\leq n$; for example, $\pi(0) = 0$, $\pi(1) = 0$, $\pi(2) = 1$, $\pi(3) = 2$, and $\pi(4) = 2$. We use the bounded sum rule to show that π is recursive. Let $F: \mathbb{N} \to \mathbb{N}$ be defined by $F(a) = 1$ whenever a is prime and $F(a) = 0$ otherwise; assuming F is recursive, we then have

$$\pi(n) = \sum_{k \le n} F(k)$$

and π is recursive by the bounded sum rule. Now let us show F to be recursive. By Example 6 in Section 8.2, the relation

$$\text{COMP}(n) \Leftrightarrow a \text{ is composite}$$

is recursive. Let K_{COMP} be the characteristic function of COMP; then

$$F(n) = \begin{cases} 0 & (n = 0) \vee (n = 1) \\ K_{\text{COMP}}(n) & n > 1, \end{cases}$$

and F is recursive by definition by cases.   □

Note: At this point, we can proceed directly to Section 8.6 for the proof of Church's theorem and Tarski's theorem. For Gödel's incompleteness theorem, however, we need the next topic, *course-of-values recursion,* and certain results from Section 8.5, on RE relations.

Course-of-Values Recursion

The basic idea of primitive recursion is that $F(n + 1)$ is calculated in terms of the previous value $F(n)$. But what if we need $F(n - 1)$ or other previous values of F to calculate $F(n + 1)$? Consider, for example, the well-known Fibonacci sequence

$$F(0) = 1$$
$$F(1) = 1$$
$$F(n + 1) = F(n) + F(n - 1) \quad (n \ge 1).$$

To handle such cases, we need a more powerful version of recursion, known as *course-of-values recursion.*

Definition 1: Let F be an $n + 1$-ary function. The *history of F* is the $n + 1$-ary function $F^{\#}$ defined by

$$F^{\#}(a_1, \ldots, a_n, 0) = 0$$
$$F^{\#}(a_1, \ldots, a_n, b + 1) = \langle F(a_1, \ldots, a_n, 0), \ldots, F(a_1, \ldots, a_n, b) \rangle_{b+1}.$$

Note that if $F^{\#}(a_1, \ldots, a_n, b + 1) = x$, we can use β to recover from x any of the calculations $F(a_1, \ldots, a_n, 0), \ldots, F(a_1, \ldots, a_n, b)$ that are required.

Theorem 3 (Course-of-values recursion): Let H be an $n + 2$-ary function and let F be the $n + 1$-ary function defined by

$$F(a_1, \ldots, a_n, b) = H(a_1, \ldots, a_n, b, F^{\#}(a_1, \ldots, a_n, b)). \tag{*}$$

If H is recursive, then F is recursive.

Proof: Given an $n + 2$-ary function H, there is a unique $n + 1$-ary function F satisfying the equation. We assume this fact and concentrate on showing that F is recursive whenever H is recursive. But first let us construct a few values of F. We use the notation $\mathbf{a} = \langle a_1, \dots, a_n \rangle$.

$$F(\mathbf{a}, 0) = H(\mathbf{a}, 0, F^{\#}(\mathbf{a}, 0))$$
$$= H(\mathbf{a}, 0, 0)$$
$$F(\mathbf{a}, 1) = H(\mathbf{a}, 1, F^{\#}(\mathbf{a}, 1))$$
$$= H(\mathbf{a}, 1, \langle F(\mathbf{a}, 0) \rangle_1)$$
$$F(\mathbf{a}, 2) = H(\mathbf{a}, 2, F^{\#}(\mathbf{a}, 2))$$
$$= H(\mathbf{a}, 2, \langle F(\mathbf{a}, 0), F(\mathbf{a}, 1) \rangle_2).$$

To prove F recursive, it suffices by composition to show that $F^{\#}$ is recursive (see (*)). The proof that $F^{\#}$ is recursive is by primitive recursion; the 2-ary recursive function $*$ (see Theorem 6 in Section 8.3) is also used. For $b > 0$,

$$F^{\#}(\mathbf{a}, b + 1) = \langle F(\mathbf{a}, 0), \dots, F(\mathbf{a}, b \doteq 1), F(\mathbf{a}, b) \rangle_{b+1}$$
$$= \langle F(\mathbf{a}, 0), \dots, F(\mathbf{a}, b \doteq 1) \rangle_b * \langle F(\mathbf{a}, b) \rangle_1$$
$$= F^{\#}(\mathbf{a}, b) * \langle F(\mathbf{a}, b) \rangle_1$$
$$= F^{\#}(\mathbf{a}, b) * \langle H(\mathbf{a}, b, F^{\#}(\mathbf{a}, b)) \rangle_1.$$

In summary, we have proved, for $b > 0$, that

$$F^{\#}(\mathbf{a}, b + 1) = F^{\#}(\mathbf{a}, b) * \langle H(\mathbf{a}, b, F^{\#}(\mathbf{a}, b)) \rangle_1.$$

This result also holds for $b = 0$ by the properties $0 * \langle a \rangle_1 = \langle a \rangle_1$, $F^{\#}(\mathbf{a}, 0) = 0$, $F^{\#}(\mathbf{a}, 1) = \langle F(\mathbf{a}, 0) \rangle_1$, and $F(\mathbf{a}, 0) = H(\mathbf{a}, 0, 0)$.

Finally, define an $n + 2$-ary function G by $G(\mathbf{a}, b, x) = x * \langle H(\mathbf{a}, b, x) \rangle_1$; the function G is recursive and

$$F^{\#}(\mathbf{a}, 0) = 0$$
$$F^{\#}(\mathbf{a}, b + 1) = G(\mathbf{a}, b, F^{\#}(\mathbf{a}, b));$$

hence $F^{\#}$ is recursive by primitive recursion as required. ∎

Example 5:

(Fibonacci sequence) Let $F(0) = F(1) = 1$ and $F(n) = F(n - 1) + F(n - 2)$ for $n > 1$. We use course-of-values recursion to show that F is recursive. Informally, we write

$$F(n) = \begin{cases} 1 & (n = 0) \vee (n = 1) \\ F(n \doteq 1) + F(n \doteq 2) & n > 1 \end{cases}$$

and then claim that F is recursive by definition by cases and by course-of-values recursion.

Formally, we need a recursive function $H(n, x)$ such that $F(n) = H(n, F^{\#}(n))$. For $n > 1$, the requirement on H is

$$F(n \doteq 1) + F(n \doteq 2) = H(n, F^{\#}(n)).$$

Now $F^{\#}(n) = \langle F(0), \ldots, F(n \dot- 2), F(n \dot- 1)\rangle_n$;   hence $F(n \dot- 2) = \beta(F^{\#}(n), n \dot- 1)$ and $F(n \dot- 1) = \beta(F^{\#}(n), n)$. This suggests

$$H(n, x) = \begin{cases} 1 & (n = 0) \vee (n = 1) \\ \beta(x, n) + \beta(x, n \dot- 1) & n > 1. \end{cases}$$

The function H is recursive by definition by cases, and for all $n \in \mathbb{N}$ we have $F(n) = H(n, F^{\#}(n))$. By course-of-values recursion, F is recursive as required. $\square$

─────────── *Primitive Recursive Functions* ───────────

What happens if we replace the operation of minimalization with the operation of primitive recursion in the definition of a recursive function?

Definition 2: An n-ary function F is *primitive recursive* if there is a finite sequence $F_1, \ldots, F_n$ of functions with $F_n = F$ and such that for $1 \leq k \leq n$, one of the following two conditions holds:

(1) F is the successor function $S(a) = a + 1$, the zero function $Z(a) = 0$, or a projection function $U_{n,k}(a_1, \ldots, a_n) = a_k$.

(2) $k > 1$ and F_k is obtained from previous functions (i.e., from among $F_1, \ldots, F_{k-1}$) by composition or by the operation of primitive recursion.

In other words, we obtain the class of primitive recursive functions by taking S, Z, and $U_{n,k}$ as starting functions and applying the operations of composition and primitive recursion to obtain new primitive recursive functions.

The primitive recursive functions are important for at least two reasons. One reason is historical: In his ground-breaking paper on formally undecidable propositions, Gödel actually worked with primitive recursive functions rather than recursive functions.

The other reason for our interest in this class of functions is that most of the recursive functions that we encounter in practice are actually primitive recursive. Indeed, it is quite difficult to give an example of a recursive function that fails to be primitive recursive. One such function is Ackermann's 2-ary function ψ, defined by

$$\psi(a, 0) = a + 1$$
$$\psi(0, b + 1) = \psi(1, b)$$
$$\psi(a + 1, b + 1) = \psi(\psi(a, b + 1), b).$$

Roughly speaking, ψ is not primitive recursive because it grows more rapidly than any primitive recursive function. In the exercises we will outline a proof that ψ is recursive but not primitive recursive.

Theorem 4, which follows, is the basic method for showing that a function is primitive recursive. We also have *induction on primitive recursive functions*, the basic technique for showing that all primitive recursive functions have some property Q (see Theorem 5, following; the proofs are left to the reader).

Theorem 4: Let F be an n-ary function. To show that F is primitive recursive, it suffices to show that F satisfies one of the following three conditions:

(1) F is a starting function (S, Z, or $U_{n,k}$ ($n \geq 1$, $1 \leq k \leq n$)).

(2) F is the composition of primitive recursive functions.

(3) F is obtained from primitive recursive functions by the operation of primitive recursion.

Theorem 5 (Induction on primitive recursive functions): Let Q be a property of functions. To prove that every primitive recursive function has property Q, it suffices to prove the following: (1) Every starting function has property Q; (2) if F is the composition of $G, H_1, \ldots, H_k$, and each of $G, H_1, \ldots, H_k$ has property Q, then F has property Q; (3) if F is obtained from G and H by primitive recursion, and both G and H have property Q, then F has property Q.

The next result is an application of induction on primitive recursive functions.

Theorem 6: Every primitive recursive function is recursive.

Outline of Proof We use induction on primitive recursive functions with Q the property of being recursive. In Theorem 1 of Section 8.1, we proved that the functions S and Z are recursive. Moreover, in Theorem 1 of this section, we proved that if F is obtained from G and H by the operation of primitive recursion, and G and H are recursive, then F is recursive. ■

We end this section with examples of primitive recursive functions and the definition of a primitive recursive relation. The theory of primitive recursive functions will be developed further in the exercises.

Theorem 7: The functions $+$, $\times$, $\dot{-}$, sg, and csg are primitive recursive.

Proof: An informal proof that addition is primitive recursive is

$$a + 0 = a$$
$$a + (b + 1) = (a + b) + 1.$$

We now proceed formally. We need a 3-ary primitive recursive function H such that $a + (b + 1) = H(a, b, a + b)$; take $H(a, b, x) = S(x) = S(U_{3,3}(a, b, x))$.

We now turn to the remaining functions. In each case we give an informal proof; a formal proof is left to the reader.

- *Multiplication*
$$a \times 0 = 0$$
$$a \times (b + 1) = (a \times b) + a.$$

- *Proper subtraction* To facilitate the proof, we use the predecessor function PD, defined by $PD(a) = a \doteq 1$. The following informal proofs show that PD and $\doteq$ are primitive recursive:

$$PD(0) = 0 \qquad\qquad a \doteq 0 = a$$
$$PD(a + 1) = a \qquad a \doteq (b + 1) = (a \doteq b) \doteq 1 = PD(a \doteq b).$$

- *Sign and cosign* $\qquad sg(0) = 0 \qquad\qquad csg\,(0) = 1$
$$sg\,(a + 1) = 1 \qquad csg\,(a + 1) = 0. \ \blacksquare$$

Definition 3: An n-ary relation R is *primitive recursive* if its characteristic function K_R is primitive recursive.

Theorem 8: $<$ is a primitive recursive relation.

Proof: $K_<(a, b) = csg((b \doteq a)) = csg((U_{2,2}(a, b) \doteq U_{2,1}(a, b)). \ \blacksquare$

──────────── *Exercises for Section 8.4* ────────────

1. Let $R = \{1^2, 2^2, 3^2, \ldots ; 1^3, 2^3, 3^3, \ldots ; 1^4, 2^4, 3^4, \ldots ; \cdots\}$. Show that R is a recursive set. This example is due to Post 1944.
2. Let $F(a, 0) = 3$, $F(a, b + 1) = G(F(a, b) + G(b))$, where G is a recursive function. Give a formal proof that F is recursive by primitive recursion. In other words, find a 3-ary recursive function H such that $F(a, b + 1) = H(a, b, F(a, b))$.
3. Let H be a 2-ary computable function. Show that F is computable, where F is defined by

$$F(0) = k \ (constant)$$
$$F(n + 1) = H(n, F(n)).$$

4. Prove Corollary 1. (*Hint:* Introduce a parameter a as follows: $F^*(a, n) = F(n)$. Then use Theorem 1 to prove that F^* is recursive; use F^* to show that F is recursive.)
5. (*Bounded sum*) Let F be an $n + 1$-ary recursive function. Show that the bounded sum of F in Theorem 2 is recursive.
6. (*Bounded sum, bounded product*) Let F be an $n + 1$-ary computable function. Show that the bounded sum and the bounded product of F are computable (see Theorem 2).
7. Show that the following functions from number theory are recursive:
   (1) $\tau(n) = $ number of positive divisors of n ($\tau(0) = 0$).
   (2) $\sigma(n) = $ sum of the positive divisors of n ($\sigma(0) = 0$).
   (*Hint:* For $1 \le k \le n$, k divides n if and only if $RM(n, k) = 0$; use the result on bounded sum.)
8. Let p_n be the nth prime; thus $p_0 = 2$, $p_1 = 3$, $p_2 = 5$, and so on. Show that the function $F(n) = p_n$ is recursive. (*Hints:* By exercise 12 in Section 8.2, PRIME(a) is recursive; use primitive recursion.)
9. Let G be a 1-ary recursive function, and define F by $F(a) = \langle G(0), \ldots, G(a) \rangle_{a+1}$. Show that F is recursive.
10. (*Simultaneous recursion*) Let k_1 and k_2 be constants, let H_1 and H_2 be 3-ary recursive functions, and define 1-ary functions F and G by

$$F(0) = k_1$$
$$G(0) = k_2$$
$$F(n + 1) = H_1(n, F(n), G(n))$$
$$G(n + 1) = H_2(n, F(n), G(n)).$$

Show that both F and G are recursive. (*Hint:* Use Cantor's pairing functions J, K, and L; let $H(n) = J(F(n), G(n))$, and use primitive recursion to show that H is recursive. Note that $F(n) = K(H(n))$ and $G(n) = L(H(n))$.)

11. Show that the following two functions are recursive:
    $F(n) = $ first $n + 1$ digits in the decimal expansion of $\sqrt{2}$.
    $G(n) = n + 1$ digit in the decimal expansion of $\sqrt{2}$.
    For example, $F(0) = 1$, $F(1) = 14$, $F(2) = 141, \ldots$; $G(0) = 1$, $G(1) = 4$, $G(2) = 1, \ldots$.

12. Let $F(0) = 1$, $F(n + 1) = F(0) + 2 \times F(1) + \cdots + (n + 1) \times F(n)$. Show that F is recursive.

13. We present four schemes for recursion. Each scheme asserts that if G and H are recursive, then F is recursive; PR4 is primitive recursion. This exercise outlines a proof that the four are equivalent.

    PR1 $\qquad\qquad F(a, 0) = a$
    $\qquad\qquad F(a, b + 1) = H(F(a, b))$
    PR2 $\qquad\qquad F(a, 0) = G(a)$
    $\qquad\qquad F(a, b + 1) = H(F(a, b))$
    PR3 $\qquad\qquad F(a, 0) = G(a)$
    $\qquad\qquad F(a, b + 1) = H(a, b, F(a, b))$
    PR4 $\qquad F(a_1, \ldots, a_n, 0) = G(a_1, \ldots, a_n)$
    $\qquad F(a_1, \ldots, a_n, b + 1) = H(a_1, \ldots, a_n, b, F(a_1, \ldots, a_n, b))$.

    (1) Show that PR3 $\Rightarrow$ PR4. (*Hint:* Let F be an $n + 1$-ary function defined by PR4, where G and H are recursive. Define F^* by $F^*(a, b) = F(J_{n,1}(a), \ldots, J_{n,n}(a), b)$. Show that $F(a_1, \ldots, a_n, b) = F^*(J_n(a_1, \ldots, a_n), b)$, and use PR3 to show that F^* is recursive.)
    (2) Show that PR2 $\Rightarrow$ PR3. (*Hint:* Let F be defined by PR3, where G and H are recursive. Define F^* by $F^*(a, b) = J_3(a, b, F(a, b))$. Show that $F(a, b) = J_{3,3}(F^*(a, b))$, and use PR2 to show that F^* is recursive.)
    (3) Show that PR1 $\Rightarrow$ PR2. (*Hint:* Let F be defined by PR2, where G and H are recursive. Define F^* by $F^*(a, 0) = a$, $F^*(a, b + 1) = H(F^*(a, b))$; F^* is recursive by PR1. Show that $F(a, b) = F^*(G(a), b)$.)
    (4) Show that PR4 $\Rightarrow$ PR1.

14. Let $F(0) = 4$, $F(1) = 6$, $F(2) = 1$, and $F(n + 1) = F(n) + 2 \times F(n - 1) + 3 \times F(n - 2)$ for $n \geq 2$. Show that F is recursive. Give complete details.

15. Let F and an $n + 1$-ary function defined by

$$F(a_1, \ldots, a_n, 0) = G(a_1, \ldots, a_n)$$
$$F(a_1, \ldots, a_n, b + 1) = H(a_1, \ldots, a_n, b, F(a_1, \ldots, a_n, L(b))),$$

where G, H, and L are recursive functions and $L(b) \leq b$ always holds. Show that F is recursive.

16. Let G be a 1-ary recursive function such that $G(a) < a$ for all $a > 0$. Let F be defined by $F(0) = 1$ and $F(a) = \beta(a, F(G(a)))$ for $a > 0$. Show that F is recursive. Give complete details.

17. (*Iteration*) Let H be a 1-ary function. Given $a \in \mathbb{N}$, the *iterates of a by H* are a, $H(a)$, $H(H(a)), \ldots$; more precisely, $H^0(a) = a$, $H^{n+1}(a) = H(H^n(a))$. Define a 2-ary function F by $F(a, 0) = a$ and $F(a, n + 1) = H(F(a, n))$. Prove the following:
    (1) If H is recursive, then F is recursive.
    (2) $F(a, n) = H^n(a)$ for all $a, n \in \mathbb{N}$.

18. Assume that F is a 2-ary recursive function that enumerates all 1-ary primitive recursive functions; in other words, given a 1-ary primitive recursive function G, there exists $e \in \mathbb{N}$ such that for all $a \in \mathbb{N}$, $G(a) = F(e, a)$. Use F to construct a 1-ary function that is recursive but not primitive recursive.

19. (*Definition by cases and primitive recursion*) Let $G_1, \ldots, G_k$ be n-ary recursive functions, let $H_1, \ldots, H_k$ be $n + 2$-ary recursive functions, and let $R_1, \ldots, R_k$ be n-ary recursive relations such that for all $a_1, \ldots, a_n \in \mathbb{N}$, exactly one of $R_1(a_1, \ldots, a_n), \ldots, R_k(a_1, \ldots, a_n)$ holds. Show that F is recursive, where F is defined by

$$F(a_1, \ldots, a_n, 0) = \begin{cases} G_1(a_1, \ldots, a_n) & R_1(a_1, \ldots, a_n) \\ \vdots & \vdots \\ G_k(a_1, \ldots, a_n) & R_k(a_1, \ldots, a_n). \end{cases}$$

$$F(a_1, \ldots, a_n, b + 1) = \begin{cases} H_1(a_1, \ldots, a_n, b, F(a_1, \ldots, a_n, b)) & R_1(a_1, \ldots, a_n) \\ \vdots & \vdots \\ H_k(a_1, \ldots, a_n, b, F(a_1, \ldots, a_n, b)) & R_k(a_1, \ldots, a_n). \end{cases}$$

20. Prove Theorem 4.

21. Prove Theorem 5.

22. This exercise develops the theory of primitive recursive functions. (*Note:* S, Z, $U_{n,k}$, $+$, $\times$, $\dot-$, sg, and csg are primitive recursive functions and $<$ is a primitive recursive relation. For hints, see Sections 8.1, 8.2, and 8.3.) Prove the following:

    (1) The sum and the product of two n-ary primitive recursive functions are primitive recursive.

    (2) The n-ary constant functions $C_{n,k}$ are primitive recursive.

    (3) If $F(a_1, \ldots, a_n) = G(b_1, \ldots, b_k)$, where G is primitive recursive and for $1 \le i \le k$, $b_i \in \{a_1, \ldots, a_n\}$ or $b_i \in \mathbb{N}$, then F is primitive recursive.

    (4) If P and Q are n-ary primitive recursive relations, then $\neg P$, $P \vee Q$, and $P \wedge Q$ are primitive recursive relations.

    (5) The 2-ary relations $\le$, $=$, $>$, $\ge$, and $\ne$ are primitive recursive.

    (6) Definition by cases holds for primitive recursive functions. Thus if $G_1, \ldots, G_k$ are n-ary primitive recursive functions and $R_1, \ldots, R_k$ are n-ary primitive recursive relations such that for all $a_1, \ldots, a_n \in \mathbb{N}$, exactly one of $R_1(a_1, \ldots, a_n), \ldots, R_k(a_1, \ldots, a_n)$ holds, then F is primitive recursive, where

$$F(a_1, \ldots, a_n) = \begin{cases} G_1(a_1, \ldots, a_n) & R_1(a_1, \ldots, a_n) \\ \vdots & \vdots \\ G_k(a_1, \ldots, a_n) & R_k(a_1, \ldots, a_n). \end{cases}$$

    (7) If F is an $n + 1$-ary primitive recursive function, then the bounded sum of F and the bounded product of F are primitive recursive.

    (8) Let R be an $n + 1$-ary primitive recursive relation. Then H_R is primitive recursive, where H_R is defined by

$$H_R(a_1, \ldots, a_n, b) = \begin{cases} 1 & \neg R(a_1, \ldots, a_n, k) \text{ for } 0 \le k \le b \\ 0 & R(a_1, \ldots, a_n, k) \text{ for some } k \le b. \end{cases}$$

    (9) (*Bounded minimalization, first version*) If R is an $n + 1$-ary primitive recursive relation, then F is primitive recursive, where F is defined by

$$F(a_1, \ldots, a_n, b) = \mu x \le b R(a_1, \ldots, a_n, x).$$

    (*Hint:* Use H_R and bounded sum. The value of $F(a_1, \ldots, a_n, b)$ is the smallest x such that $x \le b$ and $R(a_1, \ldots, a_n, b)$ holds; if there is no such x, $F(a_1, \ldots, a_n, b) = b + 1$.)

    (10) (*Bounded minimalization, second version*) If R is an $n + 1$-ary primitive recursive relation and H is an n-ary primitive recursive relation, then F is primitive recursive, where

$$F(a_1, \ldots, a_n) = \mu x \le H(a_1, \ldots, a_n) R(a_1, \ldots, a_n, x).$$

    (*Note:* If there is no $x \le H(a_1, \ldots, a_n)$ for which $R(a_1, \ldots, a_n, x)$ holds, then $F(a_1, \ldots, a_n) = H(a_1, \ldots, a_n) + 1$.)

(11) QT and RM are primitive recursive functions (see Theorem 8 in Section 8.2).

(12) (*Bounded quantifiers, first version*) If P is an $n + 1$-ary primitive recursive relation, then each of the following is primitive recursive:

(a) $Q(a_1, \ldots, a_n, b) \Leftrightarrow \exists x < bP(a_1, \ldots, a_n, x)$.

(b) $R(a_1, \ldots, a_n, b) \Leftrightarrow \forall x < bP(a_1, \ldots, a_n, x)$.

(13) (*Bounded quantifiers, second version*) If P is an $n + 1$-ary primitive recursive relation and H is an n-ary primitive recursive function, then each of the following is primitive recursive:

(a) $Q(a_1, \ldots, a_n) \Leftrightarrow \exists x < H(a_1, \ldots, a_n)P(a_1, \ldots, a_n, x)$.

(b) $R(a_1, \ldots, a_n) \Leftrightarrow \forall x < H(a_1, \ldots, a_n)P(a_1, \ldots, a_n, x)$.

(14) Cantor's pairing functions J, K, and L and Gödel's β-function are primitive recursive.

(15) For each $n \geq 1$, the function $M_n(a_1, \ldots, a_n) = \max\{a_1, \ldots, a_n\}$ is primitive recursive. (*Hint:* Induction.)

(16) For each $n \geq 1$, the n-ary coding function $\langle \cdots \rangle_n$ is primitive recursive. (*Hint:* Show there is an $n + 1$-ary primitive recursive function $H(a_0, \ldots, a_n)$ that is a bound on the number c in Theorem 3 in Section 8.3.)

(17) The 1-ary relation CD is primitive recursive.

23. (ψ *is computable*) Prove that Ackermann's function ψ is computable. (*Hint:* See Exercise 8(2) in Section 1.2.)

24. (*Rapid growth of* ψ) Ackermann's function is the 2-ary function ψ defined by

$$\psi(a, 0) = a + 1$$
$$\psi(0, b + 1) = \psi(1, b)$$
$$\psi(a + 1, b + 1) = \psi(\psi(a, b + 1), b).$$

(1) Verify the following by induction:

(a) $\psi(a, 1) = a + 2$.

(b) $\psi(a, 2) = 2 \times a + 3$.

(c) $\psi(a, 3) = 2^{a+3} - 3$.

(2) What is $\psi(a, 4)$?

25. (*Basic properties of* ψ) Prove the following properties of Ackermann's function ψ:

(1) $\psi(a, b) > a$. (Let $S(b)$ say: $\forall a(\psi(a, b) > a)$; prove $S(b)$ for all $b \in \mathbb{N}$ by induction; in the proof of $S(b + 1)$, let $T(a)$ say: $\psi(a, b + 1) > a$; prove $T(a)$ for all $a \in \mathbb{N}$ by induction.)

(2) $\psi(a + 1, b) > \psi(a, b)$. (Two cases: $b = 0$ and $b > 0$; use (1).)

(3) If $a > a'$, then $\psi(a, b) > \psi(a', b)$. (Use (2).)

(4) $\psi(a, b + 1) \geq \psi(a + 1, b)$. (Induction on a; use (1) to show $\psi(a + 1, b) \geq a + 2$.)

(5) $\psi(a, b + 1) > \psi(a, b)$. (Use (4), (2).)

(6) If $b > b'$, then $\psi(a, b) > \psi(a, b')$. (Use (5).)

(7) $\psi(a, b + 2) > \psi(2a, b)$. (Induction on a; use (1) to show $\psi(2a, b) + 1 \geq 2a + 2$.)

26. (ψ *is not primitive recursive*) This exercise outlines a proof that Ackermann's function is not primitive recursive; properties of ψ from Exercise 25 are used. An n-ary function F is *within level r* if for all $a_1, \ldots, a_n \in \mathbb{N}$, $F(a_1, \ldots, a_n) \leq \psi(x, r)$, where $x = \max\{a_1, \ldots, a_n\}$. Note that if F is within level r and $r \leq s$, then F is within level s. Prove the following:

(1) The starting functions Z, S, and $U_{n,k}$ are within level 0.

(2) If $F(a_1, \ldots, a_n) = G(H_1(a_1, \ldots, a_n), \ldots, H_k(a_1, \ldots, a_n))$, where $G, H_1, \ldots, H_k$ are all within level r, then F is within level $r + 2$. (*Hint:* $\psi(x, r + 2) \geq \psi(x + 1, r + 1) = \psi(\psi(x, r + 1), r) \geq \psi(\psi(x, r), r)$.)

(3) Let G and H be within level r and let F be defined by

$$F(a_1, \ldots, a_n, 0) = G(a_1, \ldots, a_n)$$
$$F(a_1, \ldots, a_n, b + 1) = H(a_1, \ldots, a_n, b, F(a_1, \ldots, a_n, b)).$$

(a) $F(a_1, \ldots, a_n, b) \leq \psi(x + b, r + 1)$, where $x = \max\{a_1, \ldots, a_n\}$.

(b) F is within level $r + 3$. (*Hint:* Let $x = \max\{a_1, \ldots, a_n\}$ and $y = \max\{a_1, \ldots, a_n, b\}$; note that $x + b \leq 2y$.)

(4) Every primitive recursive function is within level r for some $r \geq 0$.

(5) ψ is not primitive recursive. (*Hint:* Suppose it is; let $F(a) = \psi(a, a) + 1$.)

27. (ψ *is recursive*) This exercise outlines a proof that ψ is recursive; ψ is defined by

$$\psi(a, 0) = a + 1 \qquad\qquad \text{(Rule 1)}$$
$$\psi(0, b + 1) = \psi(1, b) \qquad\quad \text{(Rule 2)}$$
$$\psi(a + 1, b + 1) = \psi(\psi(a, b + 1), b) \quad \text{(Rule 3)}$$

We define an operator Φ: SEQ $\to$ SEQ, where SEQ is the set of all finite sequences of natural numbers (i.e., SEQ $= \mathbb{N} \cup (\mathbb{N} \times \mathbb{N}) \cup (\mathbb{N} \times \mathbb{N} \times \mathbb{N}) \cup \cdots$). On $\mathbb{N}$, Φ is just the identity function; in other words, for $a \in \mathbb{N}$, $\Phi(a) = \langle a \rangle$. For $k > 1$, Φ is defined by

$$\Phi(a_1, a_2, \ldots, a_n) = \begin{cases} \langle a_1 + 1, a_3, \ldots, a_n \rangle & a_2 = 0 \\ \langle 1, a_2 - 1, a_3, \ldots, a_n \rangle & a_1 = 0 \wedge a_2 > 0 \\ \langle a_1 - 1, a_2, a_2 - 1, a_3, \ldots, a_n \rangle & a_1 > 0 \wedge a_2 > 0 \end{cases}$$

If we interpret a_1 and a_2 as a and b in ψ, we see that Φ captures rules 1, 2, and 3 of ψ. Let Φ^n denote the nth iterate of Φ; thus, for $x \in$ SEQ, $\Phi^0(x) = x$ and $\Phi^{n+1}(x) = \Phi^n(\Phi(x))$. To illustrate, we find $\Phi^{14}(1, 2)$:

$$\Phi^1(1, 2) = \langle 0, 2, 1 \rangle \qquad\qquad \Phi^8(1, 2) = \langle 1, 1, 0, 0 \rangle$$
$$\Phi^2(1, 2) = \langle 1, 1, 1 \rangle \qquad\qquad \Phi^9(1, 2) = \langle 0, 1, 0, 0, 0 \rangle$$
$$\Phi^3(1, 2) = \langle 0, 1, 0, 1 \rangle \qquad\quad \Phi^{10}(1, 2) = \langle 1, 0, 0, 0, 0 \rangle$$
$$\Phi^4(1, 2) = \langle 1, 0, 0, 1 \rangle \qquad\quad \Phi^{11}(1, 2) = \langle 2, 0, 0, 0 \rangle$$
$$\Phi^5(1, 2) = \langle 2, 0, 1 \rangle \qquad\qquad \Phi^{12}(1, 2) = \langle 3, 0, 0 \rangle$$
$$\Phi^6(1, 2) = \langle 3, 1 \rangle \qquad\qquad\quad \Phi^{13}(1, 2) = \langle 4, 0 \rangle$$
$$\Phi^7(1, 2) = \langle 2, 1, 0 \rangle \qquad\qquad \Phi^{14}(1, 2) = \langle 5 \rangle.$$

(1) Calculate $\psi(1, 2)$ and compare the calculations with those for $\Phi^{14}(1, 2)$.

(2) Let $\Phi^r(a, b) = \langle u \rangle$. Show that for $s \geq r$, $\Phi^s(a, b) = \langle u \rangle$.

(3) Let $a, b \in \mathbb{N}$. Show that there exists $r \in \mathbb{N}^+$ such that $\Phi^r(a, b) = \langle \psi(a, b) \rangle$. (*Hint:* Let $S(b)$ say: $\forall a \exists r [\Phi^r(a, b) = \langle \psi(a, b) \rangle]$. Use induction to show that $S(b)$ holds for all $b \in \mathbb{N}$. In the proof of $S(b + 1)$, let $T(a)$ say: $\exists r [\Phi^r(a, b + 1) = \langle \psi(a, b + 1) \rangle]$. Use induction to show that $T(a)$ holds for all $a \in \mathbb{N}$.)

(4) Let $H(a, b, r) =$ code of $\Phi^r(a, b)$. For example, $\Phi^4(1, 2) = \langle 1, 0, 0, 1 \rangle$; hence $H(1, 2, 4) = \langle 1, 0, 0, 1 \rangle_4$; $\Phi^{14}(1, 2) = \langle 5 \rangle$; hence $H(1, 2, 14) = \langle 5 \rangle_1$. Our goal is to show that H is recursive. From (2) and (3), it follows that for all $a, b \in \mathbb{N}$, there exists r such that $H(a, b, r) = \langle \psi(a, b) \rangle_1$ and, moreover, if $H(a, b, r) = \langle u \rangle_1$, then $\psi(a, b) = u$. Suppose, for a moment, that H is recursive; then

$$G(a, b) = \mu r [\beta(H(a, b, r), 0) = 1]$$

is recursive and $\psi(a, b) = \beta(H(a, b, G(a, b)), 1)$; hence ψ is recursive. It remains to prove H recursive. Define 1-ary relations R_1, R_2, and R_3 by

$$R_1(c) \Leftrightarrow c = \langle a_1, \ldots, a_n \rangle_n \wedge (n \geq 2) \wedge (a_2 = 0)$$
$$R_2(c) \Leftrightarrow c = \langle a_1, \ldots, a_n \rangle_n \wedge (n \geq 2) \wedge (a_1 = 0) \wedge (a_2 > 0)$$
$$R_3(c) \Leftrightarrow c = \langle a_1, \ldots, a_n \rangle_n \wedge (n \geq 2) \wedge (a_1 > 0) \wedge (a_2 > 0).$$

Show that R_1, R_2, and R_3 are recursive. Next, show that there are recursive functions L_1, L_2, and L_3 such that for all $c \in \mathbb{N}$,

- If $R_1(c)$, then $L_1(c) = \langle a_1 + 1, a_3, \ldots, a_n \rangle_{n-1}$.
- If $R_2(c)$, then $L_2(c) = \langle 1, a_2 - 1, a_3, \ldots, a_n \rangle_n$.
- If $R_3(c)$, then $L_3(c) = \langle a_1 - 1, a_2, a_2 - 1, a_3, \ldots, a_n \rangle_{n+1}$.

Let $R(a) \Leftrightarrow R_1(a) \vee R_2(a) \vee R_3(a)$, and note that R is recursive. Finally, show that H is recursive, as follows:

$$H(a, b, 0) = \langle a, b \rangle_2$$

$$H(a, b, r + 1) = \begin{cases} L_1(H(a, b, r)) & R_1(H(a, b, r)) \\ L_2(H(a, b, r)) & R_2(H(a, b, r)) \\ L_3(H(a, b, r)) & R_3(H(a, b, r)) \\ H(a, b, r) & \text{otherwise.} \end{cases}$$

8.5

RE Relations

Recursive relations are the formal counterpart of the decidable relations. In this section we briefly study RE relations, the formal counterpart of the semidecidable relations. The definition of an RE relation is based on the following theorem (see Theorem 5 in Section 1.7):

Theorem 1: An n-ary relation R is semidecidable if and only if there is an $n + 1$-ary decidable relation Q such that for all $a_1, \ldots, a_n \in \mathbb{N}$,

$$R(a_1, \ldots, a_n) \Leftrightarrow \exists x Q(a_1, \ldots, a_n, x). \tag{$*$}$$

Definition 1: An n-ary relation R is RE ($=$ *recursively enumerable*) if there is an $n + 1$-ary recursive relation Q such that for all $a_1, \ldots, a_n \in \mathbb{N}$,

$$R(a_1, \ldots, a_n) \Leftrightarrow \exists x Q(a_1, \ldots, a_n, x).$$

Informally speaking, we obtain an RE relation by placing an existential quantifier in front of a recursive relation; note the similarity with the bounded $\exists$ rule. Our immediate goal is to show that the RE relations are indeed the formal counterpart of the semidecidable relations. To do this, it suffices to show that

(1) RE $\Rightarrow$ semidecidable.
(2) Semidecidable $\Rightarrow$ RE (assuming Church's thesis).

Both results are consequences of Theorem 1.

Corollary 1: Every RE relation is semidecidable.

Proof: Let R be an n-ary RE relation. By definition, there is an $n + 1$-ary recursive relation Q such that for all $a_1, \ldots, a_n \in \mathbb{N}$,

$$R(a_1, \ldots, a_n) \Leftrightarrow \exists x Q(a_1, \ldots, a_n, x).$$

Since Q is recursive, Q is decidable (see Corollary 1 in Section 1.8). Theorem 1 now applies and R is semidecidable as required. ∎

Corollary 2: Assume Church's thesis. Then every semidecidable relation is RE.

Proof: Let R be an n-ary semidecidable relation. By Theorem 1, there is an $n + 1$-ary decidable relation Q such that for all $a_1, \ldots, a_n \in \mathbb{N}$,

$$R(a_1, \ldots, a_n) \Leftrightarrow \exists x Q(a_1, \ldots, a_n, x).$$

By Church's thesis, Q is recursive; thus R is RE as required. ∎

It is obvious that every decidable relation is semidecidable. Moreover, if both R and $\neg R$ are semidecidable, then R is decidable (see Theorem 4 in Section 1.7). We will now prove the formal versions of these two results.

Theorem 2: Every recursive relation is RE.

Proof: Let R be an n-ary recursive relation. Define an $n + 1$-ary relation Q by $Q(a_1, \ldots, a_n, x) \Leftrightarrow R(a_1, \ldots, a_n)$. The relation Q is recursive by composition of relations. Moreover, $R(a_1, \ldots, a_n) \Leftrightarrow \exists x Q(a_1, \ldots, a_n, x)$, so R is RE as required. ∎

Note: The method of proof used in Theorem 2 is often called *superfluous quantifiers*.

Theorem 3: Let R be an n-ary relation. If both R and $\neg R$ are RE, then R is recursive.

Proof: By hypothesis, there are $n + 1$-ary recursive relations P and Q such that for all $a_1, \ldots, a_n \in \mathbb{N}$,

$$R(a_1, \ldots, a_n) \Leftrightarrow \exists x P(a_1, \ldots, a_n, x) \qquad (*)$$
$$\neg R(a_1, \ldots, a_n) \Leftrightarrow \exists x Q(a_1, \ldots, a_n, x) \qquad (**)$$

Since P and Q are recursive, $P \vee Q$ is recursive. Moreover, $P \vee Q$ is regular: Given $a_1, \ldots, a_n \in \mathbb{N}$, either $R(a_1, \ldots, a_n)$ or $\neg R(a_1, \ldots, a_n)$; thus either $(*)$ or $(**)$ applies and there exists $x \in \mathbb{N}$ such that $P(a_1, \ldots, a_n, x) \vee Q(a_1, \ldots, a_n, x)$. By minimalization, the function

$$F(a_1, \ldots, a_n) = \mu x[P(a_1, \ldots, a_n, x) \vee Q(a_1, \ldots, a_n, x)]$$

is recursive. We now prove that

$$R(a_1, \ldots, a_n) \Leftrightarrow P(a_1, \ldots, a_n, F(a_1, \ldots, a_n)). \qquad (***)$$

Assuming $(***)$ holds, it follows by composition that R is recursive as required. Let $a_1, \ldots, a_n \in \mathbb{N}$. First assume $P(a_1, \ldots, a_n, F(a_1, \ldots, a_n))$. Then $\exists x P(a_1, \ldots, a_n, x)$; hence $(*)$ applies and $R(a_1, \ldots, a_n)$. This completes the proof of $(***)$ in one direction. Now assume $R(a_1, \ldots, a_n)$. By $(*)$ and $(**)$,

(1) There exists $x \in \mathbb{N}$ such that $P(a_1, \ldots, a_n, x)$.

(2) There is no $x \in \mathbb{N}$ such that $Q(a_1, \ldots, a_n, x)$.

Let x be the smallest natural number such that $P(a_1, \ldots, a_n, x)$. Using (2), we see that $F(a_1, \ldots, a_n) = x$; hence $P(a_1, \ldots, a_n, F(a_1, \ldots, a_n))$. This completes the proof of (***) as required. ∎

Although every recursive relation is RE, the converse is false. The construction of an RE relation that is not recursive is one of the most important results in mathematical logic, and we will eventually give several proofs of this fact. In particular, in the next two sections, we will prove that THN_{NN} is RE but not recursive. For now, we will simply record the following important result.

Theorem 4 (Fundamental theorem of recursion theory): There is an RE relation that is not recursive.

We now turn to the study of closure properties for the RE relations. The class of RE relations is closed under the logical operations of existential quantification, disjunction, and conjunction; it is not, however, closed under negation (if R is an RE relation that is not recursive, then $\neg R$ cannot be RE, by Theorem 3).

Therem 5 ($\exists$ rule): Let Q be an $n + 1$-ary RE relation. Then the n-ary relation R defined by

$$R(a_1, \ldots, a_n) \Leftrightarrow \exists b Q(a_1, \ldots, a_n, b)$$

is also RE.

Proof: Since Q is RE, there is an $n + 2$-ary recursive relation P such that for all $a_1, \ldots, a_n, b \in \mathbb{N}$, $Q(a_1, \ldots, a_n, b) \Leftrightarrow \exists x P(a_1, \ldots, a_n, b, x)$. We then have

$$R(a_1, \ldots, a_n) \Leftrightarrow \exists b \exists x P(a_1, \ldots, a_n, b, x).$$
$$\Leftrightarrow \exists c P(a_1, \ldots, a_n, K(c), L(c)),$$

and R is RE as required. (To see the equivalence of the two lines, see Example 1 in Section 8.3). ∎

Note: The method of proof used in Theorem 5 is called *collapsing a quantifier.*

Theorem 6 (Conjunction/disjunction rule): If Q and R are n-ary RE relations, then $Q \vee R$ and $Q \wedge R$ are n-ary RE relations.

Proof: We will prove the result for $Q \vee R$ and leave $Q \wedge R$ to the reader. Since Q and R are RE, there exist $n + 1$-ary recursive relations Q^* and R^* such that for all $a_1, \ldots, a_n \in \mathbb{N}$,

$$Q(a_1, \ldots, a_n) \Leftrightarrow \exists x Q^*(a_1, \ldots, a_n, x)$$
$$R(a_1, \ldots, a_n) \Leftrightarrow \exists y R^*(a_1, \ldots, a_n, y).$$

We then have $(Q \vee R)(a_1, \ldots, a_n) \Leftrightarrow \exists x \exists y [Q^*(a_1, \ldots, a_n, x) \vee R^*(a_1, \ldots, a_n, y)]$. Now $\exists y [Q^*(a_1, \ldots, a_n, x) \vee R^*(a_1, \ldots, a_n, y)]$ is an $n + 1$-ary RE relation; hence, by the $\exists$-rule, $Q \vee R$ is RE as required. ∎

Theorem 7: For any n-ary function F, the following are equivalent: (1) F is recursive; (2) $\mathcal{G}_F$ is recursive; (3) $\mathcal{G}_F$ is RE.

Proof of (1) $\Rightarrow$ (3): Assume F is recursive. Now

$$\mathcal{G}_F(a_1, \ldots, a_n, b) \Leftrightarrow F(a_1, \ldots, a_n) = b;$$

hence $\mathcal{G}_F$ is recursive by composition applied to the relation $=$. From Theorem 2, it follows that $\mathcal{G}_F$ is RE.

Proof of (3) $\Rightarrow$ (2): Assume $\mathcal{G}_F$ is RE. To prove that $\mathcal{G}_F$ is recursive, it suffices to show that $\neg\mathcal{G}_F$ is RE (Theorem 3). Now

$$\neg\mathcal{G}_F(a_1, \ldots, a_n, b) \Leftrightarrow F(a_1, \ldots, a_n) \neq b$$
$$\Leftrightarrow \exists c[(F(a_1, \ldots, a_n) = c) \wedge (c \neq b)]$$
$$\Leftrightarrow \exists c[\mathcal{G}_F(a_1, \ldots, a_n, c) \wedge (c \neq b)].$$

We have written $\neg\mathcal{G}_F$ in the form $\exists c Q(a_1, \ldots, a_n, b, c)$ with Q RE; hence $\neg\mathcal{G}_F$ is RE by the $\exists$-rule. (This proof appears in Smullyan 1993.)

Proof of (2) $\Rightarrow$ (1): Assume $\mathcal{G}_F$ is recursive; since F is a function with domain $\mathbb{N}^n$, $\mathcal{G}_F$ is a regular relation. Moreover,

$$F(a_1, \ldots, a_n) = \mu b \mathcal{G}_F(a_1, \ldots, a_n, b);$$

hence F is recursive by minimalization of a regular recursive relation. ∎

Theorem 8 (Composition of an RE relation): Let Q be a k-ary RE relation and let $H_1, \ldots, H_k$ be n-ary recursive functions. Then the n-ary relation R defined by

$$R(a_1, \ldots, a_n) \Leftrightarrow Q(H_1(a_1, \ldots, a_n), \ldots, H_k(a_1, \ldots, a_n))$$

is RE.

Proof: Left as an exercise for the reader. ∎

The phrase *recursively enumerable* suggests that an RE relation is enumerated by a recursive function. This is the content of the next theorem. (Compare with Theorem 7 in Section 1.7.)

Theorem 9: A nonempty subset R of $\mathbb{N}$ is RE if and only if it is the range of a recursive function $F: \mathbb{N} \to \mathbb{N}$.

Proof: Suppose R is the range of a 1-ary recursive function F; in other words, $R = \{F(c): c \in \mathbb{N}\}$. Then $R(a) \Leftrightarrow \exists c[F(c) = a]$; hence R is RE.

Now assume R is RE, let $a_0 \in R$, and let Q be a 2-ary recursive relation such that $R(a) \Leftrightarrow \exists b Q(a, b)$ for all $a \in \mathbb{N}$. Define F by

$$F(c) = \begin{cases} K(c) & Q(K(c), L(c)) \\ a_0 & \neg Q(K(c), L(c)). \end{cases}$$

Clearly F is recursive, so it remains to show that $R = \{F(c): c \in \mathbb{N}\}$; that is, $R(a) \Leftrightarrow \exists c[F(c) = a]$ for all $a \in \mathbb{N}$.

Let $a \in \mathbb{N}$. First assume $R(a)$; we are required to find $c \in \mathbb{N}$ such that $F(c) = a$. Since $R(a)$, there exists $b \in \mathbb{N}$ such that $Q(a, b)$. Let $c = J(a, b)$, so that $K(c) = a$ and $L(c) = b$. Then $Q(K(c), L(c))$; hence $F(c) = K(c)$ by the way F is defined. But $K(c) = a$, so $F(c) = a$ as required.

Now assume there exists $c \in \mathbb{N}$ such that $F(c) = a$; we are required to prove $R(a)$. Consider two cases: First suppose $Q(K(c), L(c))$. Then $R(K(c))$, and also $F(c) = K(c)$. But $F(c) = a$; hence $K(c) = a$. But we already know that $R(K(c))$, so $R(a)$ as required. Now suppose $\neg Q(K(c), L(c))$. Then $F(c) = a_0$ by the way F is defined. But $F(c) = a$; hence $a_0 = a$. Finally, we know that $R(a_0)$; hence $R(a)$ follows. ∎

We will end this section with a summary of important properties of RE relations. The reader is referred to Chapter 9 for a deeper study of recursive functions and RE relations; in particular, see Section 9.4 on the interplay between partial recursive functions and RE relations. In the next two sections, we will use recursion theory to obtain undecidability results.

(1) R is recursive if and only if both R and $\neg R$ are RE.

(2) If R is RE but not recursive, then $\neg R$ is *not* RE.

(3) If Q is an $n + 1$-ary RE relation, then $\exists x Q$ is an n-ary RE relation.

(4) If Q and R are n-ary RE relations, then $Q \vee R$ and $Q \wedge R$ are RE.

(5) An n-ary function F is recursive if and only if $\mathscr{G}_F$ is RE.

(6) If $R(a_1, \ldots, a_n) \Leftrightarrow Q(H_1(a_1, \ldots, a_n), \ldots, H_k(a_1, \ldots, a_n))$ with Q RE and $H_1, \ldots, H_k$ recursive, then R is RE.

(7) A nonempty set $R \subseteq \mathbb{N}$ is RE if and only if R is the range of a recursive function $F: \mathbb{N} \to \mathbb{N}$.

───────────── *Exercises for Section 8.5* ─────────────

1. Let R be an RE relation that is not recursive. Show that both R and R^c are infinite.
2. Give an example of a recursive function $F: \mathbb{N} \to \mathbb{N}$ and a recursive relation R such that $F(R)$ is not recursive. (*Hint:* Assume the existence of an RE relation that is not recursive.)
3. Prove that the following are equivalent:
   (1) Every decidable relation is recursive (version of Church's thesis).
   (2) Every semidecidable relation is RE.
4. Prove Theorem 8.
5. Let $F: \mathbb{N}^n \to \mathbb{N}$ be recursive. Show that the range of F is an RE relation.
6. Let R be an infinite subset of $\mathbb{N}$. Show that R is recursive if and only if R is the range of a recursive function $F: \mathbb{N} \to \mathbb{N}$ that is strictly increasing (and therefore one-to-one).
7. Let R be an infinite RE subset of $\mathbb{N}$. Show that there is an infinite recursive set $Q \subseteq R$. (*Hint:* R is the range of a recursive function $F: \mathbb{N} \to \mathbb{N}$; use F to construct a recursive function $G: \mathbb{N} \to \mathbb{N}$ that is strictly increasing and such that $G(\mathbb{N}) \subseteq R$; Q is the range of G.)
8. Let R be an infinite recursive subset of $\mathbb{N}$. Prove the following:
   (1) There exists $Q \subseteq R$, Q is RE but not recursive.

(2) There exists $P \subseteq R$, P is not RE.
(*Hints:* By Exercise 6, there is a one-to-one recursive function $F: \mathbb{N} \to \mathbb{N}$ such that R is the range of F. Let $K \subseteq \mathbb{N}$ be RE but not recursive; take $Q(b) \Leftrightarrow \exists a[K(a) \wedge (F(a) = b)]$.)

9. Let R be an infinite RE subset of $\mathbb{N}$. Prove that there exists $Q \subseteq R$, Q is not RE. (*Hint:* Use previous exercises.)

10. Let F be an n-ary function and let R be an $n + m + 1$-ary recursive relation such that for all $a_1, \ldots, a_n, b \in \mathbb{N}$,

$$F(a_1, \ldots, a_n) = b \Leftrightarrow \exists x_1 \cdots \exists x_m R(a_1, \ldots, a_n, b, x_1, \ldots, x_m).$$

Show that F is recursive.

11. (*Bounded $\forall$ and RE relations*) Define an $n + 1$-ary relation R by

$$R(a_1, \ldots, a_n, b) \Leftrightarrow \forall x < b Q(a_1, \ldots, a_n, x).$$

Prove the following:
(1) If Q is semidecidable, then R is semidecidable.
(2) If Q is RE, then R is RE. (*Hint:* Gödel's β-function.)

12. In this exercise, we outline another proof that if $\mathcal{G}_F$ is RE, then F is recursive. If $\mathcal{G}_F$ is RE, there is a recursive relation R such that for all $a_1, \ldots, a_n \in \mathbb{N}$, $\mathcal{G}_F(a_1, \ldots, a_n, b) \Leftrightarrow \exists x R(a_1, \ldots, a_n, b, x)$. Show that $F(a_1, \ldots, a_n) = K(\mu c R(a_1, \ldots, a_n, K(c), L(c)))$. (*Hint:* First show that the relation $Q(a_1, \ldots, a_n, c) \Leftrightarrow R(a_1, \ldots, a_n, K(c), L(c))$ is regular.)

8.6

THM_Γ Is Not Recursive and TR Is Not Definable

In this section, we will effectively code the language of arithmetic and then give precise proofs of the theorems of Church and Tarski: THM_Γ is not recursive (Γ = consistent extension of NN) and TR is not definable.

Effective Coding of the Language L_{NN}

Let FOR_{NN} be the set of all formulas of L_{NN}. By an *effective coding* of L_{NN}, we mean a one-to-one function $g: FOR_{NN} \to \mathbb{N}$ satisfying conditions Gö1 and Gö2 (described in Section 7.2); if $g(A) = c$ for a formula A, we call c the *code*, or *Gödel number*, of A. An effective coding of L_{NN} was constructed in Section 7.2 using exponentiation and properties of primes. In the present case, we use the coding functions discussed in Section 8.3. Thus we have available the n-ary recursive functions $\langle \cdots \rangle_n$ (each $n \geq 1$), the recursive function β, and the 1-ary recursive relation CD. Key properties are summarized as follows:

(1) If $c = \langle a_1, \ldots, a_n \rangle_n$, then $\beta(c, 0) = n$ and $\beta(c, k) = a_k$ for $1 \leq k \leq n$.

(2) If $c = \langle a_1, \ldots, a_n \rangle_n$, then $n < c$ and $a_k < c$ for $1 \le k \le n$.

(3) If $\langle a_1, \ldots, a_n \rangle_n = \langle b_1, \ldots, b_m \rangle_m$, then $n = m$ and $a_k = b_k$ for $1 \le k \le n$.

(4) $CD(a) \Leftrightarrow a$ is the code of some n-tuple.

The first step in the construction of g is to assign a number to each symbol s of L_{NN}. This number is called the *symbol number of s* and is denoted by $SN(s)$.

$$SN(x_n) = 2n \qquad SN(\forall) = 5 \qquad SN(S) = 11 \qquad SN(<) = 17.$$
$$SN(\neg) = 1 \qquad SN(=) = 7 \qquad SN(+) = 13$$
$$SN(\vee) = 3 \qquad SN(0) = 9 \qquad SN(\times) = 15$$

In summary, the even positive integers are used as symbol numbers for variables and the odd numbers 1 through 17 are used for the remaining finite number of symbols. It is not necessary to remember the precise details.

The next step is to assign a code $g(t)$ to each term t and a code $g(A)$ to each formula A.

Code of a term
$$g(0) = \langle SN(0) \rangle_1$$
$$g(x_n) = \langle SN(x_n) \rangle_1 = \langle 2n \rangle_1$$
$$g(St) = \langle SN(S), g(t) \rangle_2$$
$$g((s + t)) = \langle SN(+), g(s), g(t) \rangle_3$$
$$g((s \times t)) = \langle SN(\times), g(s), g(t) \rangle_3$$

Code of a formula
$$g((s = t)) = \langle SN(=), g(s), g(t) \rangle_3$$
$$g((s < t)) = \langle SN(<), g(s), g(t) \rangle_3$$
$$g(\neg A) = \langle SN(\neg), g(A) \rangle_2$$
$$g((A \vee B)) = \langle SN(\vee), g(A), g(B) \rangle_3$$
$$g(\forall x_n A) = \langle SN(\forall), g(x_n), g(A) \rangle_3$$

Let us summarize this assignment of codes to terms and formulas. The code of the constant symbol 0 or a variable x_n has the form $\langle \cdot \cdot \rangle_1$; the code of a term St or a formula $\neg A$ has the form $\langle SN(\ ), \ldots \rangle_2$; the code of a term $(s + t)$ or $(s \times t)$, or of a formula $(s = t)$, $(s < t)$, $(A \vee B)$, or $\forall x A$, has the form $\langle SN(\ ), \ldots, \ldots \rangle_3$.

Example 1:

The code of the formula $\forall x_1(x_1 = S0)$ is

$$\langle SN(\forall), g(x_1), \langle SN(=), g(x_1), \langle SN(S), g(0) \rangle_2 \rangle_3 \rangle_3. \quad \square$$

Example 2:

Let $a \in \mathbb{N}$, and suppose $CD(a)$, $\beta(a, 0) = 3$, and $\beta(a, 1) = SN(\vee)$; in this case, a is a code of the form $\langle SN(\vee), \ldots, \ldots \rangle_3$. If $\beta(a, 2)$ is the code of the formula A and $\beta(a, 3)$ is the code of the formula B, then a is the code of the formula $(A \vee B)$. $\quad \square$

We now turn to the proof that g is one-to-one and satisfies Gö1 and Gö2.

Lemma 1: If s and t are distinct terms of L_{NN}, then $g(s) \neq g(t)$.

Outline of proof We will state the main ideas and leave the details to the reader. The proof is by complete induction. Let $S(n)$ say: If a term t has n occurrences of S, $+$, and $\times$, and s is any term such that $g(s) = g(t)$, then $s = t$. Prove that $S(n)$ holds for all $n \geq 0$. Use these properties:

- If $\langle a_1, \ldots, a_n \rangle_n = \langle b_1, \ldots, b_m \rangle_m$, then $n = m$ and $a_k = b_k$ for $1 \leq k \leq n$.
- If $c = \langle a_1, \ldots, a_n \rangle_n$, then $n < c$ and $a_k < c$ for $1 \leq k \leq n$. ∎

Lemma 2: If A and B are distinct formulas, then $g(A) \neq g(B)$.

Outline of proof The proof is by induction. Let $S(n)$ say: If a formula A has n occurrences of $\neg$, $\vee$, and $\forall$, and B is a formula such that $g(A) = g(B)$, then $A = B$. Prove $S(n)$ for all $n \geq 0$; use Lemma 1 for the case in which A is an atomic formula. ∎

Theorem 1 (Effective coding of the language L_{NN}): The function $g: \text{FOR}_{NN} \rightarrow \mathbb{N}$ is one-to-one and satisfies the following two conditions:

Gö1 There is an algorithm that, given a formula A of L_{NN}, computes $g(A)$.

Gö2 There is an algorithm that, given $a \in \mathbb{N}$, decides whether there is a formula A of L_{NN} such that $g(A) = a$. Moreover, if the answer is YES, the algorithm finds the formula A such that $g(A) = a$.

Outline of proof The fact that g is one-to-one follows from Lemma 2. To see that property Gö1 holds, note the following: (1) The functions $\langle \cdots \rangle_n$ are recursive and hence computable; (2) the value of $g(t)$ and that of $g(A)$ are given inductively in terms of these functions. Property Gö2 holds because we can decode using the recursive relation CD and the recursive function β. ∎

Note: An effective coding of L_{NN} is the key to showing that a set Γ of formulas of L_{NN} is decidable if and only if the 1-ary relation THM_Γ is decidable; see Lemma 1 in Section 7.5.

────────────── *Church's Theorem* ──────────────

For $\Gamma \subseteq \text{FOR}_{NN}$, THM_Γ is the set of all codes of formulas that are theorems of Γ. Church's theorem states that THM_Γ is not recursive whenever Γ is a consistent extension of NN. The proof of this result in Section 7.5 uses the 1-ary computable function DIA. To make this proof precise, we are required to show that DIA is recursive. Unfortunately, it is somewhat messy to prove DIA recursive (although this will be done in the next section), so instead we take an alternate approach based on an idea of Tarski and exploited by Smullyan in *Gödel's Incompleteness Theorems* (Oxford: Oxford University Press, 1992). The idea is that we can avoid the substitution $A_x[t]$ as follows.

Lemma 3: $\vdash_{FO} A_x[t] \leftrightarrow \forall x((x = t) \to A)$, where t is a variable-free term.

Proof: We write out the required formal proof in first-order logic; the deduction theorem is used.

(1) $\forall x((x = t) \to A)$ HYP
(2) $(t = t) \to A_x[t]$ $\forall$-ELIM
(3) $t = t$ REF AX, $\forall$-ELIM
(4) $A_x[t]$ MP

(1) $A_x[t]$ HYP
(2) $A_x[t] \to ((x = t) \to A)$ EQ AX(4)
(3) $\forall x((x = t) \to A)$ MP, GEN. ∎

Corollary 1: Let A be a formula with exactly one free variable x_1 and let $a \in \mathbb{N}$. Then $\vdash_{FO} A_{x_1}[0^a] \leftrightarrow \forall x_1(\neg(x_1 = 0^a) \vee A)$.

Lemma 4: The following 1-ary function is recursive: $NUM(a) =$ code of the term 0^a.

Proof: Use primitive recursion as follows:

$$NUM(0) = \langle SN(0) \rangle_1$$
$$NUM(a + 1) = \langle SN(S), NUM(a) \rangle_2. \quad ∎$$

Lemma 5: There is a 1-ary recursive function TS (= Tarski-Smullyan) such that if a is the code of a formula A with exactly one free variable x_1, then

$$TS(a) = \text{code of the formula } \forall x_1(\neg(x_1 = 0^a) \vee A).$$

Proof: It suffices to write an explicit formula for TS in terms of recursive functions:

$$TS(a) = \langle SN(\forall), g(x_1), \langle SN(\vee), \langle SN(\neg), \langle SN(=), g(x_1), NUM(a) \rangle_3 \rangle_2, a \rangle_3 \rangle_3. \quad ∎$$

Theorem 2 (Church): Let Γ be a consistent extension of NN. Then THM$_\Gamma$ is not recursive (i.e., Γ is recursively undecidable.)

Proof: Suppose THM$_\Gamma$ is recursive. By standard arguments, the 1-ary relation R defined by $R(a) \Leftrightarrow \neg THM_\Gamma(TS(a))$ is recursive; hence there is a formula A with exactly one free variable x_1 that expresses R. Since Γ is a consistent extension of NN, it follows by Lemma 3 in Section 7.5 that for all $a \in \mathbb{N}$, $\Gamma \vdash A_{x_1}[0^a] \Leftrightarrow \neg THM_\Gamma(TS(a))$. But by Corollary 1, we have $\vdash_{FO} A_{x_1}[0^a] \leftrightarrow \forall x_1(\neg(x_1 = 0^a) \vee A)$; hence for all $a \in \mathbb{N}$,

$$\Gamma \vdash \forall x_1(\neg(x_1 = 0^a) \vee A) \Leftrightarrow \neg THM_\Gamma(TS(a)). \quad (*)$$

Let p be the code of the formula A. Then $TS(p)$ is the code of the sentence $\forall x_1(\neg(x_1 = 0^p) \vee A)$, and by (*) we have

$$\Gamma \vdash \forall x_1(\neg(x_1 = 0^p) \vee A) \Leftrightarrow \neg THM_\Gamma(TS(p)).$$

The left side of this equivalence states that $\forall x_1(\neg(x_1 = 0^p) \vee A)$ is a theorem of Γ; the right side states that $\forall x_1(\neg(x_1 = 0^p) \vee A)$ is *not* a theorem of Γ. This is a contradiction. ∎

Let us use this result to give a precise proof of Church's theorem on the unsolvability of the validity problem for the language L_{NN} (see Sections 5.5 and 7.5). First we need the formal version of Theorem 9 in Section 5.5.

Lemma 6 (Finite extension): Let Γ be a set of formulas of L_{NN} and let Δ be obtained from Γ by adding a finite number of sentences of L_{NN}. If THM_Γ is recursive, then THM_Δ is recursive.

Proof: Let $A_1, \ldots, A_n$ be the finite number of sentences added to Γ to obtain Δ. By replacing $A_1, \ldots, A_n$ with $A_1 \wedge \cdots \wedge A_n$, we can assume that a single sentence B is added to Γ to obtain Δ. In other words, $\Delta = \Gamma \cup \{B\}$. By MP and the deduction theorem, it follows that for every formula C of L_{NN};

$$\Delta \vdash C \Leftrightarrow \Gamma \vdash (\neg B \vee C).$$

Let b be the code of B. Then THM_Δ is recursive by

$$THM_\Delta(a) \Leftrightarrow THM_\Gamma(\langle SN(\vee), \langle SN(\neg), b \rangle_2, a \rangle_3). \quad ∎$$

Theorem 3 (Church, recursive unsolvability of the validity problem for L_{NN}): Assume NN is consistent. Then LV is not recursive, where

$$LV(a) \Leftrightarrow a \text{ is the code of a logically valid formula of } L_{NN}.$$

Proof: In Lemma 6, take $\Gamma = \varnothing$ and $\Delta = $ nonlogical axioms NN1–NN9 of NN; in other words, $THM_\Delta = THM_{NN}$. Now THM_{NN} is not recursive; hence THM_Γ is not recursive (Lemma 6). But by Gödel's completeness theorem, we have $LV = THM_\Gamma$ (since $\Gamma = \varnothing$); hence LV is not recursive, as required. ∎

Tarski's Theorem

We now turn to Tarski's theorem. Recall that TR is the set of codes of sentences of L_{NN} that are true in the standard interpretation of L_{NN}.

Theorem 4 (Tarski): Assume that the standard interpretation is a model of NN. Then TR is not definable. In particular, TR is not RE.

Proof: Left as an exercise for the reader; see the proofs of Theorem 1 in Section 7.6 and Theorem 2 in this section. ∎

———————————*Exercises for Section 8.6*———————————

1. Write the code of the reflexive axiom $\forall x_1(x_1 = x_1)$.

2. Prove the following:

(1) There is a 1-ary recursive function N such that if a is the code of a formula A, then $N(a)$ is the code of the formula $\neg A$.

(2) There is a 2-ary recursive function D such that if a is the code of a formula A and b is the code of a formula B, then $D(a, b)$ is the code of the formula $A \vee B$.

(3) There is a 2-ary recursive function Q such that if a is the code of a variable x and b is the code of a formula A, then $Q(a, b)$ is the code of the formula $\forall xA$.

(4) There is a 1-ary recursive function CTN such that if a is the code of a formula of the form $A \vee A$, then CTN(a) is the code of the formula A.

(5) There is a 1-ary recursive function ASSOC such that if a is the code of a formula of the form $A \vee (B \vee C)$, then ASSOC(a) is the code of the formula $(A \vee B) \vee C$.

3. Do Exercise 1 in Section 7.5; you may assume that DIA is recursive.

4. Prove Tarski's theorem.

5. Give an example of a relation R that is not RE but is definable.

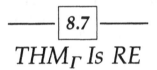

THM_Γ Is RE

In this section we prove the justly celebrated incompleteness theorem of Gödel. First, a definition.

Definition 1: For $\Gamma \subseteq \text{FOR}_{\text{NN}}$, NLA_Γ (= nonlogical axioms of Γ) is the 1-ary relation defined by

$$\text{NLA}_\Gamma(a) \Leftrightarrow a \text{ is the code of a formula } A \text{ of } L_{\text{NN}} \text{ and } A \in \Gamma.$$

If NLA_Γ is recursive, we say that Γ is *recursively axiomatized*.

Lemma 1: Every finite set of formulas of L_{NN} is recursively axiomatized. In particular, formal arithmetic NN is recursively axiomatized.

Proof: Let Γ be a finite set of formulas of L_{NN}. Then NLA_Γ $\subseteq \mathbb{N}$ is finite, and hence recursive (Theorem 5 in Section 8.2). The nonlogical axioms of NN are the nine sentences NN1–NN9, so NN is recursively axiomatized. ∎

Gödel's theorem states that every consistent recursively axiomatized extension of NN is incomplete. In broad outline, the proof proceeds as follows. Let $\Gamma \subseteq \text{FOR}_{\text{NN}}$; we will show that

(1) If Γ is recursively axiomatized, then THM_Γ is RE.

(2) If Γ is recursively axiomatized and complete, then ¬THM_Γ is RE.

Now suppose Γ is a consistent recursively axiomatized extension of NN. If Γ is also complete, then THM_Γ and ¬THM_Γ are *both* RE; hence THM_Γ is recursive, a contradiction of Church's theorem. So Γ is *not* complete, as required.

The longest, most involved step in carrying out this argument is to prove the result, due to Gödel, that THM$_\Gamma$ is RE for Γ recursively axiomatized. It is intuitively clear that THM$_\Gamma$ is semidecidable: Given $a \in \mathbb{N}$, use Gö2 to decide whether a is the code of a formula A; if so, begin listing all proofs to see whether there is a proof of A (for more detail, see Theorem 7 in Section 5.5). To prove the more precise result that THM$_\Gamma$ is RE, we need to recursively code all proofs. More precisely, we show that for Γ recursively axiomatized, the 2-ary relation PRF$_\Gamma(a, b)$ (defined in Section 7.2) is recursive; from this we obtain

$$\text{THM}_\Gamma(a) \Leftrightarrow \exists b \text{PRF}_\Gamma(a, b),$$

and THM$_\Gamma$ is RE as required. The proof that PRF$_\Gamma$ is recursive is a long and involved process, and it is easy to get lost in the details. The reader should constantly keep in mind that we are making precise the intuitively clear fact that PRF$_\Gamma$ is decidable. Another point of view is that we are giving a precise proof of Theorem 2 in Section 7.2.

Since we are coding proofs, we need results that state that certain relations concerning language and proof concepts are recursive. To begin with, we have observed that the code of a term or a formula is always the code of a 1-tuple, a 2-tuple, or a 3-tuple; it is convenient to have recursive relations to identify such codes.

Lemma 2: The following relations are recursive:

(1) $\text{CD}_1(a) \Leftrightarrow a$ is the code of a 1-tuple.
(2) $\text{CD}_2(a) \Leftrightarrow a$ is the code of a 2-tuple.
(3) $\text{CD}_3(a) \Leftrightarrow a$ is the code of a 3-tuple.

Proof: For (3), we have $\text{CD}_3(a) \Leftrightarrow \text{CD}(a) \wedge (\beta(a, 0) = 3)$. ∎

Theorem 1 (Language concepts): The following relations are recursive:

(1) $\text{VBL}(a) \Leftrightarrow a$ is the code of a variable.
(2) $\text{TRM}(a) \Leftrightarrow a$ is the code of a term.
(3) $\text{AFOR}(a) \Leftrightarrow a$ is the code of an atomic formula.
(4) $\text{FOR}(a) \Leftrightarrow a$ is the code of a formula.
(5) $\text{DFOR}(a) \Leftrightarrow a$ is the code of a formula that is a disjunction.
(6) $\text{NFOR}(a) \Leftrightarrow a$ is the code of a formula that is a negation.
(7) $\text{QFOR}(a) \Leftrightarrow a$ is the code of a formula of the form $\forall x A$.

Proof: We consider each relation in turn:

- $\text{VBL}(a) \Leftrightarrow a$ *is the code of a variable* We have

$$\text{VBL}(a) \Leftrightarrow \text{CD}_1(a) \wedge \exists n < a[(n \geq 1) \wedge (a = \langle 2n \rangle_1)].$$

Note that in applying the bounded quantifier rule, we have used the fact that $n \leq 2n < \langle 2n \rangle_1 = a$. ∎

- TRM(a) ⇔ *a is the code of a term* Let K be the characteristic function of TRM; we use definition by cases and course-of-values recursion to show that K is recursive. Introduce an auxiliary relation L defined by $L(a) ⇔ (\beta(a, 1) = SN(+)) \vee (\beta(a, 1) = SN(\times))$.

$$K(a) = \begin{cases} 0 & VBL(a) \vee (a = \langle SN(0)\rangle_1) \\ K(\beta(a, 2)) & CD_2(a) \wedge (\beta(a, 1) = SN(S)) \\ sg\,(K(\beta(a, 2)) + K(\beta(a, 3))) & CD_3(a) \wedge L(a) \\ 1 & \text{otherwise.} \end{cases}$$

Just this once, let us give a formal justification of the claim that K is recursive. By Theorem 3 in Section 8.4, we are required to find a 2-ary recursive function H such that for all $a \in \mathbb{N}$, $K(a) = H(a, K^\#(a))$; that is,

$$K(a) = H(a, \langle K(0), \ldots, K(a \dotminus 1)\rangle_a). \tag{*}$$

Let $a \in \mathbb{N}$, and suppose we are in the case $CD_3(a) \wedge L(a)$. Then (*) becomes

$$sg\,(K(\beta(a, 2)) + K(\beta(a, 3))) = H(a, \langle K(0), \ldots, K(a \dotminus 1)\rangle_a);$$

hence $H(a, x) = sg\,(\beta(x, \beta(a, 2) + 1) + \beta(x, \beta(a, 3) + 1))$ for this value of a. By considering the other possible cases for a, we eventually obtain the function H defined by

$$H(a, x) = \begin{cases} 0 & VBL(a) \vee (a = \langle SN(0)\rangle_1) \\ \beta(x, \beta(a, 2) + 1) & CD_2(a) \wedge (\beta(a, 1) = SN(S)) \\ sg\,(\beta(x, \beta(a, 2) + 1) \\ \quad + \beta(x, \beta(a, 3) + 1)) & CD_3(a) \wedge L(a) \\ 1 & \text{otherwise.} \end{cases}$$

Now H is recursive by definition by cases, and for all $a \in \mathbb{N}$ we have $K(a) = H(a, K^\#(a))$; hence K is recursive by course-of-values recursion.

- AFOR(a) ⇔ *a is the code of an atomic formula* By explicit definition,

$$AFOR(a) ⇔ CD_3(a) \wedge [(\beta(a, 1) = SN(=)) \vee (\beta(a, 1) = SN(<))]$$
$$\wedge TRM(\beta(a, 2)) \wedge TRM(\beta(a, 3)).$$

- FOR(a) ⇔ *a is the code of a formula* Let K be the characteristic function of FOR. We use definition by cases and course-of-values recursion to show that K is recursive.

$$K(a) = \begin{cases} 0 & AFOR(a) \\ K(\beta(a, 2)) & CD_2(a) \wedge (\beta(a, 1) = SN(\neg)) \\ sg\,(K(\beta(a, 2)) + K(\beta(a, 3))) & CD_3(a) \wedge (\beta(a, 1) = SN(\vee)) \\ K(\beta(a, 3)) & CD_3(a) \wedge (\beta(a, 1) = SN(\forall)) \\ & \qquad \wedge VBL(\beta(a, 2)) \\ 1 & \text{otherwise.} \end{cases}$$

- DFOR(a) ⇔ FOR(a) $\wedge$ CD$_3$(a) $\wedge$ ($\beta(a, 1) = SN(\vee)$).

- $\text{NFOR}(a) \Leftrightarrow \text{FOR}(a) \wedge \text{CD}_2(a)$.
- $\text{QFOR}(a) \Leftrightarrow \text{FOR}(a) \wedge \text{CD}_3(a) \wedge (\beta(a, 1) = \text{SN}(\forall))$. ■

The next step is to code the logical axioms. Technically, this is the most difficult part of the proof that PRF_Γ is recursive, due mainly to the complexity of substitution. Consider, for example, a substitution axiom

$$\forall xA \to A_x[t] \quad (t \text{ substitutable for } x \text{ in } A).$$

Let $\text{SUBAX}(a) \Leftrightarrow a$ is the code of a substitution axiom. To show that SUBAX is recursive, we need the following: (1) A 3-ary recursive function SUB that gives the code of $A_x[t]$ in terms of the codes of x, t, and A; (2) a 3-ary recursive relation SUBST such that $\text{SUBST}(a, b, c)$ holds precisely when a is the code of a variable x, b is the code of a term t, c is the code of a formula A, and the term t is substitutable for x in A.

Lemma 3 (Substitution function SUB): There is a 3-ary recursive function SUB such that if

$$a = \text{code of a variable } x$$
$$b = \text{code of a term } t$$
$$c = \text{code of a term } s \text{ or a formula } A,$$

then

$$\text{SUB}(a, b, c) = g(s_x[t]) = \text{code of } s_x[t] \quad (c \text{ the code of a term } s)$$
$$\text{SUB}(a, b, c) = g(A_x[t]) = \text{code of } A_x[t] \quad (c \text{ the code of a formula } A).$$

Proof: Let $a = g(x)$, $b = g(t)$, $c = g(s)$, or $c = g(A)$. We consider various cases for s and A; in each case we calculate $\text{SUB}(a, b, c)$, the code of $s_x[t]$ or $A_x[t]$, in terms of a, b, c and $\text{SUB}(a, b, x)$ for $x < c$. (See Section 4.4 for inductive definitions of $s_x[t]$ and $A_x[t]$.)

(1) *s is the constant 0 or s is a variable* $y \neq x$ In this case, $s_x[t]$ is just s, so $\text{SUB}(a, b, c) = c$.

(2) *s is the variable x* In this case, $s_x[t]$ is t, so $\text{SUB}(a, b, c) = b$.

(3) *s is Su, where u is a term* In this case, $s_x[t]$ is $Su_x[t]$ and $c = \langle \text{SN}(S), g(u) \rangle_2$. Note that $\beta(c, 2)$ is the code of the term u and $\text{SUB}(a, b, \beta(c, 2))$ is the code of $u_x[t]$. We have

$$\text{SUB}(a, b, c) = g(Su_x[t])$$
$$= \langle \text{SN}(S), g(u_x[t]) \rangle_2$$
$$= \langle \beta(c, 1), \text{SUB}(a, b, \beta(c, 2)) \rangle_2.$$

(4) *s is* $(u + v)$ *or* $(u \times v)$ In the first case, $s_x[t]$ is $(u_x[t] + v_x[t])$ and $c = \langle \text{SN}(+), g(u), g(v) \rangle_3$. We have

$$\text{SUB}(a, b, c) = g((u_x[t] + v_x[t]))$$
$$= \langle \text{SN}(+), g(u_x[t]), g(v_x[t]) \rangle_3$$
$$= \langle \beta(c, 1), \text{SUB}(a, b, \beta(c, 2)), \text{SUB}(a, b, \beta(c, 3)) \rangle_3.$$

This also covers the case in which s is $(u \times v)$.

(5) *A is an atomic formula $(u = v)$ or $(u < v)$* In the first case, $A_x[t]$ is $(u_x[t] = v_x[t])$ and $c = \langle SN(=), g(u), g(v)\rangle_3$. The calculation of $SUB(a, b, c)$ is the same as in (4). This also holds for the case $(u < v)$.

(6) *A is $\neg B$.* In this case, $A_x[t]$ is $\neg B_x[t]$ and $c = \langle SN(\neg), g(B)\rangle_2$. The calculation of $SUB(a, b, c)$ is the same as in (3).

(7) *A is $(B \vee C)$* In this case, $A_x[t]$ is $B_x[t] \vee C_x[t]$ and $c = \langle SN(\vee), g(B), g(C)\rangle_3$. The calculation of $SUB(a, b, c)$ is the same as in (4).

(8) *A is $\forall x B$* In this case, $A_x[t]$ is just A, so $SUB(a, b, c) = c$.

(9) *A is $\forall y B$ with $y \neq x$* In this case, $A_x[t]$ is $\forall y B_x[t]$, $c = \langle SN(\forall), g(y), g(B)\rangle_3$, and $\beta(c, 2) \neq a$. We have

$$\begin{aligned} SUB(a, b, c) &= g(\forall y B_x[t]) \\ &= \langle SN(\forall), g(y), g(B_x[t])\rangle_3 \\ &= \langle \beta(c, 1), \beta(c, 2), SUB(a, b, \beta(c, 3))\rangle_3. \end{aligned}$$

Now we are ready to define SUB officially. A number of cases consolidate $((4), (5), (7); (1), (8); (3), (6))$, and we have

$SUB(a, b, c)$

$$= \begin{cases} b & VBL(c) \wedge (a = c) \\ \langle \beta(c, 1), SUB(a, b, \beta(c, 2))\rangle_2 & CD_2(c) \\ \langle \beta(c, 1), \beta(c, 2), SUB(a, b, \beta(c, 3))\rangle_3 & QFOR(c) \wedge (\beta(c, 2) \neq a) \\ \langle \beta(c, 1), SUB(a, b, \beta(c, 2)), & CD_3(c) \wedge \neg QFOR(c) \\ \quad SUB(a, b, \beta(c, 3))\rangle_3 & \\ c & \text{otherwise.} \end{cases}$$

By definition by cases and course-of-values recursion, SUB is recursive as required. ∎

We note at this point that the function DIA, used in Sections 7.5 and 7.6 to prove Church's theorem and Tarski's theorem, is recursive as follows.

Corollary 1 (DIA is recursive): There is a 1-ary recursive function DIA such that if a is the code of a formula A with exactly one free variable x_1, then

$$DIA(a) = g(A_{x_1}[0^a]) = \text{code of } A_{x_1}[0^a].$$

Proof: $DIA(a) = SUB(g(x_1), NUM(a), a)$. ∎

Lemma 4 (Substitution): The following relations are recursive:

(1) $FREE(a, b) \Leftrightarrow a$ is the code of a variable x, b is the code of a formula A, and x occurs free in A.

(2) $SUBST(a, b, c) \Leftrightarrow a$ is the code of a variable x, b is the code of a term t, c is the code of a formula A, and t is substitutable for x in A.

Proof of (1): As a preliminary result, we show that the following relation is recursive:

$$\text{OCR}(a, b) \Leftrightarrow a \text{ is the code of a variable } x,$$
$$b \text{ is the code of a term } t, \text{ and}$$
$$x \text{ occurs in } t.$$

Let K be the characteristic function of OCR. We show that K is recursive by definition by cases and course-of-values recursion.

$$K(a, b) = \begin{cases} 0 & \text{VBL}(a) \wedge \text{VBL}(b) \wedge (a = b) \\ K(a, \beta(b, 2)) & \text{VBL}(a) \wedge \text{TRM}(b) \wedge \text{CD}_2(b) \\ K(a, \beta(b, 2)) \times K(a, \beta(b, 3)) & \text{VBL}(a) \wedge \text{TRM}(b) \wedge \text{CD}_3(b) \\ 1 & \text{otherwise.} \end{cases}$$

We now turn to the proof that FREE is recursive. An inductive definition that x occurs free in A is given in Section 4.4 (Definition 1). Let K be the characteristic function of FREE. We show that K is recursive by definition by cases and course-of-values recursion.

$$K(a, b) =$$
$$\begin{cases} 0 & \text{VBL}(a) \wedge \text{AFOR}(b) \\ & \wedge [\text{OCR}(a, \beta(b, 2)) \vee \text{OCR}(a, \beta(b, 3))] \\ K(a, \beta(b, 2)) & \text{VBL}(a) \wedge \text{NFOR}(b) \\ K(a, \beta(b, 2)) \times K(a, \beta(b, 3)) & \text{VBL}(a) \wedge \text{DFOR}(b) \\ K(a, \beta(b, 3)) & \text{VBL}(a) \wedge \text{QFOR}(b) \wedge (a \neq \beta(b, 2)) \\ 1 & \text{otherwise.} \end{cases}$$

Proof of (2): An inductive definition that t is substitutable for x in A is given in Section 4.4 (Definition 5). Let K be the characteristic function of SUBST. We show that K is recursive by definition by cases and course-of-values recursion. Let L be the auxiliary relation defined by $L(a, b) \Leftrightarrow \text{VBL}(a) \wedge \text{TRM}(b)$.

$$K(a, b, c) =$$
$$\begin{cases} 0 & L(a, b) \wedge \text{AFOR}(c) \\ K(a, b, \beta(c, 2)) & L(a, b) \wedge \text{NFOR}(c) \\ \text{sg}(K(a, b, \beta(c, 2)) & L(a, b) \wedge \text{DFOR}(c) \\ \quad + K(a, b, \beta(c, 3))) & \\ 0 & L(a, b) \wedge \text{QFOR}(c) \wedge \neg\text{FREE}(a, c) \\ K(a, b, \beta(c, 3)) & L(a, b) \wedge \text{QFOR}(c) \wedge \neg\text{OCR}(\beta(c, 2), b) \\ 1 & \text{otherwise.} \ \blacksquare \end{cases}$$

Theorem 2 (Logical axioms): The following relations are recursive:
(1) $\text{PAX}(a) \Leftrightarrow a$ is the code of a propositional axiom.
(2) $\text{SUBAX}(a) \Leftrightarrow a$ is the code of a substitution axiom.
(3) $\text{REFAX}(a) \Leftrightarrow a$ is the code of a reflexive axiom.
(4) $\text{EQAX}(a) \Leftrightarrow a$ is the code of an equality axiom.
(5) $\text{LAX}(a) \Leftrightarrow a$ is the code of a logical axiom.

Proof: A formula is a logical axiom if it is a propositional axiom, a substitution axiom, a reflexive axiom, or an equality axiom. Assuming that (1), (2), (3), and (4) hold, LAX is recursive by

$$\text{LAX}(a) \Leftrightarrow \text{PAX}(a) \vee \text{SUBAX}(a) \vee \text{REFAX}(a) \vee \text{EQAX}(a).$$

We recall the form of the logical axioms (written in terms of $\neg$, $\vee$):

> ***Propositional*** $\quad \neg A \vee A.$
>
> ***Substitution*** $\quad \neg \forall x A \vee A_x[t].$
>
> ***Reflexive*** $\quad \forall x_1(x_1 = x_1).$
>
> ***Equality*** $\quad \neg(x = t) \vee \neg[\neg(\neg A \vee A_x[t]) \vee \neg(\neg A_x[t] \vee A)].$

- PAX$(a) \Leftrightarrow$ DFOR$(a) \wedge (\beta(a, 2) = \langle \text{SN}(\neg), \beta(a, 3) \rangle_2)$.

- SUBAX$(d) \Leftrightarrow$ DFOR$(d) \wedge \exists a < d \, \exists b < d \, \exists c < d[\text{SUBST}(a, b, c) \wedge$
  $(\beta(d, 2) = \langle \text{SN}(\neg), \langle \text{SN}(\forall), a, c \rangle_3 \rangle_2) \wedge$
  $(\beta(d, 3) = \text{SUB}(a, b, c))]$.

- REFAX$(a) \Leftrightarrow a = \langle \text{SN}(\forall), g(x_1), \langle \text{SN}(=), g(x_1), g(x_1) \rangle_3 \rangle_3$.

- EQAX$(d) \Leftrightarrow$ DFOR$(d) \wedge \exists a < d \, \exists b < d \, \exists c < d[\text{SUBST}(a, b, c) \wedge$
  $(\beta(d, 2) = \langle \text{SN}(\neg), \langle \text{SN}(=), a, b \rangle_3 \rangle_2) \wedge$
  $(\beta(d, 3) = \langle \text{SN}(\neg), \langle \text{SN}(\vee), {*}{*}{*}, {+}{+}{+} \rangle_3 \rangle_2)]$.
  ${*}{*}{*}$ is $\langle \text{SN}(\neg), \langle \text{SN}(\vee), \langle \text{SN}(\neg), c \rangle_2, \text{SUB}(a, b, c) \rangle_3 \rangle_2$.
  ${+}{+}{+}$ is $\langle \text{SN}(\neg), \langle \text{SN}(\vee), \langle \text{SN}(\neg), \text{SUB}(a, b, c) \rangle_2, c \rangle_3 \rangle_2$. ∎

Theorem 3 (*Rules of inference*): The following relations are recursive:

(1) EXP$(a, b) \Leftrightarrow a$ is the code of a formula A and b is the code of a formula $B \vee A$.

(2) CTN$(a, b) \Leftrightarrow a$ is the code of a formula $A \vee A$ and b is the code of A.

(3) ASSOC$(a, b) \Leftrightarrow a$ is the code of a formula $A \vee (B \vee C)$ and b is the code of $(A \vee B) \vee C$.

(4) ADD$(a, b) \Leftrightarrow a$ is the code of a formula $A \vee B$, b is the code of $A \vee \forall x B$, and x is not free in A.

(5) CUT$(a, b, c) \Leftrightarrow a$ is the code of a formula $A \vee B$, b is the code of a formula $\neg A \vee C$, and c is the code of $B \vee C$.

Proof: We write an explicit definition for each relation:

- EXP$(a, b) \Leftrightarrow$ FOR$(a) \wedge$ DFOR$(b) \wedge (\beta(b, 3) = a)$.

- CTN$(a, b) \Leftrightarrow$ FOR$(b) \wedge (a = \langle \text{SN}(\vee), b, b \rangle_3)$.

- ASSOC$(a, b) \Leftrightarrow$ DFOR$(a) \wedge$ DFOR$(\beta(a, 3)) \wedge$ DFOR$(b) \wedge$
  $(\beta(b, 3) = (\beta(\beta(a, 3), 3)) \wedge$
  $(\beta(b, 2) = \langle \text{SN}(\vee), \beta(a, 2), \beta(\beta(a, 3), 2) \rangle_3)$.

- ADD$(a, b) \Leftrightarrow$ DFOR$(a) \wedge$ DFOR$(b) \wedge (\beta(a, 2) = \beta(b, 2))$
  $\wedge$ QFOR$(\beta(b, 3)) \wedge (\beta(\beta(b, 3), 3) = \beta(a, 3))$
  $\wedge \neg$FREE$(\beta(\beta(b, 3), 2), \beta(a, 2))$.

- CUT$(a, b, c) \Leftrightarrow$ DFOR$(a) \wedge$ DFOR$(b) \wedge (\beta(b, 2) = \langle \text{SN}(\neg), \beta(a, 2) \rangle_2) \wedge$
  $(c = \langle \text{SN}(\vee), \beta(a, 3), \beta(b, 3) \rangle_3)$. ∎

Theorem 4 (Gödel): Let $\Gamma \subseteq \text{FOR}_{\text{NN}}$ be recursively axiomatized. Then

(1) PRF_Γ is recursive.

(2) THM_Γ is RE.

Proof: We proceed informally at first. $\text{PRF}_\Gamma(a, b)$ holds provided that

> a = code of a formula A
>
> b = code of a finite sequence of formulas $A_1, \ldots, A_n$
>
> (i.e., $b = \langle g(A_1), \ldots, g(A_n) \rangle_n$.
>
> $A = A_n$
>
> $A_1, \ldots, A_n$ is a proof using Γ.

We use two auxiliary recursive relations as follows (AX_Γ is recursive, since NLA_Γ is recursive by hypothesis):

- $\text{AX}_\Gamma(a) \Leftrightarrow \text{LAX}(a) \vee \text{NLA}_\Gamma(a)$.
- $\text{RULE}(a, b) \Leftrightarrow \text{EXP}(a, b) \vee \text{CTN}(a, b) \vee \text{ASSOC}(a, b) \vee \text{ADD}(a, b)$.

(1) $\text{PRF}_\Gamma(a, b) \Leftrightarrow \text{FOR}(a) \wedge \text{CD}(b) \wedge (a = \beta(b, \beta(b, 0))) \wedge$
$\forall k < \beta(b, 0)[\text{FOR}(\beta(b, k + 1)) \wedge$
$\{\text{AX}_\Gamma(\beta(b, k + 1)) \vee ((k \geq 1) \wedge$
$\exists i < k\,\text{RULE}(\beta(b, i + 1), \beta(b, k + 1))) \vee ((k > 1) \wedge$
$\exists i < k \exists j < k\,\text{CUT}(\beta(b, i + 1), \beta(b, j + 1), \beta(b, k + 1)))\}]$.

(2) $\text{THM}_\Gamma(a) \Leftrightarrow \exists b\, \text{PRF}_\Gamma(a, b)$. ∎

Theorem 5 (Fundamental theorem of recursion theory):
Assume NN is consistent. Then there is an RE relation that is not recursive.

Proof: By Gödel's result, THM_{NN} is RE; by Church's theorem, THM_{NN} is not recursive. ∎

Comment Theorem 5 captures the modern interpretation of the theorems of Gödel and Church: *There is an RE set not recursive.*

The prize is within sight. First, however, we need a result about the closure of a formula. A working definition of this concept was given in Section 4.3, but for coding we need to be more precise.

Definition 2: Let A be a formula and let a be the code of A. The *closure of A*, denoted A', is the formula $\forall x_a \cdots \forall x_1 A$.

We need to show that A', as defined above, has no free variables. In other words, we need to show that if x_n occurs free in A, then $n \leq g(A)$.

Lemma 5: Let A be a formula and let a be the code of A. If x_n is any variable that occurs in A, then $n < a$.

Outline of proof We leave most of the details to the reader. The key property is: If $c = \langle a_1, \ldots, a_n \rangle_n$, then $n < c$ and $a_k < c$ for $1 \leq k \leq n$. First prove the result for terms: If $a = g(t)$ and x_n occurs in the term t, then $n < a$.

Then prove the result for formulas by induction on the number of occurrences of $\neg$, $\vee$, and $\forall$ in the formula. ∎

Lemma 6: There is a 1-ary recursive function CLS such that if a is the code of a formula A, then $\text{CLS}(a) = $ code of A'.

Proof: We first construct a 2-ary recursive function F such that if a is the code of a formula A, then $F(a, 0) = $ code of A, $F(a, 1) = $ code of $\forall x_1 A$, $F(a, 2) = $ code of $\forall x_2 \forall x_1 A$, and so on. Define F by

$$F(a, 0) = a$$
$$F(a, n + 1) = \langle \text{SN}(\forall), g(x_{n+1}), F(a, n)\rangle_3.$$

F is recursive by primitive recursion. Now suppose a is the code of a formula A. We prove by induction that for $n \geq 1$, $F(a, n) = g(\forall x_n \cdots \forall x_1 A)$.

Beginning step $F(a, 1) = \langle \text{SN}(\forall), g(x_1), F(a, 0)\rangle_3$
$$= \langle D\text{SN}(\forall), g(x_1), a\rangle_3$$
$$= g(\forall x_1 A).$$

Induction step Now let $n \geq 1$ and assume that $F(a, n) = g(\forall x_n \cdots \forall x_1 A)$. Then

$$F(a, n + 1) = \langle \text{SN}(\forall), g(x_{n+1}), F(a, n)\rangle_3$$
$$= \langle \text{SN}(\forall), g(x_{n+1}), g(\forall x_n \cdots \forall x_1 A)\rangle_3$$
$$= g(\forall x_{n+1} \forall x_n \cdots \forall x_1 A).$$

To complete the proof, define CLS by $\text{CLS}(a) = F(a, a)$. Clearly CLS is recursive. Let a be the code of a formula A. Then

$$\text{CLS}(a) = F(a, a) = g(\forall x_a \cdots \forall x_1 A) = \text{code of } A'$$

as required. ∎

Lemma 7 (Turing): Let $\Gamma \subseteq \text{FOR}_{\text{NN}}$ be recursively axiomatized and complete. Then THM_Γ is recursive.

Proof: If Γ is inconsistent, then $\text{THM}_\Gamma = \text{FOR}$; since FOR is recursive, THM_Γ is recursive as required. Thus we may assume that Γ is consistent. Since Γ is recursively axiomatized, THM_Γ is RE. Hence THM_Γ is recursive if we can show that $\neg\text{THM}_\Gamma$ is also RE (Theorem 3 in Section 8.5). Let A be a formula of L_{NN}. By the closure theorem for provability, $\Gamma \vdash A$ if and only if $\Gamma \vdash A'$. From this result and the fact that Γ is consistent and complete, we have

A is not a theorem of Γ if and only if $\Gamma \vdash \neg A'$.

Finally,

$$\neg\text{THM}_\Gamma(a) \Leftrightarrow \neg\text{FOR}(a) \vee \text{THM}_\Gamma(\langle \text{SN}(\neg), \text{CLS}(a)\rangle_2)$$
$$\Leftrightarrow \neg\text{FOR}(a) \vee \exists b \text{PRF}_\Gamma(\langle \text{SN}(\neg), \text{CLS}(a)\rangle_2, b)$$
$$\Leftrightarrow \exists b[\neg\text{FOR}(a) \vee \text{PRF}_\Gamma(\langle \text{SN}(\neg), \text{CLS}(a)\rangle_2, b)]$$

and $\neg\text{THM}_\Gamma(a)$ is RE as required. ∎

Theorem 6 (Gödel's incompleteness theorem): Every consistent recursively axiomatized extension of NN is incomplete.

Proof: Let Γ be a consistent recursively axiomatized extension of NN. Suppose Γ is also complete. By Lemma 7, THM_Γ is recursive. This contradicts Church's theorem. ∎

————————————*Exercises for Section 8.7*————————————

1. Let $SENT(a) \Leftrightarrow a$ is the code of a sentence of L_{NN}. Show that SENT is recursive.
2. An informal proof of Gödel's incompleteness theorem from Church's theorem was given in Section 7.1: Assume Γ complete, and begin listing proofs; check each proof to see whether it is a proof of A or of $\neg A$; by completeness, this procedure does not go into an infinite loop. In this way we obtain an algorithm that shows that Γ is decidable, a contradiction of Church's theorem. We will now outline a formal version of this argument. Let Γ be a consistent recursively axiomatized extension of NN. Define a 1-ary function H by

$$H(a) = \mu x[\neg FOR(a) \vee PRF_\Gamma(CLS(a), x) \vee PRF_\Gamma(\langle SN(\neg), CLS(a)\rangle_2, x)].$$

Informally, $H(a) = 0$ if a is not the code of a formula; if a is the code of a formula A, then $H(a)$ is the smallest number that is the code of a proof of A' or of $\neg A'$. H is of the form $H(a) = \mu x[R(a, x)]$, where R is a recursive relation. Assume Γ is complete. Prove the following:
   (1) R is a regular relation (so H is a recursive function).
   (2) $THM_\Gamma(a) \Leftrightarrow FOR(a) \wedge PRF_\Gamma(CLS(a), H(a))$. (This shows that $THM_\Gamma(a)$ is recursive, a contradiction of Church's theorem.)
3. (*Semantic version of Gödel's theorem*) Let Γ be a recursively axiomatized extension of NN such that the standard interpretation of L_{NN} is a model of Γ. Prove the following:
   (1) R is definable, where $R(a) \Leftrightarrow \neg THM_\Gamma(DIA(a))$ (see Section 7.6; assume THM_Γ is RE and DIA is recursive).
   (2) There is a formula A with exactly one free variable x_1 such that for all $a \in \mathbb{N}$, $\neg THM_\Gamma(DIA(a)) \Leftrightarrow A_{x_1}[0^a]$ is true in $\mathcal{N}$.
   (3) Let p be the code of A; neither $A_{x_1}[0^p]$ nor $\neg A_{x_1}[0^p]$ is a theorem of Γ.
4. (*Semantic version of Gödel's theorem*) This is a slight variation of the previous exercise, and uses Tarski's theorem that TR is not RE. Let Γ be a recursively axiomatized extension of NN such that the standard interpretation of L_{NN} is a model of Γ, and define THM_Γ^* by $THM_\Gamma^*(a) \Leftrightarrow a$ is the code of a *sentence* of L_{NN} that is a theorem of Γ. Prove the following:
   (1) $THM_\Gamma^* \subseteq TR$.
   (2) THM_Γ^* is RE. (*Hint:* See Exercise 1.)
   (3) $THM_\Gamma^* \neq TR$.
   (4) If $p \in TR$, $p \notin THM_\Gamma^*$, and p is the code of the sentence A, then neither A nor $\neg A$ is a theorem of Γ.
5. (*Post's proof of Gödel's theorem*) We will outline a proof of Gödel's theorem based on ideas of Post. Let K be an RE relation not recursive and let Γ be a recursively axiomatized extension of NN such that the standard interpretation of L_{NN} is a model of Γ. K is definable (see Corollary 1 in Section 7.6); hence there is a formula A with exactly one free variable x_1 such that for all $a \in \mathbb{N}$, $K(a) \Leftrightarrow A_{x_1}[0^a]$ is true in $\mathcal{N}$. Define a 1-ary relation R by $R(a) \Leftrightarrow \Gamma \vdash \neg A_{x_1}[0^a]$. Prove the following:
   (1) $R \subseteq K^c$.
   (2) R is RE. (*Hint:* Let p be the code of the formula A; use the recursive function SUB and the RE relation THM_Γ.)
   (3) $R \neq K^c$. (*Hint:* R is RE; K^c is not RE.)
   (4) Let $p \in K^c$, $p \notin R$; neither $A_{x_1}[0^p]$ nor $\neg A_{x_1}[0^p]$ is a theorem of Γ.

6. (*Recursive* = *expressible*; RE = *weakly expressible*) In Section 7.3, we proved that every recursive relation is expressible. This exercise outlines a proof of the converse of that result and also obtains a similar characterization of the RE relations. To simplify, we restrict attention to 1-ary relations; however, all of the results hold for n-ary relations. A 1-ary relation R is *weakly expressible* if there is a formula A with exactly one free variable x_1 such that for all $a \in \mathbb{N}$,

$$R(a) \Leftrightarrow \text{NN} \vdash A_{x_1}[0^a]. \tag{*}$$

Prove the following:

(1) Assume NN is consistent. If R is expressible, then R is weakly expressible.

(2) Assume NN is ω-consistent. If R is RE, then R is weakly expressible. (*Hint:* Let $R(a) \Leftrightarrow \exists b Q(a, b)$, where Q is recursive. Let A be a formula with exactly two free variables x_1, x_2 such that for all $a, b \in \mathbb{N}$, if $Q(a, b)$, then $\text{NN} \vdash A_{x_1, x_2}[0^a, 0^b]$; if $\neg Q(a, b)$, then $\text{NN} \vdash \neg A_{x_1, x_2}[0^a, 0^b]$. Show that $R(a) \Leftrightarrow \text{NN} \vdash \exists x_2 A_{x_1}[0^a]$.

(3) If R is weakly expressible, then R is RE. (*Hint:* Let p be the code of the formula A in (*); use the RE relation THM_{NN} and the recursive function SUB.)

(4) Assume NN is consistent. If R is expressible, then R is recursive. (*Hint:* Show that R and $\neg R$ are RE.)

9

Computability Theory

A n n-ary function $F: \mathbb{N}^n \to \mathbb{N}$ is said to be *computable* if there is an algorithm that calculates F. This definition is not precise, depending as it does on the informal notion of an algorithm. We have already discussed one way to make it precise, namely the notion of a recursive function (see Section 1.8 and Chapter 8). Every recursive function is computable; Church's thesis is the statement (or belief, if you like) that every computable function is recursive.

In this chapter, we will take an altogether different approach to the problem of giving a precise definition of computability. We describe an ideal computer called a *register machine*; given an n-ary function F, we say that F is RM-*computable* if there is a program P for the register machine with this property: Given $a_1, \ldots, a_n \in \mathbb{N}$, the program with inputs $a_1, \ldots, a_n$ eventually halts with output $F(a_1, \ldots, a_n)$. This situation certainly satisfies our intuitive notion of a computable function, so it is natural to develop a model of computation based on computers and programs.

One of our goals is to prove that the two approaches to computability (recursive and RM-computable) give the same class of functions. This fundamental result shows that the class of recursive functions is indeed a rather natural class of functions and, moreover, is evidence in support of Church's thesis.

9.1

Register Machines and RM-Computable Functions

The computer we describe is called a *register machine* (more precisely, an *unlimited register machine*) and is due to Shepherdson and Sturgis 1963. A register machine is quite primitive in terms of the number and power of its instructions. Modern-day computers have many more instructions, with apparently much greater computational power. On the other hand, a register machine differs from a real computer in that it has an unlimited memory.

There are good reasons for not putting a bound on the amount of memory. In the first place, it is difficult to put a precise bound on the memory of any computer, since it is always possible to add more tapes, disks, drives, and so on. In addition, we are interested in the theoretical limits of computation. What can a computer do *in theory*, given no limitations on time or memory?

We now describe the basic components of a register machine: *registers, instructions,* and *programs.*

Registers: There are an infinite number of registers $R_1, R_2, R_3, \ldots$. Each register contains a natural number; during a computation, however, all but a finite number of the registers contain 0. There is no bound on the size of a natural number that can be stored in a register. Registers are used in the following way: Let F be an n-ary function and let $a_1, \ldots, a_n \in \mathbb{N}$. The inputs $a_1, \ldots, a_n$ are stored in registers $R_1, \ldots, R_n$ respectively; the output $F(a_1, \ldots, a_n)$ is stored in register R_1; the remaining registers are used for storing intermediate calculations. We emphasize that register R_1 is used for both the input a_1 and the output $F(a_1, \ldots, a_n)$. We often use $\#R_k$ to denote the number in register R_k. We can visualize the registers with $a_1, \ldots, a_n$ in $R_1, \ldots, R_n$ respectively as follows:

R_1	R_2	$\cdots$	R_n	$\cdots$
a_1	a_2	$\cdots$	a_n	$\cdots$

Instructions and Programs: There are four kinds of instructions. Three of these change the number in a register; the fourth controls the sequence of steps in a computation by comparing the number in a register with 0 and acting accordingly. Instructions are described in detail below. A *program* is a finite sequence of instructions. Thus $I_1, I_2, \ldots, I_t$ denotes a program P with t instructions; instructions in a program are always numbered consecutively beginning with 1.

Increase instructions $S(k), k \in \mathbb{N}^+$ The increase instruction $S(k)$ adds 1 to the number in register R_k; the numbers in all other registers are unchanged.

Decrease instructions $D(k), k \in \mathbb{N}^+$ The decrease instruction $D(k)$ subtracts 1 from the number in register R_k; the numbers in all other registers are unchanged. If R_k contains 0, then 0 remains in R_k.

Zero instructions $Z(k), k \in \mathbb{N}^+$ The zero instruction $Z(k)$ puts the number 0 in register R_k and leaves the numbers in all other registers unchanged.

Jump instructions $J(k, q), k, q \in \mathbb{N}^+$ The jump instruction $J(k, q)$ compares the number in register R_k to 0. If $\#R_k = 0$, the next instruction executed is I_q; if $\#R_k \neq 0$, the next instruction executed is the one that follows $J(k, q)$ in the program.

Example 1:

Assume 3, 2, and 1 are in registers R_1, R_2, and R_3 respectively. We execute, in order, the instructions $Z(2)$, $S(3)$, $D(1)$, and $D(2)$:

	R_1	R_2	R_3	$\cdots$
Initial inputs	3	2	1	$\cdots$
After instruction $Z(2)$	3	0	1	$\cdots$
After instruction $S(3)$	3	0	2	$\cdots$
After instruction $D(1)$	2	0	2	$\cdots$
After instruction $D(2)$	2	0	2	$\cdots$

□

Definition 1 (Informal): Let $P = I_1, I_2, \ldots, I_t$ be a program; assume that $a_1, \ldots, a_n$ are in registers $R_1, \ldots, R_n$ respectively and that 0 is in all other registers. The *P-computation with inputs* $a_1, \ldots, a_n$ begins with instruction I_1 and then continues in sequence $I_2, I_3, \ldots$ unless a jump instruction is encountered. The *P-computation with inputs* $a_1, \ldots, a_n$ *halts* whenever one of the following conditions is satisfied:

(1) The last instruction I_t is reached and it is not a jump instruction.

(2) The last instruction I_t is reached and it is a jump instruction $J(k, q)$, but the number in register R_k is not 0 (the machine wants to execute instruction I_{t+1} next, but there is no such instruction).

(3) A jump instruction $J(k, q)$ is encountered, the number in register R_k is 0, and $q > t$ (the machine wants to execute instruction I_q next, but there is no such instruction).

If a *P*-computation does halt, the *output* is the number in register R_1.

Example 2:

The following program P never halts, regardless of input:

I_1 $Z(1)$
I_2 $S(2)$
I_3 $J(1, 2)$ □

Now that register machines, programs, and P-computations have been defined, we can say what it means to compute a function by the register machine.

Definition 2: Let P be a program for the register machine and let F be an n-ary function. The program P *computes* F if for all $a_1, \ldots, a_n \in \mathbb{N}$, the P-computation with inputs $a_1, \ldots, a_n$ halts with output $F(a_1, \ldots, a_n)$ in register R_1. Also, we say that F is RM-*computable* if there is a program P for the register machine that computes F.

Theorem 1: Every RM-computable function is computable.

Proof: A program for the register machine certainly satisfies our intuitive notion of an algorithm, so every RM-computable function is computable (in the intuitive sense). ∎

How powerful are register machines? Eventually we will prove that an n-ary function F is recursive if and only if it is RM-computable. From this result and Church's thesis, it follows that *register machines are as powerful in computational power as any computer that can ever be built*. To see why this is so, suppose we construct a hypothetical computer for calculating functions. Now a function F that can be calculated by our hypothetical computer certainly satisfies our notion of a computable function; hence, by Church's thesis, F is recursive. It follows that F is RM-computable as required (here we assume the result, which will be proved in the next section, that every recursive function is RM-computable).

The goal of the remainder of this section is to show that the starting functions $+$, $\times$, $K_<$, and $U_{n,k}$ are all RM-computable.

Theorem 2: Addition is RM-computable.

Proof: We write a program P such that the P-computation with inputs a, b in registers R_1 and R_2 respectively halts with the number $a + b$ in R_1. *Idea:* Add 1 to the number in register R_1 b times. An overview of registers R_1 and R_2 is

R_1	R_2
a	b
$a + 1$	$b - 1$
$a + 2$	$b - 2$
$\vdots$	$\vdots$
$a + b$	0

The required program, with comments, is as follows:

I_1	$J(2,5)$	Is $\#R_2 = 0$? If *Yes*, halt; $a + b$ is in R_1.
I_2	$S(1)$	Add 1 to $\#R_1$.
I_3	$D(2)$	Subtract 1 from $\#R_2$.
I_4	$J(3,1)$	Go to instruction I_1. ∎

Note: The last instruction of this program, $J(3,1)$, acts like an unconditional GO TO instruction; we assume that 0 is in R_3. In general, the technique is to have 0 in register R_k; $J(k, q)$ then tells the program to execute instruction I_q next.

Example 3:

Later, when we code computations, it will be necessary to define precisely the *P*-computation with inputs $a_1, \ldots, a_n$. In anticipation of this precise definition, let us describe, for the program *P* in Theorem 2, the *P*-computation with inputs $a = 1$ and $b = 2$. For this program, a step in a *P*-computation is denoted by

$$s = \langle r_1, r_2, r_3; j \rangle.$$

This 4-tuple tells us that the numbers r_1, r_2, and r_3 are in registers R_1, R_2, and R_3 respectively and that the next instruction is I_j (note that R_1 through R_3 are the only registers mentioned in the program *P*). The following sequence $s_1, \ldots, s_{10}$ is the *P*-computation with inputs 1, 2:

$s_1 = \langle 1, 2, 0; 1 \rangle$ $\#R_1 = 1$, $\#R_2 = 2$, $\#R_3 = 0$; next (= first) instruction is I_1.

$s_2 = \langle 1, 2, 0; 2 \rangle$ $\#R_1 = 1$, $\#R_2 = 2$, $\#R_3 = 0$; next instruction is I_2.

$s_3 = \langle 2, 2, 0; 3 \rangle$ $\#R_1 = 2$, $\#R_2 = 2$, $\#R_3 = 0$; next instruction is I_3.

$s_4 = \langle 2, 1, 0; 4 \rangle$ $\#R_1 = 2$, $\#R_2 = 2$, $\#R_3 = 0$; next instruction is I_4.

$s_5 = \langle 2, 1, 0; 1 \rangle$ $\#R_1 = 2$, $\#R_2 = 1$, $\#R_3 = 0$; next instruction is I_1.

$s_6 = \langle 2, 1, 0; 2 \rangle$ $\#R_1 = 2$, $\#R_2 = 1$, $\#R_3 = 0$; next instruction is I_2.

$s_7 = \langle 3, 1, 0; 3 \rangle$ $\#R_1 = 3$, $\#R_2 = 1$, $\#R_3 = 0$; next instruction is I_3.

$s_8 = \langle 3, 0, 0; 4 \rangle$ $\#R_1 = 3$, $\#R_2 = 0$, $\#R_3 = 0$; next instruction is I_4.

$s_9 = \langle 3, 0, 0; 1 \rangle$ $\#R_1 = 3$, $\#R_2 = 0$, $\#R_3 = 0$; next instruction is I_1.

$s_{10} = \langle 3, 0, 0; 5 \rangle$ $\#R_1 = 3$, $\#R_2 = 0$, $\#R_3 = 0$; next instruction is I_5 (does not exist).   □

Theorem 3: The characteristic function $K_<$ of $<$ is RM-computable.

Proof: We write a program *P* such that the *P*-computation with inputs a, b halts with output 0 in R_1 if $a < b$ and halts with output 1 in R_1 if $b \le a$.

Idea: Subtract 1 from each number and see which reaches 0 first. An overview of registers R_1 and R_2 is

R_1	R_2
a	b
$a - 1$	$b - 1$
$a - 2$	$b - 2$
$\vdots$	$\vdots$

The required program (with comments) is as follows:

I_1	$J(2, 6)$	Is $\#R_2 = 0$? If *Yes*, $b \le a$; go to instruction I_6.
I_2	$J(1, 8)$	Is $\#R_1 = 0$? If *Yes*, $a < b$; 0 is in R_1; halt.
I_3	$D(1)$	Subtract 1 from $\#R_1$.
I_4	$D(2)$	Subtract 1 from $\#R_2$.
I_5	$J(3, 1)$	Go to instruction I_1.
I_6	$Z(1)$	Put 0 in R_1.
I_7	$S(1)$	Add 1 to $\#R_1$ and halt. ∎

Before proceeding to the proof that the projection functions and multiplication are RM-computable, let us develop some subroutines to facilitate the writing of programs.

Lemma 1 (Subroutine MOVE(j, k; m)): Let j, k, and m be distinct positive integers. There is a program MOVE$(j, k; m)$, with 6 instructions, that transfers the number in register R_j to register R_k and puts 0 in registers R_j and R_m. The numbers in all other registers are unchanged.

Proof: The lemma states that if we start with the configuration

R_j	$\cdots$	R_k	$\cdots$	R_m	$\cdots$
a	$\cdots$	b	$\cdots$	c	$\cdots$

then the program MOVE$(j, k; m)$ halts with the configuration

R_j	$\cdots$	R_k	$\cdots$	R_m	$\cdots$
0	$\cdots$	a	$\cdots$	0	$\cdots$

The required program (with comments) is as follows:

I_1	$Z(k)$	Put 0 in R_k.
I_2	$Z(m)$	Put 0 in R_m.
I_3	$J(j, 7)$	Is $\#R_j = 0$? If *Yes*, halt.
I_4	$S(k)$	Add 1 to $\#R_k$.

I_5 $D(j)$ Subtract 1 from $\#R_j$.

I_6 $J(m, 3)$ Go to instruction I_3. ∎

Note: In this program, register R_m is used to obtain an unconditional GO TO instruction, so it is necessary to put 0 in register R_m. This means that when using the subroutine MOVE($j, k; m$) with other programs, we should choose m larger than the number of any register that occurs in any of these other programs; otherwise, a number stored in register R_m would be lost when MOVE($j, k; m$) is executed.

Theorem 4: For $n \geq 1$ and $1 \leq k \leq n$, the projection function $U_{n,k}(a_1, \ldots, a_n) = a_k$ is RM-computable.

Proof: We can assume that $k > 1$; the required program is MOVE($k, 1; n + 1$). ∎

Lemma 2 (Subroutine ADD($j, k; m$)): Let j, k, m, and $m + 1$ be distinct positive integers. There is a program ADD($j, k; m$), with 11 instructions, that adds the number in register R_j to the number in register R_k and puts 0 in registers R_m and R_{m+1}. The numbers in all other registers, including R_j, are unchanged.

Proof: The lemma states that if we start with the configuration

R_j		R_k		R_m	R_{m+1}	
a	$\cdots$	b	$\cdots$	c	d	$\cdots$

then ADD($j, k; m$) halts with the configuration

R_j		R_k		R_m	R_{m+1}	
a	$\cdots$	$a + b$	$\cdots$	0	0	$\cdots$

The idea in writing ADD($j, k; m$) is to move the number a in R_j to register R_m and then to add 1 to registers R_j and R_k a times. The required program (with comments) follows.

$\left.\begin{array}{l} I_1 \\ \vdots \\ I_6 \end{array}\right\}$ MOVE($j, m; m + 1$)   Move $\#R_j$ to R_m; 0 is now in R_j and R_{m+1}.

I_7 $J(m, 12)$ Is $\#R_m = 0$? If *Yes*, halt.

I_8 $S(j)$ Add 1 to $\#R_j$.

I_9 $S(k)$ Add 1 to $\#R_k$.

I_{10} $D(m)$ Subtract 1 from $\#R_m$.

I_{11} $J(m + 1, 7)$ Go to instruction I_7. ∎

Theorem 5: Multiplication is RM-computable.

Proof: We write a program P such that the P-computation with inputs a, b halts with $a \times b$ in R_1. *Idea:* Add a to itself b times. First we give an overview of registers $R_1, R_2,$ and R_3 and write a flow diagram:

R_1	R_2	R_3	Flow Diagram
a	b	0	(1) Is $a = 0$? If *Yes*, halt; $a \times b$ is in R_1.
0	b	a	(2) Put a in R_3, 0 in R_1.
a	$b - 1$	a	(3) Is $\#R_2 = 0$? If *Yes*, halt; $a \times b$ is in R_1.
$2 \times a$	$b - 2$	a	(4) Add a to $\#R_1$.
$\vdots$	$\vdots$		(5) Subtract 1 from $\#R_2$ and go to step 3.
$\vdots$	$\vdots$		
$b \times a$	0	a	

The required program (with comments) is as follows:

I_1 $J(1, 22)$ If $a = 0$, halt.

$\left.\begin{array}{l} I_2 \\ \vdots \\ \vdots \\ \vdots \\ I_7 \end{array}\right\}$ MOVE$(1, 3; 4)$

Obtain the configuration R_1 R_2 R_3 R_4
                          0 b a 0
Note: It is necessary to replace each jump instruction $J(k, q)$ in the program MOVE$(1, 3; 4)$ with $J(k, q + 1)$.

I_8 $J(2, 22)$ Is $\#R_2 = 0$? If *Yes*, halt.

$\left.\begin{array}{l} I_9 \\ \vdots \\ \vdots \\ \vdots \\ I_{19} \end{array}\right\}$ ADD$(3, 1; 4)$

Note: It is necessary to replace each jump instruction $J(k, q)$ in the program ADD$(3, 1; 4)$ with $J(k, q + 8)$.

I_{20} $D(2)$ Subtract 1 from $\#R_2$.

I_{21} $J(4, 8)$ Go to instruction I_8 and check whether a has been added to itself b times. ∎

The subroutine MOVE$(j, k; m)$ moves the number in register R_j to R_k, but in the process destroys the number in register R_j. It is convenient to have a subroutine that transfers the number in register R_j to register R_k while keeping the number in R_j unchanged. Such a subroutine is easily obtained by a slight modification of the subroutine ADD$(j, k; m)$.

Lemma 3 (Subroutine COPY(j, k; m)): Let j, k, m, and $m + 1$ be distinct positive integers. There is a program COPY$(j, k; m)$, with 12 instructions, that transfers the number in register R_j to register R_k and puts 0 in registers R_m and R_{m+1}. The numbers in all other registers, including R_j, are left unchanged.

Proof: The required program (with comments) is as follows:

I_1 Z(k)   Put 0 in R_k.

$\left.\begin{array}{c} I_2 \\ \vdots \\ \vdots \\ \vdots \\ I_{12} \end{array}\right\}$ ADD$(j, k; m)$ Add #R_j to #R_k.

Note: It is necessary to replace each jump instruction $J(k, q)$ in the program ADD$(j, k; m)$ with $J(k, q + 1)$. ∎

—————————*Exercises for Section 9.1*—————————

1. Show that each of the following functions is RM-computable (i.e., write a program P that computes the given function):
   (1) $F(a) = 2 \times a$.
   (2) $F(a)$ = remainder when a is divided by 3.
   (3) $F(a)$ = quotient when a is divided by 3.
2. Show that each of the following functions is RM-computable:
   (1) $a \dotdiv b = \begin{cases} a - b & b \le a \\ 0 & b > a. \end{cases}$
   (2) QT(a, b) = quotient when a is divided by $b \ne 0$ (QT$(a, 0) = 0$).
   (3) RM(a, b) = remainder when a is divided by $b \ne 0$ (RM$(a, 0) = 0$).
   (4) $sg(a) = \begin{cases} 0 & a = 0 \\ 1 & a \ne 0. \end{cases}$
   (5) $K_=(a, b) = \begin{cases} 0 & a = b \\ 1 & a \ne b. \end{cases}$
3. Write a program P such that given input $a \in \mathbb{N}$, (1) if a is even, the P-computation with input a halts with output $a/2$; (2) if a is odd, the P-computation with input a does not halt.
4. (*Zero instructions superfluous*) Assume 0 is in register R_m. Without using zero instructions Z(k), write a program ZERO$(j; m)$ that puts 0 in register R_j and leaves all other registers unchanged.

—————————— $\boxed{9.2}$ ——————————

Recursive ⇒ RM-Computable

In this section, we will show that every recursive function is RM-computable. The proof is by induction on recursive functions (see Section 1.8). We have already seen that the starting functions are RM-computable (see Section 9.1); hence it remains to consider composition and minimalization. We begin with some definitions related to the idea of executing one program followed by another program.

Definition 1: A program $P = I_1, \ldots, I_t$ is in *standard form* if for every jump instruction $J(k, q)$ we have $q \le t + 1$.

Lemma 1: Let P be a program that computes a function F. Then there is a program P^* in standard form that also computes F.

We leave the proof to the reader. As a consequence of this lemma, we can assume that all of the programs we work with are in standard form.

Definition 2: Let P and Q be programs, both in standard form; assume P has s instructions and Q has t instructions. The *join of P and Q*, written PQ, is the program $I_1, \ldots, I_s, I_{s+1}, \ldots, I_{s+t}$, where $I_1, \ldots, I_s$ is program P and $I_{s+1}, \ldots, I_{s+t}$ is program Q, except that each jump instruction $J(k, q)$ of Q is replaced with the jump instruction $J(k, q + s)$. *Idea:* In a computation, program P is executed first, followed by program Q. Note that program PQ is in standard form.

Lemma 2: The join operation is associative. More precisely, if P, Q, and R are programs in standard form, then $(PQ)R$ and $P(QR)$ are the same programs (and hence parentheses may be omitted).

Proof: Assume P has s instructions and Q has t instructions. We obtain the program $(PQ)R$ as follows: First obtain PQ by replacing each jump instruction $J(k, q)$ of Q with $J(k, q + s)$; then replace each jump instruction $J(k, q)$ of R with $J(k, q + s + t)$.

On the other hand, program $P(QR)$ is obtained as follows: First obtain QR by replacing each jump instruction $J(k, q)$ of R with $J(k, q + t)$; then replace each jump instruction $J(k, q)$ of QR with $J(k, q + s)$. This amounts to replacing each jump instruction $J(k, q)$ of Q with $J(k, q + s)$ and each jump instruction of R with $J(k, q + s + t)$; this is the same as the program $(PQ)R$. ∎

Definition 3: Let P be a program. The *width* of P, denoted $\lambda(P)$, is defined by

$$\lambda(P) = \text{\# of the largest register mentioned in program } P.$$

Thus if $S(k)$, $Z(k)$, $D(k)$, or $J(k, q)$ is an instruction of P, then $k \le \lambda(P)$. *Idea:* For $w > \max\{\lambda(P), n\}$, the number in register R_w has no influence on a P-computation with n inputs. We also note that for $m > \max\{\lambda(P), n\}$, we can safely combine program P with any of the subroutines $\text{MOVE}(j, k; m)$, $\text{ADD}(j, k; m)$, or $\text{COPY}(j, k; m)$.

Lemma 3: For each $w \ge 1$, there is a program P such that the P-computation with any inputs halts with 0 in registers $R_1, \ldots, R_w$. This program, denoted $\text{CLEAR}(w)$, has w instructions.

Proof: The required program $\text{CLEAR}(w)$ is

I_1 $Z(1)$

⋮

I_w $Z(w)$. ∎

Definition 4: Let P be a program and let n, w, z, and m be positive integers with $w > \max \{\lambda(P), n\}$ and $m > w + n$. We use

$$P[w + 1, \ldots, w + n; z; m]$$

to denote the join of the following programs:

CLEAR(w)
COPY($w + 1, 1; m$)
$\vdots$
COPY($w + n, n; m$)
P
MOVE($1, z; m$).

What does the program $P[w + 1, \ldots, w + n; z; m]$ do? Suppose P computes an n-ary function F, and assume $a_1, \ldots, a_n$ are in registers $R_{w+1}, \ldots, R_{w+n}$ respectively. A picture of the registers after CLEAR(w) is executed is as follows:

R_1	$\cdots$	R_n	$\cdots$	R_w	R_{w+1}	$\cdots$	R_{w+n}	$\cdots$	R_m	$\cdots$
0	$\cdots$	0	$\cdots$	0	a_1	$\cdots$	a_n	$\cdots$	*	$\cdots$

After the n COPY programs are executed, we have the following:

R_1	$\cdots$	R_n	$\cdots$	R_w	R_{w+1}	$\cdots$	R_{w+n}	$\cdots$	R_m	$\cdots$
a_1	$\cdots$	a_n	$\cdots$	0	a_1	$\cdots$	a_n	$\cdots$	0	$\cdots$

After P and MOVE($1, z; m$) are executed, $P[w + 1, \ldots, w + n; z; m]$ halts with $F(a_1, \ldots, a_n)$ in register R_z.

Theorem 1 (Composition): Let G be a k-ary function, let $H_1, \ldots, H_k$ be n-ary functions, and define an n-ary function F by

$$F(a_1, \ldots, a_n) = G(H_1(a_1, \ldots, a_n), \ldots, H_k(a_1, \ldots, a_n)).$$

If $G, H_1, \ldots, H_k$ are all RM-computable, then F is RM-computable.

Proof: Let $Q, P_1, \ldots, P_k$ be programs, each in standard form, that compute $G, H_1, \ldots, H_k$ respectively. We use the following idea to write a program P that computes F: To compute $F(a_1, \ldots, a_n)$, use program P_i to compute $H_i(a_1, \ldots, a_n)$ for $1 \le i \le k$ and then use program Q to compute $G(H_1(a_1, \ldots, a_n), \ldots, H_k(a_1, \ldots, a_n))$. Let

$$w = \max \{n, k, \lambda(P_1), \ldots, \lambda(P_k), \lambda(Q)\};$$

for $m > w$, register R_m does not occur in any of the programs $P_1, \ldots, P_k, Q$.

Let $m = w + n + k + 1$. Bookkeeping details and use of registers are as follows:

(1) Save inputs $a_1, \ldots, a_n$ in registers $R_{w+1}, \ldots, R_{w+n}$ respectively.
(2) For $1 \le i \le k$, use program P_i to compute $H_i(a_1, \ldots, a_n) = b_i$ and store b_i in register R_{w+n+i}.
(3) Use program Q to compute $G(b_1, \ldots, b_k) = F(a_1, \ldots, a_n)$.

R_1	$\cdots$	R_n	$\cdots$	R_w	R_{w+1}	$\cdots$	R_{w+n}	R_{w+n+1}	$\cdots$	R_{w+n+k}	R_m
a_1	$\cdots$	a_n	$\cdots$		a_1	$\cdots$	a_n	b_1	$\cdots$	b_k	0

The required program P to compute F is the join of the following programs:

MOVE$(1, w + 1; m)$	Put a_1 in R_{w+1}.
$\vdots$	$\vdots$
MOVE$(n, w + n; m)$	Put a_n in R_{w+n}.
$P_1[w + 1, \ldots, w + n; w + n + 1; m]$	Compute $H_1(a_1, \ldots, a_n)$ and store in R_{w+n+1}.
$\vdots$	$\vdots$
$P_k[w + 1, \ldots, w + n; w + n + k; m]$	Compute $H_k(a_1, \ldots, a_n)$ and store in R_{w+n+k}.
$Q[w + n + 1, \ldots, w + n + k; 1; m]$	Compute $G(H_1(a_1, \ldots, a_n), \ldots, H_k(a_1, \ldots, a_n))$ and store in R_1. ∎

Theorem 2 (Minimalization): Let F be an n-ary function that is the minimalization of an $n + 1$-ary regular function G. If G is RM-computable, then F is RM-computable.

Proof: By definition, $F(a_1, \ldots, a_n) = \mu b[G(a_1, \ldots, a_n, b) = 0]$; let P be a program in standard form that computes G and let $w = \max\{n + 1, \lambda(P)\}$. The idea for writing a program to compute F is as follows: Use P to compute $G(a_1, \ldots, a_n, b)$ for $b = 0, 1, \ldots$ until we reach the first b for which $G(a_1, \ldots, a_n, b) = 0$. A flow diagram, ignoring bookkeeping details, is as follows:

(1) Input $a_1, \ldots, a_n \in \mathbb{N}$ and let $b = 0$.
(2) Use P to compute $G(a_1, \ldots, a_n, b)$.
(3) Is $G(a_1, \ldots, a_n, b) = 0$? If *Yes*, $F(a_1, \ldots, a_n) = b$.
(4) If *No*, add 1 to b and go to (2).

The bookkeeping details are as follows: Save the inputs $a_1, \ldots, a_n$ in registers $R_{w+1}, \ldots, R_{w+n}$ respectively, and use register R_{w+n+1} for the current value of

b (initially $b = 0$). For convenience, let $m = w + n + 2$, and note that register R_m available for MOVE subroutines and so on. Registers are used as follows:

R_1	$\cdots$	R_n	$\cdots$	R_w	R_{w+1}	$\cdots$	R_{w+n}	R_{w+n+1}	R_m	$\cdots$
a_1	$\cdots$	a_n	$\cdots$		a_1	$\cdots$	a_n	b	0	$\cdots$

Let P^* be the join of the programs

   MOVE$(1, w + 1; m)$
   $\vdots$
   MOVE$(n, w + n; m)$
   $P[w + 1, \ldots, w + n, w + n + 1; 1; m]$.

Informally, P^* moves the numbers $a_1, \ldots, a_n$ in registers $R_1, \ldots, R_n$ to registers $R_{w+1}, \ldots, R_{w+n}$, computes $G(a_1, \ldots, a_n, b)$, where b is the number in register R_{w+n+1}, and then stores the result in R_1. Let t be the total number of instructions in P^*, and write P^* as $I_1, \ldots, I_k, \ldots, I_t$, where I_k is the first instruction of $P[w + 1, \ldots, w + n, w + n + 1; 1; m]$ $(k = 6n + 1)$. We now add nine instructions $I_{t+1}, \ldots, I_{t+9}$ to P^* to obtain a program that computes F:

$\left.\begin{array}{c} I_1 \\ \vdots \\ I_k \\ \vdots \\ I_t \end{array}\right\} P^*$ Program P^*; halts with $G(a_1, \ldots, a_n, b)$ in R_1 and the current value of b in R_{w+n+1}.

I_{t+1} $J(1, t + 4)$ Is $G(a_1, \ldots, a_n, b) = 0$? if *Yes*, go to instruction I_{t+4}.

I_{t+2} $S(w + n + 1)$ Add 1 to b.

I_{t+3} $J(m, k)$ Go to instruction I_k.

$\left.\begin{array}{c} I_{t+4} \\ \vdots \\ I_{t+9} \end{array}\right\}$ MOVE$(w + n + 1; 1; m)$ Put b in register R_1 and halt. *Note:* It is necessary to replace each jump instruction $J(k, q)$ of MOVE$(w + n + 1; 1; m)$ with $J(k, q + t + 3)$. ∎

Theorem 3: Every recursive function is RM-computable.

Proof: The proof is by induction on recursive functions. Starting functions are RM-computable (proved in Section 9.1); the composition of RM-computable functions is RM-computable (Theorem 1); the minimalization of a regular RM-computable function is RM-computable (Theorem 2). ∎

——————————*The Halting Problem*——————————

We will now discuss an important problem about register machines, called the *halting problem*. Consideration of this problem will lead us to a rather natural example of a 1-ary function that is not RM-computable. We note that a simple counting argument shows the existence of uncountably many 1-ary functions that are not RM-computable.

Definition 5: The *halting problem* (*for register machines*) asks whether there is an algorithm that, given a program P and a natural number a, decides whether the P-computation with input a halts. If there is such an algorithm, we say that the halting problem is *solvable*; otherwise, the halting problem is *unsolvable*.

We will prove that the halting problem is recursively unsolvable. This result was first obtained by Turing, except that his result applied to Turing machines instead of to register machines. We proceed as follows: Reduce the solvability of the halting problem to the decidability of a 2-ary relation HLT; show that the relation HLT is not recursive. This proves that the halting problem is recursively unsolvable. Church's thesis gives the stronger conclusion that the halting problem is algorithmically unsolvable. Before defining HLT, we need an effective coding of **P**, the collection of all register machine programs.

Theorem 4 (Effective coding of P): There is a one-to-one function $g: \mathbf{P} \to \mathbb{N}$ satisfying the following two conditions:

Gö1 There is an algorithm that, given a program P, computes $g(P)$.

Gö2 There is an algorithm that, given a number a, decides whether there is a program P such that $g(P) = a$. Moreover, if the answer is YES, the algorithm finds the program P such that $g(P) = a$.

If P is a program, $g(P)$ is called the *code* of P. The reader may want to construct his or her own coding function, perhaps based on the methods of Section 7.2 (the prime power method) or the methods of Chapter 8 (Gödel's β-function). A proof of Theorem 4 will be given in the next section; for now, let us assume this result.

Definition 6: HLT and K are relations on $\mathbb{N}$ defined as follows:

$$\text{HLT}(e, a) \Leftrightarrow e \text{ is the code of a program } P \text{ and}$$
$$\text{the } P\text{-computation with input } a \text{ halts.}$$
$$K(e) \Leftrightarrow \text{HLT}(e, e).$$

Note that K is the diagonalization of HLT; thus $K(e)$ if and only if e is the code of a program P and the P-computation with input e halts. The notation K for the diagonalization of HLT is standard in recursion theory and will be used for the remainder of this chapter. There is, of course, the danger that K will be interpreted as one of the Cantor pairing functions J, K, and L; we rely on the context to avoid such confusion.

The next theorem gives the connection between the 2-ary relation HLT and the halting problem. The proof is left as an exercise for the reader.

Theorem 5 (Reduction): The 2-ary relation HLT is decidable if and only if the halting problem is solvable.

Theorem 5 is the basis of the following terminology: The halting problem is *recursively solvable* or *recursively unsolvable* depending on whether HLT is recursive or not recursive. We will prove that HLT is *not* recursive, so that the halting problem is recursively unsolvable. We begin by showing that the characteristic function of K is not RM-computable (and hence not recursive).

Lemma 4 (Diagonalization): There is no program P such that for all $a \in \mathbb{N}$,

(1) If $K(a)$, then the P-computation with input a does not halt.
(2) If $\neg K(a)$, then the P-computation with input a halts.

Proof: Suppose P exists, and let e be the code of P; then from the definition of K, we have

$$K(e) \Leftrightarrow \text{the } P\text{-computation with input } e \text{ halts.}$$

But from (1) and (2) with $a = e$, we obtain

$$K(e) \Leftrightarrow \text{the } P\text{-computation with input } e \text{ does not halt,}$$

and this gives the desired contradiction. ∎

Theorem 6: The characteristic function of K is not RM-computable and not recursive.

Proof: The characteristic function of K is

$$\chi(a) = \begin{cases} 0 & K(a) \\ 1 & \neg K(a). \end{cases}$$

Suppose χ is RM-computable, and let $Q = I_1, \ldots, I_t$ be a program in standard form that computes χ. Let P be the program obtained by adding the following instruction to Q:

$$I_{t+1} = J(1, t + 1).$$

We now show that P satisfies (1) and (2) of Lemma 4; this gives the required contradiction. To check (1), assume $K(a)$. Then $\chi(a) = 0$; hence the Q-computation with input a halts with output 0. But the additional instruction I_{t+1} causes program P to go into an infinite loop. Thus the P-computation with input a does not halt. To prove (2), assume $\neg K(a)$. Then $\chi(a) = 1$; hence the Q-computation with input a halts with output 1. It is easy to see that program P with the additional instruction I_{t+1} also halts (with the same output, 1). So the P-computation with input a halts as required.

Since the characteristic function of K is not RM-computable, it cannot be recursive (Theorem 3). ∎

Corollary 1: The relation K is not recursive.

Proof: The characteristic function of K is not recursive. ∎

Theorem 7 (Turing; recursive unsolvability of the halting problem): HLT is not recursive.

Proof: $K(e) \Leftrightarrow \text{HLT}(e, e)$; if HLT were recursive, then K would be recursive. But K is not recursive, so HLT cannot be recursive, as required. ∎

Theorem 8 (Unsolvability of the halting problem): Assume Church's thesis. Then the halting problem for register machines is unsolvable.

Proof: We have: HLT is not a recursive relation (Theorem 7); HLT is not a decidable relation (Church's thesis); the halting problem is unsolvable (reduction theorem). ∎

We have proved that K and HLT are not recursive. But both are RE, and we will prove this in the next section. We will also show that there is a *specific* program P^* for the register machine such that the halting problem for this one program is unsolvable.

———————————*Exercises for Section 9.2*———————————

1. Prove Lemma 1.
2. Show that primitive recursion is RM-computable. More precisely, let G be n-ary, let H be $n + 2$-ary, and let

$$F(a_1, \ldots, a_n, 0) = G(a_1, \ldots, a_n)$$
$$F(a_1, \ldots, a_n, b + 1) = H(a_1, \ldots, a_n, b, F(a_1, \ldots, a_n, b))$$

   Show that if G and H are RM-computable, then F is RM-computable.
3. Show that HLT is semidecidable.
4. Prove Theorem 5 (reduction). (*Hint:* See Lemma 1 in Section 7.5.)
5. Prove that the number of 1-ary RM-computable functions is countable. Conclude that there are uncountably many 1-ary functions that are not RM-computable.
6. Define $F: \mathbb{N} \to \mathbb{N}$ as follows: If $K(a)$, then

$$F(a) = (\text{output of the } P\text{-computation with input } a) + 1,$$

   where P is the program with code a; if $\neg K(a)$, $F(a) = 0$. Show that F is not RM-computable.

$$\boxed{9.3}$$

Kleene Computation Relation T_n

In this section, we will prove that every RM-computable function is recursive (this is the converse of the main result in Section 9.2). Recall that in Chapter 8 we recursively coded all formulas and proofs, eventually showing that the 2-ary relation PRF_Γ is recursive. The situation in this section is similar, except that we now recursively code programs and computations, eventually showing that for each $n \geq 1$, the following $n + 2$-ary relation T_n is recursive:

$$T_n(e, a_1, \ldots, a_n, c) \Leftrightarrow e \text{ is the code of a program } P, \text{ and } c$$
$$\text{is the code of the } P\text{-computation with}$$
$$\text{inputs } a_1, \ldots, a_n \text{ that halts.}$$

The recursive relation T_n, known as the *Kleene computation relation*, is a powerful and fundamental tool in recursion theory. We also show that there is a 1-ary recursive function U such that if c is the code of a computation that halts, then

$$U(c) = \text{output of the computation.}$$

Once T_n and U are available, the proof that every RM-computable function is recursive proceeds as follows: Given an n-ary RM-computable function F, there is a number e (e = code of a program that computes F) such that for all $a_1, \ldots, a_n \in \mathbb{N}$,

$$F(a_1, \ldots, a_n) = U(\mu c T_n(e, a_1, \ldots, a_n, c)).$$

Since U and T_n are recursive, F is recursive as required. The fact that every n-ary recursive function can be expressed in terms of T_n and U is called the *Kleene normal-form theorem*. Let us also note that the recursive relation T_1 shows that HLT is RE: $\text{HLT}(e, a) \Leftrightarrow \exists c T_1(e, a, c)$.

Throughout this chapter, we will use the following recursive functions and relations for coding and decoding; proofs are given in Section 8.3.

Theorem 1 (Coding/decoding): There is a 2-ary recursive function β (Gödel's β-function), a 1-ary recursive relation CD, and for each $n \geq 1$ an n-ary recursive function $\langle \cdots \rangle_n$, such that the following hold:

(1) If $c = \langle a_1, \ldots, a_n \rangle_n$, then $\beta(c, 0) = n$ and $\beta(c, k) = a_k$ for $1 \leq k \leq n$.

(2) If $\langle a_1, \ldots, a_n \rangle_n = \langle b_1, \ldots, b_m \rangle_m$, then $n = m$ and $a_k = b_k$ for $1 \leq k \leq n$.

(3) If $c = \langle a_1, \ldots, a_n \rangle_n$, then $n < c$ and $a_k < c$ for $1 \leq k \leq n$.

(4) $\text{CD}(c) \Leftrightarrow$ there exists $n \in \mathbb{N}^+$ and $a_1, \ldots, a_n \in \mathbb{N}$ such that $c = \langle a_1, \ldots, a_n \rangle_n$.

Definition 1: Let I be an instruction. The *code of I*, denoted $\#(I)$, is the natural number obtained as follows:

$$\#(S(k)) = \langle k, 0 \rangle_2 \quad \#(D(k)) = \langle k, 2 \rangle_2$$
$$\#(Z(k)) = \langle k, 1 \rangle_2 \quad \#(J(k, q)) = \langle k, q, 0 \rangle_3.$$

Given a program $P = I_1, \ldots, I_t$, the *code of P*, denoted $\#P$, is defined by

$$\#P = \langle \#(I_1), \ldots, \#(I_t) \rangle_t.$$

Lemma 1: The following 1-ary relations are recursive:

(1) INC(a) $\Leftrightarrow$ a is the code of an increase instruction.
(2) ZERO(a) $\Leftrightarrow$ a is the code of a zero instruction.
(3) DEC(a) $\Leftrightarrow$ a is the code of a decrease instruction.
(4) JUMP(a) $\Leftrightarrow$ a is the code of a jump instruction.
(5) PGM(e) $\Leftrightarrow$ e is the code of a program.

Proof: We will prove (4) and (5) and leave the others as exercises for the reader. For (4),

$$\text{JUMP}(a) \Leftrightarrow \text{CD}(a) \wedge (\beta(a, 0) = 3) \wedge (\beta(a, 1) \geq 1)$$
$$\wedge (\beta(a, 2) \geq 1) \wedge (\beta(a, 3) = 0).$$

Before proving (5), we define an auxiliary recursive relation INST:

$$\text{INST}(a) \Leftrightarrow a \text{ is the code of an instruction}$$
$$\Leftrightarrow \text{INC}(a) \vee \text{ZERO}(a) \vee \text{DEC}(a) \vee \text{JUMP}(a).$$

$\text{PGM}(e) \Leftrightarrow \text{CD}(e) \wedge \forall k < \beta(e, 0)[\text{INST}(\beta(e, k + 1))].$ ∎

The next result states that we have an effective coding of all programs; consequently, questions about programs can be transformed into questions about relations on $\mathbb{N}$. Let **P** be the collection of all programs, and define $g: \mathbf{P} \to \mathbb{N}$ by

$$g(P) = \text{code of } P$$
$$= \langle \#(I_1), \ldots, \#(I_t) \rangle_t, \text{ where } P \text{ is } I_1, \ldots, I_t.$$

Theorem 2 (Effective coding of **P***):* The function g is one-to-one and satisfies these two conditions:

Gö1   There is an algorithm that, given a program P, calculates $g(P)$.
Gö2   There is an algorithm that, given a number e, decides whether there is a program P such that $g(P) = e$; moreover, if the answer is YES, the algorithm finds the program P.

Outline of proof Use the recursive functions $\langle \cdot \cdot \cdot \rangle_n$ ($n \geq 1$) and β and the recursive relation CD; details are left to the reader. ∎

Note: Theorem 2 was used in Section 9.2 to show that the halting problem is solvable if and only if HLT is decidable.

Lemma 2: There is a 1-ary recursive function λ^* such that if e is the code of a program P, then $\lambda^*(e) = \lambda(P)$. Informally, λ^* is a recursive function that calculates the number of the largest register mentioned in a program.

Proof: Suppose we have a program with instructions $I_1, \ldots, I_t$, and let $e = \langle \#(I_1), \ldots, \#(I_t) \rangle_t$. Informally, $\lambda^*(e)$ is the largest number that occurs among the *first* entries of $\#(I_1), \ldots, \#(I_t)$. We have

$$\lambda^*(e) = \begin{cases} \mu w[\forall j < \beta(e, 0)(\beta(\beta(e, j + 1), 1) \leq w)] & \text{PGM}(e) \\ 0 & \neg\text{PGM}(e). \end{cases}$$ ∎

We want to assign a code to each P-computation with inputs $a_1, \ldots, a_n$ that halts. Up to now, our definition of a computation has been informal, but for purposes of coding we need a precise definition. The idea, in a nutshell, is this: Let P be a program, let $a_1, \ldots, a_n \in \mathbb{N}$, and let $w = \max\{\lambda(P), n\}$. A *P-computation with inputs $a_1, \ldots, a_n$ that halts* is a finite sequence $s_1, \ldots, s_m$ of w-steps; these w-steps are unique and are related in a special way by the program P. Informally, each w-step is a record of the numbers in registers $R_1, \ldots, R_w$ and the next instruction. Therefore, let us begin with a discussion of w-steps.

Definition 2: Let $w \geq 1$. A *w-step* is a $w + 1$-tuple of natural numbers whose last coordinate is ≥ 1. A w-step s is written in the form

$$s = \langle r_1, \ldots, r_w; j \rangle.$$

The idea is that the numbers $r_1, \ldots, r_w$ are in registers $R_1, \ldots, R_w$ respectively, and that I_j is the next instruction. Observe that s tells us all of the essential information at a step of a computation. Of special interest is the w-step

$$s_1 = \langle a_1, \ldots, a_n, 0, \ldots, 0; 1 \rangle,$$

called an *initial w-step*. Informally, s_1 describes the situation at the beginning of a P-computation: The inputs $a_1, \ldots, a_n$ are in registers $R_1, \ldots, R_n$ respectively, 0 is in registers $R_{n+1}, \ldots, R_w$, and the next (= first) instruction is I_1.

Definition 3: Let $s = \langle r_1, \ldots, r_w; j \rangle$ be a w-step. The *code* of s, denoted $\#(s)$, is the number defined by

$$\#(s) = \langle r_1, \ldots, r_w, j \rangle_{w+1}.$$

For example, if a is the code of the w-step $s = \langle r_1, \ldots, r_w; j \rangle$, then $\beta(a, 0) = w + 1$, $\beta(a, k) = r_k$ for $1 \leq k \leq w$, and $\beta(a, w + 1) = j$.

Lemma 3: The following relations are recursive ($n \geq 1$):

(1) $\text{STEP}(w, a) \Leftrightarrow w \geq 1$ and a is the code of a w-step.

(2) $\text{INSTEP}_n(w, a, a_1, \ldots, a_n) \Leftrightarrow w \geq n$ and a is the code of the initial w-step

$$s_1 = \langle a_1, \ldots, a_n, 0, \ldots, 0; 1 \rangle.$$

Proof: We have explicit definitions as follows:

$\text{STEP}(w, a) \Leftrightarrow (w \geq 1) \wedge \text{CD}(a) \wedge (\beta(a, 0) = w + 1) \wedge (\beta(a, w + 1) \geq 1).$

$\text{INSTEP}_n(w, a, a_1, \ldots, a_n) \Leftrightarrow \text{STEP}(w, a) \wedge (w \geq n) \wedge (\beta(a, w + 1) = 1) \wedge$
$\qquad\qquad\qquad \forall k < n[\beta(a, k + 1) = a_{k+1}] \wedge$
$\qquad\qquad\qquad \forall k < w + 1[(k > n) \rightarrow (\beta(a, k) = 0))]. \blacksquare$

Consider the following situation: We have a program $P = I_1, \ldots, I_t$ and a w-step $s = \langle r_1, \ldots, r_w; j \rangle$ with $j \leq t$; the next instruction executed is I_j. It is convenient to have recursive functions that tell us the code of I_j (denoted CDNXI) and also the number of the register in I_j (denoted RGNXI).

Lemma 4: There are 2-ary recursive functions CDNXI(e, a) and RGNXI(e, a) such that if e is the code of a program $P = I_1, \ldots, I_t$ and a is the code of a w-step $\langle r_1, \ldots, r_w; j \rangle$ with $j \leq t$, then

CDNXI(e, a) = code of the next instruction I_j

RGNXI(e, a) = number of the register in the next instruction I_j.

Proof: Let $e = \langle \#(I_1), \ldots, \#(I_t) \rangle_t$, $a = \langle r_1, \ldots, r_w, j \rangle_{w+1}$. Then $\beta(e, j)$ is the code of I_j and $\beta(a, \beta(a, 0)) = j$; hence we have

$$\text{CDNXI}(e, a) = \beta(e, j) = \beta(e, \beta(a, \beta(a, 0)))$$
$$\text{RGNXI}(e, a) = \beta(\text{CDNXI}(e, a), 1). \blacksquare$$

Example 1:

We will illustrate the use of CDNXI and RGNXI. Suppose e is the code of a program $P = I_1, \ldots, I_t$ and a is the code of a w-step $\langle r_1, \ldots, r_w; j \rangle$ with $j \leq t$. The following hold: JUMP(CDNXI(e, a)) $\Leftrightarrow I_j$ is a jump instruction; if DEC(CDNXI(e, a)) and RGNXI$(e, a) = k$, then I_j is the decrease instruction $D(k)$. ☐

The next definition plays a key role in the precise definition of a P-computation.

Definition 4: Let $P = I_1, \ldots, I_t$ be a program, let $w \geq \lambda(P)$, and let

$$s = \langle r_1, \ldots, r_k, \ldots, r_w; j \rangle$$

be a w-step with $j \leq t$. The *P-successor of s* is the (unique) w-step s' obtained from s according to one of four cases:

S If $I_j = S(k)$, s' is obtained from s by replacing the last entry j with $j + 1$ and replacing r_k with $r_k + 1$. Thus $s' = \langle r_1, \ldots, r_k + 1, \ldots, r_w; j + 1 \rangle$. (Informally, the number in register R_k is increased by 1 and the next instruction is I_{j+1}.)

Z If $I_j = Z(k)$, s' is obtained from s by replacing the last entry j with $j + 1$ and replacing r_k with 0. (Informally, the number in register R_k becomes 0 and the next instruction is I_{j+1}.)

D If $I_j = D(k)$, s' is obtained from s by replacing the last entry j with $j + 1$ and replacing r_k with $r_k \dot{-} 1$ (i.e., if $r_k > 0$, replace r_k with $r_k - 1$, and if $r_k = 0$, do nothing to r_k).

J If $I_j = J(k, q)$, s' is obtained from s by replacing the last entry j with $j + 1$ if $r_k \neq 0$ and with q otherwise.

If $s = \langle r_1, \ldots, r_w; j \rangle$ is a w-step with $j > t$, we say that s is *P-terminal*.

Definition 5: SUCC(e, w, a, b) is the 4-ary relation defined by

$$SUCC(e, w, a, b) \Leftrightarrow e \text{ is the code of a program } P = I_1, \ldots, I_t, w \geq \lambda(P),$$
$$a \text{ is the code of a } w\text{-step } s = \langle r_1, \ldots, r_w; j \rangle \text{ with}$$
$$j \leq t, \text{ and } b \text{ is the code of the } P\text{-successor } s' \text{ of } s.$$

Lemma 5: The relation SUCC(e, w, a, b) is recursive.

Proof:

SUCC(e, w, a, b)

$\Leftrightarrow$ PGM(e) $\wedge$ (w $\geq$ $\lambda*(e)$) $\wedge$ STEP(w, a)

$\wedge$ ($\beta(a, w + 1) \leq \beta(e, 0)$) $\wedge$ STEP(w, b)

$\wedge$ [Z(e, w, a, b) $\vee$ S(e, w, a, b) $\vee$ D(e, w, a, b) $\vee$ J(e, w, a, b)].

It remains to describe S, Z, D, and J and to prove each recursive. Informally, these four relations correspond to the four cases for constructing the P-successor s' of s. We give the details for D and J and leave S and Z to the reader. Let $e = \langle \#(I_1), \ldots, \#(I_t) \rangle_t$, $a = \langle r_1, \ldots, r_w, j \rangle_{w+1}$; we use the recursive functions CDNXI(e, a) and RGNXI(e, a), defined in Lemma 4.

D(e, w, a, b)

$\Leftrightarrow$ DEC(CDNXI(e, a)) $\wedge$ [$\beta(b, w + 1) = \beta(a, w + 1) + 1$]

$\wedge$ [$\beta(b, RGNXI(e, a)) = \beta(a, RGNXI(e, a)) \dot{-} 1$]

$\wedge$ $\forall i < w[(i + 1 \neq RGNXI(e, a)) \rightarrow (\beta(b, i + 1) = \beta(a, i + 1))]$.

J(e, w, a, b)

$\Leftrightarrow$ JUMP(CDNXI(e, a)) $\wedge$ $\forall i < w[\beta(b, i + 1) = \beta(a, i + 1)]$

$\wedge$ [(($\beta(a, RGNXI(e, a)) \neq 0$) $\wedge$ ($\beta(b, w + 1) = \beta(a, w + 1) + 1$))

$\vee$ (($\beta(a, RGNXI(e, a)) = 0$) $\wedge$ ($\beta(b, w + 1) = \beta(CDNXI(e, a), 2)$))]. ∎

Definition 6: Let $P = I_1, \ldots, I_t$ be a program, let $a_1, \ldots, a_n$ be inputs, and let $w = \max\{n, \lambda(P)\}$. Given the initial w-step $s_1 = \langle a_1, \ldots, a_n, 0, \ldots, 0; 1 \rangle$, exactly one of the following holds:

(1) There is a finite sequence $s_1, \ldots, s_m$ of w-steps such that for $1 \leq i \leq m - 1$, s_{i+1} is the P-successor of s_i, and s_m is P-terminal (i.e., $\beta(\#(s_m), w + 1) > t$).

(2) There is an infinite sequence $s_1, s_2, \ldots$ of w-steps such that for all $i \geq 1$, s_{i+1} is the P-successor of s_i.

In either case, the sequence of w-steps is unique and is called the P-

computation with inputs $a_1, \ldots, a_n$. This *P*-computation with inputs $a_1, \ldots, a_n$ *halts* if (1) holds. In this case, the *output* is b_1, where $s_m = \langle b_1, \ldots, b_w; j \rangle$ and $j > t$; in other words, the output is $\beta(\#(s_m), 1)$. If $s_1, \ldots, s_m$ is a *P*-computation with inputs that halts, the *code* of this computation is

$$c = \langle \#(s_1), \ldots, \#(s_m) \rangle_m.$$

Definition 7: For each $n \geq 1$, the *Kleene computation relation* T_n is the $n + 2$-ary relation defined by

$T_n(e, a_1, \ldots, a_n, c) \Leftrightarrow e$ is the code of a program *P* and *c* is the code of the
$\qquad\qquad\qquad$ *P*-computation with inputs $a_1, \ldots, a_n$ that halts.

Theorem 3 (Kleene): For each $n \geq 1$, T_n is recursive.

Proof: Fix $n \in \mathbb{N}^+$ and define $F(e) = \max \{n, \lambda^*(e)\}$. *F* is recursive, since $F(e) = \mu w[(n \leq w) \wedge (\lambda^*(e) \leq w)]$; informally, $F(e)$ is the smallest number w such that $n \leq w$ and all registers mentioned in the program with code e are $\leq w$. T_n is recursive as follows:

$T_n(e, a_1, \ldots, a_n, c)$
$\quad \Leftrightarrow \text{PGM}(e) \wedge \text{CD}(c) \wedge [\beta(c, 0) \geq 2]$
$\qquad \wedge \forall i < \beta(c, 0)[\text{STEP}(F(e), \beta(c, i + 1))]$
$\qquad \wedge \text{INSTEP}_n(F(e), \beta(c, 1), a_1, \ldots, a_n)$
$\qquad \wedge \forall i < (\beta(c, 0) \dotminus 1)[\text{SUCC}(e, F(e), \beta(c, i + 1), \beta(c, i + 2))]$
$\qquad \wedge [\beta(\beta(c, \beta(c, 0)), F(e) + 1) > \beta(e, 0)].$ ■

Corollary 1: The relations K and HLT are RE.

Proof: We have

$$\text{HLT}(e, a) \Leftrightarrow \exists c T_1(e, a, c)$$
$$K(a) \Leftrightarrow \exists c T_1(a, a, c). \blacksquare$$

Theorem 4 (Fundamental theorem of recursion theory): There is a 1-ary relation that is RE but not recursive.

Proof: The required relation is K; K is RE by the preceding corollary and is not recursive by Corollary 1 in Section 9.2. ■

At this point, it is worthwhile to step back and compare some of the major results in Chapter 8 with results obtained in these last two sections. Let Γ be a consistent recursively axiomatized extension of NN.

Chapter 8	Chapter 9
$\text{PRF}_\Gamma(a, b)$ is recursive	$T_1(e, a, c)$ is recursive
$\text{THM}_\Gamma(a) \Leftrightarrow \exists b \text{PRF}_\Gamma(a, b)$	$\text{HLT}(e, a) \Leftrightarrow \exists c T_1(e, a, c)$
$\text{THM}_\Gamma(a)$ is RE	$\text{HLT}(e, a)$ is RE
$\text{THM}_\Gamma(a)$ is not recursive	$\text{HLT}(e, a)$ is not recursive

Theorem 5 (Output function): There is a 1-ary recursive function U such that if c is the code of a P-computation with inputs that halts, then $U(c)$ is the output.

Proof: Suppose $c = \langle \#(s_1), \ldots, \#(s_m) \rangle_m$ is the code of a P-computation with inputs that halts, and let $s_m = \langle b_1, \ldots, b_w; j \rangle$, so that the output is b_1. Now $\beta(c, \beta(c, 0)) = \#(s_m)$, so $U(c) = \beta(\beta(c, \beta(c, 0)), 1)$. ∎

The next result summarizes the relationship between T_n, U, and the function F computed by a program P.

Theorem 6: Let F be an n-ary function and let P be a program with code e that computes F. Then for all $a_1, \ldots, a_n \in \mathbb{N}$,

(1) There is a unique number c such that $T_n(e, a_1, \ldots, a_n, c)$.

(2) $F(a_1, \ldots, a_n) = U(c)$.

In other words, $F(a_1, \ldots, a_n) = U(\mu c T_n(e, a_1, \ldots, a_n, c))$.

Proof: Let $a_1, \ldots, a_n \in \mathbb{N}$. Since P computes F, the P-computation with inputs $a_1, \ldots, a_n$ halts with output $F(a_1, \ldots, a_n)$. Let c be the code of this computation. Then $T_n(e, a_1, \ldots, a_n, c)$. Moreover, a P-computation with inputs $a_1, \ldots, a_n$ is unique; hence c is also unique. This completes the proof of (1). Since $U(c)$ is the output of the computation, $F(a_1, \ldots, a_n) = U(c)$, and (2) holds as required. ∎

Corollary 2: Every RM-computable function is recursive.

Proof: Let F be an n-ary RM-computable function. There is a program P with code e that computes F. Theorem 6 now applies, and for all $a_1, \ldots, a_n \in \mathbb{N}$, $F(a_1, \ldots, a_n) = U(\mu c T_n(e, a_1, \ldots, a_n, c))$. Since U and T_n are recursive, F is recursive. ∎

We have now achieved one of the main objectives of this chapter. Corollary 2, combined with the main result in Section 9.2, gives the following.

Theorem 7: An n-ary function F is recursive if and only if it is RM-computable.

Since every recursive function is RM-computable, Theorem 6 gives the following fundamental result on the structure of recursive functions.

Theorem 8 (Kleene normal-form theorem): Let F be an n-ary recursive function. Then there exists $e \in \mathbb{N}$ such that for all $a_1, \ldots, a_n \in \mathbb{N}$,

$$F(a_1, \ldots, a_n) = U(\mu c T_n(e, a_1, \ldots, a_n, c)).$$

Comment We can prove that the Kleene relations T_n and the function U are actually primitive recursive. The Kleene normal-form theorem then tells us that every recursive function can be obtained by at most one application of the minimalization operation (and to a primitive recursive relation).

───── RE Relations and Programs ─────

There is an interesting relationship between programs for the register machine and RE relations. We will discuss the 1-ary case for now; see the exercises and Section 9.4 for the n-ary case. We begin with the following important result.

Theorem 9 (Simulation of an RE relation by a program): Let R be a 1-ary RE relation. Then there is a program P_R for the register machine such that for all $a \in \mathbb{N}$,

$$R(a) \Leftrightarrow \text{the } P_R\text{-computation with input } a \text{ halts.}$$

Proof: Let Q be a recursive relation such that $R(a) \Leftrightarrow \exists b Q(a, b)$, and let $P = I_1, \ldots, I_t$ be a program in standard form that computes the characteristic function K_Q of Q; we will use program P to write a program P_R such that the P_R-computation with input a systematically calculates $K_Q(a, 0), K_Q(a, 1), \ldots$, halting at the first $b \in \mathbb{N}$ such that $K_Q(a, b) = 0$; if no such b exists, the P_R-computation with input a continues forever. Let $w = \lambda(Q)$.

$$\left.\begin{matrix} I_1 \\ \vdots \\ I_6 \end{matrix}\right\} \text{MOVE}(1, w + 1; w + 2) \qquad \begin{matrix} \text{Move input } a \text{ to register } R_{w+1}; \\ \#R_{w+2} = 0. \end{matrix}$$

$$\left.\begin{matrix} I_7 \\ \vdots \\ \vdots \\ I_{t+6} \end{matrix}\right\} P[w + 1, w + 2; 1; w + 3] \qquad \begin{matrix} \text{Use program } P \text{ to compute } K_Q(a, b), \\ \text{where } a = \#R_{w+1}, b = \#R_{w+2}. \end{matrix}$$

I_{t+7} $J(1, t + 10)$ Halt if $K_Q(a, b) = 0$.

I_{t+8} $S(w + 2)$ Add 1 to $\#R_{w+2}$.

I_{t+9} $J(w + 3, 7)$ Go to instruction I_7. ∎

Theorem 9 suggests the following terminology: A program P *simulates* a

1-ary relation R if for all $a \in \mathbb{N}$, $R(a) \Leftrightarrow$ the P-computation with input a halts. We now have the following characterization of the 1-ary RE relations.

Theorem 10 (Characterization of RE relations): Let R be a 1-ary relation. The following are equivalent:

(1) R is RE.

(2) There is a program that simulates R.

(3) There exists $e \in \mathbb{N}$ such that for all $a \in \mathbb{N}$, $R(a) \Leftrightarrow \text{HLT}(e, a)$.

Proof: (1) $\Rightarrow$ (2) by Theorem 9, and (3) $\Rightarrow$ (1) follows from the fact that HLT is RE (also note that $R(a) \Leftrightarrow \exists c T_1(e, a, c)$). To prove (2) $\Rightarrow$ (3), assume P_R is a program that simulates R. Then for all $a \in \mathbb{N}$,

$$R(a) \Leftrightarrow \text{the } P_R\text{-computation with input } a \text{ halts.}$$

Let e be the code of P_R; then for all $a \in \mathbb{N}$,

$$R(a) \Leftrightarrow \text{HLT}(e, a). \quad \blacksquare$$

Corollary 3 (HLT enumerates): HLT is a 2-ary RE relation that enumerates all 1-ary RE relations. In other words, given a 1-ary RE relation R, there exists $e \in \mathbb{N}$ such that for all $a \in \mathbb{N}$, $R(a) \Leftrightarrow \text{HLT}(e, a)$.

Corollary 3 follows immediately from Theorem 10. Let us now use Corollary 3 and a diagonal argument to give another proof that K is not recursive (see Corollary 1 in Section 9.2).

Corollary 4: K is not recursive.

Proof: Suppose K is recursive. Then $\neg K$ is also recursive, and hence RE. Since HLT enumerates, there exists $e \in \mathbb{N}$ such that for all $a \in \mathbb{N}$, $\neg K(a) \Leftrightarrow \text{HLT}(e, a)$. For $a = e$, we obtain $\neg K(e) \Leftrightarrow \text{HLT}(e, e)$. But by the way K is defined, $K(e) \Leftrightarrow \text{HLT}(e, e)$. This is a contradiction. $\blacksquare$

We have proved that the halting problem (for register machines) is unsolvable. We now consider the following restricted version of this problem.

Definition 8: Fix a program P for the register machine. The *halting problem for P* asks whether there is an algorithm that, given $a \in \mathbb{N}$, decides whether the P-computation with input a halts. In other words, instead of asking for an algorithm that applies to *all* programs and *all* inputs $a \in \mathbb{N}$, we start with a specific program P and then ask for an algorithm that works for all inputs $a \in \mathbb{N}$ for this one program.

Corollary 5: Assume Church's thesis. Then there is a program P for the register machine such that the halting problem for P is unsolvable.

Proof: Let R be an RE relation not recursive. Since R is RE, there is a program P such that for all $a \in \mathbb{N}$,

$$R(a) \Leftrightarrow \text{the } P\text{-computation with input } a \text{ halts.}$$

But R is not recursive, and hence R is not decidable (Church's thesis); it follows from the equivalence above that the halting problem for P is unsolvable. ∎

Note: Corollary 5 plays an important role in proving the unsolvability of certain word problems. The reader who is interested in this application of recursion theory to an area of mathematics outside of logic may proceed directly to Section 9.6.

Another application of Corollary 5 is the existence of a first-order language L for which the validity problem is unsolvable; this approach is due to Turing. Most of the work is in the next theorem, which shows that for each program P, there is a first-order language that simulates P.

Theorem 11 (Simulation of a program by a first-order language): Let P be a program in standard form. There is a first-order language L and for each $a \in \mathbb{N}$ a sentence C_a of L, such that for all $a \in \mathbb{N}$,

the P-computation with input a halts $\Leftrightarrow$ the sentence C_a is logically valid.

Proof: We will outline the main ideas and leave the details to the reader. Let $I_1, \ldots, I_t$ be the instructions of P and let $w = \lambda(P)$. The nonlogical symbols of L are

R ($w + 1$-ary relation symbol)

0 (constant symbol)

S, D (1-ary function symbols).

The following notation is used: $0^0 = 0$; $0^{n+1} = S0^n$. For each instruction I_j of P $(1 \le j \le t)$, we construct a sentence A_j according to one of four cases:

(1) $I_j = S(k)$:
$$\forall x_1 \cdots \forall x_w [R(x_1, \ldots, x_k, \ldots, x_w, 0^j) \rightarrow R(x_1, \ldots, S(x_k), \ldots, x_w, 0^{j+1})].$$

(2) $I_j = D(k)$:
$$\forall x_1 \cdots \forall x_w [R(x_1, \ldots, x_k, \ldots, x_w, 0^j) \rightarrow R(x_1, \ldots, D(x_k), \ldots, x_w, 0^{j+1})].$$

(3) $I_j = Z(k)$:
$$\forall x_1 \cdots \forall x_w [R(x_1, \ldots, x_k, \ldots, x_w, 0^j) \rightarrow R(x_1, \ldots, 0, \ldots, x_w, 0^{j+1})].$$

(4) $I_j = J(k, q)$:
$$\forall x_1 \cdots \forall x_w[R(x_1, \ldots, x_w, 0^j) \to ((\neg(x_k = 0) \to R(x_1, \ldots, x_w, 0^{j+1}))$$
$$\wedge ((x_k = 0) \to R(x_1, \ldots, x_w, 0^q)))].$$

Let A_0 be $\forall x_1 \neg (S(x_1) = 0) \wedge \forall x_1(D(S(x_1)) = x_1) \wedge (D(0) = 0)$. Finally, for each $a \in \mathbb{N}$, let C_a be the sentence

$$[A_0 \wedge A_1 \wedge \cdots \wedge A_t \wedge R(0^a, 0, \ldots, 0, 0^1)] \to \exists x_1 \cdots \exists x_w R(x_1, \ldots, x_w, 0^{t+1}).$$

Now let $a \in \mathbb{N}$. We are required to show that the P-computation with input a halts if and only if the sentence C_a is logically valid. First assume the P-computation with input a halts. Then there is a finite sequence $s_1, \ldots, s_m$ of w-steps with $s_1 = \langle a, 0, \ldots, 0; 1 \rangle$ and $s_m = \langle r_1, \ldots, r_w; t + 1 \rangle$. The w-steps $s_1, \ldots, s_m$ provide us with "instructions" on how to write a formal proof of the sentence $R(0^{r_1}, \ldots, 0^{r_w}, 0^{t+1})$ using the sentences $A_0, A_1, \ldots, A_t, R(0^a, 0, \ldots, 0, 0^1)$ as nonlogical axioms. Repeated application of the substitution theorem gives

$$\{A_0, A_1, \ldots, A_t, R(0^a, 0, \ldots, 0, 0^1)\} \vdash \exists x_1 \cdots \exists x_w R(x_1, \ldots, x_w, 0^{t+1}).$$

By the deduction theorem, $\text{FO}_L \vdash C_a$, and hence C_a is logically valid as required.

Now assume C_a is logically valid. Construct an interpretation I of L as follows: The domain of I is $\mathbb{N}$; 0, D, S, and R are interpreted by

$$0_I = \text{zero}, \quad D_I(a) = a \doteq 1, \quad S_I(a) = a + 1.$$
$$R_I = \text{set of } w + 1\text{-tuples of natural numbers defined inductively by}$$
$$(1) \langle a, 0, \ldots, 0; 1 \rangle \in R_I; (2) \text{ if } s \in R_I \text{ and } s' \text{ is the P-successor of}$$
$$s, \text{ then } s' \in R_I.$$

The sentences $A_0, A_1, \ldots, A_t, R(0^a, 0, \ldots, 0, 0^1)$ are true in I. Moreover, C_a is logically valid; hence C_a is true in I. It follows, by modus ponens for interpretations, that $\exists x_1 \cdots \exists x_w R(x_1, \ldots, x_w, 0^{t+1})$ is true in I. Informally, this sentence states that there exist $r_1, \ldots, r_w \in \mathbb{N}$ such that $\langle r_1, \ldots, r_w; t + 1 \rangle \in R_I$.

Now suppose the P-computation with input a does not halt. Then $R_I = \{s_1, s_2, \ldots\}$, where $s_1 = \langle a, 0, \ldots, 0; 1 \rangle$ and for all $k \in \mathbb{N}^+$, s_{k+1} is the P-successor of s_k. Since each s_k has a P-successor, but $\langle r_1, \ldots, r_w; t + 1 \rangle$ is P-terminal, it follows that $\langle r_1, \ldots, r_w; t + 1 \rangle \notin R_I$, a contradiction. ∎

Theorem 12 (Unsolvability of the validity problem; Turing's approach): Assume Church's thesis. Then there is a first-order language for which the validity problem is unsolvable.

Proof: By Corollary 5, there is a program P for the register machine such that the halting problem for P is unsolvable. By Theorem 11, there is a first-order language L, and for each $a \in \mathbb{N}$ a sentence C_a of L, such that for all $a \in \mathbb{N}$,

the P-computation with input a halts $\Leftrightarrow$ C_a is logically valid.

Since the halting problem for P is unsolvable, the validity problem for L is unsolvable. ∎

——————————*Exercises for Section 9.3*——————————

1. Prove (1) through (3) of Lemma 1.
2. Refer to the proof of Lemma 5, on SUCC(e, w, a, b). Prove the following:
   (1) $Z(e, w, a, b)$ is recursive;   (2) $S(e, w, a, b)$ is recursive.
3. Prove the following:
   (1) Let P be a program with code e_0. The halting problem for P is solvable if and only if the 1-ary relation HLT(e_0, a) is decidable.
   (2) There is a program P with code e_0 such that the 1-ary relation HLT(e_0, a) is not recursive. (This proves that there is a program P such that the halting problem for P is recursively unsolvable.)
4. Let R be an n-ary RE relation. Show that there is a program P such that for all $a_1, \ldots, a_n \in \mathbb{N}$, $R(a_1, \ldots, a_n) \Leftrightarrow$ the P-computation with inputs $a_1, \ldots, a_n$ halts. (*Hint:* Modify the program in Theorem 9.)
5. ($n + 1$-*ary RE relation that enumerates all n-ary RE relations*) Let $R(e, a_1, \ldots, a_n) \Leftrightarrow$ HLT($e, J_n(a_1, \ldots, a_n)$). Show that R is RE and that for every n-ary RE relation Q, there exists $e \in \mathbb{N}$ such that for all $a_1, \ldots, a_n \in \mathbb{N}$, $Q(a_1, \ldots, a_n) \Leftrightarrow R(e, a_1, \ldots, a_n)$. (*Hint:* Use the fact that HLT enumerates.)
6. Prove Theorem 10 for n-ary relations.
7. (*No 2-ary recursive relation enumerates all 1-ary recursive relations*) Let R be a 2-ary relation that enumerates all 1-ary recursive relations (given any 1-ary recursive relation Q, there exists $e \in \mathbb{N}$ such that for all $a \in \mathbb{N}$, $Q(a) \Leftrightarrow R(e, a)$). Prove that R cannot be recursive. (*Hint:* Suppose R is recursive. Define a 1-ary relation Q by $Q(a) \Leftrightarrow \neg R(a, a)$ and note that Q is recursive.)
8. Let P be a program with code e_0 that computes an n-ary function F. Explain why the n-ary relation $R(a_1, \ldots, a_n, c) \Leftrightarrow T_n(e_0, a_1, \ldots, a_n, c)$ is regular (and hence $\mu c T_n(e_0, a_1, \ldots, a_n, c)$ is a total function).
9. Let R be a 1-ary RE relation. Show that there is a first-order language L and sentences $\{C_a : a \in \mathbb{N}\}$ of L, such that for all $a \in \mathbb{N}$, $R(a) \Leftrightarrow C_a$ is logically valid.
10. Let R be a 1-ary RE relation. Show that there is a program P such that for all $a \in \mathbb{N}$, $R(a) \Leftrightarrow$ the P-computation with input a halts with output 0.

——————————| **9.4** |——————————

Partial Recursive Functions

There are algorithms that, for some inputs, do not halt.

Example 1:

Consider this algorithm:

(1) Input $a \in \mathbb{N}$ and let $c = 0$.
(2) Is $a = 2c$?
(3) If *Yes*, print c and halt.
(4) If *No*, replace c with $c + 1$ and go to step 2.

If the input a is even, this algorithm eventually halts with output $a/2$, but if a is odd the algorithm continues forever. This algorithm "computes" the following function:

$$F(a) = \begin{cases} \dfrac{a}{2} & a \text{ even} \\ \text{undefined} & a \text{ odd.} \end{cases}$$ □

The function F is an example of a 1-*ary partial function*. In this section, we will extend the concepts of *computability* (in the informal sense), RM-*computability*, and *recursiveness* to n-ary partial functions. This is not just generalization for the sake of generalization; we obtain a much richer theory by widening the class of functions under consideration from the n-ary functions to the n-ary partial functions. For example, there is an enumeration theorem for partial recursive functions that does not hold for the recursive functions. In addition, this new theory leads to a much better understanding of the RE relations and provides us with powerful tools for proving unsolvability results.

Definition 1: Let $n \geq 1$. An n-*ary partial function* is a function whose domain is a subset of $\mathbb{N}^n$ and whose codomain is $\mathbb{N}$. In other words, we have

$$F : A \rightarrow \mathbb{N}, \quad \text{where } A \subseteq \mathbb{N}^n.$$

The domain of an n-ary partial function F may be all of $\mathbb{N}^n$; in this case, we say that F is *total*. By a *partial function*, we mean a function that is an n-ary partial function for some $n \geq 1$.

An important special case of an n-ary partial function is the *empty function* Λ_n. This is the n-ary partial function whose domain is $\varnothing$; in other words, the partial function that is not defined for *any* $a_1, \ldots, a_n \in \mathbb{N}$. The partial function Λ_n arises rather naturally in computability theory in connection with the existence of algorithms that do not halt for any inputs $a_1, \ldots, a_n \in \mathbb{N}$.

The notion of computability extends to partial functions as follows.

Definition 2 (Informal): An n-ary partial function F is *computable* if there is an algorithm such that for all $a_1, \ldots, a_n \in \mathbb{N}$; if $\langle a_1, \ldots, a_n \rangle \in \text{dom } F$, the algorithm with inputs $a_1, \ldots, a_n$ halts with output $F(a_1, \ldots, a_n)$; if $\langle a_1, \ldots, a_n \rangle \notin \text{dom } F$, the algorithm with inputs $a_1, \ldots, a_n$ continues forever.

The function F in Example 1 is a 1-ary partial function that is computable. Both subtraction and proper subtraction are 2-ary partial functions that are computable (proper subtraction is total but subtraction is not). The empty function Λ_n is an n-ary partial function that is computable; the required algorithm is any algorithm that, for any inputs $a_1, \ldots, a_n \in \mathbb{N}$, goes into an infinite loop. Hamlet 1974, 55, holds the pessimistic view that more computer programs are written to compute Λ_1 than all other functions.)

Example 2:

Let R be a 2-ary decidable relation and let F be the 1-ary partial function defined by

$$F(a) = \begin{cases} \mu b R(a, b) & \exists b R(a, b) \\ \text{undefined} & \neg \exists b R(a, b). \end{cases}$$

We will show that F is computable by writing an appropriate algorithm:

(1) Input $a \in \mathbb{N}$ and set $b = 0$.

(2) Use the algorithm for R to decide whether $R(a, b)$.

(3) If *Yes*, $F(a) = b$; halt.

(4) If *No*, replace b with $b + 1$ and go to step 2.   □

Partial functions arise rather naturally in the register-machine model of computability. For suppose we have a program P, and let $n \geq 1$. For all $a_1, \ldots, a_n \in \mathbb{N}$, there is a P-computation with inputs $a_1, \ldots, a_n$. If this P-computation halts, there is an output. On the other hand, the P-computation with inputs $a_1, \ldots, a_n$ may not halt. This determines an n-ary partial function F_P, as follows:

$$F_P(a_1, \ldots, a_n) = \begin{cases} \text{output} & P(a_1, \ldots, a_n) < \infty \\ \text{undefined} & P(a_1, \ldots, a_n) = \infty. \end{cases}$$

In describing F_P, we have used the following useful notation:

- $P(a_1, \ldots, a_n) < \infty$: The P-computation with inputs $a_1, \ldots, a_n$ halts.
- $P(a_1, \ldots, a_n) = \infty$: The P-computation with inputs $a_1, \ldots, a_n$ does not halt.

Definition 3: Let F be an n-ary partial function and let P be a program. Then *P computes F* if for all $a_1, \ldots, a_n \in \mathbb{N}$,

(1) If $\langle a_1, \ldots, a_n \rangle \in \text{dom } F$, then $P(a_1, \ldots, a_n) < \infty$ with output $F(a_1, \ldots, a_n)$.

(2) If $\langle a_1, \ldots, a_n \rangle \notin \text{dom } F$, then $P(a_1, \ldots, a_n) = \infty$.

In other words, P computes F if $F_P = F$. Also, an n-ary partial function F is RM-*computable* if there is some program P that computes F.

At this point, we have

C = Set of partial functions that are computable (in the intuitive sense)

RM = Set of partial functions that are RM-computable.

Moreover, **RM** ⊆ **C**, since programs satisfy our intuitive notion of an algorithm. Eventually, we will use Church's thesis to show that **RM** = **C**. But first let us introduce yet another class of functions:

R = Set of partial functions that are *partial recursive*.

This class of functions is obtained by modifying the operations of composition and minimalization.

Composition (Of partial functions): Let $H_1, \ldots, H_k$ be n-ary partial functions and let G be a k-ary partial function. The *composition of* $G, H_1, \ldots, H_k$ is the n-ary partial function F defined by

$$F(a_1, \ldots, a_n) \simeq G(H_1(a_1, \ldots, a_n), \ldots, H_k(a_1, \ldots, a_n)).$$

What does $\simeq$ mean in this situation? If $H_1, \ldots, H_k$ are all defined at $\langle a_1, \ldots, a_n \rangle$, and if G is defined at $\langle H_1(a_1, \ldots, a_n), \ldots, H_k(a_1, \ldots, a_n) \rangle$, then F is defined at $\langle a_1, \ldots, a_n \rangle$ and its value is $G(H_1(a_1, \ldots, a_n), \ldots, H_1$ $(a_1, \ldots, a_n))$. But if there is some i $(1 \leq i \leq k)$ such that H_i is not defined at $\langle a_1, \ldots, a_n \rangle$, or if G is not defined at $\langle H_1(a_1, \ldots, a_n), \ldots, H_k(a_1, \ldots, a_n) \rangle$, then F is not defined at $\langle a_1, \ldots, a_n \rangle$. We leave the proof of the following lemma to the reader.

Lemma 1: Let G be a k-ary partial function, let $H_1, \ldots, H_k$ be n-ary partial functions, and let F be the composition of $G, H_1, \ldots, H_k$. If the functions $G, H_1, \ldots, H_k$ are all computable, then F is computable.

Minimalization (Of a partial function): Let G be an $n + 1$-ary partial function. The *minimalization of G* is the n-ary partial function F defined by

$$F(a_1, \ldots, a_n) \simeq \mu b[G(a_1, \ldots, a_n, b) = 0].$$

This notation requires elaboration. Let $a_1, \ldots, a_n \in \mathbb{N}$; then $F(a_1, \ldots, a_n) = b$ provided that $G(a_1, \ldots, a_n, b) = 0$ *and* for all $k < b$, G is defined at $\langle a_1, \ldots, a_n, k \rangle$ but $G(a_1, \ldots, a_n, k) > 0$; otherwise, F is undefined at $\langle a_1, \ldots, a_n \rangle$. Thus F is undefined at $\langle a_1, \ldots, a_n \rangle$ if one of the following holds: $G(a_1, \ldots, a_n, b)$ is defined for all $b \in \mathbb{N}$ and $G(a_1, \ldots, a_n, b) > 0$; there exists $b \in \mathbb{N}$ such that G is not defined at $\langle a_1, \ldots, a_n, b \rangle$ and $G(a_1, \ldots, a_n, k) > 0$ for all $k < b$. Note that in either of these cases, a systematic search for the smallest b such that $G(a_1, \ldots, a_n, b) = 0$ would continue forever. We leave the proof of the following lemma to the reader.

Lemma 2: Let F be the minimalization of an $n + 1$-ary partial function G. If G is computable, then F is computable.

Example 3:

The function F in Example 2 is $F(a) \simeq \mu b[K_R(a, b) = 0]$, where K_R is the characteristic function of the relation R. ☐

Comment on $\simeq$ Let s and t be expressions that, if defined, give a natural number. By $s \simeq t$ we mean: Both s and t are undefined, or else both s and t are defined and are equal. Informally speaking, $\simeq$ extends the usual equality relation = and is reflexive, transitive, and symmetric.

Example 4:

Let $F(a) \simeq \mu c[2 \times c = a]$. We are defining a 1-ary partial function F. If a is even, the expression $\mu c[2 \times c = a]$ is defined and has value $a/2$; thus F is defined at a and $F(a) = a/2$. If a is odd, the expression $\mu c[2 \times c = a]$ is not defined; F is not defined at a. □

Example 5:

Let F be an n-ary partial function and let $a_1, \ldots, a_n, b \in \mathbb{N}$; the correct reading of $F(a_1, \ldots, a_n) \simeq b$ is: $\langle a_1, \ldots, a_n \rangle \in \operatorname{dom} F$ and $F(a_1, \ldots, a_n) = b$. *Reason:* The "expression" b is obviously defined, and hence $F(a_1, \ldots, a_n)$ is defined and is equal to b. □

We will now define the class **R** of partial recursive functions.

Definition 4: A partial function F is *partial recursive* if there is a finite sequence $F_1, \ldots, F_n$ of partial functions with $F_n = F$ and such that for $1 \le k \le n$, one of the following holds: F_k is a starting function; $k > 1$ and F_k is obtained by composition of partial functions from among $F_1, \ldots, F_{k-1}$; $k > 1$ and F_k is obtained by minimalization of a partial function from among $F_1, \ldots, F_{k-1}$.

Before beginning a systematic study of the partial recursive functions, let us quickly mention one of the reasons for extending recursion theory to partial functions. First, consider the following result.

Theorem 1: There is no 2-ary recursive function that enumerates all 1-ary recursive functions. In other words, there is no 2-ary recursive function H such that given a 1-ary recursive function F, there exists $e \in \mathbb{N}$ such that for all $a \in \mathbb{N}$, $F(a) = H(e, a)$.

Proof: Suppose H exists; we will use a diagonalization argument to obtain a contradiction. Let F be defined by $F(a) = H(a, a) + 1$. Since H is recursive, F is recursive. Since H enumerates, there exists $e \in \mathbb{N}$ such that for all $a \in \mathbb{N}$, $F(a) = H(e, a)$. In particular, for $a = e$, we have $F(e) = H(e, e)$. On the other hand, from the definition of F, we have $F(e) = H(e, e) + 1$. Thus

$$H(e, e) = H(e, e) + 1,$$

and this is a contradiction. ■

In our study of the partial recursive functions we will prove, among other results: (1) There *is* a 2-ary partial recursive function that enumerates all 1-ary partial recursive functions; (2) an n-ary relation R is RE if and only if it is the domain of a partial recursive function. Thus we see that the extension of recursion theory to partial functions gives a clearer insight into the RE relations and allows an enumeration theorem for partial recursive functions.

By Theorem 1, there is no enumeration theorem for the recursive functions. Why does the contradiction in Theorem 1 for recursive functions not apply in the case of the partial recursive functions? Applied to partial functions, the proof of Theorem 1 yields

$$H(e, e) \simeq H(e, e) + 1,$$

and this does not automatically give a contradiction because of the possibility that both sides may be undefined! We can summarize the situation as follows: *We can diagonalize out of the class of recursive functions, but not the class of partial recursive functions.* Now let us turn to the study of the partial recursive functions.

Theorem 2: Let F be a total function (i.e., $F: \mathbb{N}^n \to \mathbb{N}$ for some $n \geq 1$). If F is recursive, then F is partial recursive.

Proof: The conditions for a function to be recursive are given in Section 1.8. These conditions are more restrictive than those given in Definition 4; hence a recursive function is automatically partial recursive. ∎

By Theorem 2, familiar recursive functions such as QT, RM, sg, J, K, L, and so on are partial recursive. Later we will prove the converse of Theorem 2; in other words, we will prove that if $F: \mathbb{N}^n \to \mathbb{N}$ and F is partial recursive, then F is in fact recursive.

The general procedure for proving a function partial recursive is the same as in the case of the recursive functions. We leave the proof of the following lemma as an exercise for the reader.

Lemma 3: Let F be an n-ary partial function. To show that F is partial recursive, it suffices to show that F satisfies one of the following conditions: (1) F is a starting function; (2) F is the composition of partial recursive functions; (3) F is the minimalization of a partial recursive function.

The next theorem illustrates a typical use of Lemma 3; the proof is left as an exercise for the reader.

Theorem 3: Let Q be an $n + 1$-ary recursive relation. Then F is partial recursive, where

$$F(a_1, \ldots, a_n) \simeq \mu c Q(a_1, \ldots, a_n, c).$$

Induction on partial recursive functions proceeds as in the case of the recursive functions; justification is left as an exercise for the reader.

Theorem 4: Let Q be a property of partial functions. To prove that every partial recursive function has property Q, it suffices to show the following: (1) Each starting function has property Q; (2) If F is the composition of partial functions $G, H_1, \ldots, H_k$, and $G, H_1, \ldots, H_k$ all have property Q, then F has property Q; (3) if F is the minimalization of a partial function G that has property Q, then F has property Q.

We can use Theorem 4 to give a direct proof that every partial recursive function is computable (left as an exercise for the reader); instead, let us prove the following stronger result.

Theorem 5: Every partial recursive function is RM-computable.

Proof: Assume F is partial recursive. We use induction on partial recursive functions to show that F is RM-computable. The fact that the starting functions $+$, $\times$, $K_<$, and $U_{n,k}$ are RM-computable is proved in Section 9.1. Next, consider composition. Suppose F is defined by

$$F(a_1, \ldots, a_n) \simeq G(H_1(a_1, \ldots, a_n), \ldots, H_k(a_1, \ldots, a_n)),$$

where $G, H_1, \ldots, H_k$ are partial functions that are RM-computable. We ask the reader to check that the program P in the proof of Theorem 1 in Section 9.2 has the following property: If $H_1, \ldots, H_k$ are all defined at $a_1, \ldots, a_n$, and G is defined at $H_1(a_1, \ldots, a_n), \ldots, H_k(a_1, \ldots, a_n)$, then the P-computation with inputs $a_1, \ldots, a_n$ halts with output $F(a_1, \ldots, a_n)$; but if one or more of the functions $H_1, \ldots, H_k$ is not defined at $a_1, \ldots, a_n$, or if G is not defined at $H_1(a_1, \ldots, a_n), \ldots, H_k(a_1, \ldots, a_n)$, then the P-computation with inputs $a_1, \ldots, a_n$ does not halt. Thus program P shows that F is RM-computable as required.

Now suppose F is defined by

$$F(a_1, \ldots, a_n) \simeq \mu b[G(a_1, \ldots, a_n, b) = 0],$$

where G is an $n + 1$-ary partial function that is RM-computable. We ask the reader to check that the program (call it Q) that is written in the proof of Theorem 2 in Section 9.2 has the following property: If there exists $b \in \mathbb{N}$ such that $G(a_1, \ldots, a_n, b) = 0$ and $G(a_1, \ldots, a_n, k) > 0$ for all $k < b$, then the Q-computation with inputs $a_1, \ldots, a_n$ halts and the output is b; but if $G(a_1, \ldots, a_n, b) > 0$ for all $b \in \mathbb{N}$, or if there exists $b \in \mathbb{N}$ such that G is not defined at $\langle a_1, \ldots, a_n, b \rangle$ and $G(a_1, \ldots, a_n, k) > 0$ for all $k < b$, then the Q-computation with inputs $a_1, \ldots, a_n$ does not halt. So program Q shows that F is RM-computable as required. ∎

Corollary 1: Every partial recursive function is computable.

Corollary 2: A total function is recursive if and only if it is partial recursive.

Proof: It suffices to prove that each partial recursive function that is total is a recursive function. Let $F: \mathbb{N}^n \to \mathbb{N}$ be partial recursive. By Theorem 5, F is RM-computable. Since F is total, Corollary 2 in Section 9.3 applies and F is recursive as required. ∎

The next theorem summarizes the relationship between T_n, U, and the partial function F computed by a program P; we leave the proof as an exercise for the reader.

Theorem 6: Let F be an n-ary partial function and let P be a program with code e that computes F. Then for all $a_1, \ldots, a_n \in \mathbb{N}$,

(1) $\langle a_1, \ldots, a_n \rangle \in \operatorname{dom} F \Leftrightarrow \exists c T_n(e, a_1, \ldots, a_n, c)$.
(2) If $\langle a_1, \ldots, a_n \rangle \in \operatorname{dom} F$, then $F(a_1, \ldots, a_n) = U(\mu c T_n(e, a_1, \ldots, a_n, c))$.

In other words, $F(a_1, \ldots, a_n) \simeq U(\mu c T_n(e, a_1, \ldots, a_n, c))$.

Theorem 7: An n-ary partial function is partial recursive if and only if it is RM-computable. In other words, **RM = R**.

Proof: $\mathbf{R} \subseteq \mathbf{RM}$ by Theorem 5. To prove the converse, let F be an n-ary partial function that is RM-computable and let P be a program with code e that computes F. Theorem 6 applies, and for all $a_1, \ldots, a_n \in \mathbb{N}$, $F(a_1, \ldots, a_n) \simeq U(\mu c T_n(e, a_1, \ldots, a_n, c))$. The relation T_n is recursive; hence Theorem 3 applies and $\mu c T_n(e, a_1, \ldots, a_n, c)$ is a partial recursive function. Moreover, U is a recursive function; hence, by composition, F is partial recursive as required. ∎

Theorem 6 also gives the following fundamental result on the structure of partial recursive functions.

Theorem 8 (Kleene normal-form theorem): Let F be an n-ary partial recursive function. Then there exists $e \in \mathbb{N}$ such that for all $a_1, \ldots, a_n \in \mathbb{N}$,

(1) $\langle a_1, \ldots, a_n \rangle \in \operatorname{dom} F \Leftrightarrow \exists c T_n(e, a_1, \ldots, a_n, c)$.
(2) If $\langle a_1, \ldots, a_n \rangle \in \operatorname{dom} F$, then $F(a_1, \ldots, a_n) = U(\mu c T_n(e, a_1, \ldots, a_n, c))$.

In other words, $F(a_1, \ldots, a_n) \simeq U(\mu c T_n(e, a_1, \ldots, a_n, c))$.

The Kleene normal-form theorem essentially states that the function $U(\mu c T_n(e, a_1, \ldots, a_n, c))$ is "universal." To emphasize this point of view, let us introduce the following definition:

Definition 5: Let $n \geq 1$. The *universal function for n-ary partial recursive functions* is the $n + 1$-ary partial function Φ_n defined by

$$\Phi_n(e, a_1, \ldots, a_n) \simeq U(\mu c T_n(e, a_1, \ldots, a_n, c)).$$

Theorem 9 (Enumeration of partial recursive functions): For each $n \geq 1$, Φ_n is an $n + 1$-ary partial recursive function that enumerates all n-ary partial recursive functions. In other words, given an n-ary partial recursive function F, there exists $e \in \mathbb{N}$ such that for all $a_1, \ldots, a_n \in \mathbb{N}$,

$$F(a_1, \ldots, a_n) \simeq \Phi_n(e, a_1, \ldots, a_n).$$

Proof: Φ_n is partial recursive by the way in which it is defined; Φ_n enumerates by the Kleene normal-form theorem. ∎

Corollary 3: There is a 1-ary partial recursive function that is not total and, moreover, does not have an extension to a total recursive function.

Proof: $F(a) \simeq \Phi_1(a, a) + 1$ is partial recursive. Suppose there is a total recursive function $G: \mathbb{N} \to \mathbb{N}$ such that $F(a) = G(a)$ for all $a \in \text{dom } F$. Since Φ_1 enumerates, there exists $e \in \mathbb{N}$ such that for all $a \in \mathbb{N}$, $G(a) = \Phi_1(e, a)$ (replace $\simeq$ with $=$ since G is total). In particular, for $a = e$ we have $G(e) = \Phi_1(e, e)$. Since $\Phi_1(e, e)$ is defined, $F(e) = \Phi_1(e, e) + 1$ and $e \in \text{dom } F$. But $F(a) = G(a)$ for all $a \in \text{dom } F$; hence $G(e) = \Phi_1(e, e) + 1$. This contradicts the earlier calculation $G(e) = \Phi_1(e, e)$. ∎

Note: Corollary 3 shows that the class of partial recursive functions is more extensive than just the restriction of the recursive functions to subsets of $\mathbb{N}^n$.

Theorem 9 states that for each $n \geq 1$, there is a universal function Φ_n for the n-ary partial recursive functions. The next result states that in a certain sense, Φ_1 by itself is universal.

Theorem 10 (Universal function): Φ_1 is universal in the following sense: For each n-ary partial recursive function F, there exists $e \in \mathbb{N}$ such that for all $a_1, \ldots, a_n \in \mathbb{N}$,

$$F(a_1, \ldots, a_n) \simeq \Phi_1(e, J_n(a_1, \ldots, a_n)). \tag{*}$$

Proof: Let F be an n-ary partial recursive function. Define a 1-ary partial recursive function F^* by $F^*(a) \simeq F(J_{n,1}(a), \ldots, J_{n,n}(a))$. Since Φ_1 enumerates the 1-ary partial recursive functions, there exists $e \in \mathbb{N}$ such that for all $a \in \mathbb{N}$, $F^*(a) \simeq \Phi_1(e, a)$. To see that $(*)$ holds, let $a_1, \ldots, a_n \in \mathbb{N}$, let $a = J_n(a_1, \ldots, a_n)$, and use $F^*(a) \simeq \Phi_1(e, a)$. ∎

Theorem 10 suggests the existence of a *universal* RM-program P^*. Such a program has the following property: Given an RM-program P with code e and $a_1, \ldots, a_n \in \mathbb{N}$, (1) if $P(a_1, \ldots, a_n) < \infty$, then $P^*(e, a_1, \ldots, a_n) < \infty$ with the same output; (2) if $P(a_1, \ldots, a_n) = \infty$, then $P^*(e, a_1, \ldots, a_n) = \infty$. The first universal program was constructed by Turing (for Turing machines), and is referred to as a *universal Turing machine*. Informally speaking, a universal Turing machine is a list of instructions for calculating *all* computable functions.

For each $e \in \mathbb{N}$ and each $n \geq 1$, we are going to define an n-ary partial function that we denote by $\{e\}^n$. Each $\{e\}^n$ is partial recursive; moreover, the enumeration theorem implies that for each partial recursive function F, there is a number e (in fact, infinitely many numbers e) such that $\{e\}^n$ is F. This is a useful concept; informally speaking, it allows us to treat a partial recursive function as a single number. The register machine model of computability is used to define the functions $\{e\}^n$.

Definition 6: Let $e \in \mathbb{N}$ and $n \geq 1$. First assume e is the code of a program P. Then

$$\{e\}^n(a_1, \ldots, a_n) = \begin{cases} \text{output} & P(a_1, \ldots, a_n) < \infty \\ \text{undefined} & P(a_1, \ldots, a_n) = \infty. \end{cases}$$

If e is *not* the code of a program P, then $\{e\}^n$ is empty function Λ_n.

Theorem 11 (Properties of the functions $\{e\}^n$): Let $e \in \mathbb{N}$ and $n \geq 1$; then

(1) For all $a_1, \ldots, a_n \in \mathbb{N}$, $\{e\}^n(a_1, \ldots, a_n) \simeq U(\mu c T_n(e, a_1, \ldots, a_n, c))$.

(2) For all $a_1, \ldots, a_n \in \mathbb{N}$, $\{e\}^n(a_1, \ldots, a_n) \simeq \Phi_n(e, a_1, \ldots, a_n)$.

(3) $\{e\}^n$ is a partial recursive function.

Proof: If e is not the code of a program, then (1) obviously holds. Suppose e is the code of a program P. Then P computes the function $\{e\}^n$; Theorem 6 applies to give (1). Part (2) follows from (1) and the definition of Φ_n; (3) follows from (2) and the fact that Φ_n is partial recursive. ∎

The new notation $\{e\}^n$ allows a restatement of the Kleene normal-form theorem.

Theorem 12 (Kleene normal-form theorem, second version): If F is an n-ary partial recursive function, then there exists $e \in \mathbb{N}$ such that for all $a_1, \ldots, a_n \in \mathbb{N}$,

$$F(a_1, \ldots, a_n) \simeq \{e\}^n(a_1, \ldots, a_n).$$

Definition 7: Let F be an n-ary partial recursive function. The number e in Theorem 12 is called an *index* for F.

Let F be an n-ary partial recursive function and let P be a program that computes F. If e is the code of P, then e is an index of F. It is possible, however, to make trivial modifications in P to obtain an infinite number of other programs that also compute F. Each of these programs has a code, and this code is also an index for F. Thus we see that F actually has an infinite number of indices. In summary, *a program has exactly one code; a partial recursive function has infinitely many indices.*

Example 6:

Let G be a 2-ary partial recursive function and let F be defined by

$$F(a, b, e) \simeq G(\{e\}^1(a), \{e\}^2(a, b)).$$

It is not immediately obvious that F is partial recursive; as e varies, the functions $\{e\}^1$ and $\{e\}^2$ vary. We can, however, rewrite F so that it is easily seen to be partial recursive:

$$F(a, b, e) \simeq G(\Phi_1(e, a), \Phi_2(e, a, b)). \qquad \square$$

Since K is RE but not recursive, $\neg K$ is an example of a relation that is not even RE. Is there a relation that is more "natural" than $\neg K$ and that is not RE?

Definition 8: Let TOT be the set of all $e \in \mathbb{N}$ such that the function $\{e\}^1$ is a 1-ary *total* function; in other words, TOT is defined by

$$\text{TOT}(e) \Leftrightarrow \{e\}^1 \text{ is a total function.}$$

Note that we can write $\text{TOT}(e) \Leftrightarrow \forall a \exists c T_1(e, a, c)$. Can we be clever and eliminate the $\forall$? In other words, is TOT an RE relation?

Theorem 13: TOT is not RE.

Proof: Suppose TOT is RE. Clearly $\text{TOT} \neq \varnothing$, so $\text{TOT} = \{F(a): a \in \mathbb{N}\}$, where $F: \mathbb{N} \to \mathbb{N}$ is a recursive function (see Section 8.5). We now use a diagonalization argument to construct a total function H with index h such that $h \notin \{F(a): a \in \mathbb{N}\}$; since H is total, $h \in \text{TOT}$, a contradiction of $\text{TOT} = \{F(a): a \in \mathbb{N}\}$. Let

$$H(a) \simeq \{F(a)\}^1(a) + 1.$$

The function H is partial recursive by $H(a) \simeq \Phi_1(F(a), a) + 1$. To see that H is total, let $a \in \mathbb{N}$; we are required to show that H is defined at a. Now $F(a) \in \text{TOT}$, so $\{F(a)\}^1$ is a total function; it follows that H is defined at a.

Let h be an index for H; that is, $H = \{h\}^1$. Since H is total, $h \in \text{TOT}$. But $h \notin \{F(a): a \in \mathbb{N}\}$, since for all $a \in \mathbb{N}$, $\{h\}^1$ and $\{F(a)\}^1$ differ at a. ∎

—RE Relations and Partial Recursive Functions—

We will now consider the interplay between the RE relations and the partial recursive functions. The first result is an analogue of Theorem 10 in Section 9.3.

Theorem 14 (Characterization of RE relations): Let R be an n-ary relation. The following are equivalent:

(1) R is RE.

(2) R is the domain of an n-ary partial recursive function.

(3) There exists $e \in \mathbb{N}$ such that for all $a_1, \ldots, a_n \in \mathbb{N}$, $R(a_1, \ldots, a_n) \Leftrightarrow \exists c T_n(e, a_1, \ldots, a_n, c)$.

Proof: $(3) \Rightarrow (1)$ follows from the fact that T_n is recursive. To prove $(1) \Rightarrow (2)$, assume R is RE. Then there is a recursive relation Q such that for all $a_1, \ldots, a_n \in \mathbb{N}$, $R(a_1, \ldots, a_n) \Leftrightarrow \exists b Q(a_1, \ldots, a_n, b)$. The required partial recursive function is $F(a_1, \ldots, a_n) \simeq \mu b Q(a_1, \ldots, a_n, b)$.

It remains to prove that $(2) \Rightarrow (3)$. Assume R is the domain of an n-ary partial recursive function F; then for all $a_1, \ldots, a_n \in \mathbb{N}$,

$$R(a_1, \ldots, a_n) \Leftrightarrow \langle a_1, \ldots, a_n \rangle \in \text{dom } F.$$

By the Kleene normal-form theorem, there exists $e \in \mathbb{N}$ such that for all $a_1, \ldots, a_n \in \mathbb{N}$,

$$\langle a_1, \ldots, a_n \rangle \in \operatorname{dom} F \Leftrightarrow \exists c T_n(e, a_1, \ldots, a_n, c).$$

So (3) holds as required. ∎

Corollary 4 (Enumeration of RE relations): For each $n \geq 1$, $\exists c T_n(e, a_1, \ldots, a_n, c)$ is an $n + 1$-ary RE relation that enumerates all n-ary RE relations. Thus, given an n-ary RE relation R, there exists $e \in \mathbb{N}$ such that for all $a_1, \ldots, a_n \in \mathbb{N}$,

$$R(a_1, \ldots, a_n) \Leftrightarrow \exists c T_n(e, a_1, \ldots, a_n, c).$$

Proof: It is obvious that $\exists c T_n(e, a_1, \ldots, a_n, c)$ is RE; enumeration follows from Theorem 14. ∎

Definition 9: Let F be an n-ary partial function. The *graph* of F is the $n + 1$-ary relation $\mathcal{G}_F$ defined by

$$\mathcal{G}_F(a_1, \ldots, a_n, b) \Leftrightarrow \langle a_1, \ldots, a_n \rangle \in \operatorname{dom} F \quad \text{and} \quad F(a_1, \ldots, a_n) = b$$
$$\Leftrightarrow F(a_1, \ldots, a_n) \simeq b.$$

Theorem 15 (Graph theorem): Let F be an n-ary partial function. Then F is partial recursive if and only if the graph $\mathcal{G}_F$ of F is RE.

Proof: Assume F is partial recursive and let e be an index for F. For $a_1, \ldots, a_n, b \in \mathbb{N}$,

$$\mathcal{G}_F(a_1, \ldots, a_n, b) \Leftrightarrow F(a_1, \ldots, a_n) \simeq b$$
$$\Leftrightarrow \{e\}^n(a_1, \ldots, a_n) \simeq b$$
$$\Leftrightarrow U(\mu c T_n(e, a_1, \ldots, a_n, c)) \simeq b$$
$$\Leftrightarrow \exists c[T_n(e, a_1, \ldots, a_n, c) \wedge (U(c) = b)];$$

hence $\mathcal{G}_F$ is RE as required.

Now assume $\mathcal{G}_F$ is RE. Then there is an $n + 2$-ary recursive relation R such that for all $a_1, \ldots, a_n, b \in \mathbb{N}$, $\mathcal{G}_F(a_1, \ldots, a_n, b) \Leftrightarrow \exists x R(a_1, \ldots, a_n, b, x)$. From this it follows that for all $a_1, \ldots, a_n \in \mathbb{N}$,

(1) If $\langle a_1, \ldots, a_n \rangle \in \operatorname{dom} F$, then $\exists c R(a_1, \ldots, a_n, K(c), L(c))$.

(2) For any $c \in \mathbb{N}$, if $R(a_1, \ldots, a_n, K(c), L(c))$, then $\langle a_1, \ldots, a_n \rangle \in \operatorname{dom} F$ and $F(a_1, \ldots, a_n) = K(c)$.

By (1) and (2), $F(a_1, \ldots, a_n) \simeq K(\mu c R(a_1, \ldots, a_n, K(c), L(c)))$, so F is partial recursive as required (the details are left as an exercise). ∎

We will now give two applications of the graph theorem. The first states that $\mathbf{C} = \mathbf{RM} = \mathbf{R}$ (assuming Church's thesis). First, however, we need a lemma.

Lemma 4: If F is a partial function that is computable, then $\mathscr{G}_F$ is semidecidable.

Proof: Assume F is an n-ary partial function that is computable in the intuitive sense. The following algorithm shows that $\mathscr{G}_F$ is semidecidable:

(1) Input $a_1, \ldots, a_n, b \in \mathbb{N}$ and set $x = 1$.
(2) Does the algorithm for F applied to $a_1, \ldots, a_n$ halt in x steps with output b? If *Yes*, print YES and halt.
(3) Add 1 to x and go to step 2. ∎

Theorem 16: Assume Church's thesis. Then $\mathbf{C} = \mathbf{RM} = \mathbf{R}$.

Proof: We already know from Theorem 7 that $\mathbf{RM} = \mathbf{R}$. Moreover, $\mathbf{RM} \subseteq \mathbf{C}$ is obvious. Thus the proof is complete if we can show that every partial function F that is computable (in the intuitive sense) is in fact partial recursive. We have: $\mathscr{G}_F$ is semidecidable (Lemma 1); $\mathscr{G}_F$ is RE (Church's thesis—see Section 8.5); F is partial recursive (Theorem 15). ∎

Theorem 17 (Definition by cases): Let $G_1, \ldots, G_k$ be n-ary partial recursive functions and let $R_1, \ldots, R_k$ be n-ary RE relations such that for all $a_1, \ldots, a_n \in \mathbb{N}$, at most one of $R_1(a_1, \ldots, a_n), \ldots, R_k(a_1, \ldots, a_n)$ holds. Then F is partial recursive, where

$$F(a_1, \ldots, a_n) \simeq \begin{cases} G_1(a_1, \ldots, a_n) & R_1(a_1, \ldots, a_n) \\ \vdots & \vdots \\ G_k(a_1, \ldots, a_n) & R_k(a_1, \ldots, a_n) \\ \text{undefined} & \text{otherwise.} \end{cases}$$

Proof: It suffices to show that $\mathscr{G}_F$ is RE. Let $\mathbf{a} = \langle a_1, \ldots, a_n \rangle$; then

$$\mathscr{G}_F(\mathbf{a}, b) \Leftrightarrow [R_1(\mathbf{a}) \wedge \mathscr{G}_{G_1}(\mathbf{a}, b)] \vee \cdots \vee [R_k(\mathbf{a}) \wedge \mathscr{G}_{G_k}(\mathbf{a}, b)]. \quad ∎$$

Next we will introduce some notation that is standard in the literature and is used in the next section.

Definition 10: For each $e \in \mathbb{N}$, let $W_e = \text{dom}\{e\}^1$. In other words, W_e is the domain of the 1-ary partial recursive function $\{e\}^1$, so W_e is an RE set.

Let $\mathbf{RE}$ be the set of all 1-ary RE relations. We then have the following.

Theorem 18: $\mathbf{RE} = \{W_e : e \in \mathbb{N}\}$.

Proof: We have already noted that each of the sets W_e is RE; hence $\{W_e : e \in \mathbb{N}\} \subseteq \mathbf{RE}$. Conversely, let $R \in \mathbf{RE}$. Then R is the domain of a 1-ary partial recursive function F. Let e be an index for F. Then $R = W_e$, so $R \in \{W_e : e \in \mathbb{N}\}$ as required. ∎

Arithmetical Hierarchy

Results in this section are motivated by the following two observations:

- $K(e) \Leftrightarrow \exists c\, T_1(e, e, c)$; K is RE but not recursive.
- $TOT(e) \Leftrightarrow \forall a\, \exists c\, T_1(e, a, c)$; TOT is not RE.

This suggests the possibility of classifying relations in terms of the number and type of quantifiers that precede a recursive relation.

Definition 11: Let $n \geq 1$. A k-ary relation R is Σ_n if there is an $n + k$-ary recursive relation P such that for all $a_1, \ldots, a_k \in \mathbb{N}$,

$$R(a_1, \ldots, a_k) \Leftrightarrow Qx_1 \cdots Qx_n P(a_1, \ldots, a_k, x_1, \ldots, x_n);$$

each Qx_i $(1 \leq i \leq n)$ is either $\forall x_i$ or $\exists x_i$, and these quantified expressions alternate in type $(\forall, \exists)$ beginning with $\exists x_1$. For example, if P is a 4-ary recursive relation, then

$$R(a) \Leftrightarrow \exists x_1 \forall x_2 \exists x_3 P(a, x_1, x_2, x_3)$$

is a 1-ary Σ_3 relation. Note that a Σ_1 relation is the same as an RE relation. A k-ary relation R is Π_n if there is an $n + k$-ary recursive relation P such that for all $a_1, \ldots, a_k \in \mathbb{N}$,

$$R(a_1, \ldots, a_k) \Leftrightarrow Qx_1 \cdots Qx_n P(a_1, \ldots, a_k, x_1, \ldots, x_n);$$

each Qx_i $(1 \leq i \leq n)$ is either $\forall x_i$ or $\exists x_i$, and these quantified expressions alternate in type $(\forall, \exists)$, this time beginning with $\forall x_1$. For example, if P is a 4-ary recursive relation, then

$$R(a_1, a_2) \Leftrightarrow \forall x_1 \exists x_2 P(a_1, a_2, x_1, x_2)$$

is a 2-ary Π_2 relation. A k-ary relation R is both Σ_0 and Π_0 if it is a recursive relation. Finally, a k-ary relation R is in the *arithmetical hierarchy* if it is either Σ_n or Π_n for some $n \in \mathbb{N}$.

Why the emphasis on quantifiers that alternate in type? Suppose R is a 2-ary relation such that

$$R(a, b) \Leftrightarrow \exists x_1 \forall x_2 \forall x_3 \exists x_4 P(a, b, x_1, x_2, x_3, x_4).$$

By the method of collapsing quantifiers,

$$R(a, b) \Leftrightarrow \exists x_1 \forall y \exists x_4 P(a, b, x_1, K(y), L(y), x_4);$$

this gives quantifiers that alternate in type.

We will end this section with a summary of basic results about the arithmetic hierarchy; proofs are outlined in the exercises.

Example 8:

K is Σ_1 but not Π_1; TOT is Π_2 but neither Σ_1 nor Π_1.   □

Theorem 19 (Characterization): Let R be a k-ary relation.

(1) R is Σ_{n+1} if and only if there is a $k + 1$-ary Π_n relation Q such that for all $a_1, \ldots, a_k \in \mathbb{N}$, $R(a_1, \ldots, a_k) \Leftrightarrow \exists x Q(a_1, \ldots, a_k, x)$.

(2) R is Π_{n+1} if and only if there is a $k + 1$-ary Σ_n relation Q such that for all $a_1, \ldots, a_k \in \mathbb{N}$, $R(a_1, \ldots, a_k) \Leftrightarrow \forall x Q(a_1, \ldots, a_k, x)$.

Theorem 20 (Closure properties): The following hold for all $k \geq 1$:

(1) If R is a k-ary Σ_n relation, then R is both Σ_m and Π_m for all $m > n$.

(2) If R is a k-ary Π_n relation, then R is both Σ_m and Π_m for all $m > n$.

(3) If R is a k-ary Π_n relation, then $\neg R$ is a k-ary Σ_n relation.

(4) If R is a k-ary Σ_n relation, then $\neg R$ is a k-ary Π_n relation.

(5) If R is a $k + 1$-ary Σ_n relation, then $\forall x R$ is a k-ary Π_{n+1} relation.

(6) If R is a $k + 1$-ary Π_n relation, then $\exists x R$ is a k-ary Σ_{n+1} relation.

(7) If Q and R are k-ary Σ_n relations, then $Q \vee R$ and $Q \wedge R$ are k-ary Σ_n relations.

(8) If Q and R are k-ary Π_n relations, then $Q \vee R$ and $Q \wedge R$ are k-ary Π_n relations.

Theorem 21 (Enumeration): For all n and $k \geq 1$,

(1) there is a $k + 1$-ary Σ_n relation that enumerates all k-ary Σ_n relations.

(2) there is a $k + 1$-ary Π_n relation that enumerates all k-ary Π_n relations.

Theorem 22 (Hierarchy theorem): For all n and $k \geq 1$,

(1) there is a k-ary Σ_n relation that is not Π_n.

(2) there is a k-ary Π_n relation that is not Σ_n.

Exercises for Section 9.4
---

1. Prove the following (*Hint:* See Section 1.8): (1) Lemma 1; (2) Lemma 2; (3) Lemma 3.
2. Prove Theorem 3. (*Hint:* See Theorem 7 in Section 8.2.)
3. Prove Theorem 4. (*Hint:* See Section 1.8.)
4. Use Lemmas 1 and 2 and Theorem 4 to give a direct proof of Corollary 1 (every partial recursive function is computable).
5. Prove the following properties of $\approx$: (1) $s \approx s$; (2) if $s \approx t$, then $t \approx s$; (3) if $s \approx t$ and $t \approx u$, then $s \approx u$.
6. Prove Theorem 6. (*Hint:* See Theorem 6 in Section 9.3, Definition 3, and the definitions of T_n and U.
7. Let F be a partial recursive function. Show that the domain of F is RE.
8. Show that $R(e, a, b) \Leftrightarrow \{e\}^1(a) \approx b$ is RE.
9. Let R be a 1-ary RE relation that is not recursive. Show that F is not recursive, where F is defined by

$$F(a) \approx \begin{cases} 0 & R(a) \\ a^2 & \neg R(a). \end{cases}$$

10. Let R be a 1-ary RE relation. Show that F is partial recursive, where F is defined by

$$F(a) \simeq \begin{cases} 0 & R(a) \\ \text{undefined} & \neg R(a). \end{cases}$$

11. Show that there is no 3-ary recursive relation R such that given any 1-ary partial recursive function F, there exists $e \in \mathbb{N}$ such that for all $a \in \mathbb{N}$, $F(a) \simeq \mu c R(e, a, c)$ (compare with $F(a) \simeq U(\mu c T_1(e, a, c))$). (*Hint:* Suppose R exists. Let $F(a) \simeq \mu c \neg R(a, a, c)$.)

12. (*Details of graph theorem*) Prove (1) and (2) of the proof of Theorem 15; then show that $F(a_1, \ldots, a_n) \simeq K(\mu c R(a_1, \ldots, a_n, K(c), L(c)))$. (*Note:* To verify, show that both sides are defined and equal, or that both sides are undefined.)

13. Two sets $A, B \subseteq \mathbb{N}$ are *recursively inseparable* if $A \cap B = \emptyset$ and there is no recursive set R with $A \subseteq R$ and $B \subseteq R^c$. Let $A = \{e : e \in \mathbb{N}, \{e\}^1(e) = 0\}$, $B = \{e : e \in \mathbb{N}, \{e\}^1(e) = 1\}$. Show that A and B are recursively inseparable. (*Hint:* Suppose R is recursive with $A \subseteq R$ and $B \subseteq R^c$. Let e be an index of the characteristic function of R^c.)

14. Let A and B be disjoint subsets of $\mathbb{N}$. Show that A and B are recursively inseparable if and only if whenever $A \subseteq W_a$, $B \subseteq W_b$ and $W_a \cap W_b = \emptyset$, then $W_a \cup W_b \neq \mathbb{N}$.

15. Let F be a 1-ary partial recursive function. A *recursive extension* of F is a recursive function $G: \mathbb{N} \to \mathbb{N}$ such that $F(a) = G(a)$ for all $a \in \text{dom } F$. Prove the following:
    (1) Let F be a 1-ary partial recursive function whose domain is recursive. Show that F has a recursive extension.
    (2) Let $F(a) \simeq \mu c T_1(a, a, c) \times 0$. Show that the domain of F is not recursive but that nevertheless F has a recursive extension.

16. Let F be an n-ary partial function that is computable. Prove that $\text{dom } F$ is a semidecidable relation.

17. Let R be a n-ary relation. Show that R is semidecidable if and only if R is the domain of an n-ary partial function that is computable.

18. Let G be an n-ary partial function that is computable and let R be an n-ary semidecidable relation. Show that F is computable, where

$$F(a_1, \ldots, a_n) \simeq \begin{cases} G(a_1, \ldots, a_n) & R(a_1, \ldots, a_n) \\ \text{undefined} & \neg R(a_1, \ldots, a_n). \end{cases}$$

19. Define a 2-ary relation R by $R(a, e) \Leftrightarrow a \in W_e$. Show that R is a 2-ary RE relation that enumerates all 1-ary RE relations.

20. Let A be a subset of $\mathbb{N}$ that is RE. Show that $\bigcup_{a \in A} W_a$ is RE.

21. Give an example of a collection $\{R_n : n \in \mathbb{N}\}$ of recursive relations such that $\bigcup_{n \in \mathbb{N}} R_n$ is not RE.

22. Give an example of a collection $\{R_n : n \in \mathbb{N}\}$ of recursive relations such that $\bigcap_{n \in \mathbb{N}} R_n$ is not RE.

23. (*Alternate definition of minimalization*) Given a 2-ary partial function G, $\mu^* b[G(a, b) = 0]$ means: $G(a, b) = 0$, and for all $k < b$ either $G(a, k) > 0$ or G is not defined at $\langle a, k \rangle$. This exercise will show that this alternate definition of minimalization is not acceptable. Let G be defined by

$$G(a, b) \simeq \begin{cases} 0 & [(b = 0) \wedge K(a)] \vee (b = 1) \\ \text{undefined} & \text{otherwise.} \end{cases}$$

    G is partial recursive (definition by cases). Show that $F(a) \simeq \mu^* b[G(a, b) = 0]$ is the characteristic function of K; conclude that F is not a partial recursive function.

24. Show that TOT is Π_2.

25. Prove Theorem 19.

26. Prove Theorem 20. (*Hints:* For (1) and (2), add superfluous quantifiers; for (7) and (8), see the prenex rules in Section 5.6 and use contraction of quantifiers.)

27. (*Enumeration, negation, and diagonalization*) Let $\mathfrak{R}$ be a class of relations on $\mathbb{N}$ (e.g.,

$\Re$ = all recursive relations). We say that $\Re$ *admits n-ary enumeration* if there is an $n + 1$-ary relation $R_n \in \Re$ such that given any n-ary relation $Q \in \Re$, there exists $e \in \mathbb{N}$ such that for all $a_1, \ldots, a_n \in \mathbb{N}$, $Q(a_1, \ldots, a_n) \Leftrightarrow R_n(e, a_1, \ldots, a_n)$. Let $\Re$ satisfy these two properties: *negation* (if Q is a 1-ary relation with $Q \in \Re$, then $\neg Q \in \Re$); *diagonalization* (if Q is a 2-ary relation with $Q \in \Re$, then $R \in \Re$, where $R(a) \Leftrightarrow Q(a, a)$). Show that $\Re$ does not admit 1-ary enumeration.

28. Refer to Exercise 27. For each $\Re$ below, state (*Yes* or *No*) whether $\Re$ admits 1-ary enumeration. If *No*, justify your answer.
    (1) $\Re$ = primitive recursive relations.
    (2) $\Re$ = recursive relations.
    (3) $\Re$ = decidable relations.
    (4) $\Re$ = RE relations.
    (5) $\Re = \Pi_1$ relations.

29. (*Enumeration theorems for arithmetic hierarchy*) Refer to Exercise 27 for terminology. This exercise outlines a proof of Theorem 21. Prove the following:
    (1) $\exists c T_k(e, a_1, \ldots, a_k, c)$ is a $k + 1$-ary Σ_1 ($=$ RE) relation that enumerates all k-ary Σ_1 relations.
    (2) If R is a $k + 1$-ary Σ_n relation that enumerates all k-ary Σ_n relations, then $\neg R$ is a $k + 1$-ary Π_n relation that enumerates all k-ary Π_n relations.
    (3) Let $R(e, a_1, \ldots, a_k, x)$ be a $k + 2$-ary Π_n relation that enumerates all $k + 1$-ary Π_n relations. Then $R^*(e, a_1, \ldots, a_k) \Leftrightarrow \exists x R(e, a_1, \ldots, a_k, x)$ is a $k + 1$-ary Σ_{n+1} relation that enumerates all k-ary Σ_{n+1} relations.
    (4) Prove Theorem 21.

30. (*Hierarchy theorem*) Let $n \geq 1$. Use an enumeration theorem and diagonalization to construct the following:
    (1) A 1-ary Σ_n relation not Π_n (and hence not Σ_m or Π_m for $m < n$).
    (2) A 1-ary Π_n relation not Σ_n (and hence not Σ_m or Π_m for $m < n$).

31. ($\Pi_2 + \Sigma_2$ *does not imply* Π_1 *or* Σ_1) Cantor's functions J, K, and L are used in this exercise. Let Q be an RE relation not recursive and define R by

$$R(c) \Leftrightarrow \neg Q(K(c)) \wedge Q(L(c)).$$

    (1) Show that R is both Π_2 and Σ_2. (*Hint:* See Theorem 20.)
    (2) Let $b_0 \in Q$. Show that $\neg Q(a) \Leftrightarrow R(J(a, b_0))$ for all $a \in \mathbb{N}$. Conclude that R cannot be Σ_1.
    (3) Let $a_0 \notin Q$. Show that $Q(b) \Leftrightarrow R(J(a_0, b))$ for all $b \in \mathbb{N}$. Conclude that R cannot be Π_1.

32. Let $Q(a, b) \Leftrightarrow \forall x_1 \forall x_2 R(a, b, x_1, x_2)$, where R is recursive. Use contraction of quantifiers (see Section 8.5) to show that Q is Π_1.

33. Let $P(a) \Leftrightarrow \forall x_1 \exists x_2 R_1(a, x_1, x_2)$ and $Q(a) \Leftrightarrow \forall x_1 \exists x_2 R_2(a, x_1, x_2)$, where R_1 and R_2 are recursive; note that P and Q are 1-ary Π_2 relations. Show that $P \vee Q$ is a Π_2 relation (see the change-of-variable theorem and the prenex rules in Section 5.6).

34. (*Arithmetical* $\Leftrightarrow \Pi_n$ *or* Σ_n *for some* $n \in \mathbb{N}$) A k-ary relation R is said to be *arithmetical* if there is an $n + k$-ary recursive relation P such that for all $a_1, \ldots, a_k \in \mathbb{N}$,

$$R(a_1, \ldots, a_k) \Leftrightarrow Q x_1 \cdots Q x_n P(a_1, \ldots, a_k, x_1, \ldots, x_n),$$

where each $Q x_i$ is either $\forall x_i$ or $\exists x_i$. Clearly, every k-ary relation that is Σ_n or Π_n is arithmetical. Prove the converse: If R is arithmetical, then R is Π_n or Σ_n for some $n \in \mathbb{N}$. (*Hint:* Contraction of quantifiers.)

$$\boxed{9.5}$$

Parameter Theorem and Recursion Theorem

The Kleene computation relation T_n is absolutely fundamental in recursion theory. From it we obtain, among other results, the enumeration theorem for RE relations and the existence of an RE relation not recursive. In this section, we will prove another fundamental result of recursion theory, known as the *parameter theorem*. The parameter theorem is used to show that certain relations are non-recursive, non-RE, or even worse! In addition, the parameter theorem is used to prove an important and far-reaching extension of primitive recursion, known as the *recursion theorem*.

Parameter Theorem

The parameter theorem is best understood in terms of programs. Suppose we have a program P that computes a 2-ary partial recursive function $F(a, e)$. For each $e \in \mathbb{N}$, let G_e be the 1-ary partial recursive function defined by $G_e(a) \simeq F(a, e)$. Let P_e be a program that computes G_e and let $\phi(e)$ be the code of the program P_e. Note that $G_e = \{\phi(e)\}^1$, and hence $F(a, e) \simeq \{\phi(e)\}^1(a)$ for all $a \in \mathbb{N}$. We have described a function $\phi: \mathbb{N} \to \mathbb{N}$ defined by $\phi(e) = $ code of P_e. The parameter theorem tells us that we can systematically construct the programs P_e from the program P in such a way that ϕ is recursive.

Theorem 1 (Parameter theorem): Let F be an $n + 1$-ary partial recursive function. Then there is a recursive function $\phi: \mathbb{N} \to \mathbb{N}$ such that for all $a_1, \dots, a_n, e \in \mathbb{N}$,

$$F(a_1, \dots, a_n, e) \simeq \{\phi(e)\}^n(a_1, \dots, a_n).$$

Proof: Let P be a program that computes F. Thus, given the initial configuration

R_1	R_2	$\cdots$	R_n	R_{n+1}	R_{n+2}	$\cdots$
a_1	a_2	$\cdots$	a_n	e	0	$\cdots$

$(*)$

the P-computation with these inputs, if it halts at all, halts with output $F(a_1, \dots, a_n, e)$ in register R_1. For each $e \in \mathbb{N}$, we want to write a program P_e so that the P_e-computation with inputs $a_1, \dots, a_n$ is the same as the P-computation with inputs $a_1, \dots, a_n, e$. Thus for the program P_e, the initial configuration is

R_1	R_2	$\cdots$	R_n	R_{n+1}	$\cdots$
a_1	a_2	$\cdots$	a_n	0	$\cdots$

$(**)$

and we require that P_e satisfy the following:

(1) If $P(a_1, \ldots, a_n, e) < \infty$, then $P_e(a_1, \ldots, a_n) < \infty$ with output $F(a_1, \ldots, a_n, e)$.

(2) If $P(a_1, \ldots, a_n, e) = \infty$, then $P_e(a_1, \ldots, a_n) = \infty$.

The function $\phi: \mathbb{N} \to \mathbb{N}$ is then defined by $\phi(e) =$ code of program P_e. From (1) and (2), it follows that $F(a_1, \ldots, a_n, e) \simeq \{\phi(e)\}^n(a_1, \ldots, a_n)$ for all $a_1, \ldots, a_n, e \in \mathbb{N}$. Thus it remains to write the programs P_e so that (1) and (2) hold and such that ϕ is recursive. Let Q_e be the program

$I_1 \ S(n + 1)$

$\vdots$

$I_e \ S(n + 1)$.

Informally, program Q_e puts e in register R_{n+1}; in other words, configuration (∗∗) becomes (∗). Let P_e be $Q_e P$, the join of the programs Q_e and P. We leave it to the reader to check that the program P_e satisfies (1) and (2); it remains to show that ϕ is recursive.

Let $I_1, \ldots, I_t$ be the instructions of P; the code of P is $e_0 = \langle \#(I_1), \ldots, \#(I_t) \rangle_t$. Define auxiliary functions ϕ_1 and ϕ_2 by

$$\phi_1(e) = \text{code of the program } Q_e$$

$$\phi_2(e) = \text{code of the program obtained from } P \text{ by replacing}$$

$$\text{each jump instruction } J(k, q) \text{ with } J(k, q + e);$$

we then have $\phi(e) = \phi_1(e) * \phi_2(e)$, where $*$ is the concatenation function (see Theorem 6 in Section 8.3). Thus ϕ is recursive if we can show that ϕ_1 and ϕ_2 are recursive.

• $\phi_1(e) = \mu c[\mathrm{CD}(c) \wedge (\beta(c, 0) = e) \wedge \forall k < e(\beta(c, k + 1) = \langle n + 1, 0 \rangle_2)]$.

For ϕ_2, we use the recursive function G defined by

$$G(e, a) = \begin{cases} \langle \beta(a, 1), \beta(a, 2) + e, 0 \rangle_3 & \mathrm{JUMP}(a) \\ a & \neg\mathrm{JUMP}(a). \end{cases}$$

G has the following property: If a is the code of the jump instruction $J(k, q)$, then $G(e, a)$ is the code of the jump instruction $J(k, q + e)$; if a is the code of any other type of instruction, then $G(e, a) = a$.

• $\phi_2(e) = \mu c[\mathrm{CD}(c) \wedge (\beta(c, 0) = t) \wedge \forall k < t(\beta(c, k + 1) = G(e, \beta(e_0, k + 1)))]$.

■

Example 1:

We use the parameter theorem to show that there is a recursive function $\phi : \mathbb{N} \to \mathbb{N}$ such that for all $e \in \mathbb{N}$,

(1) If e is even, then $W_{\phi(e)} = \{e\}$ (i.e., the domain of $\{\phi(e)\}^1$ is just e) and $\{\phi(e)\}^1(e) = 5$.

(2) If e is odd, then $W_{\phi(e)} = \mathbb{N} - \{e\}$ (i.e., the domain of $\{\phi(e)\}^1$ is everything but e) and for all $a \neq e$, $\{\phi(e)\}^1(a) = a^2$.

Let E be the set of even natural numbers, O the set of odd natural numbers, and define F by

$$F(a, e) = \begin{cases} 5 & E(e) \wedge (a = e) \\ a^2 & O(e) \wedge (a \neq e) \\ \text{undefined} & \text{otherwise.} \end{cases}$$

F is partial recursive by definition by cases. By the parameter theorem, there is a recursive function $\phi : \mathbb{N} \to \mathbb{N}$ such that for all $a, e \in \mathbb{N}$, $F(a, e) \simeq \{\phi(e)\}^1(a)$. We verify that ϕ satisfies (2). Let e be odd; we are required to show that $W_{\phi(e)} = \mathbb{N} - \{e\}$. Let $a \in \mathbb{N} - \{e\}$. Then $O(e) \wedge (a \neq e)$, so $F(a, e) = a^2$. But $F(a, e) \simeq \{\phi(e)\}^1(a)$; hence $\{\phi(e)\}^1$ is defined at a and $a \in W_{\phi(e)}$; note also that $\{\phi(e)\}^1(a) = a^2$. Now let $a \in W_{\phi(e)}$, but suppose $a \notin \mathbb{N} - \{e\}$. Then $a = e$; hence, by the way F is defined, $F(a, e)$ is undefined. But $F(a, e) \simeq \{\phi(e)\}^1(a)$, so $\{\phi(e)\}^1$ is not defined at a and $a \notin W_{\phi(e)}$. This is a contradiction. □

Unsolvable Problems and Rice's Theorem

The parameter theorem is often used to show that certain decision problems are unsolvable. To focus attention, let us consider these two.

Definition 1: The *zero function problem* asks whether there is an algorithm that, given a 1-ary partial recursive function F, decides whether F is the zero function Z ($Z(a) = 0$ for all $a \in \mathbb{N}$). In terms of programs: Is there an algorithm that, given a program P, decides whether for all $a \in \mathbb{N}$, the P-computation with input a halts with output 0?

Definition 2: The *finite domain problem* asks whether there is an algorithm that, given a 1-ary partial recursive function F, decides whether the domain of F is finite. In terms of programs: Is there an algorithm that, given a program P, decides whether P halts for at most a finite number of inputs?

Each of these problems is an example of a decision problem. If there is no algorithm, the decision problem is said to be *unsolvable*. We proceed as follows: Formulate the problem in terms of a relation R on $\mathbb{N}$; show that R is not recursive (or not RE). For the two problems at hand, the appropriate relations on $\mathbb{N}$ are

ZERO$(e) \Leftrightarrow \{e\}^1$ is the zero function.

FIN$(e) \Leftrightarrow W_e$ is finite.

Eventually, we will prove that neither ZERO nor FIN is recursive; in other words, the zero problem and the finite-domain problem are recursively unsolvable.

We have already used diagonal arguments to show that K is not recursive and that TOT is not even RE. We now introduce a second method for showing that a relation is nonrecursive or non-RE. This method, called *reduction*, depends on the fact that we already know of the existence of a relation that is not recursive or not RE (K is often chosen as a nonrecursive relation and $\neg K$ or TOT as a non-RE relation). The reduction method for proving nonrecursiveness proceeds as follows: Given a relation R, choose a relation Q known to be nonrecursive; then show that if R were recursive, then Q would also be recursive, a contradiction of the choice of Q as a nonrecursive relation. In this situation, we say that the problem Q *reduces* to the problem R. We have seen this type of argument before in a slightly different format; in Section 9.3, we used the existence of an RE relation not recursive to show that there is a first-order language L for which the validity problem is unsolvable.

There are a number of reduction methods, of which the following is perhaps the simplest.

Definition 3: Let Q and R be 1-ary relations. We say that Q is *many-one reducible to R* (or *m-reducible to R*), written $Q \leq_m R$, if there is a recursive function $\phi: \mathbb{N} \to \mathbb{N}$ such that for all $e \in \mathbb{N}$,

$$Q(e) \Leftrightarrow R(\phi(e)).$$

Note: The phrase *many-one* is used to distinguish this type of reducibility from a related concept, called *one-one reducibility*; here, ϕ is required to be a one-to-one recursive function.

Suppose $Q \leq_m R$, so that there is a recursive function $\phi: \mathbb{N} \to \mathbb{N}$ such that $Q(e) \Leftrightarrow R(\phi(e))$. It is easy to see that the following holds: If *R is decidable, then Q is decidable.* Informally speaking, this tells us that problems about R are at least as difficult to solve as problems about Q; if we could "solve" R, then we could "solve" Q. The standard way of using $Q \leq_m R$ is as follows: Given R, choose Q such that Q is not recursive (or not RE) and then show $Q \leq_m R$; conclude that R cannot be recursive (or RE). Let us formalize these ideas.

Lemma 1: Let Q and R be 1-ary relations such that $Q \leq_m R$.

(1) If R is recursive, then Q is recursive.

(2) If R is RE, then Q is RE.

Proof of (2): Let $Q(e) \Leftrightarrow R(\phi(e))$ with $\phi: \mathbb{N} \to \mathbb{N}$ recursive. Since R is RE and ϕ is recursive, Q is RE (see Theorem 8 in Section 8.5). ∎

A basic tool for proving m-reducibility is the following consequence of the parameter theorem.

Theorem 2 (m-reducibility): Let Q be a 1-ary RE relation and let G be a 1-ary partial recursive function. Then there is a recursive function $\phi: \mathbb{N} \to \mathbb{N}$ such that for all $e \in \mathbb{N}$,

(1) If $Q(e)$, then $\phi(e)$ is an index for G (i.e., $\{\phi(e)\}^1 = G$).

(2) If $\neg Q(e)$, then $\phi(e)$ is an index for Λ_1 (i.e, $\{\phi(e)\}^1 = \Lambda_1$).

Proof: We want a recursive function $\phi: \mathbb{N} \to \mathbb{N}$ such that for all $a, e \in \mathbb{N}$,

$$\{\phi(e)\}^1(a) \simeq \begin{cases} G(a) & Q(e) \\ \text{undefined} & \neg Q(e). \end{cases}$$

In anticipation of using the parameter theorem, define a 2-ary function F by

$$F(a, e) \simeq \begin{cases} G(a) & Q(e) \\ \text{undefined} & \neg Q(e). \end{cases}$$

F is a partial recursive function by definition by cases. By the parameter theorem, there is a recursive function $\phi: \mathbb{N} \to \mathbb{N}$ such that for all $a, e \in \mathbb{N}$,

$$F(a, e) \simeq \{\phi(e)\}^1(a).$$

It remains to check (1) and (2). Suppose $Q(e)$. Then $\{\phi(e)\}^1(a) \simeq G(a)$ for all $a \in \mathbb{N}$; hence $\{\phi(e)\}^1$ is G, so $\phi(e)$ is an index for G. Now suppose $\neg Q(e)$. Then $\{\phi(e)\}^1(a)$ is undefined for all $a \in \mathbb{N}$, and hence $\phi(e)$ is an index for Λ_1. ∎

Theorem 2 is used as follows: Suppose we have an RE relation Q and we want to show that $Q \leq_m R$. Choose a 1-ary partial recursive function G such that $R(e)$ whenever e is an index of G; Theorem 2 applied to Q and G gives a recursive function $\phi: \mathbb{N} \to \mathbb{N}$ such that (1) and (2) hold; from (1), we automatically have $Q(e) \Rightarrow R(\phi(e))$.

Corollary 1: Every 1-ary RE relation is m-reducible to K.

Proof: Let Q be an arbitrary 1-ary RE relation. For G, choose any 1-ary total recursive function; note that $K(e)$ whenever e is an index of a 1-ary total recursive function. Apply Theorem 2, on m-reducibility, to Q and G to obtain a recursive function $\phi: \mathbb{N} \to \mathbb{N}$ such that for all $e \in \mathbb{N}$,

(1) If $Q(e)$, then $\phi(e)$ is an index for G.

(2) If $\neg Q(e)$, then $\phi(e)$ is an index for Λ_1.

We leave it to the reader to check that $Q(e) \Leftrightarrow K(\phi(e))$ for all $e \in \mathbb{N}$. ∎

Informally, Corollary 1 states that among semidecidable relations, K is as difficult to solve as any. But not all unsolvable problems are m-reducible to K; indeed, if $\text{TOT} \leq_m K$, then TOT would be RE, a contradiction. On the other hand, we have the following.

Corollary 2: K is m-reducible to TOT.

Proof: Let G be any 1-ary total recursive function. Apply Theorem 2, on m-reducibility, to K and G to obtain a recursive function $\phi: \mathbb{N} \to \mathbb{N}$ such that (1) if $K(e)$, then $\phi(e)$ is an index for G; (2) if $\neg K(e)$, then $\phi(e)$ is an index for Λ_1. We now have $K(e) \Leftrightarrow \text{TOT}(\phi(e))$ as required. ∎

Theorem 3: Neither TOT nor ¬TOT is RE.

Proof: We already know that TOT is not RE (see Section 9.4). By Corollary 2, there is a recursive function $\phi: \mathbb{N} \to \mathbb{N}$ such that $K(e) \Leftrightarrow$ TOT$(\phi(e))$ for all $e \in \mathbb{N}$; hence $\neg K(e) \Leftrightarrow \neg$TOT$(\phi(e))$ for all $e \in \mathbb{N}$. Since $\neg K$ is not RE, ¬TOT cannot be RE. ∎

We will now give a very general result, due to Rice, on unsolvability results. Roughly speaking, Rice's theorem states that any problem about partial recursive functions that is not obviously trivial is at least not recursive. Given a set Γ of 1-ary partial recursive functions, let R_Γ be the 1-ary relation (= subset of $\mathbb{N}$) defined by

$$R_\Gamma(e) \Leftrightarrow \{e\}^1 \in \Gamma$$
$$\Leftrightarrow e \text{ is an index for some } F \in \Gamma.$$

Let us emphasize the relationship between Γ and R_Γ. For a 1-ary partial recursive function F, (1) if $F \in \Gamma$, then $R_\Gamma(e)$ for every index e of F; (2) if $R_\Gamma(e)$ for at least one index e of F, then $F \in \Gamma$. Under very general conditions on Γ, Rice's theorem states that R_Γ is not recursive.

Theorem 4 (Rice): Let Γ be a set of 1-ary partial recursive functions that is proper ($\Gamma \neq \varnothing$ and $\Gamma \neq$ all 1-ary partial recursive functions). Then R_Γ is not recursive. Moreover, if $\Lambda_1 \in \Gamma$, then R_Γ is not RE.

Proof: We will prove that

(1) If $\Lambda_1 \notin \Gamma$, then $K \leq_m R_\Gamma$ (and hence R_Γ is not recursive).
(2) If $\Lambda_1 \in \Gamma$, then $\neg K \leq_m R_\Gamma$ (and hence R_Γ is not RE).

We will give the details for (2) and leave (1) as an exercise. Assume $\Lambda_1 \in \Gamma$. Since $\Gamma \neq$ all 1-ary partial recursive functions, there is a partial recursive function G with $G \notin \Gamma$. Apply Theorem 2, on m-reducibility, to K and G to obtain a recursive function $\phi: \mathbb{N} \to \mathbb{N}$ such that for all $e \in \mathbb{N}$, (1) if $K(e)$, then $\phi(e)$ is an index for G; (2) if $\neg K(e)$, then $\phi(e)$ is an index for Λ_1. It remains to prove that for all $e \in \mathbb{N}$,

$$\neg K(e) \Leftrightarrow R_\Gamma(\phi(e)).$$

Suppose $\neg K(e)$. Then $\phi(e)$ is an index for Λ_1; since $\Lambda_1 \in \Gamma$, $R_\Gamma(\phi(e))$. Now suppose $K(e)$. Then $\phi(e)$ is an index for G; since $G \notin \Gamma$, $\neg R_\Gamma(\phi(e))$ as required. ∎

Let us restate Rice's theorem in a slightly different way. Let $\mathbf{R}_1$ be the class of 1-ary partial recursive functions and let Q be a property of functions. If there is at least one $F \in \mathbf{R}_1$ that has property Q and at least one $F \in \mathbf{R}_1$ that does not have property Q, then R_Q, defined by

$$R_Q(e) \Leftrightarrow \{e\}^1 \text{ has property } Q,$$

is not recursive. Moreover, if Λ_1 has property Q, then R_Q is not RE.

We will now use Rice's theorem to show that the zero problem and the finite-domain problem are recursively unsolvable.

Corollary 3: ZERO is not recursive.

Proof: Let $\Gamma = \{Z\}$, where $Z: \mathbb{N} \to \mathbb{N}$ is the zero function. Clearly, $\Gamma \neq \varnothing$ and $\Gamma \neq$ all 1-ary partial recursive functions; hence, by Rice's theorem, R_Γ is not recursive. Moreover, for all $e \in \mathbb{N}$,

$$R_\Gamma(e) \Leftrightarrow \{e\}^1 \in \{Z\}$$
$$\Leftrightarrow \{e\}^1 \text{ is the zero function } Z$$
$$\Leftrightarrow \text{ZERO}(e).$$

Hence ZERO is not recursive, as required. ∎

Corollary 4: FIN is not RE.

Proof: Let

$$\Gamma = \{F: F \text{ a 1-ary partial recursive function with finite domain}\}.$$

Clearly, $\Gamma \neq \varnothing$ and $\Gamma \neq$ all 1-ary partial recursive functions. Moreover, $\Lambda_1 \in \Gamma$; hence, by Rice's theorem, FIN $= R_\Gamma$ is *not* RE. ∎

Let us sharpen the result that ZERO is not recursive.

Theorem 5: TOT $\leq_m$ ZERO; ZERO is Π_2 but not RE.

Proof: ZERO$(e) \Leftrightarrow \forall a \, \exists c[T_1(e, a, c) \wedge (U(c) = 0)]$ shows that ZERO is Π_2. To see that TOT $\leq_m$ ZERO, apply the parameter theorem to the 2-ary partial recursive function

$$F(a, e) \simeq \begin{cases} 0 & \text{HLT}(e, a) \\ \text{undefined} & \neg\text{HLT}(e, a). \end{cases}$$

There is a recursive function $\phi: \mathbb{N} \to \mathbb{N}$ such that $F(a, e) \simeq \{\phi(e)\}^1(a)$ for all $a, e \in \mathbb{N}$. The proof that TOT $\leq_m$ ZERO now proceeds as follows:

$$\text{TOT}(e) \Leftrightarrow \{e\}^1 \text{ is a total function}$$
$$\Leftrightarrow \forall a[\text{HLT}(e, a)]$$
$$\Leftrightarrow \forall a[F(a, e) = 0]$$
$$\Leftrightarrow \forall a[\{\phi(e)\}^1(a) = 0]$$
$$\Leftrightarrow \{\phi(e)\}^1 \text{ is the zero function}$$
$$\Leftrightarrow \text{ZERO}(\phi(e)).$$

Finally, since TOT $\leq_m$ ZERO and TOT is not RE, ZERO cannot be RE. ∎

To summarize, we have these statements:

(1) TOT is Π_2 but not RE; TOT$(e) \Leftrightarrow \forall a \exists c T_1(e, a, c)$.
(2) ZERO is Π_2 but not RE; ZERO$(e) \Leftrightarrow \forall a \exists c[T_1(e, a, c) \wedge (U(c) = 0)]$.
(3) FIN is Σ_2 but not RE; FIN$(e) \Leftrightarrow \exists a \forall b[b \in W_e \to (b \leq a)]$.

In the exercises, we will outline proofs that neither TOT nor ZERO is Σ_2 and that FIN is not Π_2.

——*Fixed Point Theorem and Recursion Theorem*——

We will now use the parameter theorem to prove two fundamental results about indices: the fixed point theorem and the recursion theorem.

Theorem 6 (Fixed point theorem): Let $n \geq 1$ and let $\phi: \mathbb{N} \rightarrow \mathbb{N}$ be recursive. Then there exists $e \in \mathbb{N}$ such that $\{e\}^n = \{\phi(e)\}^n$.

Proof: Consider the $n + 1$-ary function F defined by

$$F(a_1, \ldots, a_n, b) \simeq \begin{cases} \{\phi(\{b\}^1(b))\}^n(a_1, \ldots, a_n) & K(b) \\ \text{undefined} & \neg K(b). \end{cases}$$

In other words, $F(a_1, \ldots, a_n, b) \simeq \Phi_n(\phi(\Phi_1(b, b)), a_1, \ldots, a_n)$; hence F is partial recursive. (This choice of F is by no means obvious; however, as we proceed, the idea behind F will become clear.) By the parameter theorem, there is a recursive function $\psi: \mathbb{N} \rightarrow \mathbb{N}$ such that for all $a_1, \ldots, a_n, b \in \mathbb{N}$, $F(a_1, \ldots, a_n, b) \simeq \{\psi(b)\}^n(a_1, \ldots, a_n)$. Now eliminate ψ: Let f be an index for ψ, and note that $\psi(f) = \{f\}^1(f)$. Hence for all $a_1, \ldots, a_n \in \mathbb{N}$,

$$F(a_1, \ldots, a_n, f) \simeq \{\{f\}^1(f)\}^n(a_1, \ldots, a_n).$$

Our goal is an $e \in \mathbb{N}$ such that $\{\phi(e)\}^n = \{e\}^n$. Let $e = \{f\}^1(f)$, and note that e exists because ψ is total. We now have, for all $a_1, \ldots, a_n \in \mathbb{N}$,

$$F(a_1, \ldots, a_n, f) \simeq \{e\}^n(a_1, \ldots, a_n).$$

Finally, $K(f)$ holds, so $F(a_1, \ldots, a_n, f) \simeq \{\phi(\{f\}^1(f))\}^n(a_1, \ldots, a_n)$ by the definition of F; but $e = \{f\}^1(f)$, so

$$\{\phi(e)\}^n(a_1, \ldots, a_n) \simeq \{e\}^n(a_1, \ldots, a_n)$$

as required. ■

Comment: For an interesting discussion of the fixed point theorem in terms of programs, the reader is referred to Odifreddi's *Classical Recursion Theory* (p. 52) and the article "Computer Viruses: Diagonalization and Fixed Points" by W. F. Dowling, *Notices AMS* 37 (1990), 858-61.

Example 2:

There exists $e \in \mathbb{N}$ such that $\{e\}^1$ and $\{e^2\}^1$ are the same function. To see this, use the fixed point theorem with $\phi(a) = a^2$. ☐

Theorem 7 (Recursion theorem): Let H be an $n + 1$-ary partial recursive function. Then there is an n-ary partial recursive function F with index f such that for all $a_1, \ldots, a_n \in \mathbb{N}$,

$$F(a_1, \ldots, a_n) \simeq H(a_1, \ldots, a_n, f).$$

In other words, $\{f\}^n(a_1, \ldots, a_n) \simeq H(a_1, \ldots, a_n, f)$.

Proof: By the parameter theorem, applied to H, there is a recursive function $\phi: \mathbb{N} \to \mathbb{N}$ such that for all $a_1, \ldots, a_n, e \in \mathbb{N}$,

$$H(a_1, \ldots, a_n, e) \simeq \{\phi(e)\}^n(a_1, \ldots, a_n).$$

By the fixed point theorem, applied to ϕ, there exists $f \in \mathbb{N}$ such that $\{\phi(f)\}^n = \{f\}^n$. Hence for all $a_1, \ldots, a_n \in \mathbb{N}$,

$$H(a_1, \ldots, a_n, f) \simeq \{f\}^n(a_1, \ldots, a_n).$$

We also note that the required partial recursive function F is $\{f\}^n$. ∎

The recursion theorem has an interpretation in terms of programs. Let P be a program that computes an $n + 1$-ary function H. Then there is a program Q with code f such that for all $a_1, \ldots, a_n \in \mathbb{N}$, the P-computation with inputs $a_1, \ldots, a_n, f$ and the Q-computation with inputs $a_1, \ldots, a_n$ give the same result.

Corollary 5: There exists $e \in \mathbb{N}$ such that for all $a \in \mathbb{N}$, $\{e\}^1(a) = e$.

Proof: Let $H(a, b) = b$. By the recursion theorem, there is an index e such that for all $a \in \mathbb{N}$, $\{e\}^1(a) \simeq H(a, e)$. But $H(a, e) = e$, so $\{e\}^1(a) = e$ for all $a \in \mathbb{N}$ as required. ∎

Corollary 5 has an interesting interpretation in terms of programs: There is a program P with code e such that for all $a \in \mathbb{N}$, the P-computation with input a halts and the output is e, the code of P. Such a program is said to be *self-reproducing*. This is a remarkable result; to write such a program P, it would seem that we would first need to know P so that we could arrange for the output always to be the code of P!

Example 3:

(*Fibonacci sequence is recursive*) In Section 8.4, we used course-of-values recursion to show that the function $G: \mathbb{N} \to \mathbb{N}$, defined by $G(0) = G(1) = 1$ and $G(n) = G(n - 1) + G(n - 2)$ for $n > 1$, is recursive. We will now give another proof, based on the recursion theorem. Define a 2-ary function H by

$$H(n, e) \simeq \begin{cases} 1 & (n = 0) \vee (n = 1) \\ \{e\}^1(n \dotminus 1) + \{e\}^1(n \dotminus 2) & n > 1. \end{cases}$$

H is partial recursive by definition by cases. By the recursion theorem, there is a partial recursive function F with index f such that for all $n \in \mathbb{N}$, $F(n) \simeq H(n, f)$. We can now use induction to prove that for all $n \in \mathbb{N}$, $F(n) = G(n)$; since F is partial recursive and total, G is recursive as required. □

————Exercises for Section 9.5————

1. (RE *version of parameter theorem*) Let R be a 2-ary RE relation. Show that there is a recursive function $\phi: \mathbb{N} \to \mathbb{N}$ such that for all $a, e \in \mathbb{N}$, $R(a, e) \Leftrightarrow a \in W_{\phi(e)}$. (*Hint: R is the domain of a partial recursive function F; apply the parameter theorem to F.)

2. Show that there is a recursive function $\phi: \mathbb{N} \to \mathbb{N}$ such that for all $e \in \mathbb{N}$, dom $\{\phi(e)\}^1$ is just e. (*Hint:* Parameter theorem.)

3. Show that there exists $e \in \mathbb{N}$ such that the domain of $\{e\}^1$ is just e. (*Hint:* See Exercise 2 and use the fixed point theorem. Alternatively, use the recursion theorem, as in Corollary 5.)

4. Prove that the function ϕ constructed in the parameter theorem is one-to-one. Then use this fact to show that if G is a 1-ary partial recursive function, then G has infinitely many indices. (*Hint:* Apply the parameter theorem to $F(a, e) \simeq G(a)$.)

5. (*Alternate version of parameter theorem*) Prove the following version of the parameter theorem: Let F be an $n + 1$-ary partial recursive function. Then there is a recursive function $\phi: \mathbb{N} \to \mathbb{N}$ such that for all $e, a_1, \ldots, a_n \in \mathbb{N}$, $F(e, a_1, \ldots, a_n) \simeq \{\phi(e)\}^n (a_1, \ldots, a_n)$. (*Hint:* Use the parameter theorem.)

6. (*General parameter theorem*) Prove the following general version of the parameter theorem, alternate version (see Exercise 5): Let $m, n \geq 1$; let F be an $m + n$-ary partial recursive function. Then there is a recursive function $\phi: \mathbb{N}^m \to \mathbb{N}$ such that

$$F(e_1, \ldots, e_m, a_1, \ldots, a_n) \simeq \{\phi(e_1, \ldots, e_m)\}^n (a_1, \ldots, a_n).$$

(*Hint:* Let $F^*(e, a_1, \ldots, a_n) \simeq F(J_{m,1}(e), \ldots, J_{m,m}(e), a_1, \ldots, a_n)$ and apply the result in Exercise 5 to F^*.)

7. (*Uniform parameter theorem or s-m-n theorem*) Let $m, n \geq 1$. Show that there is a recursive function $\phi: \mathbb{N}^{m+1} \to \mathbb{N}$ such that

$$\{e\}^{m+n}(b_1, \ldots, b_m, a_1, \ldots, a_n) \simeq \{\phi(e, b_1, \ldots, b_m)\}^n (a_1, \ldots, a_n).$$

(*Hint:* Apply Exercise 6 to the $m + n + 1$-ary partial recursive function $\Phi_{m+n}(e, b_1, \ldots, b_m, a_1, \ldots, a_n).$)

8. Given $e, f \in \mathbb{N}$, $\{e\}^1 + \{f\}^1$ is a 1-ary partial recursive function; let $\phi(e, f)$ be an index of $\{e\}^1 + \{f\}^1$. Show that ϕ can be constructed so that it is recursive. (*Hint:* Apply the general parameter theorem in Exercise 6 to the function $F(e, f, a) \simeq \{e\}^1(a) + \{f\}^1(a)$.)

9. (*Basic properties of $\leq_m$*) Prove the following:
   (1) $\leq_m$ is reflexive and transitive.
   (2) $Q \leq_m R$ if and only if $\neg Q \leq_m \neg R$.
   (3) $\leq_m$ is not symmetric.
   (4) If R is RE but not recursive, neither $R \leq_m \neg R$ nor $\neg R \leq_m R$.
   (5) $\leq_m$ is not linear.
   (6) If Q is recursive, $R \neq \varnothing$, and $R \neq \mathbb{N}$, then $Q \leq_m R$.

10. Prove part (1) of Rice's theorem.

11. (*Alternate version of Rice's theorem*) A set $R \subseteq \mathbb{N}$ is said to *respect* functions if whenever $e \in R$ and e is an index of F, then $f \in R$ for every index f of F. Prove the following version of Rice's theorem: If $R \subseteq \mathbb{N}$, $R \neq \varnothing$, $R \neq \mathbb{N}$, and R respects functions, then R is not recursive. Moreover, if the indices of Λ_1 are in R, then R is not RE.

12. (*One-to-one problem*) In this exercise, we outline a proof that there is no algorithm that, given a 1-ary partial recursive function F, decides whether F is one-to-one. Define a 1-ary relation ONE-ONE by ONE-ONE$(e) \Leftrightarrow \{e\}^1$ is one-to-one. Prove the following: (1) ONE-ONE is not recursive; (2) $\neg$ONE-ONE is RE; (3) ONE-ONE is not RE.

13. (*Empty domain problem*) In this exercise, we outline a proof that there is no algorithm that, given a 1-ary partial recursive function F, decides whether the domain of F is empty. Define a 1-ary relation EMPTY by EMPTY$(e) \Leftrightarrow W_e = \varnothing$. Prove the following: (1) EMPTY is not RE; (2) $\neg$EMPTY is RE.

14. (*Input, output problems*) Let $a \in \mathbb{N}$ be fixed. Show that each of the following relations is RE but not recursive:
    (1) $INPUT(e) \Leftrightarrow a \in W_e$.
    (2) $OUTPUT(e) \Leftrightarrow a \in \text{ran} \{e\}^1$. (*Note:* ran F is the range of the function F.)

15. (*Verification problem for programs*) In this exercise, we outline a proof that there is no algorithm that, given two programs P_1 and P_2, decides whether P_1 and P_2 compute the same 1-ary partial function. Define a 2-ary relation by $VERIFY(e, f) \Leftrightarrow \{e\}^1 = \{f\}^1$. Show that VERIFY is not RE. (*Hint:* ZERO is not RE.)

16. Prove that each of the following relations is RE but not recursive:
    (1) $R(e) \Leftrightarrow \{e\}^1(e) = 0$.
    (2) $R(e, a) \Leftrightarrow \{e\}^1(a) = 0$.
    (3) $R(e, a) \Leftrightarrow a \in \text{ran} \{e\}^1$.

17. Let $R(e) \Leftrightarrow W_e$ is recursive; show that R is not RE.

18. (*Domain problem or verification problem for RE sets*) In this exercise, we outline a proof that there is no algorithm that, given RE sets Q and R, decides whether $Q = R$. Define a 2-ary relation by $DOMAIN(e, f) \Leftrightarrow W_e = W_f$. Show that DOMAIN is not RE. (*Hint:* Let f_0 be an index of a total recursive function; TOT is not RE.)

19. (*TOT is Π_2 but not Σ_2*) Prove the following:
    (1) TOT is Π_2.
    (2) $R \leq_m$ TOT for every 1-ary Π_2 relation R. (*Hint:* Let $R(e) \Leftrightarrow \forall a \exists b Q(e, a, b)$ be Π_2 and apply the RE parameter theorem (Exercise 1) to the RE relation $P(e, a) \Leftrightarrow \exists b Q(e, a, b)$.)
    (3) If R is a 1-ary relation such that every 1-ary Π_2 relation is m-reducible to R, then R cannot be Σ_2 (Hin hierarchy theorem (see Section 9.4), there is a 1-ary relation that is Π_2 but not Σ_2.)
    (4) TOT is not Σ_2.

20. (*ZERO is Π_2 but not Σ_2*) Prove the following:
    (1) ZERO is Π_2.
    (2) ZERO is not Σ_2. (*Hint:* TOT $\leq_m$ ZERO and TOT is not Σ_2.)

21. (*FIN is Σ_2 but not Π_2*) Prove the following:
    (1) FIN is Σ_2.
    (2) $\neg$TOT $\leq_m$ FIN. (*Hint:* Let $R(a, e) \Leftrightarrow \forall b \leq a[b \in W_e]$; prove that R is RE and then apply the RE parameter theorem (Exercise 1) to R to obtain $\phi: \mathbb{N} \to \mathbb{N}$ recursive such that $R(a, e) \Leftrightarrow a \in W_{\phi(e)}$ for all $a, e \in \mathbb{N}$. Then prove: If $TOT(e)$, then $TOT(\phi(e))$; if $\neg TOT(e)$, then $FIN(\phi(e))$.)
    (3) FIN is not Π_2. (*Hint:* $\neg$TOT is not Π_2.)

22. Let $INF(e) \Leftrightarrow W_e$ is infinite. Note that $INF = FIN^c$. Use previous exercises to prove the following: (1) INF is Π_2; (2) TOT $\leq_m$ INF; (3) INF is not Σ_2.

23. Two 1-ary relations Q, R are *many-one-equivalent* if $Q \leq_m R$ and $R \leq_m Q$; in this case, we write $Q \equiv_m R$. Prove the following:
    (1) $\equiv_m$ is an equivalence relation on $P(\mathbb{N})$.
    (2) For any $a \in \mathbb{N}$, $K \equiv_m DOM_a$, where $DOM_a(e) \Leftrightarrow a \in \text{dom} \{e\}^1$.
    (3) ZERO $\equiv_m$ TOT.
    (4) $\neg$TOT $\equiv_m$ FIN.

24. Derive the fixed point theorem from the recursion theorem.

25. (*Rice's theorem from fixed point theorem*) Prove: If Γ is a proper set of 1-ary partial recursive functions, then R_Γ is not recursive. *Hint:* Suppose R_Γ is recursive; let $e_0 \in R_\Gamma$ and $e_1 \notin R_\Gamma$, and apply the fixed point theorem to

$$\phi(a) = \begin{cases} e_0 & a \notin R_\Gamma \\ e_1 & a \in R_\Gamma. \end{cases}$$

26. (*RE versions of recursion theorem and fixed point theorem*) Prove the following:

(1) Let R be a 2-ary RE relation. Then there exists $e \in \mathbb{N}$ such that for all $a \in \mathbb{N}$, $R(a, e) \Leftrightarrow a \in W_e$.

(2) Let $\phi: \mathbb{N} \to \mathbb{N}$ be recursive. Then there exists $e \in \mathbb{N}$ such that $W_{\phi(e)} = W_e$.

27. (*Extension of fixed point theorem*) Let $\phi: \mathbb{N} \to \mathbb{N}$ be recursive and let $a \in \mathbb{N}$. Show that there exists $e > a$ such that $\{e\}^1 = \{\phi(e)\}^1$. *Hint:* Choose $c \in \mathbb{N}$ such that $\{c\}^1 \neq \{0\}^1, \ldots, \{c\}^1 \neq \{a\}^1$ and apply the fixed point theorem to

$$\psi(k) = \begin{cases} c & k \leq a \\ \phi(k) & k > a. \end{cases}$$

28. (*Primitive recursion from the recursion theorem*) Let G be an n-ary recursive function, let H be an $n + 2$-ary recursive function, let F be the $n + 1$-ary function defined by

$$F(\mathbf{a}, 0) = G(\mathbf{a}) \qquad\qquad \mathbf{a} = \langle a_1, \ldots, a_n \rangle$$

$$F(\mathbf{a}, b + 1) = H(\mathbf{a}, b, F(\mathbf{a}, b)).$$

Use the recursion theorem to show F recursive. *Hint:* Define an $n + 2$-ary partial function L by

$$L(\mathbf{a}, b, f) \simeq \begin{cases} G(\mathbf{a}) & b = 0 \\ H(\mathbf{a}, b \doteq 1, \{f\}^{n+1}(\mathbf{a}, b \doteq 1)) & b > 0. \end{cases}$$

(1) Show that L is partial recursive.

(2) By the recursion theorem, there exists $f \in \mathbb{N}$ such that $\{f\}^{n+1}(\mathbf{a}, b) \simeq L(\mathbf{a}, b, f)$. Show that $\{f\}^{n+1}(\mathbf{a}, b) = F(\mathbf{a}, b)$ and conclude that F is recursive.

29. Let $G: \mathbb{N} \to \mathbb{N}$ and $H: \mathbb{N}^3 \to \mathbb{N}$ be recursive, and define $F: \mathbb{N}^2 \to \mathbb{N}$ by

$$F(a, 0) = G(a)$$

$$F(a, b + 1) = H(a, b, F(a + 3, b)).$$

(1) Verify that F is defined for all $a, b \in \mathbb{N}$. (*Hint:* Induction on b.)

(2) Use the recursion theorem to show that F is recursive.

30. (*Ackermann's function ψ is recursive*) Ackermann's function, a recursive function not primitive recursive, is defined by

$$\psi(a, 0) = a + 1$$

$$\psi(0, b + 1) = \psi(1, b)$$

$$\psi(a + 1, b + 1) = \psi(\psi(a, b + 1), b).$$

In this exercise, we use the recursion theorem to show that ψ is recursive.

(1) Show that H is partial recursive, where H is defined by

$$H(a, b, f) \simeq \begin{cases} a + 1 & b = 0 \\ \{f\}^2(1, b \doteq 1) & a = 0, b > 0 \\ \{f\}^2(\{f\}^2(a \doteq 1, b), b \doteq 1) & a > 0, b > 0. \end{cases}$$

(2) By the recursion theorem, there exists $f \in \mathbb{N}$ such that $\{f\}^2(a, b) \simeq H(a, b, f)$. Show that $\{f\}^2(a, b) = \psi(a, b)$. (*Hint:* Double induction; first b, then a.)

31. (*Extension of fixed point theorem*) Let $n \geq 1$ and let $F: \mathbb{N} \times \mathbb{N} \to \mathbb{N}$ be recursive. Show that there is a recursive function $\phi: \mathbb{N} \to \mathbb{N}$ such that for all $b \in \mathbb{N}$, $\{\phi(b)\}^n = \{F(b, \phi(b))\}^n$. (*Hint:* Apply the general parameter theorem (Exercise 6) to an appropriate $n + 2$-ary partial recursive function G to obtain a recursive function $\psi: \mathbb{N} \times \mathbb{N} \to \mathbb{N}$ such that for all $a_1, \ldots, a_n, b, c \in \mathbb{N}$, $G(b, c, a_1, \ldots, a_n) \simeq \{\psi(b, c)\}^n(a_1, \ldots, a_n)$. Then apply the recursion theorem to ψ.

32. Show that there exists $R \subseteq \text{TOT}$; R is RE but not recursive.

9.6

Semi-Thue Systems and Word Problems

We begin with an example of a "substitution puzzle." Let $\Sigma = \{A, B, C, \ldots, Z\}$ be the 26 letters of the alphabet and let

$$\text{CS} \to \text{C}$$
$$\text{MAR} \to \text{L}$$
$$\text{MAT} \to \text{G}$$
$$\text{THE} \to \text{RO}$$

be a finite list of "substitution rules." These rules are used as follows: Suppose we have a word in which THE occurs; we may then replace this occurrence of THE with RO to obtain a new word.

Now choose two words, say MATHEMATICS and LOGIC. Can we find a finite sequence of substitutions that will allow us to obtain LOGIC from MATHEMATICS?

$$\text{MATHEMATICS} \Rightarrow \text{MATHEMATIC}$$
$$\text{MATHEMATIC} \Rightarrow \text{MAROMATIC}$$
$$\text{MAROMATIC} \Rightarrow \text{LOMATIC}$$
$$\text{LOMATIC} \Rightarrow \text{LOGIC}.$$

Next question: Can we obtain DULL from LOGIC? Certainly not!

Although this is a rather frivolous example, it turns out that many important ideas in mathematics and computer science can be expressed in terms of substitution puzzles. We will see, for example, that the RE relations can be characterized in terms of such puzzles. In addition, we have the following interesting decision problem: Given a finite set of substitution rules, is there an algorithm that, given two words W and V, decides whether V can be obtained from W by a finite sequence of substitutions? Such decision problems are called *word problems*. The first mathematician to study word problems was the Norwegian Axel Thue (pronounced "too-ay"). In this section we will prove that (1) there is a semi-Thue system Π for which the word problem is unsolvable; (2) there is a Thue system Π^+ for which the word problem is unsolvable. These results will then be used to show that there is a semigroup for which the word problem is unsolvable. This 1947 result was proved independently by Post and Markov, and is of interest for historical reasons: It is the first example of a decision problem outside of logic that was proved unsolvable using recursion theory. For more on the ubiquity of substitution puzzles, see the paper "Solvable and unsolvable problems" in *Collected Works of A. M. Turing: Pure Mathematics*, edited by J. L. Britton North Holland (Amsterdam), 1992.

First we need some definitions. By an *alphabet* Σ, we mean a nonempty but finite set of symbols. A *word* on Σ is a finite sequence of symbols in Σ. For example, $\Sigma = \{a, b\}$ is an alphabet with two symbols; *abba* and *aaab* are words on Σ. The empty word is allowed and is denoted by Λ. If Σ is an alphabet, Σ^* denotes the set of all words on Σ, including the empty word; Σ^+ denotes the set of nonempty words.

The basic operation with words is that of *juxtaposition*: If W and V are words, then WV is also a word. For example, if W is *aa* and V is *bb*, then WV is *aabb*. It is clear that the operation of juxtaposition is associative (if W, U, and V are words, then $W(UV)$ and $(WU)V$ are the same words). For any word W, we have $W\Lambda = \Lambda W = W$.

Let A and W be words: A *occurs* in W if there exist words W_1 and W_2 (one or both may be the empty word) such that $W = W_1AW_2$. For example, *aa* occurs once in *aa* and three times in *baabbaabbaa*.

Definition 1: Let Σ be an alphabet. A *production* on Σ is an ordered pair $\langle A, B \rangle$ of words on Σ; however, we will use the more convenient notation $A \rightarrow B$ to denote a production. Now suppose P is the production $A \rightarrow B$ on Σ, and let W and V be words on Σ. We say that P *produces V from W*, written $W \Rightarrow_P V$, if the following hold: (1) A occurs at least once in W; (2) the word V is obtained from W by replacing exactly one occurrence of A with B.

Example 1:

Let P be the production $a \rightarrow aa$. Then $aba \Rightarrow_P aaba$, and also $aba \Rightarrow_P abaa$.

$\square$

Definition 2: A *semi-Thue system* Π consists of an alphabet Σ and a finite nonempty set of productions on Σ, say $A_1 \rightarrow B_1, \ldots, A_k \rightarrow B_k$. Given a semi-Thue system Π and two words W and V, we say that:

(1) V is *directly derivable from* W, written $W \Rightarrow_\Pi V$, if there is a production P of Π such that $W \Rightarrow_P V$.

(2) V is *derivable from* W, written $W \vdash_\Pi V$, if $W = V$ or if there is a finite sequence $W_1, \ldots, W_k$ of words on Σ with $W_1 = W$, $W_k = V$, and $W_i \Rightarrow_\Pi W_{i+1}$ for $1 \leq i \leq k - 1$.

Example 2:

We will show that $S \vdash_\Pi (()())()$, where Π is the semi-Thue system with alphabet $\Sigma = \{S, (,)\}$ and productions

$$S \rightarrow (), \quad S \rightarrow (S), \quad S \rightarrow SS.$$

We have

$$S \Rightarrow_\Pi SS$$
$$\Rightarrow_\Pi (S)S$$
$$\Rightarrow_\Pi (SS)S$$
$$\Rightarrow_\Pi (()S)S$$
$$\Rightarrow_\Pi (()())S$$
$$\Rightarrow_\Pi (()())().$$

This is a special type of semi-Thue system, known as a *context-free grammar*. ◻

The relation $\vdash_\Pi$ is a binary relation on Σ^* that is reflexive and transitive. More generally, we have the following basic properties of $\vdash_\Pi$ (proofs left to the reader).

Lemma 1 (Basic properties of $\vdash_\Pi$): The following hold for any semi-Thue system Π and words U, V, and W:

(1) $W \vdash_\Pi W$ (reflexive).

(2) If $W \vdash_\Pi U$ and $U \vdash_\Pi V$, then $W \vdash_\Pi V$ (transitive).

(3) If $W \vdash_\Pi V$, then $UW \vdash_\Pi UV$ and $WU \vdash_\Pi VU$.

Lemma 2: Let Π be a semi-Thue system with alphabet Σ and let $\mathbf{W} \subseteq \Sigma^*$ satisfy the following:

(1) If $W \in \mathbf{W}$ and $W \Rightarrow_\Pi V$, then $V \in \mathbf{W}$ (closure).

(2) If $W \in \mathbf{W}$, and if $W \Rightarrow_\Pi V$ and $W \Rightarrow_\Pi V'$, then $V = V'$ (deterministic).

Let $W_1, \ldots, W_k$ and $V_1, \ldots, V_k$ be words with $W_i \Rightarrow_\Pi W_{i+1}$ and $V_i \Rightarrow_\Pi V_{i+1}$ for $1 \le i \le k - 1$. If $W_1 = V_1$ and $W_1 \in \mathbf{W}$, then $W_k = V_k$.

Proof: Use induction; the details are left to the reader. ∎

Definition 3: Let Π be a semi-Thue system. The *word problem for* Π asks whether there is an algorithm that, given words W and V, decides whether $W \vdash_\Pi V$. If such an algorithm exists, we say that the word problem for Π is *solvable*; otherwise the word problem for Π is *unsolvable*.

In the exercises, there are a number of examples of semi-Thue systems that are solvable. But our goal is to prove the existence of a semi-Thue system for which the word problem is unsolvable. The proof of this result uses the fact that for each program P for the register machine, there is a semi-Thue system that simulates P; since there is a program for which the halting problem is unsolvable, there is a semi-Thue system for which the word problem is unsolvable. The next result is fundamental to the rest of this section; the proof we give is from Shoenfield, *Recursion Theory* 1993.

Theorem 1 (Simulation): Let P be a program for the register machine. Then there is a semi-Thue system that simulates P. More precisely, there is a semi-Thue system $\Pi(P)$ with words V and $\{W_a: a \in \mathbb{N}\}$ such that for all $a \in \mathbb{N}$,

the P-computation with input a halts if and only if $W_a \vdash_{\Pi(P)} V$.

Proof: Let $I_1, \ldots, I_t$ be the instructions of P, let $w = \lambda(P)$, and assume P is in standard form. We first describe the alphabet Σ for the required semi-Thue system $\Pi(P)$ and the words V and $\{W_a: a \in \mathbb{N}\}$.

Symbols	Comment
x	Serves as a tally; thus xx (written x^2) is 2, x^3 is 3, and so on.
b	First symbol in certain words, including V and W_a for $a \in \mathbb{N}$.
$b_1, \ldots, b_w, b_{w+1}$	Separate the numbers in registers $R_1, \ldots, R_w$. *Example:* The word $b_1 x^3 b_2 b_3 x^2 b_4 \cdots$ tells us that 3 is in register R_1, 0 is in register R_2, 2 is in register R_3, and so on.
$c_1, \ldots, c_t, c_{t+1}$	One for each instruction $I_1, \ldots, I_t$, including c_{t+1} for "halt."
$d_1, \ldots, d_t$	Used to change c_j to c_{j+1} (next instruction).
$e_1, \ldots, e_t$	Used to change c_j to c_q (jump).

The words V and $\{W_a: a \in \mathbb{N}\}$ are

$$V = bc_{t+1}, \quad W_a = bc_1 b_1 x^a b_2 b_3 \cdots b_w b_{w+1}.$$

Informally, word W_a tells us that a is in register R_1, 0 is in registers $R_2, \ldots, R_w$, and instruction 1 is the next (= first) instruction; word V tells us that the program halts.

To better understand how symbols are used, let us associate a word with each w-step. If $s = \langle r_1, \ldots, r_w; j \rangle$ is a w-step, W_s is the word

$$W_s = bc_j b_1 x^{r_1} b_2 \cdots b_w x^{r_w} b_{w+1} \quad (x^0 \text{ means } x \text{ does not occur}).$$

Informally, the word W_s tells us that $r_1, \ldots, r_w$ are in registers $R_1, \ldots, R_w$ respectively and that the next instruction is I_j. Note that if s is the initial w-step $\langle a, 0, \ldots, 0; 1 \rangle$, then W_s is W_a.

Next we construct the productions for $\Pi(P)$. Let $s = \langle r_1, \ldots, r_w; j \rangle$ be a w-step and let s' be the P-successor of s. We require productions so that $W_s \vdash_{\Pi(P)} W_{s'}$. Also, if s is the last step of a P-computation that halts, we want productions to obtain $W_s \vdash_{\Pi(P)} V$. In summary: If a P-computation with input *a* that halts consists of *w*-steps $s_1, \ldots, s_m$, we want productions so that $W_{s_1} \vdash_{\Pi(P)} W_{s_2}, \ldots, W_{s_{m-1}} \vdash_{\Pi(P)} W_{s_m}, W_{s_m} \vdash_{\Pi(P)} V$, and hence $W_a \vdash_{\Pi(P)} V$. (We will also want: If $W_a \vdash_{\Pi(P)} V$, then the *P*-computation with input *a* halts.)

Suppose $s = \langle r_1, \ldots, r_w; j \rangle$ with $j \leq t$, and let *s'* be the *P*-successor of *s*. The productions required to obtain $W_s \vdash_{\Pi(P)} W_{s'}$ depend on instruction I_j according to one of four cases, as follows:

- I_j is $S(k)$ The required productions are

  $(1)_j \ c_j b_i \to b_i c_j \ (1 \leq i \leq k - 1)$ $(4)_j \ x d_j \to d_j x$
  $(2)_j \ c_j x \to x c_j$ $(5)_j \ b d_j \to b c_{j+1}$
  $(3)_j \ b_i d_j \to d_j b_i \ (1 \leq i \leq k - 1)$ $(6)_j \ c_j b_k \to d_j b_k x.$

Discussion In this case, we have (for $k > 1$):

$$W_s = b c_j b_1 x^{r_1} b_2 \cdots b_k x^{r_k} b_{k+1} \cdots b_w x^{r_w} b_{w+1},$$
$$W_{s'} = b c_{j+1} b_1 x^{r_1} b_2 \cdots b_k x^{r_k+1} b_{k+1} \cdots b_w x^{r_w} b_{w+1}.$$

The productions in $(1)_j$ and $(2)_j$ enable us to move c_j to the right until it is just before b_k. Production $(6)_j$ changes c_j to d_j and increases the number of *x*'s between b_k and b_{k+1} by 1. Productions $(3)_j$ and $(4)_j$ move d_j to the left until it is just after *b*. Finally, production $(5)_j$ changes d_j to c_{j+1}. In summary, we have sufficient productions to obtain $W_s \vdash_{\Pi(P)} W_{s'}$.

- I_j is $D(k)$ The required productions are $(1)_j$–$(5)_j$, as described for $S(k)$ and the two productions

  $(6)_j \ c_j b_k x \to d_j b_k$ $(7)_j \ c_j b_k b_{k+1} \to d_j b_k b_{k+1}.$

Discussion Assume c_j is just before b_k. If the number in register R_k is not 0, we have $b \cdots c_j b_k x \cdots x b_{k+1} \cdots$, and $(6)_j$ replaces c_j with d_j and diminishes the number of *x*'s by 1. If the number in register R_k is 0, we have $b \ldots c_j b_k b_{k+1} \ldots$ and $(7)_j$ replaces c_j with d_j. The reader should check that $W_s \vdash_{\Pi(P)} W_{s'}$.

- I_j is $Z(k)$ The required productions are $(1)_j$–$(5)_j$ as described for $S(k)$ and the two productions

  $(6)_j \ c_j b_k x \to c_j b_k$ $(7)_j \ c_j b_k b_{k+1} \to d_j b_k b_{k+1}.$

The reader should check that $W_s \vdash_{\Pi(P)} W_{s'}$.

- I_j is $J(k, q)$ The required productions are $(1)_j$–$(5)_j$ as described for $S(k)$ and these additional productions:

  $(6)_j \ c_j b_k x \to d_j b_k x$ $(9)_j \ x e_j \to e_j x$
  $(7)_j \ c_j b_k b_{k+1} \to e_j b_k b_{k+1}$ $(10)_j \ b e_j \to b c_q.$
  $(8)_j \ b_i e_j \to e_j b_i \ (1 \leq i \leq k - 1)$

Discussion Assume c_j is just before b_k. If the number in register R_k is not 0 (i.e., if we have $b \ldots c_j b_k x \ldots x b_{k+1} \ldots$), $(6)_j$ replaces c_j with d_j; productions $(3)_j$, $(4)_j$, and $(5)_j$ move d_j to the left until it is just after *b* and then replace d_j with c_{j+1}. On the other hand, if the number in register R_k is 0 (i.e., if we have $b \ldots c_j b_k b_{k+1} \ldots$), production $(7)_j$ replaces c_j with e_j; productions $(8)_j$, $(9)_j$, and $(10)_j$ move e_j to the left until it is just after b and then replace e_j with c_q. The reader should check that $W_s \vdash_{\Pi(P)} W_{s'}$.

Note that each instruction I_j $(1 \le j \le t)$ determines certain productions $(1)_j$, $(2)_j$, ... Finally, suppose we have a P-terminal w-step $s = \langle r_1, \ldots, r_w; t + 1 \rangle$, so that $W_s = bc_{t+1}b_1 x^{r_1} b_2 \ldots b_w x^{r_w} b_{w+1}$. The productions required for $W_s \vdash_{\Pi(P)} V$ (V is bc_{t+1}) are

$$c_{t+1}b_i \to c_{t+1} \quad (1 \le i \le w + 1); \qquad c_{t+1}x \to c_{t+1}.$$

This completes the description of the productions for the semi-Thue system $\Pi(P)$. We ask the reader to check that we have included enough productions so that the following hold:

(a) If $s = \langle r_1, \ldots, r_w; j \rangle$ is a w-step with $j \le t$ and s' is the P-successor of s, then $W_s \vdash_{\Pi(P)} W_{s'}$.

(b) If $s = \langle r_1, \ldots, r_w; t + 1 \rangle$, then $W_s \vdash_{\Pi(P)} V$.

Now let $a \in \mathbb{N}$; we must show that the P-computation with input a halts if and only if $W_a \vdash_{\Pi(P)} V$. First assume the P-computation with input a halts. Then there is a finite sequence $s_1, \ldots, s_m$ of w-steps with $s_1 = \langle a, 0, \ldots, 0; 1 \rangle$, s_{i+1} is the P-successor of s_i for $1 \le i \le m - 1$, and $s_m = \langle r_1, r_2, \ldots, r_w; t + 1 \rangle$. By (a) and (b), we have $W_{s_1} \vdash_{\Pi(P)} W_{s_2}, \ldots, W_{s_{m-1}} \vdash_{\Pi(P)} W_{s_m}, W_{s_m} \vdash_{\Pi(P)} V$. Since $W_a = W_{s_1}$, we obtain $W_a \vdash_{\Pi(P)} V$ as required.

Now assume $W_a \vdash_{\Pi(P)} V$. Let $\mathbf{W}$ be all words W such that exactly one symbol in $\{c_1, d_1, e_1, \ldots, c_t, d_t, e_t, c_{t+1}\}$ occurs in W. By looking at the productions for $\Pi(P)$, we can check that $\mathbf{W}$ satisfies conditions (1) and (2) (closure and deterministic) of Lemma 2. Suppose the P-computation with input a does not halt. Then there is an infinite sequence $s_1, s_2, \ldots$ of w-steps with $s_1 = \langle a, 0, \ldots, 0; 1 \rangle$ and s_{i+1} is the P-successor of s_i for $i = 1, 2, \ldots$. By (a), we have $W_{s_1} \vdash_{\Pi(P)} W_{s_2}, W_{s_2} \vdash_{\Pi(P)} W_{s_3}, \ldots$. It follows that there is an infinite sequence of words $W_1, W_2, \ldots$ with $W_1 = W_a$ and $W_i \Rightarrow_{\Pi(P)} W_{i+1}$ for $i = 1, 2, \ldots$. On the other hand, $W_a \vdash_{\Pi(P)} V$; hence there is a finite sequence of words $V_1, \ldots, V_k$ with $V_1 = W_a$, $V_k = V$, and $V_1 \Rightarrow_{\Pi(P)} V_2, \ldots, V_{k-1} \Rightarrow_{\Pi(P)} V_k$. Since $V_1 = W_1$ and $W_1 \in \mathbf{W}$, Lemma 2 applies and $V_k = W_k$. Now V_k is bc_{t+1}, and no production of $\Pi(P)$ applies to this word. On the other hand, $W_k \Rightarrow_{\Pi(P)} W_{k+1}$, and since $V_k = W_k$, we have $V_k \Rightarrow_{\Pi(P)} W_{k+1}$, a contradiction. ∎

Theorem 2 (Unsolvability of the word problem for semi-Thue systems): Assume Church's thesis. Then there is a semi-Thue system for which the word problem is unsolvable.

Proof: There is a program P for which the halting problem is unsolvable (see Corollary 5 in Section 9.3). Let $\Pi(P)$ be the semi-Thue system constructed in Theorem 1 (simulation). There are words V and $\{W_a : a \in \mathbb{N}\}$ such that for all $a \in \mathbb{N}$,

the P-computation with input a halts $\Leftrightarrow W_a \vdash_{\Pi(P)} V$.

An algorithm for the word problem for $\Pi(P)$ would give an algorithm for the halting problem for P. But the halting problem for P is unsolvable; hence the word problem for $\Pi(P)$ is also unsolvable. ∎

Another application of Theorem 1 is the following characterization of the 1-ary RE relations in terms of semi-Thue systems (see the exercises for the definition of *special*).

Theorem 3 (Characterization of RE relations by semi-Thue systems): A 1-ary relation R is RE if and only if there is a special semi-Thue system Π with words V and $\{W_a : a \in \mathbb{N}\}$ such that for all $a \in \mathbb{N}$, $R(a) \Leftrightarrow W_a \vdash_\Pi V$.

Outline of proof We outline the main ideas and leave the details as an exercise for the reader. By Theorem 9 in Section 9.3, there is a program P that simulates R, and by Theorem 1 there is a semi-Thue system Π that simulates P; thus there is a semi-Thue system (special) that simulates R. To prove the converse, suppose there is a semi-Thue system with words V and $\{W_a : a \in \mathbb{N}\}$ such that for all $a \in \mathbb{N}$, $R(a) \Leftrightarrow W_a \vdash_\Pi V$. Code all words and then prove that the derivability relation $\vdash_\Pi$ is RE. ∎

We have seen that there is a semi-Thue system for which the word problem is unsolvable. What happens if we look at a more restricted version of a semi-Thue system, in which each production applies in either direction?

Definition 4: A *Thue system* is a semi-Thue system Π such that for each production $A \to B$ of Π, $B \to A$ is also a production of Π; $B \to A$ is called the *inverse* of the production $A \to B$. Let Π be a Thue system; the *word problem for* Π asks whether there is an algorithm that, given words W and V, decides whether $W \vdash_\Pi V$. If no such algorithm exists, we say that the word problem for Π is *unsolvable*.

We will prove that there is even a Thue system for which the word problem is unsolvable; a key idea is Lemma 4, below. But first let us establish the symmetric property of Thue systems.

Lemma 3: Let Π be a Thue system. If $W \vdash_\Pi V$, then $V \vdash_\Pi W$ (symmetric).

Proof: Assume $W \vdash_\Pi V$. Then there exist words $W_1, \ldots, W_k$ with $W_1 = W$, $W_k = V$, and productions $P_1, \ldots, P_k$ such that for $1 \le i \le k - 1$, $W_i \Rightarrow_{P_i} W_{i+1}$. For $1 \le i \le k$, let Q_i be the inverse of P_i (if P_i is $A_i \to B_i$, Q_i is $B_i \to A_i$). Then $W_{i+1} \Rightarrow_{Q_i} W_i$ for $1 \le i \le k - 1$; hence $V \vdash_\Pi W$ as required. ∎

Definition 5: Let Π be a semi-Thue system. The Thue system *associated with* Π, denoted Π^+, is the Thue system obtained from Π by adding, for each production $A \to B$ of Π, the production $B \to A$.

Lemma 4 (Post): Let Π be a semi-Thue system with alphabet Σ, let Π^+ be the Thue system associated with Π, and let $\mathbf{W} \subseteq \Sigma^*$ satisfy the following:

   (1) If $W \in \mathbf{W}$ and $W \Rightarrow_{\Pi^+} V$, then $V \in \mathbf{W}$ (closure with respect to Π^+).
   (2) If $W \in \mathbf{W}$, and if $W \Rightarrow_\Pi V$ and $W \Rightarrow_\Pi V'$, then $V = V'$ (deterministic with respect to Π).

Let W and V be words with $W \in \mathbf{W}$ and V terminal with respect to Π (that is, no production of Π applies to V). Then $W \vdash_{\Pi^+} V$ if and only if $W \vdash_\Pi V$.

Proof: It is obvious that $W \vdash_\Pi V$ implies $W \vdash_{\Pi^+} V$. Assume $W \vdash_{\Pi^+} V$, and let $V_1, \ldots, V_k$ be a finite sequence of words with $W = V_1$, $V = V_k$, and $V_i \Rightarrow_{\Pi^+} V_{i+1}$ for $1 \leq i \leq k - 1$. Assume k is the smallest positive integer with this property. For this choice of k, we will prove that

$$V_1 \Rightarrow_\Pi V_2, \ldots, V_{k-1} \Rightarrow_\Pi V_k,$$

and hence $W \vdash_\Pi V$ as required.

First note that since $W \in \mathbf{W}$, the closure property guarantees that each $V_i \in \mathbf{W}$. Suppose there is some $i < k$ for which $V_i \Rightarrow_\Pi V_{i+1}$ fails, and assume i is the largest such positive integer. Now $V_i \Rightarrow_{\Pi^+} V_{i+1}$; hence $V_{i+1} \Rightarrow_\Pi V_i$. The word $V = V_k$ is terminal with respect to Π, so $V_{i+1} \neq V_k$; hence $i + 1 < k$. Now i is the largest positive integer for which $V_i \Rightarrow_\Pi V_{i+1}$ fails; hence $V_{i+1} \Rightarrow_\Pi V_{i+2}$. But $V_{i+1} \Rightarrow_\Pi V_i$ as well, so by the deterministic property, $V_{i+2} = V_i$. It follows that we can omit V_{i+1} in the sequence $V_1, \ldots, V_k$. But this contradicts the choice of k as the smallest positive integer for which $W = V_1$, $V = V_k$, and $V_i \Rightarrow_{\Pi^+} V_{i+1}$ for $1 \leq i \leq k - 1$. ∎

Theorem 4 (Unsolvability of the word problem for Thue systems): Assume Church's thesis. Then there is a Thue system for which the word problem is unsolvable.

Proof: There is a program P for which the halting problem is unsolvable. Let $\Pi(P)$ be the semi-Thue system constructed in Theorem 1 (simulation) and let Σ be the alphabet of $\Pi(P)$. There are words V and $\{W_a : a \in \mathbb{N}\}$ of Σ such that for all $a \in \mathbb{N}$,

the P-computation with input a halts $\Leftrightarrow W_a \vdash_{\Pi(P)} V$.

The required Thue system with an unsolvable word problem is $\Pi(P)^+$. To see this, let $\mathbf{W}$ be all words W of Σ such that exactly one symbol in $\{c_1, d_1, e_1, \ldots, c_t, d_t, e_t, c_{t+1}\}$ occurs in W. By looking at the productions for $\Pi(P)^+$ (see the proof of Theorem 1), we can check that $\mathbf{W}$ satisfies these conditions: $W_a \in \mathbf{W}$ for all $a \in \mathbb{N}$; V is terminal with respect to $\Pi(P)$; (1) and (2) of Lemma 4. It follows that for all $a \in \mathbb{N}$,

the P-computation with input a halts $\Leftrightarrow W_a \vdash_{\Pi(P)^+} V$.

Since the halting problem for P is unsolvable, the word problem for $\Pi(P)^+$ is unsolvable as required. ∎

There is a close connection between Thue systems and semigroups, and this will enable us to construct a semigroup with an unsolvable word problem. First we need several definitions.

Definition 6: A *semigroup* is an ordered pair $\langle S, \circ \rangle$, where S is a nonempty set and $\circ$ is a binary operation on S that is associative [$a \circ (b \circ c) = (a \circ b) \circ c$ for all $a, b, c \in S$].

Example 3:

The following are semigroups:

(1) $\langle \mathbb{N}^+, + \rangle$
(2) The set of all functions $F: \mathbb{N} \to \mathbb{N}$ with composition as operation
(3) The set of all words in an alphabet with juxtaposition as operation.

Definition 7: A semigroup $\langle S, \circ \rangle$ is *finitely generated* if there is a finite $G \subseteq S$ such that for all $b \in S$, there exist $a_1, \ldots, a_n \in G$ such that $b = a_1 \circ \cdots \circ a_n$; G is called a set of *generators* of S.

Example 4:

The semigroup $\langle \mathbb{Z}, + \rangle$ is finitely generated with $G = \{-1, 1\}$ a finite set of generators. On the other hand, the semigroup $\langle \mathbb{N}^+, \times \rangle$ is not finitely generated; to see this, for any finite $G \subseteq \mathbb{N}^+$, choose a prime $p \notin G$.   □

Example 5:

Let Σ be an alphabet. Then $\langle \Sigma^+, \circ \rangle$ is a semigroup with Σ a finite set of generators (the operation $\circ$ is juxtaposition of words).   □

Let $\langle S, \circ \rangle$ be a semigroup with a finite set G of generators. Then for any $c \in S$, there exist $a_1, \ldots, a_n \in G$ such that $c = a_1 \circ \cdots \circ a_n$. But this representation of c may not be unique. This suggests the following problem: Given $a_1, \ldots, a_m, b_1, \ldots, b_n \in G$, is $a_1 \circ \cdots \circ a_m = b_1 \circ \cdots \circ b_n$? This is the basic idea of the word problem for semigroups. But to formulate the problem precisely, we need the following definition.

Definition 8: Let $\langle S, \circ \rangle$ be a semigroup with a finite set $G \subseteq S$ of generators and let $\mathbf{P} = \{\langle E_i, F_i \rangle : 1 \leq i \leq k\}$ be a finite set of ordered pairs, where each E_i and each F_i is a finite product of elements of G (e.g., $E_i = c_1 \circ \cdots \circ c_j$, where $c_1, \ldots, c_j \in G$). We say that $\mathbf{P}$ is a *finite presentation* of S if for all $a_1, \ldots, a_m, b_1, \ldots, b_n \in G$,

$$a_1 \circ \cdots \circ a_m = b_1 \circ \cdots \circ b_n \quad \text{if and only if} \quad a_1 \circ \cdots a_m \Rightarrow_{\mathbf{P}} b_1 \circ \cdots \circ b_n.$$

The notation $\Rightarrow_{\mathbf{P}}$ means that $b_1 \circ \cdots \circ b_n$ can be obtained from $a_1 \circ \cdots \circ a_m$ by a finite number of applications of the following rule: For some i, $1 \leq i \leq k$, replace an occurrence of E_i with F_i (or vice versa). The *word problem* for a semigroup with a finite set G of generators and a finite presentation $\mathbf{P}$ asks whether there is an algorithm that, given $a_1, \ldots, a_m, b_1, \ldots, b_n \in G$, decides whether $a_1 \circ \cdots \circ a_m \Rightarrow_{\mathbf{P}} b_1 \circ \cdots \circ b_n$.

Example 6:

Let Π be a Thue system with alphabet Σ; we construct a semigroup $\langle S_\Pi, \circ \rangle$ with a finite set G_Π of generators and a finite presentation **P**. To begin, define a binary relation $\sim$ on Σ^+ by

$$W \sim V \Leftrightarrow W \vdash_\Pi V.$$

By Lemmas 1 and 3, $\sim$ is an equivalence relation on Σ^+. Let $S_\Pi = \{[W]: W \in \Sigma^+\}$ be the collection of all equivalence classes of $\sim$, and define a binary operation $\circ$ on S_Π by

$$[W] \circ [V] = [WV].$$

By Lemma 1(3), this operation is well defined. To check associativity, note that

$$[W] \circ ([V] \circ [U]) = [W] \circ [VU] = [WVU]$$
$$([W] \circ [V]) \circ [U] = [WV] \circ [U] = [WVU].$$

The set $G_\Pi = \{[a]: a \in \Sigma\}$ is a finite set of generators for S_Π. To see this, let $W = a_1 \cdots a_n$ be a nonempty word on Σ; we then have

$$[W] = [a_1 \cdots a_n] = [a_1] \circ \cdots \circ [a_n]. \tag{*}$$

To summarize thus far, $\langle S_\Pi, \circ, G_\Pi \rangle$ is a finitely generated semigroup. Now let $\{A_1 \to B_1, \ldots, A_k \to B_k\}$ be the set of productions of Π. Fix i, $1 \le i \le k$, and suppose $A_i = c_1 \cdots c_s$ and $B_i = d_1 \cdots d_t$; let $E_i = [c_1] \circ \cdots \circ [c_s]$ and $F_i = [d_1] \circ \cdots \circ [d_t]$. The collection $\mathbf{P} = \{\langle E_i, F_i \rangle : 1 \le i \le k\}$ is a finite presentation of $\langle S_\Pi, \circ, G_\Pi \rangle$. To prove this, it suffices to show that for all $a_1, \ldots, a_m, b_1, \ldots, b_n \in \Sigma$, the following are equivalent (the details are left to the reader):

(1) $[a_1] \circ \cdots \circ [a_m] = [b_1] \circ \cdots \circ [b_n]$.
(2) $[a_1 \cdots a_m] = [b_1 \cdots b_n]$.
(3) $a_1 \cdots a_m \sim b_1 \cdots b_n$.
(4) $a_1 \cdots a_m \vdash_\Pi b_1 \cdots b_n$.
(5) $[a_1] \circ \cdots \circ [a_m] \Rightarrow_\mathbf{P} [b_1] \circ \cdots \circ [b_n]$. $\square$

Theorem 5 (Unsolvability of the word problem for semigroups; Post, Markov): Assume Church's thesis. Then there is a semigroup for which the word problem is unsolvable.

Proof: Let Π be a Thue system with alphabet Σ for which the word problem is unsolvable; we will show that the word problem for the semigroup $\langle S_\Pi, \circ, G_\Pi, \mathbf{P} \rangle$ in Example 6 is also unsolvable. As noted in Example 6, for any words $W = a_1 \cdots a_m$ and $V = b_1 \cdots b_n$,

$$W \vdash_\Pi V \Leftrightarrow [a_1] \circ \cdots \circ [a_m] \Rightarrow_\mathbf{P} [b_1] \circ \cdots \circ [b_n].$$

Since the word problem for Π is unsolvable, the word problem for the semigroup S_Π is unsolvable. ∎

There is also a word problem for groups (much more difficult). Novikov (1955) and Boone (1959) have independently shown that there is a group for which the word problem is unsolvable.

We can simulate a Thue system with an appropriate first-order language. This result can then be used to prove Trachtenbrot's theorem that the validity problem for the first-order language with just one 2-ary function symbol is unsolvable. Proofs of these results are outlined in the exercises.

Exercises for Section 9.6

1. Prove Lemma 1.
2. Prove Lemma 2.
3. Let Π be a semi-Thue system with just one production $A \to B$. Show that the word problem for Π is solvable.
4. (*Strings of parentheses*) Refer to the semi-Thue system in Example 2. Prove the following:
   (1) $S \vdash_\Pi (())(()())$.
   (2) There is an algorithm that, given a word W on $\{(,)\}$, decides whether $S \vdash_\Pi W$. (*Hint:* Each production $A \to B$ has the property that the number of symbols in A is $<$ the number of symbols in B.)
5. Let Π be the semi-Thue system with alphabet $\Sigma = \{a, b\}$ and productions

   $aa \to a$, $ab \to b$, $ba \to a$,   and $bb \to b$.

   Let W be a word on Σ. Prove: If W ends in a, then $W \vdash_\Pi a$; if W ends in b, then $W \vdash_\Pi b$.
6. Let Π be the semi-Thue system with alphabet $\Sigma = \{a, b\}$ and productions

   $ba \to b$, $ab \to b$,   and $bb \to \Lambda$.

   Prove the following:
   (1) $baba \vdash_\Pi a$ and $baba \vdash_\Pi \Lambda$.
   (2) For any word W, one of the following holds: $W \vdash_\Pi a^n$, $W \vdash_\Pi b$, or $W \vdash_\Pi \Lambda$.
   (3) The word problem for Π is solvable. (*Hint:* Each production $A \to B$ has the property that the number of symbols in A is $>$ the number of symbols in B.)
7. Let Π be the Thue system with alphabet $\Sigma = \{a, b, c\}$ and productions

   $ab \to ba$ $ac \to ca$ $bc \to cb$
   $ba \to ab$ $ca \to ac$ $cb \to bc$.

   Prove the following:
   (1) Let W, V be words on Σ. Then $W \vdash_\Pi V$ if and only if W and V have the same number of a's, b's, and c's.
   (2) The word problem for Π is solvable.
8. Let Π be a semi-Thue system with alphabet Σ and productions $\{A_i \to B_i : 1 \le i \le n\}$. Construct a semi-Thue system Π' as follows: The alphabet of Π' is Σ; the productions are $\{B_i \to A_i : 1 \le i \le n\}$. Show that $W \vdash_\Pi V$ if and only if $V \vdash_{\Pi'} W$.
9. (*Simulation*) Let P be a program for the register machine. Show that there is a semi-Thue system $\Pi(P)$ with words V and $\{W_a : a \in \mathbb{N}\}$ such that for all $a \in \mathbb{N}$, the P-computation with input a halts $\Leftrightarrow V \vdash_\Pi W_a$. (*Hint:* See Exercise 8.)
10. (*Word problem for Π, V*) Let Π be a semi-Thue system and let V be a fixed word. The word problem for Π, V asks whether there is an algorithm that, given a word W, decides whether $V \vdash_\Pi W$. Show that there is a semi-Thue system Π and a word V such that the word problem for Π, V is unsolvable.
11. (*Simulation by a semi-Thue system with alphabet $\{a, b\}$*) Let Π be a semi-Thue system with alphabet $\Sigma = \{a_1, \ldots, a_k\}$ and productions $\{A_i \to B_i : 1 \le i \le n\}$. We construct a semi-Thue system Π^* with alphabet $\{a, b\}$ that simulates Π. For $1 \le j \le k$, let $a_j^* = ba^j b$ (a^j is a taken j times). Given a word $W = a_{i_1} \cdots a_{i_s}$ on Σ, W^* is the word on $\{a, b\}$ defined by

$$W^* = a_{i_1}^* \cdots a_{i_s}^*.$$

(*Example:* Let $\Sigma = \{a_1, a_2, a_3\}$; if W is a_3a_2, then W^* is *baaabbaab*.) The required semi-Thue system Π^* has alphabet $\{a, b\}$ and productions $\{A_i^* \rightarrow B_i^*: 1 \le i \le n\}$. Prove the following:

(1) $(W_1 W_2)^* = W_1^* W_2^*$.
(2) If $W \Rightarrow_\Pi V$, then $W^* \Rightarrow_{\Pi^*} V^*$.
(3) If $W^* \Rightarrow_{\Pi^*} V$ (W a word on Σ, V a word on $\{a, b\}$), then there is a word U on Σ such that $V = U^*$ and $W \Rightarrow_\Pi U$.
(4) If W, V are words on Σ, then $W \vdash_\Pi V$ if and only if $W^* \vdash_{\Pi^*} V^*$.

12. Show that there is a semi-Thue system with alphabet $\{a, b\}$ for which the word problem is unsolvable. (*Hint:* See Exercise 11.)

13. Let Π be a Thue system with alphabet Σ, and assume that for each symbol $a \in \Sigma$, there exists $b \in \Sigma$ such that $ab \rightarrow \Lambda$ and $\Lambda \rightarrow ab$ are productions of Π. Show that $\langle \Sigma^*/\sim, \circ \rangle$ is a group. (As in Example 6, $W \sim V \Leftrightarrow W \vdash_\Pi V$; note that $\Lambda \in \Sigma^*$.)

14. Let Π be a semi-Thue system with alphabet Σ. Show that the word problem for Π is semidecidable. In other words, show that there is an algorithm such that given $W, V \in \Sigma^*$, (1) If $W \vdash_P V$, the algorithm halts after a finite number of steps and prints YES; (2) if $W \vdash_P V$ does not hold, the algorithm continues forever.

15. A *context-free grammar* consists of:

- Σ_n = nonterminal alphabet.
- Symbol $S \in \Sigma_N$, called the *start* symbol.
- Σ_T = terminal alphabet ($\Sigma_N \cap \Sigma_T = \varnothing$).
- Finite set of productions of the form $A \rightarrow W$, where $A \in \Sigma_N$ and $W \in (\Sigma_N \cup \Sigma_T)^*$, $W \ne \Lambda$.

Note that Example 2 is a context-free grammar. If G is a context-free grammar, the *context-free language* generated by G is

$$L(G) = \{V \in \Sigma_T^*: S \vdash V\}.$$

(1) Let G be the context-free grammar with nonterminal alphabet $\Sigma_N = \{S\}$, terminal alphabet $\Sigma_T = \{a, b\}$, and productions $S \rightarrow aSb$, $S \rightarrow ab$. Show that $L(G) = \{a^n b^n: n \ge 1\}$.

(2) Let G be the context-free grammar with nonterminal alphabet $\Sigma_N = \{S\}$, terminal alphabet $\Sigma_T = \{a, b\}$, and productions

$$S \rightarrow a \qquad S \rightarrow aSa \qquad S \rightarrow \Lambda$$
$$S \rightarrow b \qquad S \rightarrow bSb$$

Show that $L(G)$ = the set of all palindromes on $\{a, b\}$.

16. (*Simulation of n-ary RE relations*) Let R be an n-ary RE relation. Show that there is a semi-Thue system Π and words V and $\{W_a: a \in \mathbb{N}\}$ such that for all $a_1, \ldots, a_n \in \mathbb{N}$, $R(a_1, \ldots, a_n) \Leftrightarrow W_{J_n(a_1, \ldots, a_n)} \vdash_\Pi V$.

17. (*Coding semi-Thue systems*) Let Π be a semi-Thue system with alphabet $\Sigma = \{a_1, \ldots, a_n\}$, and assume $A \ne \Lambda$ and $B \ne \Lambda$ for each production $A \rightarrow B$. For $1 \le i \le n$, the symbol number of a_i is $\text{SN}(a_i) = i$. If $W = b_1 \cdots b_k$ is a word on Σ, the *code* of W is $\#W = \langle \text{SN}(b_1), \ldots, \text{SN}(b_k) \rangle_k$; also let $\#(\Lambda) = 0$. Note that $\#(WV) = \#W * \#V$ ($*$ is the concatenation function). If P is the production $A \rightarrow B$, the code of P is $\#P = \langle \#A, \#B \rangle_2$.

(1) Prove that the following are recursive:
  (a) WORD(a) $\Leftrightarrow$ a is the code of a nonempty word of Π.
  (b) PROD(a) $\Leftrightarrow$ a is the code of a production of Π (PROD is a finite set).
  (c) DD(a, b) $\Leftrightarrow$ a and b are the codes of nonempty words W and V, and V is directly derivable from W.

        (d) $D(c) \Leftrightarrow c$ is the code of a derivation (i.e., $c = \langle \#W_1, \ldots, \#W_n \rangle_n$, where $n \geq 2$; $W_1, \ldots, W_n \in \Sigma^+$, and W_{i+1} is directly derivable from W_i for $1 \leq i \leq n - 1$).

     (2) Prove RE: $\text{THM}(a, b) \Leftrightarrow a$ and b are the codes of nonempty words W and V and $W \vdash_\Pi V$.

18. (*Characterization of* RE *relations by semi-Thue systems*) Call the coding in Exercise 17 the *standard* coding of a semi-Thue system. Call a semi-Thue system *special* if it satisfies these two conditions: $A \neq \Lambda$ and $B \neq \Lambda$ for each production $A \to B$; there are words $\{W_a : a \in \mathbb{N}\}$ of Π such that $F : \mathbb{N} \to \mathbb{N}$ is recursive, where F is defined by $F(a) =$ code of W_a (use the standard coding of Π). Prove the following: A 1-ary relation R is RE if and only if there is a special semi-Thue system Π and a word V such that for all $a \in \mathbb{N}$, $R(a) \Leftrightarrow W_a \vdash_\Pi V$.

19. Prove the equivalence of (1)–(5) in Example 6. (*Hint:* To prove (4) $\Leftrightarrow$ (5), first show: If $W = a_1 \cdots a_m$ and $V = b_1 \cdots b_n$ are words and P is the production $A \to B$ with $A = c_1 \cdots c_i$ and $B = d_1 \cdots d_j$, then $W \Rightarrow_P V$ if and only if $[b_1] \circ \cdots \circ [b_n]$ can be obtained from $[a_1] \circ \cdots \circ [a_m]$ by substituting $[d_1] \circ \cdots \circ [d_j]$ for $[c_1] \circ \cdots \circ [c_i]$.)

20. (*Simulation of a Thue system by a first-order language*) Let Π be a Thue system with alphabet Σ. We construct a first-order language L_Π with sentences $C_0, C_1, \ldots, C_k$ and for each $W \in \Sigma^+$ a variable-free term W^* of L_Π such that for all $W, V \in \Sigma^+$,

$$W \vdash_\Pi V \Leftrightarrow (C_0 \wedge C_1 \wedge \cdots \wedge C_k) \to (W^* = V^*) \text{ is logically valid.}$$

Let $\Sigma = \{a_1, \ldots, a_n\}$. The nonlogical symbols of L_Π are $\circ$ (2-ary function symbol) and constant symbols $a_1, \ldots, a_n$ (the symbols in Σ). The sentence C_0 is the associative law (sentence G1 of the theory of groups). For each word $W = b_1 \cdots b_n$, W^* is the variable-free term $b_1 \circ \cdots \circ b_n$ of L_Π. Let $\{A_i \to B_i : 1 \leq i \leq k\}$ be the productions of Π (assume $A_i, B_i \in \Sigma^+$); for $1 \leq i \leq k$, C_i is the atomic sentence $A_i^* = B_i^*$. We now describe the *standard interpretation* I of L_Π. The domain of I is $S_\Pi = \{[W] : W \in \Sigma^+\}$ (see Example 6), $\circ_I$ is the semigroup operation on S_Π, and the constant symbols $a_1, \ldots, a_n$ are interpreted as $[a_1], \ldots, [a_n]$ respectively. Prove the following:

     (1) The sentences $C_0, C_1, \ldots, C_k$ are true in the standard interpretation.

     (2) If $W^* = V^*$ is true in the standard interpretation, then $W \vdash_\Pi V$.

     (3) For $W, V \in \Sigma^+$,

$$W \vdash_\Pi V \Leftrightarrow (C_0 \wedge C_1 \wedge \cdots \wedge C_k) \to (W^* = V^*) \text{ is logically valid.}$$

21. (*Unsolvability of validity problem*) Let L be the first-order language whose only nonlogical symbol is a 2-ary function symbol $\circ$. Show that the validity problem for L is unsolvable. (*Hint:* See the preceding exercise and Exercise 9 in Section 5.5.)

Hilbert's Tenth Problem

What we cannot do is write out instructions once and for all and then turn over the testing of equations for solvability to a machine.

J. Robinson, *Hilbert's Tenth Problem*

In 1900, David Hilbert, one of the greatest and most influential mathematicians of the twentieth century, gave a talk at the International Congress of Mathematicians. Hilbert's talk was unusual in that it consisted of a list of problems, from all branches of mathematics, that he considered important and difficult. Hilbert regarded these problems as a challenge to mathematicians of the twentieth century. There were 23 problems in all; this chapter is concerned with the tenth problem on the list. The presentation is fairly self-contained; for sufficient background, the reader may review Section 1.7 on decidable relations, Sections 1.8 and 8.1–8.3 on recursive functions and relations, and Section 8.5 on RE relations. Topics in classical number theory, such as the division algorithm and congruences, are also used.

10.1
Overview of Hilbert's Tenth Problem

Hilbert's Tenth Problem HX($\mathbb{Z}$) *Find an algorithm that, given an arbitrary polynomial $p(x_1, \ldots, x_n)$ with integer coefficients, decides (YES or NO) whether the polynomial equation $p(x_1, \ldots, x_n) = 0$ has a solution in integers.*

A typical example of a polynomial with integer coefficients is $p(x_1, x_2) = 2x_1^2 - 3x_1x_2 + 4x_2^3$. Not allowed are $p(x_1, x_2) = \sqrt{2}x_1 - 4x_2 - 7$ and

$p(x_1, x_2) = 2^{x_1} + 2x_2$. The first is a polynomial but has an irrational coefficient $\sqrt{2}$; the second is not even a polynomial, since it involves exponentiation. The notion of a polynomial is intuitively clear, but later we will give a precise definition. Throughout this chapter, *polynomial* means *polynomial with integer coefficients* unless otherwise stated. We will now give two examples of algorithms for deciding whether certain polynomial equations have solutions in integers.

Example 1:

(*Algorithm for linear equations in two unknowns*) Consider a polynomial of the form $p(x_1, x_2) = ax_1 + bx_2 + c$ (x_1 and x_2 unknowns, a, b, c integer coefficients). From classical number theory, there is an algorithm that decides whether the equation $ax_1 + bx_2 + c = 0$ has a solution in integers:

1. Input $a, b, c \in \mathbb{Z}$ (a, b not both 0) and calculate $d = \gcd(a, b)$.
2. Does d divide c?
3. If *Yes*, print YES (the equation has a solution in integers) and halt.
4. If *No*, print NO (the equation has no solution in integers) and halt.

□

Example 2:

(*Algorithm for polynomials in one variable*) There is a simple algorithm that decides whether the polynomial equation $a_n x^n + \cdots + a_1 x + a_0 = 0$ has a solution in integers ($a_0, \ldots, a_n$ integer coefficients, x the unknown). Let $a_0, \ldots, a_n$ be given. If $a_0 = 0$, print YES (the equation has a solution in integers, namely $x = 0$). If $a_0 \neq 0$, list all divisors of a_0; there are only a finite number. Note that a solution of the equation must be a divisor of a_0. For each divisor d of a_0, ask: Does d satisfy the equation? If some divisor d of a_0 does satisfy the equation, print YES (the equation has a solution, namely $x = d$). If there is no divisor of a_0 that satisfies the equation, print NO (the equation has no solution in integers). □

These two examples show that there is an algorithm for certain types of polynomials. But this is not good enough; Hilbert asked for an algorithm that applies to *all* polynomials.

Theorem 1: Hilbert's tenth problem HX($\mathbb{Z}$) is semidecidable. In other words, there is an algorithm such that given an arbitrary polynomial $p(x_1, \ldots, x_n)$ with integer coefficients:

(1) If $p(x_1, \ldots, x_n) = 0$ has a solution in integers, the algorithm halts and prints YES.

(2) If $p(x_1, \ldots, x_n) = 0$ has no solution in integers, the algorithm continues forever.

Proof: The required algorithm is described informally as follows: Given $p(x_1, \ldots, x_n)$, systematically search for $a_1, \ldots, a_n \in \mathbb{Z}$ such that $p(a_1, \ldots, a_n) = 0$. Here is a precise description:

1. Input the polynomial $p(x_1, \ldots, x_n)$ and set $c = 0$.

2. List all possible n-tuples $\langle a_1, \ldots, a_n \rangle$ of natural numbers such that $a_1 + \cdots + a_n = c$ (there are a finite number in all).

3. For each such n-tuple $\langle a_1, \ldots, a_n \rangle$, is one of $p(\pm a_1, \ldots, \pm a_n) = 0$? (Calculate all $\pm$ combinations; there are 2^n computations in all.)

4. If *Yes*, $p(x_1, \ldots, x_n) = 0$ has a solution in integers; print YES and halt.

5. If *No*, add 1 to c and go to step 2.

This algorithm has the property that if $p(x_1, \ldots, x_n) = 0$ has a solution in integers, the algorithm eventually finds one, prints YES, and halts. To see this, suppose $a_1, \ldots, a_n \in \mathbb{Z}$ is a solution and let $c = |a_1| + \cdots + |a_n|$. Assume c is the smallest natural number obtained in this way. Then the algorithm discovers this solution when it executes steps 2 and 3 with this value for c. On the other hand, if the equation has no solution, the algorithm continues forever, looking in vain for a solution. Hence this algorithm satisfies (1) and (2) as required. ∎

By Theorem 1, Hilbert's tenth problem is semidecidable. But this does not constitute a positive solution to Hilbert's tenth problem. In fact, HX($\mathbb{Z}$) is unsolvable; more precisely, if we assume Church's thesis, then there is no algorithm that satisfies all of the requirements of Hilbert's problem. This negative solution of HX($\mathbb{Z}$) is an interesting blend of number theory and recursion theory, and is one of the major contributions of mathematical logic to twentieth-century mathematics.

The theory of recursive functions applies to natural numbers, whereas Hilbert's tenth problem allows integer solutions. We will modify HX($\mathbb{Z}$) so that recursion theory, with its emphasis on natural numbers, can be more readily applied to the problem.

Modified Hilbert's Tenth Problem HX($\mathbb{N}$) *Find an algorithm that, given an arbitrary polynomial $p(x_1, \ldots, x_n)$ with integer coefficients, decides whether the polynomial equation $p(x_1, \ldots, x_n) = 0$ has a solution in natural numbers.*

The only difference between the two problems is that in the modified version, we ask for solutions in $\mathbb{N}$ rather than in $\mathbb{Z}$; the coefficients of the polynomials are still integers. (The problem becomes trivial if both coefficients and solutions are restricted to $\mathbb{N}$!) Let us show that a negative solution to the modified HX($\mathbb{N}$) gives a negative solution to the original HX($\mathbb{Z}$).

Lemma 1: Let $p(x_1, \ldots, x_n)$ be a polynomial with integer coefficients. Then there is a polynomial $q(r_1, s_1, t_1, u_1, \ldots, r_n, s_n, t_n, u_n)$ with integer coefficients such that the following are equivalent:

(1) $p(x_1, \ldots, x_n) = 0$ has a solution in $\mathbb{N}$.
(2) $q(r_1, s_1, t_1, u_1, \ldots, r_n, s_n, t_n, u_n) = 0$ has a solution in $\mathbb{Z}$.

Proof: The polynomial q is obtained from p by replacing each variable x_k with $r_k^2 + s_k^2 + t_k^2 + u_k^2$, $1 \le k \le n$. To prove that (1) $\Rightarrow$ (2), suppose there exist $e_1, \ldots, e_n \in \mathbb{N}$ such that $p(e_1, \ldots, e_n) = 0$. By a theorem of Lagrange, each e_k is a sum of four squares: $e_k = a_k^2 + b_k^2 + c_k^2 + d_k^2$ (see the appendix on number theory). Clearly (2) holds, and moreover with solutions in $\mathbb{N}$.

To prove that (2) $\Rightarrow$ (1), suppose there exist $a_1, b_1, c_1, d_1, \ldots, a_n, b_n,$ $c_n, d_n \in \mathbb{Z}$ such that $q(a_1, b_1, c_1, d_1, \ldots, a_n, b_n, c_n, d_n) = 0$. Let $e_k = a_k^2 + b_k^2 + c_k^2 + d_k^2$ $(1 \le k \le n)$; clearly, each $e_k \in \mathbb{N}$ and $p(e_1, \ldots, e_n) = 0$ as required. ∎

Theorem 2: If there is an algorithm for Hilbert's tenth problem HX($\mathbb{Z}$), then there is an algorithm for the modified Hilbert's tenth problem HX($\mathbb{N}$).

Proof: Suppose there is an algorithm for HX($\mathbb{Z}$); the following is an algorithm for HX($\mathbb{N}$).

1. Input the polynomial $p(x_1, \ldots, x_n)$ and construct q as in Lemma 1.
2. Does $q(r_1, s_1, t_1, u_1, \ldots, r_n, s_n, t_n, u_n) = 0$ have a solution in $\mathbb{Z}$?
3. If *Yes*, print YES (by Lemma 1, $p(x_1, \ldots, x_n) = 0$ has a solution in $\mathbb{N}$).
4. If *No*, print NO (by Lemma 1, $p(x_1, \ldots, x_n) = 0$ does not have a solution in $\mathbb{N}$). ∎

We will prove that there is no algorithm for HX($\mathbb{N}$); by Theorem 2 (contrapositive form), this gives a negative solution to Hilbert's original problem HX($\mathbb{Z}$).

We now introduce a key definition. In classical number theory, we often begin with a polynomial $p(x_1, \ldots, x_n)$ and ask for the set of all solutions of $p(x_1, \ldots, x_n) = 0$. For example, given $p(x_1, x_2, x_3) = x_1^2 + x_2^2 - x_3^2$, the solutions of $p(x_1, x_2, x_3) = 0$ with nonzero entries are the Pythagorean triples $\langle 3, 4, 5 \rangle$, $\langle 5, 12, 13 \rangle$, and so on. The following definition reverses this procedure: Given a set R of solutions, find the corresponding polynomial equation.

Definition 1: An n-ary relation R is *Diophantine* if there is a polynomial $p(y_1, \ldots, y_n, x_1, \ldots, x_k)$ with $n + k$ variables $y_1, \ldots, y_n, x_1, \ldots, x_k$ $(k = 0$ allowed) and integer coefficients such that for all $a_1, \ldots, a_n \in \mathbb{N}$,

$$R(a_1, \ldots, a_n) \Leftrightarrow p(a_1, \ldots, a_n, x_1, \ldots, x_k) = 0 \text{ has a solution in } \mathbb{N}.$$

In the polynomial $p(y_1, \ldots, y_n, x_1, \ldots, x_k)$, $y_1, \ldots, y_n$ are called *parameters* and $x_1, \ldots, x_k$ *unknowns*. The idea is this: Given $a_1, \ldots, a_n \in \mathbb{N}$, replace the parameters $y_1, \ldots, y_n$ with $a_1, \ldots, a_n$ respectively to obtain a new polynomial q with unknowns $x_1, \ldots, x_k$, say $q(x_1, \ldots, x_k) = p(a_1, \ldots, a_n, x_1, \ldots, x_k)$. We then have: $R(a_1, \ldots, a_n) \Leftrightarrow q(x_1, \ldots, x_k) = 0$ has a solution in $\mathbb{N}$. (For many

Diophantine relations, unknowns $x_1, \ldots, x_k$ are not needed.) We will often write the displayed condition in the more compact form

$$R(a_1, \ldots, a_n) \Leftrightarrow \exists x_1 \cdots \exists x_k [p(a_1, \ldots, a_n, x_1, \ldots, x_k) = 0].$$

We emphasize, however, that $\exists x_1$ means "there is a natural number," and so on. Thus a more precise version of the compact form is

$$R(a_1, \ldots, a_n) \Leftrightarrow \exists b_1 \in \mathbb{N} \cdots \exists b_k \in \mathbb{N}[p(a_1, \ldots, a_n, b_1, \ldots, b_k) = 0].$$

Example 3:

The following relations are Diophantine: $\leq, <, =, >$, and $\geq$. We will give the details for $<$ and $=$ and leave the others to the reader.

- $a < b \Leftrightarrow \exists x[a + x + 1 = b] \Leftrightarrow \exists x[a + x + 1 - b = 0]$; the required polynomial is $p(y_1, y_2, x) = y_1 + x + 1 - y_2$.
- $a = b \Leftrightarrow a - b = 0$; the required polynomial is $p(y_1, y_2) = y_1 - y_2$. □

Example 4:

We will show that CP is Diophantine, where $CP(a) \Leftrightarrow a$ is composite.

$$CP(a) \Leftrightarrow a \text{ is the product of two integers, both } > 1$$
$$\Leftrightarrow \exists x_1 \exists x_2 [a = (x_1 + 2) \times (x_2 + 2)]$$
$$\Leftrightarrow \exists x_1 \exists x_2 [a - (x_1 + 2) \times (x_2 + 2) = 0].$$

The required polynomial is $p(y, x_1, x_2) = y - (x_1 + 2) \times (x_2 + 2)$. □

Now let us see how this concept is used to solve HX($\mathbb{N}$). Suppose R is a 1-ary relation that is Diophantine but not recursive. Since R is Diophantine, there is a polynomial $p(y, x_1, \ldots, x_k)$ with integer coefficients such that for all $a \in \mathbb{N}$,

$$R(a) \Leftrightarrow p(a, x_1, \ldots, x_k) = 0 \text{ has a solution in } \mathbb{N}.$$

But note: An algorithm satisfying the requirements of HX($\mathbb{N}$) would imply that R is decidable! And by Church's thesis, R would be recursive, a contradiction of the choice of R as nonrecursive. To carry out this argument, we need a 1-ary relation that is Diophantine but not recursive. Now a fundamental result of recursion theory is the existence of an RE relation that is not recursive. Thus we can achieve our goal by proving the following deep result:

Main Theorem: Every RE relation is Diophantine.

This striking theorem is due to Yuri Matiyasevich (a Russian), Julia Robinson, Martin Davis, and Hilary Putnam (Americans); Sections 10.2–10.5 will be devoted to proving the main theorem. We will end this section with a more detailed proof that the main theorem solves HX($\mathbb{N}$). The proof uses the nontrivial fact, from recursion theory, that there is an RE relation that is not recursive. Another application of the main theorem (besides settling Hilbert's tenth problem!) is a fairly easy proof of the existence of such a relation. This will be carried out in Section 10.6.

Theorem 3 (Negative solution of HX(ℕ)): Assume Church's thesis. Then there is no algorithm that decides whether an arbitrary polynomial equation with integer coefficients has a solution in natural numbers.

Proof: Let R be a 1-ary RE relation not recursive (see Section 8.7, 9.3, or 10.6). Since R is RE, R is Diophantine (main theorem). Let p be a polynomial with integer coefficients such that for all $a \in \mathbb{N}$, $R(a) \Leftrightarrow p(a, x_1, \ldots, x_k) = 0$ has a solution in $\mathbb{N}$. Now suppose there is an algorithm that decides whether a polynomial equation has a solution in $\mathbb{N}$. It easily follows that the relation R is decidable, and hence recursive (Church's thesis). This contradicts our choice of R as a nonrecursive relation. ∎

———————— *Exercises for Section 10.1* ————————

1. Construct an algorithm that, given an integer a, decides whether $x_1^2 + 3x_2^4 = a$ has a solution in $\mathbb{N}$.
2. Construct an algorithm that, given $a \in \mathbb{Z}$, decides whether $x^2 - (a^2 - 1)y^2 = 1$ has a solution in positive integers.
3. For each polynomial $p(x)$, decide whether $p(x) = 0$ has a solution in integers: (1) $p(x) = x^2 - 5x + 3$; (2) $p(x) = x^3 - 5x + 2$.
4. For each polynomial $p(x, y)$, decide whether $p(x, y) = 0$ has a solution in integers: (1) $p(x, y) = 6x - 9y - 63$; (2) $p(x, y) = 6x - 9y + 62$.
5. Show that each of the following relations is Diophantine:
   (1) $a \le b$.   (4) $\{5\}$.
   (2) $a \ge b$.   (5) $E(a) \Leftrightarrow a$ is an even natural number.
   (3) $a > b$.
6. Show that each of the following relations is Diophantine:
   (1) $R(a, b, c) \Leftrightarrow a + 2c \le 5$.
   (2) $R(a, b, c) \Leftrightarrow (b \ge 2) \wedge (c = 8)$. (*Hint:* $m = 0$ and $n = 0 \Leftrightarrow m^2 + n^2 = 0$.)
   (3) $R(a, b, c) \Leftrightarrow (a + c = 5) \vee (b \times c \ge 8)$. (*Hint:* $m = 0$ or $n = 0 \Leftrightarrow m \times n = 0$.)
7. Let Q be a 2-ary Diophantine relation. Show that R is Diophantine, where $R(a, b, c) \Leftrightarrow Q(c, a)$.
8. This problem outlines a proof that Hilbert's tenth problem HX(ℤ) is equivalent to the modified Hilbert's tenth problem HX(ℕ).
   (1) Let $p(x_1, \ldots, x_k)$ be a polynomial with integer coefficients. Show that $p(x_1, \ldots, x_k) = 0$ has a solution in $\mathbb{Z} \Leftrightarrow p(y_1 - z_1, \ldots, y_k - z_k) = 0$ has a solution in $\mathbb{N}$.
   (2) Assume there is an algorithm for HX(ℕ). Show that there is an algorithm for HX(ℤ).
   (3) Explain why there is an algorithm that solves HX(ℕ) if and only if there is an algorithm that solves HX(ℤ).
9. Let POLY be a collection of polynomials with integer coefficients (not necessarily *all* polynomials with integer coefficients). Suppose there is an algorithm that, given $p(x_1, \ldots, x_k) \in$ POLY, decides whether $p(x_1, \ldots, x_k) = 0$ has a solution in $\mathbb{N}$. Show that there is an algorithm such that, given $p(x_1, \ldots, x_k) \in$ POLY, if $p(x_1, \ldots, x_k) = 0$ has a solution in $\mathbb{N}$, the algorithm prints out a solution and halts; if $p(x_1, \ldots, x_k) = 0$ has no solution in $\mathbb{N}$, the algorithm prints NO and halts.
10. Assume Church's thesis. Show that there is a fixed polynomial $q(y, x_1, \ldots, x_k)$ with the following property: There is no algorithm that, given an arbitrary natural number a, decides whether $q(a, x_1, \ldots, x_k) = 0$ has a solution in $\mathbb{N}$.

11. (*Converse of main theorem*) We will outline a proof that every Diophantine relation is RE.

(1) Show that every Diophantine relation is semidecidable.

(2) It follows from (1) and Church's thesis (semidecidable $\Rightarrow$ RE) that every Diophantine relation is RE. However, this is a nonessential use of Church's thesis. Show directly that every Diophantine relation is RE. (*Hint:* Since R is Diophantine,

$$R(a_1, \ldots, a_n) \Leftrightarrow \exists x_1 \cdots \exists x_k [p(a_1, \ldots, a_n, x_1, \ldots, x_k) = 0]$$
$$\Leftrightarrow \exists x_1 \cdots \exists x_k [q(a_1, \ldots, a_n, x_1, \ldots, x_k) = r(a_1, \ldots, a_n, x_1, \ldots, x_k)]$$

where q, r are polynomials with coefficients in $\mathbb{N}$ (q and r are obtained from p by moving all negative coefficients of p to the right side of the equation; q and r always have values in $\mathbb{N}$). Now use Gödel's β-function to write R in the form $\exists c Q(a_1, \ldots, a_n, c)$, where Q is a recursive relation.)

10.2

Diophantine Relations and Functions

We take a three-step approach to the proof that every RE relation is Diophantine: (1) Introduce the closely related notion of a *Diophantine function*. (2) Prove that every recursive function is Diophantine. (The proof is by induction on recursive functions (see Section 1.8); we show that each starting function is Diophantine, that the composition of Diophantine functions is Diophantine, and that the minimalization of a regular Diophantine function is Diophantine.) (3) Use the result *recursive function $\Rightarrow$ Diophantine function* to show that every RE relation is Diophantine.

Definition 1: An n-ary function F is *Diophantine* if its graph $\mathcal{G}_F$ is a Diophantine relation.

Lemma 1: Let F be an n-ary function. To show that F is Diophantine, it suffices to find a polynomial $p(y_1, \ldots, y_n, y_{n+1}, x_1, \ldots, x_k)$ with integer coefficients such that for all $a_1, \ldots, a_n, b \in \mathbb{N}$,

$$F(a_1, \ldots, a_n) = b \Leftrightarrow \exists x_1 \cdots \exists x_k [p(a_1, \ldots, a_n, b, x_1, \ldots, x_k) = 0]. \quad (*)$$

Proof: By definition, $\mathcal{G}_F(a_1, \ldots, a_n, b) \Leftrightarrow F(a_1, \ldots, a_n) = b$. A polynomial p satisfying $(*)$ shows that $\mathcal{G}_F$ is a Diophantine relation; hence F is a Diophantine function as required. ∎

Example 1:

Let $C_{n,k}$ be the n-ary constant function k ($k \in \mathbb{N}$). By Lemma 1, $C_{n,k}$ is Diophantine as follows: $C_{n,k}(a_1, \ldots, a_n) = b \Leftrightarrow k - b = 0$; the required polynomial is $p(y_1, \ldots, y_n, y_{n+1}) = k - y_{n+1}$. □

Example 2:

> The 4-ary function $F(a, b, c, d) = 2 \times b + c \times d + 5$ is Diophantine as follows: $F(a, b, c, d) = e \Leftrightarrow 2 \times b + c \times d + 5 - e = 0$; the required polynomial is $p(y_1, y_2, y_3, y_4, y_5) = 2 \times y_2 + y_3 \times y_4 + 5 - y_5$.   □

Theorem 1: Each starting function is Diophantine. In other words, $+$, $\times$, the projection functions $U_{n,k}$ ($1 \le k \le n$), and $K_<$ are all Diophantine functions.

Proof: In each case, we use Lemma 1 and find the required polynomial p.

- $a + b = c \Leftrightarrow a + b - c = 0$; take $p(y_1, y_2, y_3) = y_1 + y_2 - y_3$.
- $a \times b = c \Leftrightarrow a \times b - c = 0$; take $p(y_1, y_2, y_3) = y_1 \times y_2 - y_3$.
- $U_{n,k}(a_1, \ldots, a_n) = b \Leftrightarrow a_k - b = 0$; take $p(y_1, \ldots, y_n, y_{n+1}) = y_k - y_{n+1}$.
- $K_<(a, b) = c$
  $\Leftrightarrow [(a < b) \wedge (c = 0)] \vee [(a \ge b) \wedge (c = 1)]$
  $\Leftrightarrow \exists x_1[(a + x_1 + 1 = b) \wedge (c = 0)] \vee \exists x_2[(a = b + x_2) \wedge (c = 1)]$
  $\Leftrightarrow \exists x_1 x_2[((a + x_1 + 1 - b = 0) \wedge (c = 0)) \vee ((a - b - x_2 = 0) \wedge (c - 1 = 0))]$.

The connectives $\wedge$ and $\vee$ are eliminated using familar algebraic properties: $m = 0$ and $n = 0 \Leftrightarrow m^2 + n^2 = 0$; $m = 0$ or $n = 0 \Leftrightarrow m \times n = 0$.

$$K_<(a, b) = c \Leftrightarrow \exists x_1 \exists x_2[((a + x_1 + 1 - b)^2 + c^2 = 0)$$
$$\vee ((a - b - x_2)^2 + (c - 1)^2 = 0)]$$
$$\Leftrightarrow \exists x_1 \exists x_2[((a + x_1 + 1 - b)^2 + c^2)$$
$$\times ((a - b - x_2)^2 + (c - 1)^2) = 0].$$

Take $p(y_1, y_2, y_3, x_1, x_2) = ((y_1 + x_1 + 1 - y_2)^2 + y_3^2) \times ((y_1 - y_2 - x_2)^2 + (y_3 - 1)^2)$. ■

Up to now, we have proved F Diophantine by finding an explicit polynomial (see Lemma 1). There is a second method of showing F to be Diophantine: Express $\mathcal{G}_F$ in terms of relations known to be Diophantine (Lemma 2 below). But to use this second method effectively, we need techniques for constructing new Diophantine relations from known Diophantine relations.

Lemma 2: Let F be an n-ary function. To show that F is Diophantine, it suffices to find an $n + 1$-ary Diophantine relation R such that for all $a_1, \ldots, a_n, b \in \mathbb{N}$,

$$F(a_1, \ldots, a_n) = b \Leftrightarrow R(a_1, \ldots, a_n, b).$$

Theorem 2 (Disjunction/conjunction rule): Let P and Q be n-ary Diophantine relations. Then $P \vee Q$ and $P \wedge Q$ are n-ary Diophantine relations.

Proof: The idea of the proof is illustrated in the discussion of $K_<$: Use the algebraic properties $m = 0$ and $n = 0 \Leftrightarrow m^2 + n^2 = 0$ and $m = 0$ or $n = 0$ $\Leftrightarrow m \times n = 0$. We will give the details for $P \wedge Q$ and leave $P \vee Q$ to the reader. The following notation is used: $\mathbf{a} = \langle a_1, \ldots, a_n \rangle$, $\mathbf{x} = \langle x_1, \ldots, x_k \rangle$, $\mathbf{y} = \langle y_1, \ldots, y_n \rangle$, $\mathbf{z} = \langle z_1, \ldots, z_m \rangle$. Since P and Q are Diophantine, there are polynomials $p(\mathbf{y}, \mathbf{x})$ and $q(\mathbf{y}, \mathbf{z})$ such that for all $a_1, \ldots, a_n \in \mathbb{N}$,

$$P(\mathbf{a}) \Leftrightarrow \exists x_1 \cdots \exists x_k [p(\mathbf{a}, \mathbf{x}) = 0]; \quad Q(\mathbf{a}) \Leftrightarrow \exists z_1 \cdots \exists z_m [q(\mathbf{a}, \mathbf{z}) = 0].$$

Then

$$(P \wedge Q)(\mathbf{a}) \Leftrightarrow P(\mathbf{a}) \wedge Q(\mathbf{a})$$
$$\Leftrightarrow \exists x_1 \cdots \exists x_k [p(\mathbf{a}, \mathbf{x}) = 0] \wedge \exists z_1 \cdots \exists z_m [q(\mathbf{a}, \mathbf{z}) = 0]$$
$$\Leftrightarrow \exists x_1 \cdots \exists x_k \exists z_1 \cdots \exists z_m [(p(\mathbf{a}, \mathbf{x}) = 0) \wedge (q(\mathbf{a}, \mathbf{z}) = 0)]$$
$$\Leftrightarrow \exists x_1 \cdots \exists x_k \exists z_1 \cdots \exists z_m [p(\mathbf{a}, \mathbf{x})^2 + q(\mathbf{a}, \mathbf{z})^2 = 0].$$

The required polynomial is $r(\mathbf{y}, \mathbf{x}, \mathbf{z}) = p(\mathbf{y}, \mathbf{x})^2 + q(\mathbf{y}, \mathbf{z})^2$. ∎

Theorem 3 (Variable independence): Let Q be a k-ary Diophantine relation, let G be a k-ary Diophantine function, and let $\{i_1, \ldots, i_k\} \subseteq \{1, \ldots, n\}$ (repetitions allowed). The following are Diophantine:

(1) $R(a_1, \ldots, a_n) \Leftrightarrow Q(a_{i_1}, \ldots, a_{i_k})$.
(2) $F(a_1, \ldots, a_n) = G(a_{i_1}, \ldots, a_{i_k})$.

Proof: Since Q is Diophantine, there is a polynomial $p(z_1, \ldots, z_k, x_1, \ldots, x_m)$ such that for all $b_1, \ldots, b_k \in \mathbb{N}$,

$$Q(b_1, \ldots, b_k) \Leftrightarrow \exists x_1 \cdots \exists x_m [p(b_1, \ldots, b_k, x_1, \ldots, x_m) = 0].$$

The required polynomial for R is

$$q(y_1, \ldots, y_n, x_1, \ldots, x_m) = p(y_{i_1}, \ldots, y_{i_k}, x_1, \ldots, x_m).$$

Note: q is obtained from p by replacing $z_1, \ldots, z_k$ by $y_{i_1}, \ldots, y_{i_k}$ respectively. Some of the variables $y_1, \ldots, y_n$ may not occur in q; this is allowed. Alternately, we can always include a missing variable, say y, by adding $0 \times y$ to q. Let $a_1, \ldots, a_n \in \mathbb{N}$. We have

$$R(a_1, \ldots, a_n) \Leftrightarrow Q(a_{i_1}, \ldots, a_{i_k})$$
$$\Leftrightarrow \exists x_1 \cdots \exists x_m [p(a_{i_1}, \ldots, a_{i_k}, x_1, \ldots, x_m) = 0]$$
$$\Leftrightarrow \exists x_1 \cdots \exists x_m [q(a_1, \ldots, a_n, x_1, \ldots, x_m) = 0].$$

This shows R to be Diophantine as required. To see that F is a Diophantine function, write

$$\mathscr{G}_F(a_1, \ldots, a_n, b) \Leftrightarrow F(a_1, \ldots, a_n) = b$$
$$\Leftrightarrow G(a_{i_1}, \ldots, a_{i_k}) = b$$
$$\Leftrightarrow \mathscr{G}_G(a_{i_1}, \ldots, a_{i_k}, b).$$

Since $\mathcal{G}_G$ is a Diophantine relation, part (1) of the theorem applies and $\mathcal{G}_F$ is Diophantine. ■

Theorem 3 gives us considerable flexibility in manipulating variables for both relations and functions; this result is used quite often and without explicit mention. We will illustrate with an example:

Example 3:

Let Q be a 4-ary Diophantine relation and let F be a 1-ary Diophantine function; we give the details that

$$R(a, b, c) \Leftrightarrow Q(a, c, b, a) \wedge (F(c) = b)$$

is Diophantine. Introduce auxiliary relations P_1 and P_2 by $P_1(a, b, c)$ $\Leftrightarrow Q(a, c, b, a)$ and $P_2(a, b, c) \Leftrightarrow \mathcal{G}_F(c, b)$. By Theorem 3 on variable independence, P_1 and P_2 are Diophantine. The conjunction rule now applies and R is Diophantine as required.   □

Theorem 4 ($\exists$-Rule): Let Q be an $n + 1$-ary Diophantine relation. Then the n-ary relation R defined by

$$R(a_1, \ldots, a_n) \Leftrightarrow \exists b \, Q(a_1, \ldots, a_n, b).$$

is also Diophantine.

Proof: Since Q is Diophantine, there is a polynomial $p(y_1, \ldots, y_{n+1}, x_1, \ldots, x_k)$ such that for all $a_1, \ldots, a_n, b \in \mathbb{N}$,

$$Q(a_1, \ldots, a_n, b) \Leftrightarrow \exists x_1 \cdots \exists x_k [p(a_1, \ldots, a_n, b, x_1, \ldots, x_k) = 0].$$

This same polynomial shows R to be Diophantine. For let $a_1, \ldots, a_n \in \mathbb{N}$; we leave it as an exercise for the reader to show that

$$R(a_1, \ldots, a_n) \Leftrightarrow \exists y_{n+1} \exists x_1 \cdots \exists x_k [p(a_1, \ldots, a_n, y_{n+1}, x_1, \ldots, x_k) = 0]. \blacksquare$$

Theorem 5 (*Composition of a relation*): Let Q be a k-ary Diophantine relation and let $H_1, \ldots, H_k$ be n-ary Diophantine functions. Then the n-ary relation R defined by

$$R(a_1, \ldots, a_n) \Leftrightarrow Q(H_1(a_1, \ldots, a_n), \ldots, H_k(a_1, \ldots, a_n))$$

is Diophantine.

Proof: Let $\mathbf{a} = \langle a_1, \ldots, a_n \rangle$; then

$$R(\mathbf{a}) \Leftrightarrow \exists b_1 \cdots \exists b_k [Q(b_1, \ldots, b_k) \wedge (H_1(\mathbf{a}) = b_1) \wedge \cdots \wedge (H_k(\mathbf{a}) = b_k)]$$
$$\Leftrightarrow \exists b_1 \cdots \exists b_k [Q(b_1, \ldots, b_k) \wedge \mathcal{G}_{H_1}(\mathbf{a}, b_1) \wedge \cdots \wedge \mathcal{G}_{H_k}(\mathbf{a}, b_k)].$$

By the $\exists$-rule, R is Diophantine provided that the expression in brackets $[\cdots]$

is Diophantine. By the conjunction rule, it suffices to show that each of $Q(b_1, \ldots, b_k)$, $\mathscr{G}_{H_1}(\mathbf{a}, b_1), \ldots, \mathscr{G}_{H_k}(\mathbf{a}, b_k)$ is a Diophantine relation in $a_1, \ldots,$ $a_n, b_1, \ldots, b_k$. This follows from variable independence and from the fact that $Q, \mathscr{G}_{H_1}, \ldots, \mathscr{G}_{H_k}$ are Diophantine relations. ∎

Example 4:

We show that congruence $a \equiv b \pmod{c}$ is a 3-ary Diophantine relation:

$$a \equiv b \pmod{c} \Leftrightarrow (c > 0) \wedge (c \text{ divides the difference of } a \text{ and } b)$$
$$\Leftrightarrow (c > 0) \wedge [(b \le a \wedge c \mid a - b) \vee (a \le b \wedge c \mid b - a)]$$
$$\Leftrightarrow (c > 0) \wedge [\exists d(a = b + c \times d) \vee \exists d(b = a + c \times d)].$$

□

Theorem 6 (Composition of functions): Let F be the composition of $G, H_1, \ldots, H_k$. If $G, H_1, \ldots, H_k$ are Diophantine, then F is Diophantine.

Proof:

$$F(a_1, \ldots, a_n) = b \Leftrightarrow G(H_1(a_1, \ldots, a_n), \ldots, H_k(a_1, \ldots, a_n)) = b$$
$$\Leftrightarrow \mathscr{G}_G(H_1(a_1, \ldots, a_n), \ldots, H_k(a_1, \ldots, a_n), b).$$

Since G is a Diophantine function, its graph $\mathscr{G}_G$ is a Diophantine relation. By Theorem 5 (composition of a relation), Theorem 3 (variable independence), and Lemma 2, F is a Diophantine function as required. ∎

Theorem 7 (Sum/product rule): The sum and the product of two n-ary Diophantine functions are Diophantine.

Proof: Addition and multiplication are Diophantine functions. The result now follows from Theorem 6 (composition of functions). ∎

Theorem 8: The following three functions are Diophantine:

(1) $\text{QT}(a, b) = q$, the quotient when a is divided by $b \ne 0$.
(2) $\text{RM}(a, b) = r$, the remainder when a is divided by $b \ne 0$.
(3) $\text{NEAR}(a, b) = c$, the natural number c nearest to a/b $(b \ne 0)$.

Proof: We agree that $\text{QT}(a, 0) = \text{RM}(a, 0) = \text{NEAR}(a, 0) = 0$; also that $\text{NEAR}(a, b) = c$ if $a/b = c + \frac{1}{2}$. By the division algorithm, given $a, b \in \mathbb{N}$ with $b > 0$, there exist unique $q, r \in \mathbb{N}$ such that $a = b \times q + r$ and $r < b$; by definition, $q = \text{QT}(a, b)$ and $r = \text{RM}(a, b)$. We have

$$\text{QT}(a, b) = q \Leftrightarrow [(b = 0) \wedge (q = 0)]$$
$$\vee [(b > 0) \wedge \exists r((a = b \times q + r) \wedge (r < b))];$$
$$\text{RM}(a, b) = r \Leftrightarrow [(b = 0) \wedge (r = 0)]$$
$$\vee [(b > 0) \wedge (r < b) \wedge \exists q(a = b \times q + r)].$$

The details showing that QT and RM are Diophantine functions are left to the reader. We turn to the proof that NEAR is Diophantine. For $b > 0$, we have $a = b \times q + r$ and $r < b$. Divide by b to obtain

$$\frac{a}{b} = q + \frac{r}{b} \quad \text{and} \quad \frac{r}{b} < 1.$$

Thus $\text{NEAR}(a, b) = q$ if $\frac{r}{b} \leq \frac{1}{2}$ and $\text{NEAR}(a, b) = q + 1$ if $\frac{1}{2} < \frac{r}{b}$. We have

$$
\begin{aligned}
\text{NEAR}(a, b) = c \Leftrightarrow &[(b = 0) \wedge (c = 0)] \\
&\vee [(b > 0) \wedge (c = \text{QT}(a, b)) \wedge (2 \times \text{RM}(a, b) \leq b)] \\
&\vee [(b > 0) \wedge (c = \text{QT}(a, b) + 1) \wedge (b < 2 \times \text{RM}(a, b))].
\end{aligned}
$$

To illustrate some of our tools, we give the details showing that R is Diophantine, where R is defined by $R(a, b, c) \Leftrightarrow 2 \times \text{RM}(a, b) \leq b$. Now $\leq$ is a 2-ary Diophantine relation, so by Theorem 5 (composition of a relation) it suffices to show that $G(a, b, c) = 2 \times \text{RM}(a, b)$ and $H(a, b, c) = b$ are Diophantine. This is clear for H; moreover, G is the product of two Diophantine functions and hence is Diophantine (product rule). ∎

We will now summarize the main results of this and the previous section.

Functions	$+, \times, K_<$, QT, RM, and NEAR are 2-ary Diophantine functions.
	$U_{n,k}$ $(1 \leq k \leq n)$ and $C_{n,k}$ $(k \in \mathbb{N})$ are n-ary Diophantine functions.
	Sum/product rule.
	Variable independence.
	Composition of functions.
Relations	$<, \leq, =, >$, and $\geq$ are 2-ary Diophantine relations.
	Congruence $a \equiv b \pmod{c}$ is a 3-ary Diophantine relation.
	Disjunction/conjunction rule.
	Variable independence.
	∃-rule.
	Composition of a relation.

──────── *Exercises for Section 10.2* ────────

1. Prove the disjunction rule (Theorem 2).
2. Let $\text{PS}(a) \Leftrightarrow a$ is a perfect square (i.e., $a = 1, 4, 9, \ldots$). Show that PS is a Diophantine relation.
3. Show that $a \div b$ is a Diophantine function.
4. Show that Cantor's pairing functions J, K, and L are Diophantine (see Section 8.3).
5. Show that every Diophantine function is computable. (*Hint:* It suffices to show that the graph of F is semidecidable.)

──────── **10.3** ────────

RE Relation $\Rightarrow$ Diophantine Relation (Assuming Bounded ∀-Rule)

In this section, we will outline the proof that every RE relation is Diophantine. We emphasize, however, that the proof is not complete; we still need a major result known as the *bounded* $\forall$*-rule* (proved in the next two sections). We already know that the class of Diophantine relations is closed under the logical operations $\vee$, $\wedge$, and $\exists$. This is not the case for $\neg$ or $\forall$. However, we do have:

Theorem 1 (Bounded $\forall$-rule): Let Q be an $n + 1$-ary Diophantine relation. Then R is Diophantine, where R is the $n + 1$-ary relation defined by

$$R(a_1, \ldots, a_n, b) \Leftrightarrow \forall t < b Q(a_1, \ldots, a_n, t).$$

Proof: See Sections 10.4 and 10.5. ∎

Main Theorem: Every recursive function is a Diophantine function and every RE relation is a Diophantine relation.

Proof: The proof is by induction on recursive functions. In Section 10.2, we proved that each starting function is Diophantine (Theorem 1) and that the composition of Diophantine functions is Diophantine (Theorem 6). It remains to check minimalization. Let $F(a_1, \ldots, a_n) = \mu b[G(a_1, \ldots, a_n, b) = 0]$, where G is a regular $n + 1$-ary Diophantine function. Then

$$F(a_1, \ldots, a_n) = b \Leftrightarrow (G(a_1, \ldots, a_n, b) = 0) \wedge \forall t < b[G(a_1, \ldots, a_n, t) > 0].$$

By the bounded $\forall$-rule, $\forall t < b[G(a_1, \ldots, a_n, t) > 0]$ is a Diophantine relation; it follows by standard arguments that F is a Diophantine function as required. This completes the inductive proof that every recursive function is Diophantine.

Now let R be an n-ary RE relation; $R(a_1, \ldots, a_n) \Leftrightarrow \exists b Q(a_1, \ldots, a_n, b)$, where Q is an $n + 1$-ary recursive relation. The characteristic function K_Q of Q is recursive; hence K_Q is a Diophantine function (proved above). Finally, $R(a_1, \ldots, a_n) \Leftrightarrow \exists b[K_Q(a_1, \ldots, a_n, b) = 0]$; and hence R is Diophantine by the $\exists$-rule as required. ∎

We end this section with a proof of the converse of the two results in the main theorem. Although not needed for the solution of Hilbert's tenth problem, these results, when combined with the main theorem, give an interesting characterization of the class of recursive functions and the class of RE relations.

Theorem 2: An n-ary relation is RE if and only if it is Diophantine.

Proof: We need only show that every n-ary Diophantine relation R is RE. We have:

$$R(a_1, \ldots, a_n) \Leftrightarrow p(a_1, \ldots, a_n, x_1, \ldots, x_k) = 0 \text{ has a solution in } \mathbb{N}$$
$$\Leftrightarrow q(a_1, \ldots, a_n, x_1, \ldots, x_k)$$
$$= r(a_1, \ldots, a_n, x_1, \ldots, x_k) \text{ has a solution in } \mathbb{N};$$

q and r are polynomials with coefficients in $\mathbb{N}$ obtained by moving the terms of p with negative coefficients to the right side of the equation. We do this so that

q and r have values in $\mathbb{N}$ and hence are $n + k$-ary recursive functions. Now apply Gödel's β-function:

$$R(a_1, \ldots, a_n) \Leftrightarrow \exists c[q(a_1, \ldots, a_n, \beta(c, 0), \ldots, \beta(c, k \doteq 1))$$
$$= r(a_1, \ldots, a_n, \beta(c, 0), \ldots, \beta(c, k \doteq 1))].$$

We have written R in the form $R(a_1, \ldots, a_n) \Leftrightarrow \exists c[Q(a_1, \ldots, a_n, c)]$ with Q recursive; hence R is RE as required. ∎

Theorem 3: An n-ary function is recursive if and only if it is Diophantine.

Proof: We need only show that every Diophantine function is recursive. Let F be Diophantine. Then $\mathcal{G}_F$ is Diophantine (definition); hence $\mathcal{G}_F$ is RE (Theorem 2). Finally, Theorem 7 in Section 8.5 applies and F is recursive. ∎

──────────── *Exercise for Section 10.3* ────────────

This exercise outlines a proof of the remarkable fact that given a 1-ary RE relation R (i.e., $R \subseteq \mathbb{N}$), there is a polynomial q with integer coefficients such that R is the set of nonnegative values of q. These ideas are due to Putnam.

(1) Given a polynomial $p(y, x_1, \ldots, x_k)$, define a new polynomial q by $q(y, x_1, \ldots, x_k)$
$= (y + 1)(1 - p(y, x_1, \ldots, x_k)^2) - 1$. Show the following for $a, b_1, \ldots, b_k \in \mathbb{N}$:
(a) If $q(a, b_1, \ldots, b_k) \geq 0$, then $p(a, b_1, \ldots, b_k) = 0$.
(b) If $p(a, b_1, \ldots, b_k) = 0$, then $q(a, b_1, \ldots, b_k) = a$.
(2) Let R be a 1-ary RE relation; there is a polynomial $p(y, x_1, \ldots, x_k)$ such that for all $a \in \mathbb{N}$, $R(a) \Leftrightarrow p(a, x_1, \ldots, x_k) = 0$ has a solution in $\mathbb{N}$. Let q be defined as in (1). Show that

$$R = \{q(a, b_1, \ldots, b_k): a, b_1, \ldots, b_k \in \mathbb{N} \text{ and } q(a, b_1, \ldots, b_k) \geq 0\}.$$

(3) Show RE: $PR(a) \Leftrightarrow a$ is prime. Conclude that there is a polynomial $q(y, x_1, \ldots, x_k)$ such that

$$PR = \{q(a, b_1, \ldots, b_k): a, b_1, \ldots, b_k \in \mathbb{N} \text{ and } q(a, b_1, \ldots, b_k) \geq 0\}.$$

──────────── 10.4 ────────────

The Exponential Function is Diophantine

In this section, we will prove Matiyasevich's theorem that the exponential function $EXP(b, n) = b^n$ is Diophantine ($b^0 = 1$, $b^{n+1} = b^n \times b$). This would be an interesting result in its own right, but in addition it is a key step in the proof of the bounded $\forall$-rule. We begin by reducing the problem to that of finding a Diophantine function that *behaves exponentially*.

Theorem 1 (Reduction): The exponential function $EXP(b, n)$ is Diophantine provided that we can find a 2-ary Diophantine function $F(a, n)$ such that for all $a > 1$,

$$(2a - 1)^n \leq F(a, n) \leq (2a)^n. \tag{*}$$

Proof: We proceed informally at first. By (*), $F(a, n) \approx (2a)^n$ whenever a is large. Thus for a sufficiently large,

$$\frac{F(ab, n)}{F(a, n)} \approx \frac{(2ab)^n}{(2a)^n} = b^n.$$

This suggests the following strategy: $b^n = \text{NEAR}(F(ab, n), F(a, n))$, where $a = ?$ and $?$ is a Diophantine function of b, n. We now proceed formally. Let $b > 1$ and $n > 0$; we will prove that for $a \geq nF(b, n)$,

$$b^n - \frac{1}{2} < \frac{F(ab, n)}{F(a, n)} < b^n + \frac{1}{2}. \qquad (**)$$

Suppose, for a moment, that (**) holds, and let us complete the argument that $\text{EXP}(b, n)$ is Diophantine. If we take $a = nF(b, n)$ in (**), we obtain

$$b^n - \frac{1}{2} < \frac{F(nF(b, n)b, n)}{F(nF(b, n), n)} < b^n + \frac{1}{2}.$$

Let $G(b, n) = \text{NEAR}(F(nF(b, n)b, n), F(nF(b, n), n))$; G is a Diophantine function by composition of functions and the fact that NEAR and F are Diophantine functions. Moreover, $G(b, n) = b^n$ for $b > 1$ and $n > 0$. We now have:

$$\begin{aligned}
\text{EXP}(b, n) = c &\Leftrightarrow [((n = 0) \vee (b = 1)) \wedge (c = 1)] \\
&\vee [(n > 0) \wedge (b = 0) \wedge (c = 0)] \\
&\vee [(n > 0) \wedge (b > 1) \wedge (c = G(b, n))]
\end{aligned}$$

By standard arguments, EXP is a Diophantine function as required.

We now turn to the proof of (**). Replace a with ab in (*) and then multiply the result by the reciprocals of the terms in (*) to obtain

$$\left[\frac{2ab - 1}{2a}\right]^n \leq \frac{F(ab, n)}{F(a, n)} \leq \left[\frac{2ab}{2a - 1}\right]^n$$

or

$$b^n \left[1 - \frac{1}{2ab}\right]^n \leq \frac{F(ab, n)}{F(a, n)} \leq b^n \left[1 - \frac{1}{2a}\right]^{-n}. \qquad (***)$$

The following two inequalities are used to eliminate the exponent n:

- $(1 - x)^n \geq 1 - nx$ for $0 < x < 1$
- $(1 - x)^{-n} \leq (1 - nx)^{-1}$ for $0 < x < \dfrac{1}{n}$.

The first is well known and is easily proved by induction; the second is the reciprocal of the first. We now use (***) and the two inequalities to prove

(1) $\dfrac{F(ab, n)}{F(a, n)} > b^n - \dfrac{1}{2}$ (assume $nb^{n-1} < a$).

(2) $\dfrac{F(ab, n)}{F(a, n)} < b^n + \dfrac{1}{2}$ $\left(\text{assume } nb^n + \dfrac{1}{2}n < a\right)$.

$$\frac{F(ab, n)}{F(a, n)} \geq b^n\left[1 - \frac{1}{2ab}\right]^n \geq b^n\left[1 - \frac{n}{2ab}\right] = b^n - \frac{nb^{n-1}}{2a} > b^n - \frac{1}{2}$$

$$\frac{F(ab, n)}{F(a, n)} \leq b^n\left[1 - \frac{1}{2a}\right]^{-n} \leq b^n\left[1 - \frac{n}{2a}\right]^{-1} = b^n\left[1 + \frac{n}{(2a - n)}\right] < b^n + \frac{1}{2}.$$

Combining (1) and (2), we see that (**) holds whenever $nb^n + \dfrac{n}{2} < a$. Now

$$nb^n + \frac{n}{2} = n\left(b^n + \frac{1}{2}\right)$$

$$< n(2b - 1)^n \qquad (b > 1)$$

$$\leq nF(b, n) \qquad (*).$$

Hence (**) holds whenever $nF(b, n) \leq a$. ∎

Our goal now is to construct a Diophantine function $F(a, n)$ that satisfies the inequalities in Theorem 1. For each integer $a > 1$, consider the equation

$$x^2 - (a^2 - 1)y^2 = 1.$$

We will refer to this as the *a-equation*. A Diophantine function $F(a, n)$ that satisfies $(2a - 1)^n \leq F(a, n) \leq (2a)^n$ can be expressed in terms of solutions of *a*-equations. We note that *a*-equations are special cases of Pell equations—that is, equations of the form $x^2 - cy^2 = 1$, where c is a positive integer that is not a perfect square. Each Pell equation has infinitely many solutions in natural numbers, but finding a nontrivial solution (i.e., a solution other than $x = 1$, $y = 0$) is not always easy. For example, the smallest nontrivial solution of $x^2 - 67y^2 = 1$ is $x = 48,842$, $y = 5,967$. For *a*-equations, however, the smallest nontrivial solution, namely $x = a$, $y = 1$, is easily obtained.

Perhaps the most direct way to prove the desired result is to begin by defining, for each $a > 1$, two sequences $x_n(a)$, $y_n(a)$ $(n = 0, 1, 2, \ldots)$ as follows:

$$x_{n+1}(a) = 2ax_n(a) - x_{n-1}(a) \quad (x_0(a) = 1, x_1(a) = a)$$

$$y_{n+1}(a) = 2ay_n(a) - y_{n-1}(a) \quad (y_0(a) = 0, y_1(a) = 1).$$

Note that $x_0(a) = 1$, $y_0(a) = 0$ and $x_1(a) = a$, $y_1(a) = 1$ are both solutions of the *a*-equation. We will eventually show that each pair $x_n(a)$, $y_n(a)$ is a solution of the *a*-equation and, moreover, that these pairs give all solutions in $\mathbb{N}$. We will now proceed to prove properties of $x_n(a)$ and $y_n(a)$ that will be needed later.

Theorem 2 (Inequalities for $x_n(a)$ and $y_n(a)$): The following hold for all $n \geq 0$:

I1 $1 \leq x_n(a) < x_{n+1}(a)$.

I2 $0 \leq y_n(a) < y_{n+1}(a)$.

I3 $n \leq y_n(a)$.

I4 $(2a - 1)^n \leq y_{n+1}(a) \leq (2a)^n$.

Proof: To prove I1 and I2, use induction and the recursion formulas for $x_n(a)$ and $y_n(a)$. The proof of I3 is by induction and uses I2. We leave the details to the reader. The proof of I4 is also by induction. We will skip immediately to the induction step. Let $n \geq 0$ and assume the inequalities hold for n. We have:

$$(2a - 1)^{n+1} = (2a - 1)^n(2a - 1)$$
$$\leq y_{n+1}(a)(2a - 1) \qquad \text{induction hypothesis}$$
$$= 2ay_{n+1}(a) - y_{n+1}(a)$$
$$< 2ay_{n+1}(a) - y_n(a) \qquad y_n(a) < y_{n+1}(a) \text{ by I2}$$
$$= y_{n+2}(a);$$

$$y_{n+2}(a) = 2ay_{n+1}(a) - y_n(a)$$
$$\leq 2ay_{n+1}(a)$$
$$\leq 2a(2a)^n \qquad \text{induction hypothesis}$$
$$= (2a)^{n+1}. \blacksquare$$

I4 states that $y_{n+1}(a)$ satisfies the inequalities in Theorem 1. Thus our goal now is to prove that $y_{n+1}(a)$ is a Diophantine function of n, a.

Theorem 3 (Subscript property): For $n \geq 0$, $y_n(a) \equiv n \pmod{a - 1}$.

Proof: The proof is by complete induction. Since $y_0(a) = 0$ and $y_1(a) = 1$, the result is obvious for $n = 0$ and $n = 1$. Let $n \geq 1$ and assume the result holds for all $k \leq n$.

$$y_{n+1}(a) = 2ay_n(a) - y_{n-1}(a)$$
$$\equiv 2an - (n - 1) \pmod{a - 1} \qquad \text{induction hypothesis}$$
$$\equiv n + 1 \pmod{a - 1}. \blacksquare$$

Theorem 4 (Congruence property): Let $a, b > 1$. If $a \equiv b \pmod{c}$, then $x_n(a) \equiv x_n(b) \pmod{c}$.

Proof: Again, the proof is by complete induction. For $n = 0$ and $n = 1$, the result is obvious. For $n \geq 1$, we have

$$x_{n+1}(a) = 2ax_n(a) - x_{n-1}(a)$$
$$\equiv 2ax_n(b) - x_{n-1}(b) \pmod{c} \qquad \text{induction hypothesis}$$
$$\equiv 2bx_n(b) - x_{n-1}(b) \pmod{c} \qquad a \equiv b \pmod{c}$$
$$\equiv x_{n+1}(b) \pmod{c}. \blacksquare$$

Let $d = a^2 - 1$, so that the a-equation can be written as $x^2 - dy^2 = 1$. The factorization $x^2 - dy^2 = (x + y\sqrt{d})(x - y\sqrt{d})$ suggests that numbers of the form $(x \pm y\sqrt{d})$ are useful in studying solutions of the a-equation.

Lemma 1: If p, q, r, s are integers with $p + q\sqrt{d} = r + s\sqrt{d}$, then $p = r$ and $q = s$.

Proof: This follows from the fact that $\sqrt{d}$ $(d = a^2 - 1, a > 1)$ is irrational; the details are left as an exercise for the reader. ∎

To simplify notation, we frequently omit a in $x_n(a)$, $y_n(a)$ and simply write x_n, y_n. We emphasize, however, that in the following it is assumed that $a \in \mathbb{N}$ with $a > 1$ and that $d = a^2 - 1$.

Lemma 2 (Identities for x_n and y_n): The following hold for all $n \geq 0$:

F1 $x_n \pm y_n\sqrt{d} = (a \pm \sqrt{d})^n$.
F2 $x_n - y_n\sqrt{d} = (a + \sqrt{d})^{-n}$.

Proof: The proof of F1 is by complete induction. The result is obvious for $n = 0$ and $n = 1$. Let $n \geq 1$ and assume that the identity holds for all $k \leq n$.

$$\begin{aligned}
x_{n+1} \pm y_{n+1}\sqrt{d} &= (2ax_n - x_{n-1}) \pm (2ay_n - y_{n-1})\sqrt{d} \\
&= 2a(x_n \pm y_n\sqrt{d}) - (x_{n-1} \pm y_{n-1}\sqrt{d}) \\
&= 2a(a \pm \sqrt{d})^n - (a \pm \sqrt{d})^{n-1} \\
&= (a \pm \sqrt{d})^{n-1}[2a(a \pm \sqrt{d}) - 1] \\
&= (a \pm \sqrt{d})^{n-1}[a \pm \sqrt{d}]^2 = (a \pm \sqrt{d})^{n+1}.
\end{aligned}$$

We now prove F2. By F1 and the factorization $(a + \sqrt{d})(a - \sqrt{d}) = 1$,

$$x_n - y_n\sqrt{d} = (a - \sqrt{d})^n = (a + \sqrt{d})^{-n}. ∎$$

Theorem 5 (Solutions of the a-equation): For all $n \geq 0$, x_n, y_n is a solution of the a-equation. Moreover, if $r, s \in \mathbb{N}$ is a solution of the a-equation, then there exist $n \in \mathbb{N}$ such that $r = x_n$ and $s = y_n$.

Proof: To see that x_n, y_n is a solution, use F1 and F2 as follows:

$$x_n^2 - dy_n^2 = (x_n + y_n\sqrt{d})(x_n - y_n\sqrt{d}) = (a + \sqrt{d})^n(a + \sqrt{d})^{-n} = 1.$$

Next we show that there do not exist $u, v \in \mathbb{Z}$ such that $u^2 - dv^2 = 1$ and $1 < u + v\sqrt{d} < a + \sqrt{d}$. For suppose u and v exist, in which case we have

$$(u + v\sqrt{d})(u - v\sqrt{d}) = 1 = (a + \sqrt{d})(a - \sqrt{d}). \qquad (*)$$

From $(*)$ and $1 < u + v\sqrt{d}$, we obtain $u - v\sqrt{d} < 1$; from $(*)$ and $u + v\sqrt{d} < a + \sqrt{d}$, we obtain $a - \sqrt{d} < u - v\sqrt{d}$. We now have:

$a - \sqrt{d} < u - v\sqrt{d} < 1$	just proved
$-1 < -u + v\sqrt{d} < -a + \sqrt{d}$	multiply by -1
$0 < 2v\sqrt{d} < 2\sqrt{d}$	add $1 < u + v\sqrt{d} < a + \sqrt{d}$.

It follows that $0 < v < 1$, a contradiction of $v \in \mathbb{Z}$.

Now let $r, s \in \mathbb{N}$ be a solution of the a-equation. There exists $n \in \mathbb{N}$ such that $(a + \sqrt{d})^n \le r + s\sqrt{d} < (a + \sqrt{d})^{n+1}$ (the well-ordering principle is used here). If we can prove $(a + \sqrt{d})^n = r + s\sqrt{d}$, then, by F1, we have $x_n + y_n\sqrt{d} = r + s\sqrt{d}$, and Lemma 1 then gives $x_n = r$ and $y_n = s$ as required.

Suppose not; that is, suppose $(a + \sqrt{d})^n < r + s\sqrt{d} < (a + \sqrt{d})^{n+1}$. Multiply these inequalities by $(a + \sqrt{d})^{-n}$ and use F2 to simplify the middle term:

$$1 < (r + s\sqrt{d})(x_n - y_n\sqrt{d}) < a + \sqrt{d}.$$

Now let $u = rx_n - sy_nd$ and $v = sx_n - ry_n$; hence $1 < u + v\sqrt{d} < a + \sqrt{d}$. By direct substitution, u, v is a solution of the a-equation. We have contradicted the result proved earlier. ∎

Lemma 3 (Addition formulas for x_n and y_n): The following hold for all $n \ge 0$:

A1 $x_{n\pm m} = x_nx_m \pm dy_ny_m$ ($m \le n$ for x_{n-m}).

A2 $y_{n+m} = \pm x_ny_m + x_my_n$ ($m \le n$ for y_{n-m}).

Proof:

$$\begin{aligned}
x_{n\pm m} + y_{n\pm m}\sqrt{d} &= (a + \sqrt{d})^{n\pm m} & \text{F1 of Lemma 2}\\
&= (a + \sqrt{d})^n(a + \sqrt{d})^{\pm m}\\
&= (x_n + y_n\sqrt{d})(x_m \pm y_m\sqrt{d}) & \text{F1 and F2 of Lemma 2}
\end{aligned}$$

Now multiply and use Lemma 1. ∎

Lemma 4: For $n, k \ge 1$, $y_{nk} \equiv kx_n^{k-1}y_n \pmod{y_n^2}$.

Proof: We will use the binomial theorem. We have:

$$(x_{nk} + y_{nk}\sqrt{d}) = (a + \sqrt{d})^{nk} = (x_n + y_n\sqrt{d})^k = \sum_{i=0}^{k}\binom{k}{i}x_n^{k-i}y_n^id^{i/2}.$$

By Lemma 1, $y_{nk} = \sum_{\substack{i=0\\i\text{ odd}}}^{k}\binom{k}{i}x_n^{k-i}y_n^id^{(i-1)/2}$, and hence $y_{nk} \equiv kx_n^{k-1}y_n \pmod{y_n^2}$. ∎

Theorem 6 (Solution divisibility property): Let $x, y \in \mathbb{N}$ with $y > 0$ be a solution of the a-equation. Then there is a solution $u, v \in \mathbb{N}$ of the a-equation such that $v > 1$ and $2y^2 \mid v$.

Proof: There exists $n \in \mathbb{N}$ such that $x = x_n$ and $y = y_n$. Moreover, $n \ge 1$, since $y > 0$. By Lemma 4, for all $k \ge 1$ we have $y_{nk} \equiv kx^{k-1}y \pmod{y^2}$. For $k = y$, this gives $y_{ny} \equiv 0 \pmod{y^2}$; hence $y^2 \mid y_{ny}$. Now

$$y_{2ny} = y_{ny+ny} = x_{ny}y_{ny} + x_{ny}y_{ny} = 2x_{ny}y_{ny};$$

from this and $y^2 \mid y_{ny}$ we obtain $2y^2 \mid y_{2ny}$. Thus the required solution of the a-equation is $u = x_{2ny}$, $v = y_{2ny}$. ∎

Lemma 5: Let $n \ge 1$. If $y_n \mid y_k$, then $n \mid k$.

Proof: We first show that $y_n \mid y_{nq}$ for all $q \geq 0$. This is obvious for $q = 0$. Let $q \geq 1$; by Lemma 4 we have $y_{nq} \equiv qx_n^{q-1}y_n \pmod{y_n^2}$; hence $y_n \mid y_{nq}$.

Now assume $y_n \mid y_k$. By the division algorithm, $k = nq + r$, where $0 \leq r < n$; our goal is to show that $r = 0$. (Assuming this is done, we have $k = nq$ and thus $n \mid k$ as required.) Use A1 of Lemma 3 to write

$$y_k = y_{nq+r} = x_{nq}y_r + x_r y_{nq}.$$

Now $y_n \mid y_k$ (hypothesis) and $y_n \mid y_{nq}$ (proved above), so $y_n \mid x_{nq}y_r$. We now argue that $\gcd(y_n, x_{nq}) = 1$. Suppose p is a prime such that $p \mid y_n$ and $p \mid x_{nq}$. Since $p \mid y_n$ and $y_n \mid y_{nq}$, we have $p \mid y_{nq}$. We also have $p \mid x_{nq}$ and $x_{nq}^2 - dy_{nq}^2 = 1$, and hence $p \mid 1$, which is impossible. We now have $y_n \mid x_{nq}y_r$ and $\gcd(y_n, x_{nq}) = 1$; hence $y_n \mid y_r$. Thus $y_n \leq y_r$ or $y_r = 0$. But $r < n$, so $y_n \leq y_r$ is impossible. Thus $y_r = 0$, and hence $r = 0$. ∎

Theorem 7 (Stepping down property I): Let $n \geq 1$. If $y_n^2 \mid y_k$, then $y_n \mid k$.

Proof: We can assume that $k \geq 1$. By Lemma 5, we have $n \mid k$ and thus $k = nq$ with $q \geq 1$. Apply Lemma 4 to obtain

$$y_k = y_{nq} \equiv qx_n^{q-1}y_n \pmod{y_n^2}.$$

By hypothesis, $y_n^2 \mid y_k$; hence $y_n^2 \mid qx_n^{q-1}y_n$; this gives $y_n \mid qx_n^{q-1}$. But $\gcd(x_n, y_n) = 1$ (recall that $x_n^2 - dy_n^2 = 1$), so $y_n \mid q$. Finally, $k = nq$, so $y_n \mid k$ as required. ∎

Lemma 6 (Periodicity of x_n): For $q \geq 1$, $x_{2nq \pm r} \equiv (-1)^q x_r \pmod{x_n}$ ($r \leq n$ for x_{2nq-r}).

Proof: The proof is by induction on q. For $q = 1$,

$$
\begin{aligned}
x_{2n \pm r} = x_{n+(n \pm r)} &= x_n x_{n \pm r} + dy_n y_{n \pm r} && \text{A1 of Lemma 3}\\
&\equiv dy_n y_{n \pm r} \pmod{x_n}\\
&\equiv dy_n(\pm x_n y_r + x_r y_n) \pmod{x_n} && \text{A2 of Lemma 3}\\
&\equiv dy_n^2 x_r \pmod{x_n}\\
&\equiv (x_n^2 - 1)x_r \pmod{x_n} && x_n^2 - dy_n^2 = 1\\
&\equiv -x_r \pmod{x_n}.
\end{aligned}
$$

Now assume $x_{2nq \pm r} \equiv (-1)^q x_r \pmod{x_n}$. Then

$$
\begin{aligned}
x_{2n(q+1) \pm r} = x_{2n+(2nq \pm r)} &\equiv -x_{2nq \pm r} \pmod{x_n} && \text{proved above}\\
&\equiv -(-1)^q x_r \pmod{x_n} && \text{induction hypothesis}\\
&\equiv (-1)^{q+1} x_r \pmod{x_n}. && ∎
\end{aligned}
$$

Lemma 7: For $n > 1$, the following hold:

(1) No two of the numbers $x_0, \ldots, x_n, -x_0, \ldots, -x_{n-1}$ are congruent modulo x_n.

(2) No two of the numbers $x_0, \ldots, x_{2n}$ are congruent modulo x_n.

Proof of (1): By I1, we have $1 = x_0 < \cdots < x_n$; from this it easily follows that no two of the numbers $x_0, \ldots, x_n$ are congruent modulo x_n and that no two of the numbers $-x_0, \ldots, -x_{n-1}$ are congruent modulo x_n (details left to the reader). So the proof is complete if we can show that $x_m \not\equiv -x_j$ $(\text{mod } x_n)$ for $0 \le m \le n$ and $0 \le j \le n - 1$. Suppose, for some m and j, that $x_m \equiv -x_j (\text{mod } x_n)$; note that $x_n \mid x_m + x_j$. We consider two cases: First suppose $m = n$. From $x_n \mid x_m + x_j$ and $m = n$, we obtain $x_n \mid x_j$; hence $x_n \le x_j$ and so $n \le j$, a contradiction of $j \le n - 1$. Now suppose $m < n$. From $x_n \mid x_m + x_j$, we obtain $x_n \le x_m + x_j \le 2x_{n-1}$, a contradiction of the following calculation for $n > 1$:

$$x_n = 2ax_{n-1} - x_{n-2} > 2ax_{n-1} - x_{n-1} = (2a - 1)x_{n-1} > 2x_{n-1}.$$

Proof of (2): We show that $x_{n+1}, \ldots, x_{2n}$ are congruent modulo x_n to $-x_{n-1}, \ldots, -x_0$ respectively; the required result then follows from (1). For $1 \le r \le n$, we have (by Lemma 6) $x_{n+r} = x_{2n-(n-r)} \equiv -x_{n-r} (\text{mod } x_n)$.

Theorem 8 (Stepping down property II): Let $n > 1$ and $i \le n$. If $x_k \equiv x_i (\text{mod } x_n)$, then $k \equiv \pm i (\text{mod } 2n)$.

Proof: By the division algorithm, $k = 2nq + r$ and $0 \le r < 2n$. We can assume that $k > 2n$ (Lemma 7), so $q \ge 1$. We have:

$$
\begin{aligned}
x_i &\equiv x_k (\text{mod } x_n) && \text{hypothesis} \\
&\equiv x_{2nq+r} (\text{mod } x_n) && k = 2nq + r \\
&\equiv (-1)^q x_r (\text{mod } x_n) && \text{Lemma 6.}
\end{aligned}
$$

Now consider two cases.

(1) *q is even* In this case, $x_i \equiv x_r (\text{mod } x_n)$. Now $0 \le i, r < 2n$; hence Lemma 7(2) applies and $i = r$. Thus $k = 2nq + i$, so $k \equiv i (\text{mod } 2n)$.

(2) *q is odd* In this case, $x_i \equiv -x_r (\text{mod } x_n)$. Now $0 \le i \le n$, so $0 \le r \le n - 1$ is impossible by Lemma 7(1) and we have $n \le r < 2n$. Let $s = 2n - r$, and note that $0 < s \le n$. We have:

$$
\begin{aligned}
x_i &\equiv -x_r (\text{mod } x_n) \\
&\equiv -x_{2n-s} (\text{mod } x_n) && r = 2n - s \\
&\equiv x_s (\text{mod } x_n) && \text{Lemma 6.}
\end{aligned}
$$

It follows from Lemma 7(2) that $i = s$. Thus $k = 2n(q + 1) - i$; hence $k \equiv -i (\text{mod } 2n)$. ∎

At last we have the tools to show that $y_{n+1}(a)$ is Diophantine.

Theorem 9 (Diophantine characterization of $y_{n+1}(a)$): Let $y, n, a \in \mathbb{N}$ with $a > 1$. Then $y_{n+1}(a) = y$ if and only if $n + 1 \le y$ and there exist $x, u, v, b, s, t \in \mathbb{N}$ satisfying these nine conditions:

(1) $x^2 - (a^2 - 1)y^2 = 1$.   (6) $b \equiv 1 \pmod{2y}$.
(2) $u^2 - (a^2 - 1)v^2 = 1$.   (7) $s^2 - (b^2 - 1)t^2 = 1$.
(3) $v > 1$.   (8) $t \equiv n + 1 \pmod{b - 1}$.
(4) $v \equiv 0 \pmod{y^2}$.   (9) $x \equiv s \pmod{u}$.
(5) $a \equiv b \pmod{u}$.

Proof: Assume $y_{n+1}(a) = y$. By I3, $n + 1 \le y_{n+1}(a)$; hence $n + 1 \le y$.
We now proceed to find $x, u, v, b, s, t \in \mathbb{N}$ satisfying (1)–(9). Let $x = x_{n+1}(a)$,
so that (1) holds. By Theorem 6 (solution divisibility property) there is a
solution u, v of the a-equation with $v > 1$ such that $2y^2 \mid v$. Thus (2), (3), and
(4) hold. Let $b = a + u^2(u^2 - a)$, so that (5) holds. (Since $v > 1$, the solution
u, v of the a-equation is larger than the solution $a, 1$; hence $u > a$.) To prove
(6), write

$$b = a + u^2(u^2 - a) = a + (1 + (a^2 - 1)v^2)(1 + (a^2 - 1)v^2 - a).$$

Now v was chosen so that $2y \mid v$ (i.e., $v \equiv 0 \pmod{2y}$); hence (6) holds.
Finally, let $s = x_{n+1}(b)$ and $t = y_{n+1}(b)$, so that (7) holds. All choices have
been made and it remains to check (8) and (9).

By Theorem 3 (subscript property), $y_{n+1}(b) \equiv n + 1 \pmod{b - 1}$; that is,
$t \equiv n + 1 \pmod{b - 1}$, so (8) holds. Now $a \equiv b \pmod{u}$ (see (5)), so by
Theorem 4 (congruence property), $x_{n+1}(a) \equiv x_{n+1}(b) \pmod{u}$; that is, $x \equiv s$
$\pmod{u}$, so (9) holds as required.

Now assume that $n + 1 \le y$ and that there exist $x, u, v, b, s, t \in \mathbb{N}$ satisfying
(1)–(9). We are required to show that $y_{n+1}(a) = y$. Since x, y is a solution of
the a-equation (see (1)), there exists $i \in \mathbb{N}$ such that $x = x_i(a)$ and $y = y_i(a)$.
The proof is complete if we can show that $i = n + 1$. Conditions (2)–(9) are
carefully chosen to prove this fact. Speaking quite generally, the proof that
$i = n + 1$ requires a delicate interplay among the following: another solution
u, v of the a-equation (see (2)), choice of $b > 1$, and a solution s, t of the
b-equation (see (7)).

Both $n + 1$ and i are $\le y$ ($n + 1 \le y$ by assumption; $i \le y_i(a)$ by I3, and
$y = y_i(a)$, so $i \le y$). So to prove $i = n + 1$, it suffices to show that

$$n + 1 \equiv \pm i \pmod{2y}.$$

Since u, v is a solution of the a-equation, there exists $j \in \mathbb{N}$ such that
$u = x_j(a)$ and $v = y_j(a)$. For future application of Theorem 8 (SDP II), note
that $0 \le i < j$ and $j > 1$ (use (3) and (4) to show that $0 < y < v$). Now $y^2 \mid v$
by (4), so $y \mid j$ by Theorem 7 (SDP I). From this it follows that $2y \mid 2j$; this
simply means that in proving $n + 1 \equiv \pm i \pmod{2y}$, we can use $2j$ as a
modulus instead of $2y$. Note that (6) implies $2y \mid b - 1$, so $b - 1$ can also be
used as a modulus instead of $2y$.

Since s, t is a solution of the b-equation, there exists $k \in \mathbb{N}$ such that
$s = x_k(b)$ and $t = y_k(b)$. The number k is a link between $n + 1$ and i; namely,
we will prove that

$$n + 1 \equiv k \pmod{b - 1}; \qquad k \equiv \pm i \pmod{2j}.$$

From these two congruences and $2y \mid 2j$, $2y \mid b - 1$, we obtain $n + 1 \equiv \pm i$ (mod $2y$) as required.

Proof that $n + 1 \equiv k$ (mod $b - 1$): By (8) and $t = y_k(b)$, we have $y_k(b) \equiv n + 1$ (mod $b - 1$). By Theorem 3 (subscript property), $y_k(b) \equiv k$ (mod $b - 1$), so $n + 1 \equiv k$ (mod $b - 1$) as required.

Proof that $k \equiv \pm i$ (mod $2j$): We have

$a \equiv b$ (mod u)	(5)
$x_k(a) \equiv x_k(b)$ (mod u)	Theorem 4 (congruence property)
$x_i(a) \equiv x_k(b)$ (mod u)	(9)
$x_k(a) \equiv x_i(a)$ (mod $x_j(a)$)	eliminate $x_k(b)$; $u = x_j(a)$
$k \equiv \pm i$ (mod $2j$)	SDP II ($i \leq j$ and $j > 1$). ∎

Theorem 10 (Matiyasevich): The exponential function EXP(b, n) is Diophantine.

Proof: By Theorem 1 (reduction), it suffices to find a Diophantine function $F(a, n)$ such that for $a > 1$, $(2a - 1)^n \leq F(a, n) \leq (2a)^n$. By I4, $(2a - 1)^n \leq y_{n+1}(a) \leq (2a)^n$ for $a > 1$. So define F by $F(a, n) = y_{n+1}(a)$ for $a > 1$ and $F(a, n) = 0$ for $a \leq 1$. It remains to show that F is Diophantine. We have:

$$F(a, n) = y \Leftrightarrow [(a \leq 1) \wedge (y = 0)] \vee [(a > 1) \wedge (y = y_{n+1}(a))]$$
$$\Leftrightarrow [(a \leq 1) \wedge (y = 0)] \vee [(a > 1) \wedge (n + 1 \leq y)]$$
$$\wedge \exists x \exists u \exists v \exists b \exists s \exists t (\bullet\bullet\bullet)],$$

where $\bullet\bullet\bullet$ is the conjunction of the nine Diophantine relations in the Diophantine characterization of $y_{n+1}(a)$. By standard arguments, F is Diophantine as required. ∎

Our hardest task is finished! But before proving the bounded $\forall$-rule, we still need two additional results, namely the exponential rule and the fact that $\binom{n}{k}$ is a Diophantine function.

Theorem 11 (Exponential rule): Let G and H be n-ary Diophantine functions. Then F is Diophantine, where $F(a_1, \ldots, a_n) = $ EXP$(G(a_1, \ldots, a_n), H(a_1, \ldots, a_n))$.

Proof: This follows immediately from composition of functions and the fact that EXP is Diophantine. ∎

The final result of this section is a proof that $\binom{n}{k}$ is Diophantine. To prove this result, we use the fact that the binomial coefficients satisfy

$$(b + 1)^n = \sum_{k=0}^{n} \binom{n}{k} b^k.$$

Thus if $b > \binom{n}{k}$ for $k = 0, \ldots, n$, then the binomial coefficients $\binom{n}{0}, \ldots,$

$\binom{n}{n}$ are the digits of $(b + 1)^n$ to the base b. So we begin with results on base-b expansions.

Lemma 8: Let $b > 1$ and let $a = r_n b^n + \cdots + r_0$, where $0 \le r_i < b$ for $0 \le i \le n$. Then $a < b^{n+1}$.

Proof:

$$a = r_n b^n + \cdots + r_0 \le (b - 1)b^n + \cdots + (b - 1)$$
$$= (b - 1)[b^n + \cdots + 1]$$
$$= (b - 1)[(b^{n+1} - 1)/(b - 1)] < b^{n+1}. \blacksquare$$

Theorem 12 (Base-b expansion): Let $b > 1$ and $a \ge 1$. Then there exist unique $n, r_0, \ldots, r_n \in \mathbb{N}$ such that

(1) $a = r_n b^n + \cdots + r_1 b^1 + r_0$.

(2) $0 \le r_i < b$ for $0 \le i \le n$ and $r_n \ne 0$.

Proof: The numbers $r_0, \ldots, r_n$ are called the *digits of the number a to the base b*. By repeated application of the division algorithm,

$$a = b \times q_0 + r_0 \qquad (0 \le r_0 < b)$$
$$q_0 = b \times q_1 + r_1 \qquad (0 \le r_1 < b)$$
$$q_1 = b \times q_2 + r_2 \qquad (0 \le r_2 < b)$$
$$\vdots$$
$$q_{n-2} = b \times q_{n-1} + r_{n-1} \qquad (0 \le r_{n-1} < b)$$
$$q_{n-1} = b \times 0 + r_n \qquad (0 < r_n < b),$$

where $q_0 > q_1 > \cdots > q_{n-1} > q_n = 0$. The representation of a is obtained by starting with $a = b \times q_0 + r_0$, replacing q_0 with $b \times q_1 + r_1$, and so on, eventually obtaining (1).

We now prove uniqueness. Suppose that:

(a) $a = r_n b^n + \cdots + r_1 b^1 + r_0$ $(r_n \ne 0, 0 \le r_i < b$ for $0 \le i \le n)$

(b) $a = s_k b^k + \cdots + s_1 b^1 + s_0$ $(s_k \ne 0, 0 \le s_i < b$ for $0 \le i \le k)$.

We first show that $n = k$. Suppose not; say $k < n$. From (b) and Lemma 8, we obtain $a < b^{k+1}$. But $b^{k+1} \le r_n b^n \le a$; hence $a < a$, which is impossible.

We now have $n = k$. From (a) and (b), we obtain

$$(r_n - s_n)b^n + \cdots + (r_1 - s_1)b^1 + (r_0 - s_0) = 0.$$

From this result, it follows that $b \mid (r_0 - s_0)$. But $0 \le |r_0 - s_0| < b$; hence

$r_0 = s_0$. Divide the equation by b and repeat this argument to obtain $r_1 = s_1$. Continuing this process, we eventually obtain $r_i = s_i$ for $0 \le i \le n$. ∎

Theorem 13 (Diophantine characterization of kth digit): Let $a, b, c \in \mathbb{N}$ with $b > 1$ and $a \ge 1$, let $r_0, \ldots, r_n$ (with $r_n \ne 0$) be the digits of a to the base b, and let $0 \le k \le n$. Then $r_k = c$ if and only if $c < b$ and there exist $d, e \in \mathbb{N}$ such that:

(1) $a = db^{k+1} + cb^k + e$.

(2) $d < b^{n-k}$.

(3) $e < b^k$.

Proof: First assume $r_k = c$. Since $r_k < b$, we immediately obtain $c < b$. It remains to find $d, e \in \mathbb{N}$ such that (1)–(3) hold. For $0 < k < n$, we have

$$a = (r_n b^n + \cdots + r_{k+1} b^{k+1}) + r_k b^k + (r_{k-1} b^{k-1} + \cdots + r_0)$$
$$= b^{k+1}(r_n b^{n-k-1} + \cdots + r_{k+1}) + r_k b^k + (r_{k-1} b^{k-1} + \cdots + r_0).$$

Let $d = r_n b^{n-k-1} + \cdots + r_{k+1}$, $e = r_{k-1} b^{k-1} + \cdots + r_0$. Then $a = db^{k+1} + cb^k + e$ and, moreover, $d < b^{n-k}$ and $e < b^k$ by Lemma 8; hence (1)–(3) hold as required. The cases $k = 0$ and $k = n$ are left to the reader.

Conversely, suppose that $c < b$ and that there exist $d, e \in \mathbb{N}$ such that (1)–(3) hold. Write d, e in terms of its digits to the base b (use (2), (3), and Lemma 8 to see the size of d, e), then use (1) to write a in terms of its digits to the base b. By the uniqueness of this expansion, $c = r_k$. ∎

Theorem 14 (Binomial coefficients): The 2-ary function $\binom{n}{k}$ is Diophantine.

Proof: The binomial coefficients satisfy

$$(b + 1)^n = \sum_{k=0}^{n} \binom{n}{k} b^k.$$

We want to choose a base b so that $b > \binom{n}{k}$ for $k = 0, \ldots, n$; the binomial

coefficients $\binom{n}{0}, \ldots, \binom{n}{n}$ are then the digits of $(b + 1)^n$ to the base b. Now

$2^n + 1 > \binom{n}{k}$ for $k = 0, \ldots, n$ (take $b = 1$ in the binomial theorem). Hence

for $b = 2^n + 1$, we obtain

$$(2^n + 2)^n = \sum_{k=0}^{n} \binom{n}{k}(2^n + 1)^k, \quad 0 \le \binom{n}{k} < 2^n + 1.$$

We now have (by Theorem 13)

$$\binom{n}{k} = c \Leftrightarrow [(k > n) \wedge (c = 0)] \vee [(k \le n) \wedge c \text{ is the } k\text{th digit of } (2^n + 2)^n$$

with base $2^n + 1$

$$\Leftrightarrow [(k > n) \wedge (c = 0)] \vee [(k \le n) \wedge (c < b) \wedge \exists d \exists e (\bullet\bullet\bullet)],$$

where $\bullet\bullet\bullet$ is

$$(2^n + 2)^n = (d(2^n + 1)^{k+1} + c(2^n + 1)^k + e)$$
$$\wedge (d < (2^n + 1)^{n-k}) \wedge (e < (2^n + 1)^k).$$

By standard arguments, including use of the exponential rule, $\binom{n}{k}$ is Diophantine as required. ∎

────────── *Exercises for Section 10.4* ──────────

1. Let $a > 1$, $d = a^2 - 1$.
   (1) Show that $\sqrt{d}$ is irrational.
   (2) Show that if $p, q, r, s \in \mathbb{Z}$ and $p + q\sqrt{d} = r + s\sqrt{d}$, then $p = q$ and $r = s$.
2. Prove the inequalities I1 and I2.
3. Let $0 \le a, b \le n$ and $a \equiv \pm b \pmod{2n}$. Show that $a = b$.

────── 10.5 ──────

Bounded ∀-Rule

In this section, we will prove the bounded ∀-rule. We begin with a precise definition of a polynomial.

Definition 1: Let $v_0, v_1, v_2, \ldots$ be an infinite list of variables. The *polynomials in the variables* $v_0, v_1, v_2, \ldots$ *with integer coefficients*, or simply the *polynomials*, are defined inductively as follows:

P1 Each variable v_k and each integer k is a polynomial.
P2 If p and q are polynomials, so are $(p + q)$ and $(p \times q)$.
P3 Every polynomial is obtained by a finite number of applications of P1 and P2.

For example, v_4, -7, $(-2 + v_3)$, and $((v_7 + 5) \times (-2 + v_3))$ are polynomials. For each polynomial p there exists $n \in \mathbb{N}$ such that all of the variables that occur in p are among $v_0, \ldots, v_n$; we denote this by writing $p(v_0, \ldots, v_n)$. We emphasize that some of the variables $v_0, \ldots, v_n$ may be missing.

A polynomial, as defined above, is an expression constructed from the following symbol set: $+$, $\times$, $(,)$, 0, ± 1, $\pm 2, \ldots, v_0$, v_1, $v_2, \ldots,$. But in addition, a polynomial $p(v_0, \ldots, v_n)$ calculates a function from $\mathbb{N}^{n+1}$ into $\mathbb{Z}$ in the obvious way: Given $a_0, \ldots, a_n \in \mathbb{N}$, replace the variables $v_0, \ldots, v_n$ with $a_0, \ldots, a_n$ respectively and compute (for example, if $p(v_0, v_1, v_2) = ((-7$

$+ v_0) \times v_2)$, then $p(1, 2, 3) = -18$). An inductive definition of the function from $\mathbb{N}^{n+1}$ into $\mathbb{Z}$ determined by $p(v_0, \ldots, v_n)$ proceeds as follows: Let $a_0, \ldots, a_n \in \mathbb{N}$.

- If p is k, then $p(a_0, \ldots, a_n) = k$.
- If p is v_k, then $p(a_0, \ldots, a_n) = a_k$.
- If p is $(q + r)$, then $p(a_0, \ldots, a_n) = q(a_0, \ldots, a_n) + r(a_0, \ldots, a_n)$.
- If p is $(q \times r)$, then $p(a_0, \ldots, a_n) = q(a_0, \ldots, a_n) \times r(a_0, \ldots, a_n)$.

Lemma 1: If $p(v_0, \ldots, v_n)$ is a polynomial and $a_i \equiv b_i \pmod{m}$ for $0 \le i \le n$, then $p(a_0, \ldots, a_n) \equiv p(b_0, \ldots, b_n) \pmod{m}$.

Proof: The proof is by induction on polynomials and uses these two properties of congruences:

(1) If $a \equiv b \pmod{m}$ and $c \equiv d \pmod{m}$, then $a + c \equiv b + d \pmod{m}$.
(2) If $a \equiv b \pmod{m}$ and $c \equiv d \pmod{m}$, then $a \times c \equiv b \times d \pmod{m}$.

Here are the details for the case in which p is $(q + r)$:

$$
\begin{aligned}
p(a_0, \ldots, a_n) &\equiv q(a_0, \ldots, a_n) + r(a_0, \ldots, a_n) \pmod{m} && p \text{ is } (q + r)\\
&\equiv q(b_0, \ldots, b_n) + r(b_0, \ldots, b_n) \pmod{m} &&\\
&&& \text{induction hypothesis, (1)}\\
&\equiv p(b_0, \ldots, b_n) \pmod{m} && p \text{ is } (q + r).\ \blacksquare
\end{aligned}
$$

Lemma 2: Let $p(v_0, \ldots, v_n)$ be a polynomial. Then there is an $n + 1$-ary function $|p|: \mathbb{N}^{n+1} \to \mathbb{N}$ with these properties:

(1) $|p|$ is Diophantine.
(2) If $0 \le a_i \le b_i$ for $0 \le i \le n$, then $|p(a_0, \ldots, a_n)| \le |p|(b_0, \ldots, b_n)$.

Proof: Informally, $|p|$ is obtained from p by replacing each coefficient with its absolute value. A formal proof is by induction on polynomials. Suppose p is the variable v_k. In this case $k \le n$ and $|p|$ is the function defined by $|p|(a_0, \ldots, a_n) = a_k$; in other words, $|p|$ is $U_{n+1,k+1}$. If p is the constant k, $|p|$ is the function defined by $|p|(a_0, \ldots, a_n) = |k|$; in other words, $|p|$ is $C_{n+1,|k|}$. In both cases, it is clear that $|p|$ is Diophantine and that the inequality in (2) holds.

Suppose p is $(q + r)$. By definition, $|p|$ is the sum of the two $n + 1$-ary Diophantine functions $|q|$ and $|r|$; hence $|p|$ is Diophantine. Moreover,

$$
\begin{aligned}
|p(a_0, \ldots, a_n)| &= |q(a_0, \ldots, a_n) + r(a_0, \ldots, a_n)| && p \text{ is } (q + r)\\
&\le |q(a_0, \ldots, a_n)| + |r(a_0, \ldots, a_n)| &&\\
&\le |q|(b_0, \ldots, b_n) + |r|(b_0, \ldots, b_n) && \text{induction hypothesis}\\
&= |p|(b_0, \ldots, b_n) && |p| \text{ is } |q| + |r|.
\end{aligned}
$$

The case in which p is $q \times r$ is similar and is left to the reader. $\blacksquare$

Lemma 3: If $a \mid c$, $b \mid c$, and $\gcd(a, b) = 1$, then $a \mid c/b$.

Proof: Since $\gcd(a, b) = 1$, there exist $x, y \in \mathbb{Z}$ such that $ax + by = 1$ (see the appendix on number theory). Multiply by c/b to obtain

$$ax\left(\frac{c}{b}\right) + cy = \frac{c}{b}.$$

Now $a \mid ax(c/b) + cy$; hence $a \mid c/b$ as required. ∎

Divisor Lemma: Let $s, t_1, \ldots, t_n$ be integers with $s > 0$. If $s \mid t_1 \times \cdots \times t_n$, then there exist positive integers q and k ($1 \leq k \leq n$) such that

D1 $q \mid s$;   **D2** $q \mid t_k$;   **D3** $q \geq s^{1/n}$.

Proof: For any integers $s, t_1, \ldots, t_n$ we have (see Exercise 1):

$$\gcd(s, t_1 \times \cdots \times t_n) \leq \gcd(s, t_1) \times \cdots \times \gcd(s, t_n).$$

But $s \mid t_1 \times \cdots \times t_n$, so $s = \gcd(s, t_1 \times \cdots \times t_n)$; thus we have

$$s \leq \gcd(s, t_1) \times \cdots \times \gcd(s, t_n).$$

It follows that $\gcd(s, t_k) \geq s^{1/n}$ for some k ($1 \leq k \leq n$); take $q = \gcd(s, t_k)$. ∎

Lemma 4: Let b, w, z be positive integers such that $b \leq w$ and $b!w! \mid z + 1$. Then there exist positive integers $e_0, \ldots, e_{b-1}$ satisfying these conditions:

(1) $\dbinom{z}{b} = e_0 \times e_1 \times \cdots \times e_{b-1}$.

(2) e_t and $w!$ are relatively prime, $0 \leq t < b$.
(3) $z \equiv t \pmod{e_t}$, $0 \leq t < b$.

(4) $e_0, \ldots, e_{b-1}$ are relatively prime in pairs.

Proof: Begin by writing

$$\binom{z}{b} = \frac{z}{1} \times \frac{z-1}{2} \times \cdots \times \frac{z-b+1}{b}.$$

For each $t < b$, let $e_t = (z - t)/(t + 1)$; thus $e_0 = z/1$, $e_1 = z - 1/2$,

$\ldots, e_{b-1} = (z - b + 1)/b$; this proves (1). Since $b!w! \mid z + 1$, $z = kb!w! - 1$. This allows us to write e_t as

$$e_t = \frac{z - t}{t + 1} = \frac{kb!w! - (t + 1)}{t + 1} = \frac{kb!}{t + 1} \times w! - 1. \qquad (*)$$

Since $t < b$, $t + 1 \mid b!$; from this it follows that e_t is a positive integer. Moreover, $(*)$ shows that e_t and $w!$ are relatively prime; this proves (2). Now $e_t = (z - t)/(t + 1)$; hence $e_t(t + 1) = z - t$, so e_t divides $z - t$; this proves (3).

It remains to prove (4). Let $0 \le i < j < b$; we are required to show that e_i and e_j are relatively prime. Suppose some prime p divides both e_i and e_j. Then p divides $z - i$ and $z - j$ (use (3)); hence p divides $j - i$. Now $0 < j - i < b \le w$, so p divides $w!$. But $w!$ and e_i are relatively prime, which gives a contradiction. ∎

Suppose $R(a_1, \ldots, a_n, b) \Leftrightarrow \forall t < bQ(a_1, \ldots, a_n, t)$ with Q Diophantine; our goal is to prove that R is Diophantine. Since Q is Diophantine, there is a polynomial such that for all $a_1, \ldots, a_n, b \in \mathbb{N}$,

$$R(a_1, \ldots, a_n, b) \Leftrightarrow \forall t < b \exists x_1 \cdots \exists x_k [p(a_1, \ldots, a_n, t, x_1, \ldots, x_k) = 0].$$

Thus, to prove the bounded ∀-rule, we need a Diophantine characterization of the right side of this equivalence.

Theorem 1 (Diophantine characterization for bounded ∀): Let
p be a polynomial with $n + k + 1$ variables and let $a_1, \ldots, a_n, b \in \mathbb{N}$. The following are equivalent:

BQ1 $\forall t < b \exists x_1 \cdots \exists x_k [p(a_1, \ldots, a_n, t, x_1, \ldots, x_k) = 0].$
BQ2 $b = 0$ or there exist $w, z, c_1, \ldots, c_k \in \mathbb{N}$ such that
  (1) $w > b$ and $c_i > w$ for $1 \le i \le k$.
  (2) $z > b + |p| (a_1, \ldots, a_n, b, w, \ldots, w)^{w^k} \times w^{w^{k+1}}$.
  (3) $p(a_1, \ldots, a_n, z, c_1, \ldots, c_k) \equiv 0 \left(\bmod \binom{z}{b} \right).$
  (4) $\binom{c_i}{w} \equiv 0 \left(\bmod \binom{z}{b} \right)$, $1 \le i \le k$.

Proof: We give the proof for the case $k = 2$; this will simplify notation and make the overall plan of attack more transparent. Thus we show that for $a_1, \ldots, a_n, b \in \mathbb{N}$, the following are equivalent (**a** denotes $\langle a_1, \ldots, a_n \rangle$):

BQ1 $\forall t < b \exists x \exists y [p(\mathbf{a}, t, x, y) = 0].$
BQ2 $b = 0$ or there exist $w, z, c, d \in \mathbb{N}$ such that
  (1) $w > b$ and $c, d > w$.
  (2) $z > b + |p|(\mathbf{a}, b, w, w)^{w^2} \times w^{w^3}.$

(3) $p(\mathbf{a}, z, c, d) \equiv 0 \left(\mathrm{mod} \begin{pmatrix} z \\ b \end{pmatrix} \right)$.

(4) $\begin{pmatrix} c \\ w \end{pmatrix} \equiv \begin{pmatrix} d \\ w \end{pmatrix} \equiv 0 \left(\mathrm{mod} \begin{pmatrix} z \\ b \end{pmatrix} \right)$.

Proof that BQ2 $\Rightarrow$ BQ1: If $b = 0$, then BQ1 obviously holds. So assume $b > 0$ and that there exist $w, z, c, d \in \mathbb{N}$ satisfying (1)–(4). Fix $t < b$; we are required to find $x, y \in \mathbb{N}$ such that $p(\mathbf{a}, t, x, y) = 0$. To do this, we use the divisor lemma to find $x, y \in \mathbb{N}$ and a modulus q' such that

$$p(\mathbf{a}, t, x, y) \times b! \equiv 0 \,(\mathrm{mod}\, q'), \quad 0 \le |p(\mathbf{a}, t, x, y)| \times b! < q';$$

this gives $p(\mathbf{a}, t, x, y) = 0$ as required.

It follows from $\begin{pmatrix} c \\ w \end{pmatrix} \equiv 0 \left(\mathrm{mod} \begin{pmatrix} z \\ b \end{pmatrix} \right)$ (see (4)) that

$$\frac{z \times (z - 1) \times \cdots \times (z - b + 1)}{b!} \text{ divides } \frac{c \times (c - 1) \times \cdots \times (c - w + 1)}{w!}.$$

Since $b < w$, $z \times (z - 1) \times \cdots \times (z - b + 1) \,|\, c \times (c - 1) \times \cdots \times (c - w + 1)$. Now $0 \le t < b$, so $z - t$ is one of the numbers $z, z - 1, \ldots, z - b + 1$; it follows that

$$z - t \,|\, c \times (c - 1) \times \cdots \times (c - w + 1).$$

By the divisor lemma, there exist q, x $(0 \le x < w)$ such that

D1 $q \,|\, z - t$ **D2** $q \,|\, c - x$ **D3** $q \ge (z - t)^{1/w}$.

By similar reasoning, $z - t \,|\, d \times (d - 1) \times \cdots \times (d - w + 1)$; since $q \,|\, z - t$ by D1, it follows that $q \,|\, d \times (d - 1) \times \cdots \times (d - w + 1)$. Again by the divisor lemma, there exist q', y $(0 \le y < w)$ such that

D1′ $q' \,|\, q$ **D2′** $q' \,|\, d - y$ **D3′** $q' \ge q^{1/w}$.

This completes the construction of x, y, and q'.

Next we show that $p(\mathbf{a}, t, x, y) \times b! \equiv 0 \,(\mathrm{mod}\, q')$. From (3) we have

$$z \times (z - 1) \times \cdots \times (z - b + 1) \,|\, p(\mathbf{a}, z, c, d) \times b!.$$

Now $q' \,|\, q$, $q \,|\, z - t$, and $z - t$ is one of the numbers $z, z - 1, \ldots, z - b + 1$; hence $q' \,|\, p(\mathbf{a}, z, c, d) \times b!$; that is,

$$p(\mathbf{a}, z, c, d) \times b! \equiv 0 \,(\mathrm{mod}\, q').$$

Also, $z \equiv t \,(\mathrm{mod}\, q')$, $c \equiv x \,(\mathrm{mod}\, q')$, and $d \equiv y \,(\mathrm{mod}\, q')$ (see D1, D1′, D2, and D2′), so by Lemma 1 we have $p(\mathbf{a}, z, c, d) \equiv p(\mathbf{a}, t, x, y) \,(\mathrm{mod}\, q')$. It follows that $p(\mathbf{a}, t, x, y) \times b! \equiv 0 \,(\mathrm{mod}\, q')$ as required.

It remains to show that $0 \le |p(\mathbf{a}, t, x, y)| \times b! < q'$. We have:

$$q' \ge q^{1/w} \ge (z - t)^{1/w^2} > (z - b)^{1/w^2} \qquad\qquad \text{D3′, D3}, t < b$$
$$> |p|(\mathbf{a}, b, w, w) \times w^w \qquad (2)$$
$$> |p| \,(\mathbf{a}, b, w, w) \times b! \qquad w > b$$
$$\ge |p(\mathbf{a}, t, x, y)| \times b!$$

$$\text{Lemma 2 } (t < b, x < w, y < w).$$

Proof that BQ1 $\Rightarrow$ BQ2: Assume $b > 0$; we construct $w, z, c, d \in \mathbb{N}$ such that (1)–(4) hold. By BQ1, for each $t < b$ there exist $x_t, y_t \in \mathbb{N}$ such that $p(\mathbf{a}, t, x_t, y_t) = 0$. Choose w such that $w > b$ and $w > x_0, \ldots, x_{b-1}, y_0, \ldots, y_{b-1}$. Next, choose z so that (2) holds and also such that $b!w! \mid z + 1$. We now use the Chinese remainder theorem to find $c, d \in \mathbb{N}$ such that (3) and (4) hold.

By Lemma 4, $\binom{z}{b}$ is the product of positive integers $e_0, \ldots, e_{b-1}$ satisfying these conditions:

(a) e_t and $w!$ are relatively prime, $0 \le t < b$.

(b) $e_0, \ldots, e_{b-1}$ are relatively prime in pairs.

(c) $z \equiv t \pmod{e_t}$, $0 \le t < b$.

Now apply the Chinese remainder theorem twice; there exist natural numbers $c, d > w$ such that for $0 \le t < b$,

(d) $c \equiv x_t \pmod{e_t}$,   (e) $d \equiv y_t \pmod{e_t}$.

By (c), (d), (e), and Lemma 1, we have $p(\mathbf{a}, z, c, d) \equiv p(\mathbf{a}, t, x_t, y_t) \pmod{e_t}$. But $p(\mathbf{a}, t, x_t, y_t) = 0$; hence $p(\mathbf{a}, z, c, d) \equiv 0 \pmod{e_t}$, $0 \le t < b$. Since $e_0, \ldots, e_{b-1}$ are relatively prime in pairs and their product is $\binom{z}{b}$, we have $p(\mathbf{a}, z, c, d) \equiv 0 \left(\mathrm{mod} \binom{z}{b}\right)$, so (3) holds.

We prove the first part of (4) by showing that $\binom{z}{b} \mid \binom{c}{w}$. Now $\binom{z}{b}$ is the product of $e_0, \ldots, e_{b-1}$, and these numbers are relatively prime in pairs; hence it suffices to show that each e_t divides $\binom{c}{w}$. Moreover, e_t and $w!$ are relatively prime, so by Lemma 3 it suffices to prove that $e_t \mid c \times (c - 1) \times \cdots \times (c - w + 1)$. Now $c \equiv x_t \pmod{e_t}$ by (d); hence $e_t \mid c - x_t$. Since $0 \le x_t < w$, $c - x_t$ is among the numbers $c, c - 1, \ldots, c - w + 1$. It easily follows that $e_t \mid c \times (c - 1) \times \cdots \times (c - w + 1)$ as required. ∎

Comment on proof The occurrence of $b!$ in $p(\mathbf{a}, t, x, y) \times b! \equiv 0 \pmod{q'}$ may seem strange. This could be avoided by replacing (3) by $p(\mathbf{a}, t, c, d) \equiv 0 \left(\mathrm{mod} \binom{z}{b} \times b!\right)$. But this would require proving that the factorial function is Diophantine—not an easy task. So instead we use (3) and accommodate $b!$ by including the factor w^{w^3} in (2).

Theorem 2 (Bounded ∀-rule): Let Q be an $n + 1$-ary Diophantine relation. Then R is Diophantine, where R is the $n + 1$-ary relation defined by

$$R(a_1, \ldots, a_n, b) \Leftrightarrow \forall t < b\, Q(a_1, \ldots, a_n, t).$$

Proof: Since Q is Diophantine, there is a polynomial p such that for all $a_1, \ldots, a_n, t \in \mathbb{N}$,

$$Q(a_1, \ldots, a_n, t) \Leftrightarrow \exists x_1 \cdots \exists x_k [p(a_1, \ldots, a_n, t, x_1, \ldots, x_k) = 0].$$

We can then write

$$R(a_1, \ldots, a_n, b) \Leftrightarrow \forall t < b \exists x_1 \cdots \exists x_k [p(a_1, \ldots, a_n, t, x_1, \ldots, x_k) = 0].$$

The right side of this equivalence is BQ1; replace it with BQ2 to obtain

$$R(a_1, \ldots, a_n, b) \Leftrightarrow (b = 0) \vee \exists w \exists z \exists c_1 \cdots \exists c_k [\bullet\bullet\bullet],$$

where $\bullet\bullet\bullet$ is the conjunction of the Diophantine relations (1)–(4) in BQ2. By standard arguments, including the conjunction/disjunction rule, composition, $\exists$-rule, exponential rule, $\binom{n}{k}$ a Diophantine function, and $a \equiv b \pmod{c}$ a Diophantine relation, the relation R is Diophantine as required. ∎

─────────*Exercises for Section 10.5*─────────

1. Let $s, t_1, \ldots, t_n$ be integers with $s > 0$. Use induction to prove that $\gcd(s, t_1 \times \cdots \times t_n) \leq \gcd(s, t_1) \times \cdots \times \gcd(s, t_n)$. (*Hint:* Begin by proving $\gcd(s, a \times b) \leq \gcd(s, a) \times \gcd(s, b)$. Let $c = \gcd(s, a)$ and $d = \gcd(s, b)$. Then $c = sx_1 + ay_1$, $d = sx_2 + by_2$, where $x_1, y_1, x_2, y_2 \in \mathbb{Z}$. Multiply to obtain $cd = [\cdots]s + aby_1 y_2$. Let $e = \gcd(s, a \times b)$; show that $e \mid cd$ and then conclude that $e \leq cd$.)
2. Let $b = 4$ and $w = 5$; then $b! = 24$, $w! = 120$, and $b!w! = 2880$. Let $z = 5759$. Find positive integers e_0, e_1, e_2, e_3 whose product is $\binom{5759}{4}$ and such that no two of the numbers $w!, e_0, e_1, e_2, e_3$ have a common divisor > 1.

─────────|10.6|─────────

Applications of the Main Theorem

So far we have seen two applications of the main theorem: negative solution of Hilbert's tenth problem; characterization of the class of recursive functions and the class of RE relations in terms of polynomial equations. The first application is to number theory, an area of mathematics outside of mathematical logic; the second application is to recursion theory. In this section, we will give additional applications of the main theorem to recursion theory: construction of an RE relation not recursive; proofs of the enumeration theorem for RE relations and the Kleene normal-form theorem. In Chapter 9, these results were obtained by coding programs and computations. In this section, we will code polynomials and calculations. The method of coding we use is due to J. Robinson and uses Cantor's pairing functions $J, K,$ and L. Therefore, let us begin with a summary of key properties of these three functions (proofs in Section 8.3):

(1) J is a one-to-one recursive function from $\mathbb{N} \times \mathbb{N}$ onto $\mathbb{N}$; K and L are 1-ary recursive functions.

(2) $J(a, b) = (a + b)(a + b + 1)/2 + a$.

(3) $K(J(a, b)) = a$ and $L(J(a, b)) = b$.

(4) $K(n) \leq n$ and $L(n) \leq n$.

Throughout this section, we will work with polynomials with coefficients in $\mathbb{N}$. These polynomials are adequate for the following reason: A polynomial equation $p(a, v_1, \ldots, v_k) = 0$, where p has coefficients in $\mathbb{Z}$, can always be replaced by the equation

$$q(a, v_1, \ldots, v_k) = r(a, v_1, \ldots, v_k),$$

where q and r are polynomials with coefficients in $\mathbb{N}$; q and r are obtained from p by moving the negative terms of p to the right side of the original equation.

Let $v_0, v_1, v_2, \ldots$ be an infinite list of variables. The *polynomials in the variables* $v_0, v_1, v_2, \ldots$ *with coefficients in* $\mathbb{N}$, or simply the *polynomials*, are defined inductively as in Definition 1 in Section 10.5, except that condition P1 is modified as follows: Each variable v_n and each $n \in \mathbb{N}$ is a polynomial. Thus the only difference between the two definitions is that coefficients come from $\mathbb{N}$, not from $\mathbb{Z}$. Unless otherwise stated, *polynomial* will now mean *polynomial in the variables* $v_0, v_1, v_2, \ldots$ *with coefficients in* $\mathbb{N}$. For each polynomial p, there exists $n \in \mathbb{N}$ such that the variables in p are among $v_0, \ldots, v_n$; we denote this by writing $p(v_0, \ldots, v_n)$; a polynomial $p(v_0, \ldots, v_n)$ calculates an $n + 1$-ary function in the obvious way (see Section 10.5 for details).

Polynomials are coded as follows: Define an infinite sequence $\tau_0, \tau_1, \tau_2, \ldots$, of polynomials; then show that each polynomial p occurs exactly once in the list $\tau_0, \tau_1, \tau_2, \ldots$. We then agree that if $p = \tau_k$, then the *code* of p is k.

Definition 1: Let $\tau_0, \tau_1, \tau_2, \ldots$ be the set of polynomials defined inductively as follows:

$$\tau_{4n} = v_n \qquad \tau_{4n+2} = (\tau_{K(n)} + \tau_{L(n)})$$

$$\tau_{4n+1} = n \qquad \tau_{4n+3} = (\tau_{K(n)} \times \tau_{L(n)}).$$

Since $K(n), L(n) \leq n < 4n + 2$, τ_{4n+2} and τ_{4n+3} are well defined.

Example 1:

- $\tau_0 = v_0$, $\tau_4 = v_1$, $\tau_8 = v_2, \ldots$; hence each polynomial v_n is in the list.
- $\tau_1 = 0$, $\tau_5 = 1$, $\tau_9 = 2, \ldots$; hence each polynomial n is in the list.
- $\tau_2 = (\tau_{K(0)} + \tau_{L(0)}) = (\tau_0 + \tau_0) = (v_0 + v_0)$.
- $\tau_3 = (\tau_{K(0)} \times \tau_{L(0)}) = (\tau_0 \times \tau_0) = (v_0 \times v_0)$.
- $\tau_{78} = \tau_{4 \times 19 + 2} = (\tau_{K(19)} + \tau_{L(19)}) = (\tau_4 + \tau_1) = (v_1 + 0)$. $\quad\square$

Note: For $i \neq j$, τ_i and τ_j are distinct polynomials; nevertheless, they may calculate the same function. For example, τ_4 and τ_{78} calculate the same 2-ary function.

Example 2:

Given the polynomial $((3 \times v_0) + (2 \times v_1))$, let us find $k \in \mathbb{N}$ such that $\tau_k = ((3 \times v_0) + (2 \times v_1))$. The first step is to find i and j such that $\tau_i = (3 \times v_0)$ and $\tau_j = (2 \times v_1)$; we then calculate k by $k = 4J(i, j) + 2$.

Find i $(3 \times v_0) = (\tau_{13} \times \tau_0)$; $J(13, 0) = 104$; $i = 4 \times 104 + 3 = 419$.

Find j $(2 \times v_1) = (\tau_9 \times \tau_4)$; $J(9, 4) = 100$; $j = 4 \times 100 + 3 = 403$.

Thus $((3 \times v_0) + (2 \times v_1)) = (\tau_{419} + \tau_{403})$, $J(419,403) = 338,672$,   and so $k = 4 \times 338,672 + 2 = 1,354,690$. $\square$

Lemma 1: For each polynomial p, there exists a unique $k \in \mathbb{N}$ such that $p = \tau_k$; in other words, each polynomial occurs exactly once in the list $\tau_0, \tau_1, \tau_2, \ldots$.

Proof: We prove existence and leave uniqueness to the reader. The proof is by induction on polynomials. If p is v_n, then $p = \tau_{4n}$; if p is n, then $p = \tau_{4n+1}$. Suppose p is $(q + r)$. By induction hypothesis, there exist $i, j \in \mathbb{N}$ such that $q = \tau_i$ and $r = \tau_j$. Let $n = J(i, j)$. Then $K(n) = i$ and $L(n) = j$; hence $q = \tau_{K(n)}$ and $r = \tau_{L(n)}$. We now have

$$\tau_{4n+2} = (\tau_{K(n)} + \tau_{L(n)}) = (q + r) = p.$$

The case in which p is $(q \times r)$ is similar. ■

The next result gives a bound on the variables that can occur in the kth polynomial τ_k.

Lemma 2: Let $0 \le k \le 4n + 3$. Then the variables in τ_k are among $v_0, \ldots, v_n$ and we can write $\tau_k(v_0, \ldots, v_n)$. As a special case, the variables in the polynomials $\tau_{K(n)}$ and $\tau_{L(n)}$ are among $v_0, \ldots, v_n$.

Proof: We will outline the main ideas and leave details to the reader. First prove the result for $k = 4j$ and $k = 4j + 1$ $(0 \le j \le n)$. Then proceed by induction on k. For $k = 0$, the result is covered by the case $k = 4j$. For the induction step, let $k > 0$, and assume that the result holds for all smaller values of k. It suffices to consider two cases: namely $k = 4j + 2$ and $k = 4j + 3$ $(0 \le j \le n)$. ■

Our goal is to show that there is a 2-ary RE relation SOLN that enumerates all 1-ary RE relations. Once this important result is available, it is easy to construct a 1-ary relation that is RE but not recursive.

Definition 2: SOLN is the 2-ary relation defined by

$$\text{SOLN}(n, a) \Leftrightarrow \tau_{K(n)}(a, v_1, \ldots, v_n) = \tau_{L(n)}(a, v_1, \ldots, v_n)$$

has a solution in $\mathbb{N}$.

We emphasize that SOLN is defined as follows: Given $n, a \in \mathbb{N}$, take the nth polynomial equation $\tau_{K(n)}(v_0, v_1, \ldots, v_n) = \tau_{L(n)}(v_0, v_1, \ldots, v_n)$ and replace the variable v_0 with $a \in \mathbb{N}$; SOLN(n, a) holds precisely when there exist $a_1, \ldots, a_n \in \mathbb{N}$ such that $\tau_{K(n)}(a, a_1, \ldots, a_n) = \tau_{L(n)}(a, a_1, \ldots, a_n)$. The proof that SOLN is RE is a consequence of the following result:

Theorem 1 (RE characterization of SOLN): Let $n, a \in \mathbb{N}$. The following are equivalent:

  I SOLN(n, a).

  II There exists $c \in \mathbb{N}$ such that

    (1) $\beta(c, 0) = a$.

    (2) $\beta(c, K(n)) = \beta(c, L(n))$.

    (3) $\beta(c, 4j + 1) = j, 0 \le j \le n$.

    (4) $\beta(c, 4j + 2) = \beta(c, K(j)) + \beta(c, L(j)), 0 \le j \le n$.

    (5) $\beta(c, 4j + 3) = \beta(c, K(j)) \times \beta(c, L(j)), 0 \le j \le n$.

Proof that II $\Rightarrow$ I: Let c satisfy conditions (1)–(5) in II; we are required to find $a_1, \ldots, a_n \in \mathbb{N}$ such that $\tau_{K(n)}(a, a_1, \ldots, a_n) = \tau_{L(n)}(a, a_1, \ldots, a_n)$. Let $a_j = \beta(c, 4j)$, $1 \le j \le n$. We will prove that for $0 \le k \le 4n + 3$,

$$\tau_k(a, a_1, \ldots, a_n) = \beta(c, k). \tag{$*$}$$

From $(*)$ we have

$$\begin{aligned} \tau_{K(n)}(a, a_1, \ldots, a_n) &= \beta(c, K(n)) \\ &= \beta(c, L(n)) \qquad\qquad \text{II(2)} \\ &= \tau_{L(n)}(a, a_1, \ldots, a_n), \end{aligned}$$

and SOLN(n, a) holds as required.

We first prove $(*)$ for $k = 4j$ $(1 \le j \le n)$ and $k = 4j + 1$ $(0 \le j \le n)$:

- τ_{4j} is v_j, so $\tau_{4j}(a, a_1, \ldots, a_n) = a_j = \beta(c, 4j)$ (by definition of a_j).
- τ_{4j+1} is j, so $\tau_{4j+1}(a, a_1, \ldots, a_n) = j = \beta(c, 4j + 1)$ (by II(3)).

Now proceed by induction on k. For $k = 0$, τ_0 is v_0, and we have $\tau_0(a, a_1, \ldots, a_n) = a = \beta(c, 0)$ by II(1). Next let $k > 0$ and assume that the result holds for all smaller values of k; it suffices to consider the cases $k = 4j + 2$ and $k = 4j + 3$.

$$\begin{aligned} \tau_{4j+2}(a, a_1, \ldots, a_n) & \\ = \tau_{K(j)}(a, a_1, \ldots, a_n) + \tau_{L(j)}(a, a_1, \ldots, a_n) \quad & \tau_{4j+2} \text{ is } (\tau_{K(j)} + \tau_{L(j)}) \\ = \beta(c, K(j)) + \beta(c, L(j)) \quad & \text{induction hypothesis} \\ = \beta(c, 4j + 2) \quad & \text{II(4)} \end{aligned}$$

The proof for τ_{4j+3} uses II(5) and is similar.

Proof that I ⟹ II: Assume $\text{SOLN}(n, a)$, so that there exist $a_1, \ldots, a_n \in \mathbb{N}$ such that $\tau_{K(n)}(a, a_1, \ldots, a_n) = \tau_{L(n)}(a, a_1, \ldots, a_n)$. Apply Gödel's β-function: There exists $c \in \mathbb{N}$ such that for $0 \le k \le 4n + 3$, $\beta(c, k) = \tau_k(a, a_1, \ldots, a_n)$. We will now verify that c satisfies conditions (1)–(5) in II ((4) and (5) are similar, so we omit (5)).

(1) $\quad \beta(c, 0) = \tau_0(a, a_1, \ldots, a_n) = a \qquad \tau_0$ is v_0

(2) $\quad \beta(c, K(n)) = \tau_{K(n)}(a, a_1, \ldots, a_n)$
$$= \tau_{L(n)}(a, a_1, \ldots, a_n)$$
$$= \beta(c, L(n))$$

(3) $\beta(c, 4j + 1) = \tau_{4j+1}(a, a_1, \ldots, a_n)$
$$= j \qquad\qquad\qquad \tau_{4j+1} \text{ is } j$$

(4) $\beta(c, 4j + 2) = \tau_{4j+2}(a, a_1, \ldots, a_n)$
$$= \tau_{K(j)}(a, a_1, \ldots, a_n)$$
$$+ \tau_{L(j)}(a, a_1, \ldots, a_n) \qquad \tau_{4j+2} \text{ is } (\tau_{K(j)} + \tau_{L(j)})$$
$$= \beta(c, K(j)) + \beta(c, L(j)). \quad \blacksquare$$

Theorem 2 (Enumeration of 1-ary RE relations): SOLN is a 2-ary RE relation that enumerates all 1-ary RE relations. That is, if R is any 1-ary RE relation, then there is an index $n \in \mathbb{N}$ such that for all $a \in \mathbb{N}$, $R(a) \Leftrightarrow \text{SOLN}(n, a)$.

Proof: First we prove that SOLN is RE. By Theorem 1,
$$\text{SOLN}(n, a) \Leftrightarrow \exists c[(\beta(c, 0) = a) \wedge (\beta(c, K(n)) = \beta(c, L(n)))$$
$$\wedge \ \forall j < n + 1(\bullet\bullet\bullet)],$$

where $\bullet\bullet\bullet$ is the conjunction of (3)–(5) in II; SOLN is RE because it is written in the form $\exists c Q(n, a, c)$ with Q recursive. It remains to prove that SOLN enumerates. Let R be a 1-ary RE relation. R is Diophantine (main theorem); hence there are polynomials p and q such that for all $a \in \mathbb{N}$,

$$R(a) \Leftrightarrow p(a, v_1, \ldots, v_k) = q(a, v_1, \ldots, v_k) \text{ has a solution in } \mathbb{N}.$$

By Lemma 1, there exist $i, j \in \mathbb{N}$ such that $p = \tau_i$ and $q = \tau_j$. Let $n = J(i, j)$, so that $p = \tau_{K(n)}$ and $q = \tau_{L(n)}$. By Lemma 2, all variables in $\tau_{K(n)}$ and $\tau_{L(n)}$ are among $v_0, \ldots, v_n$ and we have:

$$R(a) \Leftrightarrow \tau_{K(n)}(a, v_1, \ldots, v_n) = \tau_{K(n)}(a, v_1, \ldots, v_n) \text{ has a solution in } \mathbb{N}$$
$$\Leftrightarrow \text{SOLN}(a, n). \quad \blacksquare$$

Theorem 3 (Existence of RE relation not recursive): Let K be the 1-ary relation defined by $K(a) \Leftrightarrow \text{SOLN}(a, a)$. Then K is RE but not recursive.

Proof: The relation K is RE since SOLN is RE. But K is not recursive. For suppose K is recursive. Then $\neg K$ is also recursive; hence $\neg K$ is RE. Since SOLN enumerates the 1-ary RE relations, there exists $n \in \mathbb{N}$ such that for all $a \in \mathbb{N}$, $\neg K(a) \Leftrightarrow \text{SOLN}(n, a)$.

For $a = n$, we obtain $\neg K(n) \Leftrightarrow \mathrm{SOLN}(n, n)$. But by the definition of K, $K(n) \Leftrightarrow \mathrm{SOLN}(n, n)$, a contradiction. ∎

Comment The notation K for the diagonalization of a 2-ary RE relation that enumerates is standard in recursion theory and should not be confused with the pairing function K; the context should make it clear which use is intended.

Theorem 4 (Universal polynomial for 1-ary RE relations):
There is a polynomial $u(v_0, v_1, v_2, \ldots, v_k)$ with integer coefficients such that for every 1-ary RE relation R, there is an index $n \in \mathbb{N}$ such that for all $a \in \mathbb{N}$,

$$R(a) \Leftrightarrow u(n, a, v_2, \ldots, v_k) = 0 \text{ has a solution in } \mathbb{N}.$$

Proof: Since SOLN is RE, it is Diophantine. Let u be a polynomial such that for all $n, a \in \mathbb{N}$, $\mathrm{SOLN}(n, a) \Leftrightarrow \exists v_2 \ldots \exists v_k [u(n, a, v_2, \ldots, v_k) = 0]$. This is the required polynomial; the details are left to the reader. ∎

Theorem 5 (Enumeration of n-ary RE relations):
For each $n \geq 1$, there is an $n + 1$-ary RE relation S_n that enumerates all n-ary RE relations. Thus if R is any n-ary RE relation, there is an index $e \in \mathbb{N}$ such that for all $a_1, \ldots, a_n \in \mathbb{N}$,

$$R(a_1, \ldots, a_n) \Leftrightarrow S_n(e, a_1, \ldots, a_n).$$

Proof: Define S_n by $S_n(e, a_1, \ldots, a_n) \Leftrightarrow \mathrm{SOLN}(e, J_n(a_1, \ldots, a_n))$; since SOLN is RE and J_n is recursive, S_n is RE (Theorem 8 in Section 8.5). Given an n-ary RE relation R, define R^* by $R^*(a) \Leftrightarrow R(J_{n,1}(a), \ldots, J_{n,n}(a))$. R^* is a 1-ary RE relation; hence there is an index $e \in \mathbb{N}$ such that for all $a \in \mathbb{N}$, $R^*(a) \Leftrightarrow \mathrm{SOLN}(e, a)$. We now have:

$$\begin{aligned} R(a_1, \ldots, a_n) &\Leftrightarrow R^*(J_n(a_1, \ldots, a_n)) \\ &\Leftrightarrow \mathrm{SOLN}(e, J_n(a_1, \ldots, a_n)) \\ &\Leftrightarrow S_n(e, a_1, \ldots, a_n) \end{aligned}$$

and S_n enumerates all n-ary RE relations as required. ∎

The logic books by Bell-Machover 1977, Cohen 1987, and Manin 1977, also give complete proofs of Hilbert's tenth problem. Martin Davis 1973, has written a lively, self-contained presentation which includes a detailed discussion on the contributions of Matiyasevich, Robinson, Davis, and Putnam. Quite briefly, Davis, Putnam, and Robinson 1961, had reduced the problem to that of showing that the exponential function is Diophantine. More precisely, they proved that the decision problem for exponential Diophantine equations is unsolvable. The missing link, that the exponential function is Diophantine, was proved by Matiyasevich in 1970.

————————Exercises for Section 10.6————————

1. Explain why the 1-ary RE relations can be enumerated as $R_0, R_1, R_2, \ldots$ in such a way that the 2-ary relation $a \in R_n$ is RE.

2. Let F be an n-ary recursive function. Prove that there exists $e \in \mathbb{N}$ such that for all $a_1, \ldots, a_n \in \mathbb{N}$,

$$F(a_1, \ldots, a_n) = \mu b S_{n+1}(e, a_1, \ldots, a_n, b).$$

(*Hint:* if F is recursive, then $\mathcal{G}_F$ is RE.)

3. (*Kleene normal-form theorem*) Let $n \ge 1$. Show that there is an $n + 2$-ary recursive relation T_n and a 1-ary recursive function U such that for every n-ary recursive function F, there is an index $e \in \mathbb{N}$ such that for all $a_1, \ldots, a_n \in \mathbb{N}$,

$$F(a_1, \ldots, a_n) = U(\mu c T_n(e, a_1, \ldots, a_n, c)).$$

(*Hint:* See Exercise 2. S_{n+1} is an $n + 2$-ary RE relation that enumerates all $n + 1$-ary RE relations; there is an $n + 3$-ary recursive relation R such that for all $e, a_1, \ldots, a_n, b \in \mathbb{N}$, $S_{n+1}(e, a_1, \ldots, a_n, b) \Leftrightarrow \exists x R(e, a_1, \ldots, a_n, b, x)$. Define T_n by $T_n(e, a_1, \ldots, a_n, c) \Leftrightarrow R(e, a_1, \ldots, a_n, K(c), L(c))$. The desired function U is Cantor's function K.)

4. Define a 1-ary relation by

$$\text{SOLN*}(n) \Leftrightarrow \tau_{K(n)}(v_0, v_1, \ldots, v_n) = \tau_{L(n)}(v_0, v_1, \ldots, v_n) \text{ has a solution in } \mathbb{N}.$$

Thus, $\text{SOLN*}(n) \Leftrightarrow$ the nth polynomial equation has a solution in $\mathbb{N}$. Prove that SOLN* is RE. (*Hint:* Formulate and prove a result similar to Theorem 1.)

5. Define a 2-ary function SUB by

$$\text{SUB}(n, a) = \text{code of the polynomial } \tau_n(a, v_1, \ldots, v_n).$$

Idea: Given $n, a \in \mathbb{N}$, replace the variable v_0 in the nth polynomial $\tau_n(v_0, v_1, \ldots, v_n)$ by $a \in \mathbb{N}$ to obtain a new polynomial $\tau_n(a, v_1, \ldots, v_n)$; $\text{SUB}(n, a)$ is the code of this new polynomial. Note that $\tau_n(a, v_1, \ldots, v_n) = \tau_{\text{SUB}(n, a)}(v_0, v_1, \ldots, v_{\text{SUB}(n, a)})$. Show that SUB is recursive. *Hint:* Use definition by cases and course-of-values recursion as follows:

$$\text{SUB}(n, a) = \begin{cases} \cdots\cdots\cdots\cdots\cdots & n = 0 \\ \cdots\cdots\cdots\cdots\cdots & n = 4j, j > 0 \\ \cdots\cdots\cdots\cdots\cdots & n = 4j + 1 \\ \cdots\cdots\cdots\cdots\cdots & n = 4j + 2 \\ \cdots\cdots\cdots\cdots\cdots & n = 4j + 3. \end{cases}$$

6. Show that SOLN* is not recursive (see the preceding two exercises). *Hint:* Show that

$$\text{SOLN}(n, n) \Leftrightarrow \text{SOLN*}(J(\text{SUB}(K(n), n), \text{SUB}(L(n), n))).$$

From this it follows that if SOLN* is recursive, then K is also recursive (see Theorem 3).

Appendix: Number Theory

W e will now summarize key results of number theory that are used throughout the text. For more detail, the reader is referred to the highly recommended texts of Burton and Rosen (see References section).

Theorem 1 (Division algorithm): For all $a, b \in \mathbb{N}$ with $b > 0$, there exist unique $q, r \in \mathbb{N}$ such that $a = bq + r$ and $0 \le r < b$.

Proof of existence: Choose $q \in \mathbb{N}$ such that $bq \le a < b(q + 1)$ and let $r = a - bq$. (Informally, we obtain q by taking multiples of b until we reach a. Formally, proceed as follows: By the Archimedean order property, there exists $n \in \mathbb{N}^+$ such that $a/b < n$. By the well-ordering principle, there is a smallest such positive integer; call it $q + 1$. We then have $q \le a/b < q + 1$; hence $bq \le a < b(q + 1)$).

To see that $0 \le r < b$, subtract bq from each term of the inequality to obtain

$$0 \le a - bq < b(q + 1) - bq;$$

this simplifies to $0 \le r < b$ as required.

Proof of Uniqueness: Suppose $a = bq_1 + r_1$ with $0 \le r_1 < b$ and $a = bq_2 + r_2$ with $0 \le r_2 < b$. Then $bq_1 + r_1 = bq_2 + r_2$; hence

$$b(q_1 - q_2) = r_2 - r_1.$$

Assume $r_1 \le r_2$; it follows that $q_1 - q_2 \ge 0$. Suppose $q_1 - q_2 \ge 1$. Then $b(q_1 - q_2) \ge b$; hence $r_2 - r_1 \ge b$. This is impossible, since $0 \le r_1, r_2 < b$. Thus $q_1 = q_2$, and also $r_1 = r_2$. ∎

Definition 1: Let $a, b \in \mathbb{Z}$ with $a \neq 0$. Then a *divides* b, written $a \mid b$, if there exists $k \in \mathbb{Z}$ such that $b = ak$.

Theorem 2 (Basic properties of divisibility): Let $a, b, c, d, x, y \in \mathbb{Z}$.

(1) $a \mid a$ (assuming $a \neq 0$).
(2) If $a \mid b$ and $b \mid c$, then $a \mid c$.

(3) If $a \mid b$ and $b > 0$, then $a \leq b$.

(4) If $a \mid b$ and $a \mid c$, then $a \mid bx \pm cy$.

(5) If $a \mid b$ and $c \mid d$, then $ac \mid bd$.

(6) If $ac \mid bc$ and $c \neq 0$, then $a \mid b$.

Definition 2: An integer $p > 1$ is *prime* if the only positive divisors of p are 1 and itself. This is equivalent to the condition: There do not exist positive integers, both > 1, whose product is p. An integer $a > 1$ that is not prime is said to be *composite*. Thus $a > 1$ is composite if and only if there exist integers $b > 1$ and $c > 1$ such that $a = bc$.

Theorem 3: Every integer $a > 1$ has a prime divisor.

Proof: Use complete induction; we leave the details to the reader. ∎

Definition 3: Let $a, b \in \mathbb{Z}$, not both zero. The *greatest common divisor* of a and b, written $\gcd(a, b)$, is the unique positive integer d satisfying (1) $d \mid a$ and $d \mid b$; (2) if $e \mid a$ and $e \mid b$, then $e \leq d$. If $\gcd(a, b) = 1$, then a, b have no positive divisors > 1, in which case we say that a and b are *relatively prime*.

Theorem 4: If $\gcd(a, b) > 1$, then there is a prime p that divides both a and b.

Proof: Use Theorem 3. ∎

Theorem 5 (Fundamental theorem on gcd): Let $\gcd(a, b) = d$. Then there exist $x, y \in \mathbb{Z}$ such that $ax + by = d$.

Proof: Let $A = \{e : e \geq 1$ and $e = ax + by$ for some $x, y \in \mathbb{Z}\}$; clearly $A \neq \emptyset$. By the well-ordering principle, A has a smallest element. Call it e, and write $e = ax + by$ with $x, y \in \mathbb{Z}$. The proof is complete if we can show that $d = e$. Now $d \mid a$ and $d \mid b$ (hypothesis); from $e = ax + by$, it follows that $d \mid e$, so $d \leq e$. It remains to prove that $e \leq d$; for this, it suffices to show that e is a common divisor of a and b. We show $e \mid a$; a similar argument gives $e \mid b$. By the division algorithm, $a = qe + r$, $0 \leq r < e$. Then,

$$r = a - qe = a - q(ax + by) = (1 - qx)a + (-qy)b;$$

this shows that r is a linear combination of a, b. But e is the smallest positive integer that can be expressed as a linear combination of a, b; since $0 \leq r < e$, $r = 0$. It follows that $a = qe$, so $e \mid a$ as required. ∎

Corollary 1: If $a \mid bc$ and $\gcd(a, b) = 1$, then $a \mid c$.

Proof: Since $\gcd(a, b) = 1$, there exist $x, y \in \mathbb{Z}$ such that $1 = ax + by$. Multiply by c to obtain $c = acx + bcy$. Now $a \mid acx$ and $a \mid bcy$, so $a \mid c$ as required. ∎

Corollary 2: If p is a prime and $p \mid ab$, then $p \mid a$ or $p \mid b$.

Proof: Suppose p does not divide a. Then $\gcd(a, p) = 1$; hence $p \mid b$ by Corollary 1. ∎

Corollary 3: If p is a prime and $p \mid a_1 \times \cdots \times a_n$, then $p \mid a_k$ for some $k, 1 \le k \le n$.

Proof: Use induction and Corollary 2. ∎

Corollary 4: If $a \mid c$, $b \mid c$, and $\gcd(a, b) = 1$, then $ab \mid c$.

Proof: Since $\gcd(a, b) = 1$, there exist $x, y \in \mathbb{Z}$ such that $1 = ax + by$. Multiply by c to obtain $c = acx + bcy$. Now $ab \mid acx$ and $ab \mid bcy$; hence $ab \mid c$ as required. ∎

Corollary 5: If $a_1 \mid c, \ldots, a_n \mid c$, where $a_1, \ldots, a_n$ are relatively prime in pairs, then $a_1 \times \cdots \times a_n \mid c$.

Proof: Use induction and Corollaries 3 and 4. ∎

Divisibility can be reduced to algebra by the important concept of *congruence*.

Definition 4: Let $a, b \in \mathbb{Z}$ and let $m \in \mathbb{N}^+$; a is *congruent to b modulo m*, written $a \equiv b \pmod{m}$, if $m \mid b - a$.

Theorem 6 ($\equiv$ an equivalence relation): Let $a, b, c, m \in \mathbb{Z}$ with $m > 0$.

(1) $a \equiv a \pmod{m}$.
(2) If $a \equiv b \pmod{m}$, then $b \equiv a \pmod{m}$.
(3) If $a \equiv b \pmod{m}$ and $b \equiv c \pmod{m}$, then $a \equiv c \pmod{m}$.

Theorem 7 (Basic properties of congruence): Let $a, b, c, d, m, n \in \mathbb{Z}$ with $m, n > 0$.

(1) $a \equiv 0 \pmod{m}$ if and only if $m \mid a$.
(2) If $a \equiv b \pmod{m}$ and $0 \le a, b < m$, then $a = b$.
(3) If $a \equiv b \pmod{m}$ and $n \mid m$, then $a \equiv b \pmod{n}$.
(4) If $a \equiv b \pmod{m}$, $a \equiv b \pmod{n}$, and $\gcd(m, n) = 1$, then $a \equiv b \pmod{m \times n}$.
(5) If $a \equiv b \pmod{m}$ and $c \equiv d \pmod{m}$, then $a + c \equiv b + d \pmod{m}$.
(6) If $a \equiv b \pmod{m}$ and $c \equiv d \pmod{m}$, then $a \times c \equiv b \times d \pmod{m}$.

Theorem 8: Let $a, b, m \in \mathbb{Z}$ with $m > 0$. If $\gcd(a, m) = 1$, then $ax \equiv b$ $\pmod m$ has a solution in integers.

Proof: Since $\gcd(a, m) = 1$, there exist $c, d \in \mathbb{Z}$ such that $ac + md = 1$. Multiply by b to obtain $abc + bdm = b$. From this, it follows that $abc \equiv b \pmod m$, so the required solution is $x = bc$. ∎

Theorem 9 (Chinese remainder theorem): Let $m_1, \ldots, m_n$ be positive integers that are relatively prime in pairs and let $a_1, \ldots, a_n \in \mathbb{Z}$. Then there exists $b \in \mathbb{N}$ such that for $1 \le k \le n$, $b \equiv a_k \pmod{m_k}$. Moreover, b is unique modulo $m_1 \times \cdots \times m_n$.

Outline of proof Let $M = m_1 \times \cdots \times m_n$, and for $1 \le k \le n$, let $N_k = M/m_k$; $\gcd(N_k, m_k) = 1$. For $1 \le k \le n$, let $x_k \in \mathbb{Z}$ be such that $N_k x_k \equiv a_k \pmod{m_k}$ (see Theorem 8) and let $b = N_1 x_1 + \cdots + N_n x_n$. We leave it to the reader to verify that $b \equiv a_k \pmod{m_k}$ for $1 \le k \le n$.

To prove uniqueness, suppose $b \equiv a_k \pmod{m_k}$ and $c \equiv a_k \pmod{m_k}$ for $1 \le k \le n$. Then $b \equiv c \pmod{m_k}$, or $m_k \mid c - b$, $1 \le k \le n$. But $m_1, \ldots, m_n$ are relatively prime in pairs; hence $m_1 \times \cdots \times m_n \mid c - b$ (Corollary 5). It follows that $b \equiv c \pmod{m_1 \times \cdots \times m_n}$ as required. ∎

Theorem 10 (Unique factorization): Every integer $a > 1$ can be written as a product of primes. Moreover, this product is unique in the sense that if

$$a = p_1^{e_1} \times \cdots \times p_n^{e_n} \quad (p_1 < \cdots < p_n \text{ primes and } e_k \ge 1 \text{ for } 1 \le k \le n)$$

and

$$a = q_1^{f_1} \times \cdots \times q_m^{f_m} \quad (q_1 < \cdots < q_m \text{ primes and } f_k \ge 1 \text{ for } 1 \le k \le m),$$

then $m = n$, $q_k = q_k$ for $1 \le k \le n$, and $e_k = f_k$ for $1 \le k \le n$.

Proof: See either Burton's or Rosen's text. ∎

Theorem 11 (Lagrange): Every natural number is the sum of four squares. More precisely, given $n \in \mathbb{N}$, there exist $a, b, c, d \in \mathbb{N}$ such that $n = a^2 + b^2 + c^2 + d^2$.

Proof: See Burton's or Rosen's text. ∎

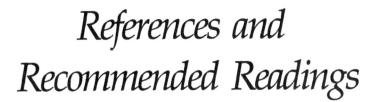

References and Recommended Readings

* Alphabetized by Subject.

──────────── *Mathematical Logic* ────────────

Andrews, Peter B. *An Introduction to Mathematical Logic and Type Theory: To Truth Through Proof.* New York: Academic Press, 1986.

Barwise, Jon and John Etchemendy. *The Language of First-Order Logic.* Stanford, CA: Center for the Study of Language and Information, 1991.

Bell, John and Moshé Machover. *A Course in Mathematical Logic.* Amsterdam: North Holland, 1977.

Boolos, George S. *The Unprovability of Consistency: An Essay in Model Logic.* Cambridge: Cambridge University Press, 1979.

Boolos, George S. and Richard C. Jeffrey. *Computability and Logic (3rd ed.).* Cambridge: Cambridge University Press, 1989.

Chang, C. C. and H. J. Keisler. *Model Theory.* Amsterdam: North Holland, 1973.

Dragalin, A. G. *Mathematical Intuitionism,* Translations of Mathematical Monographs 67. Providence, RI: Amer. Math. Soc., 1988.

Ebbinghaus, H. D., J. Flum, and W. Thomas. *Mathematical Logic.* New York: Springer-Verlag, 1984.

Enderton, Herbert B. *A Mathematical Introduction to Logic.* New York: Academic Press, 1972.

Fitting, Melvin. *First-Order Logic and Automated Theorem Proving.* New York: Springer-Verlag, 1990.

Gallier, Jean H. *Logic for Computer Science.* New York: Harper & Row, 1986.

Hamilton, A. G. *Logic for Mathematicians (revised ed.).* Cambridge: Cambridge University Press, 1988.

Kleene, Stephen C. *Introduction to Metamathematics.* New York: Van Nostrand, 1952.

Lyndon, Roger C. *Notes on Logic.* New York: Van Nostrand, 1966.

Malitz, Jerome. *Introduction to Mathematical Logic.* New York: Springer-Verlag, 1979.

Margaris, Angelo. *First-Order Mathematical Logic.* Waltham, MA: Blaisdell, 1967.

Mates, Benson. *Elementary Logic (2nd ed.).* New York: Oxford University Press, 1972.

Mendelson, Elliott. *Introduction to Mathematical Logic. (3rd ed.).* Monterey, CA: Wadsworth, 1987.

Monk, J. Donald. *Mathematical Logic.* New York: Springer-Verlag, 1976.

Nerode, Anil and Richard A. Shore. *Logic for Applications.* New York: Springer-Verlag, 1993.

Schöning, Uwe. *Logic for Computer Scientists.* Boston: Birkhäuser, 1989.

Shoenfield, J. R. *Mathematical Logic.* Reading, MA: Addison-Wesley, 1967.

Silver, Charles L. *From Symbolic Logic . . . to Mathematical Logic.* Dubuque, IA: Wm. C. Brown Publishers, 1994.

Smullyan, Raymond M. *Gödel's Incompleteness Theorems.* Oxford: Oxford University Press, 1992.

Smullyan, Raymond M. *First-Order Logic.* New York: Springer-Verlag, 1968.

Wang, Hao. *A Survey of Mathematical Logic.* Peking: Science Press, 1962.

——— *Recursion Theory and Computability* ———

Cutland, Nigel J. *Computability.* Cambridge: Cambridge University Press, 1980.

Davis, Martin D., Ron Sigal, and Elaine J. Weyuker. *Computability, Complexity, and Languages (2nd ed.).* New York: Academic Press, 1994.

Davis, Martin D. *Computability and Unsolvability.* New York: Dover, 1982.

Davis, Martin D. *The Undecidable.* Hewlett, NY: Raven Press, 1965.

Epstein, Richard L. and Walter A. Carnielli. *Computability: Computable Functions, Logic, and the Foundations of Mathematics.* Pacific Grove, CA: Wadsworth, 1989.

Hamlet, Richard G. *Introduction to Computation Theory.* New York: Intext Educational Publishers, 1974.

Hennie, Fred. *Introduction to Computability.* Reading, MA: Addison-Wesley, 1977.

Lewis, Harry R. and Christos H. Papadimitriou. *Elements of the Theory of Computation.* Englewood Cliffs, NJ: Prentice-Hall, 1981.

Kfoury, A. J., Robert N. Moll, and Michael A. Arbib. *A Programming Approach to Computability.* New York: Springer-Verlag, 1982.

Machtey, Michael and Paul Young. *An Introduction to the General Theory of Algorithms.* Amsterdam: North Holland, 1978.

Minsky, Marvin L. *Computation: Finite and Infinite Machines.* Englewood Cliffs, NJ: Prentice-Hall, 1967.

Moll, Robert N., Michael A. Arbib, and A. J. Kfoury. *An Introduction to Formal Language Theory.* New York: Springer-Verlag, 1988.

Odifreddi, P. *Classical Recursion Theory.* Amsterdam: North Holland, 1989.

Rogers, Hartley. *Theory of Recursive Functions and Effective Computability.* New York: McGraw-Hill, 1967.

Shoenfield, J. R. *Recursion Theory.* New York: Springer-Verlag, 1993.

Smullyan, Raymond M. *Recursion Theory for Metamathematicians.* Oxford: Oxford University Press, 1993.

Soare, Robert I. *Recursively Enumerable Sets and Degrees*. New York: Springer-Verlag, 1987.

Tourlakis, George J. *Computability*. Reston, VA: Reston Publishing Company, 1984.

Yasuhara, Ann. *Recursive Function Theory and Logic*. New York: Academic Press, 1971.

Set Theory

Cohen, Paul J. *Set Theory and the Continuum Hypothesis*. New York: W. A. Benjamin, 1966.

Frankel, Abraham A., Yehoshua Bar-Hillel, and Azriel Levy. *Foundations of Set Theory (2nd ed.)*. Amsterdam: North Holland, 1973.

Kunen, Kenneth. *Set Theory: An Introduction to Independence Proofs*. Amsterdam: North Holland, 1980.

Jech, Thomas. *Set Theory*. New York: Academic Press, 1978.

Hrbacek, Karel and Thomas Jech. *Introduction to Set Theory (2nd ed.)*. New York: Marcel Dekker, 1984.

Johnstone, P. T. *Notes on Logic and Set Theory*. Cambridge: Cambridge University Press, 1987.

Roitman, Judith. *Introduction to Modern Set Theory*. New York: John Wiley & Sons, 1990.

Van Dalen, D., H. C. Doets, and H. de Swart. *Sets: Naive, Axiomatic and Applied*. Oxford: Pergamon Press, 1978.

Vaught, Robert L. *Set Theory*. Boston: Birkhäuser, 1985.

Hilbert's Tenth Problem

Cohen, Daniel E. *Computability and Logic*. Chichester: Ellis Harwood, 1987.

Davis, Martin D. *Hilbert's tenth problem is unsolvable,* American Math Monthly 80(1973), 233–69.

Davis, Martin D., Yuri V. Matiyasevich, and Julia Robinson. *Hilbert's tenth problem. Diophantine equations: positive aspects of a negative solution,* Proceedings of Symposia in Pure Mathematics 28(1976). Providence, RI: Am. Math. Soc., 323–78.

Davis, Martin D., Hilary Putnam, and Julia Robinson. *The decision problem for exponential Diophantine equations,* Annals of Mathematics 74(1961), 425–36.

Jones, J. P. and Yuri V. Matiyasevich. *Proof of Recursive Unsolvability of Hilbert's Tenth Problem,* American Math Monthly 98(1991), 689–709.

Manin, Yu I. *A Course in Mathematical Logic*. New York: Springer-Verlag, 1977.

Matiyasevich, Yuri V. *Enumerable sets are Diophantine,* Soviet Math. Dokl. 11(1970), 354–57.

Matiyasevich, Yuri V. *Hilbert's Tenth Problem*. Cambridge, MA: MIT Press, 1993.

Robinson, Julia. *Hilbert's Tenth Problem*. Proceedings of Symposia in Pure Mathematics 20(1969), 191–94.

Smoryński, Craig. *Logical Number Theory I*. New York: Springer-Verlag, 1991.

———————————————— *General* ————————————————

Benacerraf, Paul and Hilary Putnam (eds.). *Philosophy of Mathematics: Selected Readings (2nd ed.)*. Cambridge: Cambridge University Press, 1983.

Carroll, Lewis. *Symbolic Logic*. New York: Dover, 1958.

Copi, Irving M. *Symbolic Logic*. New York: Macmillan, 1954.

DeLong, Howard. *A Profile of Mathematical Logic*. Reading, MA: Addison-Wesley, 1970.

Detlefsen, Michael. (ed.). *Proof, Logic and Formalization*. London: Routledge, 1992.

Hamilton, A. G. *Numbers, Sets, and Axioms*. Cambridge: Cambridge University Press, 1982.

Herken, Rolf (ed). *The Universal Turing Machine*. Oxford: Oxford University Press, 1988.

Hofstadter, Douglas R. *Gödel, Escher, Bach*. New York: Basic Books, 1979.

Kline, Morris. *Mathematics: The Loss of Certainty*. Oxford: Oxford University Press, 1980.

Sanford, David H. *If P, Then Q: Conditionals and the Foundations of Reasoning*. London: Routledge, 1989.

Shanker, S. G. (ed.). *Gödel's Theorem in Focus*. London: Routledge, 1989.

Smullyan, Raymond M. *The Lady or the Tiger?* New York: Knopf, 1982.

Smullyan, Raymond M. *Forever Undecided*. New York: Knopf, 1987.

Van Heijenoort, Jean (ed.). *From Frege to Gödel: A Source Book in Mathematical Logic 1879–1931*. Cambridge, MA: Harvard University Press, 1967.

Wang, Hao. *Popular Lectures on Mathematical Logic*. New York: Dover, 1981.

Wang, Hao. *Reflections on Kurt Gödel*. Cambridge, MA: MIT Press, 1987.

———————————————— *Research Papers* ————————————————

Gödel, Kurt. *Über formal unentscheidbare Sätze der Principia Mathematica und verwandter Systeme*. Monatshefte für Mathematik und Physik 38(1931), 173–98.

Kleene, Stephen C. *General recursive functions of natural numbers*, Math. Ann. 112(1936), 727–42.

Post, Emil. *Recursively enumerable sets of positive integers and their decision problems*, Bulletin American Math. Society 50(1944), 284–316.

Shepherdson, J. C. and H. E. Sturgis. *Computability of recursive functions*, J. Assoc. Comp. Mach. 10(1963), 217–55.

Turing, Alan M. *On computable numbers, with an application to the Entscheidungsproblem*, Proceedings of the London Mathematical Society, ser. 2, 42(1936), 230–65.

—————————— *Survey Papers* ——————————

Barwise, Jon and Larry Moss. *Hypersets,* Mathematical Intelligencer 13(1991). New York: Springer-Verlag, 31–41.

Davis, Martin D. *Unsolvable Problems,* Handbook of Math Logic, Jon Barwise (ed.). Amsterdam: North Holland, 1977, 567–94.

Van Dalen, Dirk. *Algorithms and Decision Problems: A Crash Course in Recursion Theory,* Handbook of Philosophical Logic, vol. I, D. Gabbay and F. Guenthner (eds.). Dordrecht, Holland: D. Reidel Publishing Company, 1983, 409–78.

Hilbert, David. *On the Infinite,* in *From Frege to Gödel,* 367–92.

Hodges, Wilfrid. *Elementary Predicate Logic,* Handbook of Philosophical Logic, vol. I, D. Gabbay and F. Guenthner (eds.). Dordrecht, Holland: D. Reidel Publishing Company, 1983, 1–132.

Shoenfield, J. R. *Axioms of Set Theory,* Handbook of Math Logic, Jon Barwise (ed.). Amsterdam: North Holland, 1977, 321–44.

Sundholm, Göran. *Systems of Deduction,* Handbook of Philosophical Logic, vol. I, D. Gabbay and F. Guenthner (eds.). Dordrecht, Holland: D. Reidel Publishing Company, 1983, 133–88.

Smoryński, Craig. *Hilbert's Programme,* CWI Quarterly 1(1988), 3–59.

Wang, Hao. *Gödel's and Some Other Examples of Problem Transmutation,* in *Perspectives on the History of Mathematical Logic,* Thomas Drucker. (ed.). Boston: Birkhäuser, 1991, 101–9.

—————————— *Number Theory* ——————————

Burton, David M. *Elementary Number Theory (revised ed.).* Boston: Allyn and Bacon, 1980.

Rosen, Kenneth H. *Elementary Number Theory and Its Applications (3rd ed.).* Reading, MA: Addison-Wesley, 1993.

Index